Classification of Mammals

Above the Species Level

Classification of Mammals
Above the Species Level

Malcolm C. McKenna and Susan K. Bell

WITH CONTRIBUTIONS FROM

George G. Simpson   Guy G. Musser
Rachel H. Nichols   Nancy A. Neff
Richard H. Tedford   Jeheskel Shoshani
Karl F. Koopman   Douglas M. McKenna

A Project Supported by the American Museum of Natural History

COLUMBIA UNIVERSITY PRESS ▰ NEW YORK

Columbia University Press
Publisher Since 1893
New York   Chichester, West Sussex

Library of Congress Cataloging-in-Publication Data

McKenna, Malcolm C.
Classification of mammals above the species level / Malcolm C.
McKenna and Susan K. Bell ; with contributions from George G.
Simpson . . . [et al.].
p. cm.
Includes bibliographical references (p.) and indexes.
ISBN 978-0-231-11013-6 (pbk.)

1. Mammals—Classification. I. Bell, Susan K. II. Simpson,
George Gaylord, 1902– . III. Title.
QL708.M38 1998
599' .01'2—dc21   97–30063

Classification consists in grouping things according to their
actual relationship, i.e. their consanguinity, or descent from common stocks . . .
--C. Darwin letter to G. R. Waterhouse, 26 July 1843 (see Burkhardt 1996:76)

Classification cannot be a complete and detailed expression of phylogeny,
but it should be consistent with a reasonable estimate of phylogeny.
--G. G. Simpson 1983

Classification, like all science, has in fact progressed by a perpetual struggle between
successive theoretical principles and systematic rules on the one hand,
and the facts observed on the other.
--F. A. Bather 1927

The principal advantage of vertical, and disadvantage of horizontal, classification is that
the former is more in accord with the whole conception of evolutionary descent.
--G. G. Simpson 1945

The cladists seem, unfortunately, to have swallowed a rhyming dictionary rich in classic roots of all sorts;
the resulting deposit has now fertilized a plague of toadstools, sprouting on our beautiful taxonomic lawn.
--A. J. Boucot 1979

Nomenclature ought to serve systematics and not the reverse.
--D. C. Cannatella 1991

Naming all possible monophyla runs the risk of producing a plethora of names
which only satisfies some kind of baptismal fever.
--A. Minelli 1993

Sticks and stones may break my bones,
but names will never hurt me.
--Modernized 15th century nursery rhyme (see Stevenson 1948)

CONTENTS

PREFACE

The primary purpose of the present effort is to update George Gaylord Simpson's 1945 attempt to work out in words the relationships of subtaxa of the taxon MAMMALIA. Simpson's 1945 classification has had great influence that has extended far beyond mammalogy. It has been broadly useful, erudite, and full of pithy footnotes where Simpson tucked some of his most readable and significant prose. Many classifications of mammals have been proposed since long before Linnaeus's (1758) first "official" but nonevolutionary one. Simpson's, based upon sifting a vast pile of cumulative knowledge including a distillation of all the previous classifications, has been the most complete and useful. In the libraries of mammalogists and paleomammalogists, Simpson's classification has become the most dog-eared and annotated volume. As is the fate of all such works, however, the 1945 publication was out of date even before it was issued. Not only has new information accumulated and old information been recovered or improved, but much ferment and some progress concerning the theoretical underpinnings of systematization have occurred. A new comprehensive attempt to depict higher-level mammalian systematics covering both fossil and extant mammals, to update knowledge of their temporal and geographic distribution, and to incorporate changes in systematic theory has long been needed. As Broadfield (1946:65-66) noted, classifications are always in transition.

Simpson first became seriously interested in producing a mammal classification when he joined the staff of the American Museum of Natural History in 1927. At that time the preparation of evolutionary classifications was a well-established activity at the museum. Then, as now, there was a need to maintain a continually updated working classification to serve as a systematic framework and means of communication. Because of its location, collections, support staff, and libraries, the American Museum of Natural History (AMNH) has been one of the few places in the world in which such work can be done efficiently. In 1931 Simpson provided a list of mammalian orders and subordinated families. From 1931 until late in 1942, Simpson worked, when time permitted, to extend the 1931 classification to what he regarded as the generic (and in some cases subgeneric) level. In addition to the system itself, Simpson's more comprehensive classification had a section on the principles and conventions of classification, a long review section dealing with the reasons why various taxonomic decisions had been taken, a bibliography of 960 references, and indices of scientific and vernacular names.

Simpson began service in the United States Army early in December 1942 and was unable to work on the manuscript during his wartime service, which lasted until October 1944. By March 1944, however, his completed manuscript had been edited and prepared for submission to the *Bulletin of the American Museum of Natural History* by Rachel H. Nichols, with help from J. E. Hill, G. H. H. Tate, and E. H. Colbert. Simpson's classification was issued on October 5, 1945; it was reprinted without change in 1946, 1950, 1957, and 1962.

Preparation of the present expanded work began when Simpson returned to the American Museum from his service in the U. S. Army. The American Museum of Natural History's excellent library facilities were utilized as before, and original literature and other compilations such as the *Zoological Record* and *Biological Abstracts* continued to be searched systematically, mainly by Rachel Nichols. A large card file was produced, detailing new taxonomic changes proposed since termination of compilation for the 1945 classification. In 1947 Simpson made notes on part 3 (a review of mammalian classification incorporating his reasons for various actions) of his 1945 classification, evidently then intending to produce a second edition, but these modifications were never published. No new comprehensive synthesis was attempted by Simpson after World War II, although his ideas on the principles of systematization continued to evolve and were published as a book in 1961 and as an article several years later (Simpson 1961, 1964).

After his departure from the American Museum, Simpson turned over the mammalian classification project to M. C. McKenna, who joined the staff of the American Museum in 1960. McKenna created a hierarchically arranged file of punched cards, each of which covered a single taxon of Simpson's (1945) compilation, additionally incorporating information compiled by Mrs. Nichols subsequent to 1944. In 1966 Susan K. Bell joined the AMNH staff, and she and McKenna continued compilation of information about new taxa and synonyms, and about various taxa and synonyms accidentally or deliberately omitted from Simpson's (1945) version. A cumbersome but workable mechanical system of data maintenance resulted.

Following the Museum's purchase of a Wang VS™ computer in the early 1980s, a database program was written in BASIC by Philippe Lampietti, in order to deal with the files electronically. This provided search and word processing capability and the ability to produce directly from data files a text version similar in form to Simpson's (1945) classification. For some years thereafter, the classification resided in the data processing capabilities of the Wang VS™ computer. The older card files were, for the most part, no longer accumulated but were kept for archival purposes. New information was entered directly into the computer and reasons for changes were placed in electronically linked comments files that could be viewed in connection with each taxon but which were not necessarily to be part of any eventual publication. Temporal and geographic range data were compiled by manually filling in a matrix for each taxon assigned family rank or lower. The matrix format was a contribution of Robert Evander. It allowed for the compact entry and storage of a great amount of data, provided search facilities, and, for visual reasons, was easy to use.

Gradually, however, this system too became cumbersome. The opportunity to upgrade came as central word-processing and minimal data management at the American Museum gave way to the use of decentralized personal computers capable of more rapid operations. These computers had vastly larger data storage capabilities and, in the case of the Apple Macintosh™ equipment that we purchased, were oriented toward computer graphics. Accordingly, in the early 1990s the older BASIC program used by the Wang VS™ system was replaced by a much improved program. Under the direction of the principal authors, the new program, Unitaxon™, was written in C language for the Apple Macintosh™ series of computers and their successors by Douglas M. McKenna of Mathemaesthetics, Inc. (originally incorporating 4th Dimension™, a database program produced by ACIUS, Inc.). This new program graphically presented the classification in a large-scale "cladogram format." Information screens could be called forth for each taxon to provide data on authorship, rank, synonyms, common names, extinction status, hierarchical path, notes, comments, and range. Enhanced abilities to search the files for a variety of criteria or combinations thereof were provided. Taxa could be cut from their position on the cladogram and pasted anywhere in the graphic depiction, which would be automatically redrawn to accommodate spatial requirements. New siblings or offspring could be created, ranks changed, information copied, moved or deleted, and so on. As in the former BASIC program, facilities to generate various text versions on demand, incorporating data from categories of interest chosen from a menu, were built into the program. An ability to create an index of the taxa was also provided. Updated, single, manuscript copies of the complete files or any part thereof could be printed by a laser printer in a reasonably short amount of time. By March 1997, the Macintosh™ program Unitaxon™ and associated data files representing the classification occupied about 25 megabytes on the computer hard disk. Compilation of data for the printed version of this classification was suspended March 4, 1997. The final printed version of the classification was formatted by the Unitaxon™ program. Compilation of data in the electronic version continues. A browser program that will allow its users to read the classification files (including comments not present in the printed version), to search for various types of data, and to print the data in text form is under development.

Acknowledgments

A number of people have helped us and are owed a great debt. Anthony D. Barnosky, Marc Carrasco, William A. Clemens Jr., Darrel R. Frost, Johnathan Geisler, Gina C. Gould, Bruce MacFadden, Ross MacPhee, Peter Makovicky, Jin Meng, Michael J. Novacek, Maureen O'Leary, David Polly, Guillermo Rougier, Donald E. Savage, Richard Stucky, S. David Webb, Michael O. Woodburne, André Wyss, and an anonymous reviewer went over preliminary manuscript versions of the system or its introduction and provided significant new information or important reasons for changes. Many others, including M. C. McKenna's graduate students and various unwary visitors buttonholed for the task, checked partial manuscripts dealing with particular areas of expertise. Especially notable among these helpful people were Ann R. Bleefeld, Mary R. Dawson, Eric Delson, Daryl P. Domning, Robert J. Emry, Margarita A. Erbaeva, Robert L. Evander, Larry J. Flynn, R. Ewan Fordyce, Philip D. Gingerich, Jean-Louis Hartenberger, Zofia Kielan-Jaworowska, William W. Korth, Rosendo Pascual, Martin Pickford, Peter Robinson, Gustavo J. Scillato-Yané, Nancy B. Simmons, Pascal Tassy, Elizabeth S. Vrba, Guiomar Vucetich, John H. Wahlert, and John R. Wible. We did not always follow the advice we received, but the information and thought that these colleagues provided were of great use and led to many changes that we regard as true improvements.

In the early stages of computerization and in other ways Leslie F. Marcus was of constant help. Robert Evander contributed useful ideas that led to the graphic presentation of range data made use of by Philippe K. Lampietti, who developed the program for the Wang VS computer system used in early stages of compilation. Most of the features of Lampietti's program were later supplemented, improved, or replaced in a graphics-oriented way and in a different computer language in the Macintosh™ computer system.

For help in tracking down early literature we thank Beatrice Beck of the Rancho Santa Ana Botanical Garden, Claremont, California, and Mary de Jong, Donald Clyde, Roscoe Thompson and others of the Library at the American Museum of Natural History.

Above all, we thank our respective spouses, Priscilla C. McKenna and Byron Bell, for their many years of support.

Malcolm C. McKenna
Susan K. Bell

Classification of Mammals

Above the Species Level

INTRODUCTION

Life has a hierarchical pattern because it has evolved and ramified over a very long time (Darwin 1859). It is possible to communicate features of that pattern by constructing "systems", such as cladograms, and then expressing these in words as classifications. In other words, evolutionary classifications are results of a systematization process, which involves phylogenetic analysis in addition to purely formal procedures associated with effective communication (Martin 1981; Tassy 1988:43). The systematic analysis is fundamental, but a carefully constructed classification based on that analysis is useful in communication. However, various classifications can result from the same systematic analysis (Lorch 1961). The most useful classifications are genealogically structured storage systems (Mayr 1974:101; 1981:515), whose own ramifications reflect as closely as possible what systematists hypothesize to have occurred by evolutionary modification and branching in nature (Brundin 1972; O'Hara 1991). Mayr (1968:546; see also Mayr 1974) has suggested that, "every classification is a scientific theory," but presumably he meant to restrict the characterization to evolutionary classifications.

The present effort is an attempt to update George Gaylord Simpson's 1945 classification of mammals, and also to make progress in conversion from Simpson's (and others') relatively syncretic principles of phylogenetic depiction to one that represents a somewhat more rigorous phylogenetic system (see Zimmermann 1937:1004; Hennig 1974, 1975:246; C. Patterson 1978, 1980, 1982a, 1988, 1989; Farris 1967a, 1980; Smith 1994). Simpson (see also Charig 1982) emphatically disagreed with cladistic methods that had been introduced to realize such an aim, believing that it is possible to reconcile subjectivity ("art") with science (Lamarck 1809:103; Simpson 1943:150; 1959a:294; 1961:122; 1976). We believe that "art" plays a part in making scientific progress and is not antithetic to science, but "art" should not rule at the expense of objectivity. We attempt here to arrive at a close but practical correspondence between our text and the phylogeny that we reconstruct. We realize that our work is not free of subjectivity (Burtt 1966:428; Anderson 1974). However imperfect our results may be, we continue to pursue an aim of reducing the need for subjectivity in systematics (Hull 1970). Our immediate aim is pragmatic: to make higher-level mammalian systematization better, i.e. more genealogical (Darwin 1859:420). We do not claim to produce the prose equivalent of a 1:1 map. Those who have wrestled with the many problems inherent in large hierarchical systems will appreciate the difficulty of the task and the staggering amount of work that remains to be done.

Our aim is to provide a framework that is genealogical, or at least as genealogical as possible, not primar-ily phenetic (Cain and Harrison 1960) or panphenetic (Holmes 1980:48; see also Holmes 1981). When morphological distance is to be considered, we prefer to consider it on the basis of a genealogical framework. The construction of our hierarchical system is based as much as is practical upon hypotheses of common ancestry and tree-like phylogenetic branching, founded in turn on what we take to be sound cladistic principles of phylogenetic analysis (Hennig 1950, 1965, 1966; Gaffney 1979; Nelson 1979; Eldredge and Cracraft 1980; Wiley 1975, 1981, 1986; Rieppel 1988; Schoch 1986; Panchen 1992; Forey et al. 1992; Davis and Nixon 1992; Smith 1994; Richter and Meier 1994; but see Remane 1989 for balance). We do not fully reach that goal, nor are we persuaded that we must before the system possesses utility. As will be discussed more fully below, there are major problems in adapting the Linnaean System to cladistic results. Significant problems attend paraphyly, the proliferation of hierarchical categories, and the management of inflected taxonomic names as they move up or down the hierarchy. Because our work is primarily a compilation, we are forced to retain unaddressed paraphyly in many cases where apparent paleontological lineages of identically ranked taxa have been arbitrarily chopped into segments or otherwise exclude descendants from nested taxa. This action should not be mistaken to imply approval. Where we have identified paraphyly but not yet modified the system, we have generally commented. Many taxa have yet to be analyzed adequately from a phylogenetic standpoint. In this, as well as in other higher-level phylogenetic systems, poorly understood taxa are simply carried along in their current internal disarray, pigeonholed in the system at a point at which the appropriateness of their position is judged sufficiently clear. Believing that partial solutions are preferable to none at all, we have in many cases merely assembled the components of problems needing attention.

As has often been pointed out, many systematists simply "do taxonomy" without much training in the subject or knowledge of its theoretical framework. The results have been and continue to be unfortunate. In the belief that we should investigate the theoretical underpinnings of what we are doing, and in an effort to communicate our results in a testable fashion, we have written this introduction.

Authorship

Most of the analysis on which the present version of mammalian higher systematics is based is not our own. No one can be familiar with the whole sweep of mammalian phylogeny and the entire evolving structure of its systematization. Although we have utilized our own judgment, we have nevertheless been dependent on the work of the entire community of scholars who deal with extinct and extant mammals. Methods used by that community are diverse, and the results summarized here are not always well-founded. However, our classification is not solely a passive compilation. In addition to adjustments involving matters of fact, such as corrections of authorship, dates, range, and so forth, we have made modifications that we believe improve correspondence between the storage system and the evolutionary changes and natural ramification of evolution it is designed to reflect. Other modifications incorporate our opinions about subjective synonymy. Much of what we summarize is at least approximately in its proper place in the hierarchy and thus "within view."

Authorship of this classification is complicated. Simpson' hand is still much in evidence. This is true both in many inherited fundamental features of the classification's structure and in some wording retained from his 1945 monograph. A few of the original entries have not changed since 1945, but most of the entries stemming from Simpson's work have required modification of some sort. Some of Simpson's footnotes are still as germane as they were in 1945. Those included here are identified as quotations. We are under no illusions concerning what Simpson would think about some of our actions dealing with his ideas and text, but we are enormously indebted to him and his assistants for both published and unpublished work that we incorporate. Part of the result is still very much his, and we acknowledge his responsibility for it by including him among the contributors. Needless to say, he is not responsible for aspects of our work that he surely would not have approved.

When he left the American Museum of Natural History, Simpson decided not to pursue the subject further. It was his expressed desire not to be an author of subsequent modifications of the actual classificatory nuts and bolts of the work. Not only did he believe at that time that other activities were more appropriate for him to pursue, such as study of the logic involved in principles of systematization (Simpson 1961; see also comments by Nelson and Platnick 1981), but he also had become disenchanted with some of the requirements of the International Code of Zoological Nomenclature (ICZN), notably Article 36 of the 1961 edition of the ICZN (quoted in Appendix A) that makes all family-group categories based on the type genus coordinate (International Commission on Zoological Nomenclature 1961:39). According to Article 36, the author of the earliest valid family-group taxon is deemed the official author of all taxa based on the same type genus within any particular set of family-group taxa (see also Sabrosky 1954). To convert earlier attributions of authorship to this system was regarded by Simpson as onerous.

Weekly reviews of the current literature received in the library of the AMNH have yielded most of the information we summarize. Nonetheless, important papers have surely been missed. Inasmuch as the preparation of classifications is an ongoing process, we (or our successors) would appreciate suggestions of improvements for incorporation in the ongoing electronic compilation.

Malcolm C. McKenna and Susan K. Bell have shared equally in most of the work of compilation, integration of data, and supervision of development of the underlying computer programs. We are therefore the principal authors of the present compilation, and we are ultimately responsible for its presentation. Aid beyond the amount that would require mere acknowledgment has been provided by a number of others, both in the underlying computerization and in the evolution of the compilation itself. As noted above, Simpson remains as one of the contributors because some of the synthesis as well as text is still his. Rachel H. Nichols began the compilation that led to this expanded version of the 1945 publication. Richard H. Tedford and Nancy A. Neff contributed much to the section on carnivores. Karl F. Koopman contributed much data and thought to the section on bats and also was of great help in other sections of the classification. Guy G. Musser was the major contributor to the section on muroid rodents and provided information on other rodent groups as well. Jeheskel Shoshani contributed extensively to the section dealing with tethytheres. Douglas M. McKenna is included because of his major contributions to the computerization. These will be largely invisible to readers of printed versions, but will be fully appreciated by those who make use of electronic versions of the classification.

Scope
Upper taxonomic limit

The taxon called class MAMMALIA Linnaeus, 1758, was originally diagnosed by chosen characters rather than defined by listing taxonomic content. Taxa often are casually said to be defined by characters, e.g. by Muir (1968), but this activity is really diagnosis. Recently, the difference between definition and diagnosis has been explored by Ghiselin (1966a, 1984a), Rowe (1987), de Queiroz and Gauthier (1990, 1994), Wyss and Flynn (1993), Cain (1993), and de Queiroz (1994). These authors sharpened the usual dictionary versions of these words. For further discussion of characters, see Inglis (1991).

The term "class" has several meanings: (1) a particular high-level taxon in a taxonomic system; e.g., class MAMMALIA; (2) in the general sense, any taxon built up from a conglomeration of subsumed lower-ranking taxa, the lowest of which includes only individual organisms; and (3) class as diagnosed by attributes (Colless 1985; Duarte Rodrigues 1986). Ghiselin (1984b:210) complained about conflation of the second and third concepts, but if the method of critical recognition is based upon what happened at or slightly before a particular genetically fixed cladogenetic event, the concepts for all intents and purposes can coincide (Eldredge and Salthe 1984:188; Nelson 1985).

The taxon MAMMALIA as we use it is defined as a crown group (Jefferies 1979) comprising the most proximal common ancestor of extant monotremes and therians, plus all descendants of that ancestor. The extinct multituberculates are here regarded as descendants of that ancestor, because they share the transfer of various accessory jaw bones to the skull that was present in the most proximal ancestor of therians and monotremes. They are therefore regarded as members of the crown group. A complex event like the origin of the mammalian middle ear apparatus seems to us to be more likely to be synapomorphous than convergent. Multituberculates are therefore included in MAMMALIA, but docodonts, morganucodonts, and haramiyoids (see Appendix C), which are members of MAMMALIA-FORMES and in which the transfer did not occur, are excluded from crown group MAMMALIA. Tritylodonts and other progressively more distant synapsid outgroups are excluded even from MAMMALIA-FORMES.

We therefore treat the taxonomic class MAMMALIA as definable by its subsumed taxonomic content. In this compilation the class MAMMALIA is simply a taxon at a certain hierarchical level, representing the sum of the nested taxa that it contains, actually starting with the species level (de Queiroz 1995) although we here restrict our coverage to genus-level taxa and upward. The various taxa of any given rank are here regarded as individuals (Ghiselin 1966b, 1975, 1977, 1981; Ridley 1989; Mayr 1974:101; but see Hull 1976, 1978, and his commentary *in* Ghiselin 1981; and Ereshefsky 1991), and the names of those taxonomic individuals are meant to be unique labels. Taxa are Kantian noumena, not phenomena. The full construction of such a system is called systematization (Griffiths 1974:85), but here we are primarily concerned with the expression of the system in words. It is not necessarily an essentialist program (Hull 1965:317; Mayr 1968:546; Sober 1980) to determine a practical diagnosis at each taxonomic level used. Characters are appropriate for that purpose (Gill 1871).

As a vernacular word for MAMMALIA, the word "mammifère" was used in France as early as 1791 in the *Dictionnaire Historique de la Langue Française* (J.-L. Hartenberger, personal communication). Lamarck (1802:15-17, 55) used "les mammaux." Insofar as we are aware, Good (1826) first used the plural term "mammals" in the vernacular in English, and Owen (1841:60) was apparently the first to use the term in the singular in that language (contra Gill 1902). As the name "mammal" currently signifies etymologically, extant mammals can be diagnosed as presently surviving synapsid amniotes in which the females produce milk to nourish their young (Blackburn 1991). In addition, there are many features of physiology and soft tissue anatomy that can serve to distinguish extant mammals. Unfortunately, we are uncertain of the distribution of such characters as the presence of mammary glands among extinct synapsid outgroups beyond the most recent common ancestor of extant monotremes, marsupials, and placentals. Thus, lactation is not sufficient to serve as part of the diagnosis of all MAMMALIA-FORMES. We shall never know if morganucodonts or other nonmammalian therapsids gave milk. A diagnosis employing lactation only works for a mammalian crown group based upon extant taxa (Jefferies 1979; de Queiroz 1995). Hard part proxy characters are now used in all synapsid diagnoses, and the etymology of MAMMALIA is now of secondary importance. The name has become a tag, applied to a particular taxon defined taxically.

As paleontological studies have expanded, the number of extinct taxa proposed has burgeoned. With that growth has come a necessary shift in emphasis from soft anatomical features such as the presence of hair or lactation to diagnoses based upon attributes accessible in fossils. However, many fossils are poorly known, a fact often pointed out by neontologists (usually by those whose area of expertise covers taxa lacking a good fossil record). The literature is burdened with debate about the adequacy of fossils in phylogenetic reconstruction (Doyle and Donoghue 1987; Gauthier, Kluge, and Rowe 1988a; Donoghue et al. 1989; Kluge 1990; Huelsenbeck 1991; Eernisse and Kluge 1993; Gould 1995).

If vertebrate fossils left behind in the course of Earth's long history are to be considered, features of hard parts such as bones and teeth are necessary for practical diagnoses. In fossil synapsids these proxy characters are all potentially available for study. Among mammaliaform synapsids, the departure of the articular (malleus) and quadrate (incus) bones from the jaw apparatus to the cranium has long been used in diagnosing a monophyletic MAMMALIA acceptable to many neontologists and paleontologists (Rowe 1996). Here we regard the acquisition of a functional dentary/squamosal jaw joint with a well developed dentary condyle contacting a squamosal glenoid fossa (Huxley 1869:3; 1880:658; Kermack and Mussett 1958; Kermack 1972; Kermack et al. 1973, 1981; Allin 1975:410; Kermack and Kermack 1984:49) as diagnostic of the broader taxon MAMMALIAFORMES (see also Rougier et al. 1996:37). Dentary/squamosal contact does occur in *Probainognathus*, *Diarthrognathus*, and some tritylodonts (Barghusen and Hopson 1970) that were not mammaliamorphs, but this initial type of contact does not mean the same thing as functional dentary condyle/squamosal glenoid facet contact (Gow 1981). Both of these events occurred somewhat before the transfer of the older articular/quadrate suspensorium and some of the accessory jaw apparatus away from the new jaw suspensorium to what is now exclusively the ear region of the cranium, and they were necessary precursors of that transfer (Patterson 1981a, 1981b; Kemp 1983; Miao and Lillegraven 1986). The transfer has occurred in living monotremes and therians, which are here accepted as monophyletic in that regard.

We could have worked outward beyond the closest outgroups of mammals among the nonmammalian synapsids all the way to the most primitive synapsids. In this classification we have chosen to deal mainly with a restricted monophyletic (synapsid (therapsid (cynodont (mammaliaform)))) crown group: MAMMALIA (contra Kemp 1988a, 1988b). Nonmammalian mammaliaforms are considered in Appendix C, partly for the benefit of those who expect us to classify the same organisms as Simpson (1945) did and partly because of their value in assigning character polarities in cladistic analyses of MAMMALIA. The senior author was taken to task by Patterson and Rosen (1977:159) and by Wiley (1979b) for not extending an earlier classification of mammaliaform higher categories (McKenna 1975) to the broad range of nonmammalian therapsids, but there is no compelling reason to do so. Therapsids, as they were once envisioned, were a paraphyletic group from which certain descendants were excluded. They were simply a truncated synapsid clade, the residue after subtracting one monophyletic group from another. Currently, however, the monophyletic taxon THERAPSIDA includes MAMMALIA; therefore, therapsids are alive today in the same sense that dinosaurs are alive today if birds are dinosaurs (Gauthier and Padian 1985:190; Kemp 1988a). To classify nonmammalian therapsids and mammals together is simply to classify all therapsids. Others are better qualified to do that expanded work, so we have restricted our scope. We deal mainly with the mammalian crown group (see also Appendix B), but we include information about a few close outgroups as well (Appendix C).

What one delimits as MAMMALIA among the THERAPSIDA is arbitrary if one depends upon choosing a diagnosis (Linnaeus 1758; Gill 1871; Olson 1959; Simpson 1959b, 1971a, 1971b; Reed 1960; Van Valen 1960; Crusafont Pairó 1962; MacIntyre 1967; Hopson 1967, 1994; Hopson and Crompton 1969; Kemp 1982; Ax 1985:283; Loconte 1990; Lucas 1990, 1992; Rowe and Gauthier 1992: fig. 1). Taxic definitions as well can be arbitrary and drift from their original meanings if original ones are not honored. Once an original definition is modified, as it has been for many decades with the MAMMALIA of Linnaeus, then convention, convenience, authority, and even legislation have their effects. It is important that we aim for a system of nested monophyletic taxa, to which stable and meaningful label-like names can be applied for purposes of easy communication. Linnaeus (Carl von Linné; b. 1707, d. 1778) was certainly not deliberately naming MAMMALIA as a crown group in 1758. If he had regarded it as a crown group, and if its content had not been expanded later before biologists began to consider crown groups, MAMMALIA would be a senior syn-

onym of Parker and Haswell's THERIA, proposed 139 years after Linnaeus's "official" 10th edition. See Fries (1903), Hagberg (1939), Larson (1971), Stafleu (1971), and Slaughter (1982) for details of Linnaeus's life and discussions of his methods.

The first "official" adjectival diagnosis of the class-ranked taxon MAMMALIA by Linnaeus (1758) was built upon earlier attempts, including his own, published in the 1st edition (1735: table): hirsute-bodied quadrupeds in which females are viviparous and lactiferous. In the 6th edition of the *Systema Naturae* (Linnaeus 1748), which was based upon additional characters of extant representatives then known, the diagnosis was the same: "quadrupedia, corpus pilosum, pedes quatuor, feminae viviparae, lactiferae." This diagnosis was somewhat different from Ray's (1693): "animalia, pulmone respirantia, corde ventriculis praedito duobis, vivipara," and added hair and lactation to the list. One can only wonder where Linnaeus might have placed the egg-laying monotremes in his system had he known about them.

For Linnaeus there were no extinct taxa to confuse matters. Even though fossils were then in the initial stages of impacting science, Linnaeus dealt with them as part of the mineral kingdom. For instance, Linnaeus (1740) referred to all fossil mammal bones as examples of:

> REGNUM LAPIDEUM
> CLASSIS FOSSILIA
> ORDO PETRIFICATA
> GENUS *ZOOLITHUS*

Considering the timescale accepted by Western Civilization of Linnaeus's day, one can see why it would have been unreasonable to propose that evolution played a significant role. The timescale of scientific models did not lengthen significantly until the end of the eighteenth century. We owe to comparative anatomists like Lamarck and Cuvier the transfer of fossils from the mineral to the animal and plant kingdoms, where they became representatives of "espèces perdues." Similarly, realization of the central significance of evolution is owed to Lamarck (e.g., 1809), even though evolutionary mechanisms became more fully evident only after the work of Darwin (1859).

In the present effort, we continue to classify nearly the same major groups of organisms that Simpson (1945) did, whether fossil or extant, and, in disagreement with Kemp (1988b) and Lucas (1992), we call the overarching taxon MAMMALIAFORMES Rowe, 1988, which subsumes the more familiar taxon MAMMALIA Linnaeus, 1758. The main differences between these concepts concern a few fossil taxa placed in one but not also the other (Rowe 1996). We continue to give MAMMALIA the taxonomic rank "class," but we treat it as a monophyletic cynodont synapsid clade built up from its subsumed taxonomic content and approximated by diagnosis. We do not regard

MAMMALIA as a defined grade divisible into lesser grades. See Huxley (1958, 1959), Olson (1959), Simpson (1959b, 1959c, 1960, 1971b), Reed (1960), Wood (1965), MacIntyre (1967) and Watt (1968) for divergent discussions of grades versus clades and supposed mammalian polyphyly.

Lower taxonomic limit

In 1931 Simpson produced a macrotaxonomic system of higher mammalian taxa and their names, which did not deal with taxa lower in relative rank than those of the so-called family-group (see Article 35 of the ICZN). His 1945 compilation included taxa of the genus-group (including some subgenera; see Article 42 of the ICZN) as well, as have the more recent classifications of Romer (1966), Carroll (1988), and Kalandadze and Rautian (1992). Since 1945 a number of compilations of extant mammals have dealt with taxa at the species-group level (see Article 45 of the ICZN; also see Ellerman and Morrison-Scott 1951; Walker et al. 1964; Corbet and Hill 1980; Honacki et al. 1982; Wilson and Reeder 1993). However, considerable diversity of opinion and confusion exist concerning species-level systematics (Ridley 1989; de Queiroz 1992, 1994). There continues to be a need for a hierarchical compilation of both fossil and extant mammalian taxa encompassing all currently used nested taxonomic levels, as well as of the names of ranks applied at those relative levels.

It is at the levels of individual organisms and populations represented by species-group taxa that the mechanisms of evolution operate. Moreover, it is below the species level that observable biological phenomena exist. Although we shall not be dealing primarily with taxa at the species level, for the sake of completeness we give a morphological definition for terminal species-group taxa. We believe it is practical and will hold for both fossil and extant ones. A terminal species-group taxon is a statistically diagnosable temporal cross section of an apparent lineage between branching points of the apparent lineage. The temporal cross section may be at present or at any time in the past for which contemporaneity can be demonstrated. Contemporaneity is not here interpreted to involve zero duration, but only that time interval during which the terminal species-group taxon is diagnosable. Terminal species-group taxa may be simultaneously given redundant higher ranks, notably the rank of genus but also others. Thus one encounters apparent lineages of species, which when analyzed cladistically might yield a cladogram rich in branches and therefore nested taxa of various relative ranks, yet in the apparent lineage only one evolving but unbranched population may have been present at any one time. Somewhere between the positions of doctrinaire cladists and doctrinaire stratopheneticists lies practicality.

Practicality has forced us, like Simpson, to limit coverage mainly to taxa of higher rank than species-

group taxa. The so-called genus level, the level close to or immediately above the species in the hierarchy, is a practical working level because of the arbitrary requirement of generic names brought about by the binominal Linnaean method of classifying monophyletic species (Stearn 1960). Among fossil and extant mammals there are simply too many species-level taxa and names to deal with adequately without delaying this project further. This problem is not new (see Linnaeus 1737b; Anderson 1940; Just 1953; Cain 1956; Stearn 1959).

Synonymy

Simpson was never interested in dredging for historical detail to the degree pursued here, notably in the areas of *nomina oblita* and synonyms. We have no doubt that he would have regarded our efforts excessive. We have felt more strongly than Simpson did that the fates of wholly or half forgotten taxa and synonyms, even of higher taxa outside the self-limited scope of the ICZN, are still of real (rather than merely "official") historical interest and are potentially useful. Some long-unused names are apposite for monophyletic or monotypic groups that might otherwise be named anew. If they already exist, why not use them rather than coin unsuspected synonyms and homonyms?

Among junior synonyms regulated by the ICZN, some are objective and some subjective. Objective synonymy within the family-group and at lower taxonomic levels involves nomenclatural typology: type specimens, type species, type genera. Of these, only the type specimen is concrete. If two taxa of the same rank have the same type, they are regarded as objectively synonymous. Subjectively synonymous taxa within the family-group and below have different types. Many subjective synonyms have resulted because systematists did not want to change the rank of immediately higher taxa nor to interleave more ranks than are already present, although this has often been relaxed at the genus-group and species-group levels. In some instances Simpson (1945) identified subjective synonyms as subgenera; for the cats *Felis* and *Panthera*, and for the canid *Dusicyon*, he listed their subgenera, in part, hierarchically (see section on taxonomic levels). We regard this practice as flawed, but we perpetuate it here for practical reasons, pending more detailed analysis of subjective synonyms of genus-group taxa. In the comments section of the electronic version of the classification we have provided, when possible, the type species of genera, subgenera, and their synonyms as an aid in ascertaining whether synonyms are objective or subjective.

In this printed text of the classification, objectively synonymous taxa of the genus-group are depicted in an "[= X]" format. Subjective synonyms are denoted by "[Including X]." Where matters are indefinite, we employ "[= or including X]." Above the level of the genus-group, synonymy is more difficult to codify, especially above the family-group level where even the ICZN's arbitrary rules do not apply. Subjective synonymy at higher levels is the same as at lower ones, but because higher categories are specified merely by what lower categories they include, the content of synonyms often is not identical. In order not to lose information, one might say that order X of author A is "the same as" [=] order Y of author B, but include a footnote in the text revealing that only five of the six families of order X were originally (or currently) included in order Y. An example would be the (nearly) objective synonymy of MESAXONIA with the now more familiar term PERISSODACTYLA. The same conventions were used by Simpson (1945). Common sense regarding such matters was expressed also by Brothers (1983).

Use of taxa, *incertae sedis*

Taxa of uncertain taxonomic position at one hierarchical level may nevertheless be placed with confidence within the next higher level. Such taxa are dubbed *incertae sedis*. There are, nonetheless, different definitions and methods of expression of the concept (e.g., Jussieu 1789; Simpson 1945; Nelson 1972; Wiley 1981). Monophyletic fossil taxa that are nevertheless *incertae sedis* have been needlessly dubbed adokimic taxa (Böger 1989:49; see Cannatella 1991:376; also see section on metataxa, below). We employ the *incertae sedis* concept in the following ways.

Imagine a higher-ranked nonterminal taxon Q that already subsumes two (or more) lower-ranked taxa R and S. If now a still lower taxon T with at least one known autapomorphous character (i.e., it is not an apparent ancestor; see below) is assigned to Q, but membership in either R or S is not assigned because there are no known synapomorphous characters with just R or just S, then T could be raised to the same taxonomic rank as that of R and S and listed as an additional, formal, polytomous member of Q. However, if synapomorphies are suspected but not yet adequately supported, T can be ranked provisionally two or more levels lower than the rank of Q (at least one rank below that of R or S), as "Q *incertae sedis*: taxon T." This is equivalent to saying, "we think we know that T belongs to but is not the ancestor of Q, but we do not yet think we know which of the two accepted subdivisions of Q (R or S) should subsume T." Obviously, however, a species cannot be placed in a genus *incertae sedis*, unless subgenera have been proposed and assignment of the species (obviously not the type species) to any of them is uncertain within the genus.

Simpson (1945) employed this method for some taxa (e.g., the classification of the genus *Trigenicus*), but he also used several other methods of indicating uncertainty of assignment (e.g., 1945:60, 68). We combine all these ways of dealing with uncertainties into one method of expression. We place *incertae sedis* taxa

in the lowest-ranked postulated monophyletic taxa to which we are presently confident that they belong. Thus we agree with Ax (1987:233, convention 5, derived from Wiley 1979b, 1981).

In another way that we hope promotes clarity, we depart from Simpson's several styles of listing taxa *incertae sedis*. Simpson listed such taxa at the end of the lists of more firmly allocated taxa. Thus, if Q subsumes R and S, as above, and lower ranked T is regarded as Q *incertae sedis*, Simpson would generally list them all thus:

> Taxon Q
>   Taxon R
>   Taxon S
> Taxon Q *incertae sedis*:
>   Taxon T

We depart from this in both the graphic method used in the computerized version of this classification and in its text equivalent. We simply list "*incertae sedis* taxa" in Taxon Q first, before listing the definite assignments represented by taxa R and S and their contents:

> Taxon Q
>   Taxon T
> Taxon R
> Taxon S

That Taxon T is *incertae sedis* is indicated by its position in the list and by the fact that its rank is lower than that of taxa R or S (indicated by its indentation). In the present classification, multiple cases of taxa *incertae sedis* are listed in ascending rather than descending rank, so that questions do not arise concerning whether one such taxon belongs to another.

When a list of taxa of the same rank are listed as *incertae sedis*, they are sometimes treated as sequentially interchangeable, *sedis mutabilis* (Wiley 1979b:322). This is often the case in our handling of poorly known genera. At other times when we have thought we had some inkling of the phylogenetic arrangement but have been unwilling to create new taxa and ranks to depict it, we have used a form of sequencing to suggest potential phylogenetic structure. In sequencing of *incertae sedis* taxa at the same rank, the first taxon is tentatively the sister of all those below it, the second to those below itself, and so on.

Range data

Range data are given with respect to time and space. Neither temporal range nor geographic distribution are necessary in a classification, but if used properly they add significantly to the informative value. Both kinds of range information are necessarily incomplete, but can be updated easily in the program used to compile and manipulate this classification. Although imperfect, such data are much more complete for mammals than for other vertebrate groups. At the power of resolution attempted, ranges are reasonably continuous, especially so when the availability of potentially fossiliferous rocks in which they might occur is given adequate consideration. In the case of retained paraphyletic taxa, the range is that of the artificially truncated taxon (see below).

In the present system, indication of geologic and geographic distributions is modified from the usage adopted by Simpson (1945; 1947:481). As in Simpson's (1945) classification, known geologic distributions are given in terms of periods and/or epochs. Because we consider this the time dimension of the range, we refer to early, middle and late temporal units, rather than lower, middle and upper spatial or rock-stratigraphic units. We considered, but found impractical, the idea of indicating the known time distribution for each taxon in millions of years (MY, MYBP, Ma). Although this information is available for some taxa, it has not been compiled or is unavailable for most. Simply converting period or epoch ranges into years would have been misleading; translating every known occurrence for each taxon into years would have assured that this system would not be published in our lifetimes. In the future, however, we have no doubt that indications of ranges can be much improved.

The complexity of a system that compiles local rock-stratigraphic occurrences with regional stages and broader (global) syntheses of correlation is probably not evident to the non-practitioner of stratigraphic paleontology. Keeping records in such a global biostratigraphic framework involves following many moving targets. Moreover, the timescale itself has been extensively recalibrated since the advent of radioisotopic dating (Berggren et al. 1995). Ranges of taxa given in the older literature have been adjusted accordingly in most instances. Some of the older references are no longer corrigible because the original data are either lost or were never available.

For a timescale of the Mesozoic we refer the reader to Gradstein et al. (1995). As a guide to our use of Cenozoic epochs and their subdivisions we have provided a timescale and correlation chart (fig. 1). The vertical scale of the Cenozoic chart is nonlinear because of the increased knowledge of such units with decreasing age. On average, in the early Cenozoic convenient time divisions like the North American Land Mammal Ages (NALMAs) are about four million years in duration (coincidentally, one mahayuga of the Rig Veda = 4,320,000 years). NALMAs become significantly shorter in the late Cenozoic. We have used terms familiar to a broad range of scientists as well as the various Land Mammal Ages beloved by vertebrate paleontologists. This has resulted in a certain amount of over-generalization, for which we beg the indulgence of sophisticated mammalian biostratigraphers. The South American record is shown with more than a dozen significant gaps. The chart itself has evolved enormously over the years. When new taxa were added or ranges up-dated we followed this chart, or updated the two in concert, but we have not updated the temporal range of every taxon. For various reasons, the ranges of certain rare taxa described early in the history of paleontology are difficult to update. From experience, we know that, in addition to our own errors, there exist uncorrected residual errors stemming from previous compilations.

Compilation of known stratigraphic and geographic ranges has been at the lowest taxonomic level. Known ranges of higher taxa and monotypic taxa up through but not above the so-called family level were compiled by manual summation of the generic ranges. In certain cases family-group taxa have a broader range than the sum of included taxa because of the power of resolution of identification or because the fossils yielding the information have not been assigned a generic name. Occurrences of taxa of the genus-group are recorded only where actually supported by data concerning species referred to the genus. Thus, if species of a genus are not recorded from a particular time interval, we have left the cell or cells blank in the computer data matrix that is used to generate the text concerning that genus even if it appears reasonable that the range may have been continuous. If, however, there is a questionable occurrence of a taxon at a particular time interval or place, we have sometimes recorded the occurrence of still higher parent taxa as certain for that interval and/or place. Subjective judgment has been necessary in these cases. Ranges of taxa placed *incertae sedis* have been included in the ranges of the higher parent taxa that subsume them (arbitrarily, families and lower, even though this practical regularity violates our view that "families" are not commensurate). In some cases, occurrences of unnamed lower taxa have been reported that extend the range of parent taxa. Such range extensions have been added to those of appropriate higher taxa, generally with an explanatory footnote.

For various reasons, known temporal ranges of taxa that include fossil representatives are not necessarily complete (Benton 1996). If there exists no stratigraphic sequence of rocks of a particular age within a particular area, there can be no actual record, although one might reasonably conclude that a taxon was probably present. Absences, particularly short gaps bracketed by known occurrences, can reflect sampling error of that sort and can be reasonably discounted. As a hypothetical example, one might be reasonably assured of the composition of the continental Eocene fauna of Kansas on this basis, even though there are no Eocene rocks in Kansas and therefore no Eocene fossils. If, on the other hand, a gap in the known record of a taxon should persist after long exploration and successful collecting of many other taxa throughout an otherwise normal stratigraphic section in a reasonably continuous and fossiliferous sequence, one might begin to suspect that particular taxon to be truly absent during the indicated interval, locally or even more widely. Use of negative evidence is procedurally dangerous but not always imprudent.

An additional problem with range data is caused by the paraphyly that remains in the present system, particularly among what are perceived as genera but often at higher ranks as well. By definition, a group above the species level is paraphyletic if it does not consist of an ancestor plus all of the descendants of that ancestor (Wiley 1977; Platnick 1977b:356). Nested taxa or unranked taxa can be used for the various nodes of a morphocline, but if a sequence of identically ranked taxa is created, then all but the least inclusive taxon will be paraphyletic. During the history of mammalian taxonomy a large number of genera have been created as segments of apparently anagenetic lineages, an indefinite number of which are probably real unbranched lineages (although not identifiable individually as such with certainty). In many cases boundaries delineate the application of a name, not of a biological entity. Boundaries between apparently sequential paraphyletic taxa can be arbitrary pseudoextinctions. For this reason such taxa must be (but most often have not been) treated carefully in statistical compilations (Hoffman 1985; Patterson and Smith 1987; Sepkoski 1987).

Compilers of diversity data from classifications such as this one are also warned that some of the temporal divisions used here are not at the highest possible resolution. Moreover, apparent known diversity in any particular temporal slice is artificially increased by a factor roughly proportional to the relative temporal length of that slice (McKenna 1984:474-475; see also Hoffman 1985). True diversity at a given place would be represented by the number of distinct evolutionary lineages that are simultaneously present there at a given time, not the sum total of taxa present at one time or another within a long interval of Earth's history. Additional problems for compilers of diversity are discussed below under the heading "Taxonomic levels."

We continue Simpson's and others' practice of marking extinct or phyletically extinct (Raup and Stanley 1971:98, 293) [= pseudoextinct (Van Valen 1973b:7)] taxa with a prefixed Hawthornean "†" sign. Linnaeus (1758:613), who was not concerned with evolutionary extinction, used the same symbol for organisms that he had not personally seen, such as the (mythical) *Furia infernalis* (*ibid.*:647). Krishtalka (1993:342) has suggested the use of a diagonally ascending arrow for fossil segments of a true anagenetic sequence, but we resist because of difficulties in deciding, at least on the large scale of this enterprise, if an apparently anagenetic continuous sequence is real. Moreover, diagonally ascending arrows are not part of the usual computer keyboard repertory.

We have expanded somewhat upon Simpson's (1945) use of continents as the basis for geographic distribution. Since Simpson's work the entire enterprise of plate tectonics has been integrated by vertebrate paleontologists into their schemes of historical biogeography (e.g., McKenna 1984). Yet, for practical record keeping, that profound revision need not affect the basic descriptive terms, which are based upon present geographic entities. In addition, we have included certain large islands and island groups, and we have used the oceans and seas not only as ranges for marine mammals but also to indicate smaller islands located beyond continental shelves. When occurrences of marine mammals are in epicontinental rocks of a particular landmass, we have tried to reduce ambiguity by noting which coast is involved, e.g., western versus eastern North America. Some of these geographical areas have more significance at certain times than at others.

When the range of a nonmarine taxon is limited to only part of a continent, particularly when that part belongs to a faunal zone different from that of the rest of the continent (e.g., S. As. or N. Af.), that information has been added by footnote, even though the restricted range may be caused by incomplete exploration. Also, continental allegiances may have shifted with time, as in the case of the Arabian Peninsula and India. Moreover, some boundaries used in the literature are arbitrary, both at present and at various times in the past, notably Central America and the subdivisions of Eurasia called Europe and Asia. Footnotes also indicate parts of oceans and seas, and designate island occurrences within them. No doubt, in the future the geographic framework used here will be profitably refined.

Geographic distributions used in this system and their abbreviations are as follows:

> Af., Africa
> Madagascar (not abbreviated)
> Indian O., Indian Ocean
> E. Indies, East Indies
> As., Asia
> Mediterranean, Mediterranean Sea
> Eu., Europe
> Atlantic, Atlantic Ocean
> Arctic O., Arctic Ocean
> N.A., North America
> Cent. A., Central America
> W. Indies, West Indies
> S.A., South America
> Antarctica (not abbreviated)
> New Zealand (not abbreviated)
> Aus., Australia
> New Guinea (not abbreviated)
> Pacific, Pacific Ocean

Abbreviations used for time divisions in this system are as follows:

> E., early
> M., middle
> L., late
> Trias., Triassic Period
> Juras., Jurassic Period
> Cret., Cretaceous Period
> Paleoc., Paleocene Epoch
> Eoc., Eocene Epoch

Olig., Oligocene Epoch
Mioc., Miocene Epoch
Plioc., Pliocene Epoch
Pleist., Pleistocene Epoch
R., Recent (Holocene) Epoch

Bibliographic Information

A frequent comment made to us by colleagues during the preparation of this compilation has been that a complete bibliography would be extremely useful. Although we agree, such a goal is not presently feasible; we provide references for the Introduction, Part I, and Appendixes A, B, and D only. Bibliographies of works published in mammalogy and vertebrate paleontology are sufficiently accessible, and we have provided references to them. The name and date following the name of a taxon are not bibliographic citations. Instead, they are intended to convey the author of the name and the date when it was proposed. Although page number is not necessary in the citation of authorship, we have found such information useful and have provided it when known. Some authors have buried their taxonomic creations in running text or footnotes, sometimes in long papers or books. In other cases the citation of page number helps to differentiate among several works of an author published in the same year.

References to most works in vertebrate paleontology can be found in a long sequence of bibliographies beginning with those of Hay (1902, 1929-1930) and continuing with the *Bibliography of Fossil Vertebrates* published under the editorship of Camp et al. (1940-1972), Romer et al. (1962), Bacskai et al. (1983), and Gregory et al. (1973-1994). Bibliographies appearing in the various compilations of fossil faunas (Li and Ting 1983; Maglio and Cooke 1978; Marshall et al. 1983; Marshall et al. 1984; Russell et al. 1982; Russell and Zhai 1987; Savage and Russell 1983; and Woodburne 1987) are also helpful, as are those in Piveteau (1958, 1961). Comprehensive listings of taxa and references to the literature have been provided by J. L. R. Agassiz (1848), Marschall (1873), Scudder (1882), Trouessart (1897-1905), Palmer (1904), Sherborn (1902, 1922-1933), Schulze et al. (1926-1954), Neave (1939, 1940, 1950), Conisbee (1953, 1960, 1964), Edwards and Hopwood (1966), and Edwards and Vevers (1975). Also useful are Günther (1865-1870) and the *Zoological Record 1870—*. Most large libraries have additional resources, in many cases accessible electronically.

When we have unearthed literature that has escaped the standard bibliographies, or when references are egregiously obscure, we have sometimes provided the bibliographic information in a comments section of our electronic data compilation.

PART I

History and Theory of Classification

Attempts to systematize organisms come and go; none is permanent and there are many kinds (Huxley 1869:1; Gilmour 1940, 1951; Griffiths 1974; Bouquet 1996). See Raven et al. (1971) for a discussion of the origins of taxonomy, and Gilmour (1937, 1940, 1976), Warburton (1967), Nelson (1974), Charig (1982), and Mayr (1995) for divergent discussions of the purposes of taxonomic systems. Systematization provides a general framework that places all other biological inquiry in context. Mayr (1982), Ridley (1986), Rieppel (1988), and Panchen (1992), among others, have given excellent accounts of the early development of such systems. Some systems cut across others, depending on the criteria used for their construction. Even the most useful soon become obsolete.

Most human cultures invent classifications of animals and plants, based upon some set of principles, in order to organize their knowledge of the natural world (Conklin 1962, 1980; Durkheim and Mauss 1903, 1963; Raven et al. 1971; Berlin et al. 1973; Berlin 1992). Many "folk" classifications utilize principles, such as hierarchical structuring and binominal terminal categories, that are cornerstones of scientific systematization (Berlin 1973, 1992; Ellen 1993). Binominal nomenclature, for instance, is in everyday use among certain Papuans (Bartlett 1940). The reason that so much effort is devoted to the occasional preparation of phylogenetic systems like the present one is that scientists and nonscientists alike continually need updated, structured, written compilations of their ever-increasing knowledge of phylogenetic history (O'Hara 1991; 1993). Biological classifications are, basically, written structured data storage systems from which information about systematic hypotheses is retrievable (Mayr 1968; Cracraft 1974:79; Hull 1980:437).

Phylogenetic hierarchy results from a combination of descent with modification and lineage branching (Darwin 1859; Mayr 1965; Wiley 1975). Pictures of hypothesized branching lineages, or of nested subsets within sets of taxa, or of logical hierarchies of shared derived characters, can depict this structure well (Augier 1801; Lamarck 1809, vol. 2:421; Strickland 1841:190; Lam 1936; Voss 1952; Yates 1982; Stuessy 1983; Craw 1992; Simonetta 1995; Simonetta 1995; Valentine and May 1996; Bouquet 1996). Witness the phylograms and cladograms that have illustrated an always increasing phylogenetic literature since even before the famous hypothetical branching diagram published by Darwin (1859) or the oft-republished evolutionary tree (anthropocentrically culminating in our own species) given by Haeckel (1866). To a degree, such branching is self-similar at any scale, and therefore apparently fractal (Minelli et al. 1991; Schander and Thollesson 1995), but stochastic factors keep this resemblance from being complete or fully accurate. Written classifications attempt in words to give the same information that is conveyed in diagrams. The structure of hierarchical systems of words, whether in downward classification in the Scholastic sense (Cesalpino 1583; Bicheno 1827:489; Wallace 1856:195; Cain 1959a; Stearn 1959:16; Mayr 1982:162-163; 1995, but not Mayr 1974:105), or in upward classification as we attempt here, is well suited to depict the hierarchy of nature.

We take it as self-evident that the reason hierarchical systems depict cladistically hypothesized relations so well is that evolution has occurred, and that systematic analysis of its natural results can lead to "knowledge of phylogeny" (not the other way around: see Patterson 1982b:284). The apparently anagenetic and branching lineages that we sample lie wholly behind an advancing present defined by our own existence. Their branching structure is not known with certainty in any case. However, we attempt to infer the branching structure from studies of extant organisms alone, or from both extant and fossil organisms, or from fossils alone. In agreement with Brady (1982: 290), we take it as irrelevant that "transformed cladists" (Platnick 1980) or "pattern cladists" (Nelson and Platnick 1981; Hull 1984, 1989) choose to dissociate their phylogenetic theories from evolution or just from particular evolutionary process theories, such as Darwin's, about underlying causes (see also Beatty 1982; Charig 1982; Ball 1983). Without an evolutionary context it would seem pointless to polarize characters. The challenge is to communicate in words a depiction of phylogenetic branching and descent with modification (Darwin 1859; Huxley 1880; Bock 1974:376, 1977; Cracraft 1981:459; Wiley 1975, 1981:72; Ball 1983:446; de Queiroz 1988; de Queiroz and Donoghue 1988:325).

Evolution is dynamic and implies lineage continuity, anagenesis, and branching, processes that are difficult to deal with in spoken discourse and prose. Moreover, sentences of spoken words are completely linear, like a stream of Morse code. Even written text is mainly linear, although information not contained in the meanings of the words themselves can be conveyed by indentation, capitalization, typography and other forms of emphasis or connotation. To describe the structure of a tree fully in words, one must deal with one branch at a time, sequentially until the job is done. After following a branch of a fork to its tip, one must go back to the fork and resume description along the next branch. As Simpson (1959a:295) once noted, names cannot be written *within* one another. But the same information in a picture or sculpture is accessible at a glance (Wiley 1980:79). One sees branches of branches in two or even three dimensions and we can imagine them in four. Thus, we agree with Gingerich (1980:451, 463) that "phylogeny itself is better expressed in a diagram than in a formal classification."

But diagrams are not always handy, especially if they deal with thousands of low-level or terminal taxa. Although we have, in fact, developed such a diagram, we are forced to store it electronically or in stacks of paper and can view only part of it at a time. Despite all their faults, more or less linear sequences of words, either spoken or written, but not diagrams, remain the basis of most communication regarding evolutionary modification and branching. As Clark (1956:9) wrote, "The problem is essentially one of language. Descriptions of species are verbal and words can only be used to describe discontinuous processes, since the words and the things they signify must be defined and to do that it is necessary to give them boundaries and limits."

We do not think that *things* can be defined (see Aristotle's failed attempt to define swans; Ghiselin 1966a:127; and Schander and Thollesson 1995:264) but, like Clark, we have not found a way to communicate fully our phylogenetic hypotheses using words alone, practically and with stability. The present version of mammalian higher-level systematics, like all others, is approximate, not the equivalent of a 1:1 map of phylogeny. Our effort is meant to serve principally as an elaborately organized, practical, interim housekeeping operation. It places what are hypothesized to be genealogically related evolving taxa into a hierarchical model. Recently, the relationships among taxa and their names have been explored in great detail (de Queiroz 1992; de Queiroz and Gauthier 1994).

Hierarchical Subordination
Early concepts

The word "classification" was not part of the scientific literature until the last decades of the eighteenth century. The first use of which we are aware occurs in a botanical paper by the Marquis de Condorcet (1777: 35). After that it was used in France as a noun of action denoting the act of arranging into logical classes, and as a noun to denote the result of such activity (Mayr 1968:546). Lamarck (1799:64-65; 1809) used the word, but it does not appear in Lamarck's (1786) encyclopedia article on classes.

The verb "to classify" was introduced to English in 1799 as a synonym of "to class," which had wide prior use. Adanson (1757; 1764) and other systematists had used terms like "méthode" or "natural method" for the act of classification, and "système" for the results, recalling Ray's (1682) "Methodus Plantarum" (Payer 1844; Stevens 1994). Differences between these early terms were discussed by Jussieu (1848). Of course, hierarchical classification in ordinary human affairs goes back to time immemorial; it is especially characteristic

of civic and military organizations since the time when human societies grew larger than extended family groups.

The root of "class" and "classification" is the Latin word "classis," used by Linnaeus (1735, 1758). A pertinent usage of "classis" is traceable at least to 400 B.C. and perhaps to the sixth century B.C., when semilegendary King Servius Tullius divided Roman citizens into six classes according to their wealth. This was important because in those days soldiers supplied their own equipment, of variable quality and expense. These Roman classes were hierarchically subdivided into centuries, units of 100 men:

Romans
　　Class
　　　　Century
　　　　　　individuals

In 359 B.C., Philip of Macedon organized the hoplites of his infantry into subordinated groupings, based on powers of two, as follows:

Grand phalanx 32,768 = 2^{15}
　　Division (simple phalanx) 4,096 = 2^{12}
　　　　Regiment (chiliarchia) 1,024 = 2^{10}
　　　　　　Battalion (syntagma) 256 = 2^8
　　　　　　　　Company (taxiarchia) 128 = 2^7
　　　　　　　　　　Platoon (tetrarchia) 64 = 2^6
　　　　　　　　　　　　individual hoplites

Also in the fourth century B.C., the Roman dictator Marcus Camillus introduced the following hierarchically ordered military organization:

Legion (3,000 men; after 105 B.C. 6,000 men)
　　Cohort (cohors; originally 5 per legion, later 10)
　　　　Maniple (3 per cohort)
　　　　　　Century (2 per maniple)
　　　　　　　　individual soldiers (100)

In the classification of angels ("celestial hierarchy") envisioned in a work by "Dionysius the Areopagite" (Pseudo-Dionysius), written in the fifth century A.D. and popularized in Latin by Scotus Erigena (i.e., the Irishman) c. 858 A.D., a sequence of nine orders (groups) were arranged in three hierarchies: seraphim, cherubim, thrones; dominions, principalities, powers; virtues, archangels, angels. However the structure of this hierarchy is far from clear.

Numerous other ecclesiastical and military hierarchies could be cited, drawn from the world's many cultures, but those given above will suffice (but see Croneis 1939). Their organization, and even some of their terms, were to influence formulation of later biological hierarchies. Notice that the subordinated terms above individual people are not the same as the various ranks of soldiers from private to field marshal; one cannot subdivide a colonel into two majors, although majors report to colonels. It is the power of majors that is subservient to the power of colonels. In this sense, colonels are individuals like prime numbers, indivisible, whereas categories potentially or actually comparted into aliquot subcategories are like positive factorial numbers: they are wholly subdivisible without remainder. Thus they can be either built upward or divided downward.

The distinction between specification (relational) and scalar (functional) hierarchy has been explored by Salthe (1988, 1989, 1991). Notice also that, as with Tullius' Romans, the categories are not necessarily dichotomous. Hierarchies like those described above follow the nontrivial Dirichlet box principle (pigeonhole principle of computer scientists): if p objects occupy q compartments, and p > q, then at least one compartment contains more than one object (Dantzig 1954:322). One might also note that if there is more than one example of the highest category used, then the necessity of a still higher category is implied if a complete system is desired. For additional information about early attempts to establish hierarchies, we recommend various articles contained in Papavero and Llorente-Bousquets (1993, 1994a, 1994b, 1994c) and Papavero et al. (1994).

It is more difficult to trace the notion that a classificatory level in biology can be a member of a hierarchy of taxonomic levels of upwardly increasing generality or downwardly increasing specificity, and that such a hierarchy can be further subdivided, subsumed, or interleaved. We will not do so in full detail, but we note that in biology there existed pre-Aristotelian forerunners of Linnaeus, long before the Middle Ages. Gregory (1910), in his excellent historical treatment of the early history of classification, noted that the Assyrians of the seventh century B.C. or earlier had even worked out a form of hierarchical classification.

Identification sieves

Contrary to a popular humorous misconception (Genesis 2.19, 20; Ussher 1650, 1658; Hedgpeth 1961; Mayr 1988:42; Minelli 1993:4), the practice of taxonomy is not actually humanity's oldest profession. The first human "taxonomist" claimed by these authors was the biblical Adam. However, Adam merely named organisms; there is no evidence that he attempted to classify them. Taxonomy is more than the mere giving of names; it also involves arranging them in a scientific manner. We note that Adam's work in the Fall of 4004 B.C. (Gould 1991) was conducted before Eve was approached by the world's first scientific advisor. The practical significance of early systematics can be noted, however, in the Biblical scheme of "clean" and "unclean" animals of Leviticus, XI (Cain 1993). Many human cultures have similar dichotomous schemes.

A very dim light did shine for a while in the third century A.D. with the popularizing of the Porphyrian scale ("Ramean tree" or "tree of Porphyry," actually an identification key or sieve), formulated by the Neoplatonic philosopher Porphyrius (c. 234-304). Porphyrius, a student of Plotinus (Nelson and Platnick 1981; Mayr 1982:159; Papavero and Llorente-Bousquets 1994d), has been said to have put forth the concept in an introduction to a work, "The Categories," ascribed to Aristotle (Thompson 1952:3; Owens 1960-61; Moravcsik 1967; Slaughter 1982:270). Actually, the notion dates back at least to Plato's *Sophistes* and *Politicus* (see Hopwood 1959; Panchen 1992:18). The Porphyrian scale was a branching dichotomous key having mutually exclusive adjectival differentia for its general categories. Such differentia were antonymous. The method operated by logical division (Cain 1958, 1959c; Hopwood 1959; Balme 1961, 1962; Osborne 1963; Pellegrin 1982, 1986; Ridley 1986:101; Panchen 1992:109), familiar to us in the parlor game "twenty questions." The antonymous Porphyrian identification key or sieve operated under the principle embodied in the simple Boolean formula $x^* = x$, or $x(1-x) = 0$, where x is an entity diagnosed by some adjective, * represents self multiplication or squaring, 1 is the category of everything, and 0 is the category of nothing (Boole 1854). The usual example given is the following one (e.g., Nelson and Platnick 1981:74), here supplied with Boolean abstractions given in parentheses:

Genus Substance
   Incorporeal (1-c)
   Corporeal (c)
      Genus Body
         Inanimate (1-a)
         Animate (a)
            Genus Living
               Insensible (1-s)
               Sensible (s)
                  Genus Animal
                     Irrational (1-r)
                     Rational (r)
                        Genus Man
                           Socrates
                           Plato
                           other individuals

The term "Genus" here translates, as in other Scholastic philosophy, to "general category whose rank is unassigned." However, in Scholastic philosophy there was nonetheless an overarching genus known as the "genus summum" (e.g., Linnaeus 1737a: 138). "Lower genera" could be species of "higher genera." In the oft-quoted sample given above, Socrates, Plato, and others are identified as individual instances of corporeal, animate, sensible, rational men, but here the key is only a way of seeking out by the method of logical division a particular man, e.g., Socrates, among all natural objects. Besides the conflation of phenomena with noumena, notice that in this example the method is inconsistent when individual men are finally reached. For consistency, Genus Man should have been divided into antonymous categories such as "not Socrates" (1-Soc) vs. "Socrates" (Soc), and so forth. The maximum number of additional subdivisions necessary to sieve out a particular individual, say Socrates, would depend on the number of characters necessary for a person not familiar with Socrates to single him out. The minimum number is one, a shortcut available to Xanthippe and close associates of Socrates, but not to the rest of us except by chance.

The key becomes natural only when a large number of bifurcations is reached and successive sieves are in the correct order. In that case the sequence of bifurcations might or might not represent the history of character acquisition, but that was not of interest to nonevolutionists. The chance of sieving out an individual in one random try in such a case is 1/n, where n is the number of characters necessary to characterize an individual uniquely. In this case, however, and all others with less than the correct number of sieves, the correct or full history would not be recovered. The chance of randomly getting the history right is minimally $n!^{-1}$ (the reciprocal of factorial n). Various early classifications or keys (e.g., Wilkins 1668, incorporating work by John Ray and Francis Willughby; Morison 1672; Ray 1682) and even some more modern ones (cited by Nelson and Platnick 1981) have been called Porphyrian. That of Owen (1859) is a good example. Even some "cladograms" have been erroneously constructed along Porphyrian lines; they can often be recognized by such features as single ad hoc characters at nodes, polarity problems, purported 100% consistency, and by lack of a supporting character matrix from which one might be able to assess the effects of missing data. Numerical pheneticists have also dealt with such identification schemes (e.g., DuPraw 1964).

All of the examples we have mentioned so far are representative of the method of logical division, called downward classification by Stearn (1959) and Mayr (1982:158, 1995). Given a sufficient number of steps, one can always identify something, no matter how inefficient the sieving scheme, but the branching structure of the scheme need have nothing to do with the sequence of phylogenetic branching. Although Aristotle's name is often associated with logical division, he himself believed that it did not apply to natural classification (Hopwood 1959:231; Lloyd 1961; Balme 1961: 196; Duffin 1976). See Voss (1952) for further discussion of logical division and Pankhurst (1991) for a discussion of automated identification using computers.

Upward classification

The Italian botanist Andrea Cesalpino (Andreas Caesalpinus, b. 1519, d. 1603, private physician to Pope Clement VIII) developed a hierarchical classification of plants (Cesalpino 1583; Gregory 1910:15; Bremekamp 1953a; Mayr 1982:158-162). In 1583 he remarked, "In this immense multitude of plants, I see that want which is most felt in any other unordered crowd: if such an assemblage be not arranged into brigades like an army, all must be tumult and fluctuation" (Gregory 1910:16, from Whewell 1837:280). In his classification Cesalpino arranged 840 kinds of plants in 15 classes (Adanson 1764:ix). This was a harbinger of upward classification, the combining of smaller categories into larger ones (Jussieu 1789; Darwin 1859; Stevens 1994) rather than attempting to subdivide large categories by the method of logical division (Cain 1958; Stearn 1959:16).

A century after Cesalpino, Tournefort (1694) treated plants as 22 main classes built upward from 10,146 species placed successively in 698 genera and 107 sections (Sloan 1972:39). His contemporary Ray (1693) continued to classify by means of keys (see Sloan 1972, for other examples), although as early as 1669 Ray had realized their artificiality (Nelson and Platnick 1981:76-77). Ray's classifications in the last decade of the seventeenth century mixed downward and upward classification (Stearn 1959:16; Mayr 1982:162-163, 1995), but they foreshadowed Linnaeus's binominal nomenclature by the use of a general and specific term. These terms could be used at any level, however, and the second term was still merely adjectival, e.g., "Ungulata dichela" for all artiodactyls or "Ovis domestica" for domesticated sheep. By the middle of the eighteenth century, biological hierarchical classification was in wide use, although the consistent use of binominal nomenclature throughout a work [see Article 11 (c) of the ICZN] was still to take hold.

It seems to us that the main contribution of Linnaeus's works for later zoological evolutionists was not his consistent use of binominal nomenclature for species (10th edition, 1758), nor the rules he introduced (Linnaeus 1751, 1758; Gregg 1954:30), nor even his reduction of common names of various languages to a form of Latin (Stearn 1955), but rather his massive elaboration of the essentialistic hierarchical method of biological classification. The well developed hierarchical features of Linnaeus's system were already in place in the 1st edition of his *Systema Naturae* (Linnaeus 1735). Since receiving a copy of Aristotle's *Historia Animalium* in childhood, Linnaeus had been imbued with the orderly hierarchy of nature, stemming from a faith in divine organization. Linnaeus's hierarchy was a simple one, with only six levels, lacking prefixes, suffixes, or inflexions to aid in memorizing the sequence of ranks. Such linguistic devices were not yet needed.

In the non-binominal 1st edition of his *Systema Naturae*, Linnaeus (1735) configured zoological taxa in six classes. There were five orders of class QUADRUPEDIA (= MAMMALIA of the 10th edition, 1758) and thirty-three mammalian genera (with four varieties of *HOMO* denoted by two adjectives each). The 2nd edition (Linnaeus 1740), with thirty-two genera, was similar. From 1740 onward, Linnaeus also provided each genus with a sequential number, but these numbers were not the same from edition to edition. Linnaeus's 1740 arrangement of mammals (as QUADRUPEDIA) was as follows:

REGNUM ANIMALE. CONSTAT VI. Classibus
CLASSIS I. QUADRUPEDIA
  Ordo 1. ANTHROPOMORPHA
      1. *HOMO*
          var. Europaeus albus
             Americanus rubescens
             Asiaticus fuscus
             Africanus niger
      2. *SIMIA*
      3. *BRADYPUS*
      4. *MYRMECOPHAGA* [new addition since 1735]
  Ordo 2. FERAE [lacking *Hyaena* as a separate genus]
      5. *URSUS*
      6. *LEO*
      7. *TIGRIS*
      8. *FELIS*
      9. *MUSTELA*
     10. *DIDELPHIS*
     11. *LUTRA*
     12. *PHOCA* [including *Odobaenus* of the 1st edition]
     13. *CANIS*
     14. *MELES*
     15. *ERINACEUS*
     16. *TALPA*
     17. *VESPERTILIO*
  Ordo 3. GLIRES
     18. *HYSTRIX*
     19. *LEPUS*
     20. *SCIURUS*
     21. *CASTOR*
     22. *MUS*
  Ordo 4. JUMENTA
     23. *ELEPHAS*
     24. *HIPPOPOTAMUS*
     25. *SOREX* [a member of GLIRES in the 1st edition]
     26. *EQVUS*
     27. *SUS*
  Ordo 5. PECORA
     28. *CAMELUS*
     29. *CERVUS*
     30. *CAPRA*
     31. *OVIS*
     32. *BOS*

In the 6th edition, Linnaeus (1748) increased the number of "quadruped" orders to six, renumbering them: I. ANTHROPOMORPHA; II. FERAE; III. AGRIAE; IV. GLIRES; V. JUMENTA; and VI. PECORA. There were now 34 genera, also mostly renumbered. *Myrme-*

cophaga was moved from ANTHROPOMORPHA to AGRIAE. *Manis* was added to the classification, placed in AGRIAE as well. *Leo* and *Tigris* were lumped under *Felis*. *Didelphis* was transferred from FERAE to GLIRES. *Rhinoceros* was placed in order JUMENTA and *Moschus* was added to the PECORA.

In all editions of his *Systema Naturae*, Linnaeus arranged his categories systematically as a brief series of nested sets derived from Scholastic philosophy (e.g., genus summum, genus intermedium, genus proximum, and species; see Stafleu 1969:28). Instead of placing genera within genera, however, he clearly placed genera within orders and used the set "order" as a subset of the set "class," itself a subset of the set "kingdom." The nesting was derived from logical division (downward classification of Stearn 1959, and Mayr 1982, 1995). As Linnaeus (1751) remarked (in translation), "An order is a subdivision of classes needed to avoid placing together more genera than the mind can easily follow" (Mayr 1982:175).

Linnaeus himself was for most of his life a strict creationist (e.g. 1751: "Natura non facit saltum") and essentialist, although he modified his creationist views in the 12th edition (Linnaeus 1766). He arrived at his results by going from the accepted general to the particular. However, even though only a few levels were employed, the systematization of levels of generality adopted by Linnaeus was potentially convertible to upward expression of hierarchical evolutionary relationships as well (Darwin 1859). Although Linnaeus was engaged in logical division (Cain 1958, 1959a, 1959b, 1993), his hierarchical system is serviceable whether one regards lower categories as subdivisions of higher ones, or higher ones as brigades of lower. Logical division is especially suitable for exploring the ramifications of what was proposed and then not questioned (i.e., "revealed") as the plan of God, but brigading builds up to a testable scientific model of reality. For Linnaeus, hierarchical information retrieval evidently was a matter of identification of God's plan, lacking the deeper rationale that later was to be provided by evolutionists. For discussion of the psychological factors involved, and for statistical analysis of the structure of large classifications, see especially Williams (1950) Minelli et al. (1991), and Holman (1985, 1992). For treatment of Linnaean classification in terms of set theory, see especially Woodger (1937), Gregg (1950, 1954, 1967, 1968), Parker-Rhodes (1957), Simpson (1961), Hull (1964, 1989), Buck and Hull (1966), Jardine (1969), Abe and Papavero (1992), and Papavero and Llorente-Bousquets (1993, 1994a).

In upward classification, taxa are defined by their subsumed taxonomic content. Thus order X might be defined as the sum of suborders Y and Z. Expanding this, a system can be built up from the lowest taxonomic level (species-group), combining a large number of taxa into a hierarchically multicategoried taxonomic storage system or classification.

Taxonomic levels

There has been a fairly steady increase in the number of levels of the taxonomic hierarchy since the approximately half dozen employed by Linnaeus (1735, 1740, 1748, 1758) to cover both the animate and inanimate world. As adumbrated in 1735 (text, but not in the table headings or labeled columns) and as used in 1740 and 1748, Linnaeus (1758) employed:

EMPIRE (IMPERIUM)

KINGDOM (REGNUM)

CLASS (CLASSIS)

ORDER (ORDO)

GENUS

Species

Variety

These categories were placed in an ordinal and also hierarchical (here indented) sequence that started at the most general level with the universe (IMPERIUM NATURAE) and ended with species and varieties, above the nontaxonomic level of single specimens. Each category except the highest (root) was a subset of the next higher category. Among mammals, varieties were used only for *Homo sapiens*. Families were not yet used. In 1758 Linnaeus used upper-case roman letters for names of genus-group and higher, upper-case and lower-case roman type for trivial names, and upper-case and lower-case italic type for named varieties. Modern writers use a somewhat different style, as had Linnaeus himself in earlier editions. Nevertheless, it is interesting that in the typography of the taxa of his hierarchy, Linnaeus treated taxa above the rank of species differently from taxa of the species-group. No doubt, this is related to his belief that categories are absolute, and that the various levels are qualitatively different, being recognized by essential characters (Stearn 1960). Nonetheless, a lowest-ranked species-group taxon cannot represent the sum of still lower-ranked taxa, because it is made up only of individual organisms, not taxa (Schander and Thollesson 1995).

No one has found the root term "empire" very useful; it is simply a synonym of "highest level," i.e., everything. In Boolean terms, this would be unity, or "1." Moreover, the three kingdoms, animal, vegetable, and mineral, have minimal usefulness, especially now that in post-Linnaean literature animal and plant fossils have been transferred out of the mineral kingdom. However, most people seem to be familiar with or even use the term "animal kingdom." Several of Linnaeus's (1758) six named taxonomic classes of kingdom ANIMALIA have survived the test of time fairly well, although their ranks and taxonomic content have changed somewhat. We deal here with just one of them, the taxon named MAMMALIA, ranked by Linnaeus as a taxonomic as well as a logical class. Within that class, Linnaeus (1758) recognized eight mammalian

orders, which were given mononominal proper names. These orders were subdivided into thirty-nine mononominal genera (whose proper names are given here with modern typography but original orthography). These genera were further subdivided into a larger number of species and varieties that we do not list.

Class MAMMALIA
  Order PRIMATES
    Homo, Simia, Lemur, Vespertilio
  Order BRUTA [including AGRIAE of 1748]
    *Elephas, Trichechus, Bradypus,
    Myrmecophaga, Manis*
  Order FERAE
    *Phoca, Canis, Felis, Viverra,
    Mustela, Ursus*
  Order BESTIAE
    *Sus, Dasypus, Erinaceus, Talpa,
    Sorex, Didelphis*
  Order GLIRES
    *Rhinoceros, Hystrix, Lepus, Castor,
    Mus, Sciurus*
  Order PECORA
    *Camelus, Moschus, Cervus, Capra,
    Ovis, Bos*
  Order BELLUAE [also spelled BELLUA]
    Eqvus [sic!], *Hippopotamus*
  Order CETE
    Monodon, Balaena, Physeter, Delphinus

In Linnaeus's 12th edition (1766), the taxon BESTIAE was dropped and its genera redistributed. *Rhinoceros* was removed from GLIRES, but this improvement was offset by the unwise addition to GLIRES of the bat *Noctilio*, the only new mammalian genus proposed by Linnaeus in 1766.

Class MAMMALIA
  Order PRIMATES
    Homo, Simia, Lemur, Vespertilio
  Order BRUTA
    *Elephas, Trichechus, Bradypus,
    Myrmecophaga, Manis, Dasypus*
  Order FERAE
    *Phoca, Canis, Felis, Viverra,
    Mustela, Ursus, Didelphis,
    Talpa, Sorex, Erinaceus*
  Order GLIRES
    *Hystrix, Lepus, Castor, Mus,
    Sciurus, Noctilio*
  Order PECORA
    *Camelus, Moschus, Cervus, Capra,
    Ovis, Bos*
  Order BELLUAE
    Eqvus [sic!], *Hippopotamus, Sus,
    Rhinoceros*
  Order CETE
    Monodon, Balaena, Physeter, Delphinus

In the essentialist and creationist world-view displayed in the first ten editions of his *Systema Naturae*, Linnaeus thought that taxa neither changed nor branched. Nannfeldt (1958:8) quoted an early example of Linnaeus's creed: "Species tot numeramus quot diversae formae in principio sunt creatae" ["We reckon as many species as there were diverse forms created in the beginning"]. The various categories (ranks; see Mayr 1969:413) used were regarded as having essential characters peculiar to the particular level used. Genera were created by God, and it merely remained for mortals to perceive them via their essential characters. Variation was just noise, masking a deeper essence. Likewise, extant species were *the* species created a few thousand years ago, and were immutable. Other than their mysterious creation, they had no significant history of modification. They were non-dimensional (Mayr 1949). However, in the 12th edition Linnaeus (1766) no longer insisted on "nullae speciae novae" [grammatically incorrect, but = no new species] (Osborn 1894:130; Bremekamp 1953b).

J. L. R. Agassiz (1859:233ff) and other essentialists also claimed that there are characters for "absolute" taxa ranked as genera, orders or classes as such. Even today, some would affirm that genera or suborders (as ranks) are more than sets and subsets (Boucot 1995). However, we use them strictly as subordinated levels of an ordinal sequence that can be remembered by ourselves or our computers, so that we can communicate more precisely about nested sets of organisms verbally or in prose. For us, they have no absolute value.

It is important to recognize that what we classify as taxa ranked as genera, families, and other higher categories are not commensurate with others of the same ranks (Simpson 1953:33; Gauthier, Kluge, and Rowe 1988a; de Queiroz and Donoghue 1988; Minelli 1991:186, 1993:41, 1995:305). Is a family-level category of rhinoceroses equivalent to a similarly ranked one of capybaras or of muroid rodents? What does "equivalent" mean? Van Valen (1973a) and DuBois (1988) attempted to answer these sorts of questions, but we regard their efforts as unsuccessful. Van Valen seems to have been discussing the human tendency to group entities of various hierarchical ranks into dozens, tens, sevens, etc., rather than dealing with something real "out there" in nature. Taxa are individuals (Ghiselin 1966b, 1975, 1977, 1981), but their ranking is relative and hierarchical, not absolute (Løvtrup 1987), and they are noumena, not phenomena. Compilers of diversity data based on classifications have much to learn concerning these matters.

Knowledge of taxa at the lower taxonomic levels has changed profoundly since Linnaeus's time for two reasons. First, because of new discoveries, real knowledge of animate diversity has increased at an almost exponential rate since Linnaeus's 10th and 12th editions of his *Systema Naturae*. This is especially true regarding mammals because of paleontological contributions.

Secondly, there has been what often is called taxonomic inflation [see MacLeay (1829) or Newman (1833) for early approval or complaints]. As Gill (1908:453) noted long ago, one of Linnaeus's original genera, *Vespertilio,* in which Linnaeus at first placed all bats, is now just one among the 219 currently recognized genera of the taxon CHIROPTERA. Moreover, as the details of anagenesis and branching structure have become better known since Linnaeus, many new hierarchical levels have been found useful and have either received names or have been employed without them.

Had Linnaeus not placed fossils in the mineral kingdom, presumably he would have imputed them to extant species, whether known or not yet known, each with a temporal range extending over the whole fossil record. In each case, the potential record would have been thought to start about 6,000 years ago. A hundred years later, J. L. R. Agassiz (1857, 1859) still held such a world view, and it persists today among creationists of various stripes. In contrast, the present evolutionary world-view sees the extant species of the arbitrary present (a time determined by our existence) as photographic snapshots taken of evolving and branching lineages that have a complex past and an uncertain future (Burma 1954). Likewise, fossil species are snapshots of what are hypothesized to be earlier stages of the evolution of such lineages or close relatives. There are numerous difficulties with simultaneous treatment of both fossil and extant representatives of these arrays, but we hope to show by example that these problems are not insurmountable in the construction of a useful system that deals with both. Furthermore, the vast and expanding body of paleontological evidence adds a major dimension to the systematic edifice and in a broader framework profoundly tests its phylogenetic hypotheses.

Following Cuvier's numerous works (e.g., 1817) that treated both fossil and extant taxa, the number of extinct taxa has expanded. Most mammalian taxa are now known from fossils, and most of these exclusively so. In the present classification 4075 extinct mammalian taxa are assigned generic rank, about 79% of the total of 5158 mammalian taxa given that rank in our classification.

Curiously, several centuries after Linnaeus, some authors have again advocated separate classifications of extant and extinct organisms (e.g., Hennig 1950; Bigelow 1961; Crowson 1970; Løvtrup 1973). Crowson's (1970) proposal to classify fossils separately from their extant relatives would be a special case. Crowson (1970) suggested using numbers as prefixes for taxa depending on their geological age (e.g., 5 for Jurassic, 4 for early Cretaceous, 2 for Eocene, etc.), but such taxa would be highly unstable because of geochronological vicissitudes. Griffiths (1976) even suggested that we might adopt paleotaxa, eotaxa, oligotaxa, miotaxa and pliotaxa for classifying the fossils of the various Tertiary epochs (a notion long before put forth by Gadow 1898:vii). The result would be worse than the changing of suffixes of taxa shifted in rank, dictated by the legalism of the ICZN. To his credit, Griffiths pointed out how impractical such a scheme would be. See Patterson and Rosen (1977:155-157) for further discussion of these proposals. On any scale, proliferation of ranked categories caused by the analysis and inclusion of fossils in classifications is obviously threatening to those who wish not to be bothered by fossils and the problems they cause for classifications built on extant organisms alone (Patterson and Rosen 1977; Nelson and Platnick 1981:273-274; but see Delson 1977; Huelsenbeck 1992). The present classification, however, should demonstrate that extant and fossil organisms can and need to be classified together (see also Mayr 1981:513), even though serious unresolved flaws remain.

To enrich the hierarchical structure, writers following Linnaeus have interleaved many additional categories into the Linnaean hierarchy (e.g., de Candolle 1818-1821), most notably the phylum (Haeckel 1866) and the family [Adanson 1757, 1763-1764; Jussieu 1777, 1789; Lamarck and Brisseau de Mirbel 1803; Illiger 1811 (not based on type genera); Cuvier 1817]. A number of prefixed "super," "sub," and "infra" categories like superfamilies, subgenera, and infraorders have been added as well (e.g., Bicheno 1827; Strickland 1834; Gill 1871, 1872b, 1896; Simpson 1945; Boudreaux 1987). Such prefixed cardinal names also connote ordination. Botanists utilize five categories between the ranks of genus and species: subgenus, section, subsection, series, and subseries; zoologists like Boudreaux have used sections and subsections at much higher levels in the hierarchy. Not all proposed interpolations at the various levels are listed here, but a complete list would be substantial. Many proposals have long been forgotten, such as the missus and coetus proposed by Storr (1780) or the caterva of Bleeker (1859:xvii). Moreover, like those of Boudreaux, the categories employed are often different in classifications of various different taxa. One can find discussions and incomplete lists of such categories in Dall (1878), Gill (1896, 1898), Schenk and McMasters (1936, 1948, 1956), Hargis (1956), Simpson (1961), Crowson (1970), Farris (1976), and Ax (1987).

Families as formal taxonomic levels were first introduced by botanists [Magnol 1689 (pre-Linnaean); Adanson 1757, 1763-1764; Jussieu 1777, 1789]. Jussieu (1789) utilized exactly 100 families of plants, a convenient but artificial number. These were judged useful in grouping more than a thousand taxa of plants then recognized as genera. The number of named taxa of plants given generic rank had grown steadily since compilations by Tournefort (1700-1703) listed 698 genera. Linnaeus (1737b) listed 935 genera, and later the number of Linnaean plant genera rose to 1,336. In his 1798 work Cuvier did not utilize families for vertebrates. However, following their introduction to zoo-

logical literature by Latreille (1796), families rapidly entered the taxonomic nomenclature of mammals, insects, and birds (Illiger 1811; É. Geoffroy Saint-Hilaire 1812; Kirby 1815:88; Cuvier 1817). At first, families were more or less interchangeable with orders or other categories (Croizat 1945:68), but in zoological nomenclature they gradually became the most important named level intermediate between the order and the genus (Mayr 1982:174-175).

Certain taxonomic categories came to bear prefixes suggestive of special hierarchical linkage (e.g., in the family-group, subfamilies are always subsumed in families). Others did not (e.g., tribes might alternatively have been dubbed "microfamilies" or some such term connoting subordination). Use of a prefixed category implies that the category to which the prefix applies is also used. In the Linnaean System we do not construct superfamilies directly from subfamilies without also employing families. Logically, however, above the species level there is nothing special about sub- or supercategories. They all could have received unprefixed cardinal names or simply be referred to as taxa. Such names are, after all, just labels (recognition symbols) (Mayr 1953:391). That prefixed categories did not receive unprefixed cardinal names, free of reference to another rank, seems to us to be partly a matter of practicality and memorability, and partly a function of their authors' essentialistic belief in the objective reality (beyond a construct of human language) and commensurability of various examples of such taxonomic levels as classes, orders, families, and genera (see Slaughter 1982). We employ prefixed names for the sake of stability, because they have long been in use, but we do not hesitate to allocate to an *incertae sedis* position some taxa whose names happen to be prefixed. For reasons of stability we might not wish to change their rank or to list the lower-ranked contents but not the valid but prefixed monophyletic taxon containing them.

With the proliferation of ranked categories that had increased steadily from Linnaeus's original six, came also a perceived need to encode the names of taxa themselves as signifying that the taxa for which they stood were members of some particular rank. In each particular discipline, the names of family-group taxa came to have various standardized inflected suffixes linked to the perceived rank. Thus, in zoology, a name ending in "-idae", signifies a taxon at family rank. Latreille (1796), who introduced the family category to zoology, did not use the suffix "-idae". That modification was provided later by Kirby (1815), and has not only stuck but is now legislated by the ICZN. We think of these inflective conventions as part of the "Linnaean System" but, in fact, they are arbitrary post-Linnaean additions to it, originally added for their mnemonic usefulness but now the occasion of much pedantic drudgery whenever taxonomic rank is changed or organisms are transferred from one kingdom to another (see Appendix D).

Like the number of taxa given generic rank, the number of mammalian taxa found useful at the rank of family has grown over time. Gill's (1872a) classification of mammals included 138 taxa assigned to family rank, of which 33 (24%) were extinct. Simpson's (1931) first classification of mammals contained 242 families, of which 129 (53%) were extinct; his second classification (1945:35) tabulated 257 families, of which 139 (54%) were extinct. Although we do not take such statistics very seriously, in the present classification we allocate 425 taxa of mammals to what we find useful as "family rank"; of these, 300 families (71%) are extinct. We employ 46 taxa ranked as mammalian orders.

In addition to families, other levels have become useful since Linnaeus. Whereas only orders, genera, species and varieties were used below the rank of class by Linnaeus in 1758, nearly two centuries later Simpson (1945:15) listed sixteen taxonomic levels used frequently by various taxonomists dealing with vertebrates below the rank of class:

Subclass
Infraclass
Cohort
Superorder
Order
Suborder
Infraorder
Superfamily
Family
Subfamily
Tribe
Subtribe
Genus
Subgenus
Species
Subspecies

Simpson's 1945 classification of mammals did not reach the level of species, but above that level he used all but one of the categories listed above. Notice that most of the names of ranks are prefixed. Using several formats, Simpson also formally listed or mentioned about 100 taxa given subgeneric rank within taxa listed as genera of rodents, carnivorans, proboscideans, perissodactyls (*Equus* only), and artiodactyls. In certain instances, notably for the cats *Felis* and *Panthera*, and for the canid *Dusicyon*, subgenera were listed hierarchically in an indented position. Simpson's criteria for selection of genera in which subgenera received some sort of formal treatment, whereas others did not, were not explained in 1945, although he (Simpson 1943) had recently dealt with the generically oversplit taxa of cats. It is evident in Simpson's (1943:150) theoretical work that he thought increased use of subgenera in the taxonomic hierarchy would be desirable. In most cases, however, subgenera of other authors were not recom-

mended as such but merely included by Simpson (1945) in various genera as subjective synonyms enclosed in brackets: "[ ]." One cannot always be sure whether Simpson intended subjective inclusion to indicate preservation of a taxon as a subgenus or merely abandonment of a subsumed generic name. By convention, if subgenera are employed in running text, the initially capitalized subgeneric name follows directly after the generic name but is placed in parentheses: for example, *Equus (Asinus)*.

Some of the family-group levels, such as tribes, were not used at all by Simpson in his treatments of most mammalian orders. However, tribes were found useful in certain subgroups of edentates, rodents, carnivorans, perissodactyls, and artiodactyls, where they mainly resulted from the downgrading of subfamilies.

Simpson even suggested (1945:15) that there were additional possible taxonomic categories such as infrafamilies, supertribes or supergenera, that were either then in use elsewhere, or might prove useful. But Simpson himself had no need for them at the time. Although he approved of subtribes in theory, Simpson did not use subtribes anywhere in the 1945 classification. Had he done so, calling tribes subtribes and reducing the rank of all categories above subtribe by one level of rank, then the number of taxa at family rank would have been different, a fact that compilers of numerical family diversity might profitably contemplate. The number of examples of a particular category is partly a function of how many taxonomic categories are utilized and partly a matter of where the category fits in the hierarchy (see Bleeker 1859 and Stümpke 1957 for classifications in which tribes rank above families).

In the present classification of more than 5000 mammalian taxa that are assigned generic or subgeneric rank, additional categories have proven useful in depicting in words a somewhat richer hierarchical arrangement of mammals than that found in Simpson's (1945) classification. There are now many more mammalian taxa to classify than was the case in 1945, both in real terms and because of the efforts of "splitters" and paleontological "apparent lineage choppers." Increasingly, most of these named organisms are made known from fossil materials only, sometimes very poorly represented. Moreover, the cladistic revolution in systematics has resulted in far more attention to phylogeny than was the case in the 1940s. The 25 taxonomic levels used in our classification actually fall closer to the theoretical minimum, 13 (see below for formula) than to the thousands that would be required if the classification reflected a completely pectinate (and very unlikely) sequence of taxa. The hierarchical level sequence is no more difficult (for humans) to learn than the alphabet, or probably less so in that some of the levels are easy to remember because of meaningful prefixes and suffixes. We see no particular reason why, if useful, additional categories (or simply unranked taxa) should not be proposed (or revived). Computers can remember them for us. Indeed, in the program Unitaxon™ used to process the data resulting in this classification, facilities exist to expand and keep track of the names, number and sequence of taxonomic levels indefinitely, if deemed appropriate.

In this classification of the taxon MAMMALIA, we use the following 25 subordinated named categories for taxa of rank higher than the species-group:

Class
  Subclass
    Infraclass
      Superlegion
        Legion
          Sublegion
            Infralegion
              Supercohort
                Cohort
                  Magnorder
                    Superorder
                      Grandorder
                        Mirorder
                          Order
                            Suborder
                              Infraorder
                                Parvorder
                                  Superfamily
                                    Family
                                      Subfamily
                                        Tribe
                                          Subtribe
                                            Infratribe
                                              Genus
                                                Subgenus

Ranks given the prefixed names supercohort, magnorder, grandorder, mirorder, and parvorder by McKenna (1975) have been interleaved but in less controversial positions than proposed in that paper. The intent has been twofold: to make more levels available for cases in which hierarchical structure is now richer and therefore conceivably in need of written and spoken expression by means of intermediate monophyletic subsets, and to curb inflation of taxon rank by keeping familiar names of taxa at familiar names of ranks. Linnaeus did the same when he introduced the use of orders in the eighteenth century. So did the botanists who began the use of families.

Interleaved ranks like families or infratribes are examples of the Zwischenkategorien of Hennig (1969). Obviously, a price is paid, but the "reduction in rank" alternatives, like the therapsid suborder MAMMALIA of Huene (1948:81), Nelson's (1969) and Gardiner's (1982) cohort MAMMALIA, Crowson's (1970:254, 258) order MAMMALIA, Wiley's (1979b:333-334) subdivision MAMMALIA, Hennig's (1966:187) reduction of mammalian orders to tribes, or, mirabile dictu, the genus *Mammalia* of Bigelow (1961:88) are also unpalatable to most of us working with fossil and extant

mammals (Simpson 1961:142-144; Crowson 1970:2). Generally, such reductions in rank are suggested by people who do not work on the group but view it from an armchair perspective gained from expertise on some other group, such as insects, fishes, or plants. Moreover, as the arbitrary boundaries among species-group, genus-group, family-group, and still higher taxonomic categories are crossed by rank shifts, different sets of complex and arbitrary taxonomic rules are encountered. This is true if one follows the procedures of the ICZN or presently unrequired extensions of them above family-group rank such as those suggested by Poche (1911, 1937, 1938), Stenzel (1950), Rohdendorf (1977), Rasnitsyn (1982), or especially Starobogatov (1991). For example, one wonders what the type species of Bigelow's *Mammalia* would be, or how Bigelow would have handled homonymous trivial names if all mammals were crammed into a single genus.

Our experience suggests that named hierarchical categories and their ordinal sequence become difficult to remember when there are more of them than approximately the number of letters in a Western alphabet, although the number can be expanded indefinitely by use of a computer. Since Linnaeus's time, their number has grown (e.g., see J. L. R. Agassiz 1857), and prefixes have been added as clues concerning the relative ordination of certain ranks. Generally, about twenty named categories have been found adequate by most workers to handle the internested subtaxa of a monophyletic taxon as diverse and complex as the taxon MAMMALIA. We also accept what to us is a tolerable level of paraphyly in view of our practical aims. This reduces the pressure to create new ranks. In cases in which hierarchical relations become better known than they are now, or when someone chooses to depict known hierarchical structure in greater detail than we have, it probably will become necessary to expand the number of taxonomic levels in selected parts of the classification by interpolation, just as has occurred in the past. Thus, either new levels with new names, or sublevels and superlevels of old levels, employing prefixes for the levels and suffixes for the taxonomic names themselves, may prove useful (e.g., those used by Boudreaux 1987, for "arthropods"). Or, more efficiently, formal names for levels can simply be dispensed with (D. J. Patterson 1988; Gauthier, Estes, and de Queiroz 1988; Gauthier, Kluge, and Rowe 1988a, 1988b:182-183; Gauthier et al. 1989; Lauterbach 1989; de Queiroz and Gauthier 1992). If that were done, however, an important mnemonic crutch would be lost. In cladistic literature, unnamed taxa or taxa whose rank is left unassigned are frequently used and offer relatively painless temporary solutions. This is especially true when only a few taxa are discussed and their relative ranking can be depicted graphically or, in text, by a convention such as indentation. Proper nouns for names of the ranks as well as for the names of the taxa that occupy them are memorable and can be used in conversation. If useful in prose or spoken discourse, they will endure. Reference can also be made to particular nodes in accompanying cladograms.

Contrary to claims of extravagance generated from reactions to McKenna (1975), (e.g., Patterson and Rosen 1977; Gingerich 1980:462, Panchen 1982, 1992: 158), we have not found it necessary to inflate the number beyond twenty-five here, at least within the taxon MAMMALIA as we use it. We find this number of levels useful at present because it allows the hierarchical structure of mammalian phylogeny to be expressed in more useful detail than previously, even if not completely. However, a criticism of inflation of the number of named taxonomic levels was provided by Wiley (1979b:325), commenting on McKenna's (1975) classification of class MAMMALIA. Wiley pointed out that if such a classification were to be combined with those of other vertebrates, a large number of categories intermediate between class and phylum would be required to house nested sister-groups upward from MAMMALIA to CHORDATA.

As an ichthyologist dealing with mammals, Wiley placed MAMMALIA as a subdivision of division THEROPSIDA (not a misspelling of THERAPSIDA), whose parent is superdivision AMNIOTA of the infraclass CHOANATA of the subclass SARCOPTERYGII of class EUOSTEICHTHYES. Thus, mammals, seen from the perspective of ichthyology, were to be reduced in rank in order to preserve EUOSTEICHTHYES as a class. As seen from the perspective of mammalogists, however, taxa like CHOANATA might be claimed to deserve still higher rank than they now possess. One shudders to think what might have been the consequences had Wiley been a protozoologist, but, in turn, the protozoologists might view with concern an attempt by a paleomammalogist to elevate foraminifera into the hierarchical stratosphere. As classifications move from paper to the computer, these problems of ranking will become tractable.

Wiley proposed to avoid the problem by sequencing: in a vertical list of equally indented names the first is sister to all those below it, the next is sister to those below it, and so on down the list. Polytomies in Wiley's scheme are indicated by the addition of the flag "*sedis mutabilis*." Phyletic sequencing (Nelson 1972, 1974; Cracraft 1974; Bonde 1977; Patterson and Rosen 1977; Wiley 1979b, 1981; Nelson and Platnick 1981; Schoch 1986; Ax 1987) has been proposed as a way out of the seemingly fractal dilemma of endless, approximately self-similar, subordinated ranks (Bigelow 1961; Heywood 1988).

Although we believe that sequencing on a large scale makes for difficult communication (e.g., Werdelin and Solounias 1991:70; Minelli 1993:39), in this classification we ourselves have sometimes made use of it (see section on *incertae sedis* taxa). We also use sequencing in a few cases where there are not enough named ranks available to express what we believe is fairly certain phylogenetic branching (see, for example,

ARTIODACTYLA and PROBOSCIDEA). If the phylogenetic structure were more certain than we believe it to be, we could have proposed unranked Zwischencategorien, but in order to avoid creating problems of nomenclatural instability we have not done so. Unless seen on a page in their proper relation to other names, however, names of sequenced taxa yield no clues as to their hierarchical position. Sequenced taxa are, therefore, not informative in ordinary spoken discourse or in lengthy written compilations (Heywood 1988). On a small scale, sequencing can work so long as one can see the whole sequence at once on a printed page or computer screen, or if one can simply remember a short sequence, but in spoken discourse there is no way to identify what the sequence of subordination actually is or where a discussed taxon is placed within it, although a few clues can be provided by prefixes of ranks, suffixes of taxonomic names, or analogies with the names of memorable subordinated categories of military or other familiar hierarchies. On a large scale, where the classification or the segment of a classification at hand contains more nested taxa than can be remembered easily, sequencing is increasingly impractical, both visually and in verbal communication.

Depiction of a hierarchy in prose is most often done with an additional aide-mémoire of sequential indentation, as in Simpson (1945) and examples given here. However, as ranked categories increase in number in large systems, space becomes wasted. This problem could conceivably be offset by using a different set of type for each level, or for arbitrary groups of levels. At present, however, this is accomplished reasonably well by the long-familiar convention of using italic upper- and lower-case type for genus-group and species-group names, roman upper-case and lower-case type for family-group names, and roman capital letters for all taxa higher than those of the family-group. If the text version were kept in a computer with a color monitor, one could presumably reintroduce clues about relative rank by appropriate use of the color spectrum if colors were not already in use for other purposes. Like Wiley (1979b:317), we reject the use of indentation alone as an indication of relative rank.

From time to time in the history of systematics, numbers have been tried as aids in construction of hierarchies. Ordinal numbers serving as proxies for ranks have the advantage of their easily remembered ordination, but one must not only remember the number but also what it stands for. Numbers alone lack ease of recall for most people, although we all manage a few of our own telephone numbers, personal identification numbers, and lock combinations (Amadon 1966). Ordinarily most of us do not remember Library of Congress codes for books, nor do we normally address each other by our social security numbers. Numbers also suffer from the fact that accidental transpositions, additions, or omissions of digits are less easily detected than changes in a series of letters forming a word.

Combinations of numbers and words for the names of ranks might be an improvement, but such combinations have not been generally adopted.

In the sequence of subordinated taxa given above, one might refer to the ranks of infraorder, parvorder and superfamily as 8-superfamily, 9-parvorder, and 10-infra-order, in order to remember where the rank of parvorder fits in the system (see Needham 1910; Michener 1963; Little 1964; Hull 1966, 1968; Hennig 1969; Farris 1976:281; Wiley 1979b:316, 1980:77; Løvtrup 1979; and Heppell 1991, for further exploration). One might even interleave new taxonomic levels, like new paragraphs in computer manuals (Griffiths 1976): thus the Zwischenkategorie "disorder" could be inserted between superfamily and parvorder as "8.1 disorder." This would solve the disconnectedness of Bigelow's multiple sequential classifications and would obviate need to create several dozen taxonomic levels between genus and species in case MAMMALIA were reduced to generic rank by some protozoologist who didn't have to deal with mammals on a daily basis. However, additional Zwischenkategorien are unpalatable to some (e.g., Patterson and Rosen 1977). Inasmuch as all hierarchical categories between Empire and species-group taxa are or originally were Zwischenkategorien, the term itself is not useful.

Fortunately, there is a logical relationship between the number of ranks dedicated to organizing a hierarchy and the number of terminal twigs contained within that hierarchical system. The minimum number of ranks, z, required to classify dichotomously a number of terminal taxa, n, is generally much closer to j, when $2^{j-1} < n$ and $2^j \geq n$, than it is to n-1 (the number required in the verbal equivalent of a fully pectinate cladogram of what might be a true lineage) (Anderson 1975). One can see that $j \geq 1 + \log n / \log 2$.

Therefore, the fear that proliferation of levels will get out of control (Bicheno 1827; Colless 1977; Gingerich 1977; Szalay 1994) is largely unfounded (MacLeay 1829; Strickland 1834), or at least not unbearably great, although for some people more than a dozen or so may be troublesome. For Crowson (1970:48), even seven or eight seemed quite enough. However, in addition to the long list already then in use, and still more that are long forgotten, Gaffney (1984:296) found four additional categories useful: giga-order, megaorder, hyperorder, and microorder. In 1988, Gaffney and Meylan even added two more, capaxorder and epifamily (see also Bour and Dubois 1984, 1986; Starobogatov 1991: tables 1, 2). Still others were employed by Boudreaux (1987). To depict fully a rich hierarchy of a large number of taxa in a classification, more and more cladistic nodes purportedly characterizing monophyletic subgroups can be represented by some form of wording for the names of taxonomic levels (not of the taxa themselves) as the number of taxa increases. Possibly, following suggestions of Darrel Frost and Jonathan Geisler (personal communications), one could follow the Masons' lead and use terms like "order[1]" (order, first degree) or "order[10]" (order, tenth degree).

The problem of proliferation of taxonomic levels is real and has not been completely solved in the text version of the present classification. It is dealt with effectively, however, in the computerized version of our systematization that underlies the text version. One does not need a "capacious human brain" (Gingerich 1980:463) like that of the character Ireneo Funes in Jorge Luis Borges's short story, "Funes el Memorioso" (Morrison and Morrison 1996:130). To keep track of such, one only requires a computer. One can simply store phylogenetic results in computers using the path concept of computer programmers to link unique taxon names that can be but do not have to be inflected by hierarchical rank prefixes or taxonomic name suffixes. In the computerized version of the present classification, the path to the origin of the hierarchy is made available for each taxon, and the path itself is indefinitely expandable and memorable. The computer does this by keeping track of the name of the immediate parent taxon of every taxon in the classification except the root, which has none. The procedure is an extension of the principle used in the formulation of Icelandic or Mongolian names of people. As a convenience to human users and with a bow to history, the computer additionally keeps track of the rank assigned to each taxon's proper name (see Steyskal 1988).

Potentially richer hierarchical structure exists between any parent taxon and its immediate named monophyletic taxonomic offspring if more than one independent attribute (character) has been used to diagnose that offspring (Hull 1980:436; Eldredge and Cracraft 1980). "Independent" characters are unlikely to become fixed in an evolving lineage simultaneously (Hennig 1966, 1969; Jefferies 1979:444). Thus if five uncorrelated fixed characters, A, B, C, D, and E, have been used to diagnose a monophyletic subtaxon X of taxon S, then on the basis of these characters, with adequate sampling, potentially four additional nested taxonomic levels could be created between the ranks of S and X: e.g., S(T(U(V(W(X)))))). Five levels would exist where one had been before, and each taxon occupying one of these levels would be approximately diagnosed by a single apomorphy (and its fixation) at each of five sequential cladogram nodes. The relative sequence of character fixation and presumably also acquisition would have occurred in one of 120 (= 5!) ways, e.g., CBEDA. Such a richly resolved structure would be extremely vulnerable, a situation combated by phenetic systematists since Adanson (1757, 1763-1764) but acceptable to strict cladists because it satisfies the Popperian scientific criterion of potential "falsifiability" [we would prefer a less absolute term such as "contradiction" (Gaffney 1979:99; Neff 1986:123) or "incongruence" (Sober 1988:33, 1993)].

In practice, restricting cladogram nodes to taxa recognized by single, genetically fixed characters also generates great instability because single characters are prone to reversals and parallelism. In some cases they can become unfixed. This is especially noteworthy for characters involving loss (Wheeler 1986; Neff 1986) and for molecular characters. Furthermore, character phylogenies can differ from taxon phylogenies in cases of polymorphism, in which a character arises significantly before it becomes fixed in an evolving lineage. In cases where such instability is evident, we believe that classificatory depiction (but not the underlying analysis) should retreat to a somewhat higher level of generality in order to maintain "reasonable" stability without stultifying further inquiry. In many cases in which analysis of characters is still in serious flux, we have deliberately retained unresolved polytomies in our classificatory depiction. They may reflect ambitaxa (Archibald 1994).

Various other students of hierarchical proliferation have suggested ways out of, or around, the generation of vast numbers of ranks and their associated named taxa. Generally these solutions produce about as many new problems as are solved. Thus Bigelow's (1961) or Griffiths' (1976) proposals to chop phylogenies into temporally sequential slices, each with its own independent classification, apparently have no current adherents (but see Brundin 1966:21; Tuomikoski 1967:146). Perhaps, however, our ultimate goal should be to get rid of rank names altogether (Hennig 1969:10; Griffiths 1976; Ehlers 1985:168; Ax 1987:42, 1988:58; Gauthier, Estes, and de Queiroz 1988; Gauthier, Kluge, and Rowe 1988a, 1988b; D. J. Patterson 1988; Gauthier et al. 1989; Lauterbach 1989; Minelli 1991; de Queiroz and Gauthier 1992).

Paraphyly and unresolved polytomy

Paraphyletic taxa contain a hypothetical ancestor and some but not all of the descendants of that ancestor (Hennig 1966; Ashlock 1971, 1973:434; Nelson 1971, 1973). Such taxa can be delimited arbitrarily by excluding chosen branches. If this is done, however, then by definition they are not naturally occurring entities, even though they stem from common ancestry (Farris 1991:304), because they do not contain all descendants of the ancestor. Patterson (1981b:219, 1982a; see also Løvtrup 1986 and Oosterbroek 1987) commented on their artificial limitation. Some students regard paraphyly as a sort of truncated monophyly and use the term holophyly (Ashlock 1971:65, 1973:434, 1980:443; see also Hull 1980:435 and Holmes 1980:55) for what we continue to term monophyly. We accept as axiomatic that fully monophyletic taxa (Haeckel 1874:150) are natural because of evolution alone and contain all of their descendants. Although "pattern cladists" regard evolution as irrelevant (but not necessarily nonexistent: see Beatty 1982; Platnick 1980, 1985, 1986; Farris 1985), their efforts in biological explanation are meaningless without it (Ball 1983; Dawkins 1986). They have defined paraphyly simply as lack of synapomorphy

(e.g. Patterson 1981b). This definition would apply only to putative ancestors (plesiomorphous immediate outgroups, i.e., metaspecies, discussed below). Platnick (1986) claimed that this stance is necessary in order to test evolutionary explanation.

Paraphyletic groupings and monophyletic taxa can be contrasted with polyphyletic assemblages (Farris 1975). Polyphyletic assemblages are not natural entities for a reason deeper than for paraphyletic ones: polyphyletic assemblages are not descended from an immediate hypothetical common ancestor. They are not valid taxa. In addition, like paraphyletic "groups," polyphyletic "groups" may be artificially truncated, further compromising them. Both paraphyletic and polyphyletic groupings occur in horizontal classification (Simpson 1945:17).

Simpson's family-group categories often included as subtaxa many taxa ranked as genera. These were presented in the form of lists. The taxonomic category containing the list was not further subdivided by intermediate taxonomic categories. From a cladistic standpoint, such a list of taxa all of the same rank represents either paraphyly or effectively an unresolved polytomy, and the taxa are *sedis mutabilis*. In the case of family-group taxa containing lists of genera of extinct mammals or a mixture of extinct and extant ones, Simpson's lists started with the oldest known or (usually) the most plesiomorphous members and proceeded to the extant ones, if any. This was a reflection of the fact that many paleontologists accepted (and often still accept) sequentially occurring taxa of identical rank as representing lineages, chopped into paraphyletic taxa of equal rank. These are mainly ranked as genera or species (chronospecies or simply apparent lineages) but also can occur as somewhat higher categories. Kowalevsky (1873) made such a linkage for what was believed to be a genus-level sequence leading from *Palaeotherium* to the modern horse genus *Equus* (Simpson 1953:364). Practitioners of stratophenetics make linkages at lower levels (see McKenna et al. 1977; Gould and Eldredge 1977; Wiley 1979a; Lillegraven et al. 1981: fig. 55; Schoch 1986:64; and Krishtalka 1993, for largely unanswered critiques). Many students who study extant mammals alone, or who do not take the fossil record as necessarily or literally expressing history, have found this assumption of continuity objectionable. Some object on the grounds that particular sequences of the fossil record may not mirror phylogeny accurately, although on average this might be generally expected although not recognizably so. Others aver that the method smacks of Bonnet's (1745; 1764, 2:29) infamous *echelle des êtres* of the eighteenth century (Bather 1927:lxxv; Anderson 1976, 1982; Rieppel 1988:22; O'Hara 1992: fig. 6). However, still other paleontologists have been understandably content. We take the attitude that in an unresolved dustbin one might as well arrange a mix of paraphyletic taxa (and taxa whose status is unresolved) by a descending age sequence as by any other way. If knowledge of internal hierarchical phylogenetic structure were actually available, however, one could do one of two things. One could either employ seriation based upon an hypothesis of the correct transformation series, or one could allocate some taxa to appropriate newly-created monophyletic subgroupings assigned to various lower ranks.

In this classification paraphyly remains in two main ways: (1) in some but not all cases when a number of equally-ranked supraspecific taxa, such as genera, are listed seriatim as members of a higher taxon without any intermediate taxonomic levels depicting hierarchical structure (as discussed above), and (2) when paraphyletic subtaxa result from subdividing a taxon by recognizing and naming a monophyletic subgroup that does not include the nominotypical taxon of the parent taxon and then, as a result of following Article 37 of the ICZN, feeling obliged to create a coordinate but not necessarily sister subtaxon based on the nominotypical taxon (and possibly also the remaining unassigned taxa of the parent taxon). Until further analysis provides more hierarchical structure, such paraphyletic assemblages reside uneasily in the system, and may be placed first in seriations. We do attempt to cope with paraphyletic taxa above the genus-group by holding them in abeyance when we have become convinced of their paraphyly but are not ready to suggest exactly what to do about it. For example, see our current treatment of Murinae, Equidae, and Leporidae. Readers should recognize that, in cases in which many taxa of equal rank are listed together, we do not necessarily imply evolutionary multifurcation, nor do we necessarily imply a continuous lineage; rather, these pigeonholed taxa simply await consistent analysis and proposal of a logical hierarchical taxonomic structure.

Binominal Nomenclature

Binominal (or binomial; see also Stejneger 1924, Smith 1962, Thomson 1995, and the ICZN) nomenclature haltingly became established in academic biology in the seventeenth and eighteenth centuries, although "family" and "given" names had been common enough (although not universal) in human affairs. Georg Eberhard Rumpf (Rumphius, 1627-1702) published consistent, two-part names in his *"D'Amboinische Rariteitkamer,* but whether these were binomina is debatable (Ley 1968:170). Earlier, in about 1605, Pierre Richer de Belleval (1564?-1632) had introduced a somewhat similar system of two-part names to botanical literature (Planchon 1869; Stearn 1959:6; Heller 1964: 34), but it appears to have had little or no traceable impact, partly because Belleval's classification was otherwise alphabetical and the trivial names were Greek compound words. In his *Pinax Theatri Botanici* of 1623 Bauhin used two-part names for species and genera although some of the parts themselves were compounded

(Lewis 1963; Heller 1964). Croizat (1945:59) cited binominal taxonomy used in 1657 by a botanist named Ambrosino, but we have not seen that work. John Ray (e.g., 1660) occasionally used binominal nomenclature as well, but not consistently. Still earlier, in the second century B.C., Cato (in his *De Re Rustica)* had employed two names for plants, but these were not really genera and species (Swingle 1946:220). See Gill (1898), Choate (1912), and Sprague (1950) for discussions of the early history of binominal nomenclature.

For botanists, binominal nomenclature became crystallized when Linnaeus (1749) first began the extensive use of "epithets" and then published his consistently binominal *Species Plantarum* (Linnaeus 1753b). Earlier attributions of botanical binominal nomenclature to works by Tournefort and by Lang (Langius, 1722) were refuted by Gill (1898:458, 460). For students of fungi, the important starting works and dates for "official" binominal nomenclature were much later: Persoon (1801) and Fries (1821-1832).

As early as 1737 and 1738 Linnaeus had begun to cite bibliographical references by a convenient binominal system abbreviating both the author and title of a particular work (Heller 1964:36). The usefulness of this reference technique may have influenced him to make more extensive and finally consistent use of binomina in biology. Zoological binomina were not mentioned by Linnaeus (1753a), but in botanical work (1745a, 1747, 1749) he had begun to use single words for label-like trivial botanical names and he used binominal nomenclature with increasing consistency from 1745 on (Dall 1878:14, 42; Heller 1964; Choate 1912). By 1749, Linnaeus used binomina about 90% of the time. His use of binomina in nomenclature for zoological species was not consistent until 1758. Linnaeus did not "invent" binominal nomenclature; he merely began to make it consistent. Finally, in 1758, single words were consistently used for each of the two parts of zoological binomina as well as botanical ones.

Curiously, in spite of his use of both nested categories and consistent binominal nomenclature, Linnaeus seems not to have attached much importance to either (Dall 1878; Mayr 1982:175). Nonetheless, in the 10th edition Linnaeus (1758) did actually state that he was assigning trivial names to species (Hopwood 1950a:31, 1950b), whereas in earlier editions strings of adjectives generally were used for zoological species.

As pointed out by Needham (1910), post-Linnaean modifications of the Linnaean system are often not truly binominal. Thus *"Homo sapiens Linnaeus, 1758"* is effectively quadrinominal in that it is a tag made up of four parts, the generic name, the trivial name, the author's name, and the date the name was established.

For zoologists the "official" starting date for zoological nomenclature is 1758, that of Linnaeus's 10th edition of his *Systema Naturae.* January 1, 1758 is arbitrarily used as a starting point for what has come to be called the "Linnaean system" in zoology (see Article

3 of the ICZN; also see Stejneger 1924). For a time in the nineteenth century, G. R. Gray (1841-1842), A. Agassiz (1872-1874:21, 31-36), and others championed earlier works as starting points (Dall 1878:43), but stability has been served by settling on the date of Linnaeus's 10th edition. Maverick arachnologists still follow Thorell (1869-1870) and start their binominal clock with the work of Clerck (1757). However, the publication of Clerck's work in 1757 is arbitrarily "deemed" to have been in 1758 by the ICZN, Article 3.

Linnaeus also numbered animal genera sequentially in most of his work; e.g., 1-39 for *Homo* through *Delphinus* in Class MAMMALIA, followed by *Vultur* through *Caprimulgus* (40-102) in Class AVES, in the 10th edition of his *Systema Naturae* (1758; also, see section on upward classification). Although Linnaeus abandoned numbering in some of his later botanical work, the practice lingered on in zoology (e.g., Illiger 1811; Trouessart 1897-1905) before finally dying out as the Linnaean system gradually evolved toward its present form. Occasionally, numbering systems for taxa given identical rank have reappeared (Gould 1954).

Much has been made of the consistent binominal nature of the species of the Linnaean System of the 10th and 12th editions (Linnaeus 1758, 1766). Taxa at the species-level were treated differently from those of the genus-group and other higher levels, which continued to be given single proper names. As the name of the rank "species" suggests, a species was originally somehow a specific instance of God's creation, belonging to a more general category, the genus. This was a holdover from Platonic and Scholastic idealism and from practical everyday "family" and "given" names of people. The form of the proper name of a species was consistently instituted by Linnaeus as a combination of two words, the generic name and what is often called the trivial name (or, for botanists, the epithet). *Homo* is a mononominal proper name for a genus and *Homo sapiens* is a binominal proper name for a particular species of that genus.

Binominal names for terminal twigs of keys had occurred in prior works, but with Linnaeus's publications of 1753 and 1758 their use became consistent and they became the proper names of taxa rather than two-part diagnoses composed of strings of adjectives. Moreover, the time was ripe for their introduction, although it had not been so earlier for de Belleval in the early seventeenth century nor for Rumpf at the end of that century. After the adoption of the convention of consistently binominal nomenclature, the trivial name continued in many cases to be adjectival and attempted to single out some particular distinctive "essential" quality of a particular species or perhaps to refer in latinized form to its common name. Thus names like "Felis cauda elongata, maculis superioribus orbiculatis, inferioribus virgatis," or "Ostrea major sulcata, inaequaliter utrinque ad cardinem denticulata," or "Linum ramis folique alternis lineari-lanceolatis, radice annua"

became simply *Felis pardus*, or *Ostrea edulis*, or *Linum usitatissimum* under the Linnaean scheme. The use of descriptive terms was meant to be useful in the sense of a key; the framework was still nonevolutionary. Patronyms, honoring particular persons such as other naturalists, were not at first used, but gradually entered the system as evolutionary classification superseded key-like classification. The initial letter of the trivial name is never capitalized, even for a patronym in zoological nomenclature, whereas the initial letter of a generic name is always capitalized.

Cain (1959b:310, 315) noted that the convention of binominal nomenclature forces the creation of monotypic genera for some species that could otherwise stand on their own. In these cases the generic category is redundant. Although the original reason for binominal nomenclature was to reduce strings of adjectives to manageable proportions in a system of downward logic (basically a key), there is more than the dead weight of tradition and familiarity that keeps binominal nomenclature in use. Although theoretically there is no reason why all taxa could not be mononominal, in the practical world thousands of species would have to be renamed because of homonymy if such a system were to be adopted (Amadon 1966). All taxa are individuals at various levels of generality, capable of being labeled by a single word. If memorability (for humans) were not a problem, species could be labeled by from one up to as few as five letters and/or integers, "deemed" to be a Latin word (see Article 11 of the ICZN), because $36^5 + 36^4 + 36^3 + 36^2 + 36 =$ more than the number of described organisms. The resulting names would be bizarre and would resemble automotive license plates, e.g., Q3 or Z8tja if numbers were also incorporated (see also Crowson 1970:270). Such names, like purely numerical tags, would in most cases be difficult to remember and would easily suffer undetected transpositions, additions, or omissions of the component symbols. They are definitely not what the ICZN has had in mind.

One of the principal benefits of the current system is that trivial names (epithets) can be recycled for use in distantly related genera, where homonymy can be legislated out of action. Linnaeus himself (1758) was able to recycle the trivial name *vulgaris* six times and *capensis* 29 times. Confining the effective range of homonymy to the species level is a major practical benefit gained from the use of the binominal system, but there is no such protection for higher taxa other than the abandonment of formal rules above what is deemed to be the family-group. Genera are now as common as (and often the same as) species of Linnaeus's day, so the problem of homonymy, once so serious at the species level, is now becoming common with genera. Generic homonyms are increasingly encountered. With the multiplication of generic names have come tendencies toward the lengthening of names, the greater use of anagrams, and the more extensive application of patronyms. Some authors (e.g. Dybowski 1926; Macdonald 1963) have burdened the literature with nearly unpronounceable, but sometimes perversely memorable, caconyms (Rowley 1956; Thomson 1995) that stand only a minute chance of homonymy. In spite of shortcomings, Linnaean binomina have been found to be practical.

Post-Linnaean Nomenclature and Its Management

In the early nineteenth century, evolutionary ideas had once more begun to take hold after a long hiatus following early forays by the ancient Greeks. Lamarck's (1802; 1809, 2:463; 1835-1845) evolutionary branching trees were easily expressed by hierarchical classification (Darwin 1859; Osborn 1894:14-15). A large and complicated superstructure grew out of Cesalpino's (1583), Tournefort's (1700-1703), Linnaeus's (1751), and their followers' efforts, through the work of Adanson (1763, 1764), Jussieu (1789), Lamarck (1809), Illiger (1811), de Candolle (1813), Strickland (1837, 1843), and many others. Whereas Linnaeus's own efforts at setting up a system of rules had been for all of nature (Linnaeus 1751), later interpretations and practices became centrifugal.

Like the various languages that evolved from Latin, the canons of classificatory construction came to be different in different biological disciplines, e.g., botany vs. zoology. Early on, some form of voluntary rules or even legislation was agreed to be necessary but, as in other human affairs, there was little agreement concerning which rules to adopt and to which biological disciplines they should be applied. Consequently, multiple sets of rules resulted. By the twentieth century, complex sets of rules for taxonomic procedures were evolving rapidly, but in different procedural lineages (see Dana et al. 1846; Scudder 1872; Dall 1878; Douvillé 1882; Stiles 1905; Banks and Caudell 1912; Sprague 1944; Hemming 1944; Swingle 1946; Buchanan et al. 1958; Crowson 1970; and Melville 1995 for details).

At present, separate international codes of nomenclature exist for zoology (International Commission on Zoological Nomenclature 1985), botany (Greuter et al. 1994; but see Parkinson 1990), cultivated plants (Brickell et al. 1980), bacteriology (International Code of Nomenclature of Bacteria 1966; Sneath 1992) and virology (Francki et al. 1990). The published versions of several of these codes are the result of many revisions (International Commission on Zoological Nomenclature 1902, 1961, 1964; Blanchard 1890, 1905; Briquet 1905; Arthur et al. 1907; Candolle 1867; Stearn 1953; Stafleu et al. 1972; Stafleu and Voss 1978; Greuter et al. 1994; Buchanan et al. 1958; Lapage et al. 1975). The rules for bacteriology, virology, and zoology, especially the latter, have become more legalistic than the botanical code. The zoological code goes so far as to require a judicial interpretive body (ICZN, Arts. 76-89; see also Rollins 1959; Tubbs 1992). An attempt at unification into a single International Code of Bionomenclature is under way (Hawksworth 1992, 1995; but see Strother 1995, and Bogan and Spamer 1995), but success may be distant, in part because of political infighting over the production of "standardized lists" of recognized valid taxa.

The history of development of the *International Code of Zoological Nomenclature* (ICZN, or "the Code") has been described in the introductions to the 1st and 2nd editions of the Code by N. R. Stoll (1961, 1964), and in the 3rd edition by W. D. L. Ride (1985; also see Savory 1962, Heppell 1981, Ride 1988). Among others, Bock (1994) and Melville (1995) have given exhaustive accounts. These authors trace development of the modern Code from early beginnings adopted by the British (Strickland 1843; Strickland et al. 1843) and American (Dall 1878) Associations for the Advancement of Science, in 1842 and 1877, respectively. The *Report on scientific nomenclature* of Dana et al. (1846) for the Association of American Geologists and Naturalists should also be consulted. Additional national codes followed (e.g., American Ornithologists' Union 1886, 1892), and then came a long, complicated process of internationalization that began in 1889 at the First International Congress of Zoology. The important *International rules of zoological nomenclature* embodying the results of deliberations at the Fifth International Congress of Zoology, Berlin, 1901, was published widely (Blanchard 1905). Efforts of various committees at successive Congresses of Zoology culminated two hundred years after Linnaeus's 10th edition with the adoption of the content of the 1st edition of the ICZN in 1958 (published in 1961). Two more editions followed (1964, 1985), and a 4th edition of the ICZN is presently in preparation (Savage 1990; Minelli 1995).

Because of the logical discontinuity between the names of binominal species-group categories and those of mononominal higher categories, the ICZN devotes much space to arbitrary rules governing the formation and handling of such terms. Above the species level, what is most important is that tags be unique. Taxa of species or subspecies rank must have types that are specimens (holotypes, see Article 73 of the ICZN). Taxa assigned to the genus-group have species as types (see Article 67 of the ICZN), a requirement since 1931, and taxa assigned to the family-group must now have genera as types (see Article 63 of the ICZN). All these types are for nomenclatural purposes only, although they once were thought to "typify" various concepts.

Above the family-group ranks, the ICZN has claimed no jurisdiction (see Article 1 of the ICZN), in contrast to the botanical code, which rules up through the level of order although without a judicial system. Thus, in zoology no types are at present necessary above the family-group level (but see Starobogatov 1991). Often, as in this classification, the names of taxa whose rank is placed above the family-group level are entirely capitalized. Composition of such taxa is frequently by original inclusion, often fudged a bit in subsequent use by an addition of a taxon here or a subtraction there. Ideally, however, the higher category name should apply to a monophyletic group arising at a particular phylogenetic node representing a cladogenetic event (term invented by Mickevich and Weller 1990:167).

We have discussed only in a perfunctory way the rules designed for the construction of zoological classifications, insofar as they affect mammalian taxonomy. For greater detail, consult either a current incarnation of the ICZN itself or a textbook such as those of Blackwelder (1967), Mayr (1969), Jeffrey (1973, 1974, 1989) or Mayr and Ashlock (1991). We generally follow the ICZN, because of the need to communicate via a common language, even if the chosen vehicle contains elements as arbitrary and difficult to follow as English spelling or Russian grammar. A complex legal system has been built up to keep uniform (but often arbitrary) standards in the interest of precise communication. The ICZN deals in great but depressing detail with such matters and is the Code of principal interest to neomammalogists and paleomammalogists (Savage 1990; Christoffersen 1995:440). Increasingly, attention to such matters takes time away from work of greater scientific consequence and gives taxonomists a fusty reputation (Romer 1968:7). On the other hand, to ignore such problems leads to communicative chaos.

Nonterminal Taxa and Obligatory Terminal Taxa

The Linnaean System was designed originally to deal with static species supposedly divinely created a few thousand years ago (Linnaeus 1735, 1758, etc.). For Linnaeus (except in the 12th edition of *Systema Naturae*), representatives of these species are still extant, or at least were alive within the past several thousand years (see also Lamarck 1809:75-76). Linnaeus grouped these terminal taxa hierarchically, on the basis of nested sets of shared diagnostic attributes alone, but he believed that those attributes had been distributed according to the wisdom of God. Today evolutionary biologists group them on the organizing assumption that certain shared characters are derived from common ancestry resulting from natural evolution, devoid of supernatural guidance, i.e., on the basis of descent with modification (anagenesis) and branching (cladogenesis). In cladograms and phylograms these extant species are depicted as taxa resulting from a long history. On phylograms the vertical axis always represents time, either relative or absolute. With cladograms the matter is different. For pattern cladists the internesting of nodes seems to represent logical priority only, but for others the nesting of nodes represents relative recency of cladogenetic events within the clade depicted. However, contrary to

oft-repeated denials, a vertical *relative* time axis *is* implied in cladistic methodology applied to organisms because, within nested clades, certain events are hypothesized to have happened prior to others if evolution is accepted as the cause of the pattern.

Obligatory terminal taxa must be at least species-group taxa that are further subdivisible only into perceivable individual organisms. If these terminal taxa are monotypic, however, simultaneously they can be given redundant higher ranks, notably generic rank but often a still higher one. Such redundant monotypy, often cited as exemplifying "Gregg's paradox," should not (but sometimes does) create confusion (Buck and Hull 1966; Farris 1967b, 1968, 1976; Gregg 1967; Wiley 1979b, 1980). In the present classification both paleontological and neontological terminal taxa are usually genera, but sometimes they are subgenera, for which at least a type species with a type specimen can be assumed.

Nodes of a hierarchical structure taken as depicting an evolutionary history are more complex. Monophyletic nonterminal taxa subsume all of their descendant taxa. In the informal standards of cladogram construction they are often represented by some form of named horizontal bracketing, or simply by higher taxon names placed near the nodes. Branches of nonterminal taxa must be lower in rank than their parent. Thus all monophyletic nonterminal nodes would represent taxa at various ranks higher than the terminal species-level taxa contained within them. However, this is not the case if paraphyletic taxa are allowed. We regard it as trivial that all extant terminal taxa are potentially paraphyletic because their future descendants are at present excluded. In the case of assumed anagenetic sequences without true branching, one can give both a high and a low rank, or even a sequence of ranks, to a species-level paraphyletic taxon, simply by assigning it redundant monotypic higher categories in addition to the taxon's status as a species. This is acceptable from a practical housekeeping standpoint, notably for monophyletic binomina, but is anathema to the most doctrinaire cladists (see also Gregg 1967:205).

Metataxa or Apparent Ancestors?

We do not find metataxa useful in this classification because, as we explain below, we do not employ the term with regard to taxa of the genus-group or above.

Metaspecies were named by Donoghue (1985) to cover the incompletely resolved situation in which certain sets of the constituent organisms of a species are not united by any synapomorphy and none of them shares derived characters with any other species (Kluge 1989). Donoghue treated individual organisms as terminal taxa, an erroneous procedure. Individual organisms are not taxa. Later, de Queiroz and Donoghue (1988) recommended the use of populations as terminal taxa. All this was relatively harmless, and amounted to the identification of certain infraspecific entities that might represent the ancestral condition but cannot be shown actually to do so. To us this sounds like a cladistic circumlocution for "apparent ancestor" in an attempt to avoid having a real species be necessarily paraphyletic. Perhaps "metaspecies" is a useful term covering that concept (Lee 1995).

Extending Donoghue's concept, others began to use a seemingly related, but actually quite different, concept for taxa above the species level (Gauthier 1984, 1986; Mishler and Brandon 1987; Estes et al. 1988). All these taxa, including the original metaspecies concept, were called metataxa. Mishler and Brandon even invented the term "metaphyly," to refer to the status of groups that are not known to be either paraphyletic or monophyletic. At about the same time, Böger (1989:49) independently introduced the concepts of adokimic taxa and adokimic groups. The word comes from gr. *adokimos*, "not genuine." Adokimic taxa and groups are paleontological entities without extant issue. The former are monophyletic and therefore can be regarded as fossil taxa whose position is *incertae sedis*. The latter have not been or may never be established as monophyletic. Nomina dubia and apparent ancestors are adequate categories for such entities, which do not need rebaptism.

Above the species level, if taxa are composed of immediate subtaxa then only one subtaxon can be without apomorphies; the rest must have them for otherwise they cannot be told apart. The one that does not is an apparent ancestor, a metaspecies, which as a convention can be redundantly given simultaneous monotypic higher ranks. Norell (1989:2) recommended (correctly, we believe) that metataxa not be used above the species level; the only metataxon would in that case be the metaspecies, which can "speciate," whereas higher taxa cannot.

"Metataxa" were discussed and confusingly extended by Archibald (1994). Archibald (1994:29) redefined "metataxon" as "a previously named portion of a polytomy for which positive evidence of monophyly or paraphyly is lacking or ambiguous." He then divided metataxa (as reformulated by him) into three kinds: metaspecies, mixotaxa, and ambitaxa. The latter two concepts can have a rank higher than specific rank. A mixotaxon was defined as "a metataxon at any level in the systematic hierarchy for which the constituent clades possess autapomorphies. One clade may lack autapomorphies. When the termini are organisms, more than one may lack autapomorphies." For the clade or clades possessing autapomorphies, we think that they are adequately covered by the *incertae sedis* concept that we have advocated above, and we hold that the concept is useless for individual organisms because they are not taxa. The one clade lacking autapomorphies (and synapomorphies) would be an apparent ancestor, which needs no rechristening.

An ambitaxon was defined by Archibald as "a metataxon at any level in the systematic hierarchy for which constituent clades possess autapomorphies. One clade may lack autapomorphies. When the termini are organisms, more than one may lack autapomorphies. Clades in the ambitaxon incongruously share apomorphies with one or more clades united at the same polytomy." The last sentence of Archibald's definition of an ambitaxon is the operative one and characterizes the results of most "honest" cladograms. As anyone who has ever constructed a real cladogram knows, ambiguous distributions of attributes abound. In fact, "perfect" cladograms are generally the results of special pleading. Ambitaxa resulting from the use of consensus trees (Adams 1972) are probably quite frequent in this classification, but we make no attempt to identify them as such. Inasmuch as most taxa are probably ambitaxa, we see no advantageous use for the term and will not trouble readers with it further.

Paleontological Taxa and Apparent Ancestors

If extant species are different according to fixed genetic criteria, or proxy characters indicate that they are, then, because of evolutionary continuity, branching (cladogenesis) must have occurred to have produced them. Extant sister-taxa (adelphotaxa of Ax 1987:36) represent real branching of a higher nonterminal taxon. But what about successive samples of fossils in the geological record that differ from each other in morphology by assumed anagenetic change? Such apparent lineages may span long periods of time, but lack known cladogenetic branching followed by the acquisition of autapomorphies in all immediate branches. Apparent segments of these assumed nonbranching lineages are not subject to tests of interbreeding because the samples occur at different times, all of which are in the past and not subject to direct experimentation.

This is a classic dilemma for paleontologists, who must deal with "paleontological taxa" (see Mayr et al. 1953:30 and references cited there; Krishtalka 1993). A phylogram of such an assumed lineage would lack branches and could be chopped into arbitrary paraphyletic segments, each given some particular (usually the same) rank. On the other hand, a cladogram of the same phylogenetic interpretation would contain branches at each postulated genetically fixed anagenetic novelty, leading to what Patterson and Rosen (1977) have called plesions but which here are called "apparent ancestors" (see also Willmann, 1987). Real ancestors would lack fixed autapomorphies, whereas apparent ones would only appear to lack any because of our imperfect knowledge of them. We can never be sure that we have identified a true ancestor because of the finite probability of discovery of some previously unknown autapomorphous attribute (Engelmann and Wiley 1977). Such a cautious approach is warranted but, if carried too far by inductophobes, can lead to solipsism. Noncladists simply accept certain taxa as ancestral to certain others until there is strong reason to believe otherwise. Here we attempt practicality by using the word "apparent."

Because it only takes one autapomorphous fixed character to contradict any preconceived hypothesis of ancestral status, poorly known taxa are likely to be unstable in this regard. Fossils always represent incompletely known proxy samplings of once-living organisms. Paleontologists deal continually with synecdoche. The recovery of more complete specimens could always turn out to provide previously unknown but crucial characters, although the possible number of new characters admittedly would always be less than those potentially available were the organism extant. However, this line of reasoning can be carried too far. For instance, Løvtrup (1973; 1977:21, 57, 154; 1985) stated that fossils have too few characters to test relationships worked out for extant organisms. This sweepingly incorrect conclusion was dealt with elegantly by Gauthier, Kluge, and Rowe (1988a). Some groups of organisms, of course, have more representative fossils than others. Consider, for instance, horse skeletons in comparison with scaphopod shells. Some synecdoches are more useful in systematics than others.

Neither are neontologists necessarily in possession of full information. As an extreme, one is reminded of the traditional ornithological taxonomy in which feathered skins were studied and the skeletal and many soft anatomical details of the bearer remained unknown because that part of the specimen was discarded at the time of collection. Biological disciplines sometimes have thrown away gifts in order to admire their wrappings (see also Bather 1927:lxviii). In museum collections neomammalogists as well are forced to study incompletely represented organisms in a depressing percentage of instances.

Apparent ancestors belong to higher taxa, but share no known synapomorphies with particular lower-ranked members of such taxa. They differ from some taxa placed *incertae sedis* in the parent taxon because they apparently lack autapomorphies; therefore, an available sample of an apparent ancestor could be, but cannot be proven to be, ancestral to the subdivisions of the parent nonterminal taxon or of a continuing monophyletic sequence with or without subsequent branches. At any one temporal cross section of an evolving lineage, such as the present, such samples must represent taxa at the species level, metaspecies if one prefers that term. Simultaneously, however, they can be assigned to monotypic higher categories, starting with the obligatory genus dictated by requirements of Linnaean binominal nomenclature. In general, but not with reliable clock-like behavior, such monotypic higher taxa tend to receive higher rank with increasing geologic age, but attempts to assign ranks *because* of presumed age of branch points would be intellectually disastrous (contra Ride 1988:347). The assignment of multiple ranks to monotypic taxa could be viewed as a bow to phenetics, or to Goldschmidtian saltation. It also might be thought of as tacit recognition that an unknown but larger number of nodes awaits discovery, either by recovery of unknown taxa in the extant biota or, more probably, in the fossil record.

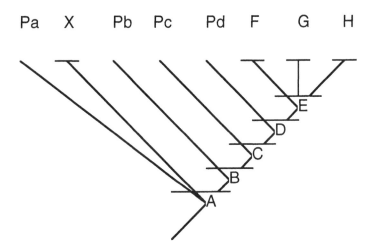

In the cladogram above, taxon A subsumes apparent ancestor Pa, monophyletic *incertae sedis* taxon X, and monophyletic taxon B with all its lower-ranked members. Insofar as known, a population of Pa could be, but is not necessarily, the most proximal known ancestor of taxon X and taxon B. Pa is a paraphyletic apparent ancestor at the species level (metaspecies), but it could be assigned additional ranks up to that of taxon B and one rank higher than that of *incertae sedis* taxon X. Paraphyletic apparent ancestors Pa, Pb, Pc, and Pd lack known autapomorphies (horizontal lines in the figure). If this were a phylogram the lines leading to them would collapse to zero length. Pb, Pc, and Pd are apparent ancestors like Pa at the species level (metaspecies) with higher ranks assignable, but the highest such rank given to Pb would normally be the same as that of taxon C, and so on. Taxa X, F, G, and H possess autapomorphies some of which would be converted to synapomorphies if cladogenesis were found to have occurred in them. In the distant future, some of today's terminal taxa would be viewed as higher-ranked taxa because new discoveries might document that they had subsequently given rise to branches.

The equivalent text systematization to the cladogram on the preceding page is:

Taxon A
    Taxon X *incertae sedis*
    Apparent ancestor Pa (metaspecies and
           possibly other monotypic categories
           based upon it)
    Taxon B
        Apparent ancestor Pb (metaspecies and
             possibly other monotypic
             categories based upon it)
        Taxon C
            Apparent ancestor Pc (metaspecies
                and possibly other mono-
                typic categories based
                upon it)
            Taxon D
                Apparent ancestor Pd
                    (metaspecies and
                    possibly other
                    monotypic categories
                    based upon it)
                Taxon E (based upon a
                    nominotypical taxon F,
                    G, or H if at or below
                    family-group rank)
                  Taxon F
                    Taxon G
                    Taxon H

From the above arrangement of words one can reconstruct the cladogram from which the pattern of words was derived. In practice, the systematization that one usually encounters in the literature is something less rigorous, such as the following, from which reconstruction of the source cladogram is no longer unambiguously possible:

Family Pc-idae (determined by node A)
    Subfamily Pa-inae
        Genus and species Pa
    Subfamily Pc-inae (determined by node B)
        Genus and species Pb
        Genus (the type genus) and species Pc
        Genus and species Pd
        Genus E (determined by node E)
            species F
            species G (type species of genus E)
            species H
Family Pc-idae, *incertae sedis*:
    Genus and species X

Proliferation of ranks in the latter hierarchy has been avoided at the expense of methodological rigor.

Nodes C and D have not been represented. Moreover, the sequence of the taxa could be scrambled if some criterion other than information about polarized morphoclines were given preference in the process of listing them. Geological first known appearance could be one such criterion. Under ordinary circumstances geological age would be expected to be approximately congruent with properly hypothesized morphoclines for which the fossil record is complete enough. However, this would be true of only a few kinds of organisms, notably those possessing preserved parts that could be interpreted to represent the whole organism adequately. Fossils are incomplete and difficult to find, and their representation in the geological record is spotty and subject to misinterpretation. Nonetheless, even though their distribution may lead to erroneous interpretation, they are not perverse. The fossil record of a particular group of organisms, if it exists at all, can be noisy in terms of information theory, but generally signal can be separated from noise. With increasing thoroughness of collection, a clearer picture of the organisms' spatial and temporal distribution, as well as of character representation, emerges (Benton 1996). The value of negative evidence also increases, however dubious or inconclusive it might still be. For instance, that is why, after more than a century of exploration, the lack of proboscideans in the comparatively well-explored Miocene of South America is strongly persuasive that they were not present on that continent at that time; their occurrence there at a later date, therefore, is probably better interpreted as part of the "Great American Biotic Interchange" (Webb 1978; Stehli and Webb 1985), not as the result of relatively ancient Gondwanic vicariance followed by cryptic nonoccurrence in well known mid-Cenozoic rocks. Negative evidence, like induction since David Hume's analysis of it in the eighteenth century, has a bad name, but, in the practical world where absolute certainty is not possible except for zealots, both induction and negative evidence can be highly efficacious if used prudently.

A complication arises in cases in which a family-group taxon, say a family, contains a number of sequentially arranged, apparent ancestors that have been given generic rank. One of the genera would have to be the type genus of the family. In the figure given above, suppose it is already Pc. Suppose further that someone now designates at node D a subfamily E-inae, composed of the type genus E and its apparent ancestor, genus and species Pd. Because Pc is already the type genus of the family, Article 37 of the ICZN requires that it would now also become the type genus of a coordinate, nominate subfamily Pc-inae, subsuming at least its type genus (and species), the apparent ancestor Pc. Here we would have the creation of an unnecessary paraphyletic taxon Pc-inae. The artificiality would be compounded if stem taxa Pa, X, and Pb were included by default, as unfortunately often is the case when family-group taxa are subdivided. The implications of Article 37 of the ICZN need to be reconsidered.

As noted above, another complication arises because of the requirements of the ICZN that a generic name is required for the naming of binominal species. Thus the naming of a monophyletic species in the Linnaean System forces the existence of two taxonomic levels, not one. Other ranks as well can be used for monotypic taxa, up to one rank below that of the higher taxon to which a lower taxon belongs. This is another difference from taxa *incertae sedis*, which in the scheme adopted here must be *two* or more ranks lower than the higher taxon to which they belong. In apparently unbranched real lineages undergoing anagenetic modification with time, arbitrarily delimited segments obviously could be assigned a number of paraphyletic arbitrary taxa based on ideas of continuity and "differentness," in turn based upon comparison with the amount of "difference" observed in extant and therefore contemporaneous taxa. For these arbitrarily created taxa, equally arbitrary higher taxa could be assigned at various nodes. Simpson (1943) attempted to arrive at a practical solution by utilizing the distribution of proxy characters "vertically" in the same way that they are used "horizontally" to recognize contemporary morphological taxa.

Noncladists have tended to accept apparent ancestors at face value, and simply call them ancestors until shown otherwise. Cladists who dread induction refuse to take that step because it is considered premature (like identifying a black sheep when one has only seen it from *this* side). Furthermore, cladists exclude from possible ancestry of a particular taxon any taxon that has autapomorphies not shared with that taxon. This is reasonable for genetically fixed morphological attributes but can lead to difficulties at the molecular level, where true reversals or paralogous mutations are known to occur, or when the effects of genes may be turned off. Both cladists and noncladists have difficulties that can be glossed over by devotees. An additional problem at the molecular level is typological thinking. All too often, molecular data are obtained from single individuals, so that polymorphism, if present in populations, remains undetected. Moreover, just as with morphological characters, ancient molecular synapomorphies can be overprinted with the passage of time to the point where few, if any, of the original synapomorphies remain (Shoshani 1986; McKenna 1987:70; Gauthier et al. 1989:350). This last situation becomes a problem in outgroup analysis.

Cladists are forced to treat evolutionary events as though they were maximally parsimonious, even though most well-constructed cladograms contain ambiguities in the form of conflicting character distributions or suffer from missing data. Consistency indices only rarely reach 100%. Although cladistic analyses must follow a general principle of parsimony, in our view nature does not always behave in such a way as to yield its secrets to such analysis. The method of analysis is merely the best available and usually works well enough on average. Even in well analyzed data sets a parsimonious analysis of a subset of data may yield different results from those obtained by a similar analysis of the entire data set. Natural processes involve random as well as nonrandom or channeled events. Real reversals in evolution can take place at the molecular level and retreats to former phenotypic conditions can occur among species by the turning off of genes or by the shifting of gene frequencies in polymorphic populations. Cladistic methods have been found to be more powerful than phenetic methods in the hypothesizing of reversals, but a one-to-one correspondence between our models and nature is not likely to be achieved. However, in genetically isolated lineages, the chance of exact convergence is increasingly unlikely with time and soon becomes statistically remote. Above the species level, where isolation and subsequent genetic divergence have reduced the level of successful actual or potential interbreeding of populations permanently to zero, reversals to exact former conditions become virtually impossible, but heterochronous convergence can nevertheless mimic reversals closely.

Practitioners of the "stratophenetic" school of mammalian paleontology (e.g., Gingerich 1979, 1980) accept various sequentially occurring populations in the fossil record as evidence of continuity. They therefore construct phylograms, a practice similar to connect-the-dots art (McKenna et al. 1977; Lillegraven et al. 1981: fig. 55). The presumed continuity is then artificially broken up into a paraphyletic sequence of presumed ancestor-descendant taxa, generally thought to be species. Among fossil mammals these taxa are often recognized by such statistical criteria as the logarithm of the area of the first lower molar. This measure is assumed by its practitioners to be sensitive enough to pick up mammalian species-level differences. The "taxa" are not assigned on the basis of morphology alone but are considered fully recognizable only if their position in the stratigraphic sequence is known as well. Two occurrences of seemingly identical morphology would be assigned to different taxa if a (presumed) sequence of other morphologies stratigraphically between them could be shown to change with time, digressing from and then returning to a condition indistinguishable from the first morphology.

The numerous problems with the stratophenetic method have been taken up in discussions by McKenna et al. (1977), Lillegraven et al. (1981: fig. 55), and Schoch (1986:64), among others. As was pointed out in those critiques, what is "continuous" at one power of resolution may be discontinuous at another. The "connectedness" of stratophenetic phylogenies may not be necessarily at or below the species level. Log areas of lower first molars of fossil mammals may indeed be the most sensitive of morphological differentiators, but at what taxonomic level? Has the species-level been reached and tested, or are the proxy differences only sensitive at some higher taxonomic level? How good a synecdoche for "biological" species are various neontological or paleontological morphospecies? [See, for instance, Simpson's (1937:19-22) treatment of successive samples of the phenacodont genus *Ectocion*, commented upon by Gazin (1956:10-13).] Another problem for the rationale of stratophenetics is that at low taxonomic levels taxa tend, with obvious exceptions such as large ungulates, many carnivorans, and *Homo sapiens*, to be geographically restricted. This is a problem because fossils are available only from places where there are outcrops of fossiliferous rock. These generally are rare or restricted in occurrence. How would one obtain a continuous record of stenotypic taxa that had lived only on eroding highlands? What are the chances that geological factors alone will permit the later collection of a continuously linked series of the real ancestors of a true anagenetic sequence? Does stratophenetics require an unjustified continuous presence of the evolving lineage in the very area in which fossilization and preservation for many millions of years will be made possible by agencies not yet operating at the time the sampled population existed?

Crown Groups

The concept of crown group was introduced by Jefferies in 1979 (see also Craske and Jefferies 1989), who described a crown group as consisting of "the latest common ancestor of the living members of a monophyletic group, plus all descendants of that ancestor" (Jefferies 1979:443). More recently the concept has been discussed by Lucas (1990, 1992), Meier and Richter (1992), Forey (1992), and Bryant (1994), among others. Crown groups add more than biological analysis to their definition, for they additionally require the restrictive convention that the defining members of the crown group be extant. Thus the biological analysis must be neontological; there can be no wholly extinct crown groups. In addition to their definition, crown groups can also be approximately diagnosed by characters.

The main advantage of the use of crown groups in taxonomy is that their extant examples provide many more characters to work with than are available from fossils. Moreover, the definition (but not necessarily the diagnosis) of a crown group is likely to remain relatively stable. If the taxon MAMMALIA is defined simply as the taxon subsuming all extant therians and monotremes, the latest common ancestor of those two subtaxa, and all descendants of that ancestor, then we have defined a taxon familiar to all neomammalogists. Only paleontologists would spend time arguing over where to place multituberculates, †*Haldanodon*, or †*Sinoconodon*. Whether or not multituberculates are members of the mammalian crown group will not affect the work of neomammalogists. When a taxon is well represented by diverse, extant, terminal subtaxa, the resulting crown group-based taxonomic definition and its

diagnosis will approximate a definition and diagnosis based additionally on fossils.

But what of situations in which the extant representation of terminal subtaxa of a large monophyletic taxon has been drastically thinned by extinction (Lucas 1990, 1992)? In such cases, the introduction of a crown group-based definition might depart widely from entrenched usage. In some instances this can be a threat to communicative stability. For example, if PROBOSCIDEA were redefined after many decades of broader application and restricted to a crown group covering only *Elephas* and *Loxodonta*, their most recent common ancestor, and all of its descendants, mastodons and deinotheres would not be considered proboscideans. We think that nearly all of our readers would object, and the world's literature confirms that viewpoint (e.g., Tassy 1988). A crown group redefinition of PROBOSCIDEA would exclude all stem groups prior to branching of the lineages represented by the two extant species of proboscideans, the Asian and African elephants. Nonetheless, living representatives of taxa thought to be extinct occasionally turn up. In the admittedly unlikely event that an extant moerithere descendant were to be discovered and recognized to represent the closest living sister group to *Elephas* and *Loxodonta*, a crown group definition and that of the traditional PROBOSCIDEA would more nearly coincide. Deinotheres and mastodons would then become respectable crown group proboscideans because of their membership among the descendants of the most immediate common ancestor of †*Moeritherium*, *Elephas*, and *Loxodonta*.

As one can see, crown groups completely coincide with traditional groups like PROBOSCIDEA if (and only if) an extant member of the most plesiomorphous subclade is known, in which case clades that would otherwise be stem groups would now be included as subclades of a broader crown group. The crown group could be extended, as MAMMALIA was after Linnaeus's time when monotremes were discovered or as PROBOSCIDEA has been because its users have not employed the crown group concept. If crown group PROBOSCIDEA were extended because of discovery of an extant moerithere, then this would satisfy both the traditionalists and the enthusiasts for crown groups. In that hypothetical case a new name could conceivably be proposed for the subsumed and lower-ranked crown group represented by *Elephas* and *Loxodonta*. Whichever course is taken, stability should be the goal. In the present classification we simply place *Elephas* and *Loxodonta* with †*Mammuthus* and †*Primelephas* in a proboscidean tribe Elephantini.

Among extant mammalian higher taxa, PROBOSCIDEA illustrates an extreme example of the traditional taxonomic definition differing greatly from what the crown group definition would be. Other taxa whose traditional and crown group definitions vary widely include HYSTRICOGNATHI, HYRACOIDEA, CETACEA, PHYLLOPHAGA, Aplodontidae, Castoridae,

Equidae (see Bryant 1994:126), Rhinocerotidae (see Bryant 1994:125; Wyss and Meng 1996), Tapiridae, Antilocapridae, Camelidae, and many more. For other taxa, crown group definitions may not differ greatly from traditional ones.

A problem with crown groups involves taxonomic history. Consider PRIMATES. In recent years there has been considerable flux concerning what to include in, or to exclude from, that taxon. Various new terms have been proposed to include extant mammals usually regarded as primates but also (since 1811) to exclude extant DERMOPTERA (*Cynocephalus*). The dermopteran *Cynocephalus* (= "*Lemur volans*" of Linnaeus) and even the chiropteran *Vespertilio* were included in Linnaeus's original concept of PRIMATES. If one were to maintain Linnaeus's exact original grouping (as monophyletic), PRIMATES would necessarily include dermopterans and bats and would not need to be restricted arbitrarily to the most immediate common ancestor of extant strepsirrhines and haplorhines and all of that ancestor's descendants. Linnaeus's PRIMATES differs from Gregory's (1910: 322) ARCHONTA only regarding tupaiids and elephant shrews.

One could exclude bats but include dermopterans in a crown group and apply the name PRIMATES to that, as is done in this classification. If one utilizes PRIMATES for strepsirrhines, haplorhines, and dermopterans, then a name is needed for the taxon that subsumes just strepsirrhines and haplorhines, their most immediate common ancestor and all its descendants. Hoffstetter has dubbed them EUPRIMATES. Thus EUPRIMATES is a tag for the most restrictive redefinition of the primate crown group. Clearly, arbitrary adjustments must be made, and some familiar terms either redefined or replaced in order to save others. In this classification DERMOPTERA is subsumed in a more traditional concept of PRIMATES than some taxonomists might be comfortable with. However, Linnaeus's *Vespertilio* has long since been expanded to the high ranking taxon CHIROPTERA. CHIROPTERA not only has been removed from the most narrowly redefined primate crown group, but also in our view probably lies outside a combined primate/dermopteran crown group and may not even be the closest outgroup. To include chiropterans in the PRIMATES (as that taxon has been restricted after Linnaeus's use of it) would in that case result in unacceptable polyphyly violating the crown group requirement of monophyly.

Another potential problem with crown groups is that polyphyletic ones will sometimes be masked. Our notions of monophyly are, after all, just hypotheses. Such a situation can be illustrated by another extreme hypothetical example. Suppose we were to define a hypothetically monophyletic (but nevertheless erroneous) crown group, "VERMINIDA," embracing extant rats, cockroaches, their most immediate common ancestor, and all of the descendants of that ancestor. Because all descendants of the ancestor were included, the group would be monophyletic! In time this group would be found to include many other extant organisms that in our ignorance we had neglected to specify originally. It would be a useless synonym of something like METAZOA.

Crown groups are evidently procedurally immune to the usual problems associated with polyphyly because, if life originated just once, then any two kinds of organisms had a common ancestor at some ancient time, and that ancestor may have spawned extant taxa not included in the original defining list. Such an expanded and redefined crown group would be procedurally valid at a higher node than the author of the taxon originally envisioned. It would have additional extant taxic contents not specified by the original author. Any polyphyletic taxon could be made whole by this means. In previous times such bizarre results were prevented by the use of paraphyletic ancestral groups, from two or more parts of which a monophyletic group could not arise.

In the future we believe that crown groups should be defined with the strict proviso that the entire extant taxonomic content stemming from descent from the immediate common ancestor should have been known to the original author and specifically enumerated. Moreover, the original author should have made his or her proposals in an evolutionary context. These strictures are not required under current formulations. By current definition, all of Linnaeus's (1758) higher taxa would have to be taken as crown groups because all are extant, but some would be unfamiliar or have unfamiliar content. Thus, BELLUAE, BESTIAE, and BRUTA each had a common ancestor at some point, and in each case that ancestor had extant descendants. However, those crown taxa would be valid only at higher levels than originally envisioned, and those now unfamiliar names would be forced to be synonyms of other, more familiar ones. BESTIAE, for instance, would become a senior synonym of THERIA, owing to the fact that its original taxonomic content comprised *Sus*, *Dasypus*, *Erinaceus*, *Talpa*, *Sorex*, and *Didelphis* (Linnaeus 1758). MAMMALIA itself would become a synonym of THERIA. As currently defined, crown groups do not automatically lend stability or practical utility to systematization. The situation is analogous to an overly zealous adherence to the rules of priority; doctrinaire solutions may not always be the best.

In some cases in this classification we have defined monophyletic, extinct, higher taxa by first listing the lower taxa comprising them and then expanding the definition to include all the extinct descendants of their most recent common ancestor. These monophyletic higher taxa are obviously not crown groups in the sense of Jefferies (1979) because they are extinct, but the method of defining them is the same except for Jefferies's artificial injection of the time we live in. As with extant groups, provision is made for future discovery of new members.

PART II

A Classification of Mammals

Class **MAMMALIA** Linnaeus, 1758:12, 14.[1]

> [= ZOOTOKA Aristotle, 330 B.C.[2]; VIVIPERA Ray, 1693:53[2]; MASTODIA Rafinesque, 1814; THRICOZOA Oken, 1847:xi; AISTHESEOZOA Oken, 1847:511; PILIFERA Bonnet, 1892[3]; MAMMALEA Kinman, 1994:37.]

[1] Crown group comprising extant monotremes and therians, and their most recent common ancestor and all its descendants.

[2] Pre-Linnaean, but included for the sake of interest.

[3] Oken coined the term TRICHOZOA with the same intent.

Subclass **PROTOTHERIA** Gill, 1872:vi.[1] Monotremes.

> [= REPTANTIA Illiger, 1811:113[2]; ORNITHODELPHIA de Blainville, 1834:82[3]; MONOTREMATA C. L. Bonaparte, 1837:5[4]; AMASTA Haeckel, 1866:cxlii; SAUROPSIDELPHIA Roger, 1887:4; ORNITHOSTOMI Cope, 1889:874; MONOTREMIFORMES Kinman, 1994:37.[5]]

[1] Proposed as a synonym of Subclass ORNITHODELPHIA; including monotremes only. Expanded by Hopson, 1970: 7.

[2] Proposed as order. REPTANTIA is also a taxon name used in the higher classification of decapods.

[3] Proposed as suborder; reduced to infraclass by Hopson, 1970:7. Originally referred to as "Didelphes Anomaux" de Blainville, 1816:251.

[4] Proposed as a order. Palmer (1904:888) gave 'Monotremes' E. Geoffroy, 1803:226 as the earliest form and use of this name; see also "Monotrema, Geoff.": Gray, 1869: 393.

[5] Proposed with a query, but not subject to the provisions of the International Code of Zoological Nomenclature if ranked above superfamily.

Order **PLATYPODA** Gill, 1872:27.[1] Platypuses.

> [= Ornithorhynques Gervais, 1854:xx.[2]]

[1] Proposed as suborder. Elevated to the rank of order by McKenna, in Stucky & McKenna, in Benton, ed., 1993:740. "Platypoda" Gray, 1865:102 was an invalid family-group name for certain mustelids.

[2] Proposed as suborder.

Family **Ornithorhynchidae** Gray, 1825:343. Platypuses.

> [= Ornithoryncina Gray, 1825:343[1]; Ornithorhynchidae Burnett, 1830:365; Ornithorhynchina C. L. Bonaparte, 1837:9.[2]]
>
> E. Cret., M. Mioc., Pleist., R.; Aus. E. Paleoc.; S.A.

[1] As tribe.

[2] As subfamily.

†*Steropodon* Archer, Flannery, Ritchie & Molnar, 1985:364.

> E. Cret.; Aus.

†*Monotrematum* Pascual, Archer, Ortiz Jaureguizar, Prado, Godthelp & Hand, in Augee, ed., 1992:3.

> E. Paleoc.; S.A.

†*Obdurodon* Woodburne & Tedford, 1975:3.

> M. Mioc.; Aus.

Ornithorhynchus Blumenbach, 1800:609-610. Duck-billed platypus.

> [= *Platypus* Shaw, 1799:385, 386[1]; *Dermipus* Wiedemann, 1800:180.]
>
> M. Mioc., Pleist., R.; Aus.

[1] Not *Platypus* Herbst, 1793, a coleopteran.

Order **TACHYGLOSSA** Gill, 1872:27.[1] Spiny anteaters, echidnas.

> [= Echidnes Gervais, 1854:xx.[2]]

[1] Proposed as suborder. Elevated to rank of order by McKenna, in Stucky & McKenna, in Benton, ed., 1993:740.

[2] Proposed as suborder.

Family **Tachyglossidae** Gill, 1872:27. Spiny anteaters, echidnas.

> [= ACULEATA É. Geoffroy Saint-Hilaire, 1795:102-103; Echidnidae Burnett, 1830:365; Echidnina C. L. Bonaparte, 1837:9[1]; Echidnida Haeckel, 1866:clvii.]

M. and/or L. Mioc. and/or Plioc., L. Pleist.-R.; Aus. L. Pleist.-R.; New Guinea.

[1] As subfamily.

Zaglossus Gill, 1877:clxxi.[1] Long-nosed spiny anteaters, long-nosed echidnas.

[= *Proechidna* Gervais, 1877:43[2]; *Acanthoglossus* Gervais, 1877:838[3]; *Bruynia* Dubois, 1882:267-270; *Bruijnia* Thomas, 1883:40; *Prozaglossus* Kerbert, 1913:166.] [Including †*Megalibgwilia* Griffiths, Wells & Barrie, 1991:90.]

M. and/or L. Mioc. and/or Plioc., L. Pleist.; Aus. L. Pleist.-R.; New Guinea.

[1] May 5, 1877.

[2] November 30, 1877.

[3] November 5, 1877. Supposedly preoccupied by *Acanthoglossa* Kraatz, 1859, a coleopteran.

Tachyglossus Illiger, 1811:114. Short-nosed spiny anteaters, short-nosed echidnas.

[= *Echidna* G. Cuvier, 1798:143[1]; *Echinopus* G. Fischer de Waldheim, 1814:691-694; *Syphomia* Rafinesque, 1815:57.[2]]

L. Pleist.-R.; Aus. R.; New Guinea.

[1] Not *Echidna* Forster, 1788, a fish.

[2] New name for *Echidna*.

Subclass **THERIIFORMES** Rowe, 1988:245, **new rank**.[1]

[= THERIIMORPHA Rowe, 1993:139.]

[1] Rowe did not assign taxon rank.

Infraclass †**ALLOTHERIA** Marsh, 1880:239.[1]

[1] Proposed as an order; used as a suborder of MARSUPIALIA by Gregory, 1910: 464; raised to subclass by Simpson, 1945:39; ranked as a "Pénéclass" by Vandebroek, 1964:155; used as an infraclass by Hopson, 1970:7.

Order †**MULTITUBERCULATA** Cope, 1884:687.[1] Multituberculates.

[= †PLAGIAULACOIDEA Simpson, 1925:9[2]; †MULTITUBERCULIFORMES Kinman, 1994:37.] [Including †PLAGIAULACIDA McKenna, 1971:736[3]; †PAULCHOFFATOIDEA Kielan-Jaworowska & Ensom, 1991:35-37.[4]]

[1] Proposed as order; used as infraclass by Vandebroek, 1964:155 and by McKenna, in Stucky & McKenna, in Benton, ed., 1993:741.

[2] Proposed as suborder to contain the ptilodontids and taeniolabidids as well as the plagiaulacids but not the haramiyids; it is effectively a synonym of †MULTITUBERCULATA and †ALLOTHERIA inasmuch as it was proposed as a balance to †TRITYLODONTOIDEA (then regarded as mammals). Later renamed as new by Simpson, 1928:31.

[3] Proposed as a suborder; used as an infraorder by McKenna, 1975:40; elevated to ordinal rank by McKenna, in Stucky & McKenna, in Benton, ed., 1993:741. Here we have eliminated the term because its content was paraphyletic.

[4] Proposed as suborder.

Family †**Plagiaulacidae** Gill, 1872:27.[1]

[= †Plagiaulacoidea Ameghino, 1894:76[2]; †Plagiaulacinae Kielan-Jaworowska & Ensom, 1991:36.] [Including †Bolodontidae Osborn, 1887:3; †Allodontidae Marsh, 1889:179; †Paulchoffatidae G. Hahn, 1969:9; †Paulchoffatiinae G. Hahn, 1971:8; †Kuehneodontinae G. Hahn, 1971:8.]

M. Juras.-E. Cret.; Eu.[3] L. Juras.-E. Cret.; N.A.[4]

[1] The †Plagiaulacidae are left here as a paraphyletic "group" (including the †Paulchoffatidae), pending adequate cladistic analysis.

[2] Evidently a family-group name.

[3] Reported but not yet described from the Middle Jurassic (Bathonian) of Europe (England).

[4] Reported but not yet described from the Early Cretaceous (Albian) of North America (Texas).

†*Paulchoffatia* Kühne, 1961:379.

L. Juras.-E. Cret.; Eu.

†*Kuehneodon* G. Hahn, 1969:35.

L. Juras.-E. Cret.; Eu.

†*Henkelodon* G. Hahn, 1977:175.

L. Juras.; Eu.

†*Meketichoffatia* Hahn, 1993:205.

L. Juras.; Eu.

†*Kielanodon* G. Hahn, 1987:182.

L. Juras.; Eu.

†*Guimarotodon* G. Hahn, 1969:83.
    L. Juras.; Eu.

†*Pseudobolodon* G. Hahn, 1977:166.
    L. Juras.; Eu.

†*Meketibolodon* G. Hahn, 1993:204.
    L. Juras.; Eu.

†*Ctenacodon* Marsh, 1879:396.
    [= †*Allodon* Marsh, 1881:511.]
    L. Juras.; N.A.

†*Psalodon* Simpson, 1926:239.
    L. Juras.; N.A.

†*Plagiaulax* Falconer, 1857:262-282.
    L. Juras.; Eu.

†*Zofiabaatar* Bakker & Carpenter, 1990:7.
    L. Juras.; N.A.

†*Galveodon* Hahn & Hahn, 1992:150.
    E. Cret.; Eu.

†*Lavocatia* Canudo & Cuenca-Bescós, 1996:218.
    E. Cret.; Eu.

†*Sunnyodon* Kielan-Jaworowska & Ensom, 1992:107-108.
    L. Juras.; Eu.

†*Gerhardodon* Kielan-Jaworowska & Ensom, 1992:103-104.
    L. Juras.; Eu.

Family †**Bolodontidae** Osborn, 1887:3.
    [Including †Eobaataridae Kielan-Jaworowska, Dashzeveg & Trofimov, 1987:7;
    †Eobaatarinae Kielan-Jaworowska & Ensom, 1991:36.]
    L. Juras.-E. Cret.; Eu. E. Cret.; As.

†*Monobaatar* Kielan-Jaworowska, Dashzeveg & Trofimov, 1987:18.
    E. Cret.; As.

†*Eobaatar* Kielan-Jaworowska, Dashzeveg & Trofimov, 1987:7.
    E. Cret.; As. E. Cret.; Eu.

†*Bolodon* Owen, 1871:54.
    [Including †*Plioprion* Cope, 1884:691.]
    L. Juras.-E. Cret.; Eu.

†*Loxaulax* Simpson, 1928:49.
    [Including †*Parendotherium* Crusafont Pairó & Adrover, 1966:30.]
    E. Cret.; Eu.

Family †**Hahnodontidae** Sigogneau-Russell, 1991:123.
    E. Cret.; Af.

†*Hahnodon* Sigogneau-Russell, 1991:123.
    E. Cret.; Af.

Family †**Albionbaataridae** Kielan-Jaworowska & Ensom, 1994:19.
    L. Juras. and/or E. Cret.; Eu.

†*Albionbaatar* Kielan-Jaworowska & Ensom, 1994:19.
    L. Juras. and/or E. Cret.; Eu.

Family †**Arginbaataridae** Hahn & Hahn, 1983:45.
    E. Cret.; As.

†*Arginbaatar* Trofimov, 1980:209.
    E. Cret.; As.[1]

[1] Mongolia.

Family †**Kogaionidae** Rădulescu & Samson, 1996:177.
L. Cret.; Eu.

†*Kogaionon* Rădulescu & Samson, 1996:177.
L. Cret.; Eu.

Suborder †**CIMOLODONTA** McKenna, 1975:40.[1]

[1] Proposed as an infraorder. Ranked as a suborder by Kielan-Jaworowska & Nessov, 1992:13. Used as an order by McKenna, in Stucky & McKenna, in Benton, ed., 1993:741.

†*Paracimexomys* Archibald, 1982:111.
E.-L. Cret.; N.A.   ?L. Cret.; Eu.[1]

[1] E. Europe (Romania).

†*Dakotamys* Eaton, 1995:770.
L. Cret.; N.A.

†*Bryceomys* Eaton, 1995:775.
L. Cret.; N.A.

†*Clemensodon* Krause, 1992:2.
L. Cret.; N.A.

†*Uzbekbaatar* Kielan-Jaworowska & Nessov, 1992:13.
L. Cret.; As.

†*Hainina* Vianey-Liaud, 1979:123.
E.-L. Paleoc.; Eu.

†*Cimexomys* Sloan & Van Valen, 1965:221.
L. Cret.-E. Paleoc.; N.A.

†*Barbatodon* Radulesco & Samson, 1986:1825.[1]
L. Cret.; Eu.

[1] (Rădulescu).

Family †**Sloanbaataridae** Kielan-Jaworowska, 1974:37.
L. Cret.; As.   L. Paleoc.; N.A.

Subfamily †**Sloanbaatarinae** Kielan-Jaworowska, 1974:37, **new rank**.
[= †Sloanbaataridae Kielan-Jaworowska, 1974:37.]
L. Cret.; As.

†*Sloanbaatar* Kielan-Jaworowska, 1970:42.
L. Cret.; As.

Subfamily †**Chulsanbaatarinae** Kielan-Jaworowska, 1974:38, **new rank**.
[= †Chulsanbaataridae Kielan-Jaworowska, 1974:38.]
L. Cret.; As.   L. Paleoc.; N.A.

†*Kamptobaatar* Kielan-Jaworowska, 1970:46.
L. Cret.; As.

†*Pentacosmodon* Jepsen, 1940:321.
L. Paleoc.; N.A.

†*Chulsanbaatar* Kielan-Jaworowska, 1974:38.
L. Cret.; As.

†*Kryptobaatar* Kielan-Jaworowska, 1970:44.
[= †*Gobibaatar* Kielan-Jaworowska, 1970:38.[1]]
[Including †*Tugrigbaatar* Kielan-Jaworowska & Dashzeveg, 1978:117.]
L. Cret.; As.

[1] †*Gobibaatar* has page priority but was selected as the junior synonym by Kielan-Jaworowska in 1980.

†*Bulganbaatar* Kielan-Jaworowska, 1974:31.
L. Cret.; As.

†*Nemegtbaatar* Kielan-Jaworowska, 1974:34.
L. Cret.; As.

†*Djadochtatherium* Simpson, 1925:1.
L. Cret.; As.

†*Catopsbaatar* Kielan-Jaworowska, 1994:134.
L. Cret.; As.

†*Tombaatar* Rougier, Novacek & Dashzeveg, 1997:4.
L. Cret.; As.

Superfamily †**Ptilodontoidea** Cope, 1887:567, **new rank**.
[= †Ptilodontidae Cope, 1887:567; †PTILODONTOIDEA Sloan & Van Valen, 1965:221.[1]]

[1] Proposed as suborder; reduced to parvorder by McKenna, 1975:40; regarded as an order by Kielan-Jaworowska, 1982:375.

Family †**Cimolodontidae** Marsh, 1889:84.
L. Cret.-L. Paleoc.; N.A.

†*Cimolodon* Marsh, 1889:84.
[= †*Nanomys* Marsh, 1889:85; †*Nanomyops* Marsh, 1892:261.]
L. Cret.; N.A.

†*Anconodon* Jepsen, 1940:289.
E.-L. Paleoc.; N.A.

Family †**Ptilodontidae** Cope, 1887:567.
[= †Chirogidae Cope, 1887:567.] [Including †Ectypodidae Sloan & Van Valen, 1965:221; †Ectypodontidae Sloan & Van Valen, 1965:221.]
L. Cret.-L. Eoc.; N.A. L. Paleoc.-E. Eoc.; Eu. E. Eoc.; As.

Subfamily †**Neoplagiaulacinae** Ameghino, 1890:176, **new rank**.
[= †Neoplagiaulacidae Ameghino, 1890:176.]
L. Cret.-L. Eoc.; N.A. L. Paleoc.-E. Eoc.; Eu. E. Eoc.; As.

†*Mesodma* Jepsen, 1940:267.
L. Cret.-L. Paleoc.; N.A.

†*Parectypodus* Jepsen, 1930:120.
E. Paleoc.-E. Eoc.; N.A.

†*Neoplagiaulax* Lemoine, 1882:1009-1011.
?L. Cret., E. Paleoc.-E. Eoc.; N.A. L. Paleoc.; Eu.

†*Cernaysia* Vianey-Liaud, 1986:143.
E. Paleoc.; N.A. L. Paleoc.; Eu.

†*Krauseia* Vianey-Liaud, 1986:155.
E. Paleoc.; N.A.

†*Xyronomys* Rigby, 1980:35.
E. Paleoc.; N.A.

†*Xanclomys* Rigby, 1980:36.
E. Paleoc.; N.A.

†*Ectypodus* Matthew & Granger, 1921:1.
[Including †*Charlesmooria* Kühne, 1969:200.]
E. Paleoc.-L. Eoc.; N.A. L. Paleoc.-E. Eoc.; Eu.

†*Mimetodon* Jepsen, 1940:265, 314-317.
E.-L. Paleoc.; N.A. ?L. Paleoc.; Eu.

†*Mesodmops* Tong & Wang, 1994:276.
E. Eoc.; As.

Subfamily †**Ptilodontinae** Cope, 1887:567.
[= †Ptilodontidae Cope, 1887:567; †Ptilodontinae Jepsen, 1940:263.]
E.-L. Paleoc.; N.A.

†*Kimbetohia* Simpson, 1936:2.
E. Paleoc.; N.A.

†*Ptilodus* Cope, 1881:921-922.
[= †*Chirox* Cope, 1884:321.]
E.-L. Paleoc.; N.A.

†*Baiotomeus* Krause, 1987:596.
E. Paleoc.; N.A.

†*Prochetodon* Jepsen, 1940:309.
E.-L. Paleoc.; N.A.

Superfamily †**Taeniolabidoidea** Granger & Simpson, 1929:603, **new rank**.
[= †Taeniolabididae Granger & Simpson, 1929:603;
†TAENIOLABIDOIDEA Sloan & Van Valen, 1965:222.[1]]
[1] Proposed as suborder; reduced to parvorder by McKenna, 1975:40;
regarded as an order by Kielan-Jaworowska, 1982:376.

†*Viridomys* Fox, 1971:935.
L. Cret.; N.A.

Family †**Cimolomyidae** Marsh, 1889:177.[1]
[= †Cimolomidae Marsh, 1889:177; †Cimolomyidae Sloan & Van Valen, 1965:222.]
L. Cret.; N.A.
[1] See the International Code of Zoological Nomenclature, Article 33(a).

†*Cimolomys* Marsh, 1889:84.
[= or including †*Allacodon* Marsh, 1889:178.[1]]
L. Cret.; N.A.
[1] †*Allacodon* may, instead, be a synonym of †*Cimolodon*.

Family †**Eucosmodontidae** Jepsen, 1940:263.
[= †Eucosmodontinae Jepsen, 1940:263; †Eucosmodontidae Sloan & Van Valen,
1965:223.] [Including †Boffiidae Hahn & Hahn, 1983:101.]
L. Cret.-E. Eoc.; N.A. E. Paleoc.-E. Eoc.; Eu.

Subfamily †**Microcosmodontinae** Holtzman & Wolberg, 1977:4.
E.-L. Paleoc.; N.A. E. Eoc.; Eu.

†*Acheronodon* Archibald, 1982:68.
E.-L. Paleoc.; N.A.

†*Microcosmodon* Jepsen, 1930:508.
E.-L. Paleoc.; N.A. E. Eoc.; Eu.

Subfamily †**Eucosmodontinae** Jepsen, 1940:263.
[Including †Boffiidae Hahn & Hahn, 1983:101.]
L. Cret.-E. Eoc.; N.A. E.-L. Paleoc.; Eu.

Tribe †**Eucosmodontini** Jepsen, 1940:263.
[= †Eucosmodontidae Jepsen, 1940:263.]
L. Cret.-E. Paleoc.; N.A.

†*Eucosmodon* Matthew & Granger, 1921:1.
E. Paleoc.; N.A.

†*Stygimys* Sloan & Van Valen, 1965:223.
L. Cret.-E. Paleoc.; N.A.

Tribe †**Boffiini** Hahn & Hahn, 1983:101, **new rank**.
[= †Boffiidae Hahn & Hahn, 1983:101.]
L. Cret.-E. Eoc.; N.A. E.-L. Paleoc.; Eu.

†*Liotomus* Cope, 1884:691, 695.
[= †*Neoctenacodon* Lemoine, 1891:266.]
E. Paleoc.; N.A. L. Paleoc.; Eu.

†*Essonodon* Simpson, 1927:2.
L. Cret.; N.A.

†*Boffius* Vianey-Liaud, 1979:119.
E. Paleoc.; Eu.

†*Neoliotomus* Jepsen, 1930:122.
L. Paleoc.-E. Eoc.; N.A.

Family †**Taeniolabididae** Granger & Simpson, 1929:603.
> [= †Polymastodontidae Cope, 1884:687;
> †Taeniolabididae Granger & Simpson, 1929:603; †Taeniolabidae Romer, 1933.]
> [Including †Dipriodontidae Marsh, 1889:85; †Tripriodontidae Marsh, 1889:86;
> †Lambdopsalidae Chow & Qi, 1978:77; †Buginbaatarinae Hahn & Hahn, 1983:220.]
> L. Cret., L. Paleoc.; As. L. Cret.-L. Paleoc.; N.A.

†*Buginbaatar* Kielan-Jaworowska & Sochava, 1969:359.
> L. Cret.; As.

†*Meniscoessus* Cope, 1882:830.
> [= †*Dipriodon* Marsh, 1889:85; †*Tripriodon* Marsh, 1889:86; †*Selenacodon*
> Marsh, 1889:86; †*Halodon* Marsh, 1889:87; †*Oracodon* Marsh, 1889:178.]
> L. Cret.; N.A.

†*Catopsalis* Cope, 1882:416-417.
> E.-L. Paleoc.; N.A.

†*Prionessus* Matthew & Granger, 1925:6.
> L. Paleoc.; As.

†*Lambdopsalis* Chow & Qi, 1978:77.
> L. Paleoc.; As.[1]
> [1] E. Asia.

†*Sphenopsalis* Matthew, Granger & Simpson, 1928:6.
> L. Paleoc.; As.

†*Taeniolabis* Cope, 1882:604.
> [= †*Polymastodon* Cope, 1882:684-685.]
> E. Paleoc.; N.A.

Suborder †**GONDWANATHERIA** Mones, 1987:237-238.[1]
> [= †Gondwanatherioidea Krause & J. F. Bonaparte, 1993:9382.[2]]
> [1] Placement among the multituberculates is increasingly supported. Krause & Bonaparte, 1990:31A assigned †GONDWANATHERIA subordinal rank within †MULTITUBERCULATA, which, in turn, was assigned ordinal rank.
> [2] A family-group name, subject to the International Code of Zoological Nomenclature and therefore invalid because of the priority of †Sudamericidae [see ICZN, Articles 23(d) and 36].

Family †**Ferugliotheriidae** J. F. Bonaparte, 1986:55.
> L. Cret.; S.A.

†*Ferugliotherium* J. F. Bonaparte, 1986:55.
> [= †*Vucetichia* J. F. Bonaparte, 1990:77.]
> L. Cret.; S.A.

Family †**Sudamericidae** Scillato-Yané & Pascual, 1984:15.
> [Including †Gondwanatheriidae J. F. Bonaparte, 1986:264.]
> L. Cret.-E. Paleoc.; S.A.

†*Gondwanatherium* J. F. Bonaparte, 1986:264.
> L. Cret.; S.A.

†*Sudamerica* Scillato-Yané & Pascual, 1984:15.
> E. Paleoc.; S.A.

Infraclass †**TRICONODONTA** Osborn, 1888:251.[1]
> [= †TRICONOTHERIA Vandebroek, 1964:155[2]; †EUTRICONODONTA Kermack, Mussett & Rigney, 1973:110[3]; †Triconodontoidea Kalandadze & Rautian, 1992:48; †TRICONODONTIFORMES Kinman, 1994:37.]
> [Including †PHASCOLOTHERIA Ameghino, 1889.[4]]
> [1] Long classified as an order. Used as an infraclass by Vandebroek, 1964:155. Elevated to rank of subclass by McKenna, 1993:740.
> [2] In part.
> [3] Proposed as a suborder. Elevated to rank of order by McKenna, in Stucky & McKenna, in Benton, ed., 1993:741.
> [4] Proposed as an order. Paper not seen, but see Ameghino's Obras Completas 6:585.

†*Hallautherium* Clemens, 1980:85.
L. Trias.; Eu.

Family †**Austrotriconodontidae** Bonaparte, 1992:100.
L. Cret.; S.A.

†*Austrotriconodon* J. F. Bonaparte, 1986:53.
L. Cret.; S.A.

Family †**Amphilestidae** Osborn, 1888:228.
[= †Amphilestinae Osborn, 1888:228; †Amphilestidae Kühne, 1958:223.] [Including †Phascolotherida Haeckel, 1866:clvii[1]; †Phascolotheridae Osborn, 1887:288.[2]]
M. and/or L. Juras., E. Cret.; As. M. Juras.; Eu. L. Juras.-E. Cret.; N.A. E. Cret.; Af.[3]

[1] Given family rank by Haeckel. Haeckel's term has long since become a *nomen oblitum*. It covered †*Phascolotherium*, †*Thylacotherium* (= †*Amphitherium*) and †*Microlestes* (= †*Thomasia*).
[2] Osborn's family, for †*Phascolotherium* and †*Tinodon*, should be suppressed, but, if ever used, its spelling should be corrected to "†Phascolotheriidae."
[3] Unnamed specimens have been reported from the early Cretaceous of Morocco.

†*Liaotherium* Zhou, Cheng & Wang, 1991:167, 173.[1]
?M. Juras.; As.

[1] (Chow, Cheng & Wang). Probably best regarded as a *nomen dubium*.

Subfamily †**Amphilestinae** Osborn, 1888:228.
[Including †Phascolotherida Haeckel, 1866:clvii[1]; †Phascolotheridae Osborn, 1887:288.[1]]
M. Juras.; Eu. L. Juras.; N.A.

[1] See footnote under †Amphilestidae.

†*Amphilestes* Owen, 1859:157-158.
M. Juras.; Eu.

†*Phascolotherium* Owen, 1838:9.
M. Juras.; Eu.

†*Aploconodon* Simpson, 1925:336.
L. Juras.; N.A.

†*Phascolodon* Simpson, 1925:334.
L. Juras.; N.A.

Subfamily †**Gobiconodontinae** Chow & Rich, 1984:226, 227.
[= †Gobiconodontidae Jenkins & Schaff, 1988:2.]
M. and/or L. Juras., E. Cret.; As. E. Cret.; N.A.

†*Klamelia* Chow & Rich, 1984:226, 227.
M. and/or L. Juras.; As.

†*Guchinodon* Trofimov, 1978:215.
E. Cret.; As.[1]

[1] Mongolia.

†*Gobiconodon* Trofimov, 1978:213.[1]
E. Cret.; As.[2] E. Cret.; N.A.

[1] *Nomen nudum* in Trofimov, 1974:20.
[2] Mongolia.

Family †**Triconodontidae** Marsh, 1887:341.
L. Juras.; Eu. L. Juras.-L. Cret.; N.A. E. Cret.; Af.

†*Dyskritodon* Sigogneau-Russell, 1995:154.
E. Cret.; Af.

†*Ichthyoconodon* Sigogneau-Russell, 1995:159.
E. Cret.; Af.

Subfamily †**Triconodontinae** Marsh, 1887:341.
[= †Triconodontidae Marsh, 1887:341; †Triconodontinae Hay, 1902:566.]
L. Juras.; Eu. L. Juras.-E. Cret.; N.A.

†*Trioracodon* Simpson, 1928:88.
L. Juras.; Eu. L. Juras.; N.A.

†*Triconodon* Owen, 1859:161.
[= †*Triacanthodon* Owen, 1871:72.]
L. Juras.; Eu.

†*Priacodon* Marsh, 1887:341.
L. Juras.; N.A.

†*Astroconodon* Patterson, 1951:31-41.
E. Cret.; N.A.

Subfamily †**Alticonodontinae** Fox, 1976:1109.
L. Cret.; N.A.

†*Alticonodon* Fox, 1969:1254.
L. Cret.; N.A.

Infraclass **HOLOTHERIA** Wible, Rougier, Novacek, McKenna & Dashzeveg, 1995:10, 11, **new rank**.[1]
[= †PATRIOTHERIA Simpson, 1971:194[2]; ORTHOTHERIA Vandebroek, 1964:155.[3]]

[1] There has been confusion of a taxon at this level with THERIA. THERIA Parker & Haswell, 1897:448 was gradually given additional paleontological content beyond the crown group (marsupials and placentals) by various authors, e.g., Gregory (1910:464), Simpson (1945), and McKenna (1975), until it subsumed essentially the same content as HOLOTHERIA used here. Hopson (1994:205, 208) referred to these animals plus monotremes (but not multituberculates) as "holotheres," but his term was diagnosed (reversed triangular molar pattern), not taxically defined, and in any case is not formal nor subject to the rules of the International Code of Zoological Nomenclature. Hopson restricted THERIA to a taxon containing MARSUPIALIA and EUTHERIA (= PLACENTALIA) alone. †HOLOTHERIA Wible, Rougier, Novacek, McKenna & Dashzeveg, 1995:10-11 was used in a formal sense, but only for the ancestor of †*Kuehneotherium* and therians, plus all its descendants. That would exclude monotremes as well as multituberculates.

[2] In part. †PATRIOTHERIA was a paraphyletic concept covering †Kuehneotheriidae, †Amphidontidae, †Spalacotheriidae, †Amphitheriidae, †Paurodontidae, †Dryolestidae, and †Aegialodontidae.

[3] In part. ORTHOTHERIA also included docodonts.

Family †**Chronoperatidae** Fox, Youzwyshyn & Krause, 1992:233.[1]
L. Paleoc.; N.A.

[1] Systematic position dubious, even at this level, but not a nonmammalian cynodont.

†*Chronoperates* Fox, Youzwyshyn & Krause, 1992:233.
L. Paleoc.; N.A.

Superlegion †**KUEHNEOTHERIA** McKenna, 1975:27.
[= †KUEHNEOTHERIDIA Tatarinov, 1994:128.]

Family †**Kuehneotheriidae** Kermack, Kermack & Mussett, 1968:412, 421.
L. Trias.-E. Juras.; Eu.

†*Kuehneotherium* D. M. Kermack, K. A. Kermack & Mussett, 1968:408.
L. Trias.-E. Juras.; Eu.[1]

[1] British Isles.

†*Kuehneon* Kretzoi, 1960:307, 308.[1]
E. Juras.; Eu.[2]

[1] Originally spelled †"*Kühneon*"; spelled †"*Kuhneon*" by Kermack, Kermack & Mussett, 1968; but see the International Code of Zoological Nomenclature, Article 32(c). The single tooth upon which the type and only species †*Kuehneon duchyense* was based, "Duchy 33", has apparently been lost. The name should be, but has not been, suppressed.

[2] Wales.

Family †**Woutersiidae** Sigogneau-Russell & Hahn, 1995:246.
L. Trias.; Eu.

†*Woutersia* Sigogneau-Russell, 1983:175, 180.
L. Trias.; Eu.

Superlegion **TRECHNOTHERIA** McKenna, 1975:27.[1]
[= YANGOTHERIA Chow & Rich, 1982:129.[2]]

[1] Proposed as superlegion. Ranked as sublegion by Chow & Rich, 1982:129; used as subclass by McKenna, in Stucky & McKenna, in Benton, ed., 1993:742.

[2] Proposed as a legion.

Legion †**SYMMETRODONTA** Simpson, 1925:560.[1]

[= †SYMMETRODONTIFORMES Kinman, 1994:37; †EUSYMMETRODONTA
Tatarinov, 1994:128; †SPALACOTHERIA Kalandadze & Rautian, 1992:49.]
[Including †YINOTHERIA Chow & Rich, 1982:129[2]; †SHUOTHERIDIA Chow &
Rich, 1982:129.[3]]

[1] Proposed as order; raised to legion by McKenna, 1975:27; used as sublegion by Chow & Rich, 1982: 129;
elevated to infraclass by Vandebroek, 1964: 155 and by McKenna, in Stucky & McKenna, in Benton, ed.,
1993: 742; called a superorder by Tatarinov, 1994: 128. In addition to the occurrences noted below, an
animal that may be a symmetrodont has been reported from the Late Cretaceous or Paleocene of India.
Another animal that may be a symmetrodont occurs in the Late Cretaceous of Portugal.
[2] Proposed as a legion.
[3] Proposed as an order.

†*Casamiquelia* Bonaparte, 1990:68.
L. Cret.; S.A.

†*Thereuodon* Sigogneau-Russell, 1989:925.
E. Cret.; Af.[1]

[1] N. Africa (Morocco).

†*Atlasodon* Sigogneau-Russell, 1991:280.
E. Cret.; Af.[1]

[1] N. Africa (Morocco).

†*Kotatherium* Datta, 1981:307-312.
E. Juras.; As.[1]

[1] S. Asia (India).

†*Trishulotherium* Yadagiri, 1984:515.
E. Juras.; As.[1]

[1] S. Asia (India).

†*Indotherium* Yadagiri, 1984:517.[1]
E. Juras.; As.[2]

[1] Preoccupied by †*Indotherium* Kretzoi, 1942: 315, a synonym of †*Aprotodon*, a rhinoceros, but a new name
has not yet been proposed.
[2] S. Asia (India).

Family †**Shuotheriidae** Chow & Rich, 1982:129.

[= †YINOTHERIA Chow & Rich, 1982:129[1]; †SHUOTHERIDIA Chow & Rich,
1982:129.[2]]
L. Juras.; As.

[1] Proposed as a legion.
[2] Proposed as an order.

†*Shuotherium* Chow & Rich, 1982:129.
L. Juras.; As.[1]

[1] E. Asia (northern Sichuan).

Order †**AMPHIDONTOIDEA** Prothero, 1981:320.[1]

[1] Proposed as sublegion; ranked as infralegion by Chow & Rich, 1982:129; elevated to rank of order by
McKenna, in Stucky & McKenna, in Benton, ed., 1993:742.

Family †**Amphidontidae** Simpson, 1925:460.
E. Juras., E. Cret.; As. L. Juras.; N.A.

†*Nakunodon* Yadagiri, 1985:415.
E. Juras.; As.[1]

[1] S. Asia (India).

†*Amphidon* Simpson, 1925:460.
L. Juras.; N.A.

†*Manchurodon* Yabe & Shikama, 1938:355.
E. Cret.; As.[1]

[1] E. Asia.

†*Gobiodon* Trofimov, 1980:210.
E. Cret.; As.[1]

[1] Mongolia.

Order †**SPALACOTHERIOIDEA** Prothero, 1981:320.[1]
[= †SPALACOTHERIIDA Prothero, 1981:321.[2]] [Including †QUIROGATHERIA
Bonaparte, 1992:104, 105.[3]]

[1] Proposed as sublegion; ranked as infralegion by Chow & Rich, 1982:129; changed to rank of order by
McKenna, in Stucky & McKenna, in Benton, ed., 1993:742.
[2] Proposed as infraclass.
[3] Described as an order. Validity questionable.

Family †**Tinodontidae** Marsh, 1887:340.
[= or including †TINODONTIDA Prothero, 1981:320.[1]] [Including †Bondesiidae
Bonaparte, 1990:66.]
M. Juras.; Eu. L. Juras., L. Cret.; N.A. L. Cret.; S.A.

[1] Proposed as infraclass.

†*Cyrtlatherium* Freeman, 1979:146.
M. Juras.; Eu.

†*Tinodon* Marsh, 1879:216.
[= †*Menacodon* Marsh, 1887:340; †*Eurylambda* Simpson, 1929:41.]
L. Juras.; N.A.

†*Mictodon* Fox, 1984:1206.[1]
L. Cret.; N.A.

[1] Assignment to †Tinodontidae is uncertain.

†*Bondesius* Bonaparte, 1990:66.
L. Cret.; S.A.

Family †**Spalacotheriidae** Marsh, 1887:340.
[= †Spalacotheridae Marsh, 1887:340; †Spalacotheriidae Lydekker, 1887:292;
†Peralestidae Osborn, 1887:289.]
L. Juras.-E. Cret.; Eu. E. Cret.; Af. E.-L. Cret.; N.A.

†*Spalacotherium* Owen, 1854:482.
[= or including †*Peralestes* Owen, 1871:33.]
L. Juras.-E. Cret.; Eu.

†*Spalacotheroides* Patterson, 1955:690.
E. Cret.; N.A.

†*Symmetrodontoides* Fox, 1976:1110.
L. Cret.; N.A.

†*Spalacotheridium* Cifelli, 1990:334.
L. Cret.; N.A.

†*Microderson* Sigogneau-Russell, 1991:281.
E. Cret.; Af.

Family †**Barbereniidae** Bonaparte, 1990:70.
L. Cret.; S.A.

†*Quirogatherium* Bonaparte, 1990:73.
L. Cret.; S.A.

†*Barberenia* Bonaparte, 1990:70.
L. Cret.; S.A.

Legion **CLADOTHERIA** McKenna, 1975:27.[1]

[1] Proposed as legion; ranked as a sublegion by Chow & Rich, 1982:129; elevated to infraclass by McKenna,
in Stucky & McKenna, in Benton, ed., 1993:742.

Sublegion †**DRYOLESTOIDEA** Butler, 1939:353.[1]
[= †DRYOLESTIFORMES Kinman, 1994:37.] [Including †PANTOTHERIA Marsh,
1880:239.[2]]

[1] Proposed as suborder; called an order by Vandebroeck, 1961, 1964; raised to rank of sublegion by
McKenna, 1975:27; used as infralegion by Chow & Rich, 1982:129; raised to rank of legion by McKenna,
in Stucky & McKenna, in Benton, ed., 1993:742.
[2] In part. Proposed as order.

Order †**DRYOLESTIDA** Prothero, 1981:321.[1]

[= †DRYOLESTIA Kalandadze & Rautian, 1992:50.]

[1] Proposed as infraclass. Rank lowered to order by McKenna, in Stucky & McKenna, in Benton (1993: 742).

Family †**Dryolestidae** Marsh, 1879:397.

[Including †Amblotheriidae Osborn, 1887:7; †Amblytheriidae Cope, 1889:876[1]; †Kurtodontidae Osborn, 1888:234; †Stylacodontidae Giebel, 1879:629; †Stylodontidae Marsh, 1879:60; †Athrodontidae Osborn, 1887:290.]

M. Juras.-E. Cret.; Eu.[2] L. Juras., L. Cret.; N.A.[3] L. Cret.; S.A.

[1] Misspelling or attempted correction.
[2] Dryolestids unassigned to genus have been found in the European Middle Jurassic (upper Bathonian).
[3] Dryolestids unassigned to genus have been found in the North American Late Cretaceous (late Campanian or early Maastrichtian).

†*Phascolestes* Owen, 1871:35; pl.1.[1]

L. Juras.; Eu.

[1] As subgenus of †*Peralestes*.

†*Miccylotyrans* Simpson, 1927:414.

L. Juras.; N.A.

†*Laolestes* Simpson, 1927:411.

L. Juras.; N.A.

†*Amblotherium* Owen, 1871:29.

[= †*Stylodon* Owen, 1866:199[1]; †*Achyrodon* Owen, 1871:37; †*Odontostylus* Trouessart, 1898:1247[2]; †*Trouessartia* Cossman, 1899:30[3]; †*Trouessartiella* Cossman, 1899:1433[4].] [Including †*Stylacodon* Marsh, 1879:60; †*Laodon* Marsh, 1887:337[5]; †*Kepolestes* Simpson, 1927:413.]

L. Juras.; Eu. L. Juras.; N.A.

[1] Not *Stylodon* Beck, 1837:46, a mollusk.
[2] Not *Odontostylus* Gray, 1840, a mollusk.
[3] New name for †*Odontostylus* Trouessart. Not *Trouessartia* Canestrini & Kramer, 1899:59, a spider.
[4] New name for †*Odontostylus* Trouessart.
[5] Objective synonym of †*Stylacodon*.

†*Dryolestes* Marsh, 1878:459.

[= †*Asthenodon* Marsh, 1887:336; †*Herpetairus* Simpson, 1927:413.]

?L. Juras.; Eu. L. Juras.; N.A.

†*Peraspalax* Owen, 1871:40.

L. Juras.; Eu.

†*Kurtodon* Osborn, 1887:1020.[1]

[= †*Athrodon* Osborn, 1887:290[2]; †*Curtodon* Zittel, 1892:102; †*Cyrtodon* Winge, 1893:118.]

L. Juras.; Eu.

[1] New name for †*Athrodon* Osborn, 1887.
[2] Not †*Athrodon* Sauvage, 1880:530, a genus of fishes.

†*Melanodon* Simpson, 1927:413.

[= †*Malthacolestes* Simpson, 1927:414; †*Laolestes* Simpson, 1927:411.]

?L. Juras., E. Cret.; Eu. L. Juras.; N.A.

†*Comotherium* Prothero, 1981:286.

L. Juras.; N.A.

†*Crusafontia* Henkel & Krebs, 1969:460.

E. Cret.; Eu.

†*Leonardus* Bonaparte, 1990:74.

L. Cret.; S.A.

†*Groebertherium* Bonaparte, 1986:51.

L. Cret.; S.A.

Family †**Paurodontidae** Marsh, 1887:341.
          [Including †Henkelotheriidae Krebs, 1991:18.]
          L. Juras.; Af. L. Juras.; Eu. L. Juras.; N.A.

    †*Brancatherulum* Dietrich, 1927:425.
        L. Juras.; Af.

    †*Paurodon* Marsh, 1887:342.
        L. Juras.; N.A.

    †*Archaeotrigon* Simpson, 1927:410.
        L. Juras.; N.A.

    †*Tathiodon* Simpson, 1927:71.
        [= †*Tanaodon* Simpson, 1927:410.[1]]
        L. Juras.; N.A.
      [1] Not †*Tanaodon* Kirk, 1927, a Chinese Devonian mollusk.

    †*Pelicopsis* Simpson, 1927:414.
        L. Juras.; N.A.

    †*Araeodon* Simpson, 1937:2.
        L. Juras.; N.A.

    †*Foxraptor* Bakker & Carpenter, 1990:4.
        L. Juras.; N.A.

    †*Henkelotherium* Krebs, 1991:18.
        L. Juras.; Eu.

    †*Euthlastus* Simpson, 1927:414.
        L. Juras.; N.A.

Family †**Donodontidae** Sigogneau-Russell, 1991:282.
        E. Cret.; Af.

    †*Donodon* Sigogneau-Russell, 1991:282.
        E. Cret.; Af.[1]
      [1] N. Africa (Morocco).

Family †**Mesungulatidae** Bonaparte, 1986:49.
        L. Cret.; S.A.

    †*Mesungulatum* Bonaparte & Soria, 1985:179.
        L. Cret.; S.A.

Family †**Reigitheriidae** Bonaparte, 1990:76.
        L. Cret.; S.A.

    †*Reigitherium* Bonaparte, 1990:76.
        L. Cret.; S.A.

Family †**Brandoniidae** Bonaparte, 1992:104, 105.
        L. Cret.; S.A.

    †*Brandonia* Bonaparte, 1990:69.
        L. Cret.; S.A.

Order †**AMPHITHERIIDA** Prothero, 1981:321.[1]
      [1] Proposed as infraclass. Rank lowered to order by McKenna, in Stucky & McKenna, in Benton, ed., 1993:742.

Family †**Amphitheriidae** Owen, 1846:29.
        M. Juras.; Eu.

    †*Amphitherium* de Blainville, 1838:417.
        [= †*Heterotherium* de Blainville, 1838:417[1]; †*Amphigonus* Agassiz, 1838:3;
        †*Thylacotherium* Valenciennes, 1838:580; †*Botheration-therium* Charlesworth,
        1838:731[2]; †*Botheratiotherium* de Blainville, 1838:735; †*Amphitylus* Osborn,
        1888:192.[3]]
        M. Juras.; Eu.
      [1] Not †*Heterotherium* Fischer de Waldheim, 1822:7, a *genus coelebs*.

[2] See the International Code of Zoological Nomenclature, Article 27.
[3] *Nomen nudum* in Osborn, 1887: 283.

Sublegion **ZATHERIA** McKenna, 1975:27.[1]
> [Including PROTOTRIBOSPHENIDA Wible, Rougier, Novacek, McKenna & Dashzeveg, 1995:19.[2]]
> [1] Used as an infralegion by Chow and Rich, 1982:129.
> [2] First proposed by Rougier, 1993 in a technically unpublished PhD dissertation.

 Family †**Arguitheriidae** Dashzeveg, 1994:2.
> E. Cret.; As.

  †*Arguitherium* Dashzeveg, 1994:2.
>   E. Cret.; As.

 Family †**Arguimuridae** Dashzeveg, 1994:2.
> E. Cret.; As.

  †*Arguimus* Dashzeveg, 1979:200.
>   E. Cret.; As.[1]
>   [1] E. Asia.

 Family †**Vincelestidae** Bonaparte, 1986:58.
> E. Cret.; S.A.

  †*Vincelestes* Bonaparte, 1986:58.
>   E. Cret.; S.A.

Infralegion †**PERAMURA** McKenna, 1975:27.[1]
> [= †PERAMURIDA McKenna, 1975:24[2]; †PERAMURIFORMES Kinman, 1994:37.]
> [1] Defined as an infraclass. Used as a suborder by Kalandadze & Yakhontov, 1992.
> [2] *Lapsus calami.*

 Family †**Peramuridae** Kretzoi, 1946:111.
> M. Juras.-E. Cret.; Eu. E. Cret.; Af.

  †*Palaeoxonodon* Freeman, 1976:1053.
>   M. Juras.; Eu.

  †*Peramus* Owen, 1871:41.
>   [= †*Leptocladus* Owen, 1871:53.]
>   L. Juras.; Eu.

  †*Pocamus* Canudo & Cuenca-Bescós, 1996:222.
>   E. Cret.; Eu.

  †*Abelodon* Brunet, Coppens, Dejax, L. Flynn, Heintz, Hell, Jacobs, Jehenne, Mouchelin, Pilbeam & Sudre, 1990:1139, 1142.
>   E. Cret.; Af.

Infralegion **TRIBOSPHENIDA** McKenna, 1975:27, **new rank**.[1]
> [1] Proposed as infraclass.

  †*Hypomylos* Sigogneau-Russell, 1992:390.
>   E. Cret.; Af.

 Family †**Necrolestidae** Ameghino, 1894:259.
> [= †Necrolestoidea Kalandadze & Rautian, 1992:50.]
> E. Mioc.; S.A.

  †*Necrolestes* Ameghino, 1891:303.
>   E. Mioc.; S.A.

Supercohort †**AEGIALODONTIA** Butler, 1978:1, **new rank**.[1]
> [= †AEGIALODONTIFORMES Kinman, 1994:37.]
> [1] Proposed as an order. Here restricted in content but raised in rank.

 Family †**Aegialodontidae** Kermack, Kermack & Mussett, 1968:421.[1]
> [= †TRIBOSPHENA Turnbull, 1971:176[2]; †TRIBOTHERIA Butler, 1978:24[3]; †AEGIALODONTIFORMES Kinman, 1994:37.]

E. Cret.; As. E. Cret.; Eu.

[1] Listed by these authors as "Aegialodontidae Kermack", but we have not found an earlier published reference although it has sometimes been incorrectly attributed to Kermack, Lees & Mussett, 1965.

[2] In part; also included †*Peramus* and therians from the Albian of Texas. Proposed as order; with ZALAMBDODONTA combined in a therian cohort TRIBOSPHENATA.

[3] In part; also included †Deltatheridiidae, †Kermackiidae, †Pappotheriidae, and †*Holoclemensia*. Proposed as an infraclass.

†*Aegialodon* Kermack, Lees & Mussett, 1965:539.

[= or including †*Kielantherium* Dashzeveg, 1975:402.]

E. Cret.; As. E. Cret.; Eu.

Supercohort **THERIA** Parker & Haswell, 1897:448, **new rank**.[1] Therians.

[= or including EUTHERIA Gill, 1872:v, vi.[2]] [Including †TRIBOTHERIA Butler, 1978:1[3]; †PAPPOTHERIDA Butler, 1978:1.[4]]

[1] THERIA has had various meanings and ranks. Here we use it for marsupials and placentals, their common ancestor, and all descendants of that ancestor. This is a crown-group definition and, following Hopson (1994), restricts its content. THERIA was originally proposed as a subclass. The rank was changed to legion by McKenna, in Stucky and McKenna, in Benton, ed., 1993:742.

[2] Not EUTHERIA: Huxley, 1880:657 and of most other authors, a term meant to include placentals only. Until recently, Gill's prior use of the term to cover both placentals and marsupials was regarded as synonymous with Huxley's homonym and Gill was credited as the author of a concept not his own (e.g., Simpson, 1945:47). Gill's concept has been said to have been modified by Huxley, but Huxley did not discuss Gill's work. The International Code of Zoological Nomenclature does not govern above the family-group, so for the sake of stability we choose to use a different term. Gill's original concept was abandoned by Gill himself in later pages of his own paper.

[3] Proposed as an infraclass.

[4] Proposed as order.

†*Anisorhizus* Ameghino, 1902:27.

E. Eoc.; S.A.

†*Tetraprothomo* Ameghino, 1907.[1]

E. Plioc.; S.A.

[1] Not Ameghino, 1884:381, a hypothetical construct.

†*Tribotherium* Sigogneau-Russell, 1991:1636.

E. Cret.; Af.[1]

[1] N. Africa (Morocco).

†*Zygiocuspis* Cifelli, 1990:344.

L. Cret.; N.A.

†*Falepetrus* Clemens & Lillegraven, 1986:56.

L. Cret.; N.A.

†*Dakotadens* Eaton, 1993:109.

L. Cret.; N.A.

†*Aethomylos* Novacek, 1976:39.

M. Eoc.; N.A.

†*Kasserinotherium* Crochet, 1986:925.

E. Eoc.; Af.

†*Paleomolops* Cifelli, 1994:117.

L. Cret.; N.A.

Subfamily †**Russellmyinae** Estravis, 1990:761, 764.

E. Eoc.; Eu.

†*Russellmys* Estravis, 1990:762.

E. Eoc.; Eu.

Family †**Pappotheriidae** Slaughter, 1965:4.

[= †PAPPOTHERIDA Butler, 1978:1.[1]] [Including †Bobolestinae Nessov, 1989:45; †Bobolestidae Nessov, 1992.]

E. Cret.; As. E. Cret.; N.A.

[1] Proposed as an order for †Pappotheriidae, which included †*Holoclemensia*. Raised to the rank of supercohort by Aplin & Archer, 1987:xxi; used at ordinal rank by Szalay, 1994.

†*Pappotherium* Slaughter, 1965:4.
E. Cret.; N.A.

†*Bobolestes* Nessov, 1985:9.
E. Cret.; As.

†*Slaughteria* Butler, 1978:12.
E. Cret.; N.A.

Family †**Holoclemensiidae** Aplin & Archer, 1987:xxviii.[1]
[= †PROTODELPHIA Archer, 1984:786.[2]]
E. Cret.; N.A.

[1] "†Holoclemensiidae" Archer, 1984:787 was proposed tentatively, as indicated by quotation marks (see the International Code of Zoological Nomenclature, Article 15).
[2] Cohort "PROTODELPHIA" was proposed tentatively by Archer, 1984:786. †PROTODELPHIA was used by Aplin & Archer, 1987:xxi as a supercohort, coordinate with MARSUPIALIA. All of this cumbersome superstructure rests on a single poorly known species, †*Holoclemensia texana*. Moreover, the tentative nature of †PROTODELPHIA in 1984 is irrelevant because the International Code of Zoological Nomenclature, Article 1, does not govern categories higher than the family group.

†*Holoclemensia* Slaughter, 1968:1306.
[= †*Clemensia* Slaughter, 1968:254.[1]]
E. Cret.; N.A.

[1] Not *Clemensia* Packard, 1864, a lepidopteran.

Family †**Kermackiidae** Butler, 1978:22.
E. Cret.; N.A.

†*Trinititherium* Butler, 1978:10.
E. Cret.; N.A.

†*Kermackia* Slaughter, 1971:133.[1]
E. Cret.; N.A.
[1] Possibly an aegialodontid.

Family †**Endotheriidae** Shikama, 1947:78.
[= †Endotherioidea Van Valen, 1967:258; †Endotheriinae Saban, 1954:429.]
E. Cret.; As.

†*Endotherium* Shikama, 1947:78.
E. Cret.; As.[1]
[1] E. Asia.

Family †**Picopsidae** Fox, 1980:1496.
E.-L. Cret.; N.A.

†*Comanchea* Jacobs, Winkler & Murry, 1989:4992.
E. Cret.; N.A.

†*Picopsis* Fox, 1980:1490.
L. Cret.; N.A.

Family †**Potamotelsidae** Nessov, 1987:201.[1]
L. Cret.; N.A.

[1] Mentioned in a figure but not elsewhere in the text. If the use of this family name is a quotation, we have not found the earlier reference. The validity is dubious.

†*Potamotelses* Fox, 1972:1483.
L. Cret.; N.A.

Family †**Plicatodontidae** Ameghino, 1904:83.
Pleist.; S.A.

†*Plicatodon* Ameghino, 1881:307.
Pleist.; S.A.

Order †**DELTATHEROIDA** Kielan-Jaworowska, 1982:373.
[= †DELTATHERALIA Marshall & Kielan-Jaworowska, 1992:372[1];
†HOLARCTIDELPHIA Szalay, 1993:233[2]; †DELTATHEROIDIFORMES Kinman, 1994:37.]
[1] In part. The taxon was defined to contain orders †DELTATHEROIDA and †SPARASSODONTA.

[2] Proposed as a cohort comprising †Deltatheroididae and †Deltatheridiidae. The content is identical to that of †DELTATHEROIDA, however, because when the latter was proposed the †Deltatheroididae had not yet been split off from the †Deltatheridiidae. Redesignated †HOLARCTIDELPHIA as a (supposedly) new cohort by Szalay, 1994:40, this time including both his order †ASIADELPHIA and order †DELTATHEROIDA.

†*Oxlestes* Nessov, 1982:236.
   L. Cret.; As.

†*Khuduklestes* Nessov, Sigogneau-Russell & D. E. Russell, 1994:84.
   Cret.; As.

Family †**Deltatheridiidae** Gregory & Simpson, 1926:6.
   [= †Deltatheridioidea Simpson, 1931:268; †Deltatheridiinae Simpson, 1945:48; †DELTATHERIDIA Van Valen, 1965:638.[1]] [= or including †Sulestinae Nessov, 1985:210.]
   L. Cret.; As. L. Cret.; N.A.
   [1] Proposed as order; raised to superorder by McKenna, 1975:41.

†*Deltatheridium* Gregory & Simpson, 1926:6.
   L. Cret.; As.

†*Sulestes* Nessov, 1985:211.[1]
   L. Cret.; As.
   [1] Sometimes spelled Nesov.

Family †**Deltatheroididae** Kielan-Jaworowska & Nessov, 1990:2.[1]
   L. Cret.; As. ?L. Cret.; N.A.
   [1] Dr. Nesov preferred to spell his name Nessov in English. Possibly the spelling of †Deltatheroididae should be "†Deltatheroideidae," by analogy with "†Titanoideidae," but "Bathyergoididae" seems to have caught hold elsewhere and the matter is of little importance. The name of this family does not date from Kielan-Jaworowska, 1982 as sometimes claimed.

†*Deltatheroides* Gregory & Simpson, 1926:11.
   L. Cret.; As. ?L. Cret.; N.A.[1]
   [1] Fragmentary remains described as †*Deltatheroides*-like.

Order †**ASIADELPHIA** Trofimov & Szalay, 1993:60A.[1]
   [1] Also said to be new in 1994.

Family †**Asiatheriidae** Trofimov & Szalay, 1994:12569.
   Cret.; As.

†*Asiatherium* Trofimov & Szalay, 1994:12569.[1]
   Cret.; As.
   [1] *Nomen nudum* in Trofimov & Szalay, 1993:60A and in Szalay, 1994:40.

Cohort **MARSUPIALIA** Illiger, 1811:75, **new rank**.[1] Marsupials.
   [= POLLICATA Illiger, 1811:66[2]; DIDELPHIA de Blainville, 1816:117[3]; FERAE Gray, 1821:308[4]; MARSUPIATA Richardson, 1837:149[5]; DIDELPHA Bonaparte, 1838:8; METATHERIA Huxley, 1880:654[6]; METADELPHIA Archer, 1984:786[7]; MARSUPIONTA W. K. Gregory, 1947:46.[8]] [Including SALIENTIA Illiger, 1811:79.[9]]
   [1] Proposed as "family" by Illiger; used as order by G. Cuvier, 1817, fide Owen, 1859:8; used as a class by Newman, 1843:27; raised to superorder by Ride, 1963:99; raised to supercohort by McKenna, 1975:27. Used as both a supercohort and an order (exclusive of †STAGODONTOIDEA) by Marshall & Kielan-Jaworowska, 1992:372. The International Code of Zoological Nomenclature does not govern such matters, but the double use of a name promotes confusion.
   [2] In part. Illiger also included various non-human primates in this unnatural assemblage, to which he gave ordinal rank. He excluded kangaroos.
   [3] Included the monotremes as "Didelphes Anomaux."
   [4] Junior homonym of FERAE Linnaeus, 1758.
   [5] Proposed as order; raised to cohort by Turnbull, 1971:176.
   [6] Proposed as infraclass. Called supercohort or infraclass by Shoshani, 1992:108.
   [7] See Archer's footnote. Cohort METADELPHIA was meant to include marsupials exclusive of †*Holoclemensia*, a genus then considered to be a marsupial, but put in its own cohort, †PROTODELPHIA.
   [8] In part. Defined as a subclass containing marsupials and monotremes.
   [9] Proposed as an order encompassing kangaroos. The International Code of Zoological Nomenclature does not deal with taxa higher than the family-group, but the amphibian taxon SALIENTIA Laurenti, 1768, antedates Illiger's term.

†*Kokopellia* Cifelli, 1993:9413.
    E. Cret.; N.A.

†*Aenigmadelphys* Cifelli & Johanson, 1994:292.
    L. Cret.; N.A.

†*Iugomortiferum* Cifelli, 1990:322.
    L. Cret.; N.A.

†*Anchistodelphys* Cifelli, 1990:325.
    L. Cret.; N.A.

†*Gashternia* Simpson, 1935:6.[1]
    L. Paleoc.; S.A.

[1] This genus was referred to the MARSUPIALIA with doubt by Simpson in 1948. Its affinities are very uncertain. Family †Gashterniidae Marshall, 1987:145, 146 was proposed for the genus and was allied with †*Groeberia* in the †Argyrolagoidea. Szalay placed †Gashterniidae in his infraorder †SIMPSONITHERIA.

†*Thylacotinga* Archer, Godthelp & Hand, 1993:193.
    E. Eoc.; Aus.

Family †**Yingabalanaridae** Archer, Every, Godthelp, Hand & Scally, 1990:194.
    E. Mioc.; Aus.

†*Yingabalanara* Archer, Every, Godthelp, Hand & Scally, 1990:194.
    E. Mioc.; Aus.

Suborder †**ARCHIMETATHERIA** Szalay, 1993:234-5, 240.

Family †**Stagodontidae** Marsh, 1889:178.[1]
    [= †Thlaeodontidae Cope, 1892:760; †Thlaeodontinae Hay, 1930:390; †Didelphodontinae Simpson, 1927:124[2]; †STAGODONTIA Archer, 1984:786; †STAGODONTOIDEA Marshall & Kielan-Jaworowska, 1992:372[3]; †STAGODONTIFORMES Kinman, 1994:37; †Stagodontinae Kalandadze & Rautian, 1992:50.]
    L. Cret.; N.A.

[1] See the International Code of Zoological Nomenclature, Articles 23(d) and 40.
[2] Not †Didelphodontinae Matthew, 1918:571.
[3] Given ordinal rank, coordinate with MARSUPIALIA.

†*Eodelphis* Matthew, 1916:482.
    L. Cret.; N.A.

†*Delphodon* Simpson, 1927:127.
    L. Cret.; N.A.

†*Didelphodon* Marsh, 1889:88.
    [= †*Didelphops* Marsh, 1889:88, inserted errata; †*Stagodon* Marsh, 1889:178; †*Thlaeodon* Cope, 1892:759; †*Ectoconodon* Osborn, 1898:171; †*Diaphorodon* Simpson, 1927:127.]
    L. Cret.; N.A.

†*Pariadens* Cifelli & Eaton, 1987:520.
    L. Cret.; N.A.

†*Boreodon* Lambe, 1902:79.
    L. Cret.; N.A.

Family †**Pediomyidae** Simpson, 1927:6.
    [= †Pediomyinae Simpson, 1927:6; †Pediomyidae Clemens, 1966:34; †PEDIOMYIDIA Archer, 1984:786.[1]] [Including †Monodelphopsinae Szalay, 1994:41.]
    L. Cret., ?M. Eoc.; N.A.[2] ?L. Cret., L. Paleoc.; S.A.[3]

[1] Named at the rank of superorder.
[2] A possible pediomyid tooth has been reported from the middle Eocene of Texas.
[3] ?Pediomyid, indet. in the Late Cretaceous of Peru.

†*Aquiladelphis* Fox, 1971:155.
    L. Cret.; N.A.

†*Pediomys* Marsh, 1889:89.
   [= †*Synconodon* Osborn, 1898:171; †*Protolambda* Osborn, 1898:172.]
   L. Cret.; N.A.

†*Monodelphopsis* Paula Couto, 1952:24.
   L. Paleoc.; S.A.

Magnorder **AUSTRALIDELPHIA** Szalay, in Archer, ed., 1982:629, **new rank**.[1]
   [= or including GONDWANADELPHIA Szalay, 1993:237, 240.[2]]
   [1] Proposed as cohort.
   [2] Proposed as an order containing MICROBIOTHERIA and DASYUROMORPHIA. Reiterated as new in Szalay, 1994:348.

Superorder **MICROBIOTHERIA** Ameghino, 1889:263, table 1, **new rank**.[1]
   [= DROMICIOPSIA Szalay, in Archer, ed., 1982:631[2]; MICROBIOTHERIIFORMES Kinman, 1994:37.]
   [1] Proposed as order; used as magnorder by McKenna, in Stucky & McKenna, in Benton, ed., 1993:742.
   [2] Proposed as an order.

Family **Microbiotheriidae** Ameghino, 1887:6.
   [= Microbiotheridae Ameghino, 1887:6; Microbiotheriinae Simpson, 1929:116; Microbiotherioidea Reig, Kirsch & Marshall, 1985:339.] [Including †Clenialitidae Ameghino, 1909:204; †Mirandatheriinae Szalay, 1994:41.]
   ?L. Cret., E. Paleoc.-E. Eoc., E. Mioc., R.; S.A.[1] M. Eoc.; Antarctica.
   [1] An edentulous jaw, of a microbiotherid or a didelphid, has been reported from the Late Cretaceous of northern Patagonia.

†*Khasia* Marshall & de Muizon, 1988:38.
   E. Paleoc.; S.A.

†*Mirandatherium* Paula Couto, 1952:503.[1]
   [= †*Mirandaia* Paula Couto, 1952:22.[2]]
   L. Paleoc.; S.A.
   [1] Replacement name for †*Mirandaia* Paula Couto, 1952, preoccupied.
   [2] Not *Mirandaia* Travassos, 1937: 360, a genus of nematodes.

†*Microbiotherium* Ameghino, 1887:6.
   [= or including †*Stylognathus* Ameghino, 1891:309; †*Eodidelphys* Ameghino, 1891:310; †*Prodidelphys* Ameghino, 1891:310; †*Hadrorhynchus* Ameghino, 1891:311; †*Proteodidelphys* Ameghino, 1898:187; †*Oligobiotherium* Ameghino, 1902:124; †*Clenia* Ameghino, 1904:260; †*Clenialites* Ameghino, 1906:422[1]; †*Microbiotheridion* Ringuelet, 1953:280.]
   E. Eoc., E. Mioc.; S.A.
   [1] Replacement name for †*Clenia* incorrectly believed to be preoccupied.

†*Eomicrobiotherium* Marshall, 1982:57.
   E. Eoc.; S.A.

†*Ideodelphys* Ameghino, 1902:43.[1]
   [= †*Ideodidelphys* Schlosser, 1923:442.]
   E. Eoc.; S.A.
   [1] Based on a species whose type specimen is an edentulous ramus; additional specimens referred by Simpson.

†*Pitheculus* Ameghino, 1894:10.
   E. Mioc.; S.A.

Dromiciops Thomas, 1894:186. Monito del monte, colocolo, llaca, kongoy-kongoy, kunuuma, huenukiki.
   R.; S.A.[1]
   [1] Chile.

Superorder **EOMETATHERIA** Simpson, 1970:38, **new rank**.[1]
   [= EOMARSUPIALIA Archer, 1984:786, 787.[2]]
   [1] Proposed tentatively as a suborder of marsupials for Dasyuroidea, Perameloidea, and Phalangeroidea as then understood. Article 15 of the International Code of Zoological Nomenclature does not apply (see Article 1). Used at ordinal rank by Archer, 1984:786 and at rank of magnorder by McKenna, in Stucky & McKenna, in Benton, ed., 1993:743.
   [2] Erroneously attributed to Simpson, 1970:38. Given rank of a cohort by Archer in his tentative classification.

Order †**YALKAPARIDONTIA** Archer, Hand & Godthelp, 1988:1528.[1]

[1] A good case of a phenetically odd beast being awarded high but nested taxonomic categories.

Family †**Yalkaparidontidae** Archer, Hand & Godthelp, 1988:1528.
M. Mioc.; Aus.

†*Yalkaparidon* Archer, Hand & Godthelp, 1988:1529.
M. Mioc.; Aus.

Order **NOTORYCTEMORPHIA** Kirsch, in Hunsaker, 1977:45.[1] Marsupial moles.
[= SYNDACTYLIFORMES Szalay, in Archer, ed., 1982:631[2];
NOTORYCTIFORMES Kinman, 1994:37.]

[1] Used as a suborder. Elevated to order by Aplin & Archer, in Archer, ed., 1987:xxi.
[2] Proposed as suborder. Assignment to SYNDACTYLI is possible.

Family **Notoryctidae** Ogilby, 1892:5. Marsupial moles.
[= Notoryctoidea Kirsch, in Hunsaker, 1977:45.]
R.; Aus.

Notoryctes Stirling, 1891:154. Marsupial mole, pouched mole.
[= *Psammoryctes* Stirling, 1889:158[1]; *Neoryctes* Stirling, 1891:186.]
R.; Aus.

[1] Not *Psammoryctes* Poeppig, 1835:252, a rodent.

Grandorder **DASYUROMORPHIA** Gill, 1872:26, **new rank**.[1]
[= CREATOPHAGA Giebel, 1859[2]; CREOPHAGA Haeckel, 1866:clvii[3]; DASYURA
Ameghino, 1889[4]; DASYURIDA Szalay, in Archer, ed., 1982:631[5];
DASYURIFORMES Kinman, 1994:37.] [Including MYRMECOBIA Ameghino,
1889.[6]]

[1] Proposed as suborder; used as order by Aplin & Archer, in Archer, ed., 1987:xxi; used as superorder by
McKenna, in Stucky & McKenna, in Benton, ed., 1993:743.
[2] Although Giebel's order CREATOPHAGA and Haeckel's order CREOPHAGA antedate Gill's suborder
DASYUROMORPHIA, Ameghino's order DASYURA, and Szalay's order DASYURIDA, priority need
not be applied above the family group level. There is, however, an obligation not to name higher
categories needlessly.
[3] Proposed as order. Not the same as †CREOPHAGA Kretzoi, 1945:62, a synonym of †CREODONTA.
[4] Original paper not seen. See Ameghino's Obras Completas 6:481.
[5] Proposed as an order.
[6] Original paper not seen. See Ameghino's Obras Completas 6:586.

Family †**Thylacinidae** C. L. Bonaparte, 1838:113.
[= †Thylacininae Bensley, 1903:91.]
E. Mioc.-E. Plioc., Pleist., R.; Aus. Plioc. and/or Pleist.; New Guinea.

†*Nimbacinus* Muirhead & Archer, 1990:203.
E.-M. Mioc.; Aus.

†*Muribacinus* Wroe, 1996:1033.
?M. Mioc.; Aus.

†*Thylacinus* Temminck, 1827:60. Tasmanian tiger, Tasmanian wolf.
[= †*Thylacynus* Temminck, 1827:23; †*Paracyon* Gray, 1827:192; †*Peralopex* Gloger,
1841:82.]
E. Mioc.-E. Plioc., Pleist., R.; Aus.[1,2] Plioc. and/or Pleist.; New Guinea.

[1] Including Tasmania in the Pleistocene.
[2] Extinct in 1936.

Family **Dasyuridae** Goldfuss, 1820:xxxiii, 447.
[= Dasyurini Goldfuss, 1820:xxxiii, 447; Dasyurina Gray, 1825:340[1]; Dasyuridae
Waterhouse, 1838[2]; Dasyurina Van der Hoeven, 1858; Dasyurida Haeckel, 1866:clvii;
Dasyuroidea Simpson, 1930:9.] [Including Myrmecobiidae Waterhouse, 1838.[2]]
?E. Mioc., M. Mioc., ?L. Mioc., E. Plioc.-R.; Aus. L. Plioc., R.; New Guinea.

[1] Proposed as tribe.
[2] Fide Waterhouse, 1841:60.

Subfamily **Dasyurinae** Goldfuss, 1820:xxxiii, 447. Native cats, pouched mice, marsupial mice, marsupial rats.

[= Dasyurini Goldfuss, 1820:xxxiii, 447; Dasyurina Gray, 1825:340[1]; Dasyurinae Thomas, 1888:253.] [Including Phascogalina Bonaparte, 1850[2]; Phascogalinae Gill, 1872:26; Antechini Murray, 1886:xv, 286, 362; Sarcophilinae Gill, 1872:26; Murexinae Archer, 1982:438; Phascolosorexinae Archer, 1982:438; Phascolosoricinae Archer, 1989:67[3]; Planigalinae Archer, 1982:439; Planigalini Krajewski, Driskell, Painter & Westerman, 1993:164[4]; Sminthopsinae Archer, 1982:439; Parantechini Archer, 1982:439; Sminthopsini Krajewski, Driskell, Painter & Westerman, 1993:164.]

?E. Mioc., M. Mioc., ?L. Mioc., E. Plioc.-R.; Aus. L. Plioc., R.; New Guinea.

[1] As tribe.
[2] Unpaged plate.
[3] Emendation of Phascolosorexinae.
[4] Probably named by earlier author(s), but we have not found the reference.

†*Wakamatha* Archer & Rich, 1979:309.
?Mioc. and/or ?Plioc. and/or ?Pleist.; Aus.

†*Dasylurinja* Archer, 1982:402.
M. Mioc.; Aus.

†*Ankotarinja* Archer, 1976:53.
M. Mioc.; Aus.

†*Keeuna* Archer, 1976:64.
M. Mioc.; Aus.

Dasyurus É. Geoffroy Saint-Hilaire, 1796:469. Native cats, tiger cats, quolls.
[Including *Nasira* Harvey, 1841:210; *Dasyurinus* Matschie, 1916:262; *Dasyurops* Matschie, 1916:262; *Notoctonus* Pocock, 1926:1082[1]; *Satanellus* Pocock, 1926:1083; *Stictophonus* Pocock, 1926:1083.[2]]
L. Plioc., Pleist., R.; Aus. R.; New Guinea.

[1] Objective synonym of *Dasyurinus*.
[2] Objective synonym of *Dasyurops*.

Planigale Troughton, 1928:282. Flat-headed marsupial mice.
E. Plioc., L. Pleist.-R.; Aus. R.; New Guinea.

Phascolosorex Matschie, 1916:263.
L. Plioc., R.; New Guinea.

Neophascogale Stein, 1933:87.
R.; New Guinea.

Antechinus MacLeay, 1841:241. Broad-footed marsupial mice, false broad-footed marsupial mice, speckled marsupial mouse.
[Including *Parantechinus* Tate, 1947:137; *Pseudantechinus* Tate, 1947:139; *Dasykaluta* Archer, 1982:435.]
Pleist., R.; Aus. R.; New Guinea.

Phascogale Temminck, 1827:xxiii, 56. Tuans, wambengers, brush-tailed marsupial mice, dibblers.
[= *Phascologale* Lenz, 1831:156-157; *Ascogale* Gloger, 1841:xxx, 83; *Phascogalea* Müller & Schlegel, 1842:149-152; *Tapoa* Lesson, 1842:190; *Phascoloictis* Matschie, 1916:263.]
Pleist., R.; Aus.

Sarcophilus I. Geoffroy Saint-Hilaire & F. Cuvier, 1837:6. Tasmanian devil.
[= *Ursinus* Boitard, 1841:290; *Diabolus* Gray, in Grey, 1841:400.]
E.-L. Plioc., M. Pleist.-R.; Aus.[1]

[1] Known from the Recent of Australia but now extinct there; extant only on Tasmania.

†*Glaucodon* Stirton, 1957:121-134.
E. Pleist.; Aus.

Dasyuroides Spencer, 1896:36. Byrne's pouched mouse, kowari, crested-tailed marsupial rat.
L. Plioc., Pleist., R.; Aus.

Dasycercus Peters, 1875:73. Crest-tailed marsupial mouse, nulgara.
[= *Chaetocercus* Krefft, 1867:434[1]; *Amperta* Cabrera, 1919:65.]
Pleist., R.; Aus.
[1] Not *Chaetocercus* G. R. Gray, 1855: 22, a genus of birds.

Sminthopsis Thomas, 1887:503. Narrow-footed marsupial mice, dunnarts,
jerboa marsupial mice, wuhl-wuhl, kultarr.
[= *Podabrus* Gould, 1845:79.[1]] [Including *Antechinomys* Krefft, 1867:434.]
Pleist., R.; Aus. R.; New Guinea.
[1] Not *Podabrus* Fischer de Waldheim, 1821, a genus of beetles.

Murexia Tate & Archbold, 1937:335, 339.
R.; New Guinea.

Myoictis Gray, 1858:112. Pouched mouse.
R.; New Guinea.

Ningaui Archer, 1975:239.
R.; Aus.

Subfamily **Myrmecobiinae** Waterhouse, 1838.
[= Myrmecobiidae Waterhouse, 1838[1]; Myrmecobida Haeckel, 1866:clvii[2];
Myrmecobiinae Gill, 1872:26; MYRMECOBIA Ameghino, 1889[3];
Ambulatoria Owen, 1841:332.]
L. Pleist.-R.; Aus.
[1] Fide Waterhouse, 1841:60; 1855:60.
[2] Family name for *Myrmecobius* and †*Plagiaulax.*
[3] Original paper not seen. See Ameghino's Obras Completas 6:586.

Myrmecobius Waterhouse, 1836:69. Marsupial anteater, numbat.
L. Pleist.-R.; Aus.

Grandorder **SYNDACTYLI** Gill, 1871:8, **new rank.**[1]
[= SYNDACTYLA Wood Jones, 1924:133.[2]]
[1] Proposed as suborder; used as superorder by McKenna, in Stucky & McKenna, in Benton, ed., 1993:743.
[2] 1923-1925. Used as a suborder. Elevated to ordinal rank by Szalay, in Archer, ed., 1982:631.

Order **PERAMELIA** Ameghino, 1889:266. Bandicoots.
[= PERAMELINA Ride, 1964:99[1]; PERAMELEMORPHIA Kirsch, 1968:420[2];
PERAMELIFORMES Szalay, in Archer, ed., 1982:631[2]; PERAMELIFORMES
Kinman, 1994:37.[3]]
[1] Used as an order, but Gray, 1825 cited as author; however, Gray used the name at a subfamilial rank.
Ranked as an infraorder by Archer, 1984:786 (but not 787), and as a semiorder by Szalay, 1993:237, 240;
1994:42, who again cited Gray, 1825 as the author.
[2] Proposed as suborder.
[3] Proposed as order.

†*Yarala* Muirhead & Filan, 1995:127.
L. Olig. and/or E. and/or M. Mioc.; Aus.

Family **Peramelidae** Gray, 1825:340.
[= Peramelina Gray, 1825:340; Peramelidae Waterhouse, 1838[1]; Peramelida Haeckel,
1866:clvii; Perameloidea Osborn, 1910:516; Syndactylina Wagner, 1855:209.[2]]
[Including Thylacomyidae Archer & Kirsch, 1977:18.]
Mioc., Plioc., ?E. and/or ?M. Pleist., L. Pleist.-R.; Aus.[3,4] R.; New Guinea.
[1] Fide Waterhouse, 1841:60.
[2] Described as a family. Not based upon a generic name.
[3] Undescribed genera are known from Miocene deposits in Australia.
[4] Including Tasmania in the Recent.

†*Ischnodon* Stirton, 1955:249.
Plioc.; Aus.

Isoodon Desmarest, 1817:409. Short-nosed bandicoots.
[= *Thylacis* of authors.[1]]
Pleist., R.; Aus.[2] R.; New Guinea.
[1] Not *Thylacis* Illiger, 1811:76, another peramelid (synonym of *Perameles*).
[2] Includes Tasmania in the Recent.

Subfamily **Peramelinae** Gray, 1825:340.
> [= Peramelina Gray, 1825:340[1]; Peramelinae Bensley, 1903:110; Peramelini Szalay, 1994:42.]
> Plioc. and/or Pleist., R.; Aus.
> [1] Proposed as tribe.

Perameles É. Geoffroy Saint-Hilaire, 1804:150. Long-nosed bandicoots.
> [= *Thylacis* Illiger, 1811:76; *Thylax* Oken, 1816:1128.]
> Plioc. and/or Pleist., R.; Aus.[1]
> [1] Includes Tasmania in the Recent.

Subfamily **Chaeropodinae** Gill, 1872:26.
> [= Choeropodinae Gill, 1872:26; Chaeropini Szalay, 1994:42.] [Including Thylacomyinae Bensley, 1903:110; Thylacomyidae Archer & Kirsch, 1977:18.]
> ?E. and/or ?M. Pleist., L. Pleist.-R.; Aus.

Macrotis Reid, 1837:31.[1] Rabbit-eared bandicoots, bilbies, pinkies.
> [= *Thylacomys* Owen, 1838:747[2]; *Thalacomys* Blyth, in Cuvier, 1840:104[3]; *Paragalia* Gray, in Grey, 1841:401; *Peragale* Thomas, 1887:397.]
> Pleist., R.; Aus.
> [1] Not *Macrotis* Dejean, 1834:186, a genus of coleopterans, *nomen nudum.*
> [2] *Nomen nudum.*
> [3] Misspelling.

Chaeropus Ogilby, 1838:25. Pig-footed bandicoot.
> [= *Choeropus* Gray, in Mitchell, 1839:131.]
> L. Pleist.-R.; Aus.[1]
> [1] Probably extinct.

Family **Peroryctidae** Archer, Godthelp, Hand & Megirian, 1989:32.[1]
> [= Peroryctidae Groves & Flannery, 1990:2.]
> R.; E. Indies.[2] R.; Aus. R.; New Guinea. R.; Pacific.[3]
> [1] This name might date from an earlier author or have been taken from the MS of Groves & Flannery, 1990, which was delayed in publication.
> [2] Moluccas.
> [3] Bismarcks.

Subfamily **Peroryctinae** Archer, Godthelp, Hand & Megirian, 1989:32, **new rank**.
> [= Peroryctidae Archer, Godthelp, Hand & Megirian, 1989:32.]
> R.; New Guinea.

Peroryctes Thomas, 1906:476. New Guinean bandicoots.
> [Including *Ornoryctes* Tate & Archbold, 1937:331, 352.]
> R.; New Guinea.

Subfamily **Echymiperinae, new**.
> R.; E. Indies.[1] R.; Aus. R.; New Guinea. R.; Pacific.[2]
> [1] Moluccas.
> [2] Bismarcks.

Echymipera Lesson, 1842:192. Spiney-haired bandicoots.
> [Including *Brachymelis* Miklouho-MacLay, 1884:713; *Anuromeles* Heller, 1897:5; *Suillomeles* Allen & Barbour, 1909:44.]
> R.; E. Indies.[1] R.; Aus. R.; New Guinea. R.; Pacific.[2]
> [1] Kei (Kai) Islands.
> [2] Bismarcks.

Rhynchomeles Thomas, 1920:430. Slender-nosed bandicoot, Seram bandicoot, Seram Island long-nosed bandicoot.
> R.; E. Indies.[1]
> [1] Known only from Seram Island; now possibly extinct.

Microperoryctes Stein, 1932:256. New Guinea mouse bandicoot.
> R.; New Guinea.

Order **DIPROTODONTIA** Owen, 1866.[1]

[= DIPROTODONTA Thomas, 1895; DIPROTODONTIFORMES Szalay, 1993:240[2]; DIPROTODONTIFORMES Kinman, 1994:37; SYNDACTYLA DIPROTODONTIA Wood Jones, 1923; DUPLICICOMMISSURALA Abbie, 1937; PHALANGERIFORMES Szalay, in Archer, ed., 1982:631.[3]] [= or including POEPHAGA Giebel, 1855:670.] [Including MACROPODA Ameghino, 1889[4]; VOMBATIFORMES Woodburne, 1984; VOMBATIMORPHIA Aplin & Archer, in Archer, 1987:xlvi; PHASCOLARCTIMORPHIA Aplin & Archer, in Archer, 1987:xlvi; PHALANGERIDA Aplin & Archer, in Archer, 1987:xxii.[5]]

[1] DIPROTODONTIA (as DIPROTODONTA) was considered to be an order by Kirsch, 1968: 420 and by Woodburne, 1984, an infraorder by Archer, 1984:786, and a semiorder by Szalay, 1993:237, 240; 1994:42.
[2] Defined as a semisuborder. Szalay did not include the Vombatoidea here.
[3] Proposed as a suborder.
[4] Original not seen. See Ameghino's Obras Completas 6:452.
[5] Originally given subordinal rank.

†*Brachalletes* de Vis, 1883:190-193.
Plioc.; Aus.

†*Koalemus* de Vis, 1889:106, pl. V.
Plioc.; Aus.

†*Sthenomerus* de Vis, 1883:11-15.
Pleist.; Aus.

Family †**Palorchestidae** Tate, 1948:238.
[= †Palorchestinae Tate, 1948:238; †Palorchestidae Archer & Bartholomai, 1978:4, 8.]
M. Mioc.-L. Pleist.; Aus.

†*Propalorchestes* Murray, 1986:196.
M. Mioc.; Aus.

†*Pitikantia* Stirton, 1967:1-44.
M. Mioc.; Aus.

†*Ngapakaldia* Stirton, 1967:1-44.
M. Mioc.; Aus.

†*Palorchestes* Owen, 1873:387.[1]
L. Mioc.-E. Pleist., ?M. Pleist., L. Pleist.; Aus.[2]

[1] Described as subgenus. Raised to generic rank by Owen, 1874.
[2] Including Tasmania in the late Pleistocene.

Family †**Wynyardiidae** Osgood, 1921:138.
[= †Wynyardioidea Kirsch, in Hunsaker, 1977:45.]
E.-M. Mioc.; Aus.

†*Wynyardia* Spencer, 1901:776-795.
E. Mioc.; Aus.[1]

[1] Tasmania.

†*Muramura* Pledge, in Archer, 1987:393.
M. Mioc.; Aus.

†*Namilamadeta* Rich & Archer, 1979:198.
M. Mioc.; Aus.

Family †**Thylacoleonidae** Gill, 1872:26.
[= †Thylacoleontidae Cope, 1889:876; †Thylacoleonioidea Szalay, 1994:44.]
M. Mioc.-E. Pleist., L. Pleist.; Aus.

†*Priscileo* Rauscher, in Archer, 1987:423.
M. Mioc.; Aus.

Subfamily †**Thylacoleoninae** Gill, 1872:26.
[= †Thylacoleonidae Gill, 1872:26; †Thylacoleoninae Murray, Wells & Plane, in Archer, 1987:435.]
E. Plioc.-E. Pleist., L. Pleist.; Aus.

†*Thylacoleo* Owen, in Gervais, 1848:192.[1]
[= †*Thylacoleon* Winge, 1893:127, 129[2]; †*Schizodon* Stutchbury, 1853[3]; †*Plectodon* Krefft, 1870; †*Prochaeris* De Vis, 1886.[4]] [Including †*Thylacopardus* Owen, 1888:99.] E. Plioc.-E. Pleist., L. Pleist.; Aus.[5]

[1] 1848-1852.
[2] Attempted emendation.
[3] Doubly preoocupied by *Schizodon* Agassiz, 1829, a fish, and *Schizodon* Waterhouse, 1842, an octodontid rodent. Invalid as well because no species was named by Stutchbury.
[4] In part.
[5] Including Tasmania in the late Pleistocene.

Subfamily †**Wakaleoninae** Murray, Wells & Plane, in Archer, 1987:435.
M.-L. Mioc.; Aus.

†*Wakaleo* Clemens & Plane, 1974:653.
M.-L. Mioc.; Aus.

Family **Tarsipedidae** Gervais & Verreaux, 1842:1.
[= Tarsipedides Gervais, 1855:277; Tarsipedina Haeckel, 1866:clvii; Tarsipedidae Gill, 1872:25; Tarsipedinae Thomas, 1888:130; Tarsipedoidea Kirsch, 1968:420.]
R.; Aus.

Tarsipes Gervais & Verreaux, 1842:40. Honey possum, noolbenger.
R.; Aus.

Superfamily **Vombatoidea** Burnett, 1830.
[= Vombatidae Burnett, 1830; Vombatoidea Kirsch, in Hunsaker, 1977:45; VOMBATIFORMES Woodburne, 1984; VOMBATIMORPHIA Aplin & Archer, 1987:xlvi[1]; VOMBATOMORPHIA Aplin & Archer, 1987:xlvi.[2]] [Including PHASCOLOMYDA Goldfuss, 1820:xxii, 444; RHIZOPHAGA Owen, 1859:52[3]; Diprotodontoidea Gill, 1872:25.]

[1] Evidently a *lapsus calami*, but used by Szalay, 1993: 240 as a semisuborder.
[2] Used by Szalay, 1994:43 as a semisuborder.
[3] Giebel, 1855: 668 had already used this term at family rank, but his taxon was not based on a genus-group name.

Family †**Ilariidae** Tedford & Woodburne, in Archer, 1987:403.
M. Mioc.; Aus.

†*Kuterintja* Pledge, in Archer, 1987:419.
M. Mioc.; Aus.

†*Ilaria* Tedford & Woodburne, in Archer, 1987:403.
M. Mioc.; Aus.

Family †**Diprotodontidae** Gill, 1872:26.
[Including †Nototheriidae Lydekker, 1887:161.]
M. Mioc.-L. Pleist.; Aus. ?E. Plioc., L. Plioc., ?E. and/or ?M. Pleist., L. Pleist.; New Guinea.

†*Alkwertatherium* Murray, 1990:54.
L. Mioc.; Aus.

Subfamily †**Zygomaturinae** Stirton, Woodburne & Plane, 1967:152.
M. Mioc.-L. Pleist.; Aus. L. Plioc., ?E. and/or ?M. Pleist., L. Pleist.; New Guinea.

†*Neohelos* Stirton, 1967:48.
M. Mioc.; Aus.

†*Raemeotherium* Rich, Archer & Tedford, 1978:86.
M. Mioc.; Aus.

†*Plaisiodon* Woodburne, 1967:53-103.
L. Mioc.; Aus.

†*Zygomaturus* Owen, 1858:49, 50.[1]
[= †*Simoprosopus* de Vis, 1907:4.]
L. Mioc.-L. Pleist.; Aus.

[1] MacLeay, 1857, published the name in a newspaper; Owen characterized it properly.

†*Kolopsis* Woodburne, 1967:53-103.
>> L. Mioc., ?E. Plioc.; Aus. L. Plioc., Pleist.; New Guinea.

†*Kolopsoides* Plane, 1967:105-128.
>> L. Plioc.; New Guinea.

†*Hulitherium* Flannery & Plane, 1986:66.
>> L. Pleist.; New Guinea.

†*Maokopia* Flannery, 1992:322.
>> L. Pleist.; New Guinea.

Subfamily †**Diprotodontinae** Gill, 1872:26.
>> [= †Diprotodontidae Gill, 1872:26; †Diprotodontinae Stirton, Woodburne & Plane, 1967:153.] [Including †Nototherida Haeckel, 1866:clvii[1]; †Nototheriinae Stirton, Woodburne & Plane, 1967:152.]
>> M. Mioc.-L. Pleist.; Aus. Plioc.; New Guinea.

> [1] Haeckel's family-group name clearly has priority and thus the matter should be taken up with the International Commission on Zoological Nomenclature. See Article 23(b) of the International Code of Zoological Nomenclature.

†*Bematherium* Tedford, 1967:217-237.
>> M. Mioc.; Aus.

†*Pyramios* Woodburne, 1967:53-103.
>> L. Mioc.; Aus.

†*Nototherium* Owen, 1845:231-236.
>> Plioc. and/or Pleist.; Aus. Plioc.; New Guinea.

†*Meniscolophus* Stirton, 1955:258.
>> Plioc.; Aus.

†*Euryzygoma* Longman, 1921:65-80.
>> Plioc.; Aus.

†*Diprotodon* Owen, 1838:xix.
>> [Including †*Diarcodon* Stephenson, 1963:615-624.]
>> E. Plioc.-L. Pleist.; Aus.

†*Euowenia* de Vis, 1891:160-165.
>> [= †*Owenia* de Vis, 1888:100.[1]]
>> E. Plioc.; Aus.

> [1] Not *Owenia* Presch, 1847, a mollusk.

†*Stenomerus* de Vis, 1907.
>> Pleist.; Aus.[1]

> [1] Tasmania.

Family **Vombatidae** Burnett, 1830. Wombats.
>> [= PHASCOLOMYDA Goldfuss, 1820:xxii, 444; Phascolomidae Gray, 1821:309; Phascolomina Gray, 1825:340[1]; Phascolomydina C. L. Bonaparte, 1837:9[2]; Phascolomyidae Owen, 1839:19; Phascolomyida Haeckel, 1866:clvii.]
>> M. Mioc., E.-L. Plioc., ?E. and/or ?M. Pleist., L. Pleist.-R.; Aus.

> [1] As tribe.
> [2] As subfamily.

†*Rhizophascolonus* Stirton, Tedford & Woodburne, 1967:454.
>> M. Mioc.; Aus.

Vombatus É. Geoffroy Saint-Hilaire, 1803:185.[1] Common wombat, naked-nosed wombat, warendj.
>> [= *Phascolomis* É. Geoffroy Saint-Hilaire, 1803:364-367[2]; †*Phascolomys* Tiedemann, 1808[3]; *Phascolomys* Haeckel, 1866:clvii; *Glirina* Haeckel, 1866:clvii.]
>> E.-L. Plioc., L. Pleist.-R.; Aus.

> [1] March, 1803.
> [2] August, 1803.
> [3] See the International Code of Zoological Nomenclature, Article 35, where Tiedemann's usage is cited.

†*Phascolonus* Owen, 1872:251, 257.[1] Giant wombat.
>> [= †*Sceparnodon* Ramsay, 1881:495.[2]]

L. Plioc., L. Pleist.; Aus.[3]

[1] Described as subgenus of *Phascolomys*; raised to generic rank by Lydekker, 1887:157-160.
[2] *Nomen nudum* in a newspaper article published in 1880.
[3] Includes Tasmania.

†*Warendja* Hope & Wilkinson, 1982:109.
　　　　　Pleist.; Aus.

†*Ramsayia* Tate, 1951:13.
　　　　　L. Plioc., L. Pleist.; Aus.

Lasiorhinus Gray, 1863:458. Hairy-nosed wombat.
　　　　　[Including *Wombatula* Troughton, 1941:145.[1]]
　　　　　L. Plioc., L. Pleist.-R.; Aus.
[1] Published as a *nomen nudum* by Iredale & Troughton, 1934: 35.

Superfamily **Phalangeroidea** Thomas, 1888:126.
　　　　　[= CARPOPHAGA Giebel, 1855:691[1]; Phalangeridae Thomas, 1888:126;
　　　　　Phalangeroidea Weber, 1928:xiii.] [Including Phascolarctoidea Woodburne, 1984.]
[1] Used as a family, but not based on a genus-group name.

Family **Phalangeridae** Thomas, 1888:126. Phalangers.
　　　　　[= Phalangistadae Gray, 1821:308; Phalangistina Gray, 1825:340[1]; Phalangistidae
　　　　　Owen, 1841:332; Phalangistida Haeckel, 1866:clvii; Phalangerini Szalay, 1994:42;
　　　　　Ailuropinae Flannery, Archer & Maynes, in Archer, 1987:503; Ailuropini Szalay,
　　　　　1994:42.] [Including Trichosurini Flynn, 1911.]
　　　　　L. Olig., M. Mioc.-R.; Aus.[2] L. Pleist.-R.; New Guinea. R.; E. Indies.[3] R.;
　　　　　Pacific.[4]
[1] As tribe.
[2] Indet. phalangerids occur in the late Oligocene Geilston Travertine of Tasmania.
[3] Sulawesi, Moluccas, Timor.
[4] Bismarcks, Solomons.

Phalanger Storr, 1780:33, 34. Cuscus.
　　　　　[= *Phalangista* G. Cuvier & É. Geoffroy Saint-Hilaire, 1795:183, 187; *Coescoes*
　　　　　Lacépède, 1799:5; *Balantia* Illiger, 1811:77-78; *Sipalus* Fischer de Waldheim,
　　　　　1813:581-582; *Cuscus* Lesson, 1826:150-160; *Ceonyx* Temminck, 1827:10-12;
　　　　　Ailurops Wagler, 1830:26.[1]] [Including *Eucuscus* Gray, 1862:316; *Strigocuscus* Gray,
　　　　　1862:319; *Spilocuscus* Gray, 1862:316.]
　　　　　M. Mioc.-R.; Aus. L. Pleist.-R.; New Guinea. R.; E. Indies.[2] R.; Pacific.[3]
[1] Not *Ailurops* Michaelles, 1830, a reptile.
[2] Sulawesi, Moluccas, Timor.
[3] Bismarcks, Solomons.

Trichosurus Lesson, 1828:333-335. Brush-tailed possums, wyulda.[1]
　　　　　[= *Cercaertus* Burmeister, 1837:814; *Psilogrammurus* Gloger, 1841:xxx, 85; *Trichurus*
　　　　　Wagner, 1843:74-83.]
　　　　　M. Mioc.-R.; Aus.[2]
[1] Not the genus *Wyulda*, which has a scale-covered tail.
[2] Includes Tasmania in the Recent.

Wyulda Alexander, 1919:31. Scaly-tailed possum.
　　　　　R.; Aus.

Family **Burramyidae** Broom, 1898:63. Pygmy possums.
　　　　　[= Burramyinae Broom, 1898:63.]
　　　　　L. Olig., M. Mioc., E. Plioc., Pleist., R.; Aus.[1] R.; New Guinea.
[1] An undescribed burramyid occurs in the late Oligocene of Tasmania.

Burramys Broom, 1895:371. Mountain pygmy possum.
　　　　　M. Mioc., E. Plioc., Pleist., R.; Aus.[1]
[1] Originally described as extinct but in 1966 discovered living at about 5,000 feet on Mt. Hotham, Victoria.
Now also known at similar altitudes in Kosciusko National Park, New South Wales.

Cercartetus Gloger, 1841:85.[1] Dormouse possums, pygmy possums.
　　　　　[= *Dromicia* Gray, in Grey, 1841:401, 407.[2]] [Including *Eudromicia* Mjoberg,
　　　　　1916:13[3]; *Dromiciella* Matschie, 1916:260; *Dromiciola* Matschie, 1916:260.]

Pleist., R.; Aus. R.; New Guinea.

[1] Published before May 1841.
[2] November 1841.
[3] January 1916.

Family **Macropodidae** Gray, 1821:308.

[= Salientia Illiger, 1811:79[1]; Halmaturini Goldfuss, 1820[2]; Halmaturidae C. L. Bonaparte, 1831:8; Halmaturida Haeckel, 1866:clvii; Macropidae Gray, 1821:308; Macropina Gray, 1825:340; Macropodae Burnett, 1830; Macropodidae Owen, 1839:19; Macropodoidea Szalay, in Archer, ed., 1982:631; Kangeroidae Gray, 1858.[3]]
[Including Hypsiprymnidae Owen, 1852[4]; Hypsiprymnodontidae Collett, 1887:906; †Protemnodontidae de Vis, 1883; Potoroidae Flannery, Archer & Plane, 1984:1087-1097.]
L. Olig. and/or E. Mioc., M. Mioc.-R.; Aus. L. Plioc., ?E. and/or ?M. Pleist., L. Pleist.-R.; New Guinea. R.; E. Indies.[5] R.; Pacific.[6]

[1] Illiger named both a "family" and an order (SALIENTIA) to encompass *Hypsiprymnus* and *Halmaturus*. The "family" was not based upon a genus-group name.
[2] See the International Code of Zoological Nomenclature, Article 37(b).
[3] There is no genus "Kangeroo," although both *Kanguroo* and *Kangurus* are synonyms of *Macropus*.
[4] Fide Palmer (1904:882).
[5] Moluccas.
[6] Bismarcks.

†*Galanarla* Flannery, Archer & Plane, 1983:296.[1]
M. Mioc.; Aus.
[1] Dated 1982 but issued March 1983.

Subfamily **Potoroinae** Gray, 1821:308. Rat kangaroos.

[= Potoridae Gray, 1821:308; Potoroinae Trouessart, 1898:1195; Potoroidae Flannery, Archer & Plane, 1984:1087-1097; Hypsiprymnidae Owen, 1852:933.]
[Including Pleopodidae Owen, 1878:574[1]; Hypsiprymnodontidae Collett, 1887:906; Hypsiprymnodontinae Thomas, 1888:4; Bettongiinae Bensley, 1903:143; †Bulungamayinae Flannery, Archer & Plane, 1983:288[2]; †Propleopinae Archer & Flannery, 1985:1331-1349; †Palaeopotoroinae Flannery & T. H. Rich, 1986:435.]
M. Mioc.-R.; Aus.

[1] Objective synonym of Hypsiprymnodontidae.
[2] As "Balungamayinae," clearly a misprint.

†*Wakiewakie* Woodburne, 1984:1063.
M. Mioc.; Aus.

†*Purtia* Case, 1984:1074.
M. Mioc.; Aus.

Tribe †**Bulungamayini** Flannery, Archer & Plane, 1983:288.[1]

[= †Bulungamayinae Flannery, Archer & Plane, 1983:288; †Bulungamayini Szalay, 1994:43.]
M. Mioc.; Aus.
[1] Dated 1982 but issued March 1983.

†*Wabularoo* Archer, 1979:299-303.
M. Mioc.; Aus.

†*Bulungamaya* Flannery, Archer & Plane, 1983:288.[1]
M. Mioc.; Aus.
[1] Dated 1982 but issued March 1983.

Tribe **Potoroini** Gray, 1821:308. Rat kangaroos.

[= Potoridae Gray, 1821:308; Potoroini Flannery, 1989:25; Potoroina Szalay, 1994:43.] [Including Pleopodidae Owen, 1879; †Propleopinae Archer & Flannery, 1985:1331-1349; Hypsiprymnodontidae Collett, 1887:906; Bettongini Flannery, 1989:25; Bettongina Szalay, 1994:43.]
M. Mioc.-R.; Aus.

Hypsiprymnodon Ramsay, 1876:33-35. Musky rat kangaroo.
[= *Pleopus* Owen, 1877:542.]
M. Mioc., E. Plioc., R.; Aus.

Bettongia Gray, 1837:584. Bettongs, tungoos.
    [Including *Bettongiops* Matschie, 1916:264.]
    M. Mioc.-R.; Aus.[1]
    [1] Includes Tasmania in the Recent.

†*Gumardee* Flannery, Archer & Plane, 1983:292.[1]
    M. Mioc.; Aus.
    [1] Dated 1982. Issued March 1983.

†*Ekaltadeta* Archer & Flannery, 1985:1332.
    M. Mioc.; Aus.

†*Propleopus* Longman, 1924:20.
    [= †*Triclis* de Vis, 1888:5-8.[1]]
    E. Plioc., L. Pleist.; Aus.
    [1] Not *Triclis* Loew, 1851, an arthropod.

Potorous Desmarest, 1804:20. Potoroo, rat kangaroo.
    [= *Hypsiprymnus* Illiger, 1811:79; *Potorus* Gray, 1821:308.] [Including *Potoroops*
    Matschie, 1916:264.]
    Pleist., R.; Aus.[1]
    [1] Includes Tasmania in the Recent.

Aepyprymnus Garrod, 1875:59. Rufous rat kangaroo.
    Pleist., R.; Aus.

Caloprymnus Thomas, 1888:114-116. Plains rat kangaroo.
    Pleist., R.; Aus.

†*Milliyowi* Flannery, T. H. Rich, Turnbull & Lundelius, 1992:4.
    E. Plioc.; Aus.

Tribe †**Palaeopotoroini** Flannery & T. H. Rich, 1986:435, **new rank**.
    [= †Palaeopotoroinae Flannery & T. H. Rich, 1986:435.]
    M. Mioc.; Aus.

†*Palaeopotorous* Flannery & T. H. Rich, 1986:435.
    M. Mioc.; Aus.

Subfamily **Macropodinae** Gray, 1821:308. Kangaroos, wallabies.
    [= Halmaturini Goldfuss, 1820; Halmaturina C. L. Bonaparte, 1837:9[1]; Macropina
    Gray, 1825:340[2]; Macropodinae Thomas, 1888:10; Macropodini Flannery, 1989:41.]
    [Including Dendrolagina Bonaparte, 1850; Dendrolagini Flannery, 1989:41;
    Protemnodontidae de Vis, 1883; Lagostrophini Flannery, 1989:30.]
    M. Mioc.-R.; Aus.[3] L. Plioc., ?E. and/or ?M. Pleist., L. Pleist.-R.; New Guinea. R.;
    E. Indies.[4] R.; Pacific.[5]
    [1] As subfamily.
    [2] As tribe.
    [3] An undescribed genus is represented in middle Miocene deposits (Kutjamarpu fauna) of Central Australia.
    [4] Moluccas.
    [5] Bismarcks.

†*Protemnodon* Owen, 1873:128.
    ?L. Mioc., E. Plioc.-L. Pleist.; Aus.[1,2] L. Plioc., ?E. and/or ?M. Pleist., L. Pleist.;
    New Guinea.
    [1] A "?protemnodont" has been recorded from late Miocene deposits in Northern Territory.
    [2] Including Tasmania in the Pleistocene.

†*Watutia* Flannery & Hoch, in Flannery, Hoch & Aplin, 1989:145.
    L. Plioc.; New Guinea.

†*Dorcopsoides* Woodburne, 1967:43.
    L. Mioc.; Aus.

†*Kurrabi* Flannery & Archer, 1984:357-383.
    E. Plioc.; Aus.

†*Troposodon* Bartholomai, 1967:21-33.
    E. Plioc.-L. Pleist.; Aus.

†*Prionotemnus* Stirton, 1955:252.
Plioc. and/or Pleist.; Aus.

†*Congruus* McNamara, 1994:111.
L. Pleist.; Aus.

Thylogale Gray, 1837:583. Scrub wallabies, pademelons.
E. Plioc., L. Pleist.-R.; Aus. L. Pleist.-R.; New Guinea. R.; Pacific.[1]
[1] Bismarcks.

Macropus Shaw, 1790:plate 33. Gray kangaroos, wallaroo, euro.
[= *Gigantomys* Link, 1794:70; *Kangurus* É. Geoffroy Saint-Hilaire & G. Cuvier, 1795:180, 188; *Kanguroo* Lacépède, 1799:6; *Halmaturus* Illiger, 1811:80; *Leptosiagon* Owen, 1874:386.[1]] [Including *Osphranter* Gould, 1842:80; *Phascolagus* Owen, 1874:262[2]; *Dendrodorcopsis* Rothschild, 1903:414[2]; *Notamacropus* Dawson & Flannery, 1985:489.[3]]
E. Plioc.-R.; Aus.[4] R.; New Guinea.
[1] Not *Leptosiagon* Trask, 1857, a worm.
[2] Synonym of *Osphranter*.
[3] Described as a subgenus.
[4] Includes Tasmania in the Recent.

Dorcopsis Schlegel & Müller, 1842:130-138. New Guinea forest wallabies.
E. Plioc., ?L. Plioc.; Aus. ?L. Plioc., R.; New Guinea. R.; E. Indies.[1]
[1] Central Halmahera, Moluccas. Extinct about 1870.

†*Baringa* Flannery & Hann, 1984:193-204.
E. Pleist.; Aus.

Megaleia Gistel, 1848:ix. Red kangaroo.
[= *Gerboides* Gervais, 1855:271; *Boriogale* Owen, 1874:247.]
Pleist., R.; Aus.

†*Bohra* Flannery & Szalay, 1982:83.
Pleist.; Aus.

Wallabia Trouessart, 1905:834. Swamp wallaby.
E.-L. Plioc., M. Pleist.-R.; Aus.

Onychogalea Gray, in Grey, 1841:402.[1] Nail-tailed wallabies.
Pleist., R.; Aus.
[1] Proposed as subgenus of *Macropus*; raised to generic rank by Gray, 1843:88.

Setonix Lesson, 1842:194. Short-tailed wallaby, quokka.
[= *Setonyx* Thomas, 1888:10.[1]]
Pleist., R.; Aus.
[1] Emendation.

†*Synaptodon* de Vis, 1889:158-160.
[= †*Synaptodus* Lydekker, 1890:52.]
Pleist.; Aus.

Petrogale Gray, 1837:583. Rock wallabies.
L. Plioc., L. Pleist.-R.; Aus.

Lagorchestes Gould, 1841:text to plate XII. Hare wallabies.
[= *Lagocheles* Owen, 1847:330.]
Pleist., R.; Aus.

†*Fissuridon* Bartholomai, 1973:365-368.
Pleist.; Aus.

Lagostrophus Thomas, 1887:544-547. Banded hare wallaby.
L. Plioc., R.; Aus.

Peradorcas Thomas, 1904:226. Little rock wallaby.
R.; Aus.

Dendrolagus Müller, 1839:63. Tree kangaroos.
E. Plioc., R.; Aus. Pleist., R.; New Guinea.

Dorcopsulus Matschie, 1916:57. New Guinea forest mountain wallabies.
R.; New Guinea.

Subfamily †**Balbarinae** Flannery, Archer & Plane, 1983:293.[1]
[= †Balbarini Szalay, 1994:43.]
L. Olig. and/or E. Mioc., M. Mioc., ?L. Mioc.; Aus.
[1] Dated 1982. Issued March 1983.

†*Balbaroo* Flannery, Archer & Plane, 1983:293.[1]
M. Mioc., ?L. Mioc.; Aus.
[1] Dated 1982. Issued March 1983.

†*Nambaroo* Flannery & Rich, 1986:421.
M. Mioc.; Aus.

†*Ganawamaya* B. N. Cooke, 1992:202.
L. Olig. and/or E. Mioc., M. Mioc.; Aus.

Subfamily †**Sthenurinae** Glauert, 1926:71.
[= †Sthenuridae Glauert, 1926:71; †Sthenurinae Raven, 1929:254; †Sthenurini Szalay, 1994:43.]
L. Mioc.-L. Plioc., M.-L. Pleist.; Aus.

†*Hadronomas* Woodburne, 1967:83.
L. Mioc.; Aus.

†*Sthenurus* Owen, 1873:128.
[Including †*Simosthenurus* Tedford, 1966:10.[1]]
E.-L. Plioc., L. Pleist.; Aus.[2]
[1] Described as subgenus of †*Sthenurus*; treated as full genus by Flannery, T. H. Rich, Turnbull & Lundelius, 1992:12.
[2] Including Tasmania in the late Pleistocene.

†*Procoptodon* Owen, 1873:784-785.
[Including †*Pachysiagon* Owen, 1874:784-785[1]; †*Simosthenurus* Tedford, 1966:10.]
M.-L. Pleist.; Aus.
[1] *Nomen nudum* in Owen, 1873.

Family **Petauridae** C. L. Bonaparte, 1838:112. Gliders.
[= Petaurina C. L. Bonaparte, 1838:112; Petaurinae Gill, 1872:25; Petauridae Kirsch, 1968:420; Petauroidea Szalay, 1994:43.] [Including Pseudocheiridae Archer, 1984:786.]
M. Mioc.-E. Pleist., ?M. Pleist., L. Pleist.-R.; Aus. R.; New Guinea.

Subfamily **Pseudocheirinae** Winge, 1893:89, 100.
[= Pseudochirini Winge, 1893:89, 100; Pseudocheirinae Turnbull & Lundelius, 1970:26; Pseudocheiridae Archer, 1984:786; Pseudocheirini Szalay, 1994:43.]
M. Mioc.-E. Pleist., L. Pleist.-R.; Aus. R.; New Guinea.

†*Pildra* Woodburne, Tedford & Archer, in Archer, 1987:643.
M. Mioc.; Aus.

†*Paljara* Woodburne, Tedford & Archer, in Archer, 1987:668.
M. Mioc.; Aus.

†*Marlu* Woodburne, Tedford & Archer, in Archer, 1987:659.
M. Mioc.; Aus.

†*Pseudokoala* Turnbull & Lundelius, 1970:26.
E. Plioc.; Aus.

Pseudocheirus Ogilby, 1837:457. Ring-tailed possums.
[= *Pseudochirus* Ogilby, 1836:26[1]; *Hepoona* Gray, in Grey, 1841:402, 407-8.]
[Including *Hemibelideus* Collett, 1884:385-387; *Pseudochirops* Matschie, 1915:86; *Pseudochirulus* Matschie, 1915:91; *Petropseudes* Thomas, 1923:250.]
E. Plioc.-E. Pleist., L. Pleist.-R.; Aus. R.; New Guinea.
[1] *Nomen nudum.*

Schoinobates Lesson, 1842:190. Great glider, great gliding possum.
>[= *Petaurista* Desmarest, 1820:268-271[1]; *Petauroides* Thomas, 1888:163-166;
>*Volucella* Bechstein, in Pennant, 1800:352.[2]]
>E. Plioc., R.; Aus.

[1] Not *Petaurista* Link, 1795, a flying squirrel.
[2] Not *Volucella* E. L. Geoffroy, 1762, an insect, nor *Voluccella* Fabricius, 1794, a genus of Diptera.

Subfamily **Petaurinae** C. L. Bonaparte, 1832:69.
>[= Petaurina C. L. Bonaparte, 1832:69; Petaurinae Gill, 1872:25;
>Petaurini Szalay, 1994:43.] [Including Dactylopsilinae Kirsch & Reig,
>in Kirsch, 1977; Dactylopsilini Szalay, 1994:43.]
>E. Plioc., Pleist., R.; Aus. R.; New Guinea.

Petaurus Shaw & Nodder, 1791:1-4. Sugar gliders, lesser gliding possums.
>[= *Ptilotus* Fischer de Waldheim, 1814:512-515; *Belideus* Waterhouse, 1839:151-152;
>*Xenochirus* Gloger, 1841:xxx, 85; *Petaurella* Matschie, 1916:261;
>*Petaurula* Matschie, 1916:261.]
>E. Plioc., Pleist., R.; Aus. R.; New Guinea.

Gymnobelideus M'Coy, 1867:287-288. Leadbeater's possum.
>[= *Gymnobelides* Marshall, 1873.[1]] [Including †*Palaeopetaurus* Broom, 1896:47.]
>Pleist., R.; Aus.

[1] Replacement name.

Dactylopsila Gray, 1858:109-111. Striped possums.
>[Including *Dactylonax* Thomas, 1910:610.]
>R.; Aus. R.; New Guinea.

Family †**Ektopodontidae** Stirton, Tedford & Woodburne, 1967:437.
>[= †Ektopodontinae Szalay, 1994:43.]
>M. Mioc., E. Plioc., E. Pleist.; Aus.[1]

[1] An unspecified genus of this family lasted until the early Pleistocene of Nelson Bay.

†*Ektopodon* Stirton, Tedford & Woodburne, 1967:438.
>M. Mioc.; Aus.

†*Chunia* Woodburne & Clemens, 1986:12.
>M. Mioc.; Aus.

†*Darcius* Rich, 1986:68.
>E. Plioc.; Aus.

Family **Phascolarctidae** Owen, 1839:19.
>[= Koladae Gray, 1821:309[1]; Koalidae Burnett, 1830:351;
>Phascolarctinae Thomas, 1888:209; Phascolarctoidea Woodburne, 1984.]
>M. Mioc., E. Plioc.-R.; Aus.

[1] Probably not based on a validly proposed genus.

†*Perikoala* Stirton, 1957:71-81.
>M. Mioc.; Aus.

†*Madakoala* Woodburne, Tedford, Archer & Pledge, in Archer, ed., 1987:295.
>M. Mioc.; Aus.

†*Litokoala* Stirton, Tedford & Woodburne, 1967:446.
>M. Mioc.; Aus.

Phascolarctos de Blainville, 1816:116.[1] Koala, native bear.
>[= *Lipurus* Goldfuss, 1817: pls. CLV Aa, Ab[2]; *Morodactylus* Goldfuss, 1820:445;
>*Kola* Gray, 1821:309[3]; *Koala* Burnett, 1830:184; *Draximenus* Anonymous, 1845:774.]
>E. Plioc.-R.; Aus.

[1] Misprint for p. 108.
[2] Not *Lipura* Illiger, 1811:95, a rodent.
[3] Gray cited Cuvier for *Kola*, but the validity is dubious because it was probably being used as a
common name. Nonetheless, this did not deter Gray from naming Koladae.

†*Koobor* Archer, 1976:389.
>Plioc.; Aus.

Family †**Pilkipildridae** Archer, Tedford & Rich, in Archer, 1987:607.
> [= †Pilkipildrinae Szalay, 1994:43.]
> M. Mioc.; Aus.

> †*Djilgaringa* Archer, Tedford & Rich, in Archer, 1987:609.
>> M. Mioc.; Aus.

> †*Pilkipildra* Archer, Tedford & Rich, in Archer, 1987:618.
>> M. Mioc.; Aus.

Family †**Miralinidae** Woodburn, Pledge & Archer, in Archer, 1987:582.
> [= †Miralininae Szalay, 1994:43.]
> M. Mioc.; Aus.

> †*Miralina* Woodburne, Pledge & Archer, in Archer, 1987:582.
>> M. Mioc.; Aus.

Family **Acrobatidae** Aplin, in Aplin & Archer, in Archer, 1987:lvii.
> [= Acrobatini Szalay, 1994:43.]
> Pleist., R.; Aus. R.; New Guinea.

> *Acrobates* Desmarest, 1817:405-406. Pygmy flying possum, feather-tail glider.
>> [= *Acrobata* Desmarest, 1820:270-271; *Cercoptenus* Gloger, 1841:xxx, 85.]
>> Pleist., R.; Aus.

> *Distoechurus* Peters, 1874:303. Pen-tailed phalanger.
>> R.; New Guinea.

Magnorder **AMERIDELPHIA** Szalay, in Archer, ed., 1982:623, **new rank**.[1]
> [= DIDELPHIDA Szalay, in Archer, ed., 1982:631.[2]] [Including †ALPHADELPHIA Marshall, Case & Woodburne, 1990:480.]

[1] Proposed as cohort.
[2] Proposed as order. The content of DIDELPHIDA is identical to that of AMERIDELPHIA and therefore the term is unnecessary.

> †*Jaskhadelphys* Marshall & de Muizon, 1988:37.[1]
>> E. Paleoc.; S.A.
>> [1] Subfamily †Jaskhadelphydae de Muizon, 1991:578, 602 was based upon this genus.

> †*Progarzonia* Ameghino, 1904:260.
>> E. Eoc.; S.A.

Subfamily †**Protodidelphinae** Marshall, 1987:145.
> [Including †Kollpaniidae Marshall, de Muizon & Sigogneau-Russell, 1995:18.[1]]
> E.-L. Paleoc., M. Eoc.; S.A.
> [1] Possibly earlier. Taxonomic position uncertain in the extreme.

> †*Kollpania* Marshall & de Muizon, 1988:39.
>> E. Paleoc.; S.A.

> †*Bobbschaefferia* Paula Couto, 1970:20.
>> [= †*Schaefferia* Paula Couto, 1952:12.[1]]
>> L. Paleoc.; S.A.
>> [1] Not *Schaefferia* Absolon, 1900: 265, a collembolan; nor *Schaefferia* Houbert, 1918: 421, a lepidopteran.

> †*Zeusdelphys* Marshall, in Archer, ed., 1987:124.
>> L. Paleoc.; S.A.

> †*Protodidelphis* Paula Couto, 1952:5.
>> L. Paleoc.; S.A.

> †*Guggenheimia* Paula Couto, 1952:11.
>> L. Paleoc.; S.A.

> †*Reigia* Pascual, 1983:266.
>> M. Eoc.; S.A.

Order **DIDELPHIMORPHIA** Gill, 1872:26.[1]

[= PEDIMANA Van der Hoeven, 1858[2]; ENTOMOPHAGA Owen, 1859:52; DIDELPHIFORMES Szalay, in Archer, ed., 1982:631[3]; DIDELPHOIDIA Kalandadze & Rautian, 1992:50[4]; DIDELPHIFORMES Kinman, 1994:37.] [Including †PERADECTIA Marshall, Case & Woodburne, 1990:480.[5]]

[1] Proposed as a suborder. Raised to ordinal rank by Marshall, Case & Woodburne, 1990: 480.
[2] Latinized version of Pédimanes Vicq d'Azyr, 1792, not Pédimanes G. Cuvier, 1795, from which Van der Hoeven removed the PRIMATES. Proposed as order.
[3] Proposed as suborder of order DIDELPHIDA Szalay, which also included suborder †BORHYAENIFORMES Szalay.
[4] Proposed as an infraorder.
[5] In part.

†*Camptomus* Marsh, 1889:87.[1]
L. Cret.; N.A.
[1] Type species based on a scapula. Validity questionable.

†*Iqualadelphis* Fox, 1987:102.
L. Cret.; N.A.

Subfamily †**Adinodontinae** Hershkovitz, 1995:163.
E. Cret.; N.A.

†*Adinodon* Hershkovitz, 1995:163.
E. Cret.; N.A.

Family **Didelphidae** Gray, 1821:308. Opossums.

[= Didelphina Gray, 1825:340; Didelphyida Haeckel, 1866:clvii; Didelphoidea Osborn, 1910:515[1]; Pedimanes de Blainville, 1834[2]; Pedimana Wagner, 1855:ix-xxvi.[3]] [Including Chironectida Haeckel, 1866:clvii; Chironectidae Anonymous, 1897[4]; †Ceciliolemuroidea Weigelt, 1933:145; †Ceciliolemuridae Weigelt, 1933:146; Monodelphidae Talice, de Mosera & Machado, 1960:149; †Peradectidae Reig, Kirsch & Marshall, 1987:75; †Peradectoidea Marshall, Case & Woodburne, 1990:480; Glironiidae Hershkovitz, 1992:209; Caluromyidae Hershkovitz, 1992:189; Caluromyidae Goin, 1993:112[5]; Marmosidae Hershkovitz, 1992:189.] L. Cret.-M. Mioc., M. Pleist.-R.; N.A. L. Cret.-E. Eoc., E. Mioc.-R.; S.A. E. Eoc., E. Olig.; Af. ?E. Eoc., M. and/or L. Eoc., E. Olig., M. Mioc.; As. E. Eoc.-M. Mioc., ?L. Mioc.; Eu. L. Olig.; Mediterranean.[6] Pleist., R.; Cent. A.

[1] Osborn's 1910 classification was actually multi-authored. A special section of the book, not listed in the table of contents, was entitled "Outline Classification of the Mammalia, Recent and Extinct, including especially the better known genera and families and those mentioned in this book." Osborn introduced it with these remarks: "This classification has been prepared under the direction of the author by W. K. Gregory and Johanna Kroeber Mosenthal. The geological range and revision of the extinct genera has been done with the cooperation of W. D. Matthew. The scheme of classification under four grand divisions, and the order throughout, from the more ancient and primitive to the more specialized forms, is that of the author."
[2] Not based upon a generic name.
[3] Used as family, but not based upon a generic name. Presumably a modification of Pedimanes de Blainville, 1834.
[4] Fide Palmer (1904:734).
[5] Goin's abstract contained this term in running text. It may have been validly proposed by someone else earlier, but we have not found a reference.
[6] Balearic Islands.

†*Bistius* Clemens & Lillegraven, 1986:58.
L. Cret.; N.A.

†*Turgidodon* Cifelli, 1990:301.
L. Cret.; N.A.

†*Varalphadon* Johanson, 1996:146.
L. Cret.; N.A.

†*Esteslestes* Novacek, Ferrusquia-Villafranca, J. Flynn, Wyss & Norell, 1991:14.
E. Eoc.; N.A.

Subfamily †**Alphadontinae** Marshall, Case & Woodburne, 1990:480.
[= †Alphadontidae Eaton, 1993:112; †ALPHADONTIA Archer, 1984:786.[1]]
L. Cret.; N.A.
[1] Proposed as superorder.

†*Alphadon* Simpson, 1927:125.
>[Including †*Protalphadon* Cifelli, 1990:297.]
>L. Cret.; N.A.

†*Albertatherium* Fox, 1971:149.
>L. Cret.; N.A.

Subfamily †**Peradectinae** Crochet, 1979:1457, 1458.
>[= †Peradectini Crochet, 1979:1457, 1458; †Peradectidae Reig, Kirsch & Marshall, 1985:339; †Peradectinae Aplin & Archer, in Archer, ed., 1987:xvii[1]; †PERADECTADONTIA Archer, 1984:786[2]; †PERADECTIFORMES Kinman, 1994:37.]
>?L. Cret., E. Paleoc.-E. Olig., ?M. Mioc.; N.A. L. Cret.-E. Paleoc.; S.A. ?E. Eoc.; Af. E. Eoc.; Eu. M. Mioc.; As.
>
>[1] Informally suggested at subfamily rank, as peradectine pediomyids, by Szalay, in Archer, ed., 1982:624-625.
>[2] Proposed as superorder.

†*Peradectes* Matthew & Granger, 1921:2.
>[Including †*Thylacodon* Matthew & Granger, 1921:2.]
>?L. Cret., E. Paleoc.-M. Eoc.; N.A. L. Cret.-E. Paleoc.; S.A. ?E. Eoc.; Af. E. Eoc.; Eu.

†*Armintodelphys* Krishtalka & Stucky, 1983:221.
>E.-M. Eoc.; N.A.

†*Mimoperadectes* Bown & Rose, 1979:90.
>L. Paleoc.-E. Eoc.; N.A.

†*Nanodelphys* McGrew, 1937:452.
>[Including †*Didelphidectes* Hough, 1961:225.]
>M. Eoc.-E. Olig., ?M. Mioc.; N.A.

†*Alloeodectes* L. S. Russell, 1984:6.
>L. Eoc., ?M. Mioc.; N.A.

†*Siamoperadectes* Ducrocq, Buffetaut, Buffetaut-Tong, Jaeger, Jongkanjanasoontorn & Suteethorn, 1992:395.
>M. Mioc.; As.[1]
>
>[1] S.E. Asia (Thailand).

Subfamily †**Herpetotheriinae** Trouessart, 1879:225.
>E. Paleoc., E. Eoc.-M. Mioc.; N.A. E. Eoc., E. Olig.; Af. ?E. Eoc., M. and/or L. Eoc., E. Olig.; As.[1,2,3] E. Eoc.-M. Mioc., ?L. Mioc.; Eu. L. Olig.; Mediterranean.[4]
>
>[1] An indet. genus has been reported from the ?early-middle Eocene of Turkey.
>[2] A true herpetotheriine marsupial has been discovered in a middle or late Eocene fissure fill deposit near Shanghai, PRC (Qi Tao, personal communication).
>[3] Kazakhstan and questionably the Arabian Peninsula in the early Oligocene.
>[4] Balearic Islands.

†*Swaindelphys* Johanson, 1996:1025.
>E. Paleoc.; N.A.

†*Garatherium* Crochet, 1984:278.
>E. Eoc.; Af.

†*Amphiperatherium* Filhol, 1879:201.
>[Including †*Oxygomphius* von Meyer, 1846:474[1]; †*Microtarsioides* Weigelt, 1933:143; †*Ceciliolemur* Weigelt, 1933:146.]
>E. Eoc.-M. Mioc., ?L. Mioc.; Eu.
>
>[1] *Nomen oblitum.*

†*Asiadidelphis* Gabunia, Shevyreva & Gabunia, 1990:101.
>E. Olig.; As.[1]
>
>[1] Kazakhstan.

†*Herpetotherium* Cope, 1873:1.
>E. Eoc.-M. Mioc.; N.A.

†*Copedelphys* Korth, 1994:383.
>L. Eoc.-E. Olig.; N.A.

†*Peratherium* Aymard, 1850:81.
>[Including †*Alacodon* Quinet, 1964:273[1]; †*Qatranitherium* Crochet, Thomas, Sen, Roger, Gheerbrant & Al-Sulaimani, 1992:540; †*Quatranitherium* Szalay, 1994:137.[2]]
>E. Eoc.-E. Mioc., ?M. Mioc.; Eu. E. Olig.; Af. ?E. Olig.; As.[3] L. Olig.; Mediterranean.[4]

> [1] Synonymy questionable. Not legally proposed?
> [2] Unjustified emendation of †*Qatranitherium*.
> [3] Asian record based on a lower tooth from Oman, Arabian Peninsula, said to be a deciduous premolar of †*Qatranitherium*, in turn said by its authors to be a peradectine marsupial.
> [4] Balearic Islands.

†*Entomacodon* Marsh, 1872:23.
>[Including †*Centracodon* Marsh, 1872:24.[1]]
>M. Eoc.; N.A.

> [1] Synonymy questionable.

Subfamily **Didelphinae** Gray, 1821:308.
>[= Didelphidae Gray, 1821:308; Didelphina Gray, 1825:340; Didelphinae Simpson, 1927:5.] [Including Chironectida Haeckel, 1866:clvii; Chironectidae Anonymous, 1897[1]; Metachirinae Hershkovitz, 1992:189; Lestodelphyinae Hershkovitz, 1992:189; Monodelphinae Hershkovitz, 1992:189; Marmosidae Hershkovitz, 1992:189; Marmosinae Hershkovitz, 1992:189; Marmosinae Goin, 1993:112.]
>E. Paleoc.-E. Eoc., M. Mioc.-R.; S.A. Pleist., R.; Cent. A. M. Pleist.-R.; N.A.

> [1] See Palmer (1904:734).

†*Pucadelphys* Marshall & de Muizon, 1988:27.
>E. Paleoc.; S.A.

†*Incadelphys* Marshall & de Muizon, 1988:31.
>E. Paleoc.; S.A.

†*Mizquedelphys* Marshall & de Muizon, 1988:33.
>E. Paleoc.; S.A.

†*Andinodelphys* Marshall & de Muizon, 1988:34.
>E. Paleoc.; S.A.

†*Marmosopsis* Paula Couto, 1962:157.
>L. Paleoc.; S.A.

†*Itaboraidelphys* Marshall & de Muizon, 1984:1297.
>L. Paleoc.; S.A.

†*Coona* Simpson, 1938:1.
>E. Eoc.; S.A.

Tribe **Monodelphini** Talice, de Mosera & Machado, 1960:149, **new rank**.
>[= Monodelphidae Talice, de Mosera & Machado, 1960:149; Monodelphinae Hershkovitz, 1992:189.] [Including Marmosini Reig, Kirsch & Marshall, 1985:340; Marmosinae Goin, 1993:112; †Zygolestini Marshall, in Archer, ed., 1987:144; Lestodelphyinae Hershkovitz, 1992:189; Thylamyinae Hershkovitz, 1992:189; Thylamyini Kirsch & Palma, 1995:422; Marmosopsini Kirsch & Palma, 1995:422.]
>M. Mioc.-R.; S.A. Pleist., R.; Cent. A. R.; N.A.

Subtribe **Monodelphina** Talice, de Mosera & Machado, 1960:149, **new rank**.
>[= Monodelphidae Talice, de Mosera & Machado, 1960:149; Monodelphinae Hershkovitz, 1992:189.] [Including Marmosini Reig, Kirsch & Marshall, 1985:340; Marmosinae Hershkovitz, 1992:189; Lestodelphyinae Hershkovitz, 1992:189; Thylamyinae Hershkovitz, 1992:189; Thylamyini Kirsch & Palma, 1995:422; Marmosopsini Kirsch & Palma, 1995:422.]
>M. Mioc.-R.; S.A. Pleist., R.; Cent. A. R.; N.A.

Marmosa Gray, 1821:308. Murine opossums.
>   [= *Grymaeomys* Burmeister, 1854:138.] [Including *Asagis* Gloger, 1842:82;
>   *Notogogus* Gloger, 1842:82; *Cuica* Liais, 1872:329; *Marmosops* Matschie, 1916:262;
>   *Stegomarmosa* Pine, 1972:279.[1]]
>   M. Mioc.-R.; S.A. Pleist., R.; Cent. A. R.; N.A.[2]
>   [1] Proposed as subgenus.
>   [2] Southern North America north to Tamaulipas and Sonora, Mexico.

Gracilinanus Gardner & Creighton, 1898:4.
>   R.; S.A.

Monodelphis Burnett, 1830:351. Short-tailed opossums.
>   [= *Peramys* Lesson, 1842:187.] [Including *Hemiurus* Gervais, 1855:101[1];
>   *Microdelphys* Burmeister, 1856:83; *Monodelphiops* Matschie, 1916:261; *Minuania*
>   Cabrera, 1919:43.]
>   ?L. Mioc., L. Pleist.-R.; S.A. R.; Cent. A.[2]
>   [1] Not *Hemiurus* Rudolphi, 1809:38, a trematode.
>   [2] Panama.

†*Thylatheridium* Reig, 1952:125.
>   L. Mioc.-E. Pleist.; S.A.

Thylamys Gray, 1843:xxiii, 101.
>   E. Plioc.-R.; S.A.

Lestodelphys Tate, 1934:154. Patagonian opossum.
>   [= *Notodelphys* Thomas, 1921:137.[1]]
>   ?L. Plioc., E. Pleist., L. Pleist.-R.; S.A.[2]
>   [1] Not *Notodelphys* Allman, 1847, a crustacean.
>   [2] Patagonia in the Recent.

Micoureus Lesson, 1842:186.
>   [= *Micoures* Reig, Kirsch & Marshall, 1987:7; *Micureus* Ringuelet, 1953:288.[1]] [= or
>   including *Caluromys* Matschie, 1916.[2]]
>   R.; Cent. A. R.; S.A.
>   [1] Misspelling.
>   [2] Not *Caluromys* Allen, 1900: 189, a caluromyine didelphid whose type species is *Didelphis philander*
>   Linnaeus.

Subtribe †**Zygolestina** Marshall, in Archer, ed., 1987:138, **new rank**.
>   [= †Zygolestini Marshall, 1987:138.]
>   E. Plioc.; S.A.

†*Zygolestes* Ameghino, 1898:243.
>   E. Plioc.; S.A.

Tribe **Metachirini** Reig, Kirsch & Marshall, 1985:341.
>   [= Metachirinae Hershkovitz, 1992:189.]
>   L. Pleist.-R.; S.A. R.; Cent. A.

Metachirus Burmeister, 1854:135.[1] Pouchless four-eyed opossum, brown four-eyed opossum.
>   L. Pleist.-R.; S.A. R.; Cent. A.
>   [1] Proposed as subgenus of *Didelphis*; raised to genus by Burmeister 1856:69.

Tribe **Didelphini** Gray, 1821:308.
>   [= Didelphidae Gray, 1821:308; Didelphina Gray, 1825:340[1]; Didelphini Crochet,
>   1979:1457.] [Including Chironectida Haeckel, 1866:clvii; Chironectidae Anonymous,
>   1897[2]; Chironectini Szalay, 1994:42.]
>   L. Mioc.-R.; S.A. M. Pleist.-R.; N.A. R.; Cent. A.
>   [1] As tribe.
>   [2] See Palmer (1904:734).

†*Hyperdidelphys* Ameghino, 1904:262.[1]
>   [= †*Hiperdidelphis* Cabrera, 1928:337; †*Hyperdidelphis* Reig, Kirsch & Marshall, in
>   Archer, 1987:5; †*Paradidelphys* Ameghino, 1904:263; †*Cladodidelphys* Ameghino,
>   1904:264; †*Paradidelphis* Reig, Kirsch & Marshall, in Archer, 1987:5.]
>   L. Mioc.-L. Plioc.; S.A.
>   [1] Ameghino consistently used "*Hyperdidelphys*" and "†*Paradidelphys*."

Lutreolina Thomas, 1910:247. Little water opossum.
    [= *Peramys* Matschie, 1916:259.[1]]
    L. Mioc.-R.; S.A.
  [1] Not *Peramys* Lesson, 1842:187, a synonym of *Monodelphis*.

Philander Tiedemann, 1808:426. Pouched four-eyed opossum, gray four-eyed opossum.
    [= *Metachirops* Matschie, 1916:262; *Holothylax* Cabrera, 1919:47.] [Including
    Metacherius Sanderson, 1949:787.]
    E. Plioc., Pleist., R.; S.A. R.; N.A.[1] R.; Cent. A.
  [1] Southern North America north to Tamaulipas, Mexico.

Chironectes Illiger, 1811:76. Yapok, water opossum.
    [= *Memina* Fischer de Waldheim, 1813:579; *Cheironectes* Gray, 1821:308; *Gamba*
    Liais, 1872:329.]
    E. Plioc.-R.; S.A. R.; N.A.[1] R.; Cent. A.
  [1] Southern North America (Oaxaca, Mexico).

†*Thylophorops* Reig, 1952:124.
    L. Plioc.; S.A.

Didelphis Linnaeus, 1758:54. Southern opossum, Virginia opossum.
    [= *Didelphys* Schreber, 1778:532[1]; *Sarigua* Muirhead, 1819:429, 505; *Thylacotherium*
    Lund, 1839:223[2]; *Dasyurotherium* Liais, 1872:331[3]; *Gambatherium* Liais, 1872:331[4];
    †*Dimerodon* Ameghino, 1889:277, 282; *Mamdidelphisus* Herrera, 1899:24[5];
    Leucodelphis Ihering, 1914:347.] [= or including *Leucodidelphys* Krumbiegel, 1941.]
    [Including *Opossum* Schmid, 1818:115.]
    M. Pleist.-R.; N.A. M. Pleist.-R.; S.A. R.; Cent. A.
  [1] Schreber's emendation has not been adopted.
  [2] Not †*Thylacotherium* Valenciennes, 1838:580, a synonym of †*Amphitherium*.
  [3] *Nomen nudum.*
  [4] New name for *Thylacotherium* Lund.
  [5] See the International Code of Zoological Nomenclature, Article 1(b)(8) and Palmer (1904:25-26).

Subfamily †**Eobrasiliinae** Marshall, in Archer, 1987:144, 150.
    E.-L. Paleoc.; S.A.

†*Tiulordia* Marshall & de Muizon, 1988:35.
    E. Paleoc.; S.A.

†*Eobrasilia* Simpson, 1947:2.
    L. Paleoc.; S.A.

†*Gaylordia* Paula Couto, 1952:16.
    [Including †*Xenodelphis* Paula Couto, 1962:160.]
    L. Paleoc.; S.A.

†*Didelphopsis* Paula Couto, 1952:7.
    L. Paleoc.; S.A.

Subfamily †**Derorhynchinae** Marshall, in Archer, 1987:144, 150.
    L. Paleoc.; S.A.

†*Minusculodelphis* Paula Couto, 1962:161.
    L. Paleoc.; S.A.

†*Derorhynchus* Paula Couto, 1952:13.
    L. Paleoc.; S.A.

Subfamily **Caluromyinae** Kirsch & Reig, in Kirsch, 1977:111.
    [= Caluromyidae Hershkovitz, 1992:189; Caluromyini Hershkovitz, 1992:209;
    Caluromyidae Goin, 1993:112.] [Including Glironiidae Hershkovitz, 1992:209;
    Glironiinae Kirsch & Palma, 1995:422; Caluromysiopsini Hershkovitz, 1992:209.]
    E. Mioc., L. Pleist.-R.; S.A. R.; N.A.[1] R.; Cent. A.
  [1] Southern North America (southern Mexico).

†*Pachybiotherium* Ameghino, 1902:123.
    E. Mioc.; S.A.

Caluromys J. A. Allen, 1900:189. Woolly opossums.
    [= *Philander* Brisson, 1762:207-214.[1]] [Including *Mallodelphys* Thomas, 1920:195.[2]]
    L. Pleist.-R.; S.A. R.; N.A.[3] R.; Cent. A.

[1] Not published in a consistently binominal work. However, see the International Code of Zoological
    Nomenclature, Article 11(c)(i) for confusion. Not *Philander* Tiedemann, 1808:426, a didelphine didelphid.
[2] Proposed as subgenus; raised to genus by Miranda-Ribeiro, 1936:328.
[3] Southern North America (Veracruz and Oaxaca, Mexico).

Caluromysiops Sanborn, 1951:474. Black-shouldered opossum.
    R.; S.A.[1]

[1] Peru, W. Brazil, ?S. Colombia.

Glironia Thomas, 1912:239. Bushy-tailed opossums.
    R.; S.A.

Family †**Sparassocynidae** Reig, 1958:249.
    [= †Sparassocyninae Reig, 1958:249; †Sparassocynidae Archer, 1984:602.]
    ?E. Eoc., L. Mioc.-L. Plioc.; S.A.[1]

[1] Unnamed teeth from the early Eocene (Casamayoran) may be the earliest sparassocynids.

†*Sparassocynus* Mercerat, 1898:59.
    [= †*Perazoyphium* Cabrera, 1928:335.]
    L. Mioc.-L. Plioc.; S.A.

Order **PAUCITUBERCULATA** Ameghino, 1894:76.[1]
    [= ASYNDACTYLIA Thomas, 1895:870; POLYDOLOPIMORPHIA Marshall,
    in Archer, ed., 1987:144[2]; †SUDAMERIDELPHIA Szalay, 1993:236, 240[3];
    GLIRIMETATHERIA Szalay, 1993:236, 240[4]; PAUCITUBERCULIFORMES
    Kinman, 1994:37.] [Including †PERADECTIA Marshall, Case & Woodburne,
    1990:480[5]; †ITABORAIFORMES Szalay, 1993:236, 240[6]; †GROEBERIDA
    Pascual, Goin & Carlini, 1994:248.[7]]

[1] Sometimes cited as Ameghino, 1892. We choose to ignore the prior use of the homonymous term
    "PAUCITUBERCULATA" at about subordinal rank by Roger, 1887:6, a name for a polyphyletic
    grouping of certain "non-multituberculate" Mesozoic mammals. The International Code of Zoological
    Nomenclature does not govern what should be done about such cases of homonymy.
[2] Proposed as a suborder. Sometimes incorrectly attributed to Ameghino, 1897. Not the same concept
    as tentative superorder "†POLYDOLOPIMORPHIA" Archer, 1984:786, meant to include only
    †Polydolopidae, †Prepidolopidae, †Glasbiidae, †Caroloameghiniidae, and †Bonapartheriidae. If
    †POLYDOPOPIMORPHIA were to prove useful, it would date from Archer, 1984, even though
    Archer's proposal was tentative (see International Code of Zoological Nomenclature, Article 1).
[3] In large part, except that Szalay also included the †SPARASSODONTA.
[4] Proposed as a suborder.
[5] In part.
[6] Proposed as an infraorder of suborder †SUDAMERIDELPHIA.
[7] Proposed as an order.

Superfamily **Caenolestoidea** Trouessart, 1898:1205.
    [= Caenolestidae Trouessart, 1898:1205; Caenolestoidea Osborn, 1910:517.]

Family †**Sternbergiidae** Szalay, 1994:42, 332, **new rank**.[1]
    [= †Sternbergiinae Szalay, 1994:332.]
    L. Paleoc.; S.A.

[1] See the International Code of Zoological Nomenclature, Article 40.

†*Carolopaulacoutoia*, **new name**.[1]
    [= †*Sternbergia* Paula Couto, 1970:30.[2]]
    L. Paleoc.; S.A.

[1] Replacement name for †*Sternbergia* Paula Couto, 1970:30 [News on the fossil marsupials
    from the Riochican of Brazil. Anais da Academia Brasileira de Ciências 42(1)], preoccupied.
    Named for Carlos de Paula Couto, the author of the preoccupied name.
[2] Not †*Sternbergia* Jordan & Gilbert, in Jordan, 1925:41, a Miocene "clupeid" fish.

Family **Caenolestidae** Trouessart, 1898:1205.[1]
    [= Caenolestidae Trouessart, 1898:1205.]
    [Including †Garzonidae Ameghino, 1891:307; †Garzoniidae Ameghino, 1891:307.]
    L. Olig.-E. Mioc., L. Mioc.-E. Plioc., R.; S.A.

[1] See International Commission on Zoological Nomenclature, Opinion 1241, 1983.

Subfamily **Caenolestinae** Trouessart, 1898:1205. Rat opossums.
[= Caenolestidae Trouessart, 1898:1205; Caenolestinae Sinclair, 1906:416; Caenolestini Winge, 1923:82, 84.]
L. Olig.-E. Mioc., R.; S.A.

†*Pseudhalmarhiphus* Ameghino, 1903:83.[1]
L. Olig.; S.A.
[1] The type and only specimen of this genus and species has been reported as lost.

†*Stilotherium* Ameghino, 1887:7.
[= †*Garzonia* Ameghino, 1891:307; †*Halmarhiphus* Ameghino, 1891:308; †*Halmatorhiphus* Winge, 1923:82.] [Including †*Parhalmarhiphus* Ameghino, 1894:356.]
E. Mioc.; S.A.

Caenolestes Thomas, 1895:367-368. Opossum-rats, rat opossums, ratónes runchos.
[= *Hyracodon* Tomes, 1863:50.[1]]
R.; S.A.[2]
[1] Not †*Hyracodon* Leidy, 1856:91, a perissodactyl.
[2] Northwestern South America.

Rhyncholestes Osgood, 1924:169. Chilean rat opossum, ratónes runchos coligruesos.
R.; S.A.[1]
[1] Chile.

Lestoros Oehser, 1934:240. Incan rat opossum, ratónes runchos Andinos.
[= *Orolestes* Thomas, 1917:3[1]; *Cryptolestes* Tate, 1934:154.[2]]
R.; S.A.[3]
[1] Not *Orolestes* MacLachlan, 1895:21, a dragonfly.
[2] Not *Cryptolestes* Ganglbaur, 1899:608, a subgenus of beetles.
[3] Peru.

Subfamily †**Pichipilinae** Marshall, 1980:40, **new rank**.
[= †Pichipilini Marshall, 1980:40.]
E. Mioc., L. Mioc.-E. Plioc.; S.A.

†*Pichipilus* Ameghino, 1890:155.
E. Mioc.; S.A.

†*Phonocdromus* Ameghino, 1894:355.
E. Mioc.; S.A.

†*Pliolestes* Reig, 1955:66.
L. Mioc.-E. Plioc.; S.A.

Family †**Palaeothentidae** Sinclair, 1906:417.[1]
[= †Epanorthidae Ameghino, 1889:268, 270[2]; †Epanorthini Winge, 1923:84; †Epanorthinae Trouessart, 1905:840; †Palaeothentinae Sinclair, 1906:417; †Palaeothentidae Osgood, 1921:143, 151.] [Including †Decastidae Ameghino, 1893:79; †Acdestinae Bown & Fleagle, 1993:17.]
L. Olig.-M. Mioc.; S.A.
[1] See the International Commission on Zoological Nomenclature, Opinion 1241, 1983.
[2] Sensu stricto.

†*Hondathentes* Dumont & Bown, in Bown & Fleagle, 1993:49.[1]
M. Mioc.; S.A.
[1] *Nomen nudum*?

Subfamily †**Acdestinae** Bown & Fleagle, 1993:17.
L. Olig.-E. Mioc.; S.A.

†*Acdestoides* Bown & Fleagle, 1993:17.
L. Olig.; S.A.

†*Acdestodon* Bown & Fleagle, 1993:18.
L. Olig.; S.A.

†*Trelewthentes* Bown & Fleagle, 1993:19.
E. Mioc.; S.A.

†*Acdestis* Ameghino, 1887:5.
  [Including †*Dipilus* Ameghino, 1890:153; †*Decastis* Ameghino, 1891:305;
  †*Callomenus* Ameghino, 1891:306.]
  E. Mioc.; S.A.

Subfamily †**Palaeothentinae** Sinclair, 1906:417.
    L. Olig.-E. Mioc.; S.A.

†*Palaeothentes* Ameghino, 1887:5.[1]
  [= †*Palaeothentes* Moreno, 1882:22[2]; †*Epanorthus* Ameghino, 1889:271.] [Including
  †*Essoprion* Ameghino, 1891:306; †*Halmadromus* Ameghino, 1891:306; †*Halmaselus*
  Ameghino, 1891:306; †*Metriodromus* Ameghino, 1894:342; †*Metaepanorthus*
  Ameghino, 1894:348; †*Paraepanorthus* Ameghino, 1894:349; †*Prepanorthus*
  Ameghino, 1894:350; †*Cladoclinus* Ameghino, 1894:358; †*Palaepanorthus*
  Ameghino, 1902:123.]
  L. Olig.-E. Mioc.; S.A.
  [1] "*Palaeothentes*" Moreno, 1882:22, was a *nomen nudum*.
  [2] *Nomen nudum.*

†*Titanothentes* Rae, Bown & Fleagle, 1996:2.
  E. Mioc.; S.A.

†*Pilchenia* Ameghino, 1903:128.
  L. Olig.; S.A.

†*Carlothentes* Bown & Fleagle, 1993:29.
  L. Olig.; S.A.

†*Propalaeothentes* Bown & Fleagle, 1993:45.
  E. Mioc.; S.A.

Family †**Abderitidae** Ameghino, 1889:268, 269.
  [= †Abderitesidae Ameghino, 1889:268, 269; †Abderitidae Ameghino, 1889[1];
  †Abderitinae Sinclair, 1906:417; †Abderitini Marshall, 1980:1, 47.]
  E. Mioc.; S.A.
  [1] Original paper not seen. See Ameghino's Obras Completas 6:456.

†*Pitheculites* Ameghino, 1902:74.
  [= †*Eomannodon* Ameghino, 1902:119; †*Micrabderites* Simpson, 1932:6.]
  E. Mioc.; S.A.

†*Abderites* Ameghino, 1887:5.
  [= †*Homunculites* Ameghino, 1902:73.]
  E. Mioc.; S.A.

Superfamily †**Polydolopoidea** Ameghino, 1897:92.
  [= †Polydolopidae Ameghino, 1897:92; †Polydolopoidea Clemens & Marshall,
  1976:79; †POLYDOLOPIMORPHIA Archer, 1984:786[1]; †POLYDOLOPIFORMES
  Kinman, 1994:37.]
  [1] Proposed as superorder. Regarded as an infraorder of suborder †SUDAMERIDELPHIA by Szalay,
  1993:236, 240 and as an infraorder by Szalay, 1994:41.

†*Rosendolops* Goin & Candela, 1996.
  E. Eoc.; S.A.

Family †**Sillustaniidae** Crochet & Sigé, 1996:629.
  E. Paleoc.; S.A.

†*Sillustania* Crochet & Sigé, 1996:629.
  E. Paleoc.; S.A.

Family †**Polydolopidae** Ameghino, 1897:496.
  [Including †Promysopidae Ameghino, 1902:36.]
  L. Paleoc.-M. Eoc., L. Eoc. and/or E. Olig., L. Olig.-E. Mioc.; S.A. M. Eoc.;
  Antarctica.

Subfamily †**Epidolopinae** Pascual & Bond, 1981:483.
    L. Paleoc.; S.A.

†*Epidolops* Paula Couto, 1952:7.
L. Paleoc.; S.A.

Subfamily †**Polydolopinae** Ameghino, 1897:496.
[= †Polydolopidae Ameghino, 1897:496; †Polydolopinae Pascual & Bond, 1981:482.] [Including †Promysopidae Ameghino, 1902:36.]
L. Paleoc.-M. Eoc., L. Eoc. and/or E. Olig.; S.A. M. Eoc.; Antarctica.

†*Polydolops* Ameghino, 1897:497.
[Including †*Pliodolops* Ameghino, 1902:41; †*Orthodolops* Ameghino, 1903:130; †*Anissodolops* Ameghino, 1903:148; †*Archaeodolops* Ameghino, 1903:150.]
L. Paleoc.-M. Eoc., L. Eoc. and/or E. Olig.; S.A.

†*Amphidolops* Ameghino, 1902:42.
[= †*Anadolops* Ameghino, 1903:186.] [Including †*Seumadia* Simpson, 1935:5.]
L. Paleoc.-E. Eoc.; S.A.

†*Eudolops* Ameghino, 1897:498.
[Including †*Promysops* Ameghino, 1902:36; †*Propolymastodon* Ameghino, 1903:100.]
E. Eoc.; S.A.

†*Pseudolops* Ameghino, 1902:40.
E. Eoc.; S.A.

†*Antarctodolops* Woodburne & Zinsmeister, 1984:913, 915.
M. Eoc.; Antarctica.

†*Eurydolops* Case, Woodburne & Chaney, 1988:508.
M. Eoc.; Antarctica.

Subfamily †**Parabderitinae** Marshall, 1980:43, **new rank**.
[= †Parabderitini Marshall, 1980:43.]
L. Olig.-E. Mioc.; S.A.

†*Parabderites* Ameghino, 1902:121-122.[1]
[Including †*Tideus* Ameghino, 1890:157[2]; †*Mannodon* Ameghino, 1893:15.[3]]
L. Olig.-E. Mioc.; S.A.

[1] *Nomen nudum* in 1901.
[2] Not *Tydeus* Koch, 1837: table 2, an arachnid. Synonymy of †*Tideus* Ameghino, 1890 with †*Parabderites* is questionable. The type specimen of †*Tideus trisulcatus* may not belong to a mammal but, if it belongs to the same taxon as the referred specimens, then stability would be threatened unless †*Tideus* and †*Mannodon* were suppressed.
[3] Objective synonym of †*Tideus* Ameghino, 1890.

Family †**Prepidolopidae** Pascual, 1980:221.
E.-M. Eoc.; S.A.

†*Prepidolops* Pascual, 1980:154.
E.-M. Eoc.; S.A.

Family †**Bonapartheriidae** Pascual, 1980:158.
E. Eoc.; S.A.

†*Bonapartherium* Pascual, 1980:159.
E. Eoc.; S.A.

Superfamily †**Argyrolagoidea** Ameghino, 1904:255.
[= †Argyrolagidae Ameghino, 1904:255; †Argyrolagoidea Simpson, 1970:3; †SIMPSONITHERIA Szalay, in Szalay, Novacek & McKenna, eds., 1993:236.[1]] [Including †Groeberioidea Clemens & Marshall, 1976:83; †Patagonioidea Pascual & Carlini, 1987:100; †GROEBERIDA Pascual, Goin & Carlini, 1994:248.]

[1] Proposed as an infraorder of suborder †GLIRIMETATHERIA. Contents the same as †Argyrolagoidea, plus †*Gashternia* (as a family).

Family †**Argyrolagidae** Ameghino, 1904:255.[1]
[= †Microtragulidae Reig, 1955:61; †Microtraguloidea Kalandadze & Rautian, 1992:52.]

L. Eoc. and/or E. Olig., L. Olig., M. Mioc.-L. Plioc.; S.A.[2]

[1] See the International Code of Zoological Nomenclature, Article 40.

[2] ?†Argyrolagidae, new genus and species, in the late Eocene and/or early Oligocene Tinguiririca fauna, Chile.

†*Proargyrolagus* Wolff, 1984:109.
L. Olig.; S.A.

†*Hondalagus* Villarroel & Marshall, 1988:465.
M. Mioc.; S.A.

†*Microtragulus* Ameghino, 1904:191.
[= †*Argyrolagus* Ameghino, 1904:255.]
L. Mioc.-L. Plioc.; S.A.

Family †**Patagoniidae** Pascual & Carlini, 1987:100.
[= †Patagonioidea Pascual & Carlini, 1987:100.]
E. Mioc.; S.A.

†*Patagonia* Pascual & Carlini, 1987:100.
E. Mioc.; S.A.

Family †**Groeberiidae** Patterson, 1952:39.
[= †Groeberioidea Clemens & Marshall, 1976:83; †GROEBERIDA Pascual, Goin & Carlini, 1994:248.[1]]
M. Eoc., L. Eoc. and/or E. Olig.; S.A.[2]

[1] Given ordinal rank by its authors.

[2] A possible groeberiid, new genus and species, has been reported from the late Eocene or early Oligocene Tinguiririca fauna, Chile.

†*Groeberia* Patterson, 1952:39.
M. Eoc.; S.A.

Superfamily †**Caroloameghinioidea** Ameghino, 1901:353.
[= †Caroloameghiniidae Ameghino, 1901:353; †Caroloameghinoidea Marshall, in Archer, 1987:144[1]; †Caroloameghiniinae Clemens, 1966:34.]

[1] Spelling corrected by Aplin & Archer, in Archer, 1987:xxi.

Family †**Glasbiidae** Clemens, 1966:24.
[= †Glasbiinae Clemens, 1966:24; †Glasbiidae Archer, 1984:786.]
L. Cret.; N.A.

†*Glasbius* Clemens, 1966:24.
L. Cret.; N.A.

Family †**Caroloameghiniidae** Ameghino, 1901:353.
[= †ITABORAIFORMES Szalay, 1993:236, 240.[1]]
E. Paleoc.-E. Eoc.; S.A.

[1] Proposed as an infraorder of suborder †SUDAMERIDELPHIA. Content identical to †Caroloameghiniidae and not meant to include †Glasbiidae, which Szalay regarded as members of his paraphyletic suborder †ARCHIMETATHERIA.

†*Roberthoffstetteria* Marshall, de Muizon & Sigé, 1983:740.
E. Paleoc.; S.A.

†*Procaroloameghinia* Marshall, 1982:711.
L. Paleoc.; S.A.

†*Robertbutleria* Marshall, in Archer, 1987:122.
L. Paleoc.; S.A.

†*Caroloameghinia* Ameghino, 1901:354.
E. Eoc.; S.A.

Order †**SPARASSODONTA** Ameghino, 1894:364.[1]
[= BORHYAENIFORMES Szalay, in Archer, ed., 1982:628, 631[2]; †BORHYAENIFORMES Kinman, 1994:37; BORHYAENIMORPHIA Archer, 1984:786.[3]]

[1] Proposed as a suborder of order SARCOBORA Ameghino, a mixture of placental and marsupial carnivores. Given ordinal rank by Aplin & Archer, in Archer, ed., 1987:xxi, and infraordinal rank in suborder †SUDAMERIDELPHIA by Szalay, 1993:236, 240.

[2] Only in part; also included Caenolestoidea (including polydolopids and groeberiids) and †Argyrolagoidea. Proposed as suborder in order DIDELPHIDA.
[3] Objectively synonymous with BORHYAENIFORMES Szalay, but proposed as a superorder.

Family †**Mayulestidae** de Muizon, 1994:208.
>   E. Paleoc.; S.A.

>   †*Mayulestes* de Muizon, 1994:208.
>>   E. Paleoc.; S.A.

Family †**Hondadelphidae** Marshall, Case & Woodburne, 1990:484.
>   [= †Hondadelphinae Szalay, 1994:41.]
>   M. Mioc.; S.A.

>   †*Hondadelphys* Marshall, 1976:405.
>>   M. Mioc.; S.A.

Family †**Borhyaenidae** Ameghino, 1894:371.
>   [= †Borhyaenoidea Simpson, 1930:9.] [Including †Acyonidae Ameghino, 1889:894[1]; †Amphiproviverridae Ameghino, 1894:333, 389; †Prothylacinidae Ameghino, 1894:377; †Hathliacynidae Ameghino, 1894:382; †Sparassodontidae Roger, 1896:16[2]; †Proborhyaenidae Ameghino, 1897:501; †Cladictidae Winge, 1923:77; †Cladosictidae F. Ameghino, 1935:131, 132[3]; †Conodonictidae F. Ameghino, 1935:131, 132[3]; †Thylacosmilidae Marshall, 1976:8.]
>   E. Paleoc.-M. Eoc., L. Olig.-L. Plioc.; S.A.

[1] This unused name has priority over †Borhyaenidae but should be suppressed for the sake of stability.
[2] There is no genus †*"Sparassodon."*
[3] Posthumously.

Subfamily †**Hathliacyninae** Ameghino, 1894:382.
>   [= †Hathlyacynidae Ameghino, 1894:382; †Hathlyacyninae Kirsch, 1977:112; †Hathliacyninae Marshall & de Muizon, 1988:25, 40; †Cladictidae Winge, 1923:77; †Cladosictinae Cabrera, 1927:273.] [Including †Acyonidae Ameghino, 1889:894; †Amphiproviverridae Ameghino, 1894:333.]
>   E. Paleoc.-M. Eoc., L. Olig.-E. Mioc., L. Mioc.-E. Plioc., ?L. Plioc.; S.A.

>   †*Allqokirus* Marshall & de Muizon, 1988:40.
>>   E. Paleoc.; S.A.

>   †*Patene* Simpson, 1935:3.
>>   [= †*Ischyrodidelphis* Paula Couto, 1952:9.]
>>   L. Paleoc.-E. Eoc.; S.A.

>   †*Palaeocladosictis* Paula Couto, 1961:331.
>>   L. Paleoc.; S.A.

>   †*Procladosictis* Ameghino, 1902:46.
>>   ?E. Eoc., M. Eoc.; S.A.

>   †*Pseudocladosictis* Ameghino, 1902:47.
>>   E. Eoc.; S.A.

>   †*Sallacyon* Villarroel & Marshall, 1982:205.
>>   [= †*Andinogale* Hoffstetter & G. Petter, 1983:207.]
>>   L. Olig.; S.A.

>   †*Notogale* Loomis, 1914:216.
>>   L. Olig.; S.A.

>   †*Cladosictis* Ameghino, 1887:7.
>>   [= †*Clasodictis* Roger, 1896:13[1]; †*Cladictis* Winge, 1923:67[2]; †*Hathliacynus* Ameghino, 1887:7; †*Hathlyacynus* Ameghino, 1894:126[3]; †*Anatherium* Ameghino, 1887:8.[3]] [Including †*Acyon* Ameghino, 1887:8.]
>>   E. Mioc.; S.A.

[1] *Lapsus.*
[2] *Lapsus?*
[3] Objective synonym of †*Hathliacynus.*

>   †*Sipalocyon* Ameghino, 1887:8.
>>   E. Mioc.; S.A.

†*Thylacodictis* Mercerat, 1891:54.
 [Including †*Amphiproviverra* Ameghino, 1891:397; †*Protoproviverra* Ameghino, 1891:312.[1]]
 E. Mioc.; S.A.
 [1] Objective synonym of †*Amphiproviverra*. Not †*Protoproviverra* Lemoine, 1891:279, a creodont.

†*Agustylus* Ameghino, 1887:7.
 E. Mioc.; S.A.

†*Ictioborus* Ameghino, 1891:315.
 E. Mioc.; S.A.

†*Amphithereutes* Ameghino, 1935:108.
 E. Mioc.; S.A.

†*Perathereutes* Ameghino, 1891:313.
 E. Mioc.; S.A.

†*Chasicostylus* Reig, 1957:29.
 L. Mioc.; S.A.

†*Notictis* Ameghino, 1889:911.
 L. Mioc.; S.A.

†*Notocynus* Mercerat, 1891:80-81.
 L. Mioc.-E. Plioc., ?L. Plioc.; S.A.

†*Borhyaenidium* Pascual & Bocchino, 1963:101.
 L. Mioc.-E. Plioc.; S.A.

Subfamily †**Proborhyaeninae** Ameghino, 1897:501.
 [= †Proborhyaenidae Ameghino, 1897:501; †Proborhyaeninae Trouessart, 1898:1211.]
 [Including †Arminiheringiidae Ameghino, 1902:44.]
 ?L. Paleoc., E. Eoc., L. Olig.; S.A.

†*Arminiheringia* Ameghino, 1902:44.
 [= †*Dilestes* Ameghino, 1902:46.]
 ?L. Paleoc., E. Eoc.; S.A.

†*Paraborhyaena* Hoffstetter & G. Petter, 1983:205.
 L. Olig.; S.A.

†*Proborhyaena* Ameghino, 1897:501.
 L. Olig.; S.A.

Subfamily †**Borhyaeninae** Ameghino, 1894:371.
 [= †Borhyaenidae Ameghino, 1894:371; †Borhyaeninae Cabrera, 1927:273, 274.]
 [Including †Sparassodontidae Roger, 1896:16[1]; †Conodonictidae F. Ameghino, 1935:131.[2]]
 ?L. Paleoc., E.-M. Eoc., L. Olig.-E. Mioc., L. Mioc.-E. Plioc.; S.A.
 [1] Not based upon a generic name.
 [2] Posthumously.

†*Nemolestes* Ameghino, 1902:48.
 ?L. Paleoc., E. Eoc.; S.A.

†*Argyrolestes* Ameghino, 1902:48.
 E. Eoc.; S.A.

†*Angelocabrerus* Simpson, 1970:2.
 E. Eoc.; S.A.

†*Pharsophorus* Ameghino, 1897:502.
 [= †*Plesiofelis* Roth, 1903:24.]
 M. Eoc., L. Olig.; S.A.

†*Borhyaena* Ameghino, 1887:8.
 [Including †*Arctodictis* Mercerat, 1891:51; †*Dynamictis* Ameghino, 1891:148.[1]]
 E. Mioc.; S.A.
 [1] Objective synonym of †*Arctodictis*.

†*Pseudoborhyaena* Ameghino, 1902:125.
E. Mioc.; S.A.

†*Acrocyon* Ameghino, 1887:8.
E. Mioc.; S.A.

†*Conodonictis* Ameghino, 1891:314.
E. Mioc.; S.A.

†*Eutemnodus* Bravard, 1858:16.[1]
[= †*Eutemnodus* Burmeister, 1885:97; †*Apera* Ameghino, 1886:13; †*Entemnodus* Trouessart, 1885:96.[2]]
L. Mioc.-E. Plioc.; S.A.

[1] Possibly a *nomen nudum* in 1858.
[2] Misprint.

†*Parahyaenodon* Ameghino, 1904:266.
E. Plioc.; S.A.

Subfamily †**Prothylacyninae** Ameghino, 1894:377.
[= †Prothylacynidae Ameghino, 1894:377; †Prothylacyninae Trouessart, 1898:121.]
E.-L. Mioc.; S.A.

†*Pseudothylacinus* Ameghino, 1902:127.
E. Mioc.; S.A.

†*Prothylacynus* Ameghino, 1891:312.
[= †*Prothylacocyon* Winge, 1923:67.] [Including †*Napodonictis* Ameghino, 1894:380.]
E. Mioc.; S.A.

†*Lycopsis* Cabrera, 1927:295.
E.-M. Mioc.; S.A.

†*Stylocynus* Mercerat, 1917:20.
L. Mioc.; S.A.

†*Pseudolycopsis* Marshall, 1976:291.
L. Mioc.; S.A.

Subfamily †**Thylacosmilinae** Riggs, 1933:65.
[= †Thylacosmilidae Marshall, 1976:8.]
M. Mioc.-L. Plioc.; S.A.

†*Thylacosmilus* Riggs, 1933:61.
[= or including †*Achlysictis* Ameghino, 1891:147; †*Hyaenodonops* Ameghino, 1908:423; †*Notosmilus* J. L. Kraglievich, 1960:55; †*Acrohyaenodon* Ameghino, 1904:267.]
M. Mioc.-L. Plioc.; S.A.

Cohort **PLACENTALIA** Owen, 1837:903, **new rank**.[1] Placentals.
[= Les monodelphes de Blainville, 1816:109; PLACENTARIA Fleming, 1822:xxxii, 2: 169; MONODELPHIA Gill, 1872:295; EUTHERIA Huxley, 1880:657[2]; PLACENTATA Turnbull, 1971:176.] [Including UNGUICULATA Linnaeus, 1766:21; FERUNGULATA Simpson, 1945:105.[3]]

[1] Rank not specified by Owen.
[2] Huxley's EUTHERIA may not have been taken from Gill, 1872. EUTHERIA Gill, 1872:v covered both marsupials and placentals and until recently lay fallow. EUTHERIA Huxley, 1880 covered "placentals" and putative "placentals," but not members of the MARSUPIALIA as well. Huxley's term was accepted for nearly a hundred years by most students (but not by Osborn, 1910:515). Moreover, re-elevation of Gill's original EUTHERIA to become a senior synonym of THERIA creates new confusion and, in any case, is not required by the International Code of Zoological Nomenclature. If the pappotheriids are therians, *incertae sedis*, then PLACENTALIA = "les monodelphes" de Blainville, 1816:109 = MONODELPHIA Gill, 1872:46 = EUTHERIA Huxley, 1880:657. EUTHERIA in its long accepted sense has been compromised and no longer has a clear meaning. In an attempt to avoid argument about the "true meaning" of EUTHERIA, we return here to Owen's PLACENTALIA, endorsed by Gill himself in preference to MONODELPHIA = "les monodelphes" de Blainville, 1816:109. PLACENTALIA has the advantage of familiarity. EUTHERIA was raised to the rank of supercohort by McKenna, 1975:27, 40; called a "division" coordinate with a "division" DIDELPHIA by Aplin and Archer, in Archer, ed., 1987:xxi; and called a "supercohort or infraclass" by Shoshani, 1992:108.
[3] Proposed as cohort.

†*Daulestes* Trofimov & Nessov, in Nessov & Trofimov, 1979:952.
    [Including †*Taslestes* Nessov, 1982:234; †*Kumlestes* Nessov, 1985:9.]
    L. Cret.; As.

†*Aspanlestes* Nessov, 1985:14.
    L. Cret.; As.

†*Sorlestes* Nessov, 1985:14.
    L. Cret.; As.

†*Bulaklestes* Nessov, 1985:16.
    L. Cret.; As.

†*Cretasorex* Nessov & Gureev, 1981:1003.[1]
    ?L. Cret.; As.[2]

[1] Nessov (Nesov) and Gureev (Gureyev) believed this almost indeterminate animal to be a Cretaceous shrew because of the pocketed coronoid process of the mandible. However, such a structure occurs in some mammals that lack a zygomatic arch, e.g. in certain apternodontids. †*Cretasorex* is apparently not referable to either *Sorex* or to *Diplomesodon*, both of which are currently resident in the Kyzyl Kum area where †*Cretasorex* was found. That the animal actually represents a shrew remains to be proven.

[2] If truly a shrew, its collection from a "paleoburrow" infilling suggests a late Cenozoic but not necessarily Recent age.

†*Kumsuperus* Nessov, 1984:62.
    L. Cret.; As.

†*Telacodon* Marsh, 1892:258.
    L. Cret.; N.A.

†*Beleutinus* Bazhanov, 1972:76.
    L. Cret.; As.

†*Obtususdon* Xu, 1977:122.
    E.-L. Paleoc.; As.

†*Wanotherium* Tang & Yan, 1976:96.
    L. Paleoc.; As.[1]

[1] E. Asia.

†*Tingamarra* Godthelp, Archer, Cifelli, Hand & Gilkeson, 1992:514.
    E. Eoc.; Aus.

†*Eutrochodon* Roth, 1903:155.
    M. Eoc.; S.A.

†*Helioseus* Sudre, 1979:100.
    M. Eoc.; Af.[1]

[1] N. Africa.

†*Eodesmatodon* Zheng & Chi, 1978:97.
    L. Eoc.; As.[1]

[1] S.E. Asia.

†*Idiogenomys* Ostrander, 1983:136.
    L. Eoc.; N.A.

†*Veratalpa* Ameghino, 1905:53.
    Mioc.; Eu.

†*Neodesmostylus* Khomenko, 1927.
    Pleist.; As.

Order †**BIBYMALAGASIA** MacPhee, 1994:201.

Subfamily †**Plesiorycteropodinae** Patterson, 1975:216.
    R.; Madagascar.

†*Plesiorycteropus* Filhol, 1895:12-14.
    [= or including †*Myoryctes* Forsyth Major, 1908;[1] †*Majoria* Thomas, 1915:57n.[2]]
    R.; Madagascar.

[1] Not †*Myoryctes* Ebert, 1863, a nematode.
[2] Replacement name for †*Myoryctes* Forsyth Major, 1908, preoccupied.

Magnorder **XENARTHRA** Cope, 1889:657, **new rank**.[1] Edentates.

> [= BRUTA Linnaeus, 1758:3[2]; MUTICI Storr, 1780[3]; EDENTATI Vicq d'Azyr, 1792:ciii; EDENTATA G. Cuvier, 1798:142[4]; OLIGODONTAE Gray, 1821:305[5]; BRUTINA Newman, 1843:148[6]; PARATHERIA Thomas, 1887:459[7]; BRADYTHERIA Haeckel, 1895:490, 521; XENARTHRIFORMES Kinman, 1994:37.[8]]

[1] Proposed as a suborder; used at rank of superorder by Pascual, Vucetich & Scillato-Yané, 1990: 628. Cope himself credited Gill, 1884:66 [actually 1883] with this taxon, but Gill's usage was merely descriptive (xenarthral edentates as opposed to nomarthral ones), not taxonomic.

[2] In part. BRUTA Linnaeus included *Elephas, Trichechus, Bradypus, Myrmecophaga,* and *Manis*.

[3] It has been suggested that Wolffer, 1780 was the correct author of this forgotten taxon.

[4] In part. Some authors list Vicq d'Azyr (1792) as the author of EDENTATA, but the International Code of Zoological Nomenclature need not be strictly applied above the family-group level.

[5] In part. Named as an order to include †*Megatherium,* an armadillo, and *Orycteropus* (as *Myrmecophaga capensis*).

[6] In part.

[7] In part. TUBULIDENTATA (i.e., *Orycteropus*) were included, although Thomas admitted that he was baffled by the tubulidentate dentition. Thomas believed that his PARATHERIA should be removed from the EUTHERIA of Huxley.

[8] Proposed as an order.

Order **CINGULATA** Illiger, 1811:110.[1]

> [= LORICATI Vicq d'Azyr, 1792:ciii; EFFODIENTIA Illiger, 1811:110[2]; LORICATA Owen, 1842:167[3]; FODIENTIA Giebel, 1855:412[4]; HICANODONTA Ameghino, 1889:758.] [Including DASYPODA Quenstedt, 1885; †GLYPTODONTIA Ameghino, 1889.]

[1] Proposed as "family". Also used by Haeckel, 1866:clviii as a "family." Used explicitly at subordinal rank by Trouessart, 1879: 2. Raised to ordinal rank by McKenna, 1975: 40.

[2] In part. Proposed as an order for "families" Cingulata (*Tolypeutes* and *Dasypus*) and Vermilinguia (*Orycteropus, Myrmecophaga,* and *Manis*). Used by Pascual, Vucetich & Scillato-Yané, 1990: 628 as grandorder.

[3] Not LORICATA Merrem, 1820:7, 34, a term for crocodiles. LORICATA has also been employed as a group of rotifers.

[4] As "family" for essentially this group, but including *Orycteropus*.

Family †**Protobradidae** Ameghino, 1902:49.

> [= †Protobradydae Ameghino, 1902:49.]
>
> E. Eoc.; S.A.

†*Protobradys* Ameghino, 1902:49.[1]

> E. Eoc.; S.A.

[1] Supposedly Casamayoran ground sloths but probably best forgotten, although Ameghino's description sets the genus off from other taxa. Ameghino even proposed a family †Protobradydae (sic) for this animal, now reported as lost.

Superfamily **Dasypodoidea** Gray, 1821:305.

> [= Dasipidae Gray, 1821:305; Dasypidae Gray, 1825:343; Dasypina Gray, 1825:343; Dasypoda Hoernes, 1886:662; Dasypodoidae Hay, 1930:415; Dasypodoidea Simpson, 1931:273.]

Family **Dasypodidae** Gray, 1821:305. Armadillos, quirquinchos, tatus.

> [= Dasipidae Gray, 1821:305; Dasypidae Gray, 1825:343; Dasypodidae C. L. Bonaparte, 1838:111; Tatusidae Burnett, 1830; Tatusiidae Lahille, 1895:8; Praopidae Ameghino, 1889:853; †Scoteopsidae Ameghino, 1894:183.] [Including Chlamydophorina Gray, 1865:381; Chlamydophoridae Gray, 1869:362, 387; Tolypeutidae Gray, 1869; †Stegotheridae Ameghino, 1889:878, 895; Scleropleuridae Lahille, 1895.]
>
> L. Paleoc.-M. Eoc., L. Olig.-R.; S.A. L. Plioc.-R.; N.A. R.; Cent. A.

Subfamily **Dasypodinae** Gray, 1821:305.

> [= Dasipidae Gray, 1821:305; Dasypidae Gray, 1825:343; Dasypina Gray, 1825:343[1]; Dasypodina C. L. Bonaparte, 1837:8[2]; Dasypodinae Gill, 1872:24; Tatusidae Burnett, 1830; Tatusiinae Gill, 1872:24; Tatusiidae Lahille, 1895:8; Praopidae Ameghino, 1889:853; †Scoteopsidae Ameghino, 1894:183.]

L. Paleoc.-E. Eoc., ?M. Eoc., E. Mioc., ?L. Plioc., E. Pleist.-R.; S.A. L. Plioc.-R.;
N.A. R.; Cent. A.

[1] As tribe.
[2] As subfamily.

Tribe †**Stegotheriini** Ameghino, 1889:878, 895.
> [= †Stegotheridae Ameghino, 1889:878, 895; †Scoteopsidae Ameghino, 1894:183;
> †Stegotherinae Trouessart, 1898:1123; †Stegotheriini Patterson & Pascual, 1968:423.]
> [Including †Astegotheriidae Ameghino, 1906:469; †Astegotheriini Vizcaíno, 1994:5.]
> L. Paleoc.-E. Eoc., ?M. Eoc., E. Mioc.; S.A.

†*Prostegotherium* Ameghino, 1902:69.
> L. Paleoc.-E. Eoc., ?M. Eoc.; S.A.

†*Pseudostegotherium* Ameghino, 1902:137.
> ?E. Eoc., E. Mioc.; S.A.

†*Astegotherium* Ameghino, 1902:67.
> E. Eoc., ?M. Eoc.; S.A.

†*Stegosimpsonia* Vizcaíno, 1994:8.
> E. Eoc.; S.A.

†*Stegotherium* Ameghino, 1887:25.
> [= or including †*Scotoeops* Ameghino, 1887:24.]
> E. Mioc.; S.A.

Tribe **Dasypodini** Gray, 1821:305.
> [= Dasipidae Gray, 1821:305; Dasypidae Gray, 1825:343; Dasypina Gray, 1825:343[1];
> Dasypodini Simpson, 1945:73; Tatusidae Burnett, 1830; Tatusiinae Gill, 1872:24;
> Tatusiidae Lahille, 1895:8; Praopidae Ameghino, 1889:853.]
> L. Plioc.-R.; N.A. ?L. Plioc., E. Pleist.-R.; S.A. R.; Cent. A.

[1] As tribe.

Dasypus Linnaeus, 1758:50-51. Long-nosed armadillos, mulitas, nine-banded armadillo.
> [= *Loricatus* Desmarest, 1804:28[1]; *Dasipus* J. E. Gray, 1821:305; *Cataphractus* Storr,
> 1762:40; *Tatus* Fermin, 1765:3; *Tatu* Frisch, 1775:5[2]; *Tatu* Blumenbach, 1779:74;
> *Tatusia* F. Cuvier, in Lesson, 1827:309-312; *Cachicamus* McMurtrie, 1831:163;
> *Cachicama* I. Geoffroy Saint-Hilaire, 1835:53; *Zonoplites* Gloger, 1841:114;
> *Mamtatusiusus* Herrera, 1899:5.[3]] [Including *Praopus* Burmeister, 1854:295-301;
> *Cryptophractus* Fitzinger, 1856:123[4]; *Hyperoambon* Peters, 1864:179-180[4]; *Muletia*
> Gray, 1874:244-246.]
> L. Plioc.-R.; N.A. ?L. Plioc., E. Pleist.-R.; S.A. R.; Cent. A.

[1] In part?
[2] Not published in a consistently binominal work.
[3] See the International Code of Zoological Nomenclature, Article 1(b)(8), and Palmer (1904:25-26).
[4] As subgenus.

†*Propraopus* Ameghino, 1881:311.
> [Including †*Pontotatus* Ameghino, 1908:427.]
> ?L. Plioc., E.-L. Pleist.; S.A.

†*Dasypodon* Castellanos, 1925:258, 284.
> Pleist.; S.A.

Subfamily **Euphractinae** Winge, 1923:304.
> [= Scleropleuridae Lahille, 1895[1]; Euphracti Winge, 1923:304; Euphractinae Pocock,
> 1924:1031; Euphractinae Weber, 1928:217; Euphractinae Yepes, 1928:30.] [Including
> Chlamydophorina Bonaparte, 1850; Chlamydophoridae Gray, 1869:387;
> †Proeuphractinae Kraglievich, 1934:98, 101.]
> E.-M. Eoc., L. Olig.-R.; S.A.

[1] See the International Code of Zoological Nomenclature, Article 37(b).

†*Coelutaetus* Ameghino, 1902:64.[1]
> E. Eoc.; S.A.

[1] Based on a single broken scute. Naive compilers of faunal diversity data will doubtless award this name a
point on their graphs; nevertheless, it can neither be synonymized nor ignored at present.

Tribe †**Utaetini** Simpson, 1945:72.
  E. Eoc.; S.A.

†*Utaetus* Ameghino, 1902:59.
  [= †*Posteutatus* Ameghino, 1902:60; †*Pareutaetus* Ameghino, 1902:62; †*Orthutaetus* Ameghino, 1902:63.]
  E. Eoc.; S.A.

Tribe †**Eutatini** Bordas, 1933:598.
  [= †Eutatinae Bordas, 1933:598; †Eutatini Simpson, 1945:72.]
  E.-M. Eoc., L. Olig.-R.; S.A.

†*Meteutatus* Ameghino, 1902:54.
  [Including †*Sadypus* Ameghino, 1902:64.]
  E.-M. Eoc., L. Olig.; S.A.

†*Anteutatus* Ameghino, 1902:58.
  E.-M. Eoc.; S.A.

†*Pseudeutatus* Ameghino, 1902:57.
  [Including †*Pachyzaedyus* Ameghino, 1902:67; †*Pachyzaedius* Ameghino, 1906:470.]
  M. Eoc., ?L. Olig.; S.A.

†*Stenotatus* Ameghino, 1891:253.
  [= or including †*Stegotheriopsis* Bordas, 1939:431.] [Including †*Prodasypus* Ameghino, 1894:172-173.]
  L. Olig.-M. Mioc.; S.A.

†*Proeutatus* Ameghino, 1891:41.
  [= or including †*Thoracotherium* Mercerat, 1891:42.]
  ?L. Olig., E.-M. Mioc.; S.A.

†*Archaeutatus* Ameghino, 1902:57.
  L. Olig.; S.A.

†*Paraeutatus* Scott, 1903:68.
  [= or including †*Proparaeutatus* Trouessart, 1905:815.]
  E. Mioc.; S.A.

†*Doellotatus* Bordas, 1932:167.
  [= †*Eutatopsis* Kraglievich, 1934:57.]
  L. Mioc.-L. Plioc.; S.A.

†*Chasicotatus* Scillato-Yané, 1977:133, 139.
  L. Mioc.; S.A.

†*Ringueletia* Reig, 1958:250.
  E.-L. Plioc.; S.A.

†*Eutatus* Gervais, 1867:279-280.
  L. Plioc.-R.; S.A.

Tribe **Euphractini** Winge, 1923:304.
  [= Scleropleuridae Lahille, 1895:8, 30[1]; Euphracti Winge, 1923:304; †Euphractinae Pocock, 1924:1031; Euphractinae Yepes, 1928:30; Euphractinae Weber, 1928:220; Euphractini Simpson, 1945:72.] [Including †Proeuphractinae Kraglievich, 1934:98, 101.]
  M. Eoc., L. Olig.-R.; S.A.

[1] See the International Code of Zoological Nomenclature, Article 37(b).

†*Isutaetus* Ameghino, 1902:65.
  M. Eoc., L. Olig.; S.A.

†*Anutaetus* Ameghino, 1902:66.
  M. Eoc., L. Olig.; S.A.

†*Hemiutaetus* Ameghino, 1902:66.
  L. Olig.; S.A.

†*Amblytatus* Ameghino, 1902:57.
  L. Olig.; S.A.

†*Eodasypus* Ameghino, 1894:429.
  L. Olig.-E. Mioc.; S.A.

†*Prozaedyus* Ameghino, 1891:41.[1]
  L. Olig.-M. Mioc.; S.A.
  [1] Sometimes incorrectly spelled "†*Prozaedius.*"

†*Vetelia* Ameghino, 1891:162-163.
  E.-L. Mioc.; S.A.

†*Proeuphractus* Ameghino, 1886:208-216.
  L. Mioc.; S.A.

†*Paleuphractus* Kraglievich, 1934:57.
  L. Mioc.-E. Plioc.; S.A.

†*Chorobates* Reig, 1958:250.
  L. Mioc.-L. Plioc.; S.A.

†*Macroeuphractus* Ameghino, 1887:19.
  [= †*Dasypotherium* Moreno, 1889:38.]
  ?L. Mioc., E.-L. Plioc.; S.A.

†*Paraeuphractus* Scillato-Yané, 1975:451.
  L. Mioc.; S.A.

†*Acantharodeia* Rovereto, 1914:108.
  [= †*Macrochorobates* Scillato-Yané, 1980:26.]
  E.-L. Plioc.; S.A.

Chaetophractus Fitzinger, 1871:268-276.[1] Hairy armadillos, peludos.
  ?E. Plioc., L. Plioc.-R.; S.A.
  [1] "In much of the literature *Chaetophractus* and *Euphractus* (together, or either one of them) are called *Dasypus*, and the genus to which the latter name is now confined is then usually called *Tatu* or *Tatusia*" (Simpson, 1945:72).

Zaedyus Ameghino, 1889:867. Pichi.
  [= *Zaedypus* Lydekker, 1890:50; *Zaedius* Lydekker, 1894:123.]
  L. Plioc.-R.; S.A.

Euphractus Wagler, 1830:36.[1] Six-banded armadillo, yellow armadillo.
  [= *Loricatus* Desmarest, 1804:28[2]; *Encoubertus* McMurtrie, 1832:104[3]; *Pseudotroctes* Gloger, 1842:112; *Scleropleura* A. Milne-Edwards, 1871:177.[4]]
  M. Pleist.-R.; S.A.
  [1] See footnote for *Chaetophractus*.
  [2] In part. Dubiously a senior synonym of *Euphractus*; if so, it should be suppressed.
  [3] Proposed as a subgenus of *Dasypus*.
  [4] "Although it is still frequently listed as a distinct genus, there is little doubt that *Scleropleura* was founded on an injured or teratological *Euphractus*" (Simpson, 1945:72).

Tribe **Chlamyphorini** Pocock, 1924:1031. Pichiciegos.
  [= Chlamydophorina Bonaparte, 1850; Chlamydophoridae Gray, 1869; Chlamyphoridae Pocock, 1924:1031; Chlamyphorinae Yepes, 1928:11; Chlamyphorini Patterson & Pascual, 1968:423.]
  Pleist., R.; S.A.

Chlamyphorus Harlan, 1825:235-246. Pichiciego menor, pichiciego mayor, fairy armadillos.
  [= *Chlamydophorus* Wagler, 1830:35.] [Including *Burmeisteria* Gray, 1865:381-382[1]; *Calyptophractus* Fitzinger, 1871:388-390.]
  Pleist., R.; S.A.
  [1] *Burmeisteria* Gray, October 1865, not †*Burmeisteria* Salter, June 1865, a trilobite. We thank John T. Thurmond for this information. Objective synonym of *Calyptophractus*.

Subfamily **Tolypeutinae** Gray, 1865:365.
  [= Tolypeutina Gray, 1865:365; Tolypeutidae Gray, 1869:385; Tolypeutinae Alberdi, Leone & Tonni, 1995:371.] [Including Priodontinae Patterson & Pascual, 1968:423.[1]]
  E. Plioc.-R.; S.A. R.; Cent. A.
  [1] Used for the taxon here called Tolypeutinae in violation of the International Code of Zoological Nomenclature, Art 23(d).

Tribe **Tolypeutini** Gray, 1865:365.
>> [= Tolypeutina Gray, 1865:365; Tolypeutini Weber, 1928:220.]
>> E. Plioc.-R.; S.A.

Tolypeutes Illiger, 1811:111. Three-banded armadillos, mataco.
>> [= *Apara* McMurtrie, 1831:163; *Tolypoides* G. Grandidier & Neveu-Lemaire, 1905:370.]
>> E. Plioc.-R.; S.A.

Tribe **Priodontini** Gray, 1873:20.
>> [= Prionodontina Gray, 1873:20[1]; Xenuri Winge, 1923:304; Cabassoinae Pocock, 1924:1031; Priodontini Weber, 1928:220.]
>> M. Pleist.-R.; S.A. R.; Cent. A.
>> [1] Not Prionodontina Gray, 1864, a viverrid family-group name.

Cabassous McMurtrie, 1831:164. Naked-tailed armadillos, eleven-banded armadillos, cabassus.
>> [= *Xenurus* Wagler, 1830:36[1]; *Arizostus* Gloger, 1841:114; *Tatoua* Gray, 1865:378; *Lysiurus* Ameghino, 1891:254.[2]] [Including *Ziphila* Gray, 1873:22.]
>> M. Pleist.-R.; S.A. R.; Cent. A.
>> [1] Not *Xenurus* Boie, 1826, a genus of birds.
>> [2] Replacement name for *Xenurus* Wagler, 1830.

Priodontes F. Cuvier, 1827:309. Giant armadillo, tatuasu.
>> [= *Priodon* McMurtrie, 1831:164; *Prionodon* Gray, 1843:xxvii[1]; *Priodonta* Gray, 1843:190; *Prionodos* Gray, 1865:374-375; *Cheloniscus* Wagler, 1830:35.]
>> R.; S.A.
>> [1] Not *Prionodon* Horsfield, 1821, a viverrid carnivore.

Family †**Peltephilidae** Ameghino, 1894:433.
>> [= †Peltephilinae Lydekker, 1894:66; †Peltophilini Winge, 1915:219; †PELTEPHILODA Bordas, 1936:4; †Peltephilinae Patterson & Pascual, 1968:423.]
>> E.-M. Eoc., L. Olig.-L. Mioc.; S.A.[1]
>> [1] †Peltephilidae indet. in the early Eocene, middle Eocene, and late Miocene.

†*Peltephilus* Ameghino, 1887:25.
>> [= †*Peltophilus* Winge, 1915:219.] [= or including †*Cochlops* Ameghino, 1889:792; †*Gephyranodus* C. Ameghino, 1891:119.]
>> L. Olig.-E. Mioc.; S.A.

†*Peltecoelus* Ameghino, 1902:138.
>> L. Olig.-E. Mioc.; S.A.

†*Parapeltecoelus* Bordas, 1938:269.
>> E. Mioc.; S.A.

†*Anantiosodon* Ameghino, 1891:41-42.
>> E. Mioc.; S.A.

†*Epipeltephilus* Ameghino, 1904:289.
>> M. Mioc.; S.A.

Superfamily †**Glyptodontoidea** Gray, 1869:387.
>> [= †Glyptodontidae Gray, 1869:387; †GLYPTODONTIA Ameghino, 1889; †Glyptodontoidea Borissiak, 1930:238, 239; †Glyptodontoidae Hay, 1930:418.] [Including †Chlamytherioidea Bordas, 1939:448.]

Family †**Pampatheriidae** Paula Couto, 1954:6.
>> [= †Pampatheriinae Paula Couto, 1954:6; †Pampatheriidae Alberdi, Leone & Tonni, 1995:371; †MESODONTIA Ameghino, 1885:131; †Chlamydotheridae Ameghino, 1889:853; †Chlamydotheriinae Lydekker, 1894:52; †Chlamydotheriini Winge, 1915:219; †Chlamytheriinae Hay, 1930:417; †Chlamytherioidea Bordas, 1939:448.]
>> E.-M. Eoc., M. Mioc.-L. Pleist.; S.A. L. Plioc.-L. Pleist.; N.A.

†*Machlydotherium* Ameghino, 1902:52.[1]
>> E.-M. Eoc.; S.A.
>> [1] Assignment to the pampatheres is uncertain. Possibly this genus should be regarded as a member of the Dasypodoidea, *incertae sedis*.

†*Kraglievichia* Castellanos, 1927:1.
　　[Including †*Plaina* Castellanos, 1937:24.]
　　M. Mioc.-L. Plioc.; S.A.

†*Vassallia* Castellanos, 1927:1, 5.
　　L. Mioc., Plioc.; S.A.

†*Pampatherium* Gervais & Ameghino, 1880:210-211.[1]
　　[= †*Chlamytherium* Lund, in Orsted, 1839[2]; †*Chlamydotherium* Lund, 1839:69.[3]]
　　[Including †*Holmesina* Simpson, 1930:3; †*Hoffstetteria* Castellanos, 1957:5.]
　　L. Plioc.-L. Pleist.; N.A. L. Plioc.-L. Pleist.; S.A.

[1] According to Palmer (1904:508), †*Pampatherium* Ameghino, 1875:528, was a *nomen nudum*. No description of †*P. typus* was given.
[2] Not 1838, and also a *lapsus calami*.
[3] Not †*Chlamydotherium* Bronn, 1838:1258, a glyptodont.

Family †**Palaeopeltidae** Ameghino, 1895:659.
　　[= †Pseudorophodontidae Hoffstetter, 1954:160.]
　　?E. Eoc., M. Eoc., L. Olig.; S.A.

†*Palaeopeltis* Ameghino, 1895:659.
　　[= †*Pseudorophodon* Hoffstetter, 1954:160.]
　　?E. Eoc., M. Eoc., L. Olig.; S.A.

Family †**Glyptodontidae** Gray, 1869:387. Glyptodonts.
　　[Including †Hoplophoridae Huxley, 1864:108[1]; †Doedicuridae Ameghino, 1889:774,
　　840; †Propalaehoplophoridae Ameghino, 1891:251; †Sclerocalyptidae Ameghino,
　　1904:137.[2]]
　　M. Eoc., L. Olig.-L. Pleist.; S.A. L. Plioc.-L. Pleist.; N.A. L. Pleist.; Cent. A.

[1] †Hoplophoridae clearly has priority over †Glyptodontidae. †Hoplophoridae is probably not truly an "unused" name, but †Glyptodontidae is so common and familiar that we retain it, in the spirit of the International Code of Zoological Nomenclature, Articles 23(b) and 79(c), rather than slavishly following Article 23(d).
[2] Possibly earlier.

†*Lomaphorelus* Ameghino, 1902:51.[1]
　　M. Eoc.; S.A.

[1] *Nomen dubium*?

Tribe †**Neothoracophorini** Castellanos, 1951:65.
　　E.-L. Pleist.; S.A.

†*Pseudoneothoracophorus* Castellanos, 1951:66.
　　Pleist.; S.A.

†*Neothoracophorus* Ameghino, 1889:790-792.[1]
　　[= †*Thoracophorus* Gervais & Ameghino, 1880:206-211.[2]]
　　E.-L. Pleist.; S.A.

[1] "It has been suggested that this genus (the validity of which is not beyond question) should be called †*Myloglyptodon* Ameghino, 1884. There is room for dispute, but it is my opinion that 'Myloglyptodon' was mentioned only as a morphological, theoretical term and was not technically introduced into zoological nomenclature. Ameghino himself evidently assumed and definitely implied this to be the case, for he never used it as if it were a real zoological name" (Simpson, 1945:75).
[2] Not *Thoracophorus* Hope, 1840, a genus of Coleoptera.

Subfamily †**Glyptatelinae** Castellanos, 1932:93.
　　[= †Glyptatelinae Castellanos, 1932:93.]
　　M. Eoc., L. Olig., M. Mioc.; S.A. E. and/or M. Pleist.; N.A.

†*Glyptatelus* Ameghino, 1897:507.
　　M. Eoc., L. Olig.; S.A.

†*Clypeotherium* Scillato-Yané, 1977:251.
　　L. Olig.; S.A.

†*Neoglyptatelus* Carlini, Vizcaino & Scillato-Yané, 1997:221.
　　M. Mioc.; S.A.

†*Pachyarmatherium* Downing & White, 1995:378.[1]
> E. and/or M. Pleist.; N.A.
> [1] Placement in the †Glyptatelinae is not certain.

Subfamily †**Propalaehoplophorinae** Ameghino, 1891:251.
> [= †Propalaehoplophoridae Ameghino, 1891:251; †Propalaeohoplophorinae Castellanos, 1932:92.]
> E.-M. Mioc.; S.A.

†*Propalaehoplophorus* Ameghino, 1887:24-25.[1]
> [= †*Propalaeohoplophorus* Castellanos, 1932:92; †*Propaleohoplophorus* Frailey, 1988:2.]
> E.-M. Mioc.; S.A.
> [1] Ameghino was consistent in his spelling. We do not know whether Castellanos and/or Frailey intended emendation.

†*Metopotoxus* Ameghino, 1895:123.[1]
> E. Mioc.; S.A.
> [1] This genus, if biologically real, should be renamed. No species was proposed by Ameghino.

†*Eucinepeltus* Ameghino, 1891:40.
> E.-M. Mioc.; S.A.

†*Asterostemma* Ameghino, 1889:822-824.
> E.-M. Mioc.; S.A.

Subfamily †**Hoplophorinae** Huxley, 1864:31.
> [= †Hoplophoridae Huxley, 1864:31; †Hoplophorinae Weber, 1928:224; †Sclerocalyptinae Trouessart, 1898:1128; †Sclerocalyptidae Ameghino, 1904:137.[1]]
> M. Mioc.-L. Pleist.; S.A.
> [1] Possibly earlier.

†*Asymmetrura* Fariña, 1981:780.
> E. Plioc.; S.A.

†*Caudaphorus* Fariña, 1981:784.
> E. Plioc.; S.A.

†*Uruguayurus* Mones, 1987:501.
> ?E. Plioc.; S.A.

Tribe †**Hoplophorini** Huxley, 1864:31, **new rank**.
> [= †Hoplophoridae Huxley, 1864:31; †Hoplophorinae Weber, 1928:224; †Sclerocalyptinae Trouessart, 1898:1128; †Sclerocalyptini Hoffstetter, in Piveteau, 1958:577.]
> M. Mioc.-L. Pleist.; S.A.

†*Hoplophorus* Lund, 1838:11.[1]
> [= †*Sclerocalyptus* Ameghino, 1891:251.[2]]
> M. Mioc., ?L. Plioc., E.-L. Pleist.; S.A.
> [1] Not *Hoplophora* Perty, 1830, an insect; nor †*Oplophorus* Milne-Edwards, 1837.
> [2] New name for †*Hoplophorus*, incorrectly thought to be preoccupied.

†*Stromaphorus* Castellanos, 1926:276.
> L. Mioc.; S.A.

†*Eosclerocalyptus* C. Ameghino, 1919:150.
> L. Mioc.; S.A.

†*Hoplophractus* Cabrera, 1939:3-35.
> L. Mioc.; S.A.

†*Trachycalyptus* Ameghino, 1908:426.
> L. Mioc., ?E. Plioc., L. Plioc.; S.A.

†*Berthawyleria* Castellanos, 1939:57.
> L. Mioc.; S.A.

†*Parahoplophorus* Castellanos, 1932:99.
> L. Mioc.; S.A.

†*Isolinia* Castellanos, 1951:96.
   Plioc. and/or Pleist.; S.A.

†*Stromaphoropsis* Kraglievich, 1932:281.
   L. Mioc.-E. Plioc.; S.A.

†*Eosclerophorus* Castellanos, 1948:3.
   L. Mioc.; S.A.

†*Trabalia* Kraglievich, 1932:308.
   Pleist.; S.A.

†*Neosclerocalyptus* Paula Couto, 1957:7.
   Pleist.; S.A.

Tribe †**Palaehoplophorini** Hoffstetter, in Piveteau, 1958:580.
   M.-L. Mioc.; S.A.

†*Palaehoplophorus* Ameghino, 1883:301-302.
   [= †*Palaeohoplophorus* Roger, 1887:21.[1]]
   M.-L. Mioc.; S.A.
  [1] And most later authors.

†*Aspidocalyptus* Cabrera, 1939:3-35.
   L. Mioc.; S.A.

†*Chlamyphractus* Castellanos, 1939:95.
   L. Mioc.; S.A.

†*Pseudoeuryurus* Ameghino, 1889:851-852.
   L. Mioc.; S.A.

†*Protoglyptodon* Ameghino, 1885:135-137.
   L. Mioc.; S.A.

Tribe †**Lomaphorini** Hoffstetter, 1958:585.
   L. Mioc.-L. Pleist.; S.A.

†*Peiranoa* Castellanos, 1946:5.
   L. Mioc.; S.A.

†*Lomaphorops* Castellanos, 1932:99.
   L. Mioc.; S.A.

†*Urotherium* Castellanos, 1926:263.
   E.-L. Plioc.; S.A.

†*Lomaphorus* Ameghino, 1889:810, 819-822.
   E.-L. Pleist.; S.A.

†*Trachycalyptoides* Saint-André, 1996.
   L. Mioc.; S.A.

Tribe †**Plohophorini** Castellanos, 1932:97.
   [= †Plohophorinae Castellanos, 1932:97; †Plohophorini Hoffstetter, in Piveteau, 1958:581.]
   L. Mioc.-L. Plioc., Pleist.; S.A.

†*Coscinocercus* Cabrera, 1939:3-35.
   L. Mioc.; S.A.

†*Phlyctaenopyga* Cabrera, 1944:42.
   L. Mioc.; S.A.

†*Plohophorops* Rusconi, 1934:106.[1]
   L. Mioc.; S.A.
  [1] Credited to Castellanos by Rusconi, but Castellanos' description was not published until 1935.

†*Plohophorus* Ameghino, 1887:17.
   L. Mioc.-L. Plioc.; S.A.

†*Pseudoplohophorus* Castellanos, 1926:269.
   L. Mioc.-E. Plioc.; S.A.

†*Teisseiria* Kraglievich, 1932:306.[1]
    L. Mioc.; S.A.
[1] Described as a subgenus of †*Stromaphoropsis.* Raised to generic rank by Castellanos, 1939:176.

†*Plohophoroides* Castellanos, 1928:1.
    L. Plioc.; S.A.

†*Zaphilus* Ameghino, 1889:828.
    Pleist.; S.A.

Tribe †**Panochthini** Castellanos, 1927:266.
    [= †Panochthinae Castellanos, 1927:266; †Panochthini Simpson, 1945:74.]
    ?L. Mioc., E. Plioc.-L. Pleist.; S.A.[1]
[1] Questionable occurrence of "†Panochthinae" in the late Miocene Epecuen Fm.

†*Nopachtus* Ameghino, 1888:16.[1]
    E. Plioc.; S.A.
[1] Not "†*Nopachthus.*"

†*Panochthus* Burmeister, 1866:190-191.
    [= †*Schistopleurum* Nodot, 1855:335-338.]
    ?L. Plioc., E.-L. Pleist.; S.A.

†*Propanochthus* Castellanos, 1925:4*n.*
    L. Plioc.; S.A.

†*Parapanochthus* Moreira, 1971:530.
    L. Pleist.; S.A.

Tribe †**Neuryurini** Hoffstetter, in Piveteau, 1958:586.
    Pleist.; S.A.

†*Neuryurus* Ameghino, 1889:840-844.[1]
    [= †*Euryurus* Gervais & Ameghino, 1880:184-187.[2]]
    Pleist.; S.A.
[1] New name for †*Euryurus* Gervais & Ameghino, 1880.
[2] Not *Euryurus* Koch, 1847, a genus of arthropods; nor *Euryurus* Von der Marck, 1864, another arthropod genus.

Subfamily †**Doedicurinae** Ameghino, 1889:774, 840.
    [= †Doedicuridae Ameghino, 1889:774, 840; †Doedicurinae Trouessart, 1898:1134; †Doedicurini Simpson, 1945:74.]
    L. Mioc.-L. Pleist.; S.A.

†*Eleutherocercus* Koken, 1888:1-28.
    ?L. Mioc., E. Plioc.; S.A.

†*Prodaedicurus* Castellanos, 1927:266.
    [Including †*Palaeodoedicurus* Castellanos, 1927:272.[1]]
    E. Plioc., ?L. Plioc.; S.A.
[1] Described as subgenus.

†*Comaphorus* Ameghino, 1886:197-199.
    L. Mioc.; S.A.

†*Castellanosia* Kraglievich, 1932:257-321.
    L. Plioc., Pleist.; S.A.

†*Xiphuroides* Castellanos, 1927:284.
    L. Plioc.; S.A.

†*Doedicurus* Burmeister, 1874:393-404.
    [= †*Daedicurus* Lydekker, 1887:122-123; †*Doedycurus* Coues, 1889:1717; †*Doedycura* Burmeister, 1872:261.[1]]
    L. Plioc.-L. Pleist.; S.A.
[1] Proposed as subgenus of †*Panochthus.*

†*Daedicuroides* Castellanos, 1941:362.
    E. and/or M. Pleist.; S.A.

†*Plaxhaplous* Ameghino, 1884:199-200.[1]
>   E.-L. Pleist.; S.A.
>   [1] Not †*Plaxhaplus*; nor †*Plaxaplus*.

Subfamily †**Glyptodontinae** Gray, 1869:387.
>   [= †Glyptodontidae Gray, 1869:387; †Glyptodontinae Trouessart, 1898:1125.[1]]
>   L. Mioc.-L. Pleist.; S.A. L. Plioc.-L. Pleist.; N.A. L. Pleist.; Cent. A.
>   [1] Or Cope, 1898:134.

Tribe †**Glyptodontini** Gray, 1869:387.
>   [= †Glyptodontidae Gray, 1869:387; †Glyptodontini Castellanos, 1951:65.]
>   L. Mioc.-L. Pleist.; S.A.

†*Glyptodontidium* Cabrera, 1944:71.
>   L. Mioc.; S.A.

†*Paraglyptodon* Castellanos, 1932:93.
>   E.-L. Plioc.; S.A.

†*Glyptodon* Owen, 1839:178.[1]
>   [= †*Lepitherium* É. Geoffroy Saint-Hilaire, 1839:127[2]; †*Pachypus* D'Alton, 1839[3];
>   †*Schistopleurum* Nodot, 1857.[4]] [Including †*Glyptocoileus* Castellanos, 1952:37[5];
>   †*Glyptopedius* Castellanos, 1953:395[5]; †*Glyptostracon* Castellanos, 1953:402;
>   †*Stromatherium* Castellanos, 1953:406.]
>   ?L. Plioc., E.-L. Pleist.; S.A.
>   [1] Not 1838.
>   [2] *Nomen nudum* in É. Geoffroy Saint-Hilaire, 1833, fide Palmer (1904:370).
>   [3] Not *Pachypus* Dejean, 1831, a coleopteran.
>   [4] Not †*Schistopleurum* Nodot, 1855, a synonym of †*Panochthus*.
>   [5] Proposed as subgenus.

†*Stromatherium* Castellanos, 1953:406.
>   Pleist.; S.A.

†*Chlamydotherium* Bronn, 1838:1256-1259.
>   Pleist.; S.A.

†*Glyptostracon* Castellanos, 1953:402.
>   Pleist.; S.A.

†*Heteroglyptodon* Roselli, 1976:137.
>   L. Pleist.; S.A.[1]
>   [1] Uruguay.

Tribe †**Glyptotheriini** Castellanos, 1953:392.
>   [= †Glyptotherini Castellanos, 1953:392.]
>   L. Plioc.-L. Pleist.; N.A. L. Pleist.; Cent. A.

†*Glyptotherium* Osborn, 1903:492.
>   [= or including †*Brachyostracon* Brown, 1912:169; †*Boreostracon* Simpson, 1929:581;
>   †*Xenoglyptodon* Meade, 1953:455.]
>   L. Plioc.-L. Pleist.; N.A. L. Pleist.; Cent. A.

Order **PILOSA** Flower, 1883:184.[1]
>   [= BRADYPODA Blumenbach, 1779[2]; ANICANODONTA Ameghino, 1889:653,
>   657.]
>   [1] Proposed as infraorder by Flower. Raised to rank of order by McKenna, 1975:40.
>   [2] Long-forgotten name proposed for *Bradypus* (as *Ignavus*) and *Myrmecophaga*, but is not resurrected here
>   for reasons of stability and because the term would tend to be associated with sloths only.

†*Trematherium* Ameghino, 1887:22.
>   E. Mioc.; S.A.

Family †**Entelopidae** Ameghino, 1889:654, 895.[1]
>   [= †Entelopsidae Ameghino, 1889:654, 895; †ENTELOPSIDA Haeckel, 1895:516[2];
>   †Entelopsoidea Pascual, 1960:127, 129; †Entelopidae Hoffstetter, 1982:420.[3]] [= or
>   including †Dideilotheridae Ameghino, 1894:183.]

E. Mioc.; S.A.

[1] †PLEIODONTA Ameghino, 1889:653-654 was a theoretical order proposed for the †Entelopidae. It was characterized (jointly with Ameghino's order PRIODONTA) by the plesiomorphic retention of enamel on the teeth and by the possession of incisors, but differed from the PRIODONTA by the plesiomorphic retention of a mandibular ascending ramus. The †PLEIODONTA and PRIODONTA were contrasted with the orders ANICANODONTA (i.e., PILOSA) and HICANODONTA (i.e., CINGULATA).
†PLEIODONTA is useless as defined, but it may be that the †Entelopsidae deserve higher taxonomic rank than given here. The †Entelopoidea might serve adequately in that case, but its placement among still higher taxa is still very much in doubt.
[2] Used as a family.
[3] Emendation.

†*Entelops* Ameghino, 1887:23-24.
E. Mioc.; S.A.

†*Delotherium* Ameghino, 1889:655.[1]
[= †*Dideilotherium* Ameghino, 1889:921.[2]]
E. Mioc.; S.A.

[1] Not †*Deilotherium* Filhol, 1882:112-113, a European Oligocene artiodactyl.
[2] Replacement name for †*Delotherium*, incorrectly thought to be preoccupied.

Suborder **VERMILINGUA** Illiger, 1811:112.[1] Anteaters.
[= VERMILINGUIA Illiger, 1811:112[2]; VERMILINGUIA Giebel, 1855:394; VERMILINGUA Gray, 1869:390; EDENTULAE Gray, 1821:305.[3]] [= or including EDENTULA Owen, 1859:52.[4]] [Including †PRIODONTA Ameghino, 1889:654.]
[1] Emended.
[2] Proposed as a "family." Used as a suborder by Gill, 1883:59.
[3] In part. Named as an order including *Myrmecophaga* and *Manis*, but not *Orycteropus*.
[4] *Nomen oblitum*. Proposed as a family.

†*Argyromanis* Ameghino, 1904:129.
E. Mioc.; S.A.

†*Orthoarthrus* Ameghino, 1904:130.
E. Mioc.; S.A.

Family **Myrmecophagidae** Gray, 1825:343. Vermilinguas.
[= Scandentia Fischer de Waldheim, 1817:372[1]; Scandentium Fischer de Waldheim, 1817:412[2]; Myrmecophagina Gray, 1825:343[3]; Myrmecophagidae Bonaparte, 1838:111; Myrmecophagini Winge, 1923:309; Myrmecophagoidea Simpson, 1931:273; Myrmecophaginae Kalandadze & Rautian, 1992:102.] [Including Tamanduina Gray, 1873; Tamanduinae Pocock, 1924:1030.]
E. Mioc.-R.; S.A. E. Pleist., R.; N.A.[4] R.; Cent. A.
[1] Not based on a generic name and not available for ranks below superfamily, but it is interesting that the tupaiids were not the first animals to be tarred with "*Scandentia*."
[2] Proposed as a family but not based on a generic name.
[3] As tribe.
[4] Southern North America (Mexico north to San Luis Potosí and Guerrero).

†*Protamandua* Ameghino, 1904:128.
E. Mioc.; S.A.

†*Neotamandua* Rovereto, 1914:98.
M.-L. Mioc.; S.A.

†*Promyrmephagus* Ameghino, 1904:128.
E. Mioc.; S.A.

Myrmecophaga Linnaeus, 1758:35. Giant anteater, jurumi.
[= *Mammyrmecophagaus* Herrera, 1899:16[1]; *Falcifer* Rehn, 1900:576.] [= or including †*Nunezia* Kraglievich, 1934:47.]
E. Plioc.-R.; S.A. E. Pleist.; N.A. R.; Cent. A.
[1] See the International Code of Zoological Nomenclature, Article 1(b)(8), and Palmer (1904:25-26).

Tamandua Gray, 1825:343.[1] Tamanduas, collared anteaters.
[= *Tamanduas* F. Cuvier, 1829:501.]
Pleist., R.; S.A. R.; N.A.[2] R.; Cent. A.
[1] Not "*Tamandua*" Frisch, 1775, which was not published in a consistently binominal work. The same applies to *Tamandua* Rafinesque, 1815. See the International Code of Zoological Nomenclature, Article 1(c)(i), for confusion, however.
[2] Southern North America (Mexico north to San Luis Potosí and Guerrero).

Family **Cyclopedidae** Pocock, 1924:1030.[1]
[= Cyclothurinae Gill, 1872:23; Cycloturini Winge, 1923:308; Cylopidae Hirschfeld, 1976:430, 432; Cyclopedidae Storch, 1981:247-289; Cyclopinae Kalandadze & Rautian, 1992:102.]
L. Mioc., R.; S.A. R.; Cent. A.
[1] See the International Code of Zoological Nomenclature, Article 40.

†*Palaeomyrmidon* Rovereto, 1914:100.
L. Mioc.; S.A.

Cyclopes Gray, 1821:305. Pygmy anteater, silky anteater, two-toed anteater.
[= *Cyclothurus* Lesson, 1842:152[1]; *Cycloturus* Sclater, 1871:546; *Mamcyclothurus* Herrera, 1899:19.[2]]
R.; Cent. A. R.; S.A.
[1] Attributed to Gray, 1825, by Lesson, but Gray's name was a *nomen nudum*.
[2] See the International Code of Zoological Nomenclature, Article 1(b)(8), and Palmer (1904:25-26).

Suborder **PHYLLOPHAGA** Owen, 1842:167.[1] Sloths.
[= TARDIGRADA Latham & Davies, 1795; BRADYPODA Haeckel, 1866:clviii[2]; PHYTOPHAGA Huxley, 1871.] [Including TARDIGRADAE Gray, 1821:404[3]; †GRAVIGRADA Owen, 1842.[4]]
[1] Owen's long-forgotten term is exact and useful. It was meant to include the arboreal as well as the terrestrial sloths. "Tardigrades" É. Geoffroy Saint-Hilaire & G. Cuvier, 1795, or TARDIGRADA Latham & Davies, 1795, or as used by G. Cuvier, 1800, is arguably a senior synonym, but it can be confused with the arthropod group Tardigrada.
[2] Used as a "family" by Haeckel, but as a suborder by Trouessart, 1879: 2. We are not sure whether Haeckel was using Blumenbach's 1779 BRADYPODA, which included *Myrmecophaga* as well as *Bradypus*.
[3] Used as an order for living sloths.
[4] Owen's †GRAVIGRADA was for the (extinct) ground sloths only. Early students of sloths believed *Bradypus* and *Choloepus* to be each other's closest relatives.

†*Diellipsodon* Berg, 1899:79.
[= †*Elipsodon* Roth, 1898:194.[1]]
M. Mioc.; S.A.
[1] Supposedly preoccupied by †*Ellipsodon* Scott, 1892, an ungulate.

†*Amphiocnus* Kraglievich, 1922:77.
L. Mioc.; S.A.

†*Pseudoglyptodon* Engelmann, 1987:287.
M. Eoc., L. Eoc. and/or E. Olig., L. Olig.; S.A.

Family †**Rathymotheriidae** Ameghino, 1904:126.[1]
L. Mioc. and/or Plioc.; S.A.
[1] *Nomen oblitum.*

†*Rathymotherium* Ameghino, 1904:126.[1]
L. Mioc. and/or Plioc.; S.A.
[1] *Nomen oblitum.*

Infraorder †**MYLODONTA, new.**[1]
[1] Definition: for the most recent common ancestor of †Mylodontoidea and †Orophodontoidea, and all its descendants.

Superfamily †**Mylodontoidea** Gill, 1872:24, **new rank.**
[= †Mylodontinae Gill, 1872:24.]

†*Pseudoprepotherium* Hoffstetter, 1961:86.
E.-M. Mioc., ?L. Mioc. and/or ?Plioc.; S.A.

Family †**Scelidotheriidae** Ameghino, 1889:665, 895.
[= †Scelidotheridae Ameghino, 1889:665, 895.] [Including †Nematheridae Mercerat, 1891:24.]
E. Mioc.-L. Pleist.; S.A.

Subfamily †**Chubutheriinae** Scillato-Yané, 1977:136.
E. Mioc.; S.A.

†*Chubutherium* Cattoi, 1962:126.
E. Mioc.; S.A.

Subfamily †**Scelidotheriinae** Ameghino, 1889:665, 895.
[= †Scelidotheridae Ameghino, 1889:665, 895; †Scelidotheriinae Ameghino, 1904:182.] [Including †Nematheridae Mercerat, 1891:24; †Nematheriinae Scillato-Yané, 1977:123.]
E. Mioc.-L. Pleist.; S.A.

†*Scelidotheriops* Ameghino, 1904:134.[1]
E. Mioc.; S.A.
[1] *Nomen oblitum.*

†*Analcitherium* Ameghino, 1891:39.
E. Mioc.; S.A.

†*Nematherium* Ameghino, 1887:22-23.
[= †*Ammotherium* Ameghino, 1891:321.] [Including †*Lymodon* Ameghino, 1891:324.]
E. Mioc.; S.A.

†*Neonematherium* Ameghino, 1904:133.
M. Mioc.; S.A.

†*Elassotherium* Cabrera, 1939:3-35.
L. Mioc.; S.A.

†*Scelidotherium* Owen, 1839:73.
[= †*Spenodon* Lund, 1839:220[1]; †*Stenodon* Ameghino, 1885:114[2]; †*Stenodontherium* Ameghino, 1889:731; †*Matschiella* Poche, 1904:47-49; †*Sphenodontherium* Trouessart, 1905:794.] [= or including †*Scelidotheridium* Kraglievich, 1934:112[3]; †*Proscelidodon* Bordas, 1935:484-491.] [Including †*Platyonyx* Lund, 1840:317[4]; †*Scelidodon* Ameghino, 1881:307; †*Catonyx* Ameghino, 1891:250.]
?L. Mioc., E. Plioc.-L. Pleist.; S.A.
[1] Lund evidently meant to write *"Sphenodon."* Not *Sphenodon* Gray, 1831, a genus of reptiles.
[2] Not *Stenodon* Rafinesque, 1818, a genus of mollusks; nor †*Stenodon* Van Beneden, 1865, a genus of cetaceans (synonym of †*Cetotheriopsis*).
[3] As subgenus.
[4] Objective synonym of †*Catonyx*. Not *Platyonyx* Schönherr, 1826, a genus of beetles.

Family †**Mylodontidae** Gill, 1872:24.
[= †Mylodontinae Gill, 1872:24; †Mylodontidae Ameghino, 1889:665, 895.]
M. Mioc.-L. Pleist.; S.A. L. Mioc., ?E. Plioc., L. Plioc.-L. Pleist.; N.A.

†*Urumacotherium* Bocquentin-Villanueva, 1983.
L. Mioc.; S.A.

Subfamily †**Mylodontinae** Gill, 1872:24.
M.-L. Mioc., E. and/or M. Pleist., L. Pleist.; S.A.

†*Glossotheriopsis* Scillato-Yané, 1976:333.
M. Mioc.; S.A.

†*Promylodon* Ameghino, 1883:298.
L. Mioc.; S.A.

†*Strabosodon* Ameghino, 1891:161.[1]
L. Mioc.; S.A.
[1] Probably best regarded as a *nomen dubium* but possibly related to †*Pleurolestodon*.

†*Megabradys* Scillato-Yané, 1981:30.
L. Mioc.; S.A.

†*Pleurolestodon* Rovereto, 1914:92.
L. Mioc.; S.A.

†*Mylodon* Owen, 1839:68.
[= †*Gnathopsis* Leidy, 1852:117; †*Grypotherium* Reinhardt, 1879:353; †*Quatriodon* Ameghino, 1881:307; †*Mesodon* Ameghino, 1882:41[1]; †*Tetrodon* Ameghino,

1882:4[2]; †*Glossotherium* of most early authors.] [Including †*Neomylodon* Ameghino, 1898:1-8; †*Iemisch* Roth, 1899:442-445.[3]]
E. and/or M. Pleist., L. Pleist.; S.A.

[1] Not *Mesodon* Rafinesque, 1819, a genus of mollusks; nor *Mesodon* Wagner, 1851, a genus of fishes.
[2] Not *Tetrodon* Linnaeus, 1766, a genus of fishes.
[3] Possibly a new name for †*Neomyodon*, sometimes considered a misnomer for a carnivoran known only from Tehuelche traditions (Palmer, 1904:349). In 1929 Kretzoi referred †*Iemisch* (as †"*Jemisch*") to †*Smilodon*. See the International Code of Zoological Nomenclature, Article 23(m).

Subfamily †**Lestodontinae** Ameghino, 1889:665, **new rank**.
[= †Lestodontidae Ameghino, 1889:665.]
L. Mioc., ?E. Plioc., L. Plioc.-L. Pleist.; N.A. L. Mioc.-L. Pleist.; S.A.

Tribe †**Thinobadistini, new**.
L. Mioc., ?E. Plioc.; N.A. L. Mioc.; S.A.

†*Thinobadistes* Hay, 1919:104-106.
L. Mioc., ?E. Plioc.; N.A.

†*Sphenotherus* Ameghino, 1891:95-99.
L. Mioc.; S.A.

Tribe †**Glossotheriini, new**.
L. Mioc.-L. Pleist.; S.A. ?E. Plioc., L. Plioc.-L. Pleist.; N.A.

†*Acremylodon* Mones, 1986:227.[1]
[= †*Stenodon* Frailey, 1986:9.[2]]
L. Mioc.; S.A.

[1] Replacement name for †*Stenodon*.
[2] Not *Stenodon* Rafinesque, 1818, a genus of mollusks; nor *Stenodon* Lesson, 1842, a genus of carnivorans; nor †*Stenodon* Van Beneden, 1865, a genus of cetaceans (synonym of †*Cetotheriopsis*); nor †*Stenodon* Ameghino, 1885, another genus of xenarthrans (synonym of †*Scelidotherium*).

†*Ranculcus* Ameghino, 1891:160.
L. Mioc.; S.A.

†*Glossotherium* Owen, 1839:57.
[= †*Eumylodon* Ameghino, 1904:136; †*Mylodon* of most early authors.] [Including †*Pseudolestodon* Gervais & Ameghino, 1880:158; †*Laniodon* Ameghino, 1881:308; †*Interodon* Ameghino, 1885:126; †*Nephotherium* Ameghino, 1886:180; †*Glossotheridium* Kraglievich, 1934; †*Oreomylodon* Hoffstetter, 1949:67.[1]]
L. Mioc.-L. Pleist.; S.A. ?E. Plioc., L. Plioc.-L. Pleist.; N.A.

[1] Proposed as subgenus.

†*Paramylodon* Brown, 1903:569.
[= †*Orycterotherium* Harlan, 1841:109-111[1]; †*Eubradys* Leidy, 1853:241[2]; †*Mylodon* sensu lato of most authors.[3]]
Pleist.; N.A.

[1] Not †*Orycterotherium* Bronn, 1838, another xenarthran.
[2] *Nomen oblitum*.
[3] But not of Owen as fixed by type.

†*Mylodonopsis* Cartelle, 1991:162.
L. Pleist.; S.A.

Tribe †**Lestodontini** Ameghino, 1889:665, **new rank**.
[= †Lestodontidae Ameghino, 1889:665.]
L. Mioc., E.-L. Pleist.; S.A.

†*Lestodon* Gervais, 1855:1114.
[= †*Pliogamphiodon* Ameghino, 1884:231.] [Including †*Prolestodon* Kraglievich, 1932:310.[1]]
L. Mioc., E.-L. Pleist.; S.A.

[1] Proposed as subgenus.

†*Lestodontidion* Roselli, 1976:125.
L. Pleist.; S.A.[1]

[1] Uruguay.

Superfamily †**Orophodontoidea** Ameghino, 1895:657.

> [= †Orophodontidae Ameghino, 1895:657; †Orophodontoidea Kraglievich & Rivas, 1951:7-28; †PARAGRAVIGRADES Hoffstetter, 1954:434.[1]]
>
> [1] Used informally.

Family †**Orophodontidae** Ameghino, 1895:657.

> [= †Orophodontinae Hoffstetter, 1954:433, 434.] [= or including †Octomylodontinae Scillato-Yané, 1977:125.] [Including †Octodontotheriinae Hoffstetter, 1954:163; †Octodontobradyinae Santos, Rancy & Ferigolo, 1993:256.]
>
> L. Olig., L. Mioc., ?E. Plioc.; S.A.

†*Proplatyarthrus* Ameghino, 1905:58, 59.[1]

> L. Olig.; S.A.
>
> [1] This genus, based on an astragalus, is probably a synonym of †*Orophodon* or †*Octodontotherium*.

†*Orophodon* Ameghino, 1895:658.

> L. Olig.; S.A.

†*Octodontotherium* Ameghino, 1895:656.

> L. Olig.; S.A.

†*Octomylodon* Ameghino, 1904:137.

> L. Mioc.; S.A.

†*Octodontobradys* Santos, Rancy & Ferigolo, 1993:256.

> L. Mioc. and/or E. Plioc.; S.A.

Infraorder **MEGATHERIA, new**.[1]

> [1] Definition: for the most recent common ancestor of Megatherioidea and Bradypodoidea, and all its descendants.

Superfamily **Megatherioidea** Gray, 1821:305.

> [= Megatheriadae Gray, 1821:305; Megatherioidea Cabrera, 1929:4; Megatherioidae Hay, 1930:403.] [Including †Megalonychoidea Simpson, 1931:272; Choloepodoidea Kalandadze & Rautian, 1992:101.[1]]
>
> [1] In part. This ill-advised taxon also included mylodonts, †Entelopidae (as †Entelopsidae), and †Ernanodontidae and was therefore useless. Emended; originally spelled "Choelopoidea."

Family †**Megatheriidae** Gray, 1821:305.

> [= †Megatheriadae Gray, 1821:305; †Megatheriidae Owen, 1842:168.] [Including †Schismotheridae Mercerat, 1891.]
>
> E. Mioc.-E. Pleist., ?M. Pleist., L. Pleist.-R.; S.A.   E. Pleist.-R.; N.A.   ?E. Pleist., L. Pleist.; Cent. A.

Subfamily †**Megatheriinae** Gray, 1821:305.

> [= †Megatheriadae Gray, 1821:305; †Megatheriinae Gill, 1872:24.] [Including †Ocnopodinae Hoffstetter, 1954:760; †Nothrotheriidae Gaudin, 1994:27A.]
>
> E. Mioc.-E. Pleist., ?M. Pleist., L. Pleist.-R.; S.A.   E. Pleist.-R.; N.A.   ?E. Pleist., L. Pleist.; Cent. A.

Tribe †**Megatheriini** Gray, 1821:305, **new rank**.

> [= †Megatheriadae Gray, 1821:305; †Megatheriini Winge, 1923:311.[1]]
>
> E.-L. Mioc., ?E. Plioc., L. Plioc., E. and/or M. Pleist., L. Pleist.; S.A.   E.-L. Pleist.; N.A.   ?E. Pleist., L. Pleist.; Cent. A.
>
> [1] Proposed as a "Gruppen" of Bradypodidae.

Subtribe †**Prepotheriina** Ameghino, 1894:161, **new rank**.

> [= †Prepotheridae Ameghino, 1894:161.] [Including †Planopsidae Scott, 1904:320; †Planopsinae Winge, 1923:331.]
>
> E. Mioc.; S.A.

†*Proprepotherium* Ameghino, 1904:131.[1]

> [= †*Proprepotheriun* Ameghino, 1904:131.[2]]
>
> E. Mioc.; S.A.
>
> [1] Emended.
> [2] Misspelling.

†*Planops* Ameghino, 1887:23.
> [Including †*Paraplanops* Ameghino, 1891:321[1]; †*Prepotheriops* Ameghino, 1904:132.[1]]
> E. Mioc.; S.A.
> [1] Synonymy questionable.

†*Prepotherium* Ameghino, 1891:157-158.
> E. Mioc.; S.A.

Subtribe †**Megatheriina** Gray, 1821:305, **new rank**.
> [= †Megatheriadae Gray, 1821:305.]
> E.-L. Mioc., ?E. Plioc., L. Plioc., E. and/or M. Pleist., L. Pleist.; S.A. E.-L. Pleist.; N.A. ?E. Pleist., L. Pleist.; Cent. A.

†*Megathericulus* Ameghino, 1904:132.
> E.-M. Mioc.; S.A.

†*Promegatherium* Ameghino, 1883:293-297.
> [Including †*Eomegatherium* Kraglievich, 1926:28, 29.]
> M.-L. Mioc.; S.A.

†*Plesiomegatherium* Roth, 1911:7-21.
> L. Mioc., ?E. Plioc.; S.A.

†*Megatheridium* Cabrera, 1928:348.
> [Including †*Pliomegatherium* Kraglievich, 1930:153-155.]
> L. Mioc.; S.A.

†*Pyramiodontherium* Rovereto, 1914:89.[1]
> [Including †*Megatheriops* C. Ameghino, 1921.]
> L. Mioc., ?L. Plioc.; S.A.
> [1] This genus might be based merely on juvenile specimens of †*Plesiomegatherium*.

†*Megatherium* G. Cuvier, 1796:303, 308.
> [= †*Essonodontherium* Ameghino, 1884:230; †*Hebetotherium* Ameghino, 1895:97-192.] [Including †*Oracanthus* Ameghino, 1885:499-504[1]; †*Neoracanthus* Ameghino, 1889:673-677[2]; †*Ocnobates* Cope, 1889:659[3]; †*Paramegatherium* Kraglievich, 1925:237.]
> L. Plioc., Pleist.; S.A.
> [1] Not *Oracanthus* Agassiz, 1837, a genus of fishes.
> [2] May 20, 1889. Replacement name for †*Oracanthus* Ameghino, 1885.
> [3] August 1889. Replacement name for †*Oracanthus* Ameghino, 1885.

†*Eremotherium* Spillmann, 1948:231, 236.
> [= †*Schaubia* Hoffstetter, 1949:1, 6[1]; †*Schaubitherium* Hoffstetter, 1950:234-235.] [= or including †*Pseuderemotherium* Paula Couto, 1954:447-463.[2]]
> E.-L. Pleist.; N.A. ?E. Pleist., L. Pleist.; Cent. A. Pleist.; S.A.
> [1] December 1949. Not †*Schaubia* Camp, Welles & Green, June 1949, a felid.
> [2] Proposed as subgenus.

†*Ocnopus* Reinhardt, 1875:234-235.[1]
> [= †*Parascelidodon* Hoffstetter, 1952:105.[2]]
> Pleist.; S.A.
> [1] †*Ocnopus* is a taxonomist's nightmare. The type species is †*Megatherium laurillardi* Lund, 1842, whose type specimen consists of two megathere teeth. Therefore, †*Ocnopus* is a megathere, valid or invalid. Referred specimens have included postcranial material, but these are not megathere bones and some or all may be nothrothere bones. Hoffstetter failed to straighten out the matter in 1954 because the type of a genus is a species, not a specimen.
> [2] Originally proposed as a subgenus of †*Scelidotherium*.

†*Perezfontanatherium* Roselli, 1976:56.
> L. Pleist.; S.A.[1]
> [1] Uruguay.

Tribe †**Nothrotheriini** Ameghino, 1920:818-819, **new rank**.
> [= †Nothrotherinae Ameghino, 1920:818-819[1]; †Nothrotherinae Kraglievich, 1923:55; †Nothrotheriinae Simpson, 1945:69; †Nothrotheriidae Gaudin, 1994:27A.]

L. Mioc.-E. Pleist., M. and/or L. Pleist., R.; S.A.   E. Pleist.-R.; N.A.

[1] Posthumus work. See Ameghino's Obras Completas 11.

†*Pronothrotherium* Ameghino, 1907:117.
[Including †*Senetia* Kraglievich, 1925:177-193.[1]]
L. Mioc., ?E. Plioc.; S.A.

[1] *Nomen dubium* in any case, based on a humerus too large to belong to the type species of †*Pronothrotherium*.

†*Xyophorus* Ameghino, 1887:23.
L. Mioc.; S.A.

†*Chasicobradys* Scillato-Yané, Carlini & Vizcaíno, 1987:211.
L. Mioc.; S.A.

†*Gilsolaresia* Roselli, 1976:148.
L. Plioc.; S.A.[1]

[1] Uruguay.

†*Diheterocnus* Kraglievich, 1928:43.
[= †*Heterocnus* Kraglievich, 1925:232.[1]]
L. Plioc.; S.A.

[1] Not *Heterocnus* Sharpe, 1895, a genus of birds.

†*Synhapalops* Kraglievich, 1930:159.[1]
L. Plioc.-E. Pleist.; S.A.

[1] Placement in the †Nothrotheriini is questionable.

†*Nothropus* Burmeister, 1882:613-620.
Pleist.; S.A.

†*Thalassocnus* de Muizon & H. G. McDonald, 1995:224.
E. Plioc.; S.A.[1]

[1] Coastal marine beds adjacent to the desert, Peru.

†*Nothrotherium* Lydekker, 1889:1299.
[= †*Coelodon* Lund, 1838:12[1]; †*Hypocoelus* Ameghino, 1891:250.]
Pleist., R.; S.A.

[1] Not *Coelodon* Serville, 1832, a genus of beetles.

†*Nothrotheriops* Hoffstetter, 1954:755.[1]
E. Pleist.-R.; N.A.

[1] Proposed as subgenus of †*Nothrotherium*; raised to generic rank by Paula Couto, 1971.

Subfamily †**Schismotheriinae** Mercerat, 1891, **new rank.**
[= †Schismotheridae Mercerat, 1891.] [Including †Metopotherini Ameghino, 1894.]
E.-L. Mioc.; S.A.

†*Hapaloides* Ameghino, 1902:131-133.[1]
E. Mioc.; S.A.

[1] Published as *nomen nudum* in 1901.

†*Schismotherium* Ameghino, 1887:21.
[Including †*Metopotherium* Ameghino, 1891:324.]
E. Mioc.; S.A.

†*Hapalops* Ameghino, 1887:22.
[= or including †*Parhapalops* Ameghino, 1891:32; †*Amphihapalops* Ameghino, 1891:319; †*Pseudhapalops* Ameghino, 1891:319; †*Geronops* Ameghino, 1891:320; †*Eugeronops* Ameghino, 1891:397; †*Stenocephalus* Mercerat, 1891:10-12.[1]]
M. Mioc.; S.A.

[1] Not *Stenocephalus* Latreille, 1825, a genus of insects.

†*Pelecyodon* Ameghino, 1891:37-38.
[Including †*Uranokurtus* Ameghino, 1894:159; †*Adiastemus* Ameghino, 1894:160.]
E. Mioc.; S.A.

†*Parapelecyodon* Scillato-Yané, 1981:684.
E. Mioc.; S.A.

†*Analcimorphus* Ameghino, 1891:34.
　　E. Mioc.; S.A.

†*Hyperleptus* Ameghino, 1891:155-157.
　　E. Mioc.; S.A.

†*Neohapalops* Kraglievich, 1923:11.
　　L. Mioc.; S.A.

Family **Megalonychidae** P. Gervais, 1855:44.[1]
　　[= Mégalonycidés P. Gervais, 1855:44[2]; Megalonycidae Ameghino, 1889:660, 695;
　　Megalonychidae Zittel, 1892:133; Megalonichidae Trouessart, 1898:1098;
　　Megalonychoidea Simpson, 1931:272.] [Including Choloepidae Pocock, 1924:1029.]
　　?M. Eoc.; Antarctica. E. Mioc., Pleist., ?R.; W. Indies.[3] E. Mioc., L. Mioc.-L.
　　Plioc., Pleist., R.; S.A. L. Mioc.-L. Pleist.; N.A. E. and/or M. Pleist., L. Pleist.-
　　R.; Cent. A.
　　[1] See the International Code of Zoological Nomenclature, Article 11(f)(iii).
　　[2] Famille des Mégalonycidés."
　　[3] Cuba, Hispaniola, Puerto Rico.

†*Imagocnus* MacPhee & Iturralde-Vinent, 1994:3.
　　E. Mioc.; W. Indies.[1]
　　[1] Cuba.

Subfamily †**Ortotheriinae** Ameghino, 1889:665.
　　[= †Ortotheridae Ameghino, 1889:665; †Ortotherinae Trouessart, 1898:1098.]
　　?M. Eoc.; Antarctica.[1] E. Mioc., L. Mioc., L. Plioc., Pleist., ?R.; S.A.[2] Pleist.; W.
　　Indies.[3]
　　[1] A megatherioid, possibly an ortotheriine megalohychid, has been reported from the La Meseta Fm. of
　　Seymour Island.
　　[2] Including Curaçao in the Pleistocene or Recent.
　　[3] Cuba.

†*Proschismotherium* Ameghino, 1902:130-131.[1]
　　E. Mioc.; S.A.
　　[1] Published as *nomen nudum* in 1901.

†*Eucholoeops* Ameghino, 1887:21-22.
　　E. Mioc.; S.A.

†*Pseudortotherium* Scillato-Yané, 1981:681.
　　E. Mioc.; S.A.

†*Megalonychotherium* Scott, 1904:278.
　　E. Mioc.; S.A.

†*Paranabradys* Scillato-Yané, 1980:194.
　　L. Mioc.; S.A.

†*Pliomorphus* Ameghino, 1885:128-130.
　　L. Mioc.; S.A.

†*Torcellia* Kraglievich, 1923:2.
　　L. Mioc.; S.A.

†*Ortotherium* Ameghino, 1885:111-113.
　　[= †*Orthotherium* Roger, 1887:17.]
　　L. Mioc.; S.A.

†*Menilaus* Ameghino, 1891:154.
　　L. Plioc.; S.A.

†*Diodomus* Ameghino, 1885:125-127.
　　[= †*Platyodon* Ameghino, 1881:308.[1]]
　　L. Plioc., Pleist.; S.A.
　　[1] Preoccupied by *Platyodon* Conrad, 1837, a mollusk, and by *Platyodon* Bravard, 1853, a lagomorph, and by
　　†*Platyodon* Gervais, 1876, *lapsus* for †*Platygnathus* Kroyer, 1841, a megathere, itself preoccupied by
　　Platygnathus Dejean, 1834, a coleopteran.

†*Habanocnus* Mayo, 1978:688.
>Pleist.; W. Indies.[1]

[1] Cuba.

†*Paulocnus* Hooijer, 1962:47.
>Pleist. and/or R.; S.A.[1]

[1] Curaçao.

Subfamily **Megalonychinae** P. Gervais, 1855:44.

[= Mégalonycidés P. Gervais, 1855:44[1]; Megalonycidae Ameghino, 1889:690, 695; Megalonychidae Zittel, 1892; Megalonichinae Trouessart, 1898:1098[2]; Megalonychinae Trouessart, 1904:793.] [Including Choloepodinae Gill, 1874:24; †Mesocninae Arredondo, 1988:2.]
L. Mioc.-L. Pleist.; N.A. L. Mioc.-E. Plioc., Pleist., R.; S.A. E. and/or M. Pleist., L. Pleist.-R.; Cent. A. Pleist., ?R.; W. Indies.[3]

[1] Famille des Mégalonycidés."
[2] *Lapsus calami.*
[3] Cuba, Hispaniola, Puerto Rico.

†*Pliometanastes* Hirschfeld & Webb, 1968:246.
>L. Mioc.; N.A.

†*Sinclairia* Ameghino, 1912:45-75.[1]
>L. Mioc.; N.A.

[1] This genus, if not a *nomen dubium*, may be a prior synonym of †*Pliometanastes*. The type specimen of the type species is almost surely from the Rattlesnake Formation, not the Mascall Formation.

†*Megalonychops* Kraglievich, 1926:30.
>L. Mioc., Pleist.; S.A.

†*Valgipes* Gervais, 1874:1-44.
>Pleist.; S.A.

†*Meizonyx* Webb, 1985:114.
>E. and/or M. Pleist.; Cent. A.

Tribe †**Megalonychini** P. Gervais, 1855:44, **new rank**.

[= †Mégalonycidés P. Gervais, 1855:44[1]; †Megalonycidae Ameghino, 1889:690, 695.]
L. Mioc.-L. Pleist.; N.A. L. Mioc.-E. Plioc., Pleist.; S.A. ?E. Pleist., L. Pleist.; Cent. A. Pleist.; W. Indies.[2]

[1] "Famille des Mégalonycidés."
[2] Cuba, Hispaniola.

Subtribe †**Megalonychina** P. Gervais, 1855:44, **new rank**.

[= †Mégalonycidés P. Gervais, 1855:44[1]; †Megalonycidae Ameghino, 1889:690, 695; †Megalonycina Kraglievich, 1925:231.]
L. Mioc.-L. Pleist.; N.A. L. Mioc.-E. Plioc., Pleist.; S.A. ?E. Pleist., L. Pleist.; Cent. A. Pleist.; W. Indies.[2]

[1] "Famille des Mégalonycidés."
[2] Cuba.

Infratribe †**Megalonychi** P. Gervais, 1855:44, **new rank**.

[= †Mégalonycidés P. Gervais, 1855:44[1]; †Megalonycidae Ameghino, 1889:690, 695.]
L. Mioc.-L. Pleist.; N.A. L. Mioc.-E. Plioc., Pleist.; S.A. ?E. Pleist., L. Pleist.; Cent. A.

[1] "Famille des Mégalonycidés."

†*Protomegalonyx* Kraglievich, 1925:180.
>L. Mioc.-E. Plioc.; S.A.

†*Megalonyx* Harlan, 1825.[1]

[= †*Aulaxodon* Harlan, 1830:284; †*Aulakodon* Scudder, 1882:39[2]; †*Aulacodon* Trouessart, 1898:1106[2]; †*Ereptodon* Leidy, 1853:241; †*Morotherium* Marsh, 1874:531-532.]
L. Mioc.-L. Pleist.; N.A. ?E. Pleist., L. Pleist.; Cent. A. Pleist.; S.A.[3]

[1] "The supposed genus '†*Megalonyx* Jefferson, 1799,' long sentimentally cherished by American paleontologists, is non-existent. Jefferson definitely did not establish such a genus in zoological nomenclature, but it can be ascribed to Harlan" (Simpson, 1945:70).
[2] Replacement name for †*Aulaxodon*, preoccupied by †*Aulacodon* Kaup, 1832, a rodent.
[3] Colombia.

Infratribe †**Megalocni** Kraglievich, 1923:54, **new rank**.
> [= †Megalocninae Kraglievich, 1923:54; †Megalocnina Kraglievich, 1925:231; †Megalocnini Varona, 1974:45.]
> Pleist.; W. Indies.[1]

[1] Cuba.

†*Megalocnus* Leidy, 1868:179-180.
> [= †*Myomorphus* Pomel, 1868:665-668.]
> Pleist.; W. Indies.[1]

[1] Cuba.

†*Neomesocnus* Arredondo, 1961:21.
> Pleist.; W. Indies.[1]

[1] Cuba.

Subtribe †**Mesocnina** Varona, 1974:46, **new rank**.[1]
> [= †Mesocnini Varona, 1974:46; †Mesocninae Arredondo, 1988:2.] [Including †Cubanocnini Varona, 1974:48.]
> Pleist.; W. Indies.[2]

[1] See the International Code of Zoological Nomenclature, Article 40.

[2] Cuba, Hispaniola.

†*Neocnus* Arredondo, 1961:29.
> [= or including †*Microcnus* Matthew, 1931:4[1]; †*Cubanocnus* Kretzoi, 1968:163.[2]]
> Pleist.; W. Indies.[3]

[1] The type species of †*Microcnus* Matthew was described in 1931. Matthew's uses of the generic name in 1915, 1918 and 1919 were instances of *nomina nuda*. Not *Microcnus* Reichenow, 1877, a subgenus of birds.

[2] Replacement for †*Microcnus* Matthew, 1931.

[3] Cuba.

†*Parocnus* Miller, 1929:28.
> [= or including †*Mesocnus* Matthew, 1931:2.[1]]
> Pleist.; W. Indies.[2]

[1] *Nomen nudum* in 1915, 1918 and 1919.

[2] Cuba, Hispaniola.

Tribe **Choloepodini** Gray, 1871:430.
> [= Choloepina Gray, 1871:430[1]; Choloepodinae Gill, 1874:24.] [Including †Acratocnini Varona, 1974:49.]
> Pleist., ?R.; W. Indies.[2] R.; Cent. A. R.; S.A.

[1] Proposed as a tribe.

[2] Cuba, Hispaniola, Puerto Rico.

Subtribe †**Acratocnina** Varona, 1974:49, **new rank**.
> [= †Acratocnini Varona, 1974:49.]
> Pleist., ?R.; W. Indies.[1]

[1] Cuba, Hispaniola, Puerto Rico.

†*Miocnus* Matthew, 1931:4.[1]
> Pleist.; W. Indies.[2]

[1] Matthew used the name in 1915, 1918 and 1919, but the type species was described in 1931.

[2] Cuba.

†*Acratocnus* Anthony, 1916:195.
> Pleist.; W. Indies.[1]

[1] Cuba, Puerto Rico.

†*Synocnus* Paula Couto, 1967:35.
> Pleist. and/or R.; W. Indies.[1]

[1] Hispaniola.

Subtribe **Choloepodina** Gray, 1871:430, **new rank**. Two-toed tree sloths.
>[= Choloepina Gray, 1871:430[1].]
>R.; Cent. A. R.; S.A.
>[1] Proposed as a tribe.

Choloepus Illiger, 1811:108. Two-toed sloths, unaus.
>R.; Cent. A. R.; S.A.

Superfamily **Bradypodoidea** Gray, 1821:304.
>[= Bradypidae Gray, 1821:304; Bradypodoidea Simpson, 1931:273.]

Family **Bradypodidae** Gray, 1821:304. Three-toed tree sloths.
>[= Bradypidae Gray, 1821:304; Bradypodidae C. L. Bonaparte, 1831:22; Bradypoda
>Haeckel, 1866:clviii[1]; Bradypodina Gray, 1871:434[2]; Bradypodini Winge, 1923:322;
>Achedae Burnett, 1830.]
>R.; Cent. A. R.; S.A.
>[1] Haeckel's "family" *Bradypoda* was effectively a suborder and was explicitly used at subordinal rank by
>Trouessart, 1879:2.
>[2] Described as a tribe.

Bradypus Linnaeus, 1758:34-35. Three-toed sloths, ais, maned sloth.
>[= *Tardigradus* Brisson, 1762:12, 21-23[1]; *Tardipes* Frisch, 1775:19[2]; *Ignavus* Frisch,
>1775:(table)[1]; *Acheus* F. Cuvier, 1825:194-195, 256; *Achaeus* Erman, 1835:22;
>*Arctopithecus* Gray, 1850:xxviii.[3]] [Including *Scaeopus* Peters, 1865:678[4];
>*Hemibradypus* R. Anthony, 1906:292[5]; *Eubradypus* Lönnberg, 1942:5; *Neobradypus*
>Lönnberg, 1942:15.]
>R.; Cent. A. R.; S.A.
>[1] Not published in a consistently binominal work. See the International Code of Zoological Nomenclature,
>Article 11(c)(i), for confusion.
>[2] Unavailable.
>[3] Not *Arctopithecus* G. Cuvier, 1817, a synonym of *Callithrix*.
>[4] As subgenus.
>[5] Replacement name for *Scaeopus*.

Magnorder **EPITHERIA** McKenna, 1975:41, **new rank**.[1]
>[1] Proposed as a cohort.

†*Asioryctes* Kielan-Jaworowska, 1975:6.[1]
>L. Cret.; As.
>[1] †Asioryctinae Kielan-Jaworowska, 1981:67 is based on this genus. †Asioryctidae Szalay, 1977:368 is
>considered a *nomen nudum* by Kielan-Jaworowska.

†*Hyotheridium* Gregory & Simpson, 1926:11.
>L. Cret.; As.

†*Deccanolestes* Prasad & Sahni, 1988:638.
>L. Cret.; As.[1]
>[1] S. Asia (India).

Superorder †**LEPTICTIDA** McKenna, 1975:41.
>[= †ICTOPSIA McKenna, 1975:41[1]; ERNOTHERIA McKenna, 1975:41[2];
>†LEPTICTIFORMES Kinman, 1994:37.] [= or including PROTEUTHERIA Romer,
>1966:209[3]; †MIXOTHERIDIA Nessov, 1985:10.[4]] [Including †KENNALESTIDA
>McKenna, 1975:41; †KENALESTIFORMES Kinman, 1994:37.]
>[1] Proposed as grandorder.
>[2] In part. Proposed as magnorder.
>[3] In part.
>[4] Described as a suborder of the paraphyletic order PROTEUTHERIA.

†*Lainodon* Gheerbrant & Astibia, 1994:1126.
>L. Cret.; Eu.

†*Labes* Sigé, in Pol, Buscalioni, Carballeira, Francés, López Martinez, Marandat, Moratalla, Sanz,
>Sigé & Villatte, 1992:296.
>L. Cret.; Eu.

†*Gallolestes* Lillegraven, 1976:438.
>L. Cret.; N.A.

†*Wania* Wang, 1995:114.
        M. Paleoc.; As.

†*Praolestes* Matthew, Granger & Simpson, 1929:3.
        L. Paleoc.; As.

Family †**Gypsonictopidae** Van Valen, 1967:259.
        [= †Gypsonictopinae Van Valen, 1967:259; †Gypsonictopidae Stucky & McKenna, in Benton, ed., 1993:746.] [Including †Kennalestidae Kielan-Jaworowska, 1981:67; †Kennalestoidea Kielan-Jaworowska & Dashzeveg, 1989:348; †Zhelestinae Nessov, 1985:15.]
        E.-L. Cret.; As. L. Cret., ?Paleoc.; N.A.

†*Prokennalestes* Kielan-Jaworowska & Dashzeveg, 1989:348.[1]
        [= or including †*Prozalambdalestes* Trofimov, in Belyaeva, Trofimov & Reshetov, 1974:20.[2]]
        E. Cret.; As.
        [1] *Nomen nudum* in Trofimov in Belyaeva, Trofimov & Reshetov, 1974:20.
        [2] *Nomen nudum.*

†*Gypsonictops* Simpson, 1927:6.
        [= †*Euangelistes* Simpson, 1929:107.]
        L. Cret.; N.A.

†*Sailestes* Nessov, 1982:237.
        L. Cret.; As.

†*Kennalestes* Kielan-Jaworowska, 1969:175.[1]
        L. Cret.; As.
        [1] Not 1968.

†*Zhelestes* Nessov, 1985:16.
        L. Cret.; As.[1]
        [1] Uzbekistan.

†*Parazhelestes* Nessov, 1993:123.
        L. Cret.; As.

†*Stilpnodon* Simpson, 1935:229.[1]
        E. Paleoc., ?L. Paleoc.; N.A.
        [1] Assignment to †Gypsonictopidae is doubtful.

Family †**Kulbeckiidae** Nessov, 1993:125.
        L. Cret.; As.

†*Kulbeckia* Nessov, 1993:125.
        L. Cret.; As.

Family †**Didymoconidae** Kretzoi, 1943:194.
        [Including †Tshelkariidae Gromova, 1960:42.]
        E. Paleoc.-L. Olig.; As.

†*Zeuctherium* Tang & Yan, 1976:94.
        E. Paleoc.; As.[1]
        [1] E. Asia.

†*Archaeoryctes* Zheng, 1979:360.[1]
        L. Paleoc., M. Eoc.; As.[2]
        [1] (Cheng).
        [2] E. Asia.

†*Hunanictis* Li, Chiu, Yan & Hsieh, 1979:77.
        E. Eoc.; As.[1]
        [1] E. Asia.

†*Kennatherium* Mellett & Szalay, 1968:1.
        M. Eoc.; As.

†*Ardynictis* Matthew & Granger, 1925:2.
        L. Eoc.; As.

†*Didymoconus* Matthew & Granger, 1924:1.
[Including †*Tshelkaria* Gromova, 1960:44.]
L. Eoc.-L. Olig.; As.

Family †**Leptictidae** Gill, 1872:19.
[= †Leptictinae Gill, 1872:19; †Leptictoidea Trouessart, 1879:59; †Leptictididae Winge, 1917:123; †Isacidae Cope, 1874:473; †Ictopsidae Schlosser, 1887:140; †Ictopsida Haeckel, 1895:578, 582.[1]]
E. Paleoc.-L. Olig.; N.A. L. Paleoc., M. Eoc.-E. Olig.; Eu. E. Olig.; As.
[1] Used as a family.

†*Leptonysson* Van Valen, 1967:235.
E. Paleoc.; N.A.

Subfamily †**Leptictinae** Gill, 1872:19.[1]
E. Paleoc.-L. Olig.; N.A. L. Paleoc.; Eu. E. Olig.; As.
[1] Redundantly renamed by Van Valen, 1967:258.

†*Prodiacodon* Matthew, 1929:171.
[= †*Palaeolestes* Matthew, 1918:576.[1]]
E. Paleoc., ?L. Paleoc.; N.A.
[1] Not †*Palaeolestes* de Vis, 1911, an extinct bird.

†*Palaeictops* Matthew, 1899:31, 35.
[Including †*Parictops* Granger, 1910:250; †*Hypictops* Gazin, 1949:222.]
?Paleoc., E.-M. Eoc.; N.A.

†*Myrmecoboides* Gidley, 1915:395.
E. Paleoc.; N.A.

†*Xenacodon* Matthew & Granger, 1921:3.
L. Paleoc.; N.A.

†*Diaphyodectes* D. E. Russell, 1964:50.
L. Paleoc.; Eu.

†*Ongghonia* Kellner & McKenna, 1996:4.
E. Olig.; As.

†*Leptictis* Leidy, 1868:315-316.
[= †*Isacus* Cope, 1873:3[1]; †*Isacis* Cope, 1873:5, 8[2]; †*Mesodectes* Cope, 1875:30.] [= or including †*Ictops* Leidy, 1868:316; †*Nanohyus* Leidy, 1869:65; †*Ictidops* Weber, 1904:381.]
L. Eoc.-L. Olig.; N.A.
[1] Palmer (1904:351) held that †*Isacus* was preoccupied by *Isaca* Walker, 1857, an insect.
[2] Misspelling or attempted correction in a different paper.

Subfamily †**Pseudorhynchocyoninae** Sigé, 1974:45.
[= †Pseudorhynchocyonidae Storch & Lister, 1985:5.]
M. Eoc.-E. Olig.; Eu.

†*Leptictidium* Tobien, 1962:27.
M.-L. Eoc.; Eu.

†*Pseudorhynchocyon* Filhol, 1892:2.
L. Eoc.-E. Olig.; Eu.

Superorder **PREPTOTHERIA** McKenna, 1975:41.[1]
[Including THERICTOIDEA Gregory, 1910:464[2]; TOKOTHERIA McKenna, 1975:41.[3]]
[1] Proposed as magnorder. Changed to rank of superorder by McKenna, in Stucky & McKenna, in Benton, ed., 1993:747.
[2] Proposed as superorder containing LIPOTYPHLA and FERAE.
[3] Proposed as a superorder.

Grandorder **ANAGALIDA** Szalay & McKenna, 1971:301.[1]
[= †ANAGALIFORMES Kinman, 1994:37.] [Including GLIRES Linnaeus, 1758:56[2]; LAGOTHERIA Kalandadze & Rautian, 1992:75[3]; †ZALAMBDOLESTA Kalandadze & Rautian, 1992:76.[4]]

[1] Emended. Proposed as order by Szalay & McKenna, 1971:301; ranked as grandorder by McKenna, 1975:41.

[2] Proposed as an order (in company with BRUTA, PECORA, BESTIAE, PRIMATES, BELLUAE, FERAE, and CETE). The original content of GLIRES included lagomorphs, rodents, and *Rhinoceros*. The characters noted were a pair of upper and lower incisors (duplicated in lagomorph upper incisors), lack of canine teeth, and possession of a diastema between the incisors and the "molars." The rhino soon went its own way taxonomically, but the notion of GLIRES persists to this day, implying that rodents and lagomorphs are sister-groups. As this is not quite certain, we here subsume the non-rhinocerotoid GLIRES within ANAGALIDA, but we do not utilize it for RODENTIA and LAGOMORPHA alone.

[3] In part. Named as a superorder and apparently equal to GLIRES plus †TAENIODONTA (as "†TAENIODONTIA").

[4] In part.

Family †**Zalambdalestidae** Gregory & Simpson, 1926:14.
> [= †Zalambdolestoidea Kalandadze & Rautian, 1992:76.]
> L. Cret.; As.

†*Zalambdalestes* Gregory & Simpson, 1926:14.
> L. Cret.; As.

†*Alymlestes* Averianov & Nessov, in Nessov, Sigogneau-Russell & D. E. Russell, 1994:67.[1]
> L. Cret.; As.
> [1] A rushed (pre-)publication, of the sort that should be avoided. Averianov & Nessov, 1995, republished the taxon as new.

†*Barunlestes* Kielan-Jaworowska, 1975:9.
> L. Cret.; As.

Family †**Anagalidae** Simpson, 1931:1.
> [= or including †Peritupaioidea Crusafont Pairó, 1966:17.[1]]
> E.-L. Paleoc., E. Olig.; As.[2]
> [1] Proposed as a superfamily but invalid because not based on a generic name.
> [2] Range may end later than early Oligocene if †*Anagalopsis* proves to be younger than that and is correctly placed in this family.

†*Yuodon* Chow, Chang, Wang & Ting, 1973:33.
> E. Paleoc.; As.[1]
> [1] E. Asia (China).

†*Linnania* Chow, Chang, Wang & Ting, 1973:31.
> E. Paleoc.; As.[1]
> [1] E. Asia (China).

†*Diacronus* Xu, 1976:243.
> E. Paleoc.; As.[1]
> [1] E. Asia.

†*Huaiyangale* Xu, 1976:175.
> E. Paleoc.; As.[1]
> [1] E. Asia.

†*Palasiodon* Tong, Zhang, Wang & Ding, 1976:19.[1]
> E. Paleoc.; As.[2]
> [1] (Tung, Chang, Wang & Ting).
> [2] E. Asia (China).

†*Eosigale* Hu, 1993:155.
> E. Paleoc.; As.[1]
> [1] E. Asia.

†*Wanogale* Xu, 1976:242.
> E. Paleoc.; As.[1]
> [1] E. Asia.

†*Anaptogale* Xu, 1976:245.
> E. Paleoc.; As.[1]
> [1] E. Asia.

†*Stenanagale* Wang, 1975:158.
> E. Paleoc.; As.[1]
> [1] E. Asia.

†*Chianshania* Xu, 1976:247.
E. Paleoc.; As.[1]
[1] E. Asia.

†*Qipania* Hu, 1993:161.
E. Paleoc.; As.[1]
[1] E. Asia.

†*Hsiuannania* Xu, 1976:179.
L. Paleoc.; As.[1]
[1] E. Asia.

†*Anagalopsis* Bohlin, 1951:28.
?E. Olig.; As.[1]
[1] E. Asia (China).

†*Anagale* Simpson, 1931:1.
E. Olig.; As.[1]
[1] Mongolia.

Family †**Pseudictopidae** Sulimski, 1969:107.
E.-L. Paleoc.; As. L. Paleoc. and/or E. Eoc.; N.A.

†*Anictops* Qiu, 1977:130.
E. Paleoc.; As.[1]
[1] E. Asia.

†*Cartictops* Ding & Tong, 1979:139.[1]
E. Paleoc.; As.[2]
[1] (Ting & Tung).
[2] E. Asia.

†*Paranictops* Qiu, 1977:135.
E. Paleoc.; As.[1]
[1] E. Asia.

†*Allictops* Qiu, 1977:138.
L. Paleoc.; As.[1]
[1] E. Asia.

†*Pseudictops* Matthew, Granger & Simpson, 1929:4.
L. Paleoc.; As.

†*Mingotherium* Schoch, 1985:5.
L. Paleoc. and/or E. Eoc.; N.A.

†*Haltictops* Ding & Tong, 1979:138.[1]
L. Paleoc.; As.[2]
[1] (Ting & Tung).
[2] E. Asia.

Mirorder **MACROSCELIDEA** Butler, 1956:479, **new rank**.[1]
[= MACROSCELIDES Peters, 1864:20[2]; MACROSCELIDIFORMES Kinman, 1994:37.]
[1] Butler coined the name, but Patterson, 1965, first gave substance to the term (as an order).
[2] Not to be confused with the genus utilizing the same spelling.

Family **Macroscelididae** Bonaparte, 1838:113. Elephant shrews.
[= Macroscelidina Bonaparte, 1838:113; Macroscelidia Haeckel, 1866:clx; Macroscelididae Mivart, 1868:143; Macroscelidoidea Gill, 1872:19; Salientia Haeckel, 1866:clx.] [Including Rhynchocyoninae Gill, 1872:19; Rynchocyonidae Gill, 1882; †Myohyracidae Andrews, 1914:171.]
L. Eoc., E. Mioc.-L. Plioc., Pleist., R.; Af.[1]
[1] An unspecified genus of Macroscelididae has been reported from the late Miocene Otavi breccias, Namibia.

Subfamily †**Metoldobotinae** Simons, Holroyd & Bown, 1991:9736.
L. Eoc.; Af.

†*Metoldobotes* Schlosser, 1910:507.[1]
    L. Eoc.; Af.
    [1] The spelling of the name of this genus is valid, despite the incorrect derivation from †*"Olbodotes."*

Subfamily †**Herodotiinae** Simons, Holroyd & Bown, 1991:9734.[1]
    [= †Herodotinae Simons, Holroyd & Bown, 1991:9734.]
    L. Eoc.; Af.
    [1] Spelling emended.

†*Herodotius* Simons, Holroyd & Bown, 1991:9734.
    L. Eoc.; Af.

Subfamily **Macroscelidinae** Bonaparte, 1838:113. Klaasneuse.
    [= Macroscelidina C. L. Bonaparte, 1838:113; Macroscelidinae Mivart, 1868:143;
    Marcoscelidini Winge, 1917:123.]
    E.-M. Mioc., ?E. Plioc., L. Plioc., Pleist., R.; Af.

†*Hiwegicyon* Butler, 1984:137.
    E. Mioc.; Af.

†*Pronasilio* Butler, 1984:135.
    M. Mioc.; Af.

†*Palaeothentoides* Stromer, 1932:185.
    Plioc.; Af.

Macroscelides Smith, 1829:435. Elephant shrew, kortoorklaasneuse.
    [= *Macroscelis* J. B. Fischer, 1830:657, 664-665.]
    L. Plioc., R.; Af.[1]
    [1] Southern Africa.

Elephantulus Thomas & Schwann, 1906:577. Long-eared elephant shrews, sakutopi, dengoli,
    short-nosed elephant shrew, klipklaasneuse, kortneusklaasneuse.
    [Including *Nasilio* Thomas & Schwann, 1906:10; *Elephantomys* Broom, 1937:758.]
    L. Plioc., Pleist., R.; Af.

Petrodromus Peters, 1846:258. Forest elephant shrews, bos-klaasneuse.
    [Including *Cercoctenus* Hollister, 1916:1[1]; *Mesoctenus* Thomas, 1918:366.[1]]
    R.; Af.
    [1] Subgenus as proposed.

Subfamily †**Myohyracinae** Andrews, 1914:171.
    [= †Myohyracidae Andrews, 1914:171; †Myohyracoidea Stromer, in Kaiser, 1926:120;
    †Myohyracinae Patterson, 1965:312.]
    E.-M. Mioc.; Af.

†*Myohyrax* Andrews, 1914:171.
    E.-M. Mioc.; Af.

†*Protypotheroides* Stromer, 1922:333.
    E. Mioc.; Af.

Subfamily **Rhynchocyoninae** Gill, 1872:119.
    [= Rhynchocyonidae Gill, 1882:119; Rhynchocynini Winge, 1917:123;
    Rhynchocyonoidea Goodman, 1975:228, 231.]
    E.-M. Mioc., E. Plioc., R.; Af.

Rhynchocyon Peters, 1847:36. Checkered-backed elephant shrews, black and rufous elephant
    shrews.
    [Including *Rhinonax* Thomas, 1918:370; †*Miorhynchocyon* Butler, 1984:120.]
    E.-M. Mioc., E. Plioc., R.; Af.

Subfamily †**Mylomygalinae** Patterson, 1965:310.
    L. Plioc.; Af.

†*Mylomygale* Broom, 1948:6-8.
    L. Plioc.; Af.[1]
    [1] S. Africa.

Mirorder **DUPLICIDENTATA** Illiger, 1811:91, **new rank.**[1]

[= SALTANTIA Newman, 1843:34[2]; DUPLICIDENTATI Alston, 1876:64.] [= or including LAGOMORPHIFORMES Kinman, 1994:37.[3]] [Including †PALAEOLOGINA Averianov, 1994:106.[4]]

[1] Proposed as a family.
[2] In part.
[3] Proposed as an order.
[4] Infraorder.

Order †**MIMOTONIDA** Li, Wilson, Dawson & Krishtalka, 1987:105.

Family †**Mimotonidae** Li, in Chow & Qi, 1978:79.[1]

[Including †Mimolagidae Szalay, 1985:120[2]; †Mimolagidae Erbajeva, 1986:159.[3]] E.-L. Paleoc., E. and/or M. Eoc., ?E. Olig.; As.

[1] According to the International Code of Zoological Nomenclature, Article 15, a name published conditionally after 1960, as this one originally was by Li, 1977:104, 117 in both the Chinese and English texts, is unavailable.
[2] Conditionally proposed; therefore, unavailable (International Code of Zoological Nomenclature, Article 15).
[3] (Erbaeva).

†*Mimotona* Li, 1977:110.

E.-L. Paleoc.; As.[1]

[1] E. Asia.

†*Anatolmylus* Averianov, 1994:401.

E. and/or M. Eoc.; As.

†*Mimolagus* Bohlin, 1951:9.

?E. Olig.; As.[1]

[1] China.

Order **LAGOMORPHA** Brandt, 1855:295.

[= NON-CLAVICULATA Cuvier, 1816[1]; LAGOMORPHI Brandt, 1855:283, 319; PLIODONTA Schlosser, 1884:133[1]; PALARODENTIA Haeckel, 1895:502.]

[1] *Nomen oblitum.*

Family **Ochotonidae** Thomas, 1897:1026. Pikas.

[= Ochotonoidea Averianov, 1994:106; Ochotoninae Kalandadze & Rautian, 1992:77; Lagomina Gray, 1825:341[1]; Lagomyidae Lilljeborg, 1866:9, 58; Lagomyinae Erbajeva, 1988:199.[2]] [Including †Desmatolaginae Burke, 1941:13; †Agispelaginae Gureev, 1953:16; †Amphilaginae Gureev, 1953:1-276; †Amphilagidae Tobien, 1974:208; †Amphilagini Kalandadze & Rautian, 1992:76; †Sinolagomyinae Gureev, 1960:8; †Prolaginae Gureev, 1960:8; †Prolagidae Erbaeva, 1988:199.[3]] L. Eoc.-R.; As. E. Olig.-L. Pleist.; Eu. ?L. Olig., E.-L. Mioc., ?E. Pleist., M. Pleist.-R.; N.A. E. Mioc.-L. Plioc., ?M. Pleist.; Af. M. Mioc., E. Plioc., E. Pleist.-R.; Mediterranean.

[1] As tribe.
[2] (Erbaeva).
[3] Casual proposal in fig. 42.

†*Desmatolagus* Matthew & Granger, 1923:8.[1]

[Including †*Agispelagus* Argyropulo, 1939:727[2]; †*Bohlinotona* de Muizon, 1977:272.] L. Eoc.-L. Mioc.; As. E. Olig.; Eu. ?L. Olig. and/or ?E. Mioc.; N.A.

[1] Some species assigned to †*Desmatolagus* may be leporids.
[2] *Nomen nudum* in Argyropulo & Pidoplichka, 1939, and in another paper published by Argyropulo in 1939.

†*Sinolagomys* Bohlin, 1937:31.

[Including †*Ochotonolagus* Gureev, 1960:10.[1]] E. Olig.-E. Mioc.; As.

[1] Synonymy questionable.

†*Piezodus* Viret, 1929:94.

L. Olig.-E. Mioc.; Eu.

†*Amphilagus* Pomel, 1854:42-43.

        L. Olig.-L. Mioc.; Eu. ?E. Mioc.; As.[1]

    [1] E. Asia (Japan).

†*Austrolagomys* Stromer, in Kaiser, 1926:127.

        E. Mioc.; Af.[1]

    [1] S. Africa.

†*Marcuinomys* Lavocat, 1951:35.

        E. Mioc.; Eu.

†*Prolagus* Pomel, 1853:43.[1]

        [= †*Anoema* König, 1825:257[2]; †*Myolagus* Hensel, 1856:689-703.]

        E. Mioc.-E. Pleist.; Eu. M. Mioc.-E. Plioc.; As.[3] L. Mioc.-L. Plioc.; Af.[4] E.

        Plioc., E. Pleist.-R.; Mediterranean.[5]

    [1] Possibly 1854:43 (see Palmer, 1904: 571).
    [2] Not *Anoema* F. Cuvier, 1809, a synonym of *Cavia*.
    [3] W. Asia.
    [4] N. Africa.
    [5] Survived until historic times on Corsica and Sardinia.

†*Ptychoprolagus* Tobien, 1975:170.

        E. Mioc.; Eu.

†*Heterolagus* Crusafont Pairó, Villalta & Truyols, 1955:146.

        E. Mioc.; Eu.

†*Lagopsis* Schlosser, 1884:13.[1]

        [= †*Opsolagus* Kretzoi, 1941.]

        E.-M. Mioc.; Eu.

    [1] "†*Lagopsis* Rafinesque, 1815, is a *nomen nudum* and does not preoccupy †*Lagopsis* Schlosser" (Simpson, 1945: 75).

†*Titanomys* von Meyer, 1843:390.

        [= †*Platyodon* Bravard, 1853:258.[1]]

        E.-L. Mioc.; Eu.

    [1] Preoccupied by *Platyodon* Conrad, 1837, a mollusk.

†*Kenyalagomys* MacInnis, 1953:2.

        E.-M. Mioc.; Af. E. Mioc.; As.

†*Oreolagus* Dice, 1917:182.

        E.-M. Mioc.; N.A.

†*Cuyamalagus* Hutchison, in Hutchison & Lindsay, 1974:10.

        E. Mioc.; N.A.

†*Oklahomalagus* Dalquest, Baskin & Schultz, 1996:118.

        L. Mioc.; N.A.

†*Albertona* López-Martínez, 1986:177, 180.

        E. Mioc.; Eu.[1]

    [1] E. Europe (Greece).

†*Eurolagus* López Martínez, 1977:8.

        M. Mioc.; Eu.

†*Bellatona* Dawson, 1961:8.

        E.-M. Mioc.; As.[1]

    [1] E. Asia.

†*Russellagus* Storer, 1970:1129.

        M. Mioc.; N.A.

†*Gymnesicolagus* Mein & Adrover, 1982:453.

        M. Mioc.; Mediterranean.[1]

    [1] Balearic Isles.

†*Hesperolagomys* Clark, Dawson & A. E. Wood, 1964:33.[1]

        M.-L. Mioc.; N.A.

    [1] Possibly a leporid.

†*Alloptox* Dawson, 1961:6.
[= †*Metochotona* Kretzoi, 1941:111.[1]]
M.-L. Mioc., ?E. Plioc.; As.
[1] *Nomen nudum.*

†*Ochotonoides* Teilhard de Chardin & Young, 1931:30.
[Including †*Prolagomys* Erbajeva, 1975:137-139.[1]]
L. Mioc.-M. Pleist.; As. L. Plioc.; Eu.[2]
[1] (Erbaeva).
[2] E. Europe.

Ochotona Link, 1795:74. Pikas, conies, mouse hares, rock rabbits, peeka (Tungus name).
[= *Lagomys* G. Cuvier, 1800: table I.[1]] [Including *Pika* Lacépède, 1799:9[2]; *Abra* Gray, 1863:11[3]; *Ogotoma* Gray, 1867:220[4]; *Conothoa* Lyon, 1904:438[2]; †*Proochotona* Khomenko, 1916; *Abrana* Strand, 1928:59[5]; *Tibetholagus* Argyropulo, in Argyropulo & Pidoplichka, 1939:727; †*Pliochotona* Kretzoi, 1941:111[6]; *Lagotona* Kretzoi, 1941:111[2]; *Buechneria* Erbajeva, 1988:170[7]; *Argyrotona* Rekovetz, in Erbajeva, 1988:186.[8]]
L. Mioc.-R.; As. L. Mioc.-L. Pleist.; Eu. L. Mioc., ?E. Pleist., M. Pleist.-R.; N.A. ?M. Pleist.; Af.
[1] *Nomen nudum* in G. Cuvier, 1798:132. Not *Lagomys* Storr, 1780, a synonym of *Marmota*.
[2] Sometimes as subgenus.
[3] Not *Abra* Leach MS, Lamarck, 1818, a mollusk.
[4] Subjective synonym of *Pika*.
[5] Replacement for *Abra* Gray, 1863.
[6] *Nomen nudum.*
[7] Corrected from *Büchneria* as proposed by Erbajeva (Erbaeva). See the International Code of Zoological Nomenclature, Articles 27 and 32 (d)(i)(2).
[8] Subgenus as defined. It is unclear whether Rekovetz published elsewhere as well.

†*Paludotona* Dawson, 1959:157-166.
L. Mioc.; Eu.

†*Pliolagomys* Erbajeva, in Agadzhanyan & Erbajeva, 1983.[1]
E.-L. Plioc.; Eu.[2]
[1] (Erbaeva).
[2] E. Europe.

Family **Leporidae** Fischer de Waldheim, 1817:409. Rabbits.
[= Leporinorum Fischer de Waldheim, 1817:409; Leporidae Gray, 1821:304; Leporina Gray, 1825:341[1]; Leporini Brandt, 1855:320[2]; Leporini Gureev, 1964:177; Leporida Haeckel, 1866:clx; Leporinae Trouessart, 1880:200; Leporoidea Averianov, 1994:106.] [Including †Palaeolagida Haeckel, 1895:502-503; †Palaeolaginae Dice, 1929:340; †Palaeolagidae Erbajeva, 1986:157, 159[3]; †Palaeolagini Kalandadze & Rautian, 1992:76; Lagidae Schulze, 1897:82; †Archaeolaginae Dice, 1929:343; †Archaeolagini Gureev, 1964:115; †Protolaginae Walker, 1931:122; †Megalaginae Walker, 1931:123; †Mytonolaginae Burke, 1941:1; †Mytonolagidae Averianov, 1996:658; Oryctolaginae Gureev, 1948:785; Oryctolagini Gureev, 1964:156; Pentalaginae Gureev, 1948:786; Pentalagini Gureev, 1964:128; †Alilepini Gureev, 1964:122; Hypsimylinae Dashzeveg & D. E. Russell, 1988:143.[4]]
?E. Eoc., M. Eoc.-L. Olig., E. and/or M. Mioc., L. Mioc.-R.; As. M. Eoc.-R.; N.A.[5] L. Eoc.-E. Olig., L. Mioc.-R.; Eu. L. Mioc.-R.; Af.[6] L. Mioc., R.; Mediterranean. Pleist., R.; S.A. L. Pleist.-R.; Cent. A. R.; E. Indies.[7] R.; Atlantic.[8]
[1] As tribe.
[2] Used at subfamilial level.
[3] Erbajeva (Erbaeva) cited Dice, 1929 as the author of this name.
[4] Rather casually proposed.
[5] Including islands of the Canadian Arctic and Greenland.
[6] Leporidae indet. reported from the L. Mioc. Mpesida and Lukeino fms., Baringo Basin, Kenya.
[7] Java (possibly by introduction), Sumatra.
[8] Subfossil deposits of the eastern Canary Islands. Recent of Faeroes and Falkland Islands by introduction. Also Recent of the West Indies, New Zealand, and Australia, and on islands of the Indian Ocean by introduction.
Recent worldwide in domestication.

†*Tsaganolagus* Li, 1978:146.
E. and/or M. Mioc.; As.

†*Shamolagus* Burke, 1941:1.
  ?E. Eoc., M. Eoc.; As. L. Eoc.-E. Olig.; Eu.

†*Mytonolagus* Burke, 1934:400.
  M. Eoc.; N.A.

†*Megalagus* Walker, 1931:234.
  [Including †*Montanolagus* Gureev, 1964:98; †*Tachylagus* Storer, 1992:230.]
  M. Eoc.-E. Mioc.; N.A.

†*Gobiolagus* Burke, 1941:5.
  M. Eoc.-E. Olig.; As.

†*Procaprolagus* Gureev, 1960:16.
  M. Eoc., E. Olig.; N.A. L. Eoc.-L. Olig.; As.

†*Lushilagus* Li, 1965:24.
  M. Eoc.; As.[1]
  [1] E. Asia.

†*Strenulagus* Tong & Lei, 1987:208.[1]
  M. Eoc.; As.[2]
  [1] (Tung & Lei).
  [2] E. Asia.

†*Hypsimylus* Zhai, 1977:174.
  L. Eoc.; As.[1]
  [1] E. Asia.

†*Chadrolagus* Gawne, 1978:1103.
  L. Eoc.; N.A.

†*Palaeolagus* Leidy, 1856:89-90.
  [Including †*Tricium* Cope, 1873:4; †*Protolagus* Walker, 1931:230.]
  L. Eoc.-L. Olig.; N.A.

†*Ordolagus* de Muizon, 1977:266.
  E.-L. Olig.; As.

†*Litolagus* Dawson, 1958:32.
  E. Olig.; N.A.

†*Gripholagomys* Green, 1972:378.[1]
  L. Olig.-E. Mioc.; N.A.
  [1] Described as an ochotonid.

†*Archaeolagus* Dice, 1917:180.
  L. Olig.-E. Mioc.; N.A.

†*Hypolagus* Dice, 1917:181.
  [= or including †*Lagotherium* Croizet, 1853:256.[1]] [Including †*Pliolagus* Kormos, 1934:66.]
  E. Mioc.-E. Pleist.; N.A. L. Mioc.-M. Pleist.; As. L. Mioc.; Mediterranean. ?L. Mioc., E. Plioc.-M. Pleist.; Eu.
  [1] This generic name long antedates †*Hypolagus* and probably should be suppressed.

†*Panolax* Cope, 1874:151.
  L. Mioc.; N.A.

†*Lepoides* White, 1987:440.
  L. Mioc.-E. Plioc.; N.A.

†*Pewelagus* White, 1984:47.
  E.-L. Plioc.; N.A.

†*Alilepus* Dice, 1931:159.
  [= †*Allolagus* Dice, 1929:342.[1]]
  L. Mioc.-E. Pleist.; As. L. Mioc.-L. Plioc.; Eu. L. Mioc.-L. Plioc.; N.A.
  [1] Not *Allolagus* Ognev, 1929: 71, a synonym of *Lepus*.

†*Trischizolagus* Radulesco & Samson, 1967:544-563.[1]
  [Including †*Hispanolagus* Janvier & Montenat, 1971:780.]

L. Mioc.; Mediterranean.[2] L. Mioc.-E. Plioc.; Eu. E. Plioc.; As.[3]

[1] (Rădulescu).
[2] Balearic Islands.
[3] Rhodes.

†*Notolagus* R. W. Wilson, 1938:98.
    [Including †*Dicea* Hibbard, 1939:509.]
    L. Mioc.-L. Plioc., ?E. Pleist.; N.A.

†*Veterilepus* Radulesco & Samson, 1967:544-563.[1]
    L. Mioc.; Eu.[2]

[1] (Rădulescu).
[2] E. Europe.

†*Pronotolagus* White, 1991:79.
    L. Mioc.; N.A.

†*Pratilepus* Hibbard, 1939:506.
    E. Plioc.; N.A.

Oryctolagus Lilljeborg, 1873:417.[1] Old World rabbits, lapins.
    [= *Cuniculus* Meyer, 1790:52.[2]]
    E. Plioc.-R.; Eu. Pleist., R.; Af.[3] R.; Atlantic.[4]

[1] Described as subgenus of *Lepus*.
[2] Not *Cuniculus* Brisson, 1762, a synonym of *Agouti*.
[3] N. Africa.
[4] Subfossil deposits of the eastern Canary Islands.
    Also Recent of North America, W. Indies, South America, New Zealand and Australia, and various islands of the Pacific, Atlantic, Indian O. and Mediterranean by introduction.
    Recent worldwide in domestication.

†*Pliopentalagus* Gureev & Konkova, in Gureev, 1964:129.
    E.-L. Plioc.; As. E. Plioc.; Eu.[1]

[1] E. Europe.

Pentalagus Lyon, 1904:428. Ryukyu rabbit, Amami rabbit.
    L. Pleist.-R.; As.[1]

[1] Ryukyu Islands, S.W. Japan.

Pronolagus Lyon, 1904:416. Rock hares, red hares, tsoarus.
    E. Plioc.-R.; Af.

†*Nekrolagus* Hibbard, 1939: table of contents.[1]
    [= †*Pediolagus* Hibbard, 1939:512.[2]]
    ?E. Plioc.; As. E.-L. Plioc.; N.A.

[1] May 1939. Replacement name for †*Pediolagus*.
[2] March 1939. Not *Pediolagus* Marelli, 1927, a synonym of *Dolichotis*.

Lepus Linnaeus, 1758:57. Hares, jack rabbits, snowshoe rabbit, arctic hare.
    [= *Chionobates* Kaup, 1829:170; *Eulepus* Acloque, 1899:52; *Mamlepus* Herrera, 1899:11.[1]] [Including *Lagos* Brookes, 1828:54; *Eulagos* Gray, 1867:222; *Macrotolagus* Mearns, 1895:698[2]; *Poecilolagus* Lyon, 1904:324, 395[2]; *Boreolepus* Barrett-Hamilton, 1911:160[2]; *Allolagus* Ognev, 1929:71[3]; *Tarimolagus* Gureev, 1947:517; *Indolagus* Gureev, 1953; *Proeulagus* Gureev, 1964:189.[2]]
    E. Plioc.-R.; Eu. L. Plioc.-R.; Af. L. Plioc.-R.; N.A.[4] E. Pleist.-R.; As. R.; E. Indies.[5] R.; Mediterranean.[6] R.; Cent. A.[7]

[1] See the International Code of Zoological Nomenclature, Article 1(b)(8), and Palmer (1904:25-26).
[2] Subgenus as used or proposed.
[3] As subgenus.
[4] Including islands of the Canadian Arctic and Greenland.
[5] Java (possibly by introduction).
[6] Balearic Islands (Mallorca, possibly by introduction), Corsica (probably by introduction), Sardinia, Sicily, and Crete.
[7] S. Mexico (Chiapas).
    Also Recent of N. Atlantic (Faeroes), W. Indies (Barbados), southern South America and Falkland Islands, New Zealand, Australia, and Indian Ocean (Reunion Island, Mauritius, etc.) by introduction.

†*Serengetilagus* Dietrich, 1941:221.
    E. Plioc.-M. Pleist.; Af. E. Plioc.; As.

†*Paranotolagus* Miller & Carranza, 1982:99.[1]
> E. Plioc.; N.A.

[1] Assignment to the Leporinae is tentative.

Sylvilagus Gray, 1867:221. Cottontails, tapetis, brush rabbits, swamp rabbit, marsh rabbit, conejos.
> [Including *Hydrolagus* Gray, 1867:221; *Tapeti* Gray, 1867:224[1]; †*Praotherium* Cope, 1871:93-94; *Limnolagus* Mearns, 1897:393[2]; *Microlagus* Trouessart, 1897:660; *Palaeotapeti* Spillmann, 1931:23; *Paludilagus* Hershkovitz, 1950:333.[3]]
> L. Plioc.-R.; N.A. Pleist., R.; S.A. L. Pleist.-R.; Cent. A.

[1] Subgenus.
[2] Replacement name for *Hydrolagus* Gray, 1867, not *Hydrolagus* Gill, 1862, a genus of fishes.
[3] Sometimes as subgenus.

†*Aztlanolagus* B. D. Russell & A. H. Harris, 1986:632.
> L. Plioc.-E. Pleist., L. Pleist.; N.A.

†*Aluralagus* Downey, 1968:171.
> L. Plioc.-E. Pleist.; N.A.

Caprolagus Blyth, 1845:247. Bristly rabbit, hispid hare.
> [Including *Sivalilagus* Gureev, 1964:137.[1]]
> L. Plioc., Pleist., R.; As.

[1] Proposed as subgenus.

Bunolagus Thomas, 1929:109. Riverine rabbit.
> M. and/or L. Pleist., R.; Af.[1]

[1] S. Africa.

Brachylagus Miller, 1900:157. Pygmy rabbit.
> L. Pleist.-R.; N.A.

Poelagus St. Leger, 1932:119. Scrub hare.
> R.; Af.

Nesolagus Forsyth Major, 1899:493. Short-eared rabbit, Sumatran rabbit.
> R.; E. Indies.[1]

[1] Sumatra.

Romerolagus Merriam, 1896:173. Volcano rabbit.
> [= *Lagomys* Herrera, 1897:34.[1]]
> R.; N.A.[2]

[1] Not *Lagomys* Storr, 1780, a synonym of *Marmota*, nor *Lagomys* G. Cuvier, 1800, a synonym of *Ochotona*.
[2] Mexico.

Mirorder **SIMPLICIDENTATA** Weber, 1904:490, 495, **new rank**.[1]
> [= CLAVICULATA Cuvier, 1816[2]; Glires Simplicidentati Lilljeborg, 1866; SIMPLICIDENTATI Alston, 1876:64; MIODONTA Schlosser, 1884:133.[3]] [= or including RODENTIFORMES Kinman, 1994:37.[4]]

[1] Proposed as suborder; used as superorder by Li, Wilson, Dawson & Krishtalka, 1987:106.
[2] *Nomen oblitum.* Contrasted with LAGOMORPHA (= NON-CLAVICULATA).
[3] *Nomen oblitum.*
[4] Proposed as an order.

Order †**MIXODONTIA** Sych, 1971:148.

Family †**Eurymylidae** Matthew, Granger & Simpson, 1929:1, 5.
> [= †Eurymyloidea Li, 1977:104.] [Including †Rhombomylidae Li, R. W. Wilson, Dawson & Krishtalka, 1987:103, 106.]
> L. Paleoc.-M. Eoc.; As.

†*Gomphos* Shevyreva, 1975:225.
> E. Eoc.; As.[1]

[1] E. Asia.

†*Zagmys* Dashzeveg, D. E. Russell & L. Flynn, 1987:135.
> E. Eoc.; As.

†*Nikolomylus* Shevyreva & Gabunia, 1986:77-82.
> M. Eoc.; As.

†*Aktashmys* Averianov, 1994:406.
E. and/or M. Eoc.; As.

Subfamily †**Eurymylinae** Matthew, Granger & Simpson, 1929:1, 5.
[= †Eurymylidae Matthew, Granger & Simpson, 1929:1, 5; †Eurymylinae Dashzeveg
& D. E. Russell, 1988:132.] [Including †Rhombomylidae Li, R. W. Wilson, Dawson
& Krishtalka, 1987:103, 106.]
L. Paleoc.-E. Eoc.; As.

†*Asiaparamys* Nessov, 1987:210.
L. Paleoc.; As.[1]
[1] Kazakhstan.

†*Kazygurtia* Nessov, 1987:209.
L. Paleoc.; As.[1]
[1] Kazakhstan.

†*Eomylus* Dashzeveg & D. E. Russell, 1988:132.
L. Paleoc.; As.

†*Eurymylus* Matthew & Granger, 1925:7.
[Including †*Baenomys* Matthew & Granger, 1925:5.]
L. Paleoc.; As.

†*Amar* Dashzeveg & D. E. Russell, 1988:139.
L. Paleoc.; As.

†*Rhombomylus* Zhai, 1978:111.[1]
[Including †*Matutinia* Li, Chiu, Yan & Hsieh, 1979:77.]
E. Eoc.; As.[2]
[1] *Nomen nudum* in Zhai, Bi & Yu, 1976:101.
[2] E. Asia.

†*Decipomys* Dashzeveg, Hartenberger, Martin & Legendre, 1997:in press.[1]
E. Eoc.; As.
[1] *Nomen nudum* in Hartenberger, 1996.

Subfamily †**Khaychininae** Dashzeveg & D. E. Russell, 1988:141.
L. Paleoc.; As.

†*Khaychina* Dashzeveg & D. E. Russell, 1988:141.
L. Paleoc.; As.

Order **RODENTIA** Bowdich, 1821:7, 51. Rodents.
[= Rodentes Vicq d'Azyr, 1792:xcvii; ROSORES Gray, 1821:302.[1]]
[1] Described as an order and including, as did RODENTIA, the lagomorphs. Curiously, later in the same
paper, Gray employed ROSORES a second time, for what is now called Daubentoniidae.

†*Heomys* Li, 1977:104.
E.-L. Paleoc.; As.[1]
[1] E. Asia.

†*Marfilomys* Ferrusquía-Villafranca, 1989:94.
M. and/or L. Eoc.; N.A.

Family †**Alagomyidae** Dashzeveg, 1990:38.
L. Paleoc.-E. Eoc.; As. L. Paleoc.; N.A.

†*Alagomys* Dashzeveg, 1990:38.
L. Paleoc.; N.A. E. Eoc.; As.

†*Tribosphenomys* Meng, Wyss, Dawson & Zhai, 1994:134.[1]
L. Paleoc.; As.[2]
[1] Assignment to the †Alagomyidae is questionable.
[2] Inner Mongolia, China.

Family †**Laredomyidae** J. A. Wilson & Westgate, 1991:257.
M. Eoc.; N.A.

†*Laredomys* J. A. Wilson & Westgate, 1991:257.
M. Eoc.; N.A.

Suborder **SCIUROMORPHA** Brandt, 1855:144, 292.
[= SCIUROMORPHI Brandt, 1855:144, 292.]

Superfamily †**Ischyromyoidea** Alston, 1876:67, 78 and plate 4.
[= †Ischyromyidae Alston, 1876:67, 78 and plate 4; †Ischyromyoidea A. E. Wood, 1937:160.]

Family †**Ischyromyidae** Alston, 1876:Plate 4, 67, 78.
[Including †Paramyida Haeckel, 1895:502; †Paramyidae Miller & Gidley, 1918:432, 439.]
L. Paleoc.-E. Olig.; N.A. E. Eoc.; Af.[1] E. Eoc.-E. Olig.; As. E. Eoc.-E. Olig., E. Mioc.; Eu. ?M. and/or ?L. Eoc.; Mediterranean.
[1] An undescribed early Eocene species of ischyromyid has been reported from Tunisia.

†*Taishanomys* Tong & Dawson, 1995:59.[1]
E. Eoc.; As.
[1] (Tung & Dawson).

†*Rapamys* R. W. Wilson, 1940:74.
M. Eoc.; N.A.

†*Anonymus* Storer, 1988:88.
M. Eoc.; N.A.

†*Hulgana* Dawson, 1968:3.
L. Eoc.; As.

Subfamily †**Microparamyinae** A. E. Wood, 1962:157.
[= †Microparamyidae Jaeger, 1988:187.[1]]
L. Paleoc.-L. Eoc.; N.A. ?E. Eoc., M. Eoc.; As. E.-M. Eoc.; Eu.
[1] *Lapsus calami*?

†*Acritoparamys* Korth, 1984:28.
L. Paleoc.-M. Eoc.; N.A. ?E. Eoc.; As.

†*Microparamys* A. E. Wood, 1959:162.
[Including †*Pantrogna* Hartenberger, 1971:104[1]; †*Sparnacomys* Hartenberger, 1971:105[2]; †*Corbarimys* Marandat, 1989:163.[2]]
L. Paleoc.-L. Eoc.; N.A. E.-M. Eoc.; Eu.
[1] Proposed as subgenus. Ranked as genus by Hartenberger, 1993 and by Hooker, 1996:148.
[2] Proposed as subgenus.

†*Decticadapis* Lemoine, 1891:289.[1]
E. Eoc.; Eu.
[1] Published as a *nomen nudum* by Lemoine, 1883.

†*Lophiparamys* Wood, 1962:167.
E. Eoc.; N.A.

†*Apatosciuravus* Korth, 1984:42.
L. Paleoc.-E. Eoc.; N.A.

†*Asiomys* Qi, 1987:17.
M. Eoc.; As.

Subfamily †**Paramyinae** Haeckel, 1895:502.
[= †Paramyida Haeckel, 1895:502; †Paramyidae Miller & Gidley, 1918:432, 439[1]; †Paramyinae Simpson, 1945:77.] [Including †Manitshini Simpson, 1941:1; †Manitshinae A. E. Wood, 1962:170; †Microparamyinae A. E. Wood, 1962:157; †Pseudoparamyinae Michaux, 1964.]
L. Paleoc.-L. Eoc., ?E. Olig.; N.A. E.-M. Eoc., ?L. Eoc.; As. E.-L. Eoc.; Eu. ?M. and/or ?L. Eoc.; Mediterranean.
[1] In part.

Tribe †**Paramyini** Haeckel, 1895:502.
[= †Paramyida Haeckel, 1895:502; †Paramyini Korth, 1984:8.]
L. Paleoc.-L. Eoc.; N.A. E.-M. Eoc., ?L. Eoc.; As. E.-M. Eoc.; Eu.

†*Paramys* Leidy, 1871:230-231.
L. Paleoc.-M. Eoc.; N.A. E.-M. Eoc.; As. E.-M. Eoc.; Eu.

†*Tapomys* A. E. Wood, 1962:154.
[Including †*Leptotomus* Matthew, 1910:50[1]; †*Uintaparamys* Kretzoi, 1968:163.[2]]
E.-L. Eoc.; N.A. ?L. Eoc.; As.
[1] Described as subgenus of †*Paramys*. Not *Leptotomus* Gahan, 1894, a carabid coleopteran.
[2] Replacement name for †*Leptotomus* Matthew, 1910.

†*Notoparamys* Korth, 1984:20.
E. Eoc.; N.A.

†*Thisbemys* A. E. Wood, 1959:163.
E.-M. Eoc.; N.A.

†*Abrosomys* Shevyreva, 1984:81.
E. Eoc.; As.

†*Quadratomus* Korth, 1984:22.
E.-M. Eoc.; N.A.

Tribe †**Pseudoparamyini** Michaux, 1964:154, **new rank**.
[= †Pseudoparamyinae Michaux, 1964:154.]
L. Paleoc.-E. Eoc.; N.A. E.-L. Eoc.; Eu. ?M. Eoc.; As. ?M. and/or ?L. Eoc.;
Mediterranean.

†*Franimys* A. E. Wood, 1962:139.
L. Paleoc.-E. Eoc.; N.A. ?M. Eoc.; As.[1]
[1] S. Asia.

†*Pseudoparamys* Michaux, 1964:154.
E.-M. Eoc.; Eu.

†*Plesiarctomys* Bravard, in Gervais, 1850:2, pl. 46.
?E. Eoc., M.-L. Eoc.; Eu. ?M. and/or ?L. Eoc.; Mediterranean.

Tribe †**Manitshini** Simpson, 1941:1.
M.-L. Eoc., ?E. Olig.; N.A.

†*Pseudotomus* Cope, 1872:467.
[Including †*Ischyrotomus* Matthew, 1910:50.[1]]
M.-L. Eoc.; N.A.
[1] Described as subgenus of †*Paramys*.

†*Manitsha* Simpson, 1941:474.
L. Eoc., ?E. Olig.; N.A.

Subfamily †**Ailuravinae** Michaux, 1968:155.
[= †Ailuraviinae Michaux, 1968:155; †Ailuravinae A. E. Wood, 1976:122.]
E.-M. Eoc., E. Olig., E. Mioc.; Eu. E.-L. Eoc.; N.A.

†*Mytonomys* A. E. Wood, 1956:753.
E.-L. Eoc.; N.A.

†*Meldimys* Michaux, 1968:155.
E. Eoc.; Eu.

†*Ailuravus* Rütimeyer, 1891:94.
[= †*Palaeomarmota* Haupt, 1921:177; †*Megachiromyoides* Weigelt, 1933:109;
†*Aeluravus* Schaub, in Stehlin & Schaub, 1951:21, 205, 353.] [Including
†*Maurimontia* Schaub, in Stehlin & Schaub, 1951:355.]
E.-M. Eoc.; Eu.

†*Eohaplomys* Stock, 1935:62.
M. Eoc.; N.A.

†*Trigonomys* Heissig, 1979:152.
E. Olig.; Eu.

†*Paracitellus* Dehm, 1950:331.
E. Olig., E. Mioc.; Eu.

Subfamily †**Ischyromyinae** Alston, 1876:67, 78, pl. 4.
　　　　　[= †Ischyromyidae Alston, 1876:67, 78, pl. 4; †Ischyromynae Trouessart, 1880:98;
　　　　　†Ischyromyinae Schlosser, 1911:425.]
　　　　　L. Eoc.-E. Olig.; N.A.

†*Titanotheriomys* Matthew, 1910:63.[1]
　　　　　L. Eoc.; N.A.
　　　　　[1] Described as subgenus.

†*Ischyromys* Leidy, 1856:89.
　　　　　[= †*Colotaxis* Cope, 1873:1; †*Gymnoptychus* Cope, 1873:5.]
　　　　　L. Eoc.-E. Olig.; N.A.

Superfamily **Aplodontoidea** Brandt, 1855:145, 151.
　　　　　[= Haploodontini Brandt, 1855:145, 151; Haploodontoidea Gill, 1872:22;
　　　　　Aplodontoidea Matthew, 1910:69.]

†*Epeiromys* Korth, 1989:406.
　　　　　E. Olig.; N.A.

†*Sinomylagaulus* Wu, 1988:251.
　　　　　M. Mioc.; As.

Family †**Allomyidae** Marsh, 1877:253.
　　　　　[Including †Prosciurinae Wilson, 1949:92.]
　　　　　M. Eoc.-M. Mioc.; N.A. E. Olig.-M. Mioc.; Eu. L. Olig.; As.

†*Spurimus* Black, 1971:207.
　　　　　M. Eoc., ?L. Eoc.; N.A.

†*Prosciurus* Matthew, 1903:213.[1]
　　　　　L. Eoc.-E. Olig.; N.A.
　　　　　[1] Proposed as subgenus; raised to generic rank by Matthew, 1909:103.

†*Pelycomys* Galbreath, 1953:54.
　　　　　L. Eoc.-E. Olig.; N.A.

†*Pseudallomys* Korth, 1992:171.
　　　　　E. Olig.; N.A.

†*Downsimus* Macdonald, 1970:27.
　　　　　E.-L. Olig.; N.A.

†*Plesispermophilus* Filhol, 1883:100.
　　　　　Olig.; Eu. L. Olig.; As.

†*Ephemeromys* Wang & Heissig, 1984:105, 106.
　　　　　E. Olig.; Eu.

†*Campestrallomys* Korth, 1989:407.
　　　　　E.-L. Olig.; N.A.

†*Sciurodon* Schlosser, 1884:73-75.
　　　　　E. Olig.-E. Mioc.; Eu.

†*Oropyctis* Korth, 1989:405.
　　　　　E. Olig.; N.A.

†*Rudiomys* Rensberger, 1983:26.
　　　　　L. Olig.; N.A.

†*Allomys* Marsh, 1877:253.
　　　　　L. Olig.-M. Mioc.; N.A.

†*Haplomys* G. S. Miller & Gidley, 1918:440.
　　　　　L. Olig.; As. L. Olig.; N.A.

†*Alwoodia* Rensberger, 1983:43.
　　　　　L. Olig.-E. Mioc.; N.A.

†*Parallomys* Rensberger, 1983:30.
　　　　　L. Olig.; Eu. E. Mioc.; N.A.

†*Ameniscomys* Dehm, 1950:326.
M. Mioc.; Eu.

Family **Aplodontidae** Brandt, 1855:145, 151.
[= Haploodontini Brandt, 1855:145, 151; Haploodontidae Lilljeborg, 1866:9, 41; Haplodontidae Alston, 1876:78; Aplodontiidae Thomas, 1897:1015; Aplodontidae Trouessart, 1897:450.] [Including †Meniscomyinae Rensberger, 1981:196; †Ansomyinae Qiu, 1987:283.]
L. Olig., M.-L. Mioc.; As. L. Olig.-E. Plioc., L. Pleist.-R.; N.A.

†*Meniscomys* Cope, 1878:5-6.
[= †*Protogaulus* Riggs, 1899:183.]
L. Olig.; N.A.

†*Niglarodon* Black, 1961:2.
[= or including †*Horatiomys* Wood, 1935:2.[1]]
L. Olig.-E. Mioc.; N.A.

[1] Possibly a senior synonym of †*Niglarodon* or of some other early aplodontid genus because the type species is based upon an aplodontid (rather than a geomyid) deciduous premolar. The name is regarded here as a *nomen dubium*. It should be suppressed.

†*Promeniscomys* Wang, 1987:33.
L. Olig.; As.

†*Ansomys* Qiu, 1987:284.
M. Mioc.; As.

†*Liodontia* Miller & Gidley, 1918:440.
M. Mioc.-E. Plioc.; N.A.

†*Tardontia* Shotwell, 1958:455.
M.-L. Mioc.; N.A.

†*Pseudaplodon* Miller, 1927:15.
L. Mioc.; As. L. Mioc.; N.A.

†*Tschalimys* Shevyreva, 1971:482.
L. Mioc.; As.

Aplodontia Richardson, 1829:334. Sewellel, mountain beaver.
[= *Apludontia* Fischer, 1830:598; *Apluodontia* Richardson, 1837:150, 159; *Haplodon* Wagler, 1830:22; *Haploodon* Brandt, 1855:150; *Hapludon* Brandt, 1855:150; *Haploudon* Coues, in Coues & J. A. Allen, 1877:556; *Haploodus* Coues, in Coues & J. A. Allen, 1877:557; *Haplodus* Coues, in Coues & J. A. Allen, 1877:557; *Haploudus* Coues, in Coues & J. A. Allen, 1877:557; *Haploudontia* Coues, 1889:2712.]
L. Pleist.-R.; N.A.

Family †**Mylagaulidae** Cope, 1881:362.
L. Olig.-E. Plioc.; N.A. L. Mioc.; As.[1]

[1] A †*Mylagaulus*-like rodent, said to be a new genus of †Promylagaulinae, is present in the late Miocene of Xinjiang, PRC.

Subfamily †**Promylagaulinae** Rensberger, 1980:1267.
L. Olig.-M. Mioc.; N.A.

†*Crucimys* Rensberger, 1980:1268.
L. Olig.; N.A.

†*Trilaccogaulus* Korth, 1992:95.
L. Olig.-M. Mioc.; N.A.

†*Promylagaulus* McGrew, 1941:5.
L. Olig.-E. Mioc.; N.A.

Subfamily †**Mylagaulinae** Cope, 1881:362, **new rank**.
[= †Mylagaulidae Cope, 1881:362.]
E. Mioc.-E. Plioc.; N.A.

RODENTIA 119

†*Mylagaulodon* Sinclair, 1903:143.
[= †*Sewelleladon* Shotwell, 1958:452.]
E. Mioc.; N.A.

†*Mesogaulus* Riggs, 1899:181.
E.-L. Mioc.; N.A.

†*Mylagaulus* Cope, 1878:384-385.
M. Mioc.-E. Plioc.; N.A.

†*Ceratogaulus* Matthew, 1902:291-294, 299.
L. Mioc.; N.A.

†*Epigaulus* Gidley, 1907:627-636.
L. Mioc.-E. Plioc.; N.A.

Infraorder †**THERIDOMYOMORPHA** A. E. Wood, 1955:173, **new rank**.[1]
[= †PARASCIUROGNATHA Lavocat, 1951:47[2]; †THERIDOMORPHES Lavocat, 1955:77[1]; †THERIDOMORPHA Thaler, 1966:11.[1]]
[1] Proposed at subordinal rank.
[2] Proposed at subordinal rank and before †THERIDOMYOMORPHA, but the rule of priority does not operate above the rank of superfamily. Lavocat himself ignored the term in 1955.

Family †**Theridomyidae** Alston, 1876:70, 88.
[= †Theridomyoidae Lavocat, 1951:47.[1]] [Including †Archaeomyidae Schlosser, 1884:327; †Nesokerodontidae Schlosser, 1884:327; †Trechomyini Winge, 1887:108, 118; †Pseudosciurini Winge, 1887:108, 118; †Pseudosciuridae Zittel, 1893:523.]
?E. Eoc., M. Eoc.-L. Olig.; Eu. M.-L. Eoc., ?E. Olig., L. Olig.; Mediterranean.[2]
[1] *Lapsus;* spelled correctly on the same page.
[2] Balearic Islands.

†*Masillamys* Tobien, 1954:18.
?E. Eoc., M. Eoc.; Eu.

Subfamily †**Pseudosciurinae** Winge, 1887:108, 118.
[= †Pseudosciurini Winge, 1887:108, 118; †Pseudosciuridae Zittel, 1893:523; †Pseudosciurinae Lavocat, 1951:73.]
M. Eoc., Olig.; Mediterranean.[1] M. Eoc.-E. Olig., ?L. Olig.; Eu.
[1] Balearic Islands.

†*Suevosciurus* Dehm, 1937:268-290.
[Including †*Microsuevosciurus* Hartenberger, 1973:15.[1]]
M. Eoc., Olig.; Mediterranean.[2] M. Eoc.-E. Olig.; Eu.
[1] Subgenus as defined.
[2] Balearic Islands.

†*Treposciurus* Schmidt-Kittler, 1970:434.
M.-L. Eoc.; Eu.

†*Pseudosciurus* Hensel, 1856:660-670.
L. Eoc., Olig.; Eu.

†*Tarnomys* Hartenberger & Schmidt-Kittler, 1976:66.
E. Olig.; Eu.

Subfamily †**Oltinomyinae** Hartenberger, 1971:1919.
[= †Oltinomyinae Bosma & Schmidt-Kittler, 1972:183.] [Including †Remyinae Hartenberger, 1973:53.]
M.-L. Eoc.; Eu.

†*Remys* Thaler, 1966:98.
M.-L. Eoc.; Eu.

†*Estellomys* Hartenberger, 1971:1918.
L. Eoc.; Eu.

†*Ectropomys* Bosma & Schmidt-Kittler, 1972:183, 185.
L. Eoc.; Eu.

†*Oltinomys* Stehlin & Schaub, 1951:363.[1]

L. Eoc.; Eu.

[1] Proposed as a subgenus of †*Theridomys*. "†*Oltinomys*" Stehlin, 1942 (not 1941): 213 was a *nomen nudum*.

†*Pairomys* Thaler, 1966:164-166.

M.-L. Eoc.; Eu.

Subfamily †**Sciuroidinae** Hartenberger, 1971:1919.

M.-L. Eoc.; Eu.

†*Sciuroides* Forsyth Major, 1873:79-86.

M.-L. Eoc.; Eu.

Subfamily †**Adelomyinae** Lavocat, 1951:73.

[= †Adelomyinea Lavocat, 1951:73.]

M. Eoc.-E. Olig.; Eu.

†*Protadelomys* Hartenberger, 1968:1818.

M.-L. Eoc.; Eu.

†*Paradelomys* Thaler, 1966:39.[1]

M. Eoc.-E. Olig.; Eu.

[1] Proposed as subgenus of †*Adelomys*.

†*Adelomys* Gervais, 1849:244.

L. Eoc.; Eu.

Subfamily †**Issiodoromyinae** Tullberg, 1899.

[= †Issidioromyidae Tullberg, 1899; †Issiodoromyidae Schlosser, 1902;
†Issiodoromyinae Lavocat, 1951:83.] [Including †Nesokerodontidae Schlosser,
1884:327; †Pseudoltinomyini Mödden, 1994:1040.]

M.-L. Eoc., Olig.; Mediterranean. M. Eoc.-L. Olig.; Eu.

†*Elfomys* Hartenberger, 1971:1919.

M. Eoc.; Mediterranean. M. Eoc.-E. Olig.; Eu.

†*Pseudoltinomys* Lavocat, 1951:53.

M.-L. Eoc., Olig.; Mediterranean.[1] M. Eoc.-E. Olig.; Eu.

[1] Mallorca.

†*Issiodoromys* Bravard, in Gervais, 1852:27.[1]

[= †*Palanoema* Pomel, 1853:39[2]; †*Palanaema* Gervais, 1859:36[3]; †*Cournomys* Zittel,
1893:525.] [Including †*Nesokerodon* Schlosser, 1884:16-20; †*Nesocerodon* Lydekker,
1885:253[3]; †*Oensingenomys* Mayo, 1987:1048.]

E.-L. Olig.; Eu.

[1] (1848-1852). Previous usages were *nomina nuda*.
[2] Spelling clearly intended by Pomel.
[3] Replacement name.

†*Bernardia* Vianey-Liaud, 1991:80.

M. Eoc.; Eu.

†*Patriotheridomys* Vianey-Liaud, 1974:200.

E. Olig.; Eu.

Subfamily †**Theridomyinae** Alston, 1876:70, 88.

[= †Theridomyidae Alston, 1876:70, 88; †Theridomyinae Miller & Gidley, 1918:442;
†Theridomyinae Lavocat, 1951:73.] [Including †Archaeomyinae Schlosser, 1884:327;
†Archaeomyidae Schlosser, 1884:327; †Archaeomyinae Lavocat, 1951:79;
†Archaeomyini Mödden, 1993:21.]

L. Eoc.-L. Olig.; Eu. L. Olig.; Mediterranean.[1]

[1] Indet. theridomyine, Mallorca, Balearic Islands.

†*Isoptychus* Pomel, 1853:81-229.

[Including †*Trechomys* Lartet, 1869:151-162; †*Quercymys* Thaler, 1966:73.[1]]

L. Eoc., Olig.; Eu.

[1] Subgenus as proposed.

†*Thalerimys* Tobien, 1972:7, 19.[1]
　　L. Eoc., ?E. Olig.; Eu.
[1] Assignment to Theridomyinae questionable.

†*Theridomys* Jourdan, 1837:483-484.
　　L. Eoc.-L. Olig.; Eu.

†*Taeniodus* Pomel, 1853:36.
　　[= †*Toeniodus* Pomel, 1853:36[1]; †*Taeniodus* Gervais, 1859:31.] [= or including
　　†*Protechimys* Schlosser, 1884:pl. IV.; †*Protechinomys* Lydekker, 1885:240-241[2];
　　†*Pararchaeomys* Schaub, in Stehlin & Schaub, 1951:364.[3]]
　　E. Olig.; Eu.
[1] The original and intended spelling, but stability would not be served by returning to it. Nevertheless, in
1987 Ginsburg & Hugueney and Mayo did so.
[2] Replacement name for †*Protechimys*.
[3] *Nomen nudum* in Stehlin, in Erni, 1942:213.

†*Blainvillimys* Gervais, 1852:4.[1]
　　[= †*Blainvillemys* Giebel, 1855:517; †*Blainvilleomys* Giebel, 1859:1087.]
　　E.-L. Olig.; Eu.
[1] 1848-1852, from Bravard (MS).

†*Archaeomys* Laizer & Parieu, 1839:206.
　　[Including †*Protechimys* Schlosser, 1884:45; †*Protechinomys* Lydekker, 1885:240[1];
　　†*Monarchaeomys* Mayo, 1983:838[2]; †*Rhombarchaeomys* Mayo, 1983:856.[2]]
　　L. Olig.; Eu.
[1] Invalid emendation.
[2] Described as subgenus.

Subfamily †**Columbomyinae** Thaler, 1966:93.
　　?E. Olig., L. Olig.; Eu.

†*Sciuromys* Schlosser, 1884:81-83.[1]
　　[Including †*Pseudosciuromys* Stehlin & Schaub, 1951:211-222.]
　　Olig.; Eu.
[1] Pagination in separate.

†*Columbomys* Thaler, 1962:2207-2209.
　　L. Olig.; Eu.

Infraorder **SCIURIDA** Carus, 1868:96, **new rank**.[1]
　　[= Sciurii Fischer de Waldheim, 1817:372; Sciuriorum Fischer de Waldheim,
　　1817:408.[1]]
[1] Proposed as family.

Family †**Reithroparamyidae** A. E. Wood, 1962:117.
　　[= †Reithroparamyinae A. E. Wood, 1962:117; †Reithroparamyidae Patterson & A. E.
　　Wood, 1982:522.]
　　E.-M. Eoc.; N.A.

†*Reithroparamys* Matthew, 1920:168.
　　[Including †*Uriscus* A. E. Wood, 1962:116.]
　　E.-M. Eoc.; N.A.

Family **Sciuridae** Fischer de Waldheim, 1817:372. Squirrels.
　　[= Sciurii Fischer de Waldheim, 1817:372; Sciuriorum Fischer de Waldheim,
　　1817:408; Sciuridae Gray, 1821:304; Sciurina Haeckel, 1866:clx[1]; Sciuroidea Gill,
　　1872:21; Sciuroidae Miller & Gidley, 1918:432.[2]] [Including Arctomydae J. E. Gray,
　　1821:303; Arctomysideae Lesson, 1841:115; Arctomyida Haeckel, 1866:clx[1];
　　Marmotinae Pocock, 1923:240; Marmotidae Weber, 1928:275.]
　　L. Eoc.-R.; N.A. Olig., ?M. Mioc., L. Mioc.; Mediterranean. E. Olig.-R.; Eu. E.
　　Mioc.-R.; Af. E. Mioc.-R.; As. R.; E. Indies. R.; Cent. A. R.; S.A.
[1] Used as a family.
[2] In part.

†*Vulcanisciurus* Lavocat, 1973:239.
　　E.-L. Mioc.; Af.

†*Sciurion* Skwara, 1986:290.
E. Mioc.; N.A.

Subfamily †**Cedromurinae** Korth & Emry, 1991:989.
L. Eoc.-E. Olig.; N.A.

†*Oligospermophilus* Korth, 1987:1248.
L. Eoc.-E. Olig.; N.A.

†*Cedromus* R. W. Wilson, 1949:29.
E. Olig.; N.A.

Subfamily **Sciurinae** Fischer de Waldheim, 1817:372.
[= Sciurii Fischer de Waldheim, 1817:372; Sciuriorum Fischer de Waldheim, 1817:408; Sciuridae Gray, 1821:304; Sciurinae Baird, 1857:240.] [Including Xeri Murray, 1866:256; Xerinae Osborn, 1910:535; Marmotinae Pocock, 1923:240; Spermophilopsinae Ognev, 1940:326, 432.]
L. Eoc.-R.; N.A. E. Olig.-R.; Eu. E. Mioc.-R.; As. M. Mioc.-E. Pleist., L. Pleist.-R.; Af. L. Mioc.; Mediterranean. R.; E. Indies. R.; Cent. A. R.; S.A.

†*Similisciurus* Stevens, 1977:21.
E. Mioc.; N.A.

†*Sinotamias* Qiu, 1991:230.
M.-L. Mioc.; As.

Sciurotamias Miller, 1901:23. Montane Chinese rock squirrel.
[Including *Rupestes* Thomas, 1922:398.]
L. Mioc., L. Plioc.-M. Pleist., ?L. Pleist., R.; As.

Tribe **Sciurini** Fischer de Waldheim, 1817:372.
[= Sciurii Fischer de Waldheim, 1817:372; Sciuriorum Fischer de Waldheim, 1817:408; Sciurina Hemprich, 1820:32; Sciurina Gray, 1825:342[1]; Sciurini Burmeister, 1854:145.]
L. Eoc.-R.; N.A. ?E. Mioc., L. Plioc., Pleist., R.; Eu. M. Mioc., M. Pleist.-R.; As. R.; E. Indies. R.; Cent. A. R.; S.A.
[1] As tribe.

†*Freudenthalia* Cuenca Bescós, 1988:90.[1]
E. Mioc.; Eu.
[1] Assignment to Sciurini is uncertain.

Rheithrosciurus Gray, 1867:273. Bornean tassel-eared squirrel.
[= *Rhithrosciurus* Lydekker, 1891:452.]
R.; E. Indies.[1]
[1] Borneo.

†*Plesiosciurus* Qiu & Liu, 1986:202.
M. Mioc.; As.

Subtribe **Sciurina** Fischer de Waldheim, 1817:372.
[= Sciurii Fischer de Waldheim, 1817:372; Sciuriorum Fischer de Waldheim, 1817:408; Sciuridae Gray, 1821:304; Sciurina Moore, 1959:198.] [Including Microsciurina Moore, 1959:199.]
L. Eoc.-R.; N.A. L. Plioc., Pleist., R.; Eu. M. Pleist.-R.; As. R.; Cent. A. R.; S.A.

†*Douglassia* Emry & Korth, 1996:775.
L. Eoc.; N.A.

†*Protosciurus* Black, 1963:138.
E. Olig.-E. Mioc.; N.A.

†*Miosciurus* Black, 1963:135.
E. Mioc.; N.A.

Sciurus Linnaeus, 1758:63. Tree squirrels, gray squirrels, fox squirrel, European red squirrel.
[= *Aphrontis* Schulze, 1893:165[1]; *Mamsciurus* Herrera, 1899:5.[2]] [Including *Guerlinguetus* Gray, 1821:304; *Macroxus* F. Cuvier, 1823:109-111; *Echinosciurus*

Trouessart, 1880:292; *Neosciurus* Trouessart, 1880:292; *Parasciurus* Trouessart, 1880:292; *Hesperosciurus* Nelson, 1899:27, 83; *Otosciurus* Nelson, 1899:28, 85; *Baiosciurus* Nelson, 1899:31-32; *Syntheosciurus* Bangs, 1902:25-27; *Tenes* Thomas, 1909:468[3]; *Notosciurus* Allen, 1914:585; *Mesosciurus* Allen, 1915:178, 212; *Hadrosciurus* Allen, 1915:178-265; *Urosciurus* Allen, 1915:178, 267; *Histriosciurus* Allen, 1915:178[4]; *Oreosciurus* Ognev, 1935:50.[5]]

M. Mioc.-R.; N.A. L. Plioc., Pleist., R.; Eu. M. Pleist.-R.; As. R.; Cent. A. R.; S.A.

[1] Proposed as subgenus.
[2] See the International Code of Zoological Nomenclature, Article 1(b)(8), and Palmer (1904:25-26).
[3] Only subgenus not known from Western Hemisphere.
[4] Described as a subgenus of *Mesosciurus*.
[5] Antedated by *Tenes* Thomas, 1909, an objective synonym.

Microsciurus J. A. Allen, 1895:332-333.
[Including *Leptosciurus* Allen, 1915:178, 199; *Simosciurus* Allen, 1915:178, 280.]
R.; Cent. A.[1] R.; S.A.[2]

[1] Panama to S. Nicaragua.
[2] Tropical South America.

Subtribe **Sciurillina** Moore, 1959:199.
R.; S.A.

Sciurillus Thomas, 1914:36. Neotropical pygmy squirrel.
R.; S.A.[1]

[1] Tropical South America.

Tribe **Xerini** Murray, 1866:256.
[= Xeri Murray, 1866:256; Xerinae Osborn, 1910:535; Xerini Simpson, 1945:79.]
[Including Spermophilopsinae Ognev, 1940:326, 432.]
E. Olig.-E. Plioc.; Eu. E. Mioc.-E. Plioc., R.; As. M. Mioc.-E. Pleist., R.; Af.

†*Heteroxerus* Schaub, in Stehlin & Schaub, 1951:358.
E. Olig.-L. Mioc.; Eu.

Atlantoxerus Forsyth Major, 1893:189.[1]
[Including †*Getuloxerus* Lavocat, 1961:47.]
E. Mioc.-E. Plioc.; As.[2] E. Mioc.-E. Plioc.; Eu.[3] M.-L. Mioc., Plioc., E. Pleist., R.; Af.[4]

[1] Proposed as subgenus of *Xerus*.
[2] Including Rhodes in the early Pliocene.
[3] Spain.
[4] N. Africa.

†*Aragoxerus* Cuenca Bescós, 1988:79.
E. Mioc.; Eu.[1]

[1] Spain.

Xerus Hemprich & Ehrenberg, 1832:pl. IX.[1] African ground squirrels.
[Including *Geosciurus* Smith, 1834:128; *Euxerus* Thomas, 1909:473.]
E. Plioc.-E. Pleist., R.; Af.

[1] Described as subgenus of *Sciurus*.

Spermophilopsis Blasius, 1884:325. Long-clawed ground squirrel.
R.; As.[1]

[1] Central Asia.

Tribe **Marmotini** Pocock, 1923:240. Northern ground squirrels.
[= Arctomydae Gray, 1821:303; Arctomina Gray, 1825:342[1]; Arctomyini Brandt, 1855:151; Marmotinae Pocock, 1923:240; Marmotidae Weber, 1928:275; Marmotini Simpson, 1945:79.]
E. Olig.-L. Mioc., E. Pleist.-R.; Eu. L. Olig.-R.; N.A. E. Mioc., L. Mioc.-E. Plioc., ?E. Pleist., M. Pleist.-R.; As. L. Mioc.; Mediterranean.[2]

[1] As tribe.
[2] Crete.

†*Palaeosciurus* Pomel, 1853:17.
    E. Olig.-M. Mioc.; Eu. E. Mioc.; As.

Subtribe **Marmotina** Pocock, 1923:240.
    [= Arctomyini Brandt, 1855:151; Arctomyida Haeckel, 1866:clx[1]; Marmotinae Pocock, 1923:240; Marmotina Moore, 1959:199.]
    L. Olig.-R.; N.A. Pleist., R.; As. Pleist., R.; Eu.
[1] Used as a family.

†*Protospermophilus* Gazin, 1930:64.[1]
    L. Olig.-L. Mioc.; N.A.
[1] Proposed as subgenus of *Citellus*; raised generic rank by Bryant, 1945:347.

†*Miospermophilus* Black, 1963:187.
    L. Olig.-M. Mioc.; N.A.

†*Palaearctomys* Douglass, 1903:182.
    L. Mioc.; N.A.

Marmota Blumenbach, 1779:79. Marmot, woodchuck.
    [= *Marmota* Frisch, 1775:9[1]; *Arctomys* Schreber, 1780:pls. ccvii-ccxi; *Lagomys* Storr, 1780:39; *Lipura* Illiger, 1811:95.[2]] [Including *Stereodectes* Cope, 1869:3; *Marmotops* Pocock, 1923:1200.]
    L. Mioc.-R.; N.A. Pleist., R.; As. Pleist., R.; Eu.
[1] *Nomen nudum.*
[2] *Lipura* was based on "*Hyrax*" *hudsonius* Schreber, actually an American marmot.

†*Arctomyoides* Bryant, 1945:361.
    L. Mioc.; N.A.

†*Paenemarmota* Hibbard & Schultz, 1948:19.
    L. Mioc.-E. Plioc.; N.A.

Subtribe **Spermophilina** Moore, 1959:199.
    [= Citellini Gromov, 1965:155.]
    E.-L. Mioc., E. Pleist.-R.; Eu. M. Mioc.-R.; N.A. L. Mioc.-E. Plioc., ?E. Pleist., M. Pleist.-R.; As. L. Mioc.; Mediterranean.[1]
[1] Crete.

†*Spermophilinus* de Bruijn & Mein, 1968:74-90.
    E.-L. Mioc.; Eu. ?L. Mioc., E. Plioc.; As.[1,2] L. Mioc.; Mediterranean.[3]
[1] E. Asia (China) in the late Miocene?
[2] W. Asia (Rhodes) in the early Pliocene.
[3] Crete.

Spermophilus F. Cuvier, 1825:255. Ground squirrels, gophers, ziesels.
    [= *Citellus* Oken, 1816:842; *Spermatophilus* Wagler, 1830:22; *Mamspermophilus* Herrera, 1899:5.[1]] [Including *Citillus* Lichtenstein, 1827:pl. 31, fig. 2[2]; *Otospermophilus* Brandt, 1844:379-380[3]; *Otocolobus* Brandt, 1844:382[4]; *Colobotis* Brandt, 1844:365[3]; *Colobates* Milne-Edwards, 1874:157[5]; *Ictidomys* Allen, 1877:821; *Xerospermophilus* Merriam, 1892:27; *Callospermophilus* Merriam, 1897:189; *Ictidomoides* Mearns, 1907:328[6]; *Urocitellus* Obolensky, 1927:192[7]; *Poliocitellus* A. H. Howell, 1938:42; *Notocitellus* A. H. Howell, 1938:44[6]; †*Pliocitellus* Hibbard, 1942:253.]
    M. Mioc.-R.; N.A. L. Mioc., ?E. Pleist., M. Pleist.-R.; As. E. Pleist.-R.; Eu.
[1] See the International Code of Zoological Nomenclature, Article 1(b)(8), and Palmer (1904:25-26).
[2] 1827-1834.
[3] Described as subgenus.
[4] Described as subgenus of *Spermophilus*. Not *Otocolobus* Brandt, 1841, a subgenus of *Felis*.
[5] 1868-1874. Objective synonym of *Colobotis* Brandt, 1844.
[6] Subgenus as defined.
[7] Sometimes as subgenus.

Ammospermophilus Merriam, 1892:27. Antelope ground squirrel.
    L. Mioc.-R.; N.A.

Cynomys Rafinesque, 1817:45. Prairie dogs.
> [= *Monax* Warden, 1819:225-228; *Cynomomus* H. L. Osborn, 1894:103; *Mamcynomisus* Herrera, 1899:22.[1]] [Including *Leucocrossuromys* Hollister, 1916:23.[2]]
> ?M. Mioc., L. Plioc.-R.; N.A.

[1] See the International Code of Zoological Nomenclature, Article 1(b)(8), and Palmer (1904:25-26).
[2] As subgenus.

Tribe **Tamiini** Weber, 1928:274. Chipmunks.
> [= Tamiidae Weber, 1928:274; Tamiina Moore, 1959:181; Tamiini Black, 1963:123.]
> L. Olig.-R.; N.A. E. Mioc.-E. Plioc., R.; Eu. M. Mioc.-R.; As.

†*Nototamias* Pratt & Morgan, 1989:95.
> L. Olig.-M. Mioc.; N.A.

Tamias Illiger, 1811:83. Eastern chipmunk, hackee, western chipmunks, borunduk, Siberian chipmunk.
> [= *Tamia* Gray, 1825:342.] [Including *Eutamias* Trouessart, 1880:86[1]; *Neotamias* Howell, 1929:11, 26.[2]]
> E. Mioc.-E. Plioc., R.; Eu. E. Mioc.-R.; N.A. M. Mioc.-R.; As.

[1] Sometimes as separate genus.
[2] Sometimes as subgenus of *Eutamias*.

Tribe **Ratufini** Moore, 1959:167, 198. Indo-Malayan giant squirrels.
> [= Ratufinae Corbet & Hill, 1992:276.]
> ?E. Mioc.; Eu. M. Mioc., R.; As.[1] R.; E. Indies.[2]

[1] S. Asia.
[2] Sumatra, Java, Borneo.

Ratufa Gray, 1867:273. Oriental giant squirrels.
> [Including *Rukaia* Gray, 1867:275; *Eosciurus* Trouessart, 1880:291.]
> ?E. Mioc.; Eu. M. Mioc., R.; As.[1] R.; E. Indies.[2]

[1] S. Asia.
[2] Sumatra, Java, Borneo.

Tribe **Nannosciurini** Forsyth Major, 1893:187-189, **new rank**. Oriental tree squirrels.
> [= Nannosciurinae Forsyth Major, 1893:187-189.] [Including Callosciurinae Pocock, 1923:239; Callosciurini Simpson, 1945:79; Hyosciurina Moore, 1959:198.]
> ?L. Mioc., L. Plioc.-M. Pleist., R.; As. R.; E. Indies.

Dremomys Heude, 1898:54.
> [Including *Zetis* Thomas, 1908:245.]
> ?L. Mioc., E.-M. Pleist., R.; As.[1,2] R.; E. Indies.[3]

[1] China in the Miocene and Pleistocene.
[2] S.E. Asia in the Recent.
[3] Borneo.

Callosciurus Gray, 1867:277.
> [Including *Baginia* Gray, 1867:279; *Erythrosciurus* Gray, 1867:285; *Heterosciurus* Trouessart, 1880:292; *Tomeutes* Thomas, 1915:385.]
> ?L. Mioc., E. Pleist., R.; As. R.; E. Indies.[1]

[1] Sumatra, Mentawi Islands, Java, Bali, Borneo, Salajar Island, Sulawesi.

Tamiops J. A. Allen, 1906:475.
> ?L. Mioc., L. Plioc.-M. Pleist., R.; As.[1,2]

[1] China in the Miocene, Pliocene and Pleistocene.
[2] S. Asia in the Recent.

Glyphotes Thomas, 1898:250-251. Sculptor squirrel.
> [Including *Hessonoglyphotes* Moore, 1959:5.]
> R.; E. Indies.[1]

[1] Borneo.

Rhinosciurus Gray, 1843:195. Ant-eating squirrel.
> R.; E. Indies.[1] R.; As.[2]

[1] Sumatra, Borneo.
[2] S.E. Asia.

Lariscus Thomas & Wroughton, 1909:389. Malayan black-striped squirrel.
          [= *Laria* Gray, 1867:276.[1]] [Including *Paralariscus* Ellerman, 1947:259.[2]]
          R.; E. Indies.[3] R.; As.[4]
          [1] Not *Laria* Scopoli, 1763, a coleopteran.
          [2] Proposed as subgenus.
          [3] Sumatra, Java, Borneo.
          [4] S.E. Asia.
Menetes Thomas, 1908:244. Siamese ground squirrel.
          R.; As.[1]
          [1] S.E. Asia.

Prosciurillus Ellerman, 1947:259. Celebesian tree squirrel.
          R.; E. Indies.[1]
          [1] Sulawesi.

Exilisciurus Moore, 1958:4. Patternless pygmy squirrel.
          R.; E. Indies.[1]
          [1] Borneo, Philippines.

Sundasciurus Moore, 1958:2.
          [Including *Aletesciurus* Moore, 1958:3.]
          R.; E. Indies.[1] R.; As.[2]
          [1] Borneo, Philippines.
          [2] S.E. Asia.
Rubrisciurus Ellerman, in Laurie & Hill, 1954:94. Celebesian giant squirrel.
          R.; E. Indies.[1]
          [1] Sulawesi.

Nannosciurus Trouessart, 1880:292. Oriental pygmy squirrel, oriental dwarf squirrel, black-eared
          pygmy squirrel.
          R.; E. Indies.[1]
          [1] Sumatra, Java, Borneo.

Hyosciurus Archbold & Tate, 1935:2. Sulawesian long-nosed squirrel.
          R.; E. Indies.[1]
          [1] Sulawesi.

Tribe **Protoxerini** Moore, 1959:198.
          L. Mioc., L. Pleist.-R.; Af.

†*Kubwaxerus* Cifelli, Ibui, Jacobs & Thorington, 1986:274.
          L. Mioc.; Af.

Heliosciurus Trouessart, 1880:292. Isindi, sun squirrel.
          L. Pleist.-R.; Af.

Protoxerus Forsyth Major, 1893:189. Oil-palm squirrels, African giant squirrels.
          [Including *Myrsilus* Thomas, 1909:470[1]; *Allosciurus* Conisbee, 1953:7.[2]]
          R.; Af.[3]
          [1] Not *Myrsilus* Stal, 1866, an insect.
          [2] Replacement for *Myrsilus* Thomas, 1909.
          [3] Tropical Africa.
Epixerus Thomas, 1909:472. African palm squirrels.
          R.; Af.[1]
          [1] W. Africa.

Tribe **Funambulini** Pocock, 1923:237. African tree squirrels.
          [= Funambulinae Pocock, 1923:237; Funambulini Simpson, 1945:79.]
          E.-L. Plioc., R.; Af. R.; As.[1]
          [1] S. Asia.

Subtribe **Funisciurina** Moore, 1959:198.
          E.-L. Plioc., R.; Af.
Paraxerus Forsyth Major, 1893:189. African bush squirrels, striped squirrels.
          [Including *Aethosciurus* Thomas, 1916:271; *Tamiscus* Thomas, 1918:33; *Montisciurus*
          Eisentraut, 1976:206.]
          E.-L. Plioc., R.; Af.

Funisciurus Trouessart, 1880:293. Common African tree squirrel.
R.; Af.

Subtribe **Funambulina** Pocock, 1923:237.
[= Funambulinae Pocock, 1923:237; Funambulina Moore, 1959:198.]
R.; As.[1]
[1] S. Asia.

Funambulus Lesson, 1832. Palm squirrel, Indian striped squirrel.
[Including *Tamiodes* Pocock, 1923:215; *Prasadsciurus* Moore & Tate, 1965:71.[1]]
R.; As.[2]
[1] Proposed as subgenus.
[2] S. Asia.

Subtribe **Myosciurina** Moore, 1959:198.
R.; Af.

Myosciurus Thomas, 1909:474. African pygmy squirrel, African dwarf squirrel.
R.; Af.[1]
[1] W. Africa.

Tribe **Tamiasciurini** Pocock, 1923:237.
[= Tamiasciurinae Pocock, 1923:237; Tamiasciurini Simpson, 1945:78.]
M. Pleist.-R.; N.A.

Tamiasciurus Trouessart, 1880:292. North American red squirrel, Douglas' squirrel, chickaree.
M. Pleist.-R.; N.A.

Subfamily **Pteromyinae** Brandt, 1855:157. Flying squirrels.
[= Pteromyini Brandt, 1855:157; Pteromidae Anderson, 1879:278; Petauristidae Miller, 1912:940; Pteromyinae G. S. Miller & Gidley, 1918:433; Petauristinae Simpson, 1945:80; Petauristini de Bruijn, van der Meulen & Katsikatsos, 1980:241, 255.]
[Including Eupetauridae Schaub, 1953:395; †Blackiini de Bruijn, van der Meulen & Katsikatsos, 1980:241, 253.]
Olig., ?M. Mioc.; Mediterranean. E. Olig., E. Mioc.-M. Pleist., R.; Eu. E.-L. Mioc., L. Plioc., M. Pleist.-R.; N.A. M.-L. Mioc., L. Plioc.-R.; As. R.; E. Indies.

†*Oligopetes* Heissig, 1979:154.
Olig.; Mediterranean. E. Olig.; Eu.

†*Petauristodon* Engesser, 1979:23.[1]
E.-L. Mioc.; N.A.
[1] Assignment to Pteromyinae is questionable.

†*Miopetaurista* Kretzoi, 1962:355.
[= †*Cryptopterus* Mein, 1970:22.]
E. Mioc.-E. Plioc.; Eu. M. Mioc.; As.[1] L. Plioc.; N.A.
[1] W. Asia (Anatolia).

†*Aliveria* de Bruijn, van der Meulen & Katsikatsos, 1980:244.
E. Mioc.; Eu.

†*Shuanggouia* Qiu & Liu, 1986:197.
M. Mioc.; As.

†*Blackia* Mein, 1970:44.
E. Mioc.-L. Plioc.; Eu. M. Mioc.; N.A.

†*Forsythia* Mein, 1970:33.
?M. Mioc.; Mediterranean. M. Mioc.; Eu.

Pteromys G. Cuvier, 1800:table 1. Old World flying squirrel.
[= *Sciuropterus* F. Cuvier, 1825:255.]
L. Plioc.-R.; As. R.; Eu.

†*Albanensia* Daxner-Höck & Mein, 1975:76.
M.-L. Mioc.; As. M.-L. Mioc.; Eu.

†*Meinia* Qiu, 1981:229.
>   M. Mioc.; As.[1]

[1] E. Asia.

Hylopetes Thomas, 1908:6.
>   [Including †*Pliopetes* Kretzoi, 1959:237-246.]
>   L. Mioc., R.; As.[1,2] E. Plioc.; Eu.[3] R.; E. Indies.[4]

[1] China in the late Miocene.
[2] S. Asia, Ceylon, Hainan Is. in the Recent.
[3] E. Europe (Macedonia).
[4] Borneo, Philippines (Palawan), Sipora Is. (west of Sumatra), Billiton (Belitong) Is.

Eoglaucomys Howell, 1915:109.
>   R.; As.[1]

[1] Western Himalayas.

Petinomys Thomas, 1908:6.
>   L. Mioc., M. Pleist., R.; As.[1,2] L. Mioc.-E. Plioc.; Eu.[3] R.; E. Indies.[4]

[1] China in the Miocene and Pleistocene.
[2] S. Asia in the Recent.
[3] E. Europe.
[4] Sumatra, Borneo, Philippines, Mentawi Is.

†*Pliopetaurista* Kretzoi, 1962:344-382.
>   L. Mioc.; As. L. Mioc.-E. Pleist.; Eu.

†*Pliosciuropterus* Sulimski, 1964:169.
>   L. Mioc.-E. Plioc.; Eu.

†*Parapetaurista* Qiu & Liu, 1986:195.
>   M. Mioc.; As.

Petaurista Link, 1795:52, 78. Taguan.
>   [= *Petauristus* Fischer de Waldheim, 1814:498-505.] [= or including *Galeolemur*
>   Lesson, 1840:261.[1]]
>   E. Pleist.-R.; As. R.; E. Indies.

[1] Synonymy questionable.

†*Petauria* Dehm, 1962:38.
>   M. Pleist.; Eu.

Glaucomys Thomas, 1908:5. American flying squirrel.
>   M. Pleist.-R.; N.A.

Aeretes G. M. Allen, 1940:745.[1]
>   E.-M. Pleist., R.; As.[2]

[1] *Nomen nudum* in 1938.
[2] China.

Eupetaurus Thomas, 1888:256-260. Woolly flying squirrel.
>   R.; As.[1]

[1] Kashmir, possibly Tibet and Yunnan.

Petaurillus Thomas, 1908:3.
>   R.; E. Indies.[1] R.; As.[2]

[1] Borneo.
[2] S.E. Asia (Malay Peninsula).

Belomys Thomas, 1908:2.
>   L. Plioc.-R.; As.[1,2]

[1] S. China in the late Pliocene and Pleistocene.
[2] S.E. Asia in the Recent.

Trogopterus Heude, 1898:46-47.
>   E.-M. Pleist., R.; As.[1]

[1] China.

Aeromys Robinson & Kloss, 1915:23.
>   R.; E. Indies.[1] R.; As.[2]

[1] Sumatra, Borneo.
[2] S.E. Asia.

Pteromyscus Thomas, 1908:3.
> R.; E. Indies.[1] R.; As.[2]
> [1] Sumatra, Borneo.
> [2] S.E. Asia.

Biswamoyopterus Saha, 1981:331.
> R.; As.[1]
> [1] S. Asia.

Olisthomys Carter, 1942:2.[1]
> R.; As.
> [1] Proposed as a subgenus of *Pteromys* (= *Petaurista*).

Iomys Thomas, 1908:1.
> R.; E. Indies.[1] R.; As.[2]
> [1] Sumatra, Borneo.
> [2] S.E. Asia.

Infraorder **CASTORIMORPHA** A. E. Wood, 1955:175, **new rank**.[1]
> [= PALMIPEDES Giebel, 1847[2]; PALMIPEDIA Van der Hoeven, 1858.[2]]
> [1] Proposed as suborder.
> [2] *Nomen oblitum.*

Family †**Eutypomyidae** Miller & Gidley, 1918:435.
> E. Eoc.-L. Olig., M. Mioc.; N.A. M. Mioc.; As. M. Mioc.; Eu.

†*Mattimys* Korth, 1984:61.
> E.-M. Eoc.; N.A.

†*Microeutypomys* Walton, 1993:262.
> M. Eoc.; N.A.

†*Janimus* Dawson, 1966:102.
> M. Eoc.; N.A.

†*Eutypomys* Matthew, 1905:21.
> M. Eoc.-L. Olig.; N.A.

†*Anchitheriomys* Roger, 1898:7-8.
> [Including †*Amblycastor* Matthew, 1918:197.]
> ?E. Olig., M. Mioc.; N.A. M. Mioc.; As. M. Mioc.; Eu.

Family **Castoridae** Hemprich, 1820:33.
> [= Castorina Hemprich, 1820:33; Castoridae Gray, 1821:302; Castorini Giebel,
> 1855:617; Castorida Haeckel, 1866:clx[1]; Castoroidea Gill, 1872:21.] [Including
> †Castoroididae Allen, 1877:419; †Chalicomyidae Miller & Gidley, 1918:435.]
> L. Eoc.-R.; As.[2] E. Olig.-R.; Eu. E. Olig.-R.; N.A.
> [1] Used as a family.
> [2] Castoridae indet. reported from the late Eocene of Kazakhstan.

†*Eocastoroides* Hibbard, 1937:244.
> E. Plioc.; N.A.

Subfamily †**Castoroidinae** Allen, 1877:419.
> [= †Castoroididae Allen, 1877:419; †Castoroidinae Trouessart, 1880:189.] [Including
> †Chloromyna Gervais, 1848:18[1]; †Chloromina Gervais, 1849:204[2]; †Trogontheriinae
> Lytschev, 1973; †Asiocastorinae Kalandadze & Rautian, 1992:95.]
> E. Olig.-M. Pleist.; Eu. L. Olig.-M. Pleist.; As. L. Olig., M. Mioc.-L. Pleist.; N.A.
> [1] 1848-1852.
> [2] †Chloromyna and †Chloromina long antedate †Castoroidinae, but the terms have not been used recently.
> Based on a synonym of †*Steneofiber*, one of these family-group names might well prove useful in the
> future, but the matter is presently of historical interest only. Possibly both †Chloromyna and †Chloromina
> should be suppressed.

†*Steneofiber* É. Geoffroy Saint-Hilaire, 1833:95.
> [Including †*Chloromys* Schlosser, 1884:39-40[1]; †*Asteneofiber* Kretzoi, 1974:418.]
> E. Olig.-M. Mioc.; Eu. E.-M. Mioc.; As.
> [1] Proposed as subgenus of †*Steneofiber* from von Meyer (MS). Not *Chloromys* Lesson, 1827: 300-301, a
> replacement name for *Cloromis* F. Cuvier, 1812, a synonym of *Dasyprocta*.

†*Neatocastor* Korth, 1996:167.
L. Olig.; N.A.

†*Asiacastor* Lytshev, in Lytshev & Aubekerova, 1971:18.
L. Olig.-L. Mioc.; As.

†*Youngofiber* Chow & Li, 1978:125.
E. Mioc.; As.

†*Trogontherium* Fischer de Waldheim, 1809:260-268.
[= †*Trongotherium* Pigeon, 1827:130[1]; †*Diabroticus* Pomel, 1848:167[2]; †*Diobroticus* Lydekker, 1891:458.[1]] [Including †*Conodontes* Laugel, 1862:715-717; †*Euroxenomys* Samson & Radulesco, 1973:441.]
E. Mioc.-M. Pleist.; Eu. M. Mioc., L. Plioc.-M. Pleist.; As. M. Mioc.; N.A.
[1] Misprint.
[2] Supposedly preoccupied by *Diabrotca* Chevrolat, 1834, a genus of Coleoptera.

†*Eucastor* Leidy, 1858:23.[1]
[Including †*Sigmogomphius* Merriam, 1896:363-370; †*Monosaulax* Stirton, 1935:416.]
M.-L. Mioc.; N.A. L. Mioc., Plioc.; As.
[1] Described as subgenus of *Castor*.

†*Schreuderia* Aldana Carrasco, 1992:102.
L. Mioc.; Eu.

†*Dipoides* Schlosser, 1902:21-23.[1]
[Including †*Castoromys* Pomel, 1854:23.[2]]
L. Mioc.-L. Plioc.; As. L. Mioc.-E. Plioc.; Eu. L. Mioc.-L. Plioc.; N.A.
[1] *Nomen nudum* in Jäger, 1835:17-18.
[2] A splendid candidate for suppression.

†*Boreofiber* Radulesco & Samson, 1972:104.[1]
E. Plioc.; Eu.[2]
[1] (Rădulescu).
[2] E. Europe.

†*Romanocastor* Radulesco & Samson, 1967:592.[1]
L. Plioc.; Eu.[2]
[1] (Rădulescu).
[2] E. Europe.

†*Zamolxifiber* Radulesco & Samson, 1967:591.[1]
L. Plioc.; Eu.[2]
[1] (Rădulescu).
[2] E. Europe.

†*Procastoroides* Barbour & Schultz, 1937:6.
E.-L. Plioc.; N.A.

†*Castoroides* Foster, 1838:80-83. Giant beaver.
L. Plioc.-L. Pleist.; N.A.

†*Paradipoides* Rinker & Hibbard, 1952:99.
M. Pleist.; N.A.

Subfamily **Castorinae** Hemprich, 1820:33.
[= Castorina Hemprich, 1820:33; Castorinae Gray, 1825:341.] [Including †Palaeocastorinae L. D. Martin, 1987:74.]
E. Olig.-R.; As. E. Olig.-R.; N.A. L. Olig.-R.; Eu.

†*Agnotocastor* Stirton, 1935:395.
E.-L. Olig.; As. E. Olig.; N.A.

Tribe †**Capacikalini** L. D. Martin, 1987:79.
L. Olig.-E. Mioc.; N.A. E. Mioc.; As.

†*Capacikala* Macdonald, 1963:197.
L. Olig.-E. Mioc.; N.A. E. Mioc.; As.

†*Pseudopalaeocastor* L. D. Martin, 1987:73-91.
 E. Mioc.; N.A.

Tribe **Castorini** Hemprich, 1820:33.
 [= Castorina Hemprich, 1820:33; Castorina Gray, 1825:341.[1]] [Including
 †Palaeocastorinae L. D. Martin, 1987:74; †Palaeocastorini L. D. Martin, 1987:74.]
 E. Olig.-R.; As. L. Olig.-R.; Eu. L. Olig.-R.; N.A.
 [1] As tribe.

Subtribe †**Euhapsina** L. D. Martin, 1987:80, **new rank**.
 [= †Euhapsini L. D. Martin, 1987:80.]
 E. Mioc.; N.A.

†*Fossorcastor* L. D. Martin, 1987:83.
 E. Mioc.; N.A.

†*Euhapsis* Peterson, 1905:179.
 E. Mioc.; N.A.

Subtribe **Castorina** Hemprich, 1820:33, **new rank**.[1]
 [= Castorina Gray, 1825:341.[2]] [Including †Palaeocastorinae L. D. Martin, 1987:74;
 †Palaeocastorini L. D. Martin, 1987:74.]
 E. Olig.-R.; As. L. Olig.-R.; Eu. L. Olig.-R.; N.A.
 [1] Hemprich's usage was at a higher rank in the family-group, but had the same suffix.
 [2] As tribe.

†*Propalaeocastor* Borisoglebskaya, 1967:130.
 E.-L. Olig.; As.

†*Palaeomys* Kaup, 1832:992.
 [= or including †*Chalicomys* Kaup, 1832:994[1]; †*Chelodus* Kaup, 1832:995;
 †*Chelodon* Gloger, 1841:105[2]; †*Chaelodus* Agassiz, 1842:7.[3]]
 L. Olig.-L. Mioc.; Eu. M.-L. Mioc.; As.[4]
 [1] "New genus" †*Chalicomys* Schlosser, 1924 was a *lapsus calami*.
 [2] Replacement for †*Chelodus* Kaup, 1832.
 [3] Misprint.
 [4] W. Asia (Anatolia).

†*Palaeocastor* Leidy, 1869:338.
 [Including †*Capatanka* Macdonald, 1963:194.]
 L. Olig.-E. Mioc.; N.A. E. Mioc.; As.

†*Hystricops* Leidy, 1858:22.[1]
 ?M. Mioc.; As. M.-L. Mioc.; N.A.
 [1] Described as subgenus of *Hystrix*.

Castor Linnaeus, 1758:58. Beavers.
 [= *Fiber* Dumeril, 1806:18, 19[1]; *Mamcastorus* Herrera, 1899:7.[2]] [Including
 †*Sinocastor* Young, 1934:57.]
 L. Mioc.-R.; As. L. Mioc.-R.; Eu. L. Mioc.-R.; N.A.
 [1] New name for *Castor* Linnaeus. Not *Fiber* G. Cuvier, 1800, a synonym of *Ondatra*.
 [2] See the International Code of Zoological Nomenclature, Article 1(b)(8), and Palmer (1904:25-26).

Family †**Rhizospalacidae** Thaler, 1966:176.
 L. Olig.; Eu.

†*Rhizospalax* Miller & Gidley, 1919:595.
 L. Olig.; Eu.

Suborder **MYOMORPHA** Brandt, 1855:152, 292.
 [= MYOMORPHI Brandt, 1855:152, 292.]

Family †**Protoptychidae** Schlosser, 1911:427.
 [= †Protoptychinae Schlosser, 1911:427; †Protoptychidae A. E. Wood, 1937:261.]
 M. Eoc.; N.A.

†*Protoptychus* Scott, 1895:269-286.
 M. Eoc.; N.A.

†*Presbymys* R. W. Wilson, 1949:7.[1]
M. Eoc.; N.A.
[1] Possibly referable to †Cylindrodontidae.

Infraorder **MYODONTA** Schaub, in Grassé & Dekeyser, 1955:1421, **new rank**.[1]
[1] Grassé & Dekeyser credited Schaub with this name, without giving a date. The term was proposed at subordinal rank for Dipodoidea and Muroidea, and was evidently taken from Schaub's MS, ultimately published as Schaub, in Piveteau, 1958:786.

Superfamily **Dipodoidea** Fischer de Waldheim, 1817:372.
[= Dipodes Fischer de Waldheim, 1817:372; Dipodum Fischer de Waldheim, 1817:407; Dipodoidea Waterhouse, 1842:203; Dipodidae Waterhouse, 1842:203; DIPODA Van der Hoeven, 1858; Dipodoidea Weber, 1904:490, 500; DIPODOMORPHA Roberts, 1951:345[1]; DIPODOMORPHA Thaler, 1966:11.[1]]
[1] Proposed as suborder.

Family †**Armintomyidae** Dawson, Krishtalka & Stucky, 1990:137.[1]
M. Eoc.; N.A.
[1] Assignment to the Dipodoidea is questionable.

†*Armintomys* Dawson, Krishtalka & Stucky, 1990:137.
M. Eoc.; N.A.

Family **Dipodidae** Fischer de Waldheim, 1817:372. Jumping mice, jerboas.
[= Dipodes Fischer de Waldheim, 1817:372; Dipodum Fischer de Waldheim, 1817:407; Dipsidae Gray, 1821:303; Dipodina Bonaparte, 1838:111; Dipodidae Waterhouse, 1842:203.] [Including Jaculini Brandt, 1855:230, 310; Jaculina Haeckel, 1866:clx[1]; Jaculidae Gill, 1872:20; Zapodidae Coues, 1875:253; Sminthidae Schulze, 1890:24; Sicistidae Weber, 1928:279[2]; Allactagidae Pavlinov & Rossolimo, 1987:151.]
E. Eoc., ?L. Eoc., E. Olig.-L. Mioc., L. Plioc.-R.; As. M. Eoc., L. Olig.-R.; N.A. L. Olig.-R.; Eu. M.-L. Mioc., L. Plioc.-M. Pleist., R.; Af.
[1] As a family.
[2] Objective synonym of Sminthidae.

†*Ulkenulastomys* Shevyreva, 1984:102.
E. Eoc.; As.

†*Blentosomys* Shevyreva, 1984:108.
E. Eoc.; As.

†*Aksyiromys* Shevyreva, 1984:99.
E. Eoc.; As.

†*Elymys* Emry & Korth, 1989:8.
M. Eoc.; N.A.

†*Simiacritomys* Kelly, 1992:113.
M. Eoc.; N.A.

Subfamily **Sicistinae** Allen, 1901:185.
[= Sminthi Brandt, 1855:164; Sminthinae Murray, 1866:360; Sminthidae Schulze, 1890:24; Sicistidae Weber, 1928:279.] [Including †Lophocricetinae Savinov, 1970:100.]
?L. Eoc., E. Olig.-L. Mioc., L. Plioc., Pleist., R.; As. L. Olig.-L. Mioc., L. Plioc.-R.; Eu. L. Olig.-L. Mioc., E. Pleist.; N.A.

†*Allosminthus* Wang, 1985:356.
E. Olig.; As.

†*Arabosminthus* Whybrow, Collinson, Daams, Gentry & McClure, 1982:111.
E. Mioc.; As.[1]
[1] S.W. Asia (Saudi Arabia).

Tribe †**Lophocricetini** Savinov, 1970:100, **new rank**.
[= †Lophocricetinae Savinov, 1970:100.]
M.-L. Mioc.; Eu. L. Mioc.; As.

†*Lophocricetus* Schlosser, 1924:41.
>[Including †*Bujoromys* Lungu, 1981; †*Sarmatosminthus* Lungu, 1981.]
>M.-L. Mioc.; Eu.[1] L. Mioc.; As.[2]
>
>[1] E. Europe.
>[2] Inner Mongolia and northern Kazakhstan.

Tribe **Sicistini** Allen, 1901:185, **new rank**.
>[= Sminthi Brandt, 1855:164; Sminthini Kalandadze & Rautian, 1992:89; Sicistinae Allen, 1901:185; Sicistidae Weber, 1928:279.]
>?L. Eoc., E. Olig.-L. Mioc., L. Plioc., Pleist., R.; As. L. Olig.-E. Mioc., L. Plioc.-R.; Eu. L. Olig.-L. Mioc., E. Pleist.; N.A.

†*Plesiosminthus* Viret, 1926:72.
>[Including †*Schaubeumys* A. E. Wood, 1935:1; †*Parasminthus* Bohlin, 1946:15; †*Sinosminthus* Wang, 1985:346; †*Heosminthus* Wang, 1985:352; †*Gobiosminthus* Huang, 1992:265; †*Shamosminthus* Huang, 1992:268.]
>?L. Eoc., E. Olig.-E. Mioc.; As. L. Olig.-E. Mioc.; Eu. L. Olig.-M. Mioc.; N.A.

†*Macrognathomys* Hall, 1930:305.
>L. Mioc.; N.A.

†*Miosicista* Korth, 1993:97.
>M. Mioc.; N.A.

†*Heterosminthus* Schaub, 1930:627.
>M.-L. Mioc.; As.[1]
>
>[1] China (Inner Mongolia).

Sicista Gray, in Griffith, Smith & Pidgeon, 1827:228. Birch mice.
>[= *Sminthus* Nordmann, in Demidoff, 1839:49.[1]]
>L. Mioc., L. Plioc., Pleist., R.; As. L. Plioc.-R.; Eu.
>
>[1] From Nathusius (MS).

†*Tyrannomys* R. A. Martin, 1989:107.
>E. Pleist.; N.A.

Subfamily **Zapodinae** Coues, 1875:253.
>[= Zapodidae Coues, 1875:253; Zapodinae Trouessart, 1880:168; Zapodini Kalandadze & Rautian, 1992:89.] [Including Eozapodini Kalandadze & Rautian, 1992:89.]
>M. Mioc.-R.; N.A. L. Mioc., R.; As. L. Mioc.-E. Plioc.; Eu.

†*Megasminthus* Klingener, 1966:3.
>M. Mioc.; N.A.

Eozapus Preble, 1899:37.[1] Chinese jumping mouse.
>[Including †*Protozapus* Bachmayer & R. W. Wilson, 1970:566.]
>L. Mioc., R.; As.[2] L. Mioc.-E. Plioc.; Eu.
>
>[1] Described as subgenus of *Zapus*.
>[2] E. Asia (China).

†*Pliozapus* R. W. Wilson, 1936:29.
>L. Mioc.; N.A.

†*Sminthozapus* Sulimski, 1962:503-512.
>E. Plioc.; Eu.[1]
>
>[1] E. Europe.

Zapus Coues, 1875:253. Jumping mice, meadow jumping mouse.
>E. Plioc.-R.; N.A.

†*Javazapus* R. A. Martin, 1989:105.
>E. Pleist.; N.A.

Napaeozapus Preble, 1899:33.[1] Woodland jumping mouse.
>M. Pleist.-R.; N.A.
>
>[1] Proposed as subgenus of *Zapus*; raised to generic rank by Miller, 1899.

Subfamily **Allactaginae** Vinogradov, 1925:578.
>[= Alactaginae Vinogradov, 1925:582; Allactaginae Vinogradov, 1930:337; Allactagidae Pavlinov & Rossolimo, 1987:151.]

M.-L. Mioc., R.; Af. M.-L. Mioc., L. Plioc.-R.; As. L. Plioc., ?E. and/or ?M. Pleist., L. Pleist.-R.; Eu.[1]

[1] E. Europe.

†*Protalactaga* Young, 1927:18.
M.-L. Mioc.; Af. M.-L. Mioc.; As.

Allactaga F. Cuvier, 1836:141. Four-toed jerboas, five-toed jerboas, earth hares.
[= *Alactaga* F. Cuvier, 1838:133; *Scirtetes* Wagner, 1841:413[1]; *Scirteta* Brandt, 1844:220-225, 230.[2]] [= or including *Beloprymnus* Gloger, 1841:106.] [Including *Scarturus* Gloger, 1841:106[3]; *Scirtomys* Brandt, 1844:220, 230[4]; *Paralactaga* Young, 1927:13[3]; *Mesoallactaga* Shenbrot, 1974:338; *Microallactaga* Shenbrot, 1974:338; *Orientallactaga* Shenbrot, 1984.[3]]
?M. Mioc., L. Mioc., L. Plioc.-R.; As. L. Plioc., Pleist., R.; Eu.[5] R.; Af.[6]

[1] New name for the "barbaric" *Allactaga*. Not *Scirtetes* Hartig, 1838, a genus of Hymenoptera.
[2] Proposed as subgenus of *Scirtetes* Wagner.
[3] Sometimes as subgenus.
[4] Proposed as subgenus of *Scirtetes*. Objective synonym of *Scarturus*.
[5] E. Europe.
[6] N. Africa (Egypt, Libya).

†*Himalayactaga* Li & Chi, 1981:246-255.
L. Mioc.; As.[1]

[1] Tibet.

†*Brachyscirtetes* Schaub, 1934:5.
L. Mioc.; As.

†*Proalactaga* Savinov, 1970:107.
L. Mioc.; As.

Pygeretmus Gloger, 1841:106. Flat-tailed jerboas.
[= *Platycercomys* Brandt, 1844:225-228, 230; *Pygerethmus* Vinogradov, 1930:337.] [Including *Alactagulus* Nehring, 1897:154[1]; †*Pliopygerethmus* Topachevsky & Skorik, 1971:113.[2]]
L. Plioc., L. Pleist.-R.; Eu.[3] R.; As.

[1] Sometimes as subgenus.
[2] In the literature the name "Topachevsky" is transliterated into English in at least four ways from the original Russian. We prefer this version.
[3] E. Europe (Ukraine and N. Crimea).

Allactodipus Kolesnikov, 1937:255.
R.; As.[1]

[1] Uzbekistan, Turkmenistan.

Subfamily **Dipodinae** Fischer de Waldheim, 1817:372.
[= Dipodes Fischer de Waldheim, 1817:372; Dipodum Fischer de Waldheim, 1817:407; Dipsidae Gray, 1821:303; Dipina Gray, 1825:342; Dipodina Bonaparte, 1838:111; Dipodidae Waterhouse, 1842:203; Dipodini Brandt, 1855:219; Dipodinae Murray, 1866:361.] [Including Jaculini Brandt, 1855:230, 310; Jaculina Haeckel, 1866:clx[1]; Jaculinae Alston, 1876:89; Cardiocraniinae Vinogradov, 1925:578; Salpingotinae Vinogradov, 1925:578.]
L. Mioc., L. Plioc., Pleist., R.; As. L. Plioc.-M. Pleist., R.; Af. L. Plioc., L. Pleist.-R.; Eu.[2]

[1] As a family.
[2] E. Europe.

Tribe **Dipodini** Fischer de Waldheim, 1817:372.
[= Dipodinae Fischer de Waldheim, 1817:372; Dipodini Brandt, 1855:219.]
L. Mioc., L. Plioc., Pleist., R.; As. L. Plioc.-M. Pleist., R.; Af. L. Plioc., L. Pleist.-R.; Eu.[1]

[1] E. Europe.

†*Scirtodipus* Savinov, 1970:114.
L. Mioc.; As.[1]

[1] Kazakhstan.

†*Sminthoides* Schlosser, 1924:34.
>> L. Mioc., L. Plioc., Pleist.; As.

Jaculus Erxleben, 1777:404. Desert jerboas, jarboua, gaurti, siebod.
>> [Including *Scirtopoda* Brandt, 1844:212-217, 230[1]; *Haltomys* Brandt, 1844:215-217.[2]]
>> L. Plioc.-M. Pleist., R.; Af. R.; As.

[1] Proposed as subgenus of *Dipus*.
[2] Proposed as a section of subgenus *Scirtopoda*.

Eremodipus Vinogradov, 1930:334.
>> R.; As.[1]

[1] Kazakhstan, Turkmenistan, Uzbekistan.

†*Plioscirtopoda* Gromov & Schevtchenko, 1969.
>> L. Plioc.; Eu.[1]

[1] E. Europe (Ukraine).

Dipus Zimmermann, 1780:358. True jerboa, feather-footed jerboa.
>> [= *Dipsus* Gray, 1821:303; *Dipodipus* Trouessart, 1910:207.]
>> Pleist., R.; As. R.; Eu.[1]

[1] E. Europe.

Stylodipus G. M. Allen, 1925:4. Feather-tailed three-toed jerboa, emuranchik.
>> [Including *Halticus* Brandt, 1844:213.[1]]
>> L. Pleist.-R.; Eu.[2] R.; As.

[1] Proposed as a section of *Scirtopoda*. Not *Halticus* Hahn, 1831, a genus of Hemiptera.
[2] E. Europe (N. Crimea, E. Ukraine, N. Caucasus).

Tribe **Cardiocraniini** Vinogradov, 1925:578. Dwarf jerboas, pygmy jerboas.
>> [= Cardiocraniinae Vinogradov, 1925:578; Cardiocraniini Pavlinov, 1980:47.]
>> [Including Salpingotinae Vinogradov, 1925:578; Salpingotini Pavlinov, 1980:49.]
>> R.; As.

Cardiocranius Satunin, 1903:582. Five-toed dwarf jerboas.
>> R.; As.[1]

[1] China, Mongolia, Kazakhstan.

Salpingotus Vinogradov, 1922:540. Three-toed dwarf jerboas.
>> [Including *Salpingotulus* Pavlinov, 1980:50[1]; *Prosalpingotus* Vorontsov & Shenbrot, 1984:739[2]; *Anguistodontus* Vorontsov & Shenbrot, 1984:741.[2]]
>> R.; As.[3]

[1] Sometimes as subgenus.
[2] As subgenus.
[3] China, Mongolia, Kazakhstan, Uzbekistan, Afghanistan, Baluchistan.

Subfamily **Paradipodinae** Pavlinov & Shenbrot, 1983:84.
>> [= Paradipodini Pavlinov & Shenbrot, 1983:84; Paradipodinae Holden, in Wilson & Reeder, eds., 1993:495.]
>> R.; As.[1]

[1] Central Asia.

Paradipus Vinogradov, 1930:333. Comb-toed jerboa.
>> R.; As.[1]

[1] Turkmenistan, Uzbekistan, Kazakhstan.

Subfamily **Euchoreutinae** Lyon, 1901:666.
>> R.; As.

Euchoreutes Sclater, 1891:610. Long-eared jerboa.
>> R.; As.[1]

[1] Mongolia, China.

Superfamily **Muroidea** Illiger, 1811:84.
>> [= Murina Illiger, 1811:84; Myoidea Gill, 1872:20; Muroidae Miller & Gidley, 1918:435.] [Including Lophiomyoidea Gill, 1872:303; Cricetinorum Fischer de Waldheim, 1817:410; Cricetoidea Thaler, 1966:137; CRICETOMORPHA Thaler, 1966:137.[1]]

[1] Proposed as suborder.

Family †**Simimyidae** A. E. Wood, 1980:55.
    [Including †Nonomyinae L. D. Martin, 1980:6.]
    M.-L. Eoc.; N.A.

  †*Simimys* R. W. Wilson, 1935:180.
    [= †*Eumysops* R. W. Wilson, 1935:26.[1]]
    M.-L. Eoc.; N.A.
    [1] Not †*Eumysops* Ameghino, 1888, a heteropsomyine rodent.

  †*Nonomys* Emry & Dawson, 1973:1003.
    [= †*Nanomys* Emry & Dawson, 1972:2.[1]] [Including †*Subsumus* A. E. Wood, 1974:101.]
    L. Eoc.; N.A.
    [1] Not †*Nanomys* Marsh, 1889, a synonym of the multituberculate genus †*Cimolodon*.

Family **Muridae** Illiger, 1811:84. Rats, mice.
    [= Murina Illiger, 1811:84[1]; Murini Fischer de Waldhein, 1817:372; Murinorum Fischer de Waldhein, 1817:410[2]; Muridae Gray, 1821:303; Musideae Lesson, 1842:134; Murini Giebel, 1855:531[2]; Murina Haeckel, 1866:clx.[2]] [Including Cricetini Fischer de Waldhein, 1817:372; Cricetinorum Fischer de Waldhein, 1817:410[2]; Cricetidae Rochebrune, 1883:66, 153; Arvicolidae Gray, 1821:303; Spalacidae Gray, 1821:303; Aspalacidae Gray, 1825:342[3]; Rattidae Burnett, 1830:350; Gerbillidae De Kay, 1842:xv, 70; Hydromysideae Lesson, 1842:125; Sigmodontes Wagner, 1843; Merionidae Burmeister, 1850:16; Lophiomyidae Gill, 1872:20; Hesperomyidae Ameghino, 1889:109; Microtidae Cope, 1891:90; Coniluridae Dahl, 1897; Nesomyidae Weber, 1904; Platacanthomyidae Miller & Gidley, 1918:437; Rhizomyidae Miller & Gidley, 1918:437; †Cricetopidae Matthew & Granger, 1923:1; †Cricetodontidae Schaub, 1925:3, 5; †Anomalomyidae Schaub, 1925:97; †Melissiodontidae Schaub, 1925:97; Myospalacidae Kretzoi, 1961; †Microtoscoptidae Kretzoi, 1969; †Trilophomyidae Kretzoi, 1969; †Tachyoryctoididae Fejfar, 1972:168, 190; Cricetomyidae Chaline, Mein & Petter, 1977:245-252; Dendromuridae Chaline, Mein & Petter, 1977:245-252; †Eucricetodontidae Ünay-Bayraktar, 1989:56; †Pseudocricetodontidae Ünay-Bayraktar, 1989:38.]
    M. Eoc.-R.; As.[4] L. Eoc.-R.; N.A.[5] ?E. Olig., L. Olig., ?M. Mioc., L. Mioc.-E. Plioc., ?L. Plioc., E. Pleist.-R.; Mediterranean. E. Olig.-R.; Eu. E. Mioc.-R.; Af. Plioc., E. Pleist.-R.; Aus. L. Plioc.-R.; S.A. L. Plioc., Pleist., R.; New Guinea. ?Pleist., R.; Madagascar. ?E. and/or ?M. Pleist., L. Pleist.-R.; Atlantic.[6] Pleist., R.; W. Indies. M. Pleist.-R.; E. Indies. L. Pleist.-R.; Pacific. R.; Indian O.[7] R.; Cent. A.[8]
    [1] Proposed as a family.
    [2] Used as a family.
    [3] Objective synonym of Spalacidae.
    [4] Including Arctic islands of Siberia.
    [5] Including Arctic islands of Canada and Greenland.
    [6] Canary Islands in the Pleistocene and Recent (now extinct). Iceland in the Recent (extant).
    [7] Christmas Island (endemics extinct; reintroduced).
    [8] Also Recent worldwide by introduction.

  †*Pappocricetodon* Tong, 1992:1.[1]
    M.-L. Eoc.; As.
    [1] (Tung).

  †*Selenomys* Matthew & Granger, 1923:5.
    E. Olig.; As.

  †*Potwarmus* Lindsay, 1988:135.
    M. Mioc.; As.

  †*Leakeymys* Lavocat, 1964.
    M. Mioc.; Af.

  †*Blancomys* Weerd, Adrover, Mein & Soria, 1977.
    L. Mioc.-E. Plioc.; Eu.

  †*Epimeriones* Daxner-Hock, 1972:143, 145.
    L. Mioc.-E. Pleist.; Eu.

Subfamily †**Cricetopinae** Matthew & Granger, 1923:1.
    [= †Cricetopidae Matthew & Granger, 1923:1; †Cricetopini Simpson, 1945:85;
    †Cricetopsinae de Bruijn & W. von Koenigswald, 1994:383.]
    Olig., E. Mioc.; As.

†*Cricetops* Matthew & Granger, 1923:1.
    Olig.; As.[1]
    [1] Mongolia, Kazakhstan.

†*Enginia* de Bruijn & von Koenigswald, 1994:383.
    E. Mioc.; As.[1]
    [1] W. Asia (Turkey).

Subfamily †**Eumyinae** Simpson, 1945:83.
    [= †Eumyini Simpson, 1945:83; †Eumyinae Stehlin & Schaub, 1951:167, 322;
    †Eumyini Chaline, Mein & Petter, 1977.]
    L. Eoc.-E. Olig.; N.A.

†*Eumys* Leidy, 1856:90.
    L. Eoc.-E. Olig.; N.A.

†*Eoeumys* L. D. Martin, 1980:16.
    L. Eoc.-E. Olig.; N.A.

†*Coloradoeumys* L. D. Martin, 1980:13.
    E. Olig.; N.A.

†*Wilsoneumys* L. D. Martin, 1980:14.
    E. Olig.; N.A.

Subfamily †**Paracricetodontinae** Mein & Freudenthal, 1971:4.
    [Including †Eucricetodontinae Mein & Freudenthal, 1971:4; †Pseudocricetodontinae
    Engesser, 1987; †Adelomyarioninae Ünay-Bayraktar, 1989:48.]
    L. Eoc.-E. Mioc.; As. ?E. Olig., L. Olig.; Mediterranean. E. Olig.-L. Mioc.; Eu. E.
    Olig.-E. Mioc.; N.A.

Tribe †**Eucricetodontini** Mein & Freudenthal, 1971:4, **new rank**.
    [= †Eucricetodontinae Mein & Freudenthal, 1971:4; †Eucricetodontidae Ünay-Bayraktar,
    1989:56.] [Including †Geringini L. D. Martin, 1980:6; †Leidymini L. D. Martin,
    1980:6; †Eumyarioninae Ünay, 1989.[1]]
    L. Eoc.-E. Mioc.; As. Olig.; Mediterranean. E. Olig.-L. Mioc.; Eu. E. Olig.-E.
    Mioc.; N.A.
    [1] (Ünay-Bayraktar).

†*Eucricetodon* Thaler, 1966:140.[1]
    [Including †*Pomelimys* Kretzoi, 1981:483.[2]]
    L. Eoc.-E. Mioc.; As. Olig.; Mediterranean. E. Olig.-E. Mioc.; Eu.
    [1] Described as subgenus of †*Cricetodon*.
    [2] Described as subgenus of †*Eucricetodon*.

†*Atavocricetodon* Freudenthal, 1996:3.
    E. Olig.; Eu.

†*Muhsinia* de Bruijn, Ünay, van den Hoek & Saraç, 1992:664.
    E. Mioc.; As.[1]
    [1] W. Asia (Anatolia).

†*Eumysodon* Argyropulo, 1939:112.
    L. Olig.; As.

†*Eumyarion* Thaler, 1966:148.[1]
    E. Mioc.; As.[2] E.-L. Mioc.; Eu.
    [1] Proposed as subgenus of †*Cricetodon*.
    [2] W. Asia (Turkey).

†*Mirabella* de Bruijn, Ünay, Saraç & Hofmeijer, 1987:120.
    E. Mioc.; As.[1] E. Mioc.; Eu.[2]
    [1] W. Asia (Turkey).
    [2] E. Europe (Greece).

†*Paciculus* Cope, 1879:2.
  E. Olig.-E. Mioc.; N.A.

†*Leidymys* A. E. Wood, 1936:3.
  [Including †*Cotimus* Black, 1961:73.]
  E. Olig.; As. E. Olig.-E. Mioc.; N.A.

†*Scottimus* A. E. Wood, 1937:255.
  E.-L. Olig.; N.A.

†*Geringia* L. D. Martin, 1980:25.
  L. Olig.; N.A.

†*Yatkolamys* L. D. Martin & Corner, 1980:2.
  E. Mioc.; N.A.

Tribe †**Pseudocricetodontini** Engesser, 1987.
  [= †Pseudocricetodontinae Engesser, 1987; †Pseudocricetodontidae Ünay-Bayraktar,
  1989:38; †Pseudocricetodontini Freudenthal, Lacomba & Sacristan, 1992:51.]
  E. Olig.-E. Mioc.; Eu. L. Olig.; Mediterranean.

Subtribe †**Pseudocricetodontina** Engesser, 1987, **new rank**.
  [= †Pseudocricetodontinae Engesser, 1987.]
  E. Olig.-E. Mioc.; Eu. L. Olig.; Mediterranean.

†*Pseudocricetodon* Thaler, 1969:191-207.
  E. Olig.-E. Mioc.; Eu. L. Olig.; Mediterranean.[1]
  [1] Mallorca, Balearic Islands.

†*Allocricetodon* Freudenthal, 1994:12.
  L. Olig.; Eu.

†*Lignitella* Ünay-Bayraktar, 1989:47.[1]
  Olig.; Eu.[2]
  [1] Described as subgenus of †*Pseudocricetodon*.
  [2] E. Europe (Turkish Thrace).

†*Kerosinia* Ünay-Bayraktar, 1989:49.
  Olig.; Eu.[1]
  [1] E. Europe (Turkish Thrace).

Subtribe †**Heterocricetodontina** Ünay-Bayraktar, 1989:60, **new rank**.
  [= †Heterocricetodontidae Ünay-Bayraktar, 1989:60; †Heterocrocetodontini Freudenthal,
  Lacomba & Sacristán, 1992:51.]
  E.-L. Olig.; Eu.

†*Heterocricetodon* Schaub, 1925:63, 96.
  E.-L. Olig.; Eu.

†*Cincamyarion* Agustí & Arbiol, 1989:266.
  L. Olig.; Eu.

Tribe †**Paracricetodontini** Mein & Freudenthal, 1971:4, **new rank**.
  [= †Paracricetodontinae Mein & Freudenthal, 1971:4.]
  E.-L. Olig.; Eu. L. Olig.; Mediterranean.[1]
  [1] Mallorca, Balearic Islands.

†*Paracricetodon* Schaub, 1925:96.
  Olig.; Mediterranean.[1] E.-L. Olig.; Eu.
  [1] Mallorca, Balearic Islands.

†*Trakymys* Ünay-Bayraktar, 1989:29.
  Olig.; Eu.[1]
  [1] E. Europe (Turkish Thrace).

†*Edirnella* Ünay-Bayraktar, 1989:34.
  Olig.; Eu.[1]
  [1] E. Europe (Turkish Thrace).

Tribe †**Adelomyarionini** Ünay-Bayraktar, 1989:48, **new rank**.
    [= †Adelomyarioninae Ünay-Bayraktar, 1989:48.]
    L. Olig.; Eu.

†*Adelomyarion* Hugueney, 1969:56.
    L. Olig.; Eu.

Subfamily †**Melissiodontinae** Schaub, 1925:97.
    [= †Melissiodontidae Schaub, 1925:97; †Melissiodontinae Stehlin & Schaub,
    1951:182, 340.]
    E. Olig.-E. Mioc.; Eu.

†*Melissiodon* Schaub, 1920:43.
    E. Olig.-E. Mioc.; Eu.

Subfamily †**Tachyoryctoidinae** Schaub, 1958:807.
    [= †Tachyoryctoididae Fejfar, 1972:168, 190; †Tachyoryctoidini Fejfar, 1972:190.]
    [Including †Argyromyini Fejfar, 1972:190.]
    L. Olig.-E. Mioc.; As.

†*Tachyoryctoides* Bohlin, 1937:43.
    [Including †*Aralomys* Argyropulo, 1939:113.]
    L. Olig.-E. Mioc.; As.[1]
    [1] China and Mongolia.

†*Argyromys* Schaub, 1958:807.
    L. Olig.; As.[1]
    [1] Kazakhstan.

Subfamily †**Microtoscoptinae** Kretzoi, 1955:355.
    [= †Microtoscoptini Repenning, 1968:60; †Microtoscoptidae Kretzoi, 1969:162.]
    L. Mioc.; As. L. Mioc.; Eu. L. Mioc.; N.A.

†*Paramicrotoscoptes* L. D. Martin, 1975:102.
    L. Mioc.; N.A.

†*Microtoscoptes* Schaub, 1934:38.
    L. Mioc.; As. L. Mioc.; Eu.[1]
    [1] E. Europe.

†*Goniodontomys* R. W. Wilson, 1937:9.
    L. Mioc.; N.A.

Subfamily †**Baranomyinae** Kretzoi, 1955:352, 355.
    [= †Baranomyini Repenning, 1968:55.]
    L. Mioc.; As.[1] E. Plioc.-E. Pleist.; Eu.
    [1] Inner Mongolia.

†*Microtodon* Miller, 1927:18.
    L. Mioc.; As.[1] L. Plioc.; Eu.
    [1] Inner Mongolia.

†*Anatolomys* Schaub, 1934:37.
    L. Mioc.; As.[1]
    [1] Inner Mongolia.

†*Baranomys* Kormos, 1933:45-54.
    [Including †*Warthamys* Kretzoi, 1969:164.]
    E. Plioc.-E. Pleist.; Eu.

Subfamily †**Trilophomyinae** Kretzoi, 1969:162, **new rank**.
    [= †Trilophomyidae Kretzoi, 1969:162.]
    E.-L. Plioc., Pleist.; Eu.

†*Trilophomys* Depéret, 1892:121-122.
    [= †*Lophiomys* Depéret, 1890:53-54.[1]]
    E.-L. Plioc., Pleist.; Eu.
    [1] Not *Lophiomys* Milne-Edwards, 1867, another muroid rodent.

Subfamily †**Gobicricetodontinae** Qiu, 1996:94.
M.-L. Mioc.; As.[1]
[1] E. Asia.

†*Gobicricetodon* Qiu, 1996:94.
M.-L. Mioc.; As.[1]
[1] Nei Mongol.

†*Plesiodipus* Young, 1927:20.
[= †*Plesiocricetodon* Schaub, 1934:24.[1]]
M.-L. Mioc.; As.[2]
[1] "Proposed because †*Plesiodipus* is not a dipodid, but such changes are not valid" (Simpson, 1945: 85).
[2] China.

Subfamily †**Cricetodontinae** Schaub, 1925:3, 5.
[= †Cricetodontidae Schaub, 1925:3, 5; †Cricetodontinae Stehlin & Schaub, 1951:159, 317.] [Including †Megacricetodontinae Jacobs & Lindsay, 1984:267; †Copemyinae Jacobs & Lindsay, 1984:267.]
E.-L. Mioc.; As. E. Mioc.-E. Plioc.; Eu. M.-L. Mioc., ?Plioc.; Af. M. and/or L. Mioc.; Mediterranean.[1] M.-L. Mioc.; N.A.
[1] Crete.

†*Tsaganocricetus* Topachevsky & Skorik, 1988.
E. Mioc.; As.

†*Aralocricetodon* Bendukidze, 1993:53.
E. Mioc.; As.

Tribe †**Megacricetodontini** Mein & Freudenthal, 1971:22.
[= †Megacricetodontinae Jacobs & Lindsay, 1984:267.]
E.-L. Mioc.; Eu. M. Mioc.; Af.[1] M. Mioc.; As.
[1] N. Africa.

†*Megacricetodon* Fahlbusch, 1964:49.[1]
[Including †*Mesocricetodon* Daxner, 1967:27-36; †*Collongomys* Mein & Freudenthal, 1971:24.[2]]
E.-L. Mioc.; Eu. M. Mioc.; As.[3]
[1] Proposed as subgenus of †*Democricetodon*.
[2] Proposed as subgenus of †*Megacricetodon*.
[3] Pakistan, China.

†*Punjabemys* Lindsay, 1988:112.
M. Mioc.; As.[1]
[1] S. Asia (Pakistan).

†*Mellalomys* Jaeger, 1977:91-125.
M. Mioc.; Af.[1]
[1] N. Africa.

Tribe †**Copemyini** Jacobs & Lindsay, 1984:267.
[= †Copemyinae Jacobs & Lindsay, 1984:267.] [Including †Democricetodontini Lindsay, 1987:483.[1]]
E.-M. Mioc.; As. E.-M. Mioc.; Eu. M. Mioc.; Af. M. and/or L. Mioc.; Mediterranean.[2] M.-L. Mioc.; N.A.
[1] Either a casual reference or previously described in a paper not seen by us.
[2] Crete.

†*Shamalina* Whybrow, Collinson, Daams, Gentry & McClure, 1982:114.
E. Mioc.; As.[1]
[1] S.W. Asia (Saudi Arabia).

†*Primus* de Bruijn, Hussain & Leinders, 1981:77.
E. and/or M. Mioc.; As.[1]
[1] S. Asia (Pakistan).

†*Democricetodon* Fahlbusch, 1964:19.
[Including †*Spanocricetodon* Li, 1977:67.]

E.-M. Mioc.; Eu. M. Mioc.; Af. M. Mioc.; As. M. and/or L. Mioc.;
Mediterranean.[1]

[1] Crete.

†*Copemys* A. E. Wood, 1936:5.
[Including †*Miochomys* Hoffmeister, 1959:697; †*Tregomys* R. L. Wilson, 1968:119;
†*Gnomomys* R. L. Wilson, 1968:120.]
M.-L. Mioc.; N.A.

†*Poamys* Matthew, 1924:86.
L. Mioc.; N.A.

†*Abelmoschomys* Baskin, 1986:287.
L. Mioc.; N.A.

Tribe †**Cricetodontini** Schaub, 1925:3, 5.
[= †Cricetodontidae Schaub, 1925:3, 5; †Cricetodontini Simpson, 1945:85;
†Cricetodontinae Stehlin & Schaub, 1951:159, 317.] [Including †Fahlbuschiini Mein
& Freudenthal, 1971:25.]
E.-L. Mioc.; As. E. Mioc.-E. Plioc.; Eu. L. Mioc., ?Plioc.; Af.[1] L. Mioc.; N.A.
[1] N. Africa.

†*Lartetomys* Mein & Freudenthal, 1971:31.
E.-M. Mioc.; Eu.

†*Fahlbuschia* Mein & Freudenthal, 1971:25.
E.-M. Mioc.; Eu.

†*Renzimys* Lacomba Andueza, 1983:1-95.
M. Mioc.; Eu.

†*Pseudofahlbuschia* Freudenthal & Daams, 1988:203.
M. Mioc.; Eu.

†*Cricetodon* Lartet, 1851:20-21.
[= †*Palaeocricetus* Argyropulo, 1938:223.[1]] [Including †*Turkomys* Tobien, 1978:210;
†*Pararuscinomys* Agustí, 1981:54.[2]]
E.-M. Mioc.; As.[3] E.-L. Mioc.; Eu.
[1] Proposed as subgenus, but the type species is synonymous with the type species of the genus.
[2] Decribed as subgenus.
[3] W. Asia.

†*Deperetomys* Mein & Freudenthal, 1971:21.[1]
E.-M. Mioc.; As.[2] M. Mioc.; Eu.
[1] Described as a subgenus of †*Cricetodon*.
[2] W. Asia (Turkey).

†*Meteamys* de Bruijn, Ünay, van den Hoek Ostende & Saraç, 1992:656.
E. Mioc.; As.[1]
[1] Earliest early Miocene of W. Asia (Anatolia).

†*Byzantinia* de Bruijn, 1976:363.
M.-L. Mioc.; As.[1] L. Mioc.; Eu.[2]
[1] W. Asia (Turkey, Samos, Rhodes).
[2] E. Europe (Greece).

†*Hispanomys* Mein & Freudenthal, 1971:19.[1]
M.-L. Mioc.; Eu.
[1] Described as a subgenus of †*Cricetodon*.

†*Zramys* Jaeger & Michaux, 1973:2477, 2480.
L. Mioc.; Af.[1]
[1] N. Africa.

†*Ruscinomys* Depéret, 1890:60-61.
[Including †*Pseudoruscinomys* Mein & Freudenthal, 1971:21.[1]]
M. Mioc.-E. Plioc.; Eu. ?Plioc.; Af.[2]
[1] Described as a subgenus of †*Cricetodon*.
[2] N. Africa.

†*Pliotomodon* Hoffmeister, 1945:186.
  L. Mioc.; N.A.

†*Neocricetodon* Schaub, 1934:26.[1]
  [= †*Epicricetodon* Kretzoi, 1941:349.[2]]
  L. Mioc.; As.
  [1] Not †*Neocricetodon* Kretzoi, in Kadic & Kretzoi, 1930:49, a *nomen nudum*; not †*Neocricetodon* Kretzoi 1954:62, a genus of cricetid rodents.
  [2] New name for †*Neocricetodon* Schaub, 1934.

Subfamily **Sigmodontinae** Wagner, 1843. New World rats, New World mice.
  [= Sigmodontes Wagner, 1843; Sigmodontinae Thomas, 1897:1019.] [Including Hesperomyinae Murray, 1866:358; Hesperomyidae Ameghino, 1889:109; Neotominae Merriam, 1894:228; Tylomyinae Reig,1984:338.]
  L. Mioc.-R.; N.A. L. Plioc.-R.; S.A. Pleist., R.; W. Indies.[1] L. Pleist.-R.; Pacific.[2]
  R.; Cent. A.
  [1] Now extinct.
  [2] E. Pacific (Galapagos).

†*Jacobsomys* Czaplewski, 1987:191.[1]
  E. Plioc.; N.A.
  [1] Tentatively assigned to the Sigmodontinae.

Baiomys True, 1894:758.
  E. Plioc.-R.; N.A. R.; Cent. A.

Scotinomys Thomas, 1913:408. Brown mice.
  R.; Cent. A.[1]
  [1] W. Panama north to S. Mexico (E. Oaxaca).

Rhagomys Thomas, 1917:192.
  R.; S.A.[1]
  [1] Brazil.

Tribe **Tylomyini** Reig, 1984:338, **new rank**.[1]
  [= Tylomyinae Reig, 1984:338.]
  R.; N.A.[2] R.; Cent. A. R.; S.A.[3]
  [1] Musser.
  [2] Mexico.
  [3] N.W. South America.

Otonyctomys Anthony, 1932:1.
  R.; Cent. A.[1]
  [1] S. Mexico (Yucatan), Belize, Guatemala.

Nyctomys de Saussure, 1860:106.[1] Vesper rat.
  R.; N.A.[2] R.; Cent. A.
  [1] Proposed as subgenus of *Hesperomys* Murray, 1866, not *Hesperomys* Waterhouse, 1839; raised to generic rank by Bangs, 1902.
  [2] S. North America (Mexico north to Jalisco).

Tylomys Peters, 1866:404. Climbing rats.
  R.; N.A.[1] R.; Cent. A. R.; S.A.[2]
  [1] Mexico.
  [2] Colombia and Ecuador.

Ototylomys Merriam, 1901:561.
  R.; N.A.[1] R.; Cent. A.[2]
  [1] Isolated record from Guerro, Mexico.
  [2] S. Mexico (Yucatan and Chiapas) south to Costa Rica.

Tribe **Neotomini** Merriam, 1894:228, **new rank**.[1]
  [= Neotominae Merriam, 1894:228.]
  L. Mioc.-R.; N.A. R.; Cent. A.
  [1] Musser.

†*Repomys* May, 1981:220.
  L. Mioc.-L. Plioc.; N.A.

†*Galushamys* Jacobs, 1977:516.
    L. Mioc.; N.A.

Neotoma Say & Ord, 1825:345. Wood rats, pack rats.
    [= *Nectoma* Agassiz, 1842:22.[1]] [Including *Teonoma* Gray, 1843:117[2]; *Teanopus*
    Merriam, 1903:81[2]; *Homodontomys* Goldman, 1910:86; †*Parahodomys* Gidley &
    Gazin, 1933:356; †*Paraneotoma* Hibbard, 1967:125.[3]]
    L. Mioc.-R.; N.A. R.; Cent. A.[4]
    [1] Misprint.
    [2] Subgenus.
    [3] As subgenus.
    [4] Central America south to Nicaragua.

Xenomys Merriam, 1892:160.
    R.; N.A.[1]
    [1] Mexico.

Hodomys Merriam, 1894:232.
    R.; N.A.[1]
    [1] Mexico.

Nelsonia Merriam, 1897:277.
    R.; N.A.[1]
    [1] Mexico.

Tribe **Peromyscini** Hershkovitz, 1966:734.
    [= or including Hesperomyinae Murray, 1866:358; Hesperomyini Simpson, 1945:83.]
    L. Mioc.-R.; N.A. R.; Cent. A. R.; S.A.[1]
    [1] N.W. South America.

†*Bensonomys* Gazin, 1942:489.
    L. Mioc.-E. Plioc.; N.A.

†*Paronychomys* Jacobs, 1977:514.
    L. Mioc.; N.A.

Peromyscus Gloger, 1841:95. White-footed mice, deer mice.
    [= *Sitomys* Fitzinger, 1867:97; *Vesperimus* Coues, 1874:178[1]; *Trinodontomys*
    Rhoads, 1894:256-257.[2]] [= or including *Hesperomys* Murray, 1866:358.[3]] [Including
    Haplomylomys Osgood, 1904:53.[4]]
    L. Mioc.-R.; N.A. R.; Cent. A.
    [1] As subgenus.
    [2] As subgenus of *Sitomys*.
    [3] Not *Hesperomys* Waterhouse, 1839, a synonym of the phyllotine genus *Calomys* Waterhouse, 1837.
    [4] Subgenus as described.

†*Symmetrodontomys* Hibbard, 1941:354.
    E. Plioc.; N.A.

Reithrodontomys Giglioli, 1873:160. Harvest mice.
    [Including *Aporodon* Howell, 1914:63[1]; †*Cudahyomys* Hibbard, 1944:725.]
    E. Plioc.-R.; N.A. R.; Cent. A. R.; S.A.[2]
    [1] Subgenus.
    [2] N.W. South America.

Onychomys Baird, 1857:458.[1] Grasshopper mice.
    E. Plioc.-R.; N.A.
    [1] Described as subgenus of *Hesperomys* Waterhouse, 1839, a synonym of *Calomys* Waterhouse, 1837;
    raised to generic rank by Bailey, 1888:442.

†*Cimarronomys* Hibbard, 1953:403.
    E. Plioc.; N.A.

Neotomodon Merriam, 1898:127.
    Pleist., R.; N.A.[1]
    [1] Mexico.

Ochrotomys Osgood, 1909:222.[1] Golden mouse.
    L. Pleist.-R.; N.A.
    [1] Described as subgenus of *Peromyscus*.

Podomys Osgood, 1909:226.[1]

       L. Pleist.-R.; N.A.

[1] Described as subgenus of *Peromyscus*.

Isthmomys Hooper & Musser, 1964:12.[1]

       R.; Cent. A.[2] R.; S.A.[3]

[1] Described as subgenus of *Peromyscus*; raised to generic rank by Carleton, 1980.
[2] Panama.
[3] N.W. Colombia.

Megadontomys Merriam, 1898:115.[1]

       R.; N.A.[2]

[1] Described as subgenus of *Peromyscus*; raised to generic rank by Bangs, 1902:27.
[2] Mexico.

Habromys Hooper & Musser, 1964:12.[1]

       R.; N.A.[2] R.; Cent. A.[3]

[1] Described as subgenus of *Peromyscus*; raised to generic rank by Carleton, 1980.
[2] Mexico (Oaxaca and Veracruz).
[3] S. Mexico (Chiapas), Guatemala, El Salvador.

Osgoodomys Hooper & Musser, 1964:12.[1]

       R.; N.A.[2]

[1] Described as subgenus of *Peromyscus*; raised to generic rank by Carleton, 1980.
[2] Mexico.

Tribe **Oryzomyini** Vorontzov, 1959.

       ?L. Mioc., L. Pleist.-R.; N.A. Pleist., R.; W. Indies.[1] ?E. Pleist., M. Pleist.-R.; S.A. R.; Cent. A. R.; Pacific.[2]

[1] Now extinct.
[2] E. Pacific (Galapagos).

†*Megalomys* Trouessart, 1881:357.[1]

       [= †*Moschomys* Trouessart, 1903:387[2]; †*Moschophoromys* Elliot, 1904:270.]
       Pleist., R.; W. Indies.[3] Pleist.; S.A.[4]

[1] Described as subgenus of *Hesperomys*; raised to generic rank by Allen, 1902:21.
[2] Replacement name for †*Megalomys* Trouessart, 1881, thought to be preoccupied by †*Megamys* d'Orbigny & Laurillard, 1842, emended as †*Megalomys* Trouessart, 1903. Not *Moschomys* Billberg, 1827, a synonym of *Ondatra*.
[3] Lesser Antilles (Martinique and Santa Lucia). Extinct since the late 19th Century.
[4] Curaçao.

Oryzomys Baird, 1857:458.[1] Rice rats.

       [Including *Micronectomys* Hershkovitz, 1948:55[2]; *Macruroryzomys* Hershkovitz, 1948:56.[2]]
       ?L. Mioc., L. Pleist.-R.; N.A. Pleist., R.; S.A. R.; Cent. A. R.; W. Indies.[3] R.; Pacific.[4]

[1] Proposed as subgenus of *Hesperomys*; raised to generic rank by Coues, 1890:4164.
[2] *Nomen nudum*.
[3] Saint Vincent (presumably extinct); Jamaica.
[4] E. Pacific (Galapagos).

Nesoryzomys Heller, 1904:241.

       R.; Pacific.[1]

[1] E. Pacific (Galapagos).

Melanomys Thomas, 1902:248.

       R.; Cent. A. R.; S.A.

Sigmodontomys J. A. Allen, 1897:38.

       R.; Cent. A.[1] R.; S.A.[2]

[1] Honduras to Panama.
[2] Venezuela, Colombia, Ecuador.

Nectomys Peters, 1861:151.

       [Including *Cercomys* F. Cuvier, 1829 (unpaged).[1]]
       Pleist., R.; S.A.

[1] A composite based on a skin of *Nectomys*, a skull of *Proechimys*, and a mandible of *Echimys*. *Cercomys* is a senior synonym and should be suppressed.

Amphinectomys Malygin, Aniskin, Isaev & Milishnikov, 1994:198.
  R.; S.A.

Oligoryzomys Bangs, 1900:94.[1]
  Pleist., R.; S.A. R.; Cent. A.
  [1] Described as subgenus of *Oryzomys*.

Neacomys Thomas, 1900:153.
  R.; Cent. A.[1] R.; S.A.[2]
  [1] Panama.
  [2] Venezuela, Colombia, Ecuador, Peru, Brazil, Guianas.

Zygodontomys J. A. Allen, 1897:38.
  Pleist., R.; S.A.[1,2] R.; Cent. A.
  [1] Aruba in the Pleistocene.
  [2] Northern South America, including Trinidad and Tobago, in the Recent.

Lundomys Voss & Carleton, 1993:5.
  L. Pleist.-R.; S.A.

Holochilus Brandt, 1835:428.[1]
  [= *Holochyse* Lesson, 1842:137.]
  M. Pleist.-R.; S.A.
  [1] Described as subgenus of *Mus*.

Pseudoryzomys Hershkovitz, 1959:8.
  Pleist., R.; S.A.[1]
  [1] Brazil in the Pleist. and Recent; Paraguay, E. Bolivia, and N. Argentina also in the Recent.

Microakodontomys Hershkovitz, 1993:2.
  R.; S.A.[1]
  [1] Brazil.

Oecomys Thomas, 1906:444.[1]
  R.; Cent. A. R.; S.A.
  [1] Described as subgenus of *Oryzomys*.

Microryzomys Thomas, 1917:1.
  [Including *Thallomyscus* Thomas, 1926:612.[1]]
  R.; S.A.
  [1] Described as subgenus of *Oryzomys*.

Scolomys Anthony, 1924:1.
  R.; S.A.[1]
  [1] Ecuador and Peru.

Tribe **Thomasomyini** Steadman & Ray, 1982:15.[1]
  Pleist., R.; S.A. L. Pleist.-R.; Pacific.[2] R.; Cent. A.[3]
  [1] Used informally by Hershkovitz, 1962:21; 1966:125.
  [2] E. Pacific (Galapagos). Extinct before the arrival of humans.
  [3] Panama.

†*Megaoryzomys* Lenglet & Coppois, 1979:635.
  L. Pleist.-R.; Pacific.[1]
  [1] E. Pacific (Galapagos). Extinct before the arrival of humans.

Chilomys Thomas, 1897:500.
  R.; S.A.[1]
  [1] Venezuela, Colombia, Ecuador.

Abrawayaomys Cunha & Cruz, 1979:2.
  R.; S.A.[1]
  [1] Brazil and Argentina.

Delomys Thomas, 1917:196.
  Pleist., R.; S.A.[1]
  [1] Brazil.

Thomasomys Coues, 1884:1275.
  [Including *Erioryzomys* Bangs, 1900:96-97; *Inomys* Thomas, 1917:197.]

R.; S.A.[1]

[1] Venezuela, Colombia, Ecuador, Peru and Bolivia.

Wilfredomys Avila Pires, 1960:3.
R.; S.A.[1]

[1] N.E. Brazil, Cent. Uruguay and N.E. Argentina.

Aepeomys Thomas, 1898:452.
R.; S.A.[1]

[1] N.W. South America.

Phaenomys Thomas, 1917:196.
R.; S.A.[1]

[1] Brazil.

Rhipidomys Tschudi, 1844:252.[1] Climbing mice.
Pleist., R.; S.A. R.; Cent. A.[2]

[1] Proposed as subgenus of *Hesperomys*; raised to generic rank by Winge, 1888:54-57.
[2] Panama.

Tribe **Wiedomyini** Reig, 1980:265.
E. Pleist., R.; S.A.

†*Cholomys* Reig, 1980:266.
E. Pleist.; S.A.[1]

[1] Argentina.

Wiedomys Hershkovitz, 1959:5.
R.; S.A.[1]

[1] Brazil.

Tribe **Akodontini** Vorontzov, 1959:134-137.
L. Plioc.-R.; S.A.

†*Dankomys* Reig, 1978:169.
L. Plioc.-E. Pleist.; S.A.

Akodon Meyen, 1833:599. South American field mice, grass mice.
[= *Acodon* Agassiz, 1846:5; *Axodon* Giebel, 1855:48.[1]] [Including *Habrothrix* Wagner, 1843:516-523[2]; *Microxus* Thomas, 1909:237[3]; *Chalcomys* Thomas, 1916:338; *Deltamys* Thomas, 1917:98[3]; *Hypsimys* Thomas, 1918:190[4]; *Thaptomys* Thomas, 1918:339[3]; *Plectomys* Borchert & Hansen, 1983:229-240.[5]]
L. Plioc.-R.; S.A.[6]

[1] Emendation.
[2] Proposed as subgenus.
[3] Sometimes as subgenus.
[4] Subgenus.
[5] *Nomen nudum.*
[6] Including Trinidad and Tobago.

Bolomys Thomas, 1916:339.
[Including *Cabreramys* Massoia & Fornes, 1967:418.]
L. Plioc.-R.; S.A.

Podoxymys Anthony, 1929:4.
R.; S.A.[1]

[1] Northern South America.

Thalpomys Thomas, 1916:339.
R.; S.A.[1]

[1] Brazil.

Abrothrix Waterhouse, 1837:21.
R.; S.A.[1]

[1] Argentina and Chile.

Chroeomys Thomas, 1916:340.
R.; S.A.[1]

[1] Peru, Bolivia, N. Chile, N. Argentina.

Chelemys Thomas, 1903:242.[1]
    R.; S.A.[2]
[1] Described as subgenus of *Akodon*.
[2] Chile and Argentina.

Notiomys Thomas, in Milne-Edwards, 1890:23.[1] Mole mouse.
    R.; S.A.[2]
[1] Proposed as subgenus of *Hesperomys*; raised to generic rank by Thomas, 1897:1020.
[2] Southern Argentina.

Pearsonomys Patterson, 1992:132.
    R.; S.A.[1]
[1] Chile.

Geoxus Thomas, 1919:209. Mole mouse, raton topo.
    R.; S.A.[1]
[1] Chile, S. Argentina.

Blarinomys Thomas, 1896:310. Brazilian shrew mouse.
    Pleist., R.; S.A.[1]
[1] Brazil.

Juscelinomys Moojen, 1965:281.
    Pleist., R.; S.A.[1]
[1] Brazil.

Oxymycterus Waterhouse, 1837:21.[1] Burrowing mice.
    [= *Oxymicterus* Tomes, 1861:285.[2]]
    M. Pleist., ?L. Pleist., R.; S.A.
[1] Proposed as subgenus of *Mus*.
[2] As genus.

Lenoxus Thomas, 1909:236. Andean rat.
    R.; S.A.[1]
[1] Peru, Bolivia.

Tribe **Phyllotini** Vorontzov, 1959.
    [Including Reithrodontini Vorontzov, 1959.]
    L. Plioc.-R.; S.A.

Calomys Waterhouse, 1837:21.[1]
    [= *Hesperomys* Waterhouse, 1839:74-77; *Callomys* Gray, 1843:112.[2]] [Including †*Necromys* Ameghino, 1889:120-121.]
    Pleist., R.; S.A.[3]
[1] Proposed as subgenus of *Mus*; raised to generic rank by Gray, 1843:112.
[2] Not *Callomys* d'Orbigny & Geoffroy Saint-Hilaire, 1830, a chinchillid.
[3] Includes Curaçao.

Eligmodontia F. Cuvier, 1837:169.
    [= *Heligmodontia* Agassiz, 1846:136, 175; *Elimodon* Fitzinger, 1867:463; *Eligmodon* Thomas, 1896:307.]
    L. Pleist.-R.; S.A.[1]
[1] Argentina, Chile, W. Bolivia, S. Peru.

Graomys Thomas, 1916:141.
    [= †*Bothriomys* Ameghino, 1889:118.[1]] [Including *Andalgalomys* Williams & Mares, 1978:197.]
    L. Plioc.-R.; S.A.
[1] *Nomen oblitum.*

Salinomys Braun & Mares, 1995:505.
    R.; S.A.

Phyllotis Waterhouse, 1837:27.[1] Leaf-eared mice, pericotes.
    [Including *Paralomys* Thomas, 1926:315.]
    Pleist., R.; S.A.
[1] Proposed as subgenus of *Mus*; raised to generic rank by Fitzinger, 1867:83-84.

Loxodontomys Osgood, 1947:172.[1]
        R.; S.A.[2]
[1] Proposed as subgenus of *Phyllotis*.
[2] S. Andes of Chile and Argentina.

Auliscomys Osgood, 1915:190.[1]
        [Including *Maresomys* Braun, 1993:40.]
        Plioc. and/or Pleist., R.; S.A.
[1] Proposed as a subgenus of *Phyllotis*; raised to generic rank by Thomas, 1926.

Galenomys Thomas, 1916:143.[1]
        R.; S.A.[2]
[1] Described as subgenus of *Euneomys*; raised to generic rank by Thomas, 1926.
[2] Bolivia, S. Peru, N. Chile.

Chinchillula Thomas, 1898:280.
        R.; S.A.[1]
[1] Peru, Bolivia, Chile, Argentina.

Punomys Osgood, 1943:369.
        R.; S.A.[1]
[1] Peru.

Andinomys Thomas, 1902:116.
        Pleist., R.; S.A.[1]
[1] Peru, Bolivia, Chile, Argentina.

Irenomys Thomas, 1919:201.
        R.; S.A.[1]
[1] Chile, Argentina.

Euneomys Coues, 1874:185.[1]
        [Including *Chelemyscus* Thomas, 1925:585.[2]]
        R.; S.A.[3]
[1] Proposed as subgenus of *Reithrodon*; raised to generic rank by Thomas, 1901:254.
[2] Doubtfully assigned.
[3] Chile, Argentina.

Neotomys Thomas, 1894:346.[1]
        R.; S.A.[2]
[1] Not *Neotomys* Wallace, 1876, misprint for *Nectomys* Peters, 1861.
[2] Peru, Bolivia, Chile, Argentina.

Reithrodon Waterhouse, 1837:29.
        [= *Rithrodon* Agassiz, 1846:327; *Rheitrodon* Roger, 1887:102; *Rhithrodon* Flower
        & Lydekker, 1891:464.] [Including †*Ptyssophorus* Ameghino, 1889:111-112;
        †*Tretomys* Ameghino, 1889:119-120; †*Proreithrodon* Ameghino, 1908:424.]
        L. Plioc.-R.; S.A.[1]
[1] Chile, Argentina, Uruguay.

Tribe **Sigmodontini** Wagner, 1843:509.
        [= Sigmodontes Wagner, 1843:509; Sigmodontini Vorontzov, 1959;
        Sigmodontini Hershkovitz, 1966:747.]
        E. Plioc.-R.; N.A. R.; Cent. A. R.; S.A.

†*Prosigmodon* Jacobs & Lindsay, 1981:425.
        E. Plioc.; N.A.

Sigmodon Say & Ord, 1825:352. Cotton rats.
        [Including *Lasiomys* Burmeister, 1854:15-17; *Deilemys* de Saussure, 1860:98-101[1];
        Sigmomys Thomas, 1901:150-151.[2]]
        E. Plioc.-R.; N.A. R.; Cent. A. R.; S.A.
[1] Described as subgenus of *Hesperomys*.
[2] Sometimes as subgenus.

Tribe **Scapteromyini** Massoia, 1979.
        L. Plioc.-R.; S.A.

Scapteromys Waterhouse, 1837:20.[1] Water rat, rata acuatica.
    L. Plioc.-R.; S.A.
[1] Proposed as subgenus of *Mus*; raised to generic rank by Fitzinger, 1867:79-80.

Kunsia Hershkovitz, 1966:112.
    Pleist., R.; S.A.[1]
[1] Brazil in the Pleist. and Recent; also N.E. Argentina and N.E. Bolivia in the Recent.

Bibimys Massoia, 1979:2.
    Pleist., R.; S.A.[1]
[1] Brazil in the Pleist. and Recent; Argentina in the Recent.

Tribe **Ichthyomyini** Vorontzov, 1959:134-137.
    R.; N.A.[1] R.; Cent. A. R.; S.A.
[1] Mexico.

Neusticomys Anthony, 1921:2.
    [Including *Daptomys* Anthony, 1929:1.]
    R.; S.A.[1]
[1] Guianas, Venezuela, Colombia, Ecuador, Peru.

Rheomys Thomas, 1906:421.
    [Including *Neorheomys* Goodwin, 1959:3.]
    R.; N.A.[1] R.; Cent. A.
[1] S. Mexico.

Anotomys Thomas, 1906:86.
    R.; S.A.[1]
[1] Ecuador.

Chibchanomys Voss, 1988:321.
    R.; S.A.[1]
[1] Venezuela, Colombia, Peru.

Ichthyomys Thomas, 1893:286. Fish-eating rats.
    R.; Cent. A.[1] R.; S.A.[2]
[1] Panama.
[2] Venezuela, Colombia, Ecuador, Peru.

Subfamily **Calomyscinae** Vorontsov, Kartavtseva & Potapova, 1978:17.
    [= Calomyscini Vorontsov, Kartavtseva & Potapova, 1978:17; Calomyscinae Musser
    & Carleton, in Wilson & Reeder, eds., 1993:535.]
    L. Mioc., ?E. Plioc.; Eu. E. Plioc., R.; As.

Calomyscus Thomas, 1905:23. Mouse-like hamster.
    L. Mioc., ?E. Plioc.; Eu. E. Plioc., R.; As.[1]
[1] Rhodes in the early Pliocene.

Subfamily **Cricetinae** Fischer de Waldheim, 1817:372. Hamsters.
    [= Cricetini Fischer de Waldheim, 1817:372; Cricetina Gray, 1825:342[1]; Cricetinae
    Murray, 1866:358; Cricetidae Rochebrune, 1883:66, 153; Cricetini Winge, 1887:109[2];
    Cricetini Simpson, 1945:85.[1]] [Including †Ischymomyini Topachevsky, 1992:29.]
    M. Mioc.; Af.[3] M. Mioc.-R.; Eu. L. Mioc.-R.; As. E. Plioc., R.; Mediterranean.
[1] As tribe.
[2] As subfamily.
[3] N. Africa.

Cricetus Leske, 1779:168. Common hamster, krietsch, black-bellied hamster.
    [= *Hamster* Lacépède, 1799:10; *Heliomys* Gray, 1873:417.]
    M. Mioc.; Af.[1] L. Mioc.-R.; Eu. E. Plioc., M. Pleist., R.; As.[2,3]
[1] N. Africa (Egypt).
[2] W. Asia (Rhodes) in the early Pliocene.
[3] W. Asia (Israel) in the middle Pleistocene.

†*Rotundomys* Mein, 1965:106.
    [= or including †*Cricetulodon* Hartenberger, 1966:493.]
    M.-L. Mioc., Plioc.; Eu.

†*Microtocricetus* Fahlbusch & Mayr, 1975:79.
    [Including †*Sarmatomys* Topachevsky & Skorik, 1988:37-45.]
    M. Mioc.; Eu.

†*Ischymomys* Zazhigin, in Gromov, 1971.
    L. Mioc.; As.

†*Collimys* Daxner-Hock, 1972:145.
    L. Mioc.; Eu.

†*Pannonicola* Kretzoi, 1965:131-139.
    L. Mioc.; Eu.

†*Hattomys* Freudenthal, 1985:31.
    ?L. Mioc.; Eu.

†*Nannocricetus* Schaub, 1934:29.
    L. Mioc., ?L. Plioc., ?M. Pleist.; As.

†*Karstocricetus* Kordos, 1987:75.
    L. Mioc.; Eu.[1]
    [1] E. Europe.

†*Kowalskia* Fahlbusch, 1969:103.
    L. Mioc.; As. L. Mioc.-L. Plioc.; Eu. E. Plioc.; Mediterranean.

Cricetulus Milne-Edwards, 1867:375. Rat-like hamsters.
    [Including *Urocricetus* Satunin, 1903:573-575; †*Cricetinus* Zdansky, 1928:54;
    †*Allocricetus* Schaub, 1930:32; †*Moldavimus* Samson & Radulesco, 1973:214.[1]]
    L. Mioc.-R.; Eu. E. Plioc.-R.; As.[2]
    [1] Subgenus as defined.
    [2] Including Rhodes in the early Pliocene.

†*Hypsocricetus* Daxner-Höch, 1992:351.
    L. Mioc.; Eu.[1]
    [1] E. Europe (Macedonia).

†*Sinocricetus* Schaub, 1930:44.
    L. Mioc.; As.[1]
    [1] Inner Mongolia.

†*Paracricetulus* Young, 1927:30.
    E. Plioc.; As.

†*Chuanocricetus* Zheng, 1993:60.
    L. Plioc.; As.[1]
    [1] China.

†*Amblycricetus* Zheng, 1993:64.
    L. Plioc.; As.[1]
    [1] China.

Mesocricetus Nehring, 1898:494.[1] Golden hamster.
    [= *Mediocricetus* Nehring, 1898:494; *Semicricetus* Nehring, 1898:494.]
    E. Plioc., M. Pleist.-R.; As.[2,3] L. Pleist.-R.; Eu.[4] R.; Mediterranean.
    [1] Proposed as subgenus of *Cricetus*; raised to generic rank by Nehring, 1902:57-60.
    [2] Including Rhodes in the early Pliocene.
    [3] W. Asia.
    [4] E. Europe.

†*Rhinocricetus* Kretzoi, 1956.
    E.-M. Pleist.; Eu.

Phodopus Miller, 1910:498. Small desert hamsters, Djungarian hamsters, mai-tsang-or.
    [Including *Cricetiscus* Thomas, 1917:456.]
    Pleist.; Eu. M. and/or L. Pleist., R.; As.

Tscherskia Ognev, 1914:102.
    [= *Asiocricetus* Kishida, 1929:148.]
    L. Plioc.; Eu.[1] R.; As.
    [1] E. Europe.

Cansumys G. M. Allen, 1928:244.

    R.; As.[1]

    [1] China.

Allocricetulus Argyropulo, 1932:242.

    R.; As.[1]

    [1] Mongolia, China.

Subfamily **Arvicolinae** Gray, 1821:303. Voles, lemmings, muskrats.

    [= Arvicolidae Gray, 1821:303; Arvicolina C. L. Bonaparte, 1837:9[1]; Arvicolinae Pavlinov & Rossolimo, 1987:174.] [Including Lemnina Gray, 1825:342; Lemminae Kretzoi, 1969:185; Ellobiinae Gill, 1872:20; Microtidae Cope, 1891; Microtinae Miller, 1896:8; †Braminae Miller & Gidley, 1918:438; Prometheomyinae Kretzoi, 1955; †Microtoscoptidae Kretzoi, 1969; Dicrostonychinae Chaline, 1973; Ondatrinae Repenning, in Honacki, Kinman & Koeppl, eds., 1982:484.]

    L. Mioc.-R.; As.[2] L. Mioc.-R.; Eu. E. Plioc.-R.; N.A.[3] M. Pleist., R.; Af.[4] M. Pleist.-R.; Mediterranean. R.; Cent. A.[5]

    [1] As subfamily.

    [2] Including Arctic islands of Siberia.

    [3] Including Arctic islands of Canada and Greenland.

    [4] N. Africa.

    [5] Also Recent of South America by introduction.

†*Aratomys* Zazhigin, in Gromov, 1972.

    L. Mioc.; As.

†*Loupomys* W. von Koenigswald & L. D. Martin, 1984:118.

    L. Plioc.; N.A.

†*Nebraskomys* Hibbard, 1957:43.

    E.-L. Plioc.; N.A.

†*Atopomys* Patton, 1965:466.

    M. Pleist.; N.A.

†*Visternomys* Radulesco & Samson, 1986:71.[1]

    M. Pleist.; Eu.[2]

    [1] (Rădulescu).

    [2] E. Europe.

Tribe **Arvicolini** Gray, 1821:303.

    [= Arvicolidae Gray, 1821:303; Arvicolini Giebel, 1855:602; Arvicolini Kretzoi, 1955:353-355; Arvicolina Pavlinov & Rossolimo, 1987:187.] [Including Microti Miller, 1896:8; Microtini Simpson, 1945:87; Lagurini Kretzoi, 1955:355; Lagurina Pavlinov & Rossolimo, 1987:185; Pitymyini Repenning, 1983.]

    L. Mioc.-R.; Eu. E. Plioc.-R.; As. E. Plioc.-R.; N.A. M. Pleist.-R.; Mediterranean. R.; Af.[1] R.; Cent. A.

    [1] N. Africa.

†*Mimomys* Forsyth Major, 1902:431.

    [Including †*Microtomys* Méhely, 1914:210; †*Kislangia* Kretzoi, 1954:247[1]; †*Promimomys* Kretzoi, 1955:90[2]; †*Cseria* Kretzoi, 1959:242[3]; †*Katamys* Kretzoi, 1962[4]; †*Hintonia* Kretzoi, 1969:185[5]; †*Tianschanomys* Savinov, 1974:48[5]; †*Borsodia* Janossy & van der Meulen, 1975:389[5]; †*Pusillomimus* Rabeder, 1981:292; †*Tcharinomys* Savinov & Tut'kova, 1987.]

    L. Mioc.-L. Pleist.; Eu. E. Plioc.-M. Pleist.; As.

    [1] Proposed as genus; as subgenus in Michaux,1971.

    [2] Proposed as genus.

    [3] Proposed as genus; as subgenus in Kowalski, 1960, and Kretzoi, 1964:185.

    [4] As subgenus in Kretzoi, 1964:185.

    [5] Proposed as subgenus.

†*Prosomys* Shotwell, 1956:732.

    [Including †*Polonomys* Kretzoi, 1959.]

    E.-L. Plioc.; Eu. E. Plioc.; N.A. L. Plioc.; As.

†*Cosomys* R. W. Wilson, 1932:150.

    Plioc.; N.A.

†*Ophiomys* Hibbard & Zakrzewski, 1967:258.
  E.-L. Plioc.; N.A.

†*Ogmodontomys* Hibbard, 1941:362.
  E.-L. Plioc.; N.A.

†*Kilarcola* Kotlia, 1985:82.[1]
  L. Plioc.-E. Pleist.; As.[2]
  [1] *Nomen nudum* in Kotlia, 1984 (unpublished PhD thesis).
  [2] S. Asia (Kashmir).

†*Hibbardomys* Zakrzewski, 1984:204.
  L. Plioc.; N.A.

†*Cromeromys* Zazhigin, 1980:108.
  L. Plioc.; As. E. Pleist.; N.A.

Arvicola Lacépède, 1799:10. Water vole.
  [= *Alviceola* de Blainville, 1817:287; *Hemiotomys* de Sélys-Longchamps, 1836:7[1]; *Ochetomys* Fitzinger, 1867:103.] [= or including *Paludicola* Blasius, 1857:333[2]; *Praticola* Fatio, 1867:36.[3]]
  E. Pleist.-R.; As. E. Pleist.-R.; Eu.
  [1] Described as subgenus of *Arvicola*.
  [2] Described as subgenus of *Arvicola*. Not *Paludicola* Wagler, 1830, a genus of amphibians, nor *Paludicola* Hodgson, 1837, a genus of birds.
  [3] Described as subgenus of *Arvicola*. Not *Praticola* Swainson, 1837, a genus of birds.

†*Tyrrhenicola* Forsyth Major, 1905:504, 505.[1]
  M. Pleist.-R.; Mediterranean.[2]
  [1] Proposed as subgenus of *Arvicola*.
  [2] Corsica, Sardinia.

Microtus Schrank, 1798:72. Voles, meadow mice.
  [= *Arvalomys* Chaline, 1974:450.] [Including *Mynomes* Rafinesque, 1817:45[1]; *Psammomys* Le Conte, 1830:132[2]; *Ammomys* Bonaparte, 1831:20[3]; *Pitymys* McMurtrie, 1831:434[4]; *Pinemys* Lesson, 1836:436; *Neodon* Hodgson, 1849:203[1]; *Chilotus* Baird, 1857:516[5]; *Pedomys* Baird, 1857:517[6]; *Agricola* Blasius, 1857:334[1]; *Phaiomys* Blyth, 1863:89[7]; *Bicunedens* Hodgson, 1863:11; *Sylvicola* Fatio, 1867:63[8]; *Terricola* Fatio, 1867:73[9]; †*Isodelta* Cope, 1871:87-88; *Micrurus* Forsyth Major, 1877:124[10]; *Lasiopodomys* Lataste, 1887:268[11]; *Campicola* Schulze, 1890:24[12]; *Aulacomys* Rhoads, 1894:182-185[1]; *Tetramerodon* Rhoads, 1894:282-283; *Orthriomys* Merriam, 1898:106-107[5]; *Herpetomys* Merriam, 1898:107-108; *Euarvicola* Acloque, 1899:49; *Stenocranius* Kastschenko, 1901:167[1]; *Steneocranius* Trouessart, 1904:457[13]; *Alexandromys* Ognev, 1914:109[1]; *Arbusticola* Shidlovsky, 1919:21; *Campicoloma* Strand, 1928:61[14]; †*Allophaiomys* Kormos, 1932:324[15]; *Sumeriomys* Argyropulo, 1933:180[5]; *Lemmimicrotus* Tokuda, 1941:68; *Pallasiinus* Kretzoi, 1964:136[1]; *Iberomys* Chaline, 1972[5]; *Suranomys* Chaline, 1972[5]; *Meridiopitymys* Chaline, 1974; *Parapitymys* Chaline, 1978; *Oecomicrotus* Rabeder, 1981:300; †*Tibericola* W. von Koenigswald, Fejfar & Tchernov, 1992:3.[16]]
  L. Plioc.-R.; As. L. Plioc.-R.; Eu. L. Plioc.-R.; N.A. R.; Af.[17] R.; Mediterranean.[18] R.; Cent. A.[19]
  [1] Subgenus.
  [2] Not *Psammomys* Cretzschmar, 1828, a gerbil.
  [3] New name for *Psammomys*.
  [4] New name for *Psammomys*. Subgenus.
  [5] Sometimes as subgenus.
  [6] Sometimes as subgenus of *Microtus*; sometimes as subgenus of *Pitymys*.
  [7] Sometimes as separate genus.
  [8] Objective synonym of *Agricola*. Not *Sylvicola* Harris, 1782, a dipteran, nor *Sylvicola* Humphrey, 1797, a genus of mollusks.
  [9] Sometimes as subgenus. Not *Terricola* Fleming, 1828, a genus of mollusks.
  [10] Not *Micrurus* Ehrenberg, 1831, a worm.
  [11] Sometimes as subgenus; sometimes as separate genus.
  [12] Not *Campicola* Swainson, 1827, a bird.
  [13] Misspelling?
  [14] Replacement for *Campicola* Schulze, 1890.
  [15] Proposed as a genus; later used as subgenus.
  [16] Described as subgenus.

[17] N. Africa (Libya).
[18] Sicily.
[19] S. Mexico and Guatemala.

Lemmiscus Thomas, 1912:401.[1] Sagebrush vole.
  M. Pleist.-R.; N.A.
  [1] Described as subgenus of *Lagurus*; raised to generic rank by Davis, 1939.

Blanfordimys Argyropulo, 1933:182.[1]
  R.; As.
  [1] Proposed as subgenus of *Microtus*.

Chionomys Miller, 1908:97.[1]
  M. Pleist.-R.; Eu. R.; As.
  [1] Sometimes considered a subgenus of *Microtus*.

Proedromys Thomas, 1911:4.
  ?Pleist., R.; As.[1]
  [1] China.

Volemys Zagorodnyuk, 1990:28.
  R.; As.

†*Villanyia* Kretzoi, 1956:188.
  [Including †*Kulundomys* Zazhigin, 1980:1-156.[1]]
  L. Plioc.; As. L. Plioc.-E. Pleist.; Eu.
  [1] Proposed as subgenus.

Lagurus Gloger, 1841:97. Steppe vole.
  [= *Eremiomys* Polyakov, 1881:34; *Eremomys* Heude, 1898:61.] [Including †*Prolagurus* Kormos, 1938; †*Lagurodon* Kretzoi, 1956:162; †*Laguropsis* Kretzoi, 1956.]
  L. Plioc.-R.; Eu. E. Pleist.-R.; As.

Eolagurus Argyropulo, 1946:44. Yellow steppe lemming.
  M. Pleist.; Eu.[1] L. Pleist.-R.; As.
  [1] E. Europe.

†*Jordanomys* Haas, 1966.
  E. Pleist.; As.[1]
  [1] W. Asia (Israel).

†*Kalymnomys* W. von Koenigswald, Fejfar & Tchernov, 1992:15.
  E.-M. Pleist.; As.[1]
  [1] W. Asia.

†*Nemausia* Chaline & Laborier, 1981:633.
  L. Pleist.; Eu.

†*Huananomys* Zheng, 1992:147.
  [= †*Hexianomys* Zheng & Li, 1990:431-442.[1]]
  E.-M. Pleist.; As.[2]
  [1] *Nomen nudum.*
  [2] China.

Tribe **Ondatrini** Gray, 1825:341.
  [= Ondatrina Gray, 1825:341[1]; Ondatrini Kretzoi, 1955:353, 355; Ondatrinae Repenning, in Honacki, Kinman & Koeppl, eds., 1982:484; Fibrini Mehely, 1914.]
  E. Plioc.-R.; N.A.[2]
  [1] Proposed as tribe.
  [2] Also Recent of South America, Europe, and Asia by introduction.

†*Pliopotamys* Hibbard, 1937:249.
  [= †*Neondatra* Hibbard, 1937:251.]
  E. Plioc.; N.A.

Ondatra Link, 1795:76. Muskrat.
  [= *Ondatra* Lacépède, 1799:9; *Fiber* G. Cuvier, 1800: table 1; *Mussascus* Oken, 1816:886; *Simotes* Fischer de Waldheim, 1817:444; *Moschomys* Billberg, 1828: before 1.] [Including †*Anaptogonia* Cope, 1871:91[1]; †*Sycium* Cope, 1899:203.[2]]

L. Plioc.-R.; N.A.[3]

[1] As subgenus of *Arvicola*.
[2] Objective synonym of †*Anaptogonia* Cope, 1871.
[3] Also Recent of South America (S. Argentina), Europe, and Asia by introduction.

Tribe **Clethrionomyini** Hooper & Hart, 1962:64.
[= Clethrionomyi I. Gromov, in I. Gromov & Polyakov, 1977:138; Clethrionomyina Pavlinov & Rossolimo, 1987:175.] [Including †Pliomyini Kretzoi, 1969; †Pliomyini Chaline, 1975:115; †Dolomyinae Chaline, 1975:115; Alticoli Gromov, in Gromov, Polyakov & Polevki, 1977:126; Phenacomyini Zagorodnyuk, 1990:27.]
E. Plioc.-R.; Eu.   E. Plioc.-R.; N.A.   L. Plioc.-R.; As.

†*Pliolemmus* Hibbard, 1937:247.
E. Plioc.; N.A.

†*Pliophenacomys* Hibbard, 1937:248.[1]
E. Plioc.-E. Pleist.; N.A.

[1] Proposed as subgenus of *Phenacomys*.

†*Guildayomys* Zakrzewski, 1984:201.
L. Plioc.; N.A.

Phenacomys Merriam, 1889:32. Heather voles.
[Including *Arborimus* Taylor, 1915:119[1]; †*Propliophenacomys* Martin, 1975:105; *Paraphenacomys* Repenning & Grady, 1988:8.[2]]
E. Pleist., L. Pleist.-R.; N.A.

[1] Described as subgenus; sometimes used as separate genus.
[2] Described as subgenus.

†*Dolomys* Nehring, 1898:13-16.
E. Plioc.-M. Pleist.; Eu.

†*Pliomys* Méhely, 1914.
[Including †*Apistomys* Méhely, 1914; †*Propliomys* Kretzoi, 1959:137-246; †*Laugaritiomys* Fejfar, 1961:257-273.]
E. Plioc.-L. Pleist.; Eu.   L. Plioc.-M. Pleist.; As.

Dinaromys Kretzoi, 1955:351.
E. Pleist.-R.; Eu.

Hyperacrius Miller, 1896:54.[1]
E. Pleist., R.; As.[2,3]

[1] Described as subgenus of *Microtus*.
[2] E. Asia (China) in the early Pleistocene.
[3] S. Asia (Kashmir and Pakistan) in the Recent.

Alticola Blanford, 1881:96.
[Including *Platycranius* Kastschenko, 1901:201[1]; *Aschizomys* Miller, 1898:368-371.[1]]
Pleist., R.; As.

[1] Subgenus.

Eothenomys Miller, 1896:45.
[Including *Anteliomys* Miller, 1896:9, 47-49; *Caryomys* Thomas, 1911:4.]
L. Plioc.-M. Pleist., R.; As.

Clethrionomys Tilesius, 1850:28. Red-backed voles.
[= *Evotomys* Coues, 1874:186; *Euotomys* Schulze, 1900:203; *Eotomys* Forsyth-Major, 1902:107.] [Including *Craseomys* Miller, 1900:87[1]; †*Acrorhizomys* Topachevsky, 1914; *Neoaschizomys* Tokuda, 1935:242; *Glareomys* Razorenova, 1952:23.]
L. Plioc., M. Pleist.-R.; As.   E. Pleist.-R.; Eu.   M. Pleist.-R.; N.A.

[1] Sometimes as subgenus.

Phaulomys Thomas, 1905:493.
L. Pleist.-R.; As.[1]

[1] E. Asia (Japan).

Tribe **Prometheomyini** Kretzoi, 1955:355.
[= Prometheomyinae Kretzoi, 1955:355; Prometheomyini Hooper & Hart, 1962.]
E. Plioc.-E. Pleist.; Eu.   L. Pleist.-R.; As.[1]

[1] W. Asia.

†*Stachomys* Kowalski, 1960:461.
  [= †*Leukaristomys* Fejfar, 1961:60-62.]
  E. Plioc.-E. Pleist.; Eu.[1]
  [1] E. Europe (Western Russia).

Prometheomys Satunin, 1901:572. Prometheus mouse.
  L. Pleist.-R.; As.[1]
  [1] W. Asia (Caucasus and N.E. Turkey).

Tribe **Ellobiini** Gill, 1872:20.
  [= Ellobiinae Gill, 1872:20[1]; Ellobii Weber, 1928:284; Ellobiini Simpson, 1945:88.]
  [Including †Braminae Miller & Gidley, 1918:438.]
  E. Plioc.-R.; Eu. L. Plioc.-R.; As. M. Pleist.; Af.[2]
  [1] Not Ellobiinae Adams, 1858, a subfamily of mollusks.
  [2] N. Africa.

†*Ungaromys* Kormos, 1932:336.
  [= †*Betfiamys* Terzea, 1973:421-426.] [Including †*Germanomys* Heller, 1936:130-131.]
  E. Plioc.-M. Pleist.; Eu. L. Plioc.; As.

Ellobius Fischer de Waldheim, 1814:72.[1] Mole voles, mole lemmings.
  [= *Chthonergus* Nordmann, in Demidoff, 1839:37; *Chtonoergus* Keyserling & Blasius, 1840:vii, 12, 32; *Lemmomys* Lesson, 1842:123.] [Including *Myospalax* Blyth, 1846:141[2]; †*Bramus* Pomel, 1892:1159-1163; *Afganomys* Topachevsky, 1965:98[3]; *Afghanomys* Baryshnikov & Baranova, 1983:103.]
  L. Plioc.-R.; As. L. Plioc.-R.; Eu.[4] M. Pleist.; Af.[5]
  [1] Not preoccupied by *Ellobium* Bolton, 1798, a genus of mollusks.
  [2] Not *Myospalax* Laxmann, 1773, a genus of myospalacine murids, nor *Myospalax* Hermann, 1783, a synonym of *Spalax*, a spalacine murid.
  [3] Subgenus.
  [4] E. Europe.
  [5] N. Africa.

Tribe **Lemmini** Gray, 1825:342. True lemmings.
  [= Lemnina Gray, 1825:342[1]; Lemmi Miller, 1896:8; Lemminae Kretzoi, 1969:185; Lemmini Simpson, 1945:86; Myodini Kretzoi, 1969:185.] [Including Synaptomyini W. von Koenigswald & L. D. Martin, 1984:126.]
  E. Plioc.-R.; Eu. L. Plioc., E. and/or M. Pleist., L. Pleist.-R.; As. L. Plioc.-R.; N.A.
  [1] Proposed as tribe.

Synaptomys Baird, 1857:558. Bog lemming.
  [Including *Mictomys* True, 1894:242-243[1]; †*Metaxyomys* Zakrzewski, 1972:2[1]; †*Praesynaptomys* Kowalski, 1977:299[2]; †*Plioctomys* Suchov, 1976:121-139; †*Kentuckomys* W. von Koenigswald & L. D. Martin, 1984:132.]
  E.-L. Plioc.; Eu. L. Plioc.; As. L. Plioc.-R.; N.A.
  [1] Sometimes as subgenus.
  [2] Described as subgenus.

Lemmus Link, 1795:75. Lemmings.
  [= *Hypudaeus* Illiger, 1811:87; *Myodes* Pallas, 1811:173[1]; *Lemnus* Rochebrune, 1843:216; †*Mirus* Brunner, 1938:33[2]; †*Miromus* Brunner, 1951:32.[3]] [Including *Brachyurus* Fischer de Waldheim, 1813:14, 24.]
  Pleist., R.; As. E. Pleist.-R.; Eu. E. Pleist.-R.; N.A.
  [1] Synonymy questionable. Possibly a synonym of *Clethrionomys*.
  [2] Not *Mirus* Albers, 1850, a mollusk.
  [3] Replacement name for *Mirus* Brunner, 1938.

Myopus Miller, 1910:497.
  L. Pleist.-R.; As. R.; Eu.

Tribe **Neofibrini** Hooper & Hart, 1962:64.
  E. Pleist.-R.; N.A.

†*Proneofiber* Hibbard & Dalquest, 1973:271.
    E. Pleist.; N.A.[1]
    [1] Texas.

Neofiber True, 1884:34. Round-tailed muskrat, water rat.
    [Including †*Schistodelta* Cope, 1899:206.]
    M. Pleist.-R.; N.A.

Tribe **Dicrostonychini** Kretzoi, 1955:355.
    [= Dicrostonyxini Gromov, 1972:16; Dicrostonyxini Gromov & Polyakov, 1977:177;
    Dicrostonychinae Chaline, 1973; Dicrostonychina Pavlinov & Rossolimo, 1987:183.]
    E. Pleist.-R.; As.[1] E. Pleist.-R.; Eu. E. Pleist.-R.; N.A.[2]
    [1] Including Arctic islands of Siberia.
    [2] Including Arctic islands of Canada and Greenland.

†*Predicrostonyx* Guthrie & Matthews, 1971:484.
    E.-M. Pleist.; As. E.-M. Pleist.; Eu. E. Pleist.; N.A.

Dicrostonyx Gloger, 1841:97. Collared lemming.
    [Including *Cuniculus* Wagler, 1830:21[1]; *Myolemmus* Pomel, 1854:27-28;
    Misothermus Hensel, 1855:492[2]; *Borioikon* Polyakov, 1881:34[3]; *Borioicon* Büchner,
    1889:127; *Tylonyx* Schulze, 1897:83.]
    E. Pleist.-R.; Eu. M. Pleist.-R.; As.[4] M. Pleist.-R.; N.A.[5]
    [1] Not *Cuniculus* Brisson, 1762, a synonym of *Agouti*, nor *Cuniculus* Gronovius, 1763, nor *Cuniculus* Meyer,
    1790, a synonym of *Oryctolagus*.
    [2] Sometimes as subgenus.
    [3] Objective synonym of *Misothermus*.
    [4] Including Arctic islands of Siberia.
    [5] Including Arctic islands of Canada and Greenland.

Subfamily †**Afrocricetodontinae** Lavocat, 1973:199.
    E.-L. Mioc.; Af.

†*Notocricetodon* Lavocat, 1973:226.
    E.-M. Mioc.; Af.[1]
    [1] Kenya.

†*Afrocricetodon* Lavocat, 1973:199.
    E. Mioc.; Af.[1]
    [1] Kenya.

†*Protarsomys* Lavocat, 1973:213.
    E.-L. Mioc.; Af.

Subfamily **Lophiomyinae** Milne-Edwards, 1867:114.
    [= Lophiomyidae Gill, 1872:20; Lophiomyoidea Gill, 1872:303; Lophiomyini Stehlin
    & Schaub, 1951:167.]
    L. Mioc.; Eu. Plioc. and/or E. Pleist., R.; Af. R.; As.[1]
    [1] S.W. Asia, now extinct.

†*Microlophiomys* Topachevsky & Skorik, 1984:58.
    L. Mioc.; Eu.[1]
    [1] E. Europe (Ukraine).

†*Protolophiomys* Aguilar & Thaler, 1987:859.
    L. Mioc.; Eu.[1]
    [1] Spain.

Lophiomys Milne-Edwards, 1867:46. Maned rat, crested rat, yaidado.
    [= *Phractomys* Peters, 1867:195; *Phragmomys* Peters, 1867.[1]]
    Plioc. and/or E. Pleist., R.; Af.[2,3] R.; As.[4]
    [1] Apparently a different spelling of *Phractomys*.
    [2] N. Africa (Morocco) in the Pliocene and/or early Pleistocene.
    [3] E. Africa in the Recent.
    [4] S.W. Asia (Israel), now extinct.

Subfamily **Nesomyinae** Forsyth Major, 1897:718. Malagasy mice, Malagasy rats.
[= Nesomyidae Weber, 1904.] [Including Eliuri Ellerman, 1941:75; Gymnuromyinae Ellerman, 1941:487; Brachyuromyes Ellerman, 1941:491; Brachytarsomyes Ellerman, 1941:555.]
?L. Mioc.; Af. ?Pleist., R.; Madagascar.

Brachytarsomys Günther, 1875:79.
?Pleist., R.; Madagascar.

Macrotarsomys A. Milne-Edwards & A. Grandidier, 1898:179.
?Pleist., R.; Madagascar.

Nesomys Peters, 1870:54.
[Including *Hallomys* Jentink, 1879:107-109.]
R.; Madagascar.

Gymnuromys Forsyth Major, 1896:324. Voalavoanala.
R.; Madagascar.

Eliurus Milne-Edwards, 1885:1.
R.; Madagascar.

Brachyuromys Forsyth Major, 1896:322. Volane andrivo, ramirohitra.
?L. Mioc.; Af. R.; Madagascar.

Hypogeomys A. Grandidier, 1869:338. Voalavo voutsoutse, votsotsa.
R.; Madagascar.

Subfamily **Cricetomyinae** Roberts, 1951:434. Pouched rats.
[= Cricetomyidae Chaline, Mein & Petter, 1977:245-252.] [Including Saccostomurinae Roberts, 1951:434, 436.]
E. Plioc.-R.; Af.

Saccostomus Peters, 1846:258. African pouched rats.
[= *Eosaccomys* Palmer, 1903:873.]
E. Plioc.-R.; Af.

Beamys Thomas, 1909:107. Long-tailed pouched rats.
R.; Af.[1]
[1] Zambia, Malawi, Tanzania, Kenya.

Cricetomys Waterhouse, 1840:2.[1] African giant pouched rats.
R.; Af.
[1] Proposed as subgenus of *Mus*; raised to generic rank by Lesson, 1842:120.

Subfamily **Delanymyinae** Denys, Michaux, Catzeflis, Ducrocq & Chevret, 1995:186.
L. Mioc.-L. Plioc., ?Pleist., R.; Af.

†*Stenodontomys* Pocock, 1987:86.
L. Mioc.-L. Plioc., ?Pleist.; Af.[1]
[1] Southern Africa.

Delanymys Hayman, 1962:129. Climbing swamp mouse.
Plioc. and/or Pleist., R.; Af.[1]
[1] Uganda, Zaire.

Subfamily **Mystromyinae** Vorontsov, 1966:437.
[= Mystromyini Vorontsov, 1966:437; Mystromyinae Lavocat, 1973:237.]
?L. Mioc., E. Plioc.-R.; Af.[1]
[1] S. Africa.

Mystromys Wagner, 1841:421. White-tailed rat, witstertrotte.
?L. Mioc., E. Plioc.-R.; Af.[1]
[1] S. Africa.

†*Proodontomys* Pocock, 1987:82.
L. Plioc.-E. Pleist.; Af.[1]
[1] S. Africa.

Subfamily **Petromyscinae** Roberts, 1951:434.
>   M. Mioc., ?L. Mioc., Plioc. and/or Pleist., R.; Af.[1,2]
>
>   [1] Petromyscinae indet. in the middle Miocene Otavi breccias, Berg Aukas I, Namibia.
>   [2] Southern Africa.

Petromyscus Thomas, 1926:179. Rock mice.
>   ?L. Mioc., Plioc. and/or Pleist., R.; Af.[1]
>
>   [1] Southern Africa.

Subfamily **Gerbillinae** Gray, 1825:342. Jirds, gerbils, sand rats.
>   [= Gerbillina Gray, 1825:342; Gerbillidae De Kay, 1842:xv, 70; Gerbillinae Alston, 1876:81.] [Including Merionina Brandt, 1844:231; Merionidinae Schmidtlein, 1893:401; Merioninae Heptner, 1933:10; Rhombomyinae Heptner, 1933:11; †Myocricetodontinae Lavocat, 1961; Taterillinae Chaline, Mein & F. Petter, 1977.] E. Mioc.-R.; As. M. Mioc.-R.; Af. L. Mioc.; Mediterranean. L. Mioc.-E. Plioc., R.; Eu.[1]
>
>   [1] E. Europe in the Recent.

†*Pseudomeriones* Schaub, 1934:31.
>   L. Mioc.-E. Plioc.; As.[1] L. Mioc.; Eu.[2]
>
>   [1] Including Rhodes in the early Pliocene.
>   [2] S. Europe (Spain and Greece).

Tribe †**Myocricetodontini** Lavocat, 1961:38, **new rank**.
>   [= †Myocricetodontinae Lavocat, 1961:38.]
>   E.-L. Mioc.; As.[1] M. Mioc.-E. Plioc., ?L. Plioc. and/or ?E. Pleist.; Af. L. Mioc.-E. Plioc.; Eu.
>
>   [1] †Myocricetodontinae genus and sp. indet. in the early Miocene of S. Asia (Pakistan).

†*Myocricetodon* Lavocat, 1952:190.
>   M. Mioc., L. Mioc. and/or E. Plioc.; Af.[1] M.-L. Mioc.; As.[2,3] L. Mioc.-E. Plioc.; Eu.
>
>   [1] N. Africa.
>   [2] S. Asia in the middle Miocene.
>   [3] S.W. Asia in the late Miocene.

†*Paradakkamys* Lindsay, 1988:132.
>   M.-L. Mioc.; As.[1]
>
>   [1] S. Asia (Pakistan).

†*Dakkamyoides* Lindsay, 1988:123.
>   M.-L. Mioc., Plioc. and/or E. Pleist.; Af.[1] M. Mioc.; As.[2]
>
>   [1] Southern Africa.
>   [2] S. Asia (Pakistan).

†*Dakkamys* Jaeger, 1977:91-125.
>   M. Mioc., ?L. Mioc.; Af.[1] M.-L. Mioc.; As.
>
>   [1] N. Africa.

†*Eulmus* Ameur-Chehbeur, 1991:510.
>   L. Mioc.-E. Plioc.; Af.[1]
>
>   [1] N. Africa (Algeria).

Tribe **Gerbillini** Gray, 1825:342.
>   [= Gerbillina Gray, 1825:342[1]; Gerbillini Pavlinov, 1982.] [Including Rhombomyinae Heptner, 1933:11; Rhombomyini Pavlinov & Rossolimo, 1987:212.]
>   L. Mioc., ?E. Plioc., L. Plioc.-R.; As. L. Plioc.-R.; Af. R.; Eu.[2]
>
>   [1] Proposed as tribe.
>   [2] E. Europe.

†*Abudhabia* de Bruijn & Whybrow, 1994:412.
>   L. Mioc.; As.

Subtribe **Gerbillina** Gray, 1825:342.[1]
>   [= Gerbillina Pavlinov, 1982.]
>   L. Mioc., Plioc., E. Pleist.-R.; As. L. Plioc.-R.; Af.
>
>   [1] Proposed as tribe.

Gerbillus Desmarest, 1804:22. Gerbils, pygmy gerbils, sand rats.
    [Including *Dipodillus* Lataste, 1881:506[1]; *Endecapleura* Lataste, 1882:127[2];
    Hendecapleura Thomas, 1883:28[3]; *Monodia* Heim de Balsac, 1943:287; *Petteromys*
    Pavlinov, 1982:30.]
    ?L. Mioc., Plioc., E. Pleist.-R.; As.[4,5] L. Plioc.-R.; Af.
    [1] Described as subgenus of *Gerbillus*.
    [2] Described as subgenus.
    [3] Emendation of *Endecapleura*. Sometimes as subgenus.
    [4] *Gerbillus*? in the late Miocene of Rhodes.
    [5] Pliocene records, mostly from China, may be examples of *Meriones*, not *Gerbillus*.

Microdillus Thomas, 1910:197.
    R.; Af.[1]
    [1] Somalia.

Subtribe **Merionina** Brandt, 1844:231.
    [= Merionides Giebel, 1855:580; Merioninae Heptner, 1933:10.] [Including
    Rhombomyinae Heptner, 1933:11; Rhombomyini Pavlinov & Rossolimo, 1987:212;
    Rhombomyina Pavlinov, Dubrovskii, Rossollimo & Potapova, 1990:101.]
    L. Plioc.-R.; Af.[1] L. Plioc.-R.; As. R.; Eu.[2]
    [1] N. Africa.
    [2] E. Europe.

†*Mascaramys* Tong, 1986.[1]
    L. Plioc.-M. Pleist.; Af.[2]
    [1] (Tung).
    [2] N. Africa.

Meriones Illiger, 1811:82. Tamarisk gerbils, jirds, sand rats.
    [= *Meriaeus* Billberg, 1828[1]; *Idomeneus* Schulze, 1900:201.] [Including *Cheliones*
    Thomas, 1919:265[2]; *Pallasiomys* Heptner, 1933:150[2]; *Parameriones* Heptner,
    1937:190.[2]]
    E. Pleist.-R.; Af.[3] E. Pleist.-R.; As.[4] R.; Eu.[5]
    [1] New name for *Meriones* Illiger.
    [2] Subgenus.
    [3] N. Africa.
    [4] W. Asia.
    [5] E. Europe.

Rhombomys Wagner, 1841:421. Great gerbil.
    [= *Amphiaulacomys* Lataste, 1882:11.] [Including †*Pliorhombomys* Fokanov,
    1976:122.[1]]
    L. Plioc.-R.; As.
    [1] *Nomen nudum* in 1964.

Psammomys Cretschmar, in Rüppell, 1828:56. Fat sand rats, diurnal sand rats.
    [Including *Parameriones* Tchernov & Chetboun, 1984:560.[1]]
    L. Pleist.-R.; As.[2] R.; Af.[3]
    [1] Not *Parameriones* Heptner, 1937, a synonym of *Meriones*.
    [2] W. Asia.
    [3] N. Africa and coastal Sudan.

Sekeetamys Ellerman, 1947:271. Bushy-tailed jird.
    R.; Af.[1] R.; As.[2]
    [1] N. Africa (eastern Egypt).
    [2] S.W. Asia (Sinai, Negev Desert, Jordan, Saudi Arabia).

Brachiones Thomas, 1925:548. Przewalski's gerbil.
    R.; As.[1]
    [1] China.

Subtribe **Desmodilliscina** Pavlinov, 1982:30.
    L. Pleist.-R.; Af.

Desmodilliscus Wettstein, 1916:153.
    L. Pleist.-R.; Af.

Subtribe **Pachyuromyina** Pavlinov, 1982:30.
R.; Af.[1]
[1] N. Africa.

Pachyuromys Lataste, 1880:313. Fat-tailed gerbil.
R.; Af.[1]
[1] N. Africa.

Tribe **Taterillini** Chaline, Mein & F. Petter, 1977:250, **new rank**.
[= Taterillinae Chaline, Mein & F. Petter, 1977:250.]
L. Mioc.-E. Pleist., ?M. Pleist., L. Pleist.-R.; Af. L. Mioc.-E. Plioc., L. Pleist.-R.;
As. L. Mioc.; Mediterranean.

†*Protatera* Jaeger, 1977:88.
L. Mioc.-E. Plioc.; Af.[1] L. Mioc.-E. Plioc.; As.[2] L. Mioc.; Mediterranean.[3]
[1] N. Africa.
[2] Afghanistan.
[3] Balearic Islands (Ibiza).

Subtribe **Taterillina** Chaline, Mein & F. Petter, 1977:250. Naked-soled gerbils.
[= Taterillinae Chaline, Mein & F. Petter, 1977:250; Taterillina Pavlinov, 1982.]
E. Plioc.-E. Pleist., L. Pleist.-R.; Af. L. Pleist.-R.; As.

Tatera Lataste, 1882:126.[1] Large naked-soled gerbil.
[Including *Gerbilliscus* Thomas, 1897:433[2]; *Taterona* Wroughton, 1917:40.[3]]
E. Plioc.-E. Pleist., L. Pleist.-R.; Af. L. Pleist.-R.; As.
[1] Proposed as subgenus of *Gerbillus*; raised to generic rank by Thomas, 1902:441-442.
[2] Subgenus of *Tatera*. Described as subgenus of *Gerbillus*; raised to generic rank by Thomas, 1902:441-442.
[3] Subgenus.

Taterillus Thomas, 1910:222. Small naked-soled gerbil.
[Including *Taterina* Wettstein, 1916:152.[1]]
?L. Plioc., L. Pleist.-R.; Af.
[1] Described as subgenus of *Gerbillus*.

Subtribe **Gerbillurina** Pavlinov, 1982:29.
Pleist., R.; Af.[1]
[1] Southern Africa.

Desmodillus Thomas & Schwann, 1904:6. Cape short-eared gerbil.
Pleist., R.; Af.[1]
[1] Southern Africa.

Gerbillurus Shortridge, 1942:161.
[Including *Progerbillurus* Pavlinov, 1982:23[1]; *Paratatera* Petter, 1983:266.[1]]
R.; Af.[2]
[1] Considered a subgenus of *Gerbillurus* by Meester et al., 1986.
[2] Southern Africa.

Tribe **Ammodillini** Pavlinov, 1981.
R.; Af.

Ammodillus Thomas, 1904:102. Ammodille.
R.; Af.[1]
[1] Ethiopia, Somalia.

Subfamily **Dendromurinae** G. M. Allen, 1939:82.
[= Dendromyinae Alston, 1876:82; Dendromuridae Chaline, Mein & F. Petter,
1977:245-252.] [Including Deomyinae Lydekker, 1889.]
M. Mioc.-R.; Af. L. Mioc.; As.[1] L. Mioc.; Eu.
[1] S.W. Asia.

†*Ternania* Tong & Jaeger, 1993:62.
M. Mioc.; Af.

†*Senoussimys* Ameur, 1984:167-175.
L. Mioc.; Af.[1]
[1] N. Africa (Algeria).

Dendromus Smith, 1829:438. African climbing mice, tree mice.
    [= *Dendromys* Smuts, 1832:32.] [Including *Chortomys* Thomas, 1916:238[1]; *Poemys* Thomas, 1916:238.[1]]
    L. Mioc.-R.; Af. L. Mioc.; As.[2] L. Mioc.; Eu.
  [1] Subgenus as defined.
  [2] S.W. Asia (Arabian Peninsula).

Steatomys Peters, 1846:258. Fat mice.
    ?M. and/or ?L. Mioc., L. Plioc.-R.; Af.

Malacothrix Wagner, 1843:496.[1] Large-eared mouse, gerbil mouse.
    [= *Otomys* A. Smith, 1834:147-148.[2]]
    L. Plioc.-R.; Af.
  [1] Replacement name for *Otomys* A. Smith, 1834.
  [2] Not *Otomys* F. Cuvier, 1824, another rodent.

Megadendromus Dieterlen & Rupp, 1978:129.
    R.; Af.[1]
  [1] Ethiopia.

Dendroprionomys F. Petter, 1966:129.
    R.; Af.[1]
  [1] Congo.

Prionomys Dollman, 1910:226. Dollman's tree mouse.
    R.; Af.

Deomys Thomas, 1888:130. Link mice.
    R.; Af.

Leimacomys Matschie, 1893:107. Togo mice.
    [= *Limacomys* Lydekker, 1894:31.]
    R.; Af.[1]
  [1] Togo.

Subfamily **Murinae** Illiger, 1811:84. Old World rats, Old World mice.
    [= Murina Illiger, 1811:84[1]; Murina Gray, 1825:341[2]; Murinae Murray, 1866:xv.]
    [Including Hydromina Gray, 1825:341[3]; Hydromyinae Alston, 1876:80; Hydromyini Lee, Baverstock & Watts, in Keast, ed., 1981:1530; Rattidae Burnett, 1830:350; Phloeomyinae Alston, 1876:81; Rhynchomyinae Thomas, 1897:1017; Otomyinae Thomas, 1897:1017; Otomyini Tullberg, 1899; Pseudomyinae Simpson, 1961:433; Conilurini Lee, Baverstock & Watts, in Keast, ed., 1981:1530; Uromyini Lee, Baverstock & Watts, in Keast, ed., 1981:1530; Anisomyini Lidicker & Brylski, 1987:635.]
    M. Mioc.-R.; Af.[4] M. Mioc.-R.; As. M. Mioc.-R.; Eu. L. Mioc., ?Plioc., E. Pleist.-R.; Mediterranean.[5,6,7] Plioc., E. Pleist.-R.; Aus. L. Plioc., Pleist., R.; New Guinea.[8] ?E. and/or ?M. Pleist., L. Pleist.-R.; Atlantic.[9] M. Pleist.-R.; E. Indies. R.; Indian O.[10] R.; Pacific.[11]
  [1] Proposed as family.
  [2] As tribe.
  [3] Proposed as tribe.
  [4] N. Africa in the middle Miocene.
  [5] Crete in the L. Miocene.
  [6] Sicily and Sardinia in the Pliocene?
  [7] Crete, Sicily, Sardinia, and Corsica in the Pleistocene and Recent.
  [8] Indet. "hydromyine" in the late Pliocene.
  [9] Canary Islands in the Pleistocene and Recent (now extinct); Iceland in the Recent (extant).
  [10] Christmas Island (endemics extinct; reintroduced).
  [11] W. Pacific.
    Also Recent worldwide by introduction.

Acomys I. Geoffroy Saint-Hilaire, 1838:126.[1] Spiny mice.
    [= *Acanthomys* Lesson, 1842:135.[2]] [Including *Peracomys* F. Petter & Roche, 1981:381[3]; *Subacomys* Denys, Gautun, Tranier & Volobouev, 1994:217.[3]]
    E. Plioc.-E. Pleist., L. Pleist.-R.; Af Plioc.; Eu. L. Pleist.-R.; As.[4] R.; Mediterranean.[5]
  [1] Molecular and morphological evidence suggests that *Acomys*, *Uranomys* and *Lophuromys* form a monophyletic group.

[2] Not *Acanthomys* Gray, 1867, a synonym of *Rattus*, nor *Acanthomys* Tokuda, 1941, a synonym of *Tokudaia*.
[3] Subgenus as described.
[4] W. Asia.
[5] Crete.

Uranomys Dollman, 1909:551.[1]
R.; Af.

[1] Molecular and morphological evidence suggests that *Acomys*, *Uranomys*, and *Lophuromys* form a monophyletic group.

Lophuromys Peters, 1874:234.[1]
[= *Lasiomys* Peters, 1866:409.[2]] [Including *Neanthomys* Toschi, 1947:4; *Kivumys* Dieterlen, 1987:193.[3]]
R.; Af.

[1] Molecular and morphological evidence suggests that *Acomys*, *Uranomys*, and *Lophuromys* form a monophyletic group.
[2] Not *Lasiomys* Burmeister, 1854, a synonym of *Sigmodon*.
[3] Subgenus.

†*Huaxiamys* Wu & L. Flynn, 1992:19.[1]
L. Mioc.-E. Plioc.; As.

[1] Adequate diagnosis not provided. Strict application of Article 13(a)(i) of the International Code of Zoological Nomenclature would probably require rejection of this name.

†*Antemus* Jacobs, 1978:3.
M. Mioc.; As.[1]

[1] S. Asia (Pakistan).

†*Progonomys* Schaub, 1938:19.
[Including †*Karnimata* Jacobs, 1978:51; †*Proceromys* Brandy, 1979:81-82.]
M.-L. Mioc.; Af.[1] M. Mioc.-E. Plioc.; As. M.-L. Mioc.; Eu. L. Mioc.; Mediterranean.[2]

[1] N. Africa.
[2] Crete.

†*Huerzelerimys* Mein, Martín Suárez & Agustí, 1993:49.
L. Mioc.; Eu.

†*Ratchaburimys* Chaimanee, Suteethorn, Triamwichanon & Jaeger, 1996:156.
E. Pleist.; As.

Millardia Thomas, 1911:998.
[Including *Grypomys* Thomas, 1911:999; *Guyia* Thomas, 1917:201; *Millardomys* Sody, 1941:261.]
?E. Pleist., R.; As.[1]

[1] Afghanistan, S. Asia, and Burma.

Cremnomys Wroughton, 1912:340.
[Including *Madromys* Sody, 1941:260.]
R.; As.[1]

[1] S. Asia (India, Sri Lanka).

Diomys Thomas, 1917:203.
R.; As.[1]

[1] N.E. India, Nepal, and N. Burma.

Micromys Dehne, 1841:1.
L. Mioc.-E. Plioc., Pleist., R.; As. L. Mioc.-R.; Eu.

Vandeleuria Gray, 1842:265.
E. Pleist., R.; As.[1]

[1] S. and S.E. Asia in the Recent.

Vernaya Anthony, 1941:110.
[= *Octopodomys* Sody, 1941:261.]
L. Plioc.-M. Pleist., R.; As.[1]

[1] S. China, Burma.

Chiropodomys Peters, 1868:448.
[Including *Insulaemus* Taylor, 1934:401, 469.]

L. Plioc., R.; As.[1,2] R.; E. Indies.[3]

[1] S. China in the Late Pliocene.
[2] S.E. Asia. in the Recent.
[3] Sumatra, Nias Island, Mentawai Islands, Java, Bali, Natuna Islands, Borneo, Philippines.

Hapalomys Blyth, 1859:296.

L. Plioc., L. Pleist.-R.; As.[1,2]

[1] S. China in the Pliocene and Pleistocene.
[2] S.E. Asia in the Recent.

Apodemus Kaup, 1829:154.

[Including *Sylvaemus* Ognev, in Ognev & Vorobiev, 1924:143[1]; *Nemomys* Thomas, 1924:889; *Alsomys* Dukelski, 1928:42[1]; *Silvimus* Ognev & Heptner, 1929:96; *Petromys* Martino, 1935:85[2]; *Karstomys* Martino, 1939:88.[3]]

L. Mioc.-R.; As. L. Mioc.-R.; Eu. E.-L. Plioc., L. Pleist.-R.; Af.[4] ?Plioc., R.; Mediterranean.[5,6] R.; Atlantic.[7]

[1] Subgenus.
[2] Not *Petromys* Smith, 1834, an unjustifiable emendation of *Petromus* A. Smith, 1831.
[3] New name for *Petromys* Martino, 1935. Subgenus as proposed.
[4] N. Africa.
[5] Sicily and Sardinia in the ?Pliocene.
[6] Crete, Sicily, and Corsica in the Recent.
[7] N. Atlantic (Iceland).

†*Rhagapodemus* Kretzoi, 1959:137-246.

L. Mioc.-E. Plioc.; As. L. Mioc.-L. Plioc., Pleist.; Eu.[1] ?Plioc.; Mediterranean.[2]

[1] Including Rhodes in the early Pliocene.
[2] Sardinia.

†*Rhagamys* Forsyth Major, 1905:503.

E. Pleist.-R.; Mediterranean.[1]

[1] Sardinia and Corsica; to subrecent only.

Tokudaia Kuroda, 1943:61.

[= *Acanthomys* Tokuda, 1941:93[1]; *Tokudamys* Johnson, 1946:169.]

L. Pleist.-R.; As.[2]

[1] Not *Acanthomys* Lesson, 1842, a synonym of *Acomys*, nor *Acanthomys* Gray, 1867, a synonym of *Rattus*.
[2] Ryukyu Islands (Okinawa and Le-jima in the late Pleistocene; Okinawa and Amami-oshima in the Recent).

†*Microtia* Freudenthal, 1976:3.

L. Mioc.; Eu.

†*Paraethomys* F. Petter, 1968:54.

L. Mioc.-L. Pleist.; Af.[1] L. Mioc.-E. Plioc.; As.[2,3] L. Mioc.-L. Plioc.; Eu.[4]

[1] N. Africa.
[2] Turkey in the L. Mioc.
[3] Rhodes in the E. Plioc.
[4] Spain, France.

†*Castillomys* Michaux, 1969:5.

[Including †*Occitanomys* Michaux, 1969:8; †*Valerymys* Michaux, 1969:11; †*Centralomys* de Giuli, 1989:204[1]; †*Hansdebruijnia* Storch & Dahlmann, 1995:130.[1]]

L. Mioc.-E. Plioc.; As. L. Mioc.-E. Pleist.; Eu. E. Plioc.; Af.[2]

[1] Subgenus as defined.
[2] N. Africa.

†*Orientalomys* de Bruijn & van der Meulen, 1975:317.

[Including †*Euxinomys* Sen, 1975:317-324.]

L. Mioc.-L. Plioc.; As. E.-L. Plioc.; Eu.[1]

[1] E. Europe.

†*Chardinomys* Jacobs & Li, 1982:256.

[Including †*Teilhardomys* Cordy, 1977.[1]]

E. Plioc.-E. Pleist.; As.[2]

[1] PhD thesis.
[2] China.

†*Parapodemus* Schaub, 1938:14.

L. Mioc.; As.[1] L. Mioc.-L. Plioc., Pleist.; Eu.

[1] China, Pakistan, Afghanistan.

†*Castromys* Martín Suárez & Freudenthal, 1994:12.
>   L. Mioc.; Eu.

†*Stephanomys* Schaub, 1938:21.
>   L. Mioc.-E. Pleist.; Eu. E. Plioc.; Af.[1]

[1] N. Africa.

†*Anthracomys* Schaub, 1938:17.
>   L. Mioc.; Af.[1] L. Mioc.-E. Plioc.; Eu.

[i] N. Africa.

†*Beremendimys* Kretzoi, 1956:41.
>   E. Pleist.; Eu.[1]

[1] E. Europe.

Mus Linnaeus, 1758:59. House mice.
>   [= *Musculus* Rafinesque, 1814:13; *Drymomys* Tschudi, 1844:178-180; *Mammus* Herrera, 1899:24.[1]] [Including *Leggada* Gray, 1837:586; *Nannomys* Peters, 1876:480-481[2]; *Pseudoconomys* Rhoads, 1896:531-532; *Pyromys* Thomas, 1911:996[2]; *Leggadilla* Thomas, 1914:682; *Coelomys* Thomas, 1915:414[2]; *Oromys* Robinson & Kloss, 1916:270[3]; *Tautatus* Kloss, 1917:279; *Mycteromys* Robinson & Kloss, 1918:57; *Hylenomys* Thomas, 1925:667; †*Budamys* Kretzoi & Vertes, 1965:76; *Gatamiya* Deraniyagala, 1966:214.]
>   L. Mioc.-R.; As. L. Plioc.-R.; Af. L. Plioc.-R.; Eu. L. Pleist.; Mediterranean.[4] R.; E. Indies.[5]

[1] See the International Code of Zoological Nomenclature, Article 1(b)(8), and Palmer (1904:25-26).
[2] Subgenus.
[3] Objective synonym of *Mycteromys*. Not *Oromys* Leidy, 1853: 241, *nomen nudum*, a synonym of †*Neochoerus*.
[4] Crete.
[5] Sumatra and Java.
Also Recent worldwide by introduction.

Muriculus Thomas, 1903:314.
>   R.; Af.[1]

[1] Ethiopia.

†*Saidomys* James & Slaughter, 1974:342.
>   L. Mioc., L. Plioc.; Af.[1,2] E. Plioc.; As.[3]

[1] N. Africa (Egypt) and possibly E. Africa (Tanzania) in the late Miocene.
[2] Ethiopia in the late Pliocene.
[3] Afghanistan.

†*Yunomys* Qiu & Storch, 1990:470.
>   L. Mioc.; As.

†*Parapelomys* Jacobs, 1978:61.
>   L. Mioc.-L. Plioc.; As.

Hadromys Thomas, 1911:999.
>   E. Pleist., R.; As.[1,2]

[1] Pakistan in the early Pleistocene.
[2] N.E. India and S. China (Yunnan) in the Recent.

Golunda Gray, 1837:586.
>   ?L. Plioc.; Af.[1] L. Plioc., Pleist., R.; As.[2]

[1] Ethiopia.
[2] S. Asia.

Pelomys Peters, 1852:275.[1] Groove-toothed rat.
>   [Including *Komemys* de Beaux, 1924:207.[2]]
>   E. Plioc.; As.[3] L. Plioc., R.; Af.

[1] Proposed as subgenus of *Mus*; raised to generic rank by Peters the same year in another paper.
[2] Sometimes as subgenus.
[3] Rhodes and Afganistan.

Desmomys Thomas, 1910:284.
>   R.; Af.[1]

[1] Ethiopia.

Mylomys Thomas, 1906:224.
    R.; Af.

†*Dilatomys* Sen, 1983:33, 50.
    E. Plioc.; As.[1]
    [1] Afghanistan.

Arvicanthis Lesson, 1842:147.
    [= *Rattus* Donovan (?), 1827: unpaged plate;[1] *Arviacanthis* Beddard, 1902:473.]
    [Including *Isomys* Sundevall, 1842:219-220.]
    E. Plioc., L. Pleist.-R.; As.[2] L. Plioc.-M. Pleist., ?L. Pleist., R.; Af.
    [1] Not *Rattus* Fischer de Waldheim, 1803, nor *Rattus* Frisch, 1775.
    [2] S.W. Asia (Arabia) in the Recent.

Lemniscomys Trouessart, 1881:124.[1] Single-striped mouse.
    L. Plioc., ?E. Pleist., M. Pleist.-R.; Af.
    [1] Proposed as subgenus of *Mus*.

Rhabdomys Thomas, 1916:69. Striped mouse.
    L. Plioc., ?Pleist., R.; Af.

†*Euryotomys* T. N. Pocock, 1976:58.
    E. Plioc.; Af.[1]
    [1] S. Africa.

Otomys F. Cuvier, 1824:255, pl. 60. Vlei rats.
    [= *Euryotis* Brants, 1827:93-99.] [Including *Oreomys* Heuglin, 1877:76-77;
    Oreinomys Trouessart, 1881:111[1]; *Myotomys* Thomas, 1918:204, 206; *Anchotomys*
    Thomas, 1918:204, 208[2]; *Lamotomys* Thomas, 1918:208[2]; *Palaeotomys* Broom,
    1937:764; *Metotomys* Broom, 1937:765; †*Prototomys* Broom, 1948:1-38.]
    L. Plioc.-R.; Af.
    [1] Replacement for *Oreomys* Heuglin, 1877, supposedly preoccupied by *Orenomys* Aymard, 1855, an
      hystricid.
    [2] Described as subgenus.

Parotomys Thomas, 1918:204. Karoo rats.
    [Including *Liotomys* Thomas, 1918:204, 205.]
    M. Pleist.-R.; Af.[1]
    [1] S. Africa.

Dasymys Peters, 1875:12.
    L. Plioc.-E. Pleist., R.; Af.

Lamottemys F. Petter, 1986:98.
    R.; Af.[1]
    [1] Cameroon.

Oenomys Thomas, 1904:416.
    L. Plioc., R.; Af.

Thamnomys Thomas, 1907:121.
    L. Pleist.-R.; Af.

Thallomys Thomas, 1920:141. Tree rats.
    E.-L. Plioc., L. Pleist.-R.; Af.

Grammomys Thomas, 1915:150. Woodland mice.
    L. Plioc.-R.; Af.

Aethomys Thomas, 1915:477.[1]
    [Including *Micaelamys* Ellerman, 1941:170.[2]]
    E. Plioc.-R.; Af.
    [1] Described as subgenus of *Epimys*.
    [2] Described as subgenus of *Rattus*; now regarded as subgenus of *Aethomys*.

Stochomys Thomas, 1926:176.
    R.; Af.

Dephomys Thomas, 1926:177.
    R.; Af.

Hybomys Thomas, 1910:85.
>[Including *Typomys* Thomas, 1911:382.[1]]
>R.; Af.

[1] Subgenus.

Heimyscus Misonne, 1969:125.
>R.; Af.

†*Canariomys* Crusafont Pairó & F. Petter, 1964:607.
>Pleist., R.; Atlantic.[1]

[1] Canary Islands, up until about 2000 yrs. B.P.

†*Malpaisomys* Hutterer, López-Martínez & Michaux, 1988:246.
>L. Pleist.-R.; Atlantic.[1]

[1] Canary Islands.

Zelotomys Osgood, 1910:7. Desert rat.
>[Including *Ochromys* Thomas, 1920:142.]
>L. Plioc.-R.; Af.

Colomys Thomas & Wroughton, 1907:379.
>[= *Nilopegamys* Osgood, 1928:185.]
>Pleist., R.; Af.[1]

[1] Zaire basin.

Nilopegamys Osgood, 1908:185. Ethiopian water mouse.
>R.; Af.[1]

[1] Ethiopia.

Stenocephalemys Frick, 1914:7.
>R.; Af.[1]

[1] Ethiopia.

Praomys Thomas, 1915:477.
>L. Plioc.-R.; Af.

Mastomys Thomas, 1915:477. Multimammate rats.
>L. Plioc.-R.; Af. L. Pleist.; As.[1]

[1] S.W. Asia (Israel).

Myomys Thomas, 1915:477.
>[= *Myomyscus* Shortridge, 1942:93.[1]]
>R.; Af. R.; As.[2]

[1] Described as subgenus of *Myomys*.
[2] S.W. Asia (Yemen and Saudia Arabia).

Hylomyscus Thomas, 1926:174.
>R.; Af.

Malacomys Milne-Edwards, 1877:10.
>R.; Af.

†*Kritimys* Kuss & Misonne, 1968:62.
>M. Pleist.; Mediterranean.[1]

[1] Crete.

Rattus Fischer de Waldheim, 1803:128. Rats.
>[= *Rattus* Frisch, 1775: unpaged plate[1]] [Including *Acanthomys* Gray, 1867:598-599[2]; *Epimys* Trouessart, 1881:117; *Christomys* Sody, 1941:260; *Cironomys* Sody, 1941:260; *Geromys* Sody, 1941:260; *Mollicomys* Sody, 1941:260; *Pullomys* Sody, 1941:260; *Octomys* Sody, 1941:261[3]; *Togomys* Dieterlen, 1986:12.]
>E. Pleist.-R.; As. E. Pleist.-R.; Aus. M. Pleist.-R.; E. Indies. R.; Indian O.[4] R.; New Guinea. R.; Pacific.[5]

[1] Invalid because not published in a consistently binominal work. See the International Code of Zoological Nomenclature, Article 11(c).
[2] Not *Acanthomys* Lesson, 1842, a synonym of *Acomys*, nor *Acanthomys* Tokuda, 1941, a synonym of *Tokudaia*.
[3] Not *Octomys* Thomas, 1920, an octodontid.
[4] Christmas Island (endemic species extinct; genus introduced).
[5] W. Pacific.
  Also Recent worldwide by introduction.

Nesokia Gray, 1842:264. Pest rat.
> [= *Spalacomys* Peters, 1861:139; *Nesocia* Blanford, 1891:421.] [Including
> *Erythronesokia* Khajuria, 1981.[1]]
> Plioc., ?E. Pleist., R.; As. L. Pleist.-R.; Af.[2]
> [1] Sometimes as subgenus.
> [2] N. Africa (Egypt and N. Sudan).

Bandicota Gray, 1873:418. Bandicoot rats.
> [Including *Gunomys* Thomas, 1907:203.]
> R.; As.[1]
> [1] Also Recent of E. Indies (Sumatra and Java) by introduction.

Berylmys Ellerman, 1947:261.[1]
> E.-M. Pleist., R.; As.[2,3] R.; E. Indies.[4]
> [1] Described as a subgenus of *Rattus*; elevated to generic rank by Musser & Newcomb, 1983.
> [2] S. China in the Pleistocene.
> [3] S.E. Asia in the Recent.
> [4] N.W. Sumatra.

Diplothrix Thomas, 1916:404.
> L. Pleist.-R.; As.[1]
> [1] Japan (Ryukyu Islands).

Sundamys Musser & Newcomb, 1983:401.
> R.; E. Indies.[1] R.; As.[2]
> [1] Palawan, Borneo, Sumatra, Java.
> [2] S.E. Asia (Malay Peninsula).

Palawanomys Musser & Newcomb, 1983:335.
> R.; E. Indies.[1]
> [1] Philippines (Palawan).

Kadarsanomys Musser, 1981:5.
> R.; E. Indies.[1]
> [1] Java in the subrecent and Recent.

Tryphomys Miller, 1910:399.
> R.; E. Indies.[1]
> [1] Philippines (Luzon).

Abditomys Musser, 1982:3.
> R.; E. Indies.[1]
> [1] Philippines (Luzon).

Bullimus Mearns, 1905:450.
> R.; E. Indies.[1]
> [1] Philippines.

Tarsomys Mearns, 1905:453.
> R.; E. Indies.[1]
> [1] Philippines (Mindanao).

Limnomys Mearns, 1905:451.
> R.; E. Indies.[1]
> [1] Philippines (Mindanao).

Taeromys Sody, 1941:260.
> [Including *Arcuomys* Sody, 1941:260.]
> R.; E. Indies.[1]
> [1] Sulawesi in the subrecent and Recent.

Paruromys Ellerman, in Laurie & Hill, 1954:117.
> R.; E. Indies.[1]
> [1] Sulawesi in the subrecent and Recent.

†*Hooijeromys* Musser, 1981:69, 71, 99.
> L. Pleist.; E. Indies.[1]
> [1] Lesser Sunda Islands (Flores).

Papagomys Sody, 1941:262.

R.; E. Indies.[1]

[1] Lesser Sunda Islands (Flores) in the subrecent and Recent.

Komodomys Musser & Boeadi, 1980:397.

R.; E. Indies.[1]

[1] Rintja, Padar, and Flores Islands in the subrecent and Recent.

Bunomys Thomas, 1910:508.

[Including *Frateromys* Sody, 1941:260.]

R.; E. Indies.[1]

[1] Sulawesi in the subrecent and Recent.

Paulamys Musser, in Musser, van de Weerd & Strasser, 1986:1.

[= †*Floresomys* Musser, 1981.[1]]

R.; E. Indies.[2]

[1] Not †*Floresomys* Fries, Hibbard & Dunkle, 1955, a geomyoid.
[2] Lesser Sunda Islands (Flores).

Stenomys Thomas, 1910:507.

[Including *Nesoromys* Thomas, 1920:425.]

R.; E. Indies.[1] R.; New Guinea.

[1] Moluccas (Seram).

Maxomys Sody, 1936:55.

R.; E. Indies.[1] R.; As.[2]

[1] Borneo, Indonesian islands (including Sulawesi in the subrecent) and Philippines.
[2] S.E. Asia.

Leopoldamys Ellerman, 1947:267.

L. Plioc.-R.; As.[1] R.; E. Indies.[2]

[1] Late Pliocene to Recent of China and Recent of S.E. Asia (N.E. India to the Malay Peninsula).
[2] Mentawai Islands, Sumatra (subrecent to Recent), Borneo, smaller islands of the Sunda Shelf, and Java (subrecent to Recent).

†*Wushanomys* Zheng, 1993:188.

L. Plioc.-E. Pleist.; As.[1]

[1] China.

†*Qianomys* Zheng, 1993:195.

M. Pleist.; As.[1]

[1] China.

Niviventer Marshall, 1976:402.

L. Plioc.-R.; As.[1,2] R.; E. Indies.[3]

[1] China in the Pliocene and Pleistocene.
[2] S. and S.E. Asia in the Recent.
[3] Borneo, Sumatra, Java, Bali and smaller islands on the Sunda Shelf only.

Chiromyscus Thomas, 1925:503.

R.; As.[1]

[1] S.E. Asia.

Dacnomys Thomas, 1916:404.

R.; As.[1]

[1] Nepal, Sikkim, Assam, Laos, S. China.

Srilankamys Musser, 1981:268.

R.; As.[1]

[1] S. Asia (Sri Lanka).

Lenothrix Miller, 1903:466.

R.; E. Indies.[1] R.; As.[2]

[1] Borneo, Tuangku Island.
[2] S.E. Asia (Malaya).

Pithecheir F. Cuvier, 1838:447.

[= *Pithecochirus* Gloger, 1841:93; *Pithechirus* Agassiz, 1842:26; *Pitechirus* Kaup, 1844:76[1]; *Pitcheir* Schinz, 1845:260; *Pithechir* Jentink, 1892:122-126.] [Including *Pithecheirops* Emmons, 1993:752.]

R.; E. Indies.[2] R.; As.[3]

[1] Misprint.
[2] Java, Borneo.
[3] S.E. Asia (Malay Peninsula).

Eropeplus Miller & Hollister, 1921:94.
   R.; E. Indies.[1]
[1] Sulawesi.

Lenomys Thomas, 1898:409.
   R.; E. Indies.[1]
[1] Sulawesi in the subrecent and Recent.

Margaretamys Musser, 1981:275.
   R.; E. Indies.[1]
[1] Sulawesi.

Melasmothrix Miller & Hollister, 1921:93.
   R.; E. Indies.[1]
[1] Sulawesi.

Tateomys Musser, 1969:2.
   R.; E. Indies.[1]
[1] Sulawesi.

Haeromys Thomas, 1911:207.
   R.; E. Indies.[1]
[1] Borneo, Sulawesi, Philippines (Palawan).

Anonymomys Musser, 1981:300.
   R.; E. Indies.[1]
[1] Philippines (Mindoro).

Echiothrix Gray, 1867:599.
   [= *Craurothrix* Thomas, 1896:246.[1]]
   R.; E. Indies.[2]
[1] Replacement name.
[2] Sulawesi.

Phloeomys Waterhouse, 1839:108.[1]
   R.; E. Indies.[2]
[1] Proposed as subgenus of *Mus*; raised to generic rank by Gray, 1850:20.
[2] Philippines (Luzon, Marinduque, Catanduanes).

Crateromys Thomas, 1895:163.
   R.; E. Indies.[1]
[1] Philippines (Luzon, Ilin, Dinagat).

Batomys Thomas, 1895:162-163.
   [Including *Mindanaomys* Sanborn, 1953:287.]
   R.; E. Indies.[1]
[1] Philippines (Luzon, Biliran, Leyte, Mindanao).

Carpomys Thomas, 1895:161.
   R.; E. Indies.[1]
[1] Philippines (Luzon).

Apomys Mearns, 1905:455.
   R.; E. Indies.[1]
[1] Philippines (Luzon, Catanduanes, Mindoro, Leyte, Dinagat, Mindanao, Negros).

Rhynchomys Thomas, 1895:160. Shrew rat.
   R.; E. Indies.[1]
[1] Philippines (Luzon).

Crunomys Thomas, 1897:393.
   R.; E. Indies.[1]
[1] Philippines (Luzon, Leyte, Mindanao), Sulawesi.

Archboldomys Musser, 1982:30.
    R.; E. Indies.[1]
    [1] Philippines (Luzon).

Chrotomys Thomas, 1895:161.
    R.; E. Indies.[1]
    [1] Philippines (Luzon, Mindoro).

Celaenomys Thomas, 1898:390.
    R.; E. Indies.[1]
    [1] Philippines (Luzon).

Pseudomys Gray, 1832:39. Broad-toothed rat, Australian false mice, Australian field mice.
    [= †*Paraleporillus* Martínez & Lidicker, 1971:775.] [Including *Mastacomys*
    Thomas, 1882:413; *Thetomys* Thomas, 1910:606; *Gyomys* Thomas, 1910:607.]
    Plioc., E. Pleist.-R.; Aus.[1] R.; New Guinea.
    [1] Including Tasmania in the Recent.

Notomys Lesson, 1842:129. Hopping mice.
    [Including *Podanomalus* Waite, 1898:117-121; *Thylacomys* Waite, 1898:121-124[1];
    Ascopharynx Waite, 1900:223.[2]]
    Pleist., R.; Aus.[3]
    [1] Not *Thylacomys* Owen, a marsupial.
    [2] Replacement name for *Thylacomys* Waite.
    [3] Including Tasmania.

Conilurus Ogilby, 1838:124. Brush-tailed tree rat, white-footed tree rat, rabbit rat.
    [= *Hapalotis* Lichtenstein, 1829.[1]]
    E. Pleist.-R.; Aus. R.; New Guinea.
    [1] Not *Hapalotis* Hübner, 1816, a genus of lepidopterans.

Leporillus Thomas, 1906:83. Stick-nest rats, tillikins.
    E. Pleist.-R.; Aus.

Mesembriomys Palmer, 1906:97. Golden-backed tree rats, black-footed tree rats.
    [= *Ammomys* Thomas, 1906:83.[1]]
    ?Pleist., R.; Aus.
    [1] Not *Ammomys* Bonaparte, 1831, another rodent.

Zyzomys Thomas, 1909:372. Rock rats.
    [Including *Laomys* Thomas, 1909:373.]
    ?Pleist., R.; Aus.

Leggadina Thomas, 1910:606.[1]
    Pleist., R.; Aus.
    [1] Described as a subgenus of *Pseudomys* Gray, 1832.

Xeromys Thomas, 1889:248.
    R.; Aus.

Hydromys É. Geoffroy Saint-Hilaire, 1804:353.
    [Including *Baiyankomys* Hinton, 1943:552.]
    Pleist., R.; Aus. R.; E. Indies.[1] R.; New Guinea. R.; Pacific.[2]
    [1] Obi Island, Moluccas.
    [2] W. Pacific (New Britain).

Crossomys Thomas, 1907:70.
    R.; New Guinea.

Microhydromys Tate & Archbold, 1941:2.
    R.; New Guinea.

Parahydromys Poche, 1906:326.[1]
    [= *Limnomys* Thomas, 1906:325[2]; *Drosomys* Thomas, 1906:199.[1]]
    R.; New Guinea.
    [1] Replacement name for *Limnomys* Thomas, 1906.
    [2] Not *Limnomys* Mearns, 1905, another murine genus.

Paraleptomys Tate & Archbold, 1941:1.
    R.; New Guinea.

Mayermys Laurie & Hill, 1954:133.
  R.; New Guinea.

Neohydromys Laurie, 1952:311.
  R.; New Guinea.

Pseudohydromys Rümmler, 1934:47.
  R.; New Guinea.

Leptomys Thomas, 1897:610.
  R.; New Guinea.

Lorentzimys Jentink, 1911:166.
  R.; New Guinea.

Melomys Thomas, 1922:261.
  [Including *Paramelomys* Rümmler, 1936:248; *Mammelomys* Menzies, 1996:383; *Protochromys* Menzies, 1996:416.]
  Pleist., R.; Aus. R.; E. Indies.[1] R.; New Guinea. R.; Pacific.[2]
  [1] Timor (subrecent), Moluccas, Seram, Obi Island, Talaud Island.
  [2] W. Pacific (Bismarcks, Solomons).

Pogonomelomys Rümmler, 1936:248.[1]
  [Including *Abelomelomys* Menzies, 1990:133.]
  R.; E. Indies.[2] R.; New Guinea.
  [1] Described as subgenus of *Melomys*; elevated to generic rank by Tate & Archbold, 1941.
  [2] Including Timor in the subrecent.

Coccymys Menzies, 1990:132.
  R.; New Guinea.

Uromys Peters, 1867:343.
  [= *Gymnomys* Gray, 1867:597-598.] [Including *Cyromys* Thomas, 1910:507[1]; *Melanomys* Winter, 1983.[2]]
  Pleist., R.; New Guinea. R.; E. Indies.[3] R.; Aus.[4] R.; Pacific.[5]
  [1] Subgenus.
  [2] Not *Melanomys* Thomas, 1902, a synonym of *Oryzomys*, a sigmodontine murid.
  [3] E. Moluccas.
  [4] Queensland.
  [5] W. Pacific (New Britain, Guadalcanal).

Solomys Thomas, 1922:261.
  [Including *Unicomys* Troughton, 1935:259.]
  R.; Pacific.[1]
  [1] W. Pacific (Solomon Islands).

Xenuromys Tate & Archbold, 1941:3.
  R.; New Guinea.

Hyomys Thomas, 1904:198.
  Pleist., R.; New Guinea.

Anisomys Thomas, 1904:199.
  Pleist., R.; New Guinea.

Mallomys Thomas, 1898:1.
  [= *Dendrosminthus* de Vis, 1907:10.]
  Pleist., R.; New Guinea.

Macruromys Stein, 1933:94.
  R.; New Guinea.

†*Spelaeomys* Hooijer, 1957:306.
  R.; E. Indies.[1]
  [1] Lesser Sunda Islands (Flores) in the subrecent.

†*Coryphomys* Schaub, 1937:2.
  R.; E. Indies.[1]
  [1] Timor in the subrecent.

Chiruromys Thomas, 1888:237.
  R.; New Guinea.

Pogonomys Milne-Edwards, 1877:1081.
>R.; Aus. R.; New Guinea. R.; Pacific.[1]
>
>[1] W. Pacific (Bismarcks).

Subfamily **Platacanthomyinae** Alston, 1876:81.
>[= Platacanthomyidae Miller & Gidley, 1918:437;
>Platacanthomyini Stehlin & Schaub, 1951:330.]
>E.-L. Mioc.; Eu. M.-L. Mioc., L. Plioc.-R.; As.

†*Neocometes* Schaub & Zapfe, 1953:198.
>E.-L. Mioc.; Eu. M. Mioc.; As.

Typhlomys Milne-Edwards, 1877:9. Blind tree mouse.
>L. Mioc., L. Plioc.-R.; As.[1]
>
>[1] N. Vietnam and S. China.

Platacanthomys Blyth, 1859:288. Malabar spiny mouse.
>[= *Platacanthomus* Marschall, 1873:10; *Platyacanthomys* Coues, 1890:4536.]
>L. Mioc., R.; As.[1,2]
>
>[1] China (Yunnan) in the late Miocene.
>[2] S. Asia (S.W. India) in the Recent.

Subfamily **Myospalacinae** Lilljeborg, 1866:25.
>[= Myospalacini Lilljeborg, 1866:25; Myospalacinae Miller & Gidley, 1918:438;
>Myospalacidae Kretzoi, 1961:124; Siphneinae Gill, 1872:20; Myotalpinae Miller,
>1896:8; †Mesosiphneinae Zheng, in Tomida, Li & Setoguchi, 1994:62.]
>[Including †Prosiphneinae Leroy, 1940:167-193.]
>L. Mioc.-R.; As.

†*Prosiphneus* Teilhard de Chardin, 1926:48.
>[= †*Myotalpavus* Miller, 1927:16.] [= or including †*Youngia* Zheng, in Tomida, Li
>& Setoguchi, 1994:62.[1]] [Including †*Mesosiphneus* Kretzoi, 1961:12; †*Episiphneus*
>Kretzoi, 1961:12; †*Pliosiphneus* Zheng, in Tomida, Li & Setoguchi, 1994:60;
>†*Chardina* Zheng, in Tomida, Li & Setoguchi, 1994:60.]
>L. Mioc.-E. Pleist.; As.
>
>[1] Invalid because no type species was chosen.

Myospalax Laxmann, 1773:134.[1] Zokors, mole rats, ha-whei, ha-lao, hsia-lao, ti-pai.
>[Including *Myotalpa* Kerr, 1792:516-17, 520; *Siphneus* Brants, 1827:19;
>*Aspalomys* Gervais, 1841:56[2]; *Eospalax* G. M. Allen, 1938:i, vii;
>*Zokor* Ellerman, 1941:541[3]; †*Allosiphneus* Kretzoi, 1961:128.]
>L. Plioc. and/or E. Pleist., M. Pleist.-R.; As.
>
>[1] *Nomen nudum* in Laxmann, 1769:74-77 (no type species).
>[2] Attributed to Laxmann by Gervais.
>[3] Objective synonym of *Eospalax*.

Subfamily **Spalacinae** Gray, 1821:303.
>[= Spalacidae Gray, 1821:303; Spalasina Reichenbach, 1836:50;
>Spalacini Giebel, 1855:517; Spalacinae Thomas, 1896; Spalacini Fejfar, 1972:
>168, 190; Aspalacidae Gray, 1825:342; Aspalacina Gray, 1825:342[1];
>Aspalacina C. L. Bonaparte, 1838:113.[2]]
>E. Mioc., L. Mioc.-R.; Eu. M. Mioc.-E. Plioc., ?L. Plioc. and/or ?E. Pleist.,
>M. Pleist.-R.; As.[3] ?Pleist., R.; Af.[4]
>
>[1] As tribe.
>[2] As subfamily.
>[3] Spalacinae indet. in the late Miocene Songshan fauna of China.
>[4] N. Africa.

†*Heramys* Hofmeijer & de Bruijn, 1985:186.[1]
>E. Mioc.; Eu.[2]
>
>[1] *Nomen nudum* in de Bruijn, 1984:419.
>[2] E. Europe (Island of Evia, Greece).

†*Pliospalax* Kormos, 1932:193-200.
> M. Mioc., E. Plioc.; As.[1] L. Mioc.-L. Plioc.; Eu.[2]
>
> [1] W. Asia (Turkey in the middle Miocene and early Pliocene; Rhodes in the early Pliocene).
> [2] E. Europe.

Spalax Güldenstaedt, 1770:410. Russian mole rat, blind mole rats, blindmause.
> [= *Glis* Erxleben, 1777:358[1]; *Myospalax* Hermann, 1783:83[2]; *Talpoides* Lacépède, 1799:10; *Aspalax* Desmarest, 1804:24; *Anotis* Rafinesque, 1815:58[3]; *Ommatostergus* Nordmann, in Keyserling & Blasius, 1840:vii, 31; *Macrospalax* Méhely, 1909:23.]
> ?L. Plioc. and/or ?E. Pleist.; As. L. Plioc.-R.; Eu.
>
> [1] Not *Glis* Brisson, 1762, a synonym of *Myoxus*.
> [2] Not *Myospalax* Laxmann, 1769, a genus of myospalacine murids, nor *Myospalax* Blyth, 1846:141, a synonym of *Ellobius*, an arvicoline murid.
> [3] New name for *Talpoides*.

Nannospalax Palmer, 1903:873.[1] Lesser mole rat, Palestine mole rat.
> [= *Microspalax* Méhely, 1909[2]; *Ujhelyiana* Strand, 1922:142.[3]]
> [Including *Mesospalax* Méhely, 1909:22.[4]]
> E. Plioc., E.-M. Pleist., ?L. Pleist., R.; Eu.[5] ?Pleist., R.; Af.[6] M. Pleist.-R.; As.[7]
>
> [1] A replacement name for *Microspalax* Nehring.
> [2] Not *Microspalax* Megnin & Trouessart, 1885, an arachnid genus. *Microspalax* Nehring, 1898:168, proposed as a subgenus of *Spalax*, was a *nomen nudum* in addition to being preoccupied.
> [3] Replacement name for *Microspalax*.
> [4] *Mesospalax* Nehring, 1897, was a *nomen nudum*.
> [5] E. Europe.
> [6] N. Africa.
> [7] W. Asia.

Subfamily †**Anomalomyinae** Schaub, 1925:97.
> [= †Anomalomyidae Schaub, 1925:97; †Anomalomyinae Stehlin & Schaub, 1951:172, 326; †Anomalomyini Fejfar, 1972:168, 190.]
> [Including †Prospalacinae Topachevsky, 1969:123.]
> E. Mioc.-E. Pleist.; Eu.

†*Anomalomys* Gaillard, 1900:191-192.
> [= †*Miospalax* Stromer, 1928:24.]
> E. Mioc.-L. Plioc.; Eu.

†*Anomalospalax* Kordos, 1985:27.
> L. Mioc.; Eu.

†*Prospalax* Méhely, 1908:243-258.
> [Including †*Pterospalax* Kretzoi, 1971:113[1]; †*Allospalax* Kretzoi, 1971:115.[1]]
> L. Mioc.-E. Pleist.; Eu.

Subfamily **Rhizomyinae** Winge, 1887:109.
> [= Rhizomyini Winge, 1887:109; Rhizomyinae Thomas, 1897:1021; Rhizomyidae Miller & Gidley, 1918:437.] [Including Tachyoryctinae Miller & Gidley, 1918:437; Tachyoryctinae Chaline, Mein & Petter, 1977:250.]
> E. Mioc.-R.; As. M.-L. Mioc., L. Plioc.-R.; Af. R.; E. Indies.

Tribe **Tachyoryctini** Miller & Gidley, 1918:437, **new rank**.[1]
> [= Tachyoryctinae Miller & Gidley, 1918:437; Tachyoryctinae Chaline, Mein & Petter, 1977:250.]
> E. Mioc.-E. Plioc.; As. M.-L. Mioc., L. Plioc.-R.; Af.
>
> [1] Musser.

†*Prokanisamys* de Bruijn, Hussain & Leinders, 1981:72.
> E.-M. Mioc.; As.

†*Kanisamys* A. E. Wood, 1937:66.
> M.-L. Mioc.; As.[1]
>
> [1] S. Asia (Pakistan and India).

†*Eicooryctes* L. Flynn, 1982:350.
> L. Mioc.; As.[1]
>
> [1] S. Asia (Pakistan).

†*Protachyoryctes* Hinton, 1933:620-622.
L. Mioc.; As.[1]

[1] S. Asia (Pakistan).

†*Pronakalimys* Tong & Jaeger, 1993:56.
M. Mioc.; Af.[1]

[1] E. Africa (Kenya).

†*Nakalimys* L. Flynn & Sabatier, 1984:161.
L. Mioc.; Af.[1]

[1] E. Africa (Kenya).

†*Rhizomyides* Bohlin, 1946:68.
[= †*Rhizomyoides* Black, 1972:249.]
L. Mioc.-E. Plioc.; As.[1]

[1] Afghanistan, Pakistan, India.

Tachyoryctes Rüppell, 1835:35. African mole rats.
[= *Chrysomys* Gray, 1843:150.]
L. Plioc.-R.; Af.

Tribe **Rhizomyini** Winge, 1887:109.
L. Mioc.-R.; As. R.; E. Indies.[1]

[1] Sumatra.

†*Brachyrhizomys* Teilhard de Chardin, 1942:1-101.
L. Mioc.-E. Plioc., M. Pleist.; As.[1]

[1] Pakistan, India, China.

†*Anepsirhizomys* L. Flynn, 1982:375.
E. Plioc.; As.[1]

[1] Pakistan.

†*Pararhizomys* Teilhard de Chardin & Young, 1931:11.
E. Plioc., ?L. Plioc.; As.[1,2]

[1] China in the early Pliocene.
[2] Mongolia in the Pliocene.

Rhizomys Gray, 1831:95. Bamboo rats.
[Including *Nyctocleptes* Temminck, 1832:1-8.]
L. Plioc.-R.; As.[1] R.; E. Indies.[2]

[1] China in the Pliocene, Pleistocene and Recent; India (Assam) and S.E. Asia in the Recent.
[2] Sumatra.

Cannomys Thomas, 1915:57. Lesser bamboo rat, bay bamboo rat.
R.; As.[1]

[1] Nepal, Bhutan, Bangladesh, N.E. India, Burma, Yunnan, Thailand, Cambodia.

Infraorder **GLIRIMORPHA** Thaler, 1966:11, 101, **new rank**.[1]

[1] Proposed as suborder.

Family **Myoxidae** Gray, 1821:303.[1] Dormice.
[= Glirini Muirhead, 1819[2]; Gliridae Thomas, 1897:1016; Gliroidea Simpson, 1945:91; Myosidae Gray, 1821:303; Myoxina Gray, 1825:342; Myoxidae Waterhouse, 1839:184; Myoxini Giebel, 1855:621; Myoxida Haeckel, 1866:clx; Myoxoidea Gill, 1872:21.] [Including Leithiidae Lydekker, 1895:862; Muscardinidae Palmer, 1899:413; Seleviniidae Bashanov & Belosludov, 1939:3-8.]
E. Eoc., E. Mioc.-E. Plioc., E. Pleist.-R.; As. E. Eoc.-R.; Eu. M. Eoc., L. Olig.-R.; Mediterranean. M.-L. Mioc., L. Plioc.-R.; Af.

[1] The currently fashionable return to Myoxidae and other family-group names based upon it, in preference to Gliridae and its coordinate names, may be a violation of the International Code of Zoological Nomenclature, Article 40, depending on the vague unwritten definition of "general acceptance." If *Glis* Brisson were to be validated by the International Commission on Zoological Nomenclature, then Muirhead's Glirini, not Gliridae Thomas, would be the basis for Gliridae.
[2] See Palmer (1904:859).

†*Miniglis* Vianey-Liaud, 1994:151.
L. Eoc.; Eu.

†*Tenuiglis* Vianey-Liaud, 1994:150.
    L. Eoc.-E. Olig.; Eu.

†*Caenomys* Lydekker, 1885:225.[1]
    E. Mioc.; Eu.
    [1] From Bravard (MS.).

†*Brachymys* von Meyer, 1847:456.
    [= †*Micromys* von Meyer, 1846:475.[1]]
    Mioc.; Eu.
    [1] Not *Micromys* Dehne, 1841, a murid.

†*Szechenyia* Kretzoi, 1980:348.[1]
    L. Mioc.; Eu.
    [1] ("1978").

†*Myolidus* Alvarez-Sierra, in Alvarez-Sierra & García-Moreno, 1986:157.
    L. Mioc.; Eu.

Subfamily **Graphiurinae** Winge, 1887:109, 123.
    [= Graphiurini Winge, 1887:109, 123; Graphiurinae Palmer, 1899:413;
    Graphiuridae Miller & Gidley, 1918:440.]
    L. Mioc.; Eu. E. Pleist.-R.; Af.

†*Graphiurops* Bachmayer & R. W. Wilson, 1980:367.
    L. Mioc.; Eu.

†*Graphiglis* Kretzoi, 1984:217.
    L. Mioc.; Eu.

Graphiurus Smuts, 1832:32-33. African dormice.
    [= *Graphidurus* Brandt, 1855:300.] [Including *Claviglis* Jentink, 1888:41-42;
    Gliriscus Thomas & Hinton, 1925:232; *Aethoglis* Allen, 1936:292.]
    L. Plioc.-R.; Af.

Subfamily †**Gliravinae** Schaub, 1958:775.
    E. Eoc.; As. E. Eoc.-E. Mioc.; Eu.

†*Chaibulakomys* Shevyreva, 1992:114.
    E. Eoc.; As.

†*Eogliravus* Hartenberger, 1971:112.
    E.-M. Eoc.; Eu.

†*Gliravus* Schaub, in Stehlin & Schaub, 1951:368.[1]
    L. Eoc.-E. Mioc.; Eu.
    [1] *Nomen nudum* in Stehlin, in Erni, 1942 (not 1941): 213.

Subfamily †**Glamyinae** Vianey-Liaud, 1994:122.
    M. Eoc., L. Olig.; Mediterranean. M. Eoc.-E. Olig.; Eu.

†*Glamys* Vianey-Liaud, 1989:220.
    M. Eoc., L. Olig.; Mediterranean.[1] M. Eoc.-E. Olig.; Eu.
    [1] Balearic Islands (Mallorca).

Subfamily **Leithiinae** Lydekker, 1896:862.
    [= †Leithiidae Lydekker, 1895:862; †Leithiinae Trouessart, 1897:446.]
    [Including Seleviniidae Bashanov & Belosludov, 1939:3-8;
    Dryomyinae de Bruijn, 1966:483; Myomiminae Daams, 1981:98.]
    L. Eoc.-R.; Eu. L. Olig.-M. Mioc., E. Plioc.-R.; Mediterranean.
    E. Mioc.-E. Plioc., E. Pleist.-R.; As. M.-L. Mioc., M. Pleist.-R.; Af.[1]
    [1] N. Africa.

Tribe **Leithiini** Lydekker, 1895:866.
    [= Leithiidae Lydekker, 1895:866; Leithiini Wahlert, Sawitzke & Holden, 1993:26.]
    [Including Dryomyini Pavlinov & Rossolimo, 1987:145.]
    L. Eoc.-R.; Eu. L. Olig.-E. Mioc., E. Plioc.-R.; Mediterranean.
    E.-M. Mioc., E. Plioc., M. Pleist.-R.; As. M.-L. Mioc., M. Pleist.-R.; Af.[1]
    [1] North Africa.

†*Bransatoglis* Hugueney, 1967:92.
> [= †*Branssatoglis* Hugueney, 1967:92; †*Bransatoglis* Hugueney, 1969:141.[1]]
> [Including †*Oligodyromys* Bahlo, 1975:122.]
> L. Eoc.-M. Mioc.; Eu. L. Olig.; Mediterranean.[2] E. Mioc.; As.[3]
> [1] Emendation.
> [2] Balearic Islands (Mallorca).
> [3] W. Asia (Turkey).

†*Microdyromys* de Bruijn, 1966:68.
> [Including †*Afrodryomys* Jaeger, 1975.]
> E. Olig.-L. Mioc.; Eu. E. Mioc.; Mediterranean.[1] M.-L. Mioc.; Af.[2] M. Mioc.; As.
> [1] Sardinia.
> [2] N. Africa.

†*Anthracoglis* Engesser, 1983:765.
> L. Mioc.; Eu.

Eliomys Wagner, 1840:175. Garden dormice, lerots.
> [Including *Bifa* Lataste, 1885:61-63; †*Hypnomys* Bate, 1919:210[1]; †*Tyrrhenoglis* Engesser, 1976:784[1]; †*Maltamys* Maempel & de Bruijn, 1982:118[1]; †*Eivissia* Alcover & Agusti, 1985:52.[2]]
> M. Mioc.-R.; Eu. E. Plioc., M. Pleist.-R.; As.[3] E. Plioc.-R.; Mediterranean.[4] M. Pleist.-R.; Af.[5]
> [1] Sometimes as subgenus.
> [2] Described as subgenus.
> [3] W. Asia.
> [4] Balearic Islands, Corsica, Sardinia, Sicily, Malta.
> [5] N. Africa.

Dryomys Thomas, 1906:348. Tree dormouse, lerotin.
> [= *Dyromys* Thomas, 1907:406.] [= or including *Chaetocauda* Wang, 1985:67.]
> M. Pleist.-R.; As.[1] M. Pleist.-R.; Eu.
> [1] W. Asia in the Pleistocene.

†*Tempestia* van de Weerd, 1976:1-217.
> M.-L. Mioc.; Eu.

†*Prodryomys* Mayr, 1979:1-380.[1]
> M. Mioc.; Eu.
> [1] PhD thesis.

†*Leithia* Lydekker, 1895:862.[1]
> Pleist.; Mediterranean.[2]
> [1] Dec. 3, 1895, not 1896, nor Adams, 1863.
> [2] Malta, Sicily.

Tribe **Seleviniini** Bashanov & Belosludov, 1939:3, **new rank.**
> [= Seleviniidae Bashanov & Belosludov, 1939:3; Seleviniinae Ognev, 1947:551.]
> ?E. Olig., L. Olig.-L. Plioc., M. and/or L. Pleist., R.; Eu. L. Olig.-M. Mioc.; Mediterranean. E. Mioc.-E. Plioc., E. Pleist.-R.; As.

Subtribe **Seleviniina** Bashanov & Belosludov, 1939:3, **new rank.**
> [= Seleviniidae Bashanov & Belosludov, 1939:3-8.]
> E. Plioc.; Eu.[1] R.; As.
> [1] E. Europe (Poland).

†*Plioselevinia* Sulimski, 1962:503-512.
> E. Plioc.; Eu.[1]
> [1] E. Europe (Poland).

Selevinia Belosludov & Bashanov, 1939:81. Desert dormouse.
> [= *Salevinia* Argyropulo & Vinogradov, 1939:81.[1]]
> R.; As.
> [1] Misprint; spelled "*Selevinia*" in the remainder of the paper. This name was based on a different type species name than that used by Belosludov & Bashanov. Argyropulo is spelled "Argiropulo" in this paper.

Subtribe **Myomimina** Daams, 1981:98, **new rank.**
> [= Myomiminae Daams, 1981:98; Myomimini Pavlinov & Rossolimo, 1987:144.]

?E. Olig., L. Olig.-L. Plioc., M. and/or L. Pleist., R.; Eu. L. Olig.-M. Mioc.; Mediterranean. E. Mioc.-E. Plioc., E. Pleist.-R.; As.

†*Vasseuromys* Baudelot & de Bonis, 1966:341.
> L. Olig.; Mediterranean.[1] E. Mioc.; As.[2] E. Mioc., L. Mioc.; Eu.
>
> [1] Balearic Islands (Mallorca).
> [2] W. Asia (Anatolia).

†*Peridyromys* Schaub, in Stehlin & Schaub, 1951:368.
> ?E. Olig., L. Olig.-E. Mioc.; Eu. E.-M. Mioc.; Mediterranean.[1,2]
>
> [1] Sardinia in the early Miocene.
> [2] Balearic Islands (Mallorca) in the middle Miocene.

†*Quercomys* Lacomba & Martínez-Salanova, 1988:106, 108.
> E. Mioc.; Eu.

†*Miodyromys* Kretzoi, 1943:272.
> [Including †*Pseudodryomys* de Bruijn, 1966:64.]
> L. Olig.-M. Mioc.; Eu. ?M. Mioc.; As. M. Mioc.; Mediterranean.[1]
>
> [1] Endemic insular fauna of Murchas, Granada Basin, Spain.

Myomimus Ognev, 1924:115. Mouse-like dormouse.
> [Including †*Philistomys* Bate, 1937:399.]
> E. Mioc.-E. Plioc., E. Pleist.-R.; As.[1,2] M.-L. Mioc., M. and/or L. Pleist., R.; Eu.[3]
>
> [1] W. Asia (Anatolia) in the early Miocene.
> [2] Including Rhodes in the early Pliocene.
> [3] E. Europe in the Pleistocene and Recent.

†*Nievella* Daams, 1976:156.
> E. Mioc.; Eu.

†*Altomiramys* Díaz Molina & López-Martínez, 1979:158.
> E. Mioc.; Eu.

†*Armantomys* de Bruijn, 1966:61.
> E.-M. Mioc.; Eu.

†*Praearmantomys* de Bruijn, 1966:58-78.
> E.-M. Mioc.; Eu.

†*Carbomys* Mein & Adrover, 1982:455.
> M. Mioc.; Mediterranean.[1]
>
> [1] Balearic Islands (Mallorca).

†*Margaritamys* Mein & Adrover, 1982:457.
> M. Mioc.; Mediterranean.[1]
>
> [1] Balearic Islands (Mallorca).

†*Ramys* García Moreno & López-Martínez, 1986:337, 341.
> L. Mioc.; Eu.

†*Dryomimus* Kretzoi, 1959:240.
> E.-L. Plioc.; Eu.

Subfamily **Myoxinae** Gray, 1821:303.
> [= Glirini Muirhead, in Brewster, 1819:433; Glirinae Thomas, 1897:1016; Myosidae Gray, 1821:303; Myoxina Gray, 1825:342; Myoxinae Huxley, 1872:369.]
> [Including Muscardininae Palmer, 1904:859n; Glirulinae de Bruijn, 1966:373.]
> ?E. Olig., L. Olig.-R.; Eu. E. Mioc., M. Pleist.-R.; As.
> E. Mioc., L. Mioc., R.; Mediterranean.

Glirulus Thomas, 1906:347. Japanese dormouse, yamane.
> [Including †*Amphidyromys* Heller, 1936:125[1]; †*Paraglirulus* Engesser, 1972:208.[2]]
> E. Mioc., M. Pleist.-R.; As.[3] E. Mioc.-E. Pleist.; Eu.
>
> [1] Described as genus; used as subgenus of †*Glirulus* by Kowalski, 1963.
> [2] Described as genus; reduced to subgeneric rank by van der Meulen & de Bruijn, 1982.
> [3] E. Asia (Japan).

Tribe **Myoxini** Gray, 1821:303.
> [= Glirini Muirhead, in Brewster, 1819:433; Myosidae Gray, 1821:303; Myoxina Gray, 1825:342[1]; Myoxini Giebel, 1855:621.]

Olig., E. Mioc.-R.; Eu. E. Mioc., R.; As. E. Mioc., R.; Mediterranean.
[1] Proposed as tribe.

†*Stertomys* Daams & Freudenthal, 1985:22. Giant dormouse.
?L. Mioc.; Eu.

Myoxus Zimmermann, 1780:351. Common dormouse, edible dormouse, siebenschläfer, loir.
[= *Glis* Brisson, 1762:113[1]; *Myorus* Reichenbach, 1835:7.] [= or including *Elius* Schulze, 1900:200.[2]]
Olig., E. Mioc.-R.; Eu. E. Mioc., R.; As.[3] E. Mioc., R.; Mediterranean.[4,5]
[1] See the International Code of Zoological Nomenclature, Articles 11(c)(i) and 40, for confusing rules. Brisson, 1762 was not consistently binominal.
[2] Proposed as subgenus.
[3] W. Asia (Anatolia) in the early Miocene.
[4] Sardinia in the early Miocene.
[5] Sicily, Elba, Sardinia and Corsica in the Recent.

Tribe **Muscardinini** Palmer, 1899:413, **new rank**.
[= Muscardinidae Palmer, 1899:413; Muscardininae Palmer, 1904:859.]
L. Olig.-R.; Eu. E. Mioc., R.; As.[1] L. Mioc., R.; Mediterranean.
[1] W. Asia.

†*Glirudinus* de Bruijn, 1966:73.
[Including †*Muscardinulus* Thaler, 1966:110.[1]]
L. Olig.-E. Mioc.; Eu. E. Mioc.; As.[2]
[1] Described as subgenus of *Muscardinus*.
[2] W. Asia (Anatolia).

†*Myoglis* Baudelot, 1966:760.
[= or including †*Pentaglis* Kretzoi, 1943:271.[1]]
E.-M. Mioc.; Eu.
[1] *Nomen dubium.*

†*Heteromyoxus* Dehm, 1938:338.
E. Mioc.; Eu.

†*Paraglis* Baudelot, 1970:303.
E.-M. Mioc.; Eu.

†*Eomuscardinus* Hartenberger, 1966:597.[1]
M.-L. Mioc.; Eu.
[1] Proposed as subgenus of *Muscardinus*.

Muscardinus Kaup, 1829:139. Hazel mouse.
M. Mioc.-R.; Eu. L. Mioc., R.; Mediterranean.[1,2] R.; As.[3]
[1] Balearic Islands (Menorca) in the late Miocene.
[2] Sicily in the Recent.
[3] W. Asia.

Infraorder **GEOMORPHA** Thaler, 1966:11, **new rank**.[1]
[= SCIUROSPALACOIDES Brandt, 1855.[2]]
[1] Proposed as suborder.
[2] *Nomen oblitum.*

†*Griphomys* R. W. Wilson, 1940:93.
M.-L. Eoc.; N.A.

†*Meliakrouniomys* Harris & A. E. Wood, 1969:3.
L. Eoc.; N.A.

Superfamily †**Eomyoidea** Winge, 1887:109, 122.
[= †Eomyini Winge, 1887:109, 122[1]; †Eomyoidea Wahlert, 1978:15.]
[1] Although this paper bears 1888 as its date, author's separates were distributed in December, 1887, fide Palmer (1904:740).

Family †**Eomyidae** Winge, 1887:109, 122.
[= †Eomyini Winge, 1887:109, 122[1]; †Eomyidae Depéret & Douxami, 1902:69.]
[Including †Adjidaumidae Miller & Gidley, 1918:434.]

E. Eoc., E.-L. Olig., M. Mioc.-E. Plioc., ?L. Plioc.; As.[2] M. Eoc.-E. Plioc.; N.A.
E. Olig.-E. Pleist., ?M. and/or ?L. Pleist.; Eu. L. Olig.; Mediterranean.[3]

[1] Although this paper bears 1888 as its date, author's separates were distributed in December 1887, fide
   Palmer (1904:740).
[2] †Eomyidae genus and sp. nov. in the early Oligocene of Central Asia.
[3] Mallorca, Ballearic Islands.

†*Zetamys* L. D. Martin, 1974:6.[1]
   L. Olig.; N.A.

[1] Assignment to †Eomyidae is questionable.

Subfamily †**Eomyinae** Winge, 1887:109, 122.
   [= †Eomyini Winge, 1887:109, 122[1]; †Eomyidae Depéret & Douxami, 1902:69.]
   [Including †Adjidaumidae Miller & Gidley, 1918:434.]
   M. Eoc.-E. Plioc.; N.A. E. Olig.-E. Pleist., ?M. and/or ?L. Pleist.; Eu. L. Olig., M.
   Mioc.-E. Plioc., ?L. Plioc.; As. L. Olig.; Mediterranean.[2]

[1] Although this paper bears 1888 as its date, author's separates were distributed in December 1887, fide
   Palmer (1904:740).
[2] Mallorca, Balearic Islands.

Tribe †**Eomyini** Winge, 1887:109, 122.
   M. Eoc.-E. Plioc.; N.A. E. Olig.-E. Pleist., ?M. and/or ?L. Pleist.; Eu. L. Olig., M.
   Mioc.-E. Plioc., ?L. Plioc.; As. L. Olig.; Mediterranean.[1]

[1] Mallorca, Balearic Islands.

†*Protadjidaumo* Burke, 1934:394.
   M. Eoc.; N.A.

†*Aguafriamys* J. A. Wilson & Runckel, 1991:9.
   M. Eoc.; N.A.

†*Centimanomys* Galbreath, 1955:75.
   L. Eoc.; N.A.

†*Cupressimus* Storer, 1978:26.
   L. Eoc.; N.A.

†*Viejadjidaumo* A. E. Wood, 1974:63.
   L. Eoc.; N.A.

†*Paradjidaumo* Burke, 1934:396.
   M. Eoc.-E. Olig., ?E. Mioc.; N.A.

†*Aulolithomys* Black, 1965:35.
   L. Eoc.; N.A.

†*Eomys* Schlosser, 1884:84-85.
   [= or including †*Omegodus* Pomel, 1853:37[1]; †*Decticus* Aymard, 1853:250.[1]]
   [Including †*Adjidaumo* Hay, 1899:253.]
   M. Eoc.-E. Olig.; N.A. E. Olig.-E. Mioc.; Eu. L. Olig.; As. L. Olig.;
   Mediterranean.[2]

[1] For nomenclatural purposes best forgotten. Stability of †*Eomys* is long-established.
[2] Mallorca, Balearic Islands.

†*Orelladjidaumo* Korth, 1989:42.
   E. Olig.; N.A.

†*Metadjidaumo* Setoguchi, 1978:32.
   E. Olig.; N.A.

†*Leptodontomys* Shotwell, 1956:731.
   L. Olig.-L. Mioc.; N.A. M.-L. Mioc., Plioc.; As.[1] M. Mioc.-E. Pleist.; Eu.
[1] China.

†*Pseudotheridomys* Schlosser, 1926:379.
   L. Olig., M. Mioc.; As. L. Olig.-L. Mioc.; Eu. E.-M. Mioc.; N.A.

†*Ligerimys* Schaub, in Stehlin & Schaub, 1951:357.
   E.-M. Mioc.; Eu.

†*Eomyodon* Engesser, 1987:969.
   L. Olig.; As. L. Olig.-E. Mioc.; Eu.

†*Rhodanomys* Depéret & Douxami, 1902:69.
L. Olig.-E. Mioc.; Eu.

†*Ritteneria* Schaub, in Stehlin & Schaub, 1951:357.
E. Mioc.; Eu.

†*Pentabuneomys* Engesser, 1990:117.
E. Mioc.; Eu.

†*Apeomys* Fahlbusch, 1968:233.
E. Mioc.; Eu.

†*Eomyops* Engesser, 1979:27.
M.-L. Mioc., E. Pleist.; Eu.

†*Keramidomys* Hartenberger, 1966:602.[1]
M. Mioc., E. Plioc.; As.[2] M. Mioc.-E. Plioc.; Eu.
[1] Proposed as subgenus of †*Pseudotheridomys*; elevated to generic rank by Hugueney & Mein, 1968:187.
[2] W. Asia (Rhodes).

†*Estramomys* Janossy, 1969:35.
E. Plioc.-E. Pleist.; Eu.[1]
[1] E. Europe.

†*Meteomys* Kretzoi, 1952:88.
?Pleist.; Eu.[1]
[1] E. Europe.

†*Pseudadjidaumo* Lindsay, 1972:34.
M. Mioc.; N.A.

†*Ronquillomys* Jacobs, 1977:507.
L. Mioc.-E. Plioc.; N.A.

†*Kansasimys* A. E. Wood, 1936:392-394.
L. Mioc.-E. Plioc.; N.A.

†*Comancheomys* Dalquest, 1983:11.
L. Mioc.; N.A.

Tribe †**Namatomyini** Korth, 1992:78.
M.-L. Eoc.; N.A.

†*Metanoiamys* Chiment & Korth, 1996:116.
M. Eoc.; N.A.

†*Namatomys* Black, 1965:32.
L. Eoc.; N.A.

†*Montanamus* Ostrander, 1983:140.
L. Eoc.; N.A.

†*Paranamatomys* Korth, 1992:78.
L. Eoc.; N.A.

Subfamily †**Yoderimyinae** A. E. Wood, 1955:519.
E. Eoc.; As. L. Eoc., E. Mioc.; N.A.

†*Zaisaneomys* Shevyreva, 1993:166.
E. Eoc.; As.

†*Yoderimys* A. E. Wood, 1955:520.
L. Eoc.; N.A.

†*Zemiodontomys* Emry & Korth, 1993:1053.
L. Eoc.; N.A.

†*Litoyoderimys* Emry & Korth, 1993:1055.
L. Eoc.; N.A.

†*Arikareeomys* Korth, 1992:113.
E. Mioc.; N.A.

Superfamily **Geomyoidea** Bonaparte, 1845:5.
> [= Pseudotomina Gray, 1825:342[1]; Geomina Bonaparte, 1845:5; Geomyina Bonaparte, 1850; Geomyinae Baird, 1857:366; Geomyoidea Weber, 1904:490.] [Including Saccomyina Waterhouse, 1848:8; Saccomyoidea Gill, 1872:21; †Eomyoidea Wahlert, 1978:15; †Heliscomyidae Korth, Wahlert & Emry, 1991:247.]
> [1] *Lapsus calami.*

†*Floresomys* Fries, Hibbard & Dunkle, 1955:16.
> M. and/or L. Eoc.; N.A.[1]
> [1] Mexico.

†*Texomys* Slaughter, 1981:111.
> E.-M. Mioc.; N.A. E. Mioc.; Cent. A.

†*Jimomys* Wahlert, 1976:1.
> E.-M. Mioc.; N.A.

†*Heliscomys* Cope, 1873:3.
> [Including †*Apletotomeus* Reeder, 1960:512; †*Syphyriomys* Korth, 1995:193.[1]]
> M. Eoc.-E. Mioc.; N.A.
> [1] Subgenus as defined.

Family †**Florentiamyidae** A. E. Wood, 1936:41.
> [= †Florentiamyinae A. E. Wood, 1936:41; †Florentiamyidae Wahlert, 1983:4.]
> E. Olig.-M. Mioc.; N.A.[1]
> [1] Undescribed florentiamyids have been reported from the middle Miocene (Barstovian) of North America.

†*Ecclesimus* Korth, 1989:36.
> E. Olig.; N.A.

†*Kirkomys* Wahlert, 1984:2.
> E.-L. Olig.; N.A.

†*Hitonkala* Macdonald, 1963:186.
> L. Olig.-E. Mioc.; N.A.

†*Sanctimus* Macdonald, 1970:43.
> L. Olig.-E. Mioc.; N.A.

†*Florentiamys* A. E. Wood, 1936:41.
> L. Olig.-E. Mioc.; N.A.

†*Fanimus* Korth, Bailey & Hunt, 1990:40.
> E. Mioc.; N.A.

Family **Geomyidae** Bonaparte, 1845:5.
> [= Pseudotomina Gray, 1825:342[1]; Pseudostomidae Gervais, 1853:245; Geomina Bonaparte, 1845:5; Geomyina Bonaparte, 1850: unpaged plate; Geomyinae Baird, 1857:366; Geomyidae Gill, 1872:21.] [= or including Sciurospalacini Giebel, 1855:528[2]; Sciurospalacida Haeckel, 1866:clx.[2]] [Including Saccomyna Gray, 1843:xxiv, 120[3]; Saccomyidae Baird, 1857:236, 365; Saccomyinae Baird, 1857:405; Dipodomyna Gervais, 1853:245; Dipodomyina Gray, 1868:200; Heteromyina Gray, 1868:200; Heteromyidae Allen & Chapman, 1893:233; †Heliscomyidae Korth, Wahlert & Emry, 1991:247.]
> E. Olig.-R.; N.A. L. Pleist.-R.; Cent. A. R.; S.A.[4]
> [1] *Lapsus calami.*
> [2] Used as a family. Not based upon a generic name.
> [3] Objective synonym of Heteromyidae.
> [4] Northern South America.

†*Diplolophus* Troxell, 1923:157.[1]
> [= †*Gidleumys* A. E. Wood, 1936.]
> E. Olig.; N.A.
> [1] Assignment to Geomyidae is questionable.

†*Schizodontomys* Rensberger, 1973:60.
> E.-M. Mioc.; N.A.

†*Lignimus* Storer, 1970:1127.
M. Mioc.; N.A.

Subfamily †**Entoptychinae** Miller & Gidley, 1918:434.
[= †Entoptychidae Wahlert, 1985:17.]
L. Olig.-L. Mioc.; N.A.

Tribe †**Pleurolicini** Rensberger, 1973:11, **new rank**.
[= †Pleurolicinae Rensberger, 1973:11.]
L. Olig.-E. Mioc., ?M. Mioc.; N.A.

†*Pleurolicus* Cope, 1878:4-5.
[= †*Grangerimus* A. E. Wood, 1936:13.]
L. Olig.-E. Mioc.; N.A.

†*Ziamys* Gawne, 1975:18.
E. Mioc., ?M. Mioc.; N.A.

Tribe †**Entoptychini** Miller & Gidley, 1918:434, **new rank**.
[= †Entoptychinae Miller & Gidley, 1918:434.]
E.-L. Mioc.; N.A.

†*Gregorymys* A. E. Wood, 1936:9.
E.-L. Mioc.; N.A.

†*Entoptychus* Cope, 1878:2-4.
[= †*Palustrimus* A. E. Wood, 1935:368-372.]
E. Mioc.; N.A.

Subfamily **Geomyinae** Bonaparte, 1845:5. Pocket gophers.
[= Pseudotomina Gray, 1825:342[1]; Geomina Bonaparte, 1845:5; Geomyina Bonaparte, 1850: unpaged plate; Geomyinae Baird, 1857:366.]
E. Olig.-R.; N.A. L. Pleist.-R.; Cent. A.
[1] *Lapsus calami.*

†*Tenudomys* Rensberger, 1973:78.
E.-L. Olig.; N.A.

†*Phelosaccomys* Korth & Reynolds, 1994:91.
M.-L. Mioc.; N.A.

Tribe †**Dikkomyini** R. J. Russell, 1968:515.
E.-L. Mioc.; N.A.

†*Dikkomys* A. E. Wood, 1936:26.
E. Mioc.; N.A.

†*Parapliosaccomys* Shotwell, 1967:34.
M.-L. Mioc.; N.A.

†*Mojavemys* Lindsay, 1972:59.
M. Mioc.; N.A.[1]
[1] California, Oregon.

†*Pliosaccomys* R. W. Wilson, 1936:20.
L. Mioc.; N.A.

Tribe **Geomyini** Bonaparte, 1845:5. Tuzas, taltuzas.
[= Pseudotomina Gray, 1825:342[1]; Geomina Bonaparte, 1845:5; Geomyina Bonaparte, 1850: unpaged plate.; Geomyinae Baird, 1857:366; Geomyini R. J. Russell, 1968:521.]
L. Mioc.-R.; N.A. L. Pleist.-R.; Cent. A.
[1] *Lapsus calami* for Pseudostomina, based on *Pseudostoma* Say, 1823. Proposed as tribe.

†*Pliogeomys* Hibbard, 1954:353.
L. Mioc.-L. Plioc.; N.A.

Geomys Rafinesque, 1817:45. Eastern pocket gophers.
[= *Diplostoma* Rafinesque, 1817:44-45; *Saccophorus* Kuhl, 1820:65-66; *Pseudostoma* Say, 1823:406-407; *Ascomys* Lichtenstein, 1825:20; *Mamgeomysus* Herrera,

1899:28.[1]] [Including †*Nerterogeomys* Gazin, 1942:507; †*Parageomys* Hibbard, 1944:735; †*Progeomys* Dalquest, 1983:13.]
L. Mioc.-R.; N.A.
[1] See the International Code of Zoological Nomenclature, Article 1(b)(8), and Palmer (1904:25-26).

Pappogeomys Merriam, 1895:145.
[Including *Cratogeomys* Merriam, 1895:23, 25, 150[1]; *Craterogeomys* J. A. Allen, 1895:690[2]; *Platygeomys* Merriam, 1895:23, 26, 162.]
E. Plioc.-R.; N.A.
[1] As subgenus. (Sometimes as separate genus.)
[2] Misprint.

Orthogeomys Merriam, 1895:23, 26, 172. Giant pocket gophers, hispid pocket gophers.
[Including *Heterogeomys* Merriam, 1895:23, 26, 179[1]; *Macrogeomys* Merriam, 1895:23, 26, 185.[1]]
L. Pleist.-R.; N.A. L. Pleist.-R.; Cent. A.
[1] Subgenus.

Zygogeomys Merriam, 1895:195. Michoacán pocket gopher.
[= *Gygogeomys* J. A. Allen, 1895:242.[1]]
R.; N.A.[2]
[1] Misprint.
[2] Michoacan, Mexico.

Tribe **Thomomyini** R. J. Russell, 1968:518.
E. Plioc.-R.; N.A.

Thomomys Wied-Neuwied, 1839:377-384.[1] Western pocket gophers, smooth-toothed pocket gophers.
[= *Tomomys* Brandt, 1855:188.] [Including *Megascapheus* Elliot, 1903:190[2]; †*Plesiothomomys* Gidley & Gazin, 1933:343-357.[3]]
E. Plioc.-R.; N.A.
[1] Prince Maximilian zu Wied-Neuwied.
[2] Subgenus.
[3] Sometimes as subgenus.

Subfamily **Heteromyinae** Gray, 1868:201.
[= Heteromyina Gray, 1868:201; Heteromyinae Alston, 1876:88; Heteromyidae Allen & Chapman, 1893:233.]
E. Olig.-R.; N.A. R.; Cent. A. R.; S.A.[1]
[1] Northern South America.

†*Proheteromys* A. E. Wood, 1932:45.
[Including †*Akmaiomys* Reeder, 1960:517.]
E. Olig.-M. Mioc.; N.A.

Tribe †**Harrymyini** Wahlert, 1991:3, **new rank**.
[= †Harrymyinae Wahlert, 1991:3.]
E.-M. Mioc.; N.A.

†*Harrymys* Munthe, 1988:67.
E.-M. Mioc.; N.A.

Tribe **Heteromyini** Gray, 1868:200. Spiny pocket mice.
[= Heteromyina Gray, 1868:200.[1]]
E.-L. Mioc., L. Pleist.-R.; N.A. R.; Cent. A. R.; S.A.[2]
[1] Proposed as tribe.
[2] Northern South America.

†*Peridiomys* Matthew, 1924:85.
E.-M. Mioc.; N.A.

†*Diprionomys* Kellogg, 1910:433.
M.-L. Mioc.; N.A.

Liomys Merriam, 1902:44. Spiny pocket mice.
[Including *Schaeferia* Lehmann & Schaefer, 1979:232.[1]]

L. Pleist.-R.; N.A.[2] R.; Cent. A.

[1] Described as subgenus.
[2] Southern North America (southern Texas and Mexico).

Heteromys Desmarest, 1817:181. Forest spiny pocket mice.
[= *Saccomys* F. Cuvier, 1823:419-428; *Dasynotus* Wagler, 1830:21.[1]] [Including *Xylomys* Merriam, 1902:43-44.[2]]
R.; N.A.[3] R.; Cent. A. R.; S.A.[4]

[1] Replacement name.
[2] Subgenus as defined.
[3] Southernmost North America (Oaxaca, Veracruz and ?Guerrero, Mexico).
[4] Northern South America (including Trinidad and Tobago).

Tribe **Perognathini** Coues, 1875:277, **new rank**. Pocket mice.
[= Peragnathi Brandt, 1855:305[1]; Perognathidinae Coues, 1875:277; Perognathinae A. E. Wood, 1935:88.]
E. Mioc.-R.; N.A.

[1] Proposed as a "section" based on "*Peragnathus*." The name should be suppressed.

†*Mookomys* A. E. Wood, 1931:4.
E.-M. Mioc.; N.A.

†*Cupidinimus* A. E. Wood, 1935:118.
[Including †*Perognathoides* A. E. Wood, 1935:92.]
E.-L. Mioc.; N.A.

†*Trogomys* Reeder, 1960:121.
E. Mioc.; N.A.

Perognathus Wied-Neuwied, 1839:368.[1] Pocket mice.
[= *Peragnathus* Brandt, 1855:305.[2]] [Including *Cricetodipus* Peale, 1848:52-53; *Abromys* Gray, 1868:202; *Otognosis* Coues, 1875:305.]
E. Mioc.-R.; N.A.

[1] Prince Maximilian zu Wied-Neuwied.
[2] Unjustified emendation or *lapsus calami*.

†*Stratimus* Korth, Bailey & Hunt, 1990:27.
E. Mioc.; N.A.

†*Oregonomys* J. E. Martin, 1984:105.
L. Mioc.-L. Plioc.; N.A.

Chaetodipus Merriam, 1889:5.[1]
[Including *Burtognathus* Hoffmeister, 1986.[2]]
R.; N.A.

[1] Proposed as subgenus of *Perognathus*.
[2] Subgenus as defined.

Tribe **Dipodomyini** Gervais, 1853:245. Kangaroo rats.
[= Dipodomyna Gervais, 1853:245; Dipodomyina Gray, 1868:200[1]; Dipodomyinae Coues, 1875:277.] [= or including Macrocolini Brandt, 1855:231.[2]]
L. Mioc.-R.; N.A.

[1] Proposed as tribe.
[2] Brandt used this term at subfamilial rank. It should be suppressed.

†*Eodipodomys* Voorhies, 1975:163.
L. Mioc.; N.A.

†*Prodipodomys* Hibbard, 1939:458.
[Including †*Etadonomys* Hibbard, 1943:185.]
L. Mioc.-M. Pleist.; N.A.

Dipodomys Gray, 1841:521. Kangaroo rats.
[= *Dipodamys* Agassiz, 1842:10[1]; *Mamdipodomysus* Herrera, 1899:24.[2]] [= or including *Macrocolus* Wagner, 1846:172-177.[3]] [Including *Perodipus* Fitzinger, 1867:126; *Dipodops* Merriam, 1890:72.[4]]
L. Plioc.-R.; N.A.

[1] Misprint.
[2] See the International Code of Zoological Nomenclature, Article 1(b)(8), and Palmer (1904:25-26).

[3] *Nomen nudum* in Wagner, 1844.
[4] Objective synonym of *Perodipus*.

Microdipodops Merriam, 1891:115. Pygmy kangaroo rats, kangaroo mice.
  L. Pleist.-R.; N.A.

Suborder **ANOMALUROMORPHA** Bugge, 1974:48.

Superfamily **Pedetoidea** Gray, 1825:342.
  [= Pedestina Gray, 1825:342; Pedetoidae Ellerman, 1940:33; Pedetoidea Lavocat,
  1973:197.]

Family **Pedetidae** Gray, 1825:342.
  [= Halamydae Gray, 1821:303[1]; Pedestina Gray, 1825:342[2]; Pedetidae Owen,
  1847:242; Pedetini Brandt, 1855:219; Pedetini Winge, 1924:20; Helamyina Degland,
  1854:98.] [Including †Megapedetinae MacInnes, 1957:1.]
  E.-M. Mioc., E. Plioc.-R.; Af. E.-M. Mioc.; As.
  [1] Based not on a generic name but on a common name for *Pedetes*.
  [2] Proposed as a tribe.

†*Megapedetes* MacInnes, 1957:1.
  E.-M. Mioc.; Af. E.-M. Mioc.; As.[1]
  [1] W. Asia (Israel, Saudi Arabia, Turkey).

†*Diatomys* Li, 1974:43.
  M. Mioc.; As.

Pedetes Illiger, 1811:81. Springhaas, springhares, African jumping hares.
  [= *Helamys* F. Cuvier, 1817:202-203; *Helamis* F. Cuvier, 1821:341-344; *Pedestes*
  Gray, 1821:303[1]; *Gerbua* F. Cuvier, 1825:254.] [= or including *Yerbua* Forster,
  1778:108-119.[2]]
  E. Plioc.-R.; Af.
  [1] Attributed to Illiger.
  [2] In part?

Family †**Parapedetidae**, new.
  E.-M. Mioc.; Af.

†*Parapedetes* Stromer, in Kaiser, 1926:128, 143.
  E.-M. Mioc.; Af.

Superfamily **Anomaluroidea** Gervais, in d'Orbigny, 1849:203.
  [= Anomalurina Gervais, 1849:203; Anomaluroidea Gill, 1872:21.]

Family †**Zegdoumyidae** Vianey-Liaud, Jaeger, Hartenberger & Mahboubi, 1994:98.
  E. and/or M. Eoc.; Af.[1]
  [1] N. Africa.

†*Zegdoumys* Vianey-Liaud, Jaeger, Hartenberger & Mahboubi, 1994:100.
  E. and/or M. Eoc.; Af.[1]
  [1] N. Africa (Algeria and Tunisia).

†*Glibia* Vianey-Liaud, Jaeger, Hartenberger & Mahboubi, 1994:103.
  E. and/or M. Eoc.; Af.[1]
  [1] N. Africa (Algeria).

†*Glibemys* Vianey-Liaud, Jaeger, Hartenberger & Mahboubi, 1994:108.
  E. and/or M. Eoc.; Af.[1]
  [1] N. Africa (Algeria).

Family **Anomaluridae** Gervais, in d'Orbigny, 1849:203. Scaly-tailed squirrels.
  [= Anomalurina Gervais, 1849:203; Anomaluri Brandt, 1855:298; Anomaluridae Gill,
  1872:21.]
  L. Eoc.-E. Olig., E.-M. Mioc., E. Plioc., R.; Af.[1]
  [1] Anomaluridae nov. reported from Lothagam-l, early Pliocene of Kenya.

†*Nementchamys* Jaeger, Denys & Coiffait, 1985:580.
  L. Eoc.; Af.[1]
  [1] N. Africa (Algeria).

Subfamily **Anomalurinae** Gervais, in d'Orbigny, 1849:203.
[= Anomalurina Gervais, 1849:203; Anomalurinae Alston, 1875:95.]
E. Olig., E.-M. Mioc., R.; Af.

†*Paranomalurus* Lavocat, 1973:173.
E. Olig., E.-M. Mioc.; Af.

Anomalurus Waterhouse, 1843:124. Scaly-tailed flying squirrels.
[= *Aroaethrus* Waterhouse, 1843:124n.[1]] [Including *Anomalurodon* Matschie, 1914:350;[2] *Anomalurella* Matschie, 1914:351;[2] *Anomalurops* Matschie, 1914:351.[2]]
M. Mioc., R.; Af.
[1] Provisional replacement name.
[2] Subgenus as defined.

Subfamily **Zenkerellinae** Matschie, 1898:26.
[Including Idiuridae Miller & Gidley, 1918:442; Idiurinae Miller & Gidley, 1918:442.]
E. Mioc., R.; Af.

Zenkerella Matschie, 1898:23.[1] Flightless scaly-tailed squirrel.
[= *Aethurus* de Winton, 1898:1.[2]]
E. Mioc., R.; Af.
[1] May 17, 1898.
[2] May 20, 1898.

Idiurus Matschie, 1894:194. Pigmy scaly-tailed squirrels, small African flying squirrels.
R.; Af.[1]
[1] Tropical Africa.

Suborder **SCIURAVIDA, new**.[1]
[1] Definition: for the most recent common ancestor of †Ivanantoniidae, †Sciuravidae, †Chapattimyidae, †Cylindrodontidae, Ctenodactylidae, and all its descendants.

†*Prolapsus* A. E. Wood, 1973:22.[1]
M. Eoc.; N.A.
[1] Possibly referable to †Sciuravidae.

†*Pipestoneomys* Donohoe, 1956:264.[1]
L. Eoc.-E. Olig.; N.A.
[1] Possibly referable to †Sciuravidae.

†*Khodzhentia* Averianov, 1996:649.
E. Eoc.; As.

†*Hohomys* Hu, 1995:25.
E. Eoc.; As.

Family †**Ivanantoniidae** Shevyreva, 1989:71.[1]
E. Eoc.; As.
[1] *Nomen nudum* in Shevyreva, 1983:57.

†*Ivanantonia* Shevyreva, 1989:71.[1]
E. Eoc.; As.
[1] *Nomen nudum* in Shevyreva, 1983:57.

Family †**Sciuravidae** Miller & Gidley, 1918:442.
E.-M. Eoc., ?L. Eoc.; N.A. M. Eoc., ?L. Eoc.; As.

†*Pauromys* Troxell, 1923:155.
E.-M. Eoc.; N.A.

†*Knightomys* Gazin, 1961:193.
E.-M. Eoc.; N.A.

†*Tillomys* Marsh, 1872:219.
M. Eoc.; N.A.

†*Taxymys* Marsh, 1872:219.
[= †*Tachymys* Osborn, Scott & Speir, 1878:138; †*Toxymys* Zittel, 1893:522.[1]]
M. Eoc.; N.A.
[1] Misprint.

†*Sciuravus* Marsh, 1871:122.
> [= †*Colonymys* Marsh, 1872:220.]
> E.-M. Eoc.; N.A. ?L. Eoc.; As.

†*Zelomys* Wang & Li, 1990:182.
> M. Eoc.; As.

†*Guanajuatomys* Black & Stevens, 1973:2.
> M. and/or L. Eoc.; N.A.[1]
> [1] Mexico.

Family †**Chapattimyidae** Hussain, de Bruijn & Leinders, 1978:80.
> E.-M. Eoc., E. Olig., E. Mioc.; As. ?M. Eoc.; Eu.[1] L. Eoc.; Af.
> [1] Possible indet. chapattimyid in the middle Eocene of Spain.

†*Protophiomys* Jaeger, Denys & Coiffait, 1985:569.
> L. Eoc.; Af.

†*Esesempomys* Shevyreva, 1989:70.[1]
> E. Eoc.; As.
> [1] *Nomen nudum* in Shevyreva, 1983:57.

†*Fallomus* L. Flynn, Jacobs & Cheema, 1986:13.
> E. Mioc.; As.

Subfamily †**Cocomyinae** de Bruijn, Hussain & Leinders, 1982:252.
> [= †Cocomyidae de Bruijn, Hussain & Leinders, 1982:252[1]; †Cocomyidae Dawson, Li & Qi, 1984:147; †Cocomyinae Dashzeveg, 1990:18.] [Including †Tamquammyidae Shevyreva, 1983:57; †Orogomyidae Dashzeveg, 1990:40.]
> E.-M. Eoc.; As.
> [1] The priority of this name, attributed to Dawson, Li & Qi, 1982 (in press), over †Tamquammyidae Shevyreva, 1983, and †Cocomyidae Dawson, Li & Qi, 1984, is questionable, but it is retained here (as †Cocomyinae) for the purpose of stability, pending eventual adjudication by the International Commission on Zoological Nomenclature.

†*Orogomys* Dashzeveg, 1990:40.
> E. Eoc.; As.

†*Bumbanomys* Shevyreva, 1989:67.[1]
> E. Eoc.; As.
> [1] *Nomen nudum* in Shevyreva, 1983:57.

†*Adolomys* Shevyreva, 1989:65.[1]
> E. Eoc.; As.
> [1] *Nomen nudum* in Shevyreva, 1983:57.

†*Tsagankhushumys* Shevyreva, 1989:62.
> [= †*Tsaganchuschumys* Shevyreva, 1983:57.[1]]
> E. Eoc.; As.
> [1] *Nomen nudum.*

†*Cocomys* Dawson, Li & Qi, 1984:139.
> E. Eoc.; As.[1]
> [1] E. Asia.

†*Sharomys* Dashzeveg, 1990:20.
> E. Eoc.; As.

†*Kharomys* Dashzeveg, 1990:25.
> E. Eoc.; As.

†*Tsagamys* Dashzeveg, 1990:27.
> E. Eoc.; As.

†*Alaymys* Averianov, 1993:151.[1]
> E. Eoc.; As.
> [1] *Nomen nudum* in Averianov, 1991:150, 216.

†*Ulanomys* Dashzeveg, 1990:28.
> E. Eoc.; As.

†*Tsinlingomys* Li, 1963:151.[1]

    M. Eoc.; As.

[1] In the English summary beginning on p. 156 of the same paper, the orthography is consistently given as †"*Tsinlinomys.*"

†*Tamquammys* Shevyreva, 1971:745.

    M. Eoc.; As.

Subfamily †**Chapattimyinae** Hussain, de Bruijn & Leinders, 1978:80.

    [= †Chapattimyidae Hussain, de Bruijn & Leinders, 1978:80; †Chapattimyinae Hartenberger, 1982:22.[1]]

    E.-M. Eoc.; As.

[1] Informally proposed.

†*Geitonomys* Shevyreva, 1984:95.

    E. Eoc.; As.

†*Bolosomys* Shevyreva, 1984:92.

    E. Eoc.; As.

†*Chkhikvadzomys* Shevyreva, 1984:89.

    E. Eoc.; As.

†*Gumbatomys* Hartenberger, 1982:27.

    E. Eoc.; As.[1]

[1] S. Asia.

†*Birbalomys* Sahni & Khare, 1973:34.

    [= †*Metkamys* Sahni & Srivastava, 1976:924.] [Including †*Basalomys* Hartenberger, 1982:26.[1]]

    E.-M. Eoc.; As.[2]

[1] Described as subgenus of †*Birbalomys.*

[2] S. Asia.

†*Chapattimys* Hussain, de Bruijn & Leinders, 1978:87.

    E.-M. Eoc.; As.[1]

[1] S. Asia.

Subfamily †**Yuomyinae** Dawson, Li & Qi, 1984:147.

    [= †Yuomyidae Dawson, Li & Qi, 1984:147.] [Including †Advenimurinae Dashzeveg, 1990:30.]

    E.-M. Eoc., E. Olig.; As.

†*Bandaomys* Tong & Dawson, 1995:56.[1]

    E. Eoc.; As.

[1] (Tung & Dawson).

†*Advenimus* Dawson, 1964:4.

    [= or including †*Saykanomys* Shevyreva, 1972:134.]

    ?E. Eoc., M. Eoc.; As.

†*Petrokozlovia* Shevyreva, 1972:136.

    E.-M. Eoc.; As.

†*Yuomys* Li, 1975:58.

    M. Eoc.; As.[1]

[1] E. Asia.

†*Euboromys* Dashzeveg & McKenna, 1991:527.

    [= †*Boromys* Dashzeveg, 1990:31.[1]]

    M. Eoc.; As.

[1] Not †*Boromys* Miller, 1916: 8, a heteropsomyine echimyid rodent from the Quaternary of Cuba.

†*Dianomys* Wang, 1984:38.[1]

    E. Olig.; As.[2]

[1] The Kingdom of Dian is now Yunnan.

[2] South China (Yunnan).

Subfamily †**Baluchimyinae** L. Flynn, Jacobs & Cheema, 1986:18.

    E. Mioc.; As.

†*Lindsaya* L. Flynn, Jacobs & Cheema, 1986:31.
E. Mioc.; As.

†*Baluchimys* L. Flynn, Jacobs & Cheema, 1986:22.
E. Mioc.; As.

†*Lophibaluchia* L. Flynn, Jacobs & Cheema, 1986:34.
E. Mioc.; As.

†*Hodsahibia* L. Flynn, Jacobs & Cheema, 1986:38.
E. Mioc.; As.

†*Asterattus* Flynn & Cheema, in Tomida, Li & Setoguchi, 1994:122.
E. Mioc.; As.

†*Zindapiria* Flynn & Cheema, in Tomida, Li & Setoguchi, 1994:124.
E. Mioc.; As.

Family †**Cylindrodontidae** Miller & Gidley, 1918:440.
E.-L. Eoc., L. Olig.; N.A. L. Eoc.-E. Mioc.; As.

†*Dawsonomys* Gazin, 1961:194.
E. Eoc.; N.A.

†*Downsimys* L. Flynn, Jacobs & Cheema, 1986:40.[1]
E. Mioc.; As.

[1] Assignment uncertain. Not a junior homonym of †*Downsimus* Macdonald, 1970, an allomyid rodent.

Subfamily †**Cylindrodontinae** Miller & Gidley, 1918:440.
[= †Cylindrodontidae Miller & Gidley, 1918:440; †Cylindrodontinae Simpson, 1945:77.]
M.-L. Eoc.; N.A. L. Eoc.-L. Olig.; As.

†*Pseudocylindrodon* Burke, 1935:1.
M.-L. Eoc.; N.A. Olig.; As.

†*Ardynomys* Matthew & Granger, 1925:5.
L. Eoc.-L. Olig.; As. L. Eoc.; N.A.

†*Cylindrodon* Douglass, 1901:15.
L. Eoc.; N.A.

Subfamily †**Jaywilsonomyinae** A. E. Wood, 1974:52.
M.-L. Eoc., L. Olig.; N.A. E. Olig.; As.

†*Mysops* Leidy, 1871:231.
[= †*Syllophodus* Cope, 1881:375.[1]]
M. Eoc.; N.A.

[1] Invalid replacement. †*Mysops* (as "†*Myops*") was supposedly preoccupied by *Myops* Schiner, 1868, a dipteran.

†*Pareumys* Peterson, 1919:66.
M. Eoc.; N.A.

†*Anomoemys* Wang, 1986:289.
M. Eoc.; N.A. E. Olig.; As.

†*Jaywilsonomys* Ferrusquía-Villafranca & A. E. Wood, 1969:6.
L. Eoc.; N.A.

†*Sespemys* R. W. Wilson, 1934:13.
L. Olig.; N.A.

Family **Ctenodactylidae** Gervais, 1853:245. Gundis.
[= Ctenodactylinae Alston, 1876:90; Ctenodactyloidei Tullberg, 1899:387;
Ctenodactylina Gervais, 1853:245; Ctenodactylidae Zittel, 1893:542; Ctenodactyloidea
Simpson, 1945:99.] [Including Pectinatoridae Murray, 1866:xv; †Tataromyidae
Bohlin, 1946:133; †Tataromyinae Lavocat, 1961:52; †Pellegriniidae Schaub, 1950;
†Sayimyinae Lavocat, 1961:58; †Karakoromyinae Wang, 1994:40; †Distylomyidae
Wang, 1988:35; †Distylomyinae Wang, in Tomida, Li & Setoguchi, 1994:41.]

E. Olig.-E. Plioc.; As. Olig., E. Mioc., Pleist.; Mediterranean.[1] M. Mioc.-E. Pleist., R.; Af.

[1] An unnamed taxon has been reported from the Oligocene of Paguera, Mallorca, Balearic Islands.

†*Karakoromys* Matthew & Granger, 1923:6.
[= †*Terrarboreus* Shevyreva, 1971:81; †*Woodromys* Shevyreva, 1968:155[1]; †*Woodomys* Shevyreva, 1971:83.]
E. Olig., ?L. Olig.; As.

[1] *Nomen nudum.* Probably also a misprint.

†*Muratkhanomys* Shevyreva, 1994:151.
E. Olig.; As.

†*Roborovskia* Shevyreva, 1994:154.
E. Olig.; As.

†*Tataromys* Matthew & Granger, 1923:5.
[= †*Leptotataromys* Bohlin, 1946:107.]
L. Olig.-E. Mioc.; As.

†*Yindirtemys* Bohlin, 1946:108.
L. Olig.-E. Mioc.; As.

†*Prodistylomys* Wang & Qi, 1989:28.
L. Olig.; As.

†*Distylomys* Wang, 1988:35.
L. Olig., M. Mioc.; As.

†*Pireddamys* de Bruijn & Rümke, 1974:52.
E. Mioc.; Mediterranean.[1]

[1] Sardinia.

†*Metasayimys* Lavocat, 1961:59.
[= †*Dubiomys* Lavocat, 1961:66.]
E.-M. Mioc.; As.[1] L. Mioc.; Af.

[1] S.W. Asia (Israel, Saudi Arabia).

†*Sayimys* A. E. Wood, 1937:64-76.
E. Mioc.-E. Plioc.; As. Plioc.; Af.

†*Akzharomys* Shevyreva, 1994:160.
E. Mioc.; As.

†*Sardomys* de Bruijn & Rümke, 1974:48.
E. Mioc.; Mediterranean.[1]

[1] Sardinia.

†*Africanomys* Lavocat, 1961:52.
M.-L. Mioc.; Af.

†*Irhoudia* Jaeger, 1971:119.
L. Mioc.-E. Pleist.; Af.[1]

[1] N. Africa.

†*Testouromys* Robinson & Black, 1973:446.
L. Mioc.; Af.

Pectinator Blyth, 1856:294-296.
[= *Petrobates* Heuglin, 1860:413.]
?E. Plioc.; As. R.; Af.

†*Pellegrinia* de Gregorio, 1886:234.
[= †*Pellegrina* de Gregorio, 1886:234-241; †*Pellegrinia* Zittel, 1893:542.]
Pleist.; Mediterranean.[1]

[1] Sicily.

Massouteria Lataste, 1885:21-22.
R.; Af.

Felovia Lataste, 1886:287.[1]
>R.; Af.

[1] Proposed as subgenus of *Massouteria*; recognized as valid genus by Thomas, 1913:31.

Ctenodactylus Gray, 1830:10-11.
>R.; Af.[1]

[1] N. Africa.

Suborder **HYSTRICOGNATHA** Woods, 1976:268.

[= HYSTRICHOMORPHI Brandt, 1855:235, 294; HYSTRICHOMORPHA Brandt, 1855:294; HYSTRICOMORPHA Simpson, 1945:93.]
[Including HYSTRICOGNATHIFORMES Bryant & McKenna, 1996:30, 34.]

Family †**Tsaganomyidae** Matthew & Granger, 1923:4.

[= †Tsaganomyinae Matthew & Granger, 1923:4; †Tsaganomyidae A. E. Wood, in Patterson & A. E. Wood, 1982:515.]
E.-L. Olig., ?E. Mioc.; As.

†*Tsaganomys* Matthew & Granger, 1923:2.[1]

[= †*Cyclomylus* Matthew & Granger, 1923:5; †*Pseudotsaganomys* Vinogradov & Gambarian, 1952:18-22; †*Morosomys* Shevyreva, 1972:138; †*Sepulkomys* Shevyreva, 1972:139; †*Beatomus* Shevyreva, 1972:143.]
E.-L. Olig., ?E. Mioc.; As.

[1] †*Tsaganomys* is a widespread genus represented by many ontogenetic stages often mistaken for new taxa. Thus specimens of "†*Cyclomylus*," etc. are just young individuals of †*Tsaganomys*.

Infraorder **HYSTRICOGNATHI** Tullberg, 1899:69.[1]

[Including ACULEATA Illiger, 1811:89[2]; NOTOTROGOMORPHA Schaub, 1953:395[3]; CAVIOMORPHA A. E. Wood, 1955:180.[4]; PHIOMORPHA Lavocat, 1967:773.]

[1] We utilize HYSTRICOGNATHI in its taxic, not its descriptive, sense.
[2] ACULEATA Geoffroy Saint-Hilaire, 1795, was used earlier as a synonym of Tachyglossidae, but this is not technically preoccupation.
[3] Although Schaub's term has priority, it is unfamiliar. Moreover, the Rule of Priority does not apply above family group rank.
[4] Proposed as suborder. Cited as "Wood and Patterson (in press)," a reference to A. E. Wood & Patterson, 1959, where it was used at the rank of suborder. Used as infraorder by Patterson & A. E. Wood, 1982:380. Kraglievich, 1930, has also been cited as the author, but we have not yet found such a reference.

†*Luribayomys* Hoffstetter & Lavocat, 1976:64.
>L. Olig.; S.A.

Family **Hystricidae** Fischer de Waldheim, 1817:372. Old World porcupines.

[= Histricini Fischer de Waldheim, 1817:372; Hystricinorum Fischer de Waldheim, 1817:411; Hystricidae Gray, 1821:304; Hystrices Giebel, 1855:471; Hystricida Haeckel, 1866:clx; Hystricoidea Gill, 1872:22; Hystrioidae Miller & Gidley, 1918:444[1]; HYSTRICOTA Kalandadze & Rautian, 1992:92.[2]]
?Olig., E. Mioc.-L. Plioc., Pleist., R.; Eu.[3] M.-L. Mioc., ?E. Plioc., L. Plioc.-R.; As. L. Mioc.-R.; Af. Pleist., R.; E. Indies. R.; Mediterranean.[4]

[1] In part.
[2] In part. Named as a parvorder for Hystricoidea, Anomaluroidea, Pedetidae, †Theridomyidae, Bathyergidae, Ctenodactyloidea, Thryonomyidae, Gliroidea, and Platacanthomyidae, but not the kitchen sink.
[3] Recent of Italy and some parts of E. Europe possibly by introduction.
[4] Sicily, possibly by introduction.

Subfamily **Hystricinae** Fischer de Waldheim, 1817:372.

[= Hystricini Fischer de Waldheim, 1817:372; Hystricina C. L. Bonaparte, 1838:113[1]; Hystricinae Murray, 1866:xiv, 351.]
?Olig., E. Mioc.-L. Plioc., Pleist., R.; Eu.[2] L. Mioc.-R.; Af. L. Mioc., ?E. Plioc., L. Plioc.-R.; As. Pleist., R.; E. Indies. R.; Mediterranean.[3]

[1] As subfamily.
[2] Recent of Italy and some parts of E. Europe possibly by introduction.
[3] Sicily, possibly by introduction.

Hystrix Linnaeus, 1758:56.[1] Old World porcupines, Indonesian porcupines, landak.
>   [= *Histrix* Cuvier, 1798:130; *Oedocephalus* Gray, 1866:308-309.] [Including *Acanthion* F. Cuvier, 1822:424[2]; †*Orenomys* Aymard, 1855:507[3]; *Thecurus* Lyon, 1907:582.[2]]
>   ?Olig., E. Mioc.-L. Plioc., Pleist., R.; Eu.[4] L. Mioc.-R.; Af. L. Mioc., ?E. Plioc., L. Plioc.-R.; As. Pleist., R.; E. Indies.[5,6] R.; Mediterranean.[7]
>   [1] Some of the fossil species are certainly *Hystrix* only in the broad sense.
>   [2] As subgenus.
>   [3] Said to be a *nomen nudum* (Palmer 1904).
>   [4] Recent of Italy, Albania, and N. Greece, possibly by introduction.
>   [5] Java in the Pleistocene.
>   [6] Sumatra, Borneo, Java, Sulawesi, Lesser Sundas, and Philippines in the Recent.
>   [7] Sicily, possibly by introduction.

†*Miohystrix* Kretzoi, 1951:389, 407.
>   L. Mioc.; Eu.[1]
>   [1] E. Europe.

†*Xenohystrix* Greenwood, 1955:81.
>   E.-L. Plioc., Pleist.; Af.

Subfamily **Atherurinae** Lyon, 1907:576, 584.
>   M.-L. Mioc., E. Pleist.-R.; As. R.; Af. R.; E. Indies.

†*Sivacanthion* Colbert, 1933:3.
>   M.-L. Mioc.; As.

Atherurus F. Cuvier, 1829:483. Brush-tailed porcupines.
>   E. Pleist.-R.; As.[1,2] R.; Af.[3] R.; E. Indies.[4]
>   [1] China in the Pleistocene.
>   [2] China and S.E. Asia in the Recent.
>   [3] Tropical Africa.
>   [4] Sumatra.

Trichys Günther, 1877:739. Long-tailed porcupines.
>   M. Pleist., R.; As.[1,2] R.; E. Indies.[3]
>   [1] S. China in the Pleistocene.
>   [2] S.E. Asia in the Recent.
>   [3] Sumatra, Borneo.

Family **Erethizontidae** Bonaparte, 1845:5. New World porcupines, tree porcupines.
>   [= Erethyzonina Bonaparte, 1845:5; Erethizontidae Thomas, 1897:1025; Erethizontoidea Simpson, 1945:94; ERETHIZONTOMORPHIA Bugge, 1974:56.[1]] [Including Cercolabina Gray, 1843:xxiv, 123; Cercolabidae Ameghino, 1889[2]; Coendidae Trouessart, 1897:619.]
>   L. Olig.-L. Mioc., L. Pleist.-R.; S.A. L. Plioc.-R.; N.A. L. Pleist.-R.; Cent. A.
>   [1] Proposed as suborder.
>   [2] Not seen. See Ameghino's Obras Completas 6:218.

Subfamily **Erethizontinae** Bonaparte, 1845:5.
>   [= Erethyzonina Bonaparte, 1845:5; Erethizontinae Thomas, 1897:1025.] [Including Cercolabina Gray, 1843:xxiv, 123; Sphingurinae Alston, 1876:93; †Steiromyinae Ameghino, 1902:109; Coendinae Pocock, 1922:422.[1]]
>   L. Olig.-L. Mioc., L. Pleist.-R.; S.A. L. Plioc.-R.; N.A. L. Pleist.-R.; Cent. A.
>   [1] Objective synonym of Sphingurinae.

†*Protosteiromys* A. E. Wood & Patterson, 1959:377.
>   L. Olig.; S.A.

†*Eosteiromys* Ameghino, 1902:110-111.[1]
>   E. Mioc., ?M. Mioc.; S.A.
>   [1] *Nomen nudum* in 1901.

†*Hypsosteiromys* Patterson, 1958:2.
>   E. Mioc.; S.A.

†*Parasteiromys* Ameghino, 1904:101.
>   E. Mioc.; S.A.

†*Steiromys* Ameghino, 1887:9-10.
  E. Mioc.; S.A.

†*Disteiromys* Ameghino, 1904:102.
  M. Mioc.; S.A.

†*Neosteiromys* Rovereto, 1914:75.
  ?M. Mioc., L. Mioc.; S.A.

Erethizon F. Cuvier, 1822:426. North American porcupine.
  [= *Eretizon* Cuvier, 1825:178-179, 256; *Erethison* Cuvier, 1829:484; *Erithizon*
  Burnett, 1830:350; *Eretison* McMurtrie, 1831:154; *Erythizon* Alston, 1876:94.]
  L. Plioc.-R.; N.A.

Coendou Lacépède, 1799:11. South American tree porcupines, prehensile-tailed porcupines,
  coandú, cuandu, couiy.
  [= *Coendus* E. Geoffroy Saint-Hilaire, 1803:157; *Coandu* G. Fischer, 1814:102-105;
  Coendu Lesson, 1827:290-291; *Cuandu* Liais, 1872:532, 550; *Sinethere* F. Cuvier,
  1822:427; *Sinoetherus* F. Cuvier, 1825:256; *Laboura* Billberg, 1828:before 1[1];
  Synetheres G. Cuvier, 1829:216; *Cercolabes* Brandt, 1835:55-58[2]; *Mamsynetheresus*
  Herrera, 1899:16.[3]] [Including *Sphiggure* F. Cuvier, 1822:427; *Sphiggurus* F. Cuvier,
  1825:256[4]; *Echinoprocta* Gray, 1865:321.[5]]
  ?L. Pleist., R.; N.A.[6] L. Pleist.-R.; Cent. A. L. Pleist.-R.; S.A.[7]

[1] Replacement name.
[2] New name for the "barbarous" *Coendu* Lacépède, 1799 (Palmer, 1903:171).
[3] See the International Code of Zoological Nomenclature, Article 1(b)(8), and Palmer (1904:25-26).
[4] Also as *Sphingura* Wagler, 1830:18-19; *Spigurus* Swainson, 1835:390; *Spiggurus* Gray, 1847:45; and
  Sphingurus Waterhouse, 1848:409.
[5] Proposed as subgenus of *Erethizon*.
[6] Mexico.
[7] Including Trinidad in the Recent.

Subfamily **Chaetomyinae** Thomas, 1897:1026.
  R.; S.A.

Chaetomys Gray, 1843:21. Thin-spined porcupine.
  [= *Plectrochoerus* Pictet, 1843:225-227.]
  R.; S.A.[1]

[1] Brazil.

Family †**Myophiomyidae** Lavocat, 1973:103.
  E. Olig., E.-M. Mioc.; Af.

Subfamily †**Phiocricetomyinae** Lavocat, 1973:103.
  E. Olig.; Af.

†*Phiocricetomys* A. E. Wood, 1968:73.
  E. Olig.; Af.

Subfamily †**Myophiomyinae** Lavocat, 1973:103.
  E.-M. Mioc.; Af.

†*Phiomyoides* Stromer, in Kaiser, 1926:136, 143.
  E. Mioc.; Af.

†*Elmerimys* Lavocat, 1973:105.
  E.-M. Mioc.; Af.

†*Myophiomys* Lavocat, 1973:103.
  E. Mioc.; Af.

Family †**Diamantomyidae** Schaub, 1958:786.
  L. Eoc.-E. Olig., E.-M. Mioc.; Af. L. Eoc. and/or E. Olig.; As.[1]
  [1] Arabian Peninsula.

Subfamily †**Metaphiomyinae** Lavocat, 1973:52.
  L. Eoc.-E. Olig.; Af. L. Eoc. and/or E. Olig.; As.[1]
  [1] Arabian Peninsula.

†*Metaphiomys* Osborn, 1908:270.
>   L. Eoc.-E. Olig.; Af.[1] L. Eoc. or E. Olig.; As.[2]
>   [1] N. Africa.
>   [2] S.W. Asia (Oman).

Subfamily †**Diamantomyinae** Schaub, 1958:786.
>   [= †Diamantomyidae Schaub, 1958:786; †Diamantomyinae Lavocat, 1973:52.]
>   E.-M. Mioc.; Af.

†*Diamantomys* Stromer, 1922:334.
>   E.-M. Mioc.; Af.

†*Pomonomys* Stromer, 1922:334-335.
>   E. Mioc.; Af.

Family †**Phiomyidae** A. E. Wood, 1955:172.
>   L. Eoc.-E. Olig., E. Mioc.; Af. L. Eoc. or E. Olig., M. Mioc.; As.[1,2]
>   Olig.; Mediterranean.[3]
>   [1] S.W. Asia in the late Eocene or early Oligocene.
>   [2] An unnamed phiomyid has been reported in the middle Miocene of W. Asia (Anatolia).
>   [3] An unnamed phiomyid has been reported from the Oligocene of Mallorca.

†*Phiomys* Osborn, 1908:269.
>   L. Eoc.-E. Olig., E. Mioc.; Af. L. Eoc. or E. Olig.; As.[1]
>   [1] S.W. Asia (Oman).

†*Andrewsimys* Lavocat, 1973:259.
>   E. Mioc.; Af.

Family †**Kenyamyidae** Lavocat, 1973:93.
>   E. Mioc.; Af.

†*Kenyamys* Lavocat, 1973:93.
>   E. Mioc.; Af.

†*Simonimys* Lavocat, 1973:99.
>   E. Mioc.; Af.

Family **Petromuridae** Tullberg, 1899:147.
>   [= Muriformia Van der Hoeven, 1858:684[1]; Petromyidae Tullberg, 1899:147;
>   Petromyinae Ellerman, 1940:149; Petromurinae Lavocat, 1973:50;
>   Petromuridae Swanepoel, Smithers & Rautenbach, 1980:161.]
>   Pleist., R.; Af.
>   [1] In part. Proposed as a family but not based on a generic name.

Petromus A. Smith, 1831:10.[1] Rock rat.
>   [= *Petromys* Smith, 1834:146-147.[2]]
>   Pleist., R.; Af.
>   [1] Misprint for page 2.
>   [2] Unjustifiable emendation.

Family **Thryonomyidae** Pocock, 1922:423.
>   [= Thrynomyidae Pocock, 1922:423; Thryonomyoidea A. E. Wood, 1955:183;
>   Aulacodina C. L. Bonaparte, 1845:5; Ulacodidae Brandt, 1855:251.[1]]
>   L. Eoc.-E. Olig., E.-L. Mioc., L. Plioc.-R.; Af. Olig.; Mediterranean.
>   M.-L. Mioc.; As.
>   [1] Based upon *Aulacodus*.

†*Paraphiomys* C. W. Andrews, 1914:178.
>   [= †*Neosciuromys* Stromer, 1922:333; †*Phthinylla* Hopwood, 1929:4.]
>   [Including †*Apodecter* Hopwood, 1929:3.]
>   E. Olig., E.-L. Mioc.; Af. M. Mioc.; As.[1]
>   [1] S.W. Asia (Saudi Arabia).

†*Gaudeamus* A. E. Wood, 1968:68.
>   L. Eoc.; Af.[1]
>   [1] N. Africa (Egypt).

†*Sacaresia* Hugueney & Adrover, 1991:208.

      Olig.; Mediterranean.[1]

    [1] Mallorca.

†*Epiphiomys* Lavocat, 1973:46.

      E. Mioc.; Af.

†*Paraulacodus* Hinton, 1933:621.

      M.-L. Mioc.; Af. M.-L. Mioc.; As.

†*Kochalia* de Bruijn, 1986:125.

    [= †*Kirtharia* de Bruijn & Hussain, 1985:157.[1]]

      M. Mioc.; As.

    [1] De Bruijn replaced †*Kirtharia* with †*Kochalia* because he incorrectly believed †*Kirtharia* to be a junior homonym of †*Khirtharia* Pilgrim, 1940, an artiodactyl. This is a clear violation of the International Code of Zoological Nomenclature, Article 56(b). The matter should possibly be adjudicated.

Thryonomys Fitzinger, 1867:141. Cane rats.

    [= *Aulacodus* Temminck, 1827:245-248;[1] *Triaulacodus* Lydekker, 1896:91, 240n.[2]]

    [Including *Choeromys* Thomas, 1922:390.]

      L. Plioc.-R.; Af.

    [1] Not *Aulacodus* Eschscholtz, 1822, a genus of Coleoptera.
    [2] New name for *Aulacodus* Temminck.

Parvorder **BATHYERGOMORPHI** Tullberg, 1899:71, **new rank**.[1]

    [= Bathyergidae Waterhouse, 1841:81; Bathyergoidea Osborn, 1910:538; Bathyergoidae Miller & Gidley, 1918:443; BATHYERGOMORPHA Roberts, 1951:345.[2]]

    [1] Proposed as a "subtribe" but with rank higher than family.
    [2] Proposed as suborder.

†*Paracryptomys* Lavocat, 1973:147.

      E. Mioc.; Af.

Family **Bathyergidae** Waterhouse, 1841:81. Mole-rats, sand-rats.

    [= Orycterideae Lesson, 1842:120; Bathyergini Winge, 1924:57.] [Including Georychi Wiegmann, 1832; Georychina Gravenhorst, 1843: facing 502; Georychini Brandt, 1855:202; Georrhychida Haeckel, 1866:clx[1]; Georychidae (Anon.), 1897:14.]

      E. Mioc.-R.; Af.[2] Pleist.; As.[3]

    [1] In part. Used as a family.
    [2] An unspecified bathyergid has been reported in the late Miocene of Africa.
    [3] W. Asia (Israel).

Subfamily **Bathyerginae** Waterhouse, 1841:81.

    [= Bathyergidae Waterhouse, 1841:81; Bathyerginae Alston, 1876:87; Bathyergini Winge, 1924:57; Bathyerginae Landry, 1957:74, 80.] [Including Georychi Wiegmann, 1832[1]; Georychini Brandt, 1855:202; Georychinae Roberts, 1951:382.]

      E.-M. Mioc., E. Plioc.-R.; Af. Pleist.; As.[2]

    [1] A fine candidate for suppression.
    [2] W. Asia (Israel).

†*Proheliophobius* Lavocat, 1973:139.

      E. Mioc.; Af.

†*Richardus* Lavocat, 1988:1301.

      M. Mioc.; Af.

†*Gypsorhychus* Broom, 1934:471-480.

      E. Pleist.; Af.[1]

    [1] S. Africa.

Cryptomys Gray, 1864:124.[1] Common mole-rats.

    [Including *Coetomys* Gray, 1864:124-125; *Typhloryctes* Fitzinger, 1867:502-503.[2]]

      E. Plioc.-R.; Af. Pleist.; As.[3]

    [1] Proposed as subgenus of *Georychus*.
    [2] Objective synonym of *Coetomys*.
    [3] W. Asia (Israel).

Georychus Illiger, 1811:87. Cape mole-rat.
[= *Georrhychus* Minding, 1829:80; *Georhychus* Wagner, 1843:369-375; *Fossor* Lichtenstein, 1844:31-32.]
M. Pleist.-R.; Af.[1]
[1] S. Africa.

Bathyergus Illiger, 1811:86. Cape sand-rats.
[= *Orycterus* F. Cuvier, 1829:481-482.]
M. Pleist.-R.; Af.[1]
[1] Southern Africa.

Heliophobius Peters, 1846:259. Sand-rat.
[= *Myoscalops* Thomas, 1890:448[1]; *Heliphobius* Beddard, 1902:481.[2]]
R.; Af.
[1] New name for *Heliophobius* Peters, 1846, thought to be preoccupied by *Heliophobus* Boisduval, 1829:69, a genus of Lepidoptera.
[2] Misprint.

Subfamily **Heterocephalinae** Landry, 1957:74, 80.
E. Plioc.-M. Pleist., R.; Af.

Heterocephalus Rüppell, 1842:99-101. Naked mole-rat.
[Including *Fornarina* Thomas, 1904:336.[1]]
E. Plioc.-M. Pleist., R.; Af.
[1] See Conisbee (1953:32) for erudite comment.

Family †**Bathyergoididae** Lavocat, 1973:109.
E. Mioc.; Af.

†*Bathyergoides* Stromer, 1923:263.
E. Mioc.; Af.

Parvorder **CAVIIDA** Bryant & McKenna, 1995:30, 34.
[= Caviae Fischer de Waldheim, 1817:372; Caviarum Fischer de Waldheim, 1817:409.[1]]
[1] As "family."

Superfamily **Cavioidea** Fischer de Waldheim, 1817:372.
[= Sub-ungulata Illiger, 1811:92[1]; Caviae Fischer de Waldheim, 1817:372; Caviarum Fischer de Waldheim, 1817:409[1]; Caviadae Gray, 1821:304; Cavioidea Kraglievich, 1930:60.] [Including Dinomyoidea Schaub, 1953:396.]
[1] As a "family."

Family **Agoutidae** Gray, 1821:304.
[= Coelogenyidae Gervais, 1849:204; Coelogenina Gervais, 1849:204; Coelogenyidae Burmeister, 1854:227; Cuniculidae Miller & Gidley, 1918:446.] [Including Dasyporcina Gray, 1825:341; Dasyproctina Bonaparte, 1838:112; Dasyproctidae H. Smith, 1842:307; Dasyproctida Haeckel, 1866:clx; †Cephalomyidae Ameghino, 1897:493.]
L. Eoc. and/or E. Olig., L. Olig.-M. Mioc., R.; S.A. R.; N.A.[1] R.; Cent. A. R.; W. Indies.[2]
[1] Southern North America (tropical Mexico).
[2] Lesser Antilles, Cuba, and Cayman Islands by introduction.

†*Scotamys* Loomis, 1914:192.
L. Olig.; S.A.

Subfamily **Dasyproctinae** Gray, 1825:341.
[= Dasyporcina Gray, 1825:341; Dasyproctina C. L. Bonaparte, 1837:9[1]; Dasyproctidae H. Smith, 1842:307; Dasyproctini Brandt, 1855:280.] [Including †Cephalomyidae Ameghino, 1897:493.]
L. Eoc. and/or E. Olig., L. Olig.-M. Mioc., R.; S.A.[2] R.; Cent. A. R.; W. Indies.[3]
[1] As subfamily.
[2] ?Dasyproctidae, new genus and species, from the late Eocene and/or early Oligocene Tinguiriirica Fauna, Chile.
[3] Lesser Antilles, Cuba, and Cayman Islands by introduction.

†*Palmiramys* Kraglievich, 1932:314.
    L. Olig.; S.A.

†*Litodontomys* Loomis, 1914:193.
    L. Olig.; S.A.

†*Incamys* Hoffstetter & Lavocat, 1970:172.
    L. Olig.; S.A.[1]
[1] Bolivia.

†*Branisamys* Hoffstetter & Lavocat, 1970:172.
    [= †*Villarroelomys* Hartenberger, 1975:427.]
    L. Olig.; S.A.[1]
[1] Bolivia.

†*Cephalomys* Ameghino, 1897:18.
    L. Olig.; S.A.

†*Cephalomyopsis* Vucetich, 1985:243.
    E. Mioc.; S.A.

†*Neoreomys* Ameghino, 1887:10-11.
    E.-M. Mioc.; S.A.

†*Megastus* Roth, 1898:193-194.
    [= †*Magestus* Ameghino, 1899:7.[1]]
    M. Mioc.; S.A.

[1] Pagination of separate; original pagination not determined. New name for †*Megastus* Roth thought to be preoccupied by *Megastes* Guénée, 1854, and by *Megastes* Boisduval, 1870, both lepidopterans.

†*Alloiomys* Vucetich, 1977:216.
    M. Mioc.; S.A.

Myoprocta Thomas, 1903:464. Acushi, acuchi, acuschy.
    R.; S.A.[1]
[1] Tropical South America.

Dasyprocta Illiger, 1811:93. Agoutis, agutis.
    [= *Mamdasyproctaus* Herrera, 1899:29.[1]] [= or including *Cloromis* F. Cuvier, 1812:290-291; *Chloromys* (F. Cuvier) Rafinesque, 1815:56; *Chloromys* Lesson, 1827:300-301.[2]]
    R.; Cent. A. R.; W. Indies.[3] R.; S.A.

[1] See the International Code of Zoological Nomenclature, Article 1(b)(8), and Palmer (1904:25-26).
[2] Not †*Chloromys* (von Meyer, MS.) Schlosser, 1884, a subgenus of †*Steneofiber*.
[3] Lesser Antilles, Cuba, and Cayman Islands by introduction.

Subfamily **Agoutinae** Gray, 1821:304.
    [= Agoutidae Gray, 1821:304; Agoutinae Hall & Kelson, 1959:787; Coelogenina Gervais, 1849:204; Coelogenyidae Burmeister, 1854:227; Coelogenyidae Pocock, 1922:424; Cuniculidae Miller & Gidley, 1918:446; Cuniculinae Simpson, 1945:96.]
    R.; N.A.[1] R.; Cent. A. R.; S.A.
[1] Southern North America (tropical Mexico).

Agouti Lacépède, 1799:9.[1] Pacas.
    [= *Cuniculus* Brisson, 1762:13, 98-104[2]; *Coelogenus* F. Cuvier, 1807:203[3]; *Coelogenys* Illiger, 1811:92; *Caelogenus* Fleming, 1822:192; *Caelogenys* Agassiz, 1842:5; *Caelogonus* Anonymous, 1845:747; *Genyscoelus* Liais, 1872:537[4]; *Paca* Fisher, 1814:85; *Osteopera* Harlan, 1825:126; *Mamcoelogenysus* Herrera, 1899:120.[5]] [Including *Stictomys* Thomas, 1924:238.]
    R.; N.A.[6] R.; Cent. A. R.; S.A.

[1] Note that *Agouti* is not applicable to the agoutis (agutis) but to the pacas.
[2] Not published in a consistently binominal work. However, see the International Code of Zoological Nomenclature, Article 11(c) (i), for confusion. Not a senior homonym of *Cuniculus* Meyer, 1790, a synonym of *Oryctolagus*, nor of *Cuniculus* Wagler, 1830, a synonym of *Dicrostonyx*.
[3] *Coelogenys* of most authors.
[4] Substitute name suggested, but never used, for *Coelogenus* Cuvier, 1807.
[5] See the International Code of Zoological Nomenclature, Article 1(b)(8), and Palmer (1904:25-26).
[6] Southern North America (tropical Mexico north to San Luis Potosí).

Family †**Eocardiidae** Ameghino, 1891:145.
  [= †Eocardidae Ameghino, 1891:145.]
  L. Olig.-L. Mioc.; S.A.

Subfamily †**Eocardiinae** Ameghino, 1891:145, **new rank**.
  [= †Eocardidae Ameghino, 1891:145.]
  L. Olig.-L. Mioc.; S.A.

†*Chubutomys* A. E. Wood & Patterson, 1959:373.
  L. Olig.; S.A.

†*Phanomys* Ameghino, 1887:13-14.
  E. Mioc.; S.A.

†*Schistomys* Ameghino, 1887:13.
  [Including †*Procardia* Ameghino, 1891:16.[1]]
  E. Mioc.; S.A.
  [1] Proposed as subgenus of †*Eocardia*; raised to generic rank by Ameghino, 1894.

†*Eocardia* Ameghino, 1887:65-66.
  [Including †*Dicardia* Ameghino, 1891:16[1]; †*Tricardia* Ameghino, 1891:16-17.[1]]
  E.-M. Mioc.; S.A.
  [1] As subgenus.

†*Matiamys* Vucetich, 1984:57.
  M. Mioc.; S.A.

†*Neophanomys* Rovereto, 1914:60.
  L. Mioc.; S.A.

Subfamily †**Luantinae** A. E. Wood & Patterson, 1959:365.
  L. Olig.-E. Mioc.; S.A.

†*Asteromys* Ameghino, 1897:495.
  L. Olig.; S.A.

†*Luantus* Ameghino, 1899:7.[1]
  E. Mioc.; S.A.
  [1] Pagination of separate. "Usually cited as '*Luanthus*,' but the original spelling was intentionally *Luantus*" (Simpson 1945:94).

Family **Dinomyidae** Peters, 1873:552.
  [= Dinomyes Peters, 1873:552; Dinomyina Troschel, 1874:132; Dinomyidae Alston, 1876:96; Dinomyoidea Schaub, 1953:396.] [Including †Eumegamyidae Kraglievich, 1926:121.]
  E. Mioc.-L. Plioc., Pleist., R.; S.A.

†*Pseudodiodomus* Paula Couto, 1979:105.
  L. Mioc.; S.A.

†*Agnomys* Kraglievich, 1940:490.[1]
  L. Mioc. and/or E. Plioc.; S.A.
  [1] Best considered a *nomen dubium*.

Subfamily †**Eumegamyinae** Kraglievich, 1926:121.
  [= †Eumegamyidae Kraglievich, 1926:121; †Eumegamyinae Kraglievich, 1932:179; †Eumegamyini Mones, 1981:610.] [Including †Phoberomyinae Kraglievich, 1926:127; †Gyriabrinae Kraglievich, 1930:220; †Tetrastylinae Kraglievich, 1931:255; †Tetrastylusinae Kraglievich, 1940:717.]
  L. Mioc.-L. Plioc., Pleist.; S.A.

†*Doellomys* Alvarez, 1947:60.
  L. Mioc.; S.A.

†*Gyriabrus* Ameghino, 1891:246.
  [= †*Gyrabrius* Lydekker, 1892:33.]
  L. Mioc.; S.A.

†*Briaromys* Ameghino, 1889:904-905.
  L. Mioc.; S.A.

†*Tetrastylus* Ameghino, 1886:46.

    [= †*Loxomylus* Burmeister, 1891:384[1]; †*Loxopygus* Burmeister, 1891: pl. 7.[2]]
    [Including †*Tetrastylopsis* Kraglievich, 1931:393; †*Protelicomys* Kraglievich, 1931:393.[3]]
    L. Mioc.-E. Plioc., ?L. Plioc., Pleist.; S.A.

    [1] Not †*Loxomylus* Cope, 1869: 186, a fossil rodent from Anguilla, West Indies.
    [2] Misprint?
    [3] Subgenus as proposed.

†*Phoberomys* Kraglievich, 1926:127.[1]

    L. Mioc. and/or E. Plioc.; S.A.

    [1] The largest known rodent, about the size of a small ox.

†*Colpostemma* Ameghino, 1891:141.

    [= †*Calpostemma* Zittel, 1893:549.]
    L. Mioc.; S.A.

†*Orthomys* Ameghino, 1881:306.

    L. Mioc.; S.A.

†*Eumegamys* Kraglievich, 1926:122.

    [= †*Megamys* Ameghino, 1886:41.[1]] [Including †*Isostylomys* Kraglievich, 1926:125; †*Carlesia* Kraglievich, 1926:126; †*Diaphoromys* Kraglievich, 1931:392; †*Rusconia* Kraglievich, 1931:392; †*Protomegamys* Kraglievich, 1932:212.]
    L. Mioc.-L. Plioc.; S.A.

    [1] And almost all authors but not †*Megamys* d'Orbigny & Laurillard, 1837: pl. 8, figs. 4-8; 1842:110-112, which Kraglievich has shown to be a litoptern.

†*Pseudosigmomys* Kraglievich, 1931:394.

    L. Mioc. and/or Plioc.; S.A.

†*Pentastylodon* Alvarez, 1947:62.

    L. Mioc.; S.A.

†*Eumegamysops* Alvarez, 1947:61.[1]

    L. Mioc.; S.A.

    [1] Not Kraglievich, 1940:750, a *nomen nudum* for a different animal.

†*Telicomys* Kraglievich, 1926:126.

    L. Mioc.-L. Plioc.; S.A.

†*Perumys* Kretzoi & Voros, 1989:111.

    L. Plioc.; S.A.

†*Artigasia* Francis & Mones, 1966:92.

    L. Plioc.; S.A.[1]

    [1] Uruguay.

Subfamily †**Potamarchinae** Kraglievich, 1926:129.

    [= †Potamarchidae Kraglievich, 1926:129; †Potamarchinae Simpson, 1945:96.]
    E.-L. Mioc.; S.A.

†*Scleromys* Ameghino, 1887:11.

    [Including †*Lomomys* Ameghino, 1891:15.]
    E.-M. Mioc.; S.A.

†*Olenopsis* Ameghino, 1889:145.

    [Including †*Paranomys* Ameghino, 1889:901[1]; †*Paranamys* Kraglievich, 1934:73; †*Drytomomys* Anthony, 1922:2.]
    E.-L. Mioc.; S.A.

    [1] From Scalabrini (MS).

†*Simplimus* Ameghino, 1904:104.

    E.-M. Mioc.; S.A.

†*Eusigmomys* Ameghino, 1905:75.

    [= †*Sigmomys* Ameghino, 1904:103.[1]]
    M. Mioc.; S.A.

    [1] Not *Sigmomys* Thomas, 1901:150, a muroid rodent.

†*Potamarchus* Burmeister, 1885:154-157.
[= or including †*Discolomys* Ameghino, 1889:148.]
L. Mioc.; S.A.

Subfamily **Dinomyinae** Peters, 1873:552.
[= Dinomyes Peters, 1873:552; Dinomyina Troschel, 1874:132; Dinomyidae Alston, 1876:96; Dinomyinae Schaub, 1953:396; Dinomyini Mones, 1981:609.]
L. Mioc., R.; S.A.

†*Telodontomys* Kraglievich, 1934:72.
L. Mioc.; S.A.

Dinomys Peters, 1873:551-552. Long-tailed paca, false paca, pacarana.
R.; S.A.

Family **Caviidae** Fischer de Waldheim, 1817:372.
[= Caviae Fischer de Waldheim, 1817:372; Caviarum Fischer de Waldheim, 1817:409[1]; Caviadae Gray, 1821:304; Caviina Gray, 1825:341; Caviidae Waterhouse, 1839:91.]
M. Mioc.-R.; S.A.
[1] As "family."

†*Strata* Ameghino, 1886:70-71.[1]
L. Mioc.; S.A.
[1] Assignment to Caviidae is questionable.

Subfamily **Caviinae** Fischer de Waldheim, 1817:372. Guinea pigs, cavies, cuys.
[= Caviae Fischer de Waldheim, 1817:372; Caviarum Fischer de Waldheim, 1817:409[1]; Caviadae Gray, 1821:304; Caviina Gray, 1825:341[2]; Cavini Brandt, 1855:280[3]; Caviinae Murray, 1866:xiv, 350.]
L. Mioc.-R.; S.A.
[1] As "family."
[2] As tribe.
[3] Also used by Giebel, 1855.

†*Neoprocavia* Ameghino, 1889:235-236.[1]
[= †*Procavia* Ameghino, 1885:66, 68.[2]]
L. Mioc.; S.A.
[1] New name for †*Procavia* Ameghino.
[2] Not *Procavia* Storr, 1780, a genus of hyraxes.

†*Allocavia* Pascual, 1962:169-174.
L. Mioc.; S.A.

†*Palaeocavia* Ameghino, 1889:231-233.
L. Mioc.-L. Plioc.; S.A.

†*Neocavia* Kraglievich, 1932:164.
L. Mioc. and/or E. Plioc.; S.A.

†*Dolicavia* C. Ameghino, 1916:283-284.
L. Plioc.-E. Pleist.; S.A.

†*Macrocavia* Rusconi, 1933:16.
L. Plioc.; S.A.

†*Caviops* Ameghino, 1908.
L. Plioc.; S.A.

†*Pascualia* Ortega Hinojosa, 1963.
L. Plioc.; S.A.

Galea Meyen, 1833:597. Cuy.
Pleist., R.; S.A.

Microcavia Gervais & Ameghino, 1880:50-55. Mountain guinea pigs.
[Including *Caviella* Osgood, 1915:194; *Monticavia* Thomas, 1916:303; *Nanocavia* Thomas, 1925:418.]
M. Pleist.-R.; S.A.

Cavia Pallas, 1766:30-33. Common guinea pigs, domestic guinea pigs, cavies, perea, preá.
> [= *Cauia* Storr, 1780:39, table B; *Anoema* F. Cuvier, 1809:394; *Cobaya* G. Cuvier, 1817:481-482; *Coiza* Billberg, 1828:45[1]; *Cobaia* Aymard, 1854:393; *Mamcaviaus* Herrera, 1899:13.[2]] [= or including †*Moco* Lund, 1840:191; †*Perea* Lund, 1840:191[3]; *Prea* Liais, 1872:540-545.[4]]
> M. Pleist.-R.; S.A.

[1] Replacement name.
[2] See the International Code of Zoological Nomenclature, Article 1(b)(8), and Palmer (1904:25-26).
[3] Defined as subgenus but no type species designated.
[4] Possibly a misspelling of *Perea* Lund, 1840.

Kerodon F. Cuvier, 1825:151. Moco.
> [= *Kerodons* F. Cuvier, 1829:493; *Cerodon* Wagler, 1830:18; †*Ceratodon* Wagler, 1830:18.]
> L. Pleist.-R.; S.A.

Subfamily **Dolichotinae** Pocock, 1922:426.
> M. Mioc.-L. Plioc., E. and/or M. Pleist., L. Pleist.-R.; S.A.

†*Prodolichotis* Kraglievich, 1932:158.
> M. Mioc.-L. Plioc.; S.A.

†*Orthomyctera* Ameghino, 1889:218-221.
> [Including †*Orocavia* Kraglievich, 1932:163.[1]]
> L. Mioc.-L. Plioc.; S.A.

[1] Described as subgenus of †*Orthomyctera*.

†*Pliodolichotis* Kraglievich, 1927:596.
> L. Mioc.; S.A.

†*Propediolagus* Ortega Hinojosa, 1963:21-28.
> ?L. Plioc.; S.A.

Dolichotis Desmarest, 1820:360. Maras, Patagonian hares.
> [= *Mara* d'Orbigny, 1829:220.] [Including *Pediolagus* Marelli, 1927:5[1]; *Weyenberghia* Kraglievich, 1927:578[2]; *Paradolichotis* Kraglievich, 1927:594[1]; *Lagospedius* Marelli, 1928:103.]
> E. and/or M. Pleist., L. Pleist.-R.; S.A.

[1] As subgenus.
[2] Proposed as subgenus. Preoccupied by an insect homonym. Objective synonym of *Pediolagus* Marelli, 1927.

Subfamily †**Cardiomyinae** Kraglievich, 1930:61.
> L. Mioc.-L. Plioc.; S.A.

†*Procardiomys* Pascual, 1961:61-71.
> L. Mioc.; S.A.

†*Parodimys* Kraglievich, 1932:175.
> L. Mioc.; S.A.

†*Cardiomys* Ameghino, 1885:57.
> [Including †*Caviodon* Ameghino, 1885:65-66; †*Diocartherium* Ameghino, 1888:10; †*Lelongia* Kraglievich, 1930:178[1]; †*Paracaviodon* Kraglievich, 1932:169, 179[1]; †*Pseudocardiomys* Kraglievich, 1932:169.[2]]
> L. Mioc.-L. Plioc.; S.A.

[1] Described as subgenus of †*Caviodon*.
[2] Described as subgenus of †*Cardiomys*.

Family **Hydrochoeridae** Gray, 1825:341.
> [= Hydrocharina Gray, 1825:341; Hydrochoeridae Gill, 1872:22; Hydrochaeridae Cabrera, 1961:583.]
> L. Mioc.-R.; S.A. E. Plioc.-L. Pleist.; N.A. L. Pleist.-R.; Cent. A.

†*Porcellusignum* Casamiquela, in Angulo, 1982:52-53.[1]
> L. Mioc.; S.A.

[1] Assignment to Hydrochoeridae is questionable.

Subfamily †**Cardiatheriinae** Kraglievich, 1930:241.
>> [= †Cardiotheriinae Kraglievich, 1930:241; †Cardiatheriinae Simpson, 1945:95.]
>> L. Mioc.-E. Plioc., ?L. Plioc.; S.A.

†*Procardiatherium* Ameghino, 1885:55-59.
>> L. Mioc.; S.A.

†*Cardiatherium* Ameghino, 1883:270-274.[1]
>> [= or including †*Plexochoerus* Ameghino, 1886:58.]
>> L. Mioc., ?E. Plioc.; S.A.

[1] "Almost always spelled *Cardiotherium* by authors (including Ameghino in later papers), but the first spelling does not appear to be a misprint" (Simpson 1945:95).

†*Anchimys* Ameghino, 1886:71-72.
>> [Including †*Cardiodon* Ameghino, 1885:61-65[1]; †*Eucardiodon* Ameghino, 1891:241.]
>> L. Mioc.; S.A.

[1] Objective synonym of †*Eucardiodon*. Not †*Cardiodon* Owen, 1841, a Jurassic brachiosaur.

†*Phugatherium* Ameghino, 1887:6-7.
>> [= †*Phregatherium* Lydekker, 1888:37[1]; †*Neoanchimys* Pascual & Bondesio, 1961:101.]
>> E. Plioc.; S.A.

[1] Misprint.

†*Xenocardia* Pascual & Bondesio, 1963:45.
>> L. Mioc.; S.A.

†*Anchimysops* Kraglievich, 1927:597.
>> E. Plioc., ?L. Plioc.; S.A.

†*Kiyutherium* Francis & Mones, 1965:47.
>> L. Mioc.-E. Plioc.; S.A.

Subfamily †**Anatochoerinae** Mones & Vucetich, in Mones, 1991:32.
>> L. Mioc., L. Plioc.; S.A.

†*Contracavia* Burmeister, 1885:158.
>> L. Mioc.; S.A.

†*Anatochoerus* Vucetich & Mones, in Mones, 1991:32.
>> L. Mioc.; S.A.

†*Hydrochoeropsis* Kraglievich, 1930:159.
>> L. Plioc.; S.A.

Subfamily **Hydrochoerinae** Gray, 1825:341.
>> [= Hydrocharina Gray, 1825:341[1]; Hydrochaerina Bonaparte, 1850; Hydrobii Brandt, 1855:281[2]; Hydrochoerinae Weber, 1928:290.] [Including †Protohydrochoerinae Kraglievich, 1930:240.]
>> E. Plioc.-L. Pleist.; N.A. E. Plioc.-R.; S.A. L. Pleist.-R.; Cent. A.

[1] Proposed as tribe.
[2] Not based on a generic name.

†*Protohydrochoerus* Rovereto, 1914:140.
>> E.-L. Plioc.; S.A.

†*Chapalmatherium* F. Ameghino, 1908:423.[1]
>> E.-L. Plioc.; S.A.

[1] Originally described as a litoptern.

†*Nothydrochoerus* Rusconi, 1935:1.
>> L. Plioc.; S.A.

†*Neochoerus* Hay, 1926:5.
>> [= †*Oromys* Leidy, 1853:241.[1]] [= or including †*Palaeohydrochoerus* Spillmann, 1942:379.[2]] [Including †*Pliohydrochoerus* Kraglievich, 1930:30[3]; †*Prohydrochoerus* Spillmann, 1941:196-201.]
>> E. Plioc.-L. Pleist.; N.A. E.-L. Pleist.; S.A. L. Pleist.; Cent. A.

[1] *Nomen nudum?*
[2] *Nomen nudum.*
[3] Subgenus as defined.

Hydrochaeris Brunnich, 1772:24. Capybara, carpincho, capivara, capigua, ronsoco, samani, poncho.

[= *Hydrochoerus* Brisson, 1762:80-81[1]; *Hydrochaerus* Erxleben, 1777:191-195; *Hydrochoenus* Gray, 1821:304[2]; *Hydrocherus* F. Cuvier, 1829:492; *Hydrochoerus* Wagler, 1830:18; *Hydrochoeris* Allen, 1916:568; *Hydrocheirus* Hollande & Batisse, 1959:1; *Capibara* Moussy, 1860:13; *Capiguara* Liais, 1872:545.[3]] [Including †*Xenohydrochoerus* Rusconi, 1934:5.]

Pleist.; N.A. E. and/or M. Pleist., L. Pleist.-R.; S.A. L. Pleist.-R.; Cent. A.

[1] Not published in a consistently binominal work. However, see the International Code of Zoological Nomenclature, Article 11(c)(i), for confusion.
[2] Misprint?
[3] Replacement for *Hydrochoerus* Brisson, 1762.

Superfamily **Octodontoidea** Waterhouse, 1839:172.

[= Octodontidae Waterhouse, 1839:172; Octodontoidea Simpson, 1945:97.] [= or including Muriformes Giebel, 1855:486[1]; Muriformidae Ameghino, 1887:10.[2]]

[1] Proposed as a family essentially for this group (with the addition of certain African genera) but not based on a generic name.
[2] Not based on a generic name.

†*Morenella* Palmer, 1903:873.[1]

[= †*Morenia* Ameghino, 1886:51.[2]]

L. Mioc.; S.A.

[1] Assignment to Octodontoidea is questionable.
[2] Not †*Morenia* Gray, 1870, a reptile.

†*Plataeomys* Ameghino, 1881:306.[1]

[= or including †*Pseudoplataeomys* Kraglievich, 1934:77.]

L. Mioc.-L. Plioc., E. and/or M. Pleist.; S.A.

[1] Assignment to Octodontoidea is questionable.

†*Dicolpomys* Winge, 1887:99-101.[1]

Pleist.; S.A.

[1] Assignment to Octodontoidea is questionable.

†*Caviocricetus* Vucetich & Verzi, 1996:297.

E. Mioc.; S.A.

Family **Octodontidae** Waterhouse, 1839:172.

[= Octodontina Waterhouse, 1858:147.] [Including Psammoryctina Wagner, 1841:312; Psammoryctida Haeckel, 1866:clx[1]; Ctenomyidae Gervais, 1849:203; Spalacopodidae Lilljeborg, 1866:44; †Paradoximyina Ameghino, 1886:79, 222; †Paradoxomydae Ameghino, 1889:122; †Acaremyidae A. E. Wood, 1949:4.]

L. Olig.-R.; S.A.

[1] Used as a family.

†*Phtoramys* Ameghino, 1887:4-5.

[= †*Phloromys* Lydekker, 1888:36.[1]]

L. Mioc.-E. Plioc.; S.A.

[1] Misprint.

†*Tarijamys* Takai, Arozqueta, Mizuno, Yoshida & Kondo, 1984:35.

M. Pleist.; S.A.

Subfamily †**Acaremyinae** Ameghino, 1902:111.

[= †Acaremyidae A. E. Wood, 1949:4.]

L. Olig.-M. Mioc.; S.A.

†*Migraveramus* Patterson & A. E. Wood, 1982:380.

L. Olig.; S.A.

†*Platypittamys* A. E. Wood, 1949:5.

L. Olig.; S.A.

†*Acaremys* Ameghino, 1887:9.

E. Mioc.; S.A.

†*Sciamys* Ameghino, 1887:9.
      E. Mioc.; S.A.

†*Massoiamys* Vucetich, 1978:331.
      M. Mioc.; S.A.

Subfamily **Octodontinae** Waterhouse, 1839:172, **new rank**.
      [= Octodontidae Waterhouse, 1839:172.] [Including Psammoryctina Wagner,
      1841:312; Psammoryctida Haeckel, 1866:clx[1]; Spalacopodidae Lilljeborg, 1866:44;
      †Paradoximyina Ameghino, 1886:79, 222; †Paradoxomydae Ameghino, 1889:122.]
      L. Mioc.-R.; S.A.
    [1] Used as a family.

†*Chasicomys* Pascual, 1967:269.
      L. Mioc.; S.A.

Tribe **Octodontini** Waterhouse, 1839:172, **new rank**.
      [= Octodontidae Waterhouse, 1839:172.] [Including Psammoryctina Wagner,
      1841:312; Psammoryctida Haeckel, 1866:clx[1]; Spalacopidae Lilljeborg, 1866:44;
      †Paradoximyina Ameghino, 1886:77, 222; †Paradoxomydae Ameghino, 1889:122.]
      L. Mioc.-E. Pleist., R.; S.A.
    [1] Used as a family.

†*Paradoxomys* Ameghino, 1885:68.
      L. Mioc.; S.A.

Pithanotomys Ameghino, 1887:5.
      [Including *Schizodon* Waterhouse, 1842:89-91[1]; *Aconaemys* Ameghino, 1891:245;
      Acondemys Sclater, 1899:280.[2]]
      E. Plioc.-E. Pleist., R.; S.A.[3]
    [1] Objective synonym of *Aconaemys*. Not *Schizodon* Agassiz, 1829, a genus of fishes.
    [2] Misprint.
    [3] Argentina, Chile. Described as fossil before being recognized as extant.

†*Abalosia* Reig & Quintana, 1991:294.
      L. Plioc.; S.A.

Octomys Thomas, 1920:117.
      R.; S.A.[1]
    [1] Argentina.

Tympanoctomys Yepes, 1941:569.
      R.; S.A.[1]
    [1] Argentina.

Spalacopus Wagler, 1832:1219.
      [= *Psammoryctes* Poeppig, 1835:252.] [= or including *Poephagomys* F. Cuvier,
      1834:321-326.]
      R.; S.A.[1]
    [1] Chile.

Octodontomys Palmer, 1903:873. Chozchori, soco.
      [= *Neoctodon* Thomas, 1902:227.[1]]
      R.; S.A.[2]
    [1] Not *Neoctodon* Bedel, 1892, a coleopteran.
    [2] Argentina, Bolivia, Chile.

Octodon Bennett, 1832:46-47. Degu.
      [= *Dendroleius* Meyen, 1833: table xliv; *Dendrobius* Meyen, 1833:600.[1]]
      R.; S.A.[2]
    [1] Misprint.
    [2] Chile.

Tribe **Ctenomyini** Lesson, 1842:105, **new rank**.
      [= Ctenomysideae Lesson, 1842:105; Ctenomyinae Gervais, 1849:203.]
      L. Mioc.-R.; S.A.

†*Palaeoctodon* Rovereto, 1914:221.
> [Including †*Proctenomys* Pascual, Pisano & Ortega, 1965:21.]
> L. Mioc.-E. Plioc.; S.A.

†*Xenodontomys* Kraglievich, 1927:592.
> E. Plioc.; S.A.

†*Eucoelophorus* Ameghino, 1909:425.
> [= †*Eucelophorus* Ameghino, 1909:425; †*Eucoelophorus* Rovereto, 1914:198.[1]]
> E. Plioc.-E. Pleist.; S.A.

[1] Emendation.

Ctenomys de Blainville, 1826:62. Tucu-tuco, tunduque, anguya-tutu, oculto.
> [= †*Dicoelophorus* Ameghino, 1888:6.] [Including †*Paractenomys* Ameghino,
> 1908:425; *Haptomys* Thomas, 1916:305[1]; *Chacomys* Osgood, 1946:47.[1]]
> L. Plioc.-R.; S.A.

[1] Proposed as subgenus.

†*Praectenomys* Villarroel, 1975:496.
> L. Plioc.; S.A.

†*Actenomys* Burmeister, 1888:179.[1]
> E. Plioc.; S.A.

[1] Generally, but incorrectly, called †*Dicoelophorus* Ameghino, 1888, a synonym of *Ctenomys*.

†*Megactenomys* Rusconi, 1930:1.
> L. Plioc.; S.A.

Family **Echimyidae** Gray, 1825:341. Spiny rats, porcupine rats, urares.
> [= Echymina Gray, 1825:341; Echymyda Pictet, 1843:202; Echymyidae C. L.
> Bonaparte, 1850: unpaged chart.; Echimyina Waterhouse, 1858:147; Echimyidae Miller
> & Gidley, 1918:445; Loncherini Giebel, 1847:93; Loncheridae Burmeister, 1850:17;
> Echinomyidae Ameghino, 1889:122, 131.]
> L. Olig.-R.; S.A. Pleist., ?R.; W. Indies. R.; Cent. A.[1]

[1] Also widespread in the Recent of North America, Europe, N. Asia, and E. Africa by introduction.

†*Paulacoutomys* Vucetich, de Souza Cunha & Ferraz de Alvarenga, 1993:247.[1]
> L. Olig.; S.A.

[1] Assignment to the Echimyidae is questionable.

†*Willidewu* Vucetich & Verzi, 1991:67.
> E. Mioc.; S.A.

†*Maruchito* Vucetich, Mazzoni & Pardiñas, 1993:370.
> M. Mioc.; S.A.

Subfamily **Heteropsomyinae** Anthony, 1917:189.
> [Including Eumysopidae Rusconi, 1935:2; Eumysopinae Patton & Reig, 1989:76.]
> L. Olig.-L. Plioc., ?E. and/or ?M. Pleist., L. Pleist.-R.; S.A. Pleist., ?R.; W. Indies.
> R.; Cent. A.

†*Sallamys* Hoffstetter & Lavocat, 1970:172.
> L. Olig.; S.A.

†*Protadelphomys* Ameghino, 1902:112-113.[1]
> [= †*Gaimanomys* Vucetich & Bond, 1984:106.]
> E. Mioc.; S.A.

[1] Proposed as subgenus of †*Adelphomys*. †*Protadelphomys* Ameghino, 1901, was a *nomen nudum*.

†*Protacaremys* Ameghino, 1902:111-112.[1]
> [= †*Eoctodon* Ameghino, 1902:115; †*Archaeocardia* Ameghino, 1902.]
> E.-M. Mioc.; S.A.

[1] Proposed as subgenus of †*Acaremys*. †*Protacaremys* Ameghino, 1901, was a *nomen nudum*.

†*Acarechimys* Patterson, in J. L. Kraglievich, 1965:258.[1]
> E.-M. Mioc.; S.A.

[1] Quoted by Kraglievich from a MS and credited to Patterson; validated by Patterson, in Patterson & A. E. Wood, 1982:529.

†*Chasichimys* Pascual, 1967:262.
>   L. Mioc.; S.A.[1]
>   [1] Argentina.

†*Pattersomys* Pascual, 1967:265.
>   L. Mioc.; S.A.[1]
>   [1] Argentina.

†*Eumysops* Ameghino, 1888:5-6.
>   [= or including †*Therydomyops* Ameghino, 1906:417, 483; †*Proaguti* Ameghino, 1908:424.]
>   ?L. Mioc., E. Plioc., ?L. Plioc.; S.A.

Thrichomys Trouessart, 1881:179.[1] Punare.
>   [= *Nelomys* Lund, 1841:241, etc.[2]; *Thricomys* Trouessart, 1897:606.[3]]
>   L. Mioc.-L. Plioc., Pleist., R.; S.A.
>   [1] Proposed as subgenus of *Echimys*; raised to generic rank by Thomas, 1897:1025.
>   [2] Not *Nelomys* F. Cuvier, 1837, a synonym of *Echimys* G. Cuvier, 1809.
>   [3] Misprint. Date of "*Thricomys*" given as "1880". Sometimes also as "*Trichomys*."

†*Pampamys* Verzi, Vucetich & Montalvo, 1995:191.
>   L. Mioc.; S.A.

†*Palaeoechimys* Spillmann, 1949:29.
>   Plioc. and/or Pleist.; S.A.[1]
>   [1] Peru.

Proechimys J. A. Allen, 1899:264. Casiragua.
>   [Including *Trinomys* Thomas, 1921:140.[1]]
>   Pleist., R.; S.A. R.; Cent. A.
>   [1] As subgenus.

Euryzygomatomys Goeldi, 1901:179.
>   Pleist., R.; S.A.

†*Boromys* Miller, 1916:7.
>   Pleist.; W. Indies.[1]
>   [1] Cuba and Isle of Pines.

†*Heteropsomys* Anthony, 1916:202.
>   [= or including †*Homopsomys* Anthony, 1917:187.]
>   Pleist.; W. Indies.[1]
>   [1] Puerto Rico and Vieques Island.

†*Brotomys* Miller, 1916:6.
>   Pleist. and/or R.; W. Indies.[1]
>   [1] Hispaniola and La Gonave Island.

Carterodon Waterhouse, 1848:351.
>   Pleist., R.; S.A.[1]
>   [1] Brazil. Found first as fossil and later discovered to be still living (Simpson 1945:98).

†*Puertoricomys* Woods, 1989:748.
>   Pleist.; W. Indies.[1]
>   [1] Puerto Rico.

Clyomys Thomas, 1916:300.
>   L. Pleist.-R.; S.A.[1]
>   [1] Brazil.

Lonchothrix Thomas, 1920:113.
>   R.; S.A.[1]
>   [1] Brazil.

Hoplomys J. A. Allen, 1908:649. Armored rat.
>   R.; Cent. A. R.; S.A.

Mesomys Wagner, 1845:145.
>   R.; S.A.

Subfamily †**Adelphomyinae** Patterson & Pascual, 1968:5.
    L. Olig.-M. Mioc.; S.A.

†*Xylechimys* Patterson & Pascual, 1968:9.
    L. Olig.; S.A.

†*Deseadomys* A. E. Wood & Patterson, 1959:303.
    L. Olig.; S.A.

†*Paradelphomys* Patterson & Pascual, 1968:9.
    E. Mioc.; S.A.[1]
    [1] Argentina.

†*Adelphomys* Ameghino, 1887:10.
    E. Mioc.; S.A.

†*Stichomys* Ameghino, 1887:10.
    E.-M. Mioc.; S.A.

Subfamily **Myocastorinae** Ameghino, 1902.
    [= Myiopotamyina Bonaparte, 1850: unpaged chart; Myopotamini Tullberg, 1900:128;
    Myocastoridae Ameghino, 1902[1]; Myocastorinae Patterson & Pascual, 1968:6.]
    E. Mioc., L. Mioc.-R.; S.A.[2]
    [1] Obras Completas 13:456.
    [2] Also widespread in the Recent of North America, Europe, N. Asia, and E. Africa by introduction.

†*Prospaniomys* Ameghino, 1902:113-114.
    E. Mioc.; S.A.

†*Spaniomys* Ameghino, 1877:10.
    [= †*Gyrignophus* Ameghino, 1891:14; †*Graphimys* Ameghino, 1891:14.]
    E. Mioc.; S.A.

†*Strophostephanos* Ameghino, 1891:142-143.
    L. Mioc. and/or E. Plioc.; S.A.

†*Haplostropha* Ameghino, 1891:140.
    L. Mioc.; S.A.

†*Proatherura* Ameghino, 1906:413.[1]
    E. Plioc.; S.A.
    [1] Or 1904?

†*Tribodon* Ameghino, 1887:7-8.
    [= †*Trilodon* Flower & Lydekker, 1891:484.[1]]
    E. Plioc.; S.A.
    [1] Misprint.

†*Isomyopotamus* Rovereto, 1914:132.
    E.-L. Plioc.; S.A.

Myocastor Kerr, 1792:225. Nutria, copyu.
    [= *Myopotamus* Commerson, in É. Geoffroy Saint-Hilaire, 1805:81; *Potamys*
    Larranhaga, 1823:83; *Mastonotus* Wesmael, 1841:61; *Guillinomys* Lesson, 1842:126.]
    [Including *Tramyocastor* Rusconi, 1936:1-4.]
    L. Plioc.-R.; S.A.[1]
    [1] Also widespread in the Recent of North America, Europe, N. Asia, and E. Africa by introduction.

†*Paramyocastor* Ameghino, 1904:103.
    L. Plioc.; S.A.

†*Matyoscor* Ameghino, 1902:241.
    Pleist.; S.A.

Subfamily **Dactylomyinae** Tate, 1935:295.
    Pleist., R.; S.A.

Kannabateomys Jentink, 1891:105-110.
    [= *Cannabateomys* Lydekker, 1892:32.]
    Pleist., R.; S.A.

Dactylomys I. Geoffroy Saint-Hilaire, 1838:201. Coro-coro.
[Including *Lachnomys* Thomas, 1916:296, 299.]
R.; S.A.

Olallamys Emmons, 1988:421.
[= *Thrinacodus* Günther, 1879:144-145.[1]]
R.; S.A.
[1] Not *Thrinacodus* St. John and Worthen, 1875, a Carboniferous shark.

Subfamily **Echimyinae** Gray, 1825:341.
[= Echymina Gray, 1825:341[1]; Echymydina C. L. Bonaparte, 1837:9[2]; Echimyinae Murray, 1866:xiv, 350; Loncherini Giebel, 1847:93; Loncherinae Thomas, 1897:1024.]
Pleist., R.; S.A. R.; Cent. A. ?R.; W. Indies.
[1] Proposed as tribe.
[2] As subfamily.

Echimys G. Cuvier, 1809:394. Spiny rats.
[= *Loncheres* Illiger, 1811:90; *Echinomys* Wagner, 1840:203.] [Including *Nelomys* F. Cuvier, 1837; *Phyllomys* Lund, 1839:225-226, 233.]
Pleist., R.; S.A.

Isothrix Wagner, 1845:145-146.
[= *Lasiuromys* Deville, 1852:357-361.]
Pleist., R.; S.A.

Diplomys Thomas, 1916:240.
R.; Cent. A.[1] R.; S.A.
[1] Panama.

Makalata Husson, 1978:445.
?R.; W. Indies.[1] R.; S.A.
[1] Recent of West Indies (Martinique)?, possibly based on a cataloging error.

Family **Capromyidae** Smith, 1842:308.
[Including Plagiodontinae Ellerman, 1940:25; †Isolobodontinae Woods, 1989:76; †Hexolobodontinae Woods, 1989:76.]
E. Mioc., ?E. and/or ?M. Pleist., L. Pleist.-R.; W. Indies.[1] M. Mioc.; S.A.[2]
[1] Greater Antilles and Bahamas.
[2] Unnamed capromyids have been reported from Quebrada Honda, Bolivia.

†*Zazamys* MacPhee & Iturralde-Vinent, 1995:3.
E. Mioc.; W. Indies.[1]
[1] Cuba.

†*Macrocapromys* Arredondo, 1958:10.
Pleist.; W. Indies.[1]
[1] Cuba.

†*Hexolobodon* Miller, 1929:20.
Pleist.; W. Indies.[1]
[1] Hispaniola and La Gonave Island.

†*Quemisia* Miller, 1929:22.
Pleist.; W. Indies.[1]
[1] Hispaniola.

Capromys Desmarest, 1822:185-188. Hutias, jutias.
[= *Isodon* Say, 1822:333.] [= or including *Procapromys* Chapman, 1901:322.]
[Including *Mysateles* Lesson, 1842:124; *Mesocapromys* Varona, 1970:8[1]; *Paracapromys* Kratochvíl, Rodriguez & Barus, 1978:15[2]; *Leptocapromys* Kratochvíl, Rodriguez & Barus, 1978:15[3]; †*Palaeocapromys* Varona, 1979:4[1]; *Pygmaeocapromys* Varona, 1979:5[1]; †*Brachycapromys* Varona, 1979:15[1]; †*Stenocapromys* Varona, 1979:17.[1]]
Pleist., R.; W. Indies.[4]
[1] Described as subgenus of *Capromys*.
[2] Described as subgenus of *Mesocapromys*.

[3] Described as subgenus of *Mysateles*.
[4] Isle of Pines and Cuba.

†*Aphaetreus* Miller, 1922:3.
      Pleist.; W. Indies.[1]
   [1] Hispaniola.

†*Isolobodon* J. A. Allen, 1916:19.
      [= †*Ithydontia* Miller, 1922:4.]
      Pleist., R.; W. Indies.[1]
   [1] Hispaniola; Puerto Rico, and the Virgin Islands by introduction.

Geocapromys Chapman, 1901:314.[1] Jamaican hutia, Bahaman hutia.
      Pleist., R.; W. Indies.[2]
   [1] Proposed as subgenus of *Capromys*; considered a distinct genus by Miller, 1929.
   [2] Jamaica, Little Swan Island (recently extinct), and Bahamas.

Plagiodontia F. Cuvier, 1836:347.
      [Including †*Hyperplagiodontia* Rimoli, 1977:33.]
      Pleist., R.; W. Indies.[1]
   [1] Hispaniola and La Gonave Island.

†*Rhizoplagiodontia* Woods, 1989:62.
      L. Pleist.-R.; W. Indies.[1]
   [1] Hispaniola (Haiti).

Family †**Heptaxodontidae** Anthony, 1917:186.[1]
      [= †Heptaxodontinae Anthony, 1917:186; †Heptaxodontidae Miller & Gidley,
      1918:477.] [Including †Amblyrhizinae Schaub, 1951:96; †Clidomyinae Woods,
      1989:753.]
      L. Mioc.; S.A. Pleist.; W. Indies.
   [1] See the International Code of Zoological Nomenclature, Article 40.

†*Tetrastylomys* Kraglievich, 1926:128.
      L. Mioc.; S.A.

†*Pentastylomys* Kraglievich, 1926:128.
      L. Mioc.; S.A.

†*Elasmodontomys* Anthony, 1916:199.
      [= †*Heptaxodon* Anthony, 1917:183.]
      Pleist.; W. Indies.[1]
   [1] Puerto Rico.

†*Clidomys* Anthony, 1920:469.
      [= or including †*Speoxenus* Anthony, 1920:473; †*Spirodontomys* Anthony, 1920:473;
      †*Alterodon* Anthony, 1920:474.]
      Pleist.; W. Indies.[1]
   [1] Jamaica.

†*Amblyrhiza* Cope, 1868:313.
      [= †*Loxomylus* Cope, 1869:186.]
      Pleist.; W. Indies.[1]
   [1] N. Lesser Antilles.

Superfamily **Chinchilloidea** Bennett, 1833:58.
      [= Chinchillidae Bennett, 1833:58; Chinchilloidea Kraglievich, 1940:756; Eriomyina
      Van der Hoeven, 1858.]

Family **Chinchillidae** Bennett, 1833:58.
      [= Chinchillina Waterhouse, 1858:147; Eriomyidae Burmeister, 1854:188; Eriomyina
      Van der Hoeven, 1858.] [Including Lagostomi Wiegmann, 1832[1]; Lagostomidae
      Bonaparte, 1838:113; Lagostomida Haeckel, 1866:clx[2]; Viscacidae Ameghino,
      1904:252; Viscacciidae Rovereto, 1914:36.]
      L. Olig.-M. Mioc., ?L. Plioc., E. Pleist.-R.; S.A.
   [1] Used as a family. Antedates the family name Chinchillidae and therefore should be suppressed as a senior
   synonym.
   [2] Used as a family.

†*Eoviscaccia* Vucetich, 1989:234.
　　　　L. Olig.; S.A.

Subfamily **Lagostominae** Wiegmann, 1832.
　　　　[= Lagostomi Wiegmann, 1832[1]; Lagostomina C. L. Bonaparte, 1837:9[2]; Lagostomida Haeckel, 1866:clx[1]; Lagostominae Pocock, 1922:425.]
　　　　E.-M. Mioc., ?L. Plioc., E. Pleist.-R.; S.A.
　　　　[1] Used as a family.
　　　　[2] As subfamily.

†*Pliolagostomus* Ameghino, 1887:12.
　　　　E.-M. Mioc.; S.A.

†*Prolagostomus* Ameghino, 1887:11.
　　　　[= †*Scotaeumys* Ameghino, 1887:12; †*Sphaeramys* Ameghino, 1889:13; †*Sphaeromys* Ameghino, 1889:169.]
　　　　E.-M. Mioc.; S.A.

Lagostomus Brookes, 1828:133-134. Plains vizcacha, vizcachon.
　　　　[Including †*Lagostomopsis* Kraglievich, 1926:45-88.[1]]
　　　　?L. Plioc., E. Pleist.-R.; S.A.
　　　　[1] As subgenus.

Subfamily **Chinchillinae** Bennett, 1833:58.
　　　　[= Chinchillidae Bennett, 1833:58; Eriomyina Van der Hoeven, 1858; Chinchillinae Pocock, 1922:425.] [Including Viscacidae Ameghino, 1904:252; Viscacciidae Rovereto, 1914:36.]
　　　　R.; S.A.

Lagidium Meyen, 1833:576. Mountain vizcacha, chinchillon.
　　　　[= *Viscacia* Rafinesque, 1815:56[1]; *Viscaccia* Oken, 1816:835-837.[2]]
　　　　R.; S.A.
　　　　[1] *Nomen nudum.*
　　　　[2] Invalid.

Chinchilla Bennett, 1829:1. Chinchilla.
　　　　[= *Eriomys* Lichtenstein, 1829:table XXVIII.]
　　　　R.; S.A.

Family †**Neoepiblemidae** Kraglievich, 1926:128.
　　　　[= †Neoepibleminae Simpson, 1945:96.] [Including †Perimyidae Landry, 1957:59.]
　　　　E. Mioc., L. Mioc., Plioc.; S.A.

†*Perimys* Ameghino, 1887:12.
　　　　[= †*Sphiggomys* Ameghino, 1887:12; †*Sphodromys* Ameghino, 1887:13; †*Sphingomys* Lydekker, 1892:33.]
　　　　E. Mioc.; S.A.

†*Neoepiblema* Ameghino, 1889:208.[1]
　　　　[= †*Epiblema* Ameghino, 1886:44-45.[2]] [Including †*Euphilus* Ameghino, 1889:903-904.]
　　　　L. Mioc., Plioc.; S.A.
　　　　[1] New name for †*Epiblema* Ameghino, 1886.
　　　　[2] Not *Epiblema* Hübner, 1816, a lepidopteran.

†*Dabbenea* Kraglievich, 1926:127.
　　　　[Including †*Prodabbenea* Kraglievich, 1940:96.[1]]
　　　　L. Mioc.; S.A.
　　　　[1] As subgenus.

Family **Abrocomidae** Miller & Gidley, 1918:447.
　　　　L. Mioc., R.; S.A.

†*Protabrocoma* Kraglievich, 1927:591.
　　　　L. Mioc.; S.A.

Abrocoma Waterhouse, 1837:30-32. Rat chinchilla.
　　　　[= *Habrocoma* Wagner, 1842:5-8.]
　　　　R.; S.A.

Grandorder **FERAE** Linnaeus, 1758:37.[1]

> [= RAPACIA Newman, 1843:34[2]; FERINA Newman, 1843:148; CARNASSIA
> Haeckel, 1895:490, 573; SARCOTHERIA Haeckel, 1895:490, 573;
> CARNIVORAMORPHA Kalandadze & Rautian, 1992:53.[3]] [= or including
> CARNARIA Haeckel, 1866:cxlvii.[4]]

> [1] One of Linnaeus' eight major mammalian subdivisions but not formally assigned a rank. Used as an order
> by Gregory, 1910:465; as a superorder by Simpson, 1945:105; and as a grandorder by McKenna, 1975:41.
> Also used at ordinal rank by Gray, 1821:308, but the content was wholly composed of marsupials.
> [2] Included other carnivorous animals, e.g., starfish!
> [3] In part. LIPOTYPHLA and †Deltatheridiidae were included as well.
> [4] Proposed as an order for true Carnivora, but it also contained †*Palaeonyctis* and †*Arctocyon*.

Order †**CIMOLESTA** McKenna, 1975:41.

> [= †CIMOLESTIFORMES Kinman, 1994:37.]

†*Alostera* Fox, 1989:29.

> L. Cret.; N.A.

†*Avitotherium* Cifelli, 1990:353.

> L. Cret.; N.A.

†*Ravenictis* Fox & Youzwyshyn, 1994:388.

> E. Paleoc.; N.A.

†*Pararyctes* Van Valen, 1966:57.

> E. Paleoc.-E. Eoc.; N.A.

Subfamily †**Wyolestinae** Gingerich, 1981:528.

> E.-M. Eoc.; As. E. Eoc.; N.A.

†*Wyolestes* Gingerich, 1981:529.[1]

> E. Eoc.; N.A.

> [1] Originally described as a didymoconid.

†*Hsiangolestes* Zheng & Huang, 1984:202.

> E. Eoc.; As.[1]

> [1] S.E. Asia (Hunan).

†*Mongoloryctes* Van Valen, 1966:68.

> M. Eoc.; As.

Family †**Palaeoryctidae** Winge, 1917:161.

> [= †Palaeoryctae Winge, 1917:161; †Palaeoryctidae Simpson, 1931:268;
> †Palaeoryctinae Van Valen, 1966:110; †Palaeoryctoidea Van Valen, 1963:40.[1]]
> ?L. Cret.; As.[2] ?L. Cret.; Eu.[3] E. Paleoc.-E. Eoc., ?M. Eoc.; N.A.[4] L. Paleoc.-E.
> Eoc.; Af.[5]

> [1] Also Van Valen, 1966:110.
> [2] Based on undescribed material from Khulsan, Mongolia.
> [3] A possible palaeoryctid is represented by unnamed specimens from the Calizas de Lychnus Fm.,
> Maastrichtian of Burgos, Spain.
> [4] Occurrence in middle Eocene (Uintan) of North America is based on an unnamed taxon.
> [5] N. Africa.

†*Palaeoryctes* Matthew, 1913:309.

> E. Paleoc.-E. Eoc.; N.A. L. Paleoc.-E. Eoc.; Af.[1]

> [1] N. Africa (Morocco).

†*Aaptoryctes* Gingerich, 1982:38.

> L. Paleoc.; N.A.

†*Eoryctes* Thewissen & Gingerich, 1989:460.[1]

> E. Eoc.; N.A.

> [1] Not †*Eoryctes* Romer, 1966:381L, a *nomen nudum* based on an unpublished MS by McKenna dealing with
> a new Bridgerian apternodontid.

Suborder †**DIDELPHODONTA** McKenna, 1975:41.

Family †**Cimolestidae** Marsh, 1889:89.
[= †Cimolestinae Trouessart, 1905:842.] [Including †Didelphodontinae Matthew, 1918:571; †Procerberinae Sloan & Van Valen, 1965:225.]
?L. Cret., L. Paleoc.-M. Eoc.; Eu.[1] L. Cret.-M. Eoc.; N.A. ?E. Paleoc.; S.A. L. Paleoc.-E. Eoc.; Af.[2] E. Eoc.; As.

[1] A single tooth of a possible cimolestid-like mammal has been reported from the Late Cretaceous (Campanian) of France.
[2] Indet. didelphodontines has been reported from the early Eocene (Ypresian) of N. Africa (Morocco).

†*Cimolestes* Marsh, 1889:89.
[Including †*Nyssodon* Simpson, 1927:124; †*Puercolestes* Reynolds, 1936:204.[1]]
L. Cret.-E. Paleoc.; N.A. ?E. Paleoc.; S.A. L. Paleoc.; Af.

[1] Synonymy questionable.

†*Procerberus* Sloan & Van Valen, 1965:225.
E. Paleoc.; N.A.

†*Gelastops* Simpson, 1935:227.
[= †*Emperodon* Simpson, 1935:229.]
E.-L. Paleoc.; N.A.

†*Avunculus* Van Valen, 1966:16.
L. Paleoc.; N.A.

†*Acmeodon* Matthew & Granger, 1921:3.
E.-L. Paleoc.; N.A.

†*Paleotomus* Van Valen, 1967:250.
L. Paleoc.; N.A.

†*Tinerhodon* Gheerbrant, 1995:88.
L. Paleoc.; Af.

†*Aboletylestes* D. E. Russell, 1964:75.
L. Paleoc.; Af. L. Paleoc.; Eu.

†*Protentomodon* Simpson, 1928:3.
L. Paleoc.; N.A.

†*Ilerdoryctes* Marandat, 1989:165.
E. Eoc.; Eu.

†*Tsaganius* D. E. Russell & Dashzeveg, 1986:284.
E. Eoc.; As.

†*Didelphodus* Cope, 1882:522.
[= †*Didelphyodus* Winge, 1923:221.] [Including †*Phenacops* Matthew, 1909:535.]
E.-M. Eoc.; Eu. E.-M. Eoc.; N.A.

†*Naranius* D. E. Russell & Dashzeveg, 1986:280.
E. Eoc.; As.

Suborder †**APATOTHERIA** Scott & Jepsen, 1936:26.[1]

[1] Proposed as order (tentatively). Used as suborder by McKenna, 1975:41.

Family †**Apatemyidae** Matthew, 1909:543.
[= †Apatemyoidea Saban, 1954:429.[1]] [= or including †Heterohyins Gervais, 1859.[2]]
E. Paleoc., E.-L. Eoc.; Eu.[3] E. Paleoc.-L. Olig.; N.A.

[1] Orthography emended by McKenna, 1960.
[2] See the International Code of Zoological Nomenclature, Article 11(f)(ii). As "Tribu des Heterohyins".
[3] An apatemyid has been reported from the early Paleocene Hainin fauna.

Subfamily †**Apatemyinae** Matthew, 1909:543, **new rank**.
[= †Apatemyidae Matthew, 1909:543.]
E. Paleoc.-L. Olig.; N.A. E.-L. Eoc.; Eu.

†*Jepsenella* Simpson, 1940:186.
E.-L. Paleoc.; N.A.

†*Apatemys* Marsh, 1872:221-222.

[= or including †*Teilhardella* Jepsen, 1930:126.] [Including †*Labidolemur* Matthew & Granger, 1921:4; †*Stehlinius* Matthew, 1921:2[1]; †*Stehlinella* Matthew, 1929:171.[2]]

L. Paleoc.-M. Eoc.; N.A. E. Eoc.; Eu.

[1] Not †*Stehlinia* Revilliod, 1919, a bat.
[2] New name for †*Stehlinius*.

†*Heterohyus* Gervais, 1848:7.

[= or including †*Necrosorex* Filhol, 1890:174-175; †*Heterochiromys* Stehlin, 1916:1299-1552; †*Amphichiromys* Stehlin, 1916:1299-1552; †*Gardonyus* Sigé, 1975:667[1]; †*Chardinyus* Sigé, 1975:667.[1]] [Including †*Gervaisyus* Sigé, 1990:8.[2]]

E.-L. Eoc.; Eu.

[1] Proposed as subgenus.
[2] Sigé in Sudre, Sigé, Remy, Marandat, Hartenberger, Godinot & Crochet. Proposed as subgenus.

†*Eochiromys* Teilhard de Chardin, 1927:14.

E. Eoc.; Eu.

†*Sinclairella* Jepsen, 1934:291.

L. Eoc.-L. Olig.; N.A.

Subfamily †**Unuchiniinae** Van Valen, 1966:260.[1]

E.-L. Paleoc.; N.A.

[1] Not Van Valen and McKenna as sometimes stated.

†*Unuchinia* Simpson, 1937:78.

[= †*Apator* Simpson, 1936:16.[1]]

E.-L. Paleoc.; N.A.

[1] Not *Apator* Semenow, 1898, a coleopteran.

Suborder †**TAENIODONTA** Cope, 1876:39.[1]

[= †GANODONTA Wortman, 1896:259[2]; †STYLINODONTIA Haeckel, 1895:502.]

[1] Proposed as order. Used as suborder by McKenna, 1975:41.
[2] Proposed or used as suborder.

Family †**Stylinodontidae** Marsh, 1875:221.

[= †Taeniodontidae Szalay, 1977:368.[1]] [Including †Calamodontidae Cope, 1876:39; †Ectoganidae Cope, 1876:39; †Hemiganidae Cope, 1888:310; †Conoryctidae Wortman, 1896:260; †Onychodectini Winge, 1923:123; †Onychodectinae Matthew, 1937:238; †Conoryctellini Schoch, 1982:470.]

E. Paleoc.-M. Eoc.; N.A.

[1] Not based upon a generic name.

†*Onychodectes* Cope, 1888:317.

E. Paleoc.; N.A.

†*Conoryctella* Gazin, 1939:276.

E. Paleoc.; N.A.

Subfamily †**Conoryctinae** Wortman, 1896:260.

[= †Conoryctidae Wortman, 1896:260; †Conoryctinae Schlosser, 1911:414; †Conoryctini Winge, 1917:123.]

E. Paleoc.; N.A.

†*Huerfanodon* Schoch & Lucas, 1981:683.

E. Paleoc.; N.A.

†*Conoryctes* Cope, 1881:829.

[Including †*Hexodon* Cope, 1884:794.[1]]

E. Paleoc.; N.A.

[1] Not †*Hexodon* Olivier, 1789, a coleopteran.

Subfamily †**Stylinodontinae** Marsh, 1875:221.

[= †Stylinodontidae Marsh, 1875:221; †Stylinodontinae Schlosser, 1911:414; †Stylinodontini Winge, 1917:106.] [Including †Calamodontidae Cope, 1876:39; †Ectoganidae Cope, 1876:39; †Ectoganini Schoch, 1983:205; †Hemiganidae Cope,

1888:310; †Psittacotheriinae Matthew, 1937:255; †Psittacotheriini Schoch, 1982:470; †Wortmaniinae Schoch, 1982:470.]
E. Paleoc.-M. Eoc.; N.A.

†*Schochia* Lucas & Williamson, 1993:175.
E. Paleoc.; N.A.

†*Wortmania* Hay, 1899:593.
E. Paleoc.; N.A.

†*Psittacotherium* Cope, 1882:156.
[= †*Hemiganus* Cope, 1882:831.[1]]
E. Paleoc.; N.A.
[1] Not the animal discussed as †*Hemiganus* by Wortman in 1897.

†*Ectoganus* Cope, 1874:592.
[= †*Calamodon* Cope, 1874:593; †*Dryptodon* Marsh, 1876:403; †*Conicodon* Cope, 1894:594.] [= or including †*Lampadophorus* Patterson, 1949:41.]
L. Paleoc.-E. Eoc.; N.A.

†*Stylinodon* Marsh, 1874:531.
E.-M. Eoc.; N.A.

Suborder †**TILLODONTIA** Marsh, 1875:221.[1]
[= †TILLODONTA Trouessart, 1879:2; †TILLOTHERIDA Haeckel, 1895:502; ESTHONYCHIDA Haeckel, 1895:502[2]; †TILLODONTIFORMES Kinman, 1994:37.[3]]
[1] Proposed as order. Reduced to suborder by McKenna, in Stucky & McKenna, in Benton, ed., 1993:751.
[2] In Part. Suborder proposed for †*Esthonyx* and †*Platychoerops*.
[3] Proposed with a query but not subject to the provisions of the International Code of Zoological Nomenclature.

Family †**Tillotheriidae** Marsh, 1875:221.
[= †Anchippodontidae Gill, 1872:11[1]; †Tillotheridae Marsh, 1875:221; †Tillotheriini Winge, 1917:123.] [Including †Esthonychidae Cope, 1883:80; †Esthonychini Winge, 1917:123; †Trogosinae Gazin, 1953:33; †Plethorodontidae Huang & Zheng, 1987:20.]
E.-L. Paleoc., ?E. Eoc., M.-L. Eoc.; As. E. Paleoc.-M. Eoc.; N.A. E.-M. Eoc.; Eu.
[1] Validity doubtful. Based on "†*Anchippodus*" Leidy, 1868, a *nomen dubium*, from the Shark River Marl, New Jersey.

Subfamily †**Tillotheriinae** Marsh, 1875:221, **new rank**.
[= †Tillotheriidae Marsh, 1875:221.]
E.-L. Paleoc., ?E. Eoc., M.-L. Eoc.; As. L. Paleoc.-M. Eoc.; N.A. E.-M. Eoc.; Eu.

†*Anchilestes* Chiu & Li, 1977:94.[1]
E. Paleoc.; As.[2]
[1] (Qiu & Li).
[2] E. Asia.

†*Lofochaius* Chow, Chang, Wang & Ting, 1973:31.
E. Paleoc.; As.[1]
[1] E. Asia (China).

†*Franchaius* Baudry, 1992:216.
E. Eoc.; Eu.

†*Plethorodon* Huang & Zheng, 1987:20.
E. Paleoc.; As.[1]
[1] E. Asia.

†*Meiostylodon* Wang, 1975:161.
E. Paleoc.; As.[1]
[1] E. Asia.

†*Esthonyx* Cope, 1874:6-7.
[Including †*Plesiesthonyx* Lemoine, 1891:276; †*Azygonyx* Gingerich, 1989:25.]
L. Paleoc.-E. Eoc.; N.A. E.-M. Eoc.; Eu.

†*Interogale* Huang & Zheng, 1983:59.
    L. Paleoc.; As.[1]
  [1] E. Asia.

†*Megalesthonyx* Rose, 1972:2.
    E. Eoc.; N.A.

†*Basalina* Dehm & Oettingen-Spielberg, 1958:8.
    ?E. Eoc., M. Eoc.; As.[1]
  [1] S. Asia (Pakistan).

†*Trogosus* Leidy, 1871:113-115.
    [= or including †*Anchippodus* Leidy, 1868:232.[1]] [Including †*Tillotherium* Marsh, 1873:485-486; †*Kuanchuanius* Chow, 1963:97.]
    E.-M. Eoc.; N.A. M. Eoc.; As.[2]
  [1] †*Anchippodus* was a *nomen nudum*, or nearly so. It should be suppressed.
  [2] E. Asia.

†*Tillodon* Gazin, 1953:48.
    M. Eoc.; N.A.

†*Adapidium* Young, 1937:434.
    L. Eoc.; As.[1]
  [1] E. Asia.

†*Chungchienia* Chow, 1963:1889-1893.
    L. Eoc.; As.[1]
  [1] E. Asia.

Subfamily †**Deltatheriinae** Van Valen, 1988:39.
    E. Paleoc., ?L. Paleoc.; N.A.

†*Deltatherium* Cope, 1881:337.
    [= †*Lipodectes* Cope, 1881:1019.]
    E. Paleoc., ?L. Paleoc.; N.A.

Suborder †**PANTODONTA** Cope, 1873:40, 67.[1]
    [= †CORYPHODONTIA Marsh, 1884:193.] [Including †TALIGRADA Cope, 1883:406; †EUPANTODONTA Lucas, 1993:186.[2]]
  [1] Proposed as suborder of PROBOSCIDEA; used as suborder of †CIMOLESTA by McKenna, 1975:41.
  [2] Proposed as a suborder of order †PANTODONTA, for pantodonts exclusive of the †Bemalambdidae.

Family †**Wangliidae** Van Valen, 1988:38.
    [Including †Alcidedorbignyidae de Muizon & Marshall, 1992:501.]
    E. Paleoc.; S.A. L. Paleoc.; As.

†*Alcidedorbignya* de Muizon & Marshall, 1987:205.
    E. Paleoc.; S.A.

†*Wanglia* Van Valen, 1988:38.
    L. Paleoc.; As.

Superfamily †**Bemalambdoidea** Chow, Chang, Wang & Ting, 1973:33.[1]
    [= †Bemalambdidae Chow, Chang, Wang & Ting, 1973:33; †Bemalambdoidea Tong, 1979:377.[2]]
  [1] (Zhou, Zhang, Wang & Ding).
  [2] (Tung).

Family †**Harpyodidae** Wang, 1979:366.
    E.-L. Paleoc.; As.[1]
  [1] E. Asia.

†*Harpyodus* Chiu & Li, 1977:96.[1]
    E.-L. Paleoc.; As.[2]
  [1] (Qiu & Li).
  [2] E. Asia.

†*Dysnoetodon* Zhang, 1980:126.[1]
E. Paleoc.; As.[2]
[1] (Chang).
[2] E. Asia.

Family †**Bemalambdidae** Chow, Chang, Wang & Ting, 1973:33.[1]
E. Paleoc.; As.[2]
[1] (Zhou, Zhang, Wang & Ding).
[2] E. Asia.

†*Hypsilolambda* Wang, 1975:154.
E. Paleoc.; As.[1]
[1] E. Asia.

†*Bemalambda* Chow, Chang, Wang & Ting, 1973:34.[1]
E. Paleoc.; As.[2]
[1] (Zhou, Zhang, Wang & Ding).
[2] E. Asia (China).

Superfamily †**Pantolambdoidea** Cope, 1883:406.
[= †Pantolambdidae Cope, 1883:406; †Pantolambdoidea Simons, 1960:15.]

Family †**Pastoralodontidae** Chow & Qi, 1978:80.
E.-L. Paleoc.; As.[1]
[1] E. Asia.

†*Altilambda* Chow & Wang, 1978:86.
E.-L. Paleoc.; As.[1]
[1] E. Asia (China).

†*Pastoralodon* Chow & Qi, 1978:80.
[= or including †*Convallisodon* Chow & Qi, 1978:81.]
L. Paleoc.; As.[1]
[1] E. Asia.

Family †**Titanoideidae** Patterson, 1934:72.
[= †Titanoidinae Patterson, 1934:72; †Titanoideidae Scott, 1937:468, 479.]
E. Paleoc.-E. Eoc.; N.A.[1]
[1] Possible unnamed new genus in the early Eocene (Wasatchian) of Baja California, Mexico.

†*Titanoides* Gidley, 1917:431.
[Including †*Sparactolambda* Patterson, 1939:352.]
E.-L. Paleoc.; N.A.

Family †**Pantolambdidae** Cope, 1883:406.
[= †Pantolambdina Haeckel, 1895:530, 542; †Pantolambdinae Kalandadze & Rautian, 1992:60.]
E.-L. Paleoc.; N.A.

†*Pantolambda* Cope, 1882:418.
E. Paleoc.; N.A.

†*Caenolambda* Gazin, 1956:48.
L. Paleoc.; N.A.

Family †**Barylambdidae** Patterson, 1937:230.
[= †Barylambdinae Patterson, 1937:230; †Barylambdidae Patterson, 1939:361, 372.]
L. Paleoc.-E. Eoc.; N.A.

†*Barylambda* Patterson, 1937:229.
[Including †*Leptolambda* Patterson & Simons, 1958:1-8.]
L. Paleoc.-E. Eoc.; N.A.

†*Haplolambda* Patterson, 1939:365.
L. Paleoc., ?E. Eoc.; N.A.

†*Ignatiolambda* Simons, 1960:29.
L. Paleoc.; N.A.

Family †**Cyriacotheriidae** Rose & Krause, 1982:27.
>> L. Paleoc.; N.A.

>> †*Cyriacotherium* Rose & Krause, 1982:28.
>>> L. Paleoc.; N.A.

Family †**Pantolambdodontidae** Granger & Gregory, 1934:6.
>> [Including †Archaeolambdidae Flerov, 1952:44.]
>> L. Paleoc.-L. Eoc.; As.

>> †*Archaeolambda* Flerov, 1952:44.
>>> [Including †*Dilambda* Tong, 1978:88[1]; †*Nanlingilambda* Tong, 1979:379[1];
>>> †*Oroklambda* Dashzeveg, 1980:113.]
>>> L. Paleoc.-M. Eoc.; As.[2]
>>> [1] (Tung).
>>> [2] E. Asia.

>> †*Pantolambdodon* Granger & Gregory, 1934:7.
>>> M.-L. Eoc.; As.

Superfamily †**Coryphodontoidea** Marsh, 1876:428.
>> [= †Coryphodontidae Marsh, 1876:428; †Coryphodontoidea Simons, 1960.]

Family †**Coryphodontidae** Marsh, 1876:428.
>> [= †Bathmodontidae Cope, 1872:279; †Coryphodontinae Kalandadze & Rautian,
>> 1992:60.]
>> L. Paleoc.-E. Eoc.; Eu. L. Paleoc.-E. Eoc., ?M. Eoc.; N.A. E.-L. Eoc.; As.

>> †*Coryphodon* Owen, 1845:607-609.
>>> [= or including †*Bathmodon* Cope, 1872:417-420; †*Metalophodon* Cope, 1872:542;
>>> †*Ectacodon* Cope, 1882:165, 167; †*Manteodon* Cope, 1882:165, 167.]
>>> L. Paleoc.-E. Eoc.; Eu. L. Paleoc.-E. Eoc., ?M. Eoc.; N.A. E. Eoc.; As.

>> †*Heterocoryphodon* Lucas & Tong, 1987:362.[1]
>>> E.-M. Eoc.; As.
>>> [1] (Lucas & Tung).

>> †*Asiocoryphodon* Xu, 1976:185.
>>> E.-M. Eoc.; As.[1]
>>> [1] E. Asia.

>> †*Metacoryphodon* Chow & Qi, 1982:304.[1]
>>> M. Eoc.; As.
>>> [1] Not Qi (1987: 23).

>> †*Hypercoryphodon* Osborn & Granger, 1932:1.
>>> M. Eoc.; As.

>> †*Eudinoceras* Osborn, 1924:2.
>>> M.-L. Eoc.; As.

Suborder †**PANTOLESTA** McKenna, 1975:41.[1]
>> [Including †PTOLEMAIIDA Simons & Bown, 1995:3269.[2]]
>> [1] Used as order by Marandat, 1989:164.
>> [2] Proposed as order.

>> †*Simidectes* Stock, 1933:481.[1]
>>> [= †*Pleurocyon* Peterson, 1919:52[2]; †*Sespecyon* Stock, 1933:482[3]; †*Petersonella*
>>> Kraglievich, 1948:162.[4]]
>>> M. Eoc.; N.A.
>>> [1] Proposed as a subgenus of †*Pleurocyon*; raised to generic rank by Van Valen, 1965:394.
>>> [2] Not †*Pleurocyon* Mercerat, 1917:13, a junior synonym of †*Theriodictis* Mercerat, 1891, a canid.
>>> [3] *Nomen nudum.*
>>> [4] New name for †*Pleurocyon* Peterson.

Family †**Pantolestidae** Cope, 1884:719.
>> [= †Pantolestoidea Cope, 1887:378; †Pantolestida Haeckel, 1895:530, 554.] [Including
>> †Dyspternidae Kretzoi, 1943:195; †Cymaprimadontidae J. Clark, 1968:242;
>> †Todralestidae Gheerbrant, 1991:1249.]

E. Paleoc.-E. Olig.; N.A.[1] L. Paleoc.-E. Eoc.; Af. L. Paleoc.-L. Olig.; Eu. E.-M. Eoc., ?L. Eoc., E. Olig.; As.[2]

[1] An unnamed pantolestid occurs in the early Oligocene (Whitneyan) of South Dakota.
[2] An unnamed pantolestid occurs in the early Eocene of Mongolia.

Subfamily †**Pentacodontinae** Simpson, 1937:123.
[= †Pentacodontidae Van Valen, 1967:230.]
E. Paleoc.-E. Eoc.; N.A.

†*Coriphagus* Douglass, 1908:17.
[= †*Pantomimus* Van Valen, 1967:228.] [Including †*Mixoclaenus* Matthew & Granger, 1921:7.]
E. Paleoc.; N.A.

†*Aphronorus* Simpson, 1935:230.
E.-L. Paleoc.; N.A.

†*Pentacodon* Scott, 1892:296-297.
E. Paleoc.; N.A.

†*Bisonalveus* Gazin, 1956:17.
L. Paleoc.; N.A.

†*Amaramnis* Gazin, 1962:24.
E. Eoc.; N.A.

Subfamily †**Pantolestinae** Cope, 1884:719.
[= †Pantolestidae Cope, 1884:719; †Pantolestinae Simpson, 1937:121.] [Including †Cymaprimadontidae J. Clark, 1968:242; †Todralestidae Gheerbrant, 1991:1249.]
E. Paleoc.-L. Eoc.; N.A. L. Paleoc.-E. Eoc.; Af. L. Paleoc.-L. Eoc.; Eu. M. Eoc., ?L. Eoc., E. Olig.; As.

†*Propalaeosinopa* Simpson, 1927:2.
[Including †*Bessoecetor* Simpson, 1936:9.]
E.-L. Paleoc.; N.A. ?L. Paleoc.; Eu.

†*Palaeosinopa* Matthew, 1901:22.[1]
[= or including †*Niphredil* Van Valen, 1978:66.]
L. Paleoc.-E. Eoc.; N.A. E. Eoc.; Eu.
[1] *Nomen nudum* in Matthew, 1899:31.

†*Thelysia* Gingerich, 1982:44.
L. Paleoc.; N.A.

†*Pagonomus* D. E. Russell, 1964:65.
L. Paleoc.; Eu.

†*Todralestes* Gheerbrant, 1991:1249.
L. Paleoc.-E. Eoc.; Af.

†*Pantolestes* Cope, 1872:2.
[= or including †*Passalacodon* Marsh, 1872:16; †*Anisacodon* Marsh, 1872:17.]
M. Eoc.; As. M. Eoc.; N.A.

†*Bogdia* Dashzeveg & D. E. Russell, 1985:872.
M. Eoc.; As.

†*Chadronia* Cook, 1954:388.
[Including †*Cymaprimadon* J. Clark, 1968:243.]
?L. Eoc.; As. L. Eoc.; N.A.

†*Buxolestes* Jaeger, 1970:66.
M. Eoc.; Eu.

†*Bouffinomus* Mathis, 1989:35.
L. Eoc.; Eu.

†*Galethylax* Gervais, 1850:132-133.
L. Eoc.; Eu.

†*Oboia* Gabunia, 1989:177.
E. Olig.; As.

Subfamily †**Dyspterninae** Kretzoi, 1943:195.
                [= †Dyspternidae Kretzoi, 1943:195; †Dyspterninae D. E. Russell & Godinot,
                1988:319, 321.] [Including †Kochictidae Kretzoi, 1933:194.]
                E. Eoc., L. Eoc.-L. Olig.; Eu. E. Olig.; As.

    †*Fordonia* Marandat, 1989:164.
                E. Eoc.; Eu.

    †*Cryptopithecus* Schlosser, 1890:65 [451].
                [Including †*Opsiclaenodon* Butler, 1946:691; †*Androconus* Quinet, 1965:2.]
                L. Eoc.-E. Olig.; Eu.

    †*Dyspterna* Hopwood, 1927:174-176.
                L. Eoc.-E. Olig.; Eu.

    †*Kochictis* Kretzoi, 1943:10-17, 190.
                L. Olig.; Eu.

    †*Gobipithecus* Dashzeveg & D. E. Russell, 1992:648.
                E. Olig.; As.

Family †**Paroxyclaenidae** Weitzel, 1933:86.
                [= †Paroxyclaenidae Kretzoi, 1943:194.]
                Eoc.; As. E.-L. Eoc.; Eu.

    †*Kiinkerishella* Gabunia & Birjukov, 1978:489.
                Eoc.; As.

Subfamily †**Paroxyclaeninae** Weitzel, 1933:86.
                E.-L. Eoc.; Eu.

    †*Paroxyclaenus* Teilhard de Chardin, 1922:85-89.
                ?E. Eoc., M.-L. Eoc.; Eu.

    †*Spaniella* Crusafont Pairó & D. E. Russell, 1967:759.
                E. Eoc.; Eu.

    †*Vulpavoides* Matthes, 1952:229.
                [= or including †*Russellites* Van Valen, 1965:392.]
                M. Eoc.; Eu.

    †*Pugiodens* Matthes, 1952:232.
                M. Eoc.; Eu.

    †*Kopidodon* Weitzel, 1933:83.
                M. Eoc.; Eu.

Subfamily †**Merialinae** D. E. Russell & Godinot, 1988:321.
                E. Eoc., L. Eoc.; Eu.

    †*Merialus* D. E. Russell & Godinot, 1988:322.
                E. Eoc.; Eu.

    †*Euhookeria* D. E. Russell & Godinot, 1988:321.
                L. Eoc.; Eu.

Family †**Ptolemaiidae** Osborn, 1908:267.
                L. Eoc.-E. Olig.; Af.

    †*Ptolemaia* Osborn, 1908:267.
                L. Eoc.; Af.

    †*Qarunavus* Simons & Gingerich, 1974:162.
                L. Eoc.; Af.[1]
                [1] N. Africa (Egypt).

    †*Cleopatrodon* Bown & Simons, 1987:315.
                L. Eoc.-E. Olig.; Af.[1]
                [1] N. Africa (Egypt).

Suborder **PHOLIDOTA** Weber, 1904:412-413, 420, **new rank**.[1]

[= REPENTIA Newman, 1843:34, 50[2]; Manides Gervais, 1854:xx[3]; SCUTATA Murray, 1866:xiii[4]; SQUAMATA Huxley, 1872:287[5]; NOMARTHRA Cope, 1889:657; MANITHERIA Haeckel, 1895:490, 519[2]; PHOLIDOTHERIA Haeckel, 1895:516, 519; SQUAMIGERA Gill, 1910:56; LEPIDOTA Lane, 1910[6]; PHOLIDOTIFORMES Kinman, 1994:37.] [Including PALAEANODONTA Matthew, 1918:620[7]; †AFREDENTATA Szalay & Schrenk, 1994:48A; †PALAEANODONTIFORMES Kinman, 1994:37.[8]]

[1] Proposed as an order. Not PHOLIDOTA Merrem, 1820:5.
[2] In part.
[3] Proposed as sous-ordre.
[4] Also included †*Macrotherium*.
[5] Not SQUAMATA Oppel, 1811:14.
[6] Not LEPIDOTA Vogt, 1851.
[7] Proposed as a suborder (of EDENTATA, sensu XENARTHRA). Used at rank of superorder by Pascual, Vucetich & Scillato-Yané, 1990:628.
[8] Proposed with a query, but not subject to the provisions of the International Code of Zoological Nomenclature.

†*Melaniella* Fox, 1984:1335.
L. Paleoc.; N.A.

†*Tubulodon* Jepsen, 1932:255.
E. Eoc.; N.A.

Family †**Epoicotheriidae** Simpson, 1927:285.
L. Paleoc.-L. Eoc.; N.A.   ?M. Eoc.; As.[1]   ?E. Olig.; Eu.

[1] A poorly known epoicothere-like mammal has been reported from the Eocene of Yunnan.

†*Amelotabes* Rose, 1978:659.
L. Paleoc.; N.A.

†*Alocodontulum* Rose, Bown & Simons, 1978:1162.
[= †*Alocodon* Rose, Bown & Simons, 1977:1.[1]]
E. Eoc.; N.A.

[1] Not †*Alocodon* Thulborn, 1973, a Jurassic ornithopod.

†*Pentapassalus* Gazin, 1952:32.
E. Eoc.; N.A.

†*Dipassalus* Rose, Krishtalka & Stucky, 1991:69.[1]
E.-M. Eoc.; N.A.

[1] Assigned to †Epoicotheriidae with doubt by the original authors.

†*Tetrapassalus* Simpson, 1959:2.
M. Eoc.; N.A.

†*Epoicotherium* Simpson, 1927:285.
[= †*Xenotherium* Douglass, 1906:204[1]; †*Pseudochrysochloris* Turnbull & Reed, 1967:623.]
L. Eoc.; N.A.   ?E. Olig.; Eu.

[1] Not †*Xenotherium* Ameghino, 1904:24, a dubiously valid notoungulate.

†*Xenocranium* Colbert, 1942:3.
L. Eoc.-E. Olig.; N.A.

Family †**Metacheiromyidae** Wortman, 1903:347 (197).
L. Paleoc.-M. Eoc.; N.A.

†*Propalaeanodon* Rose, 1979:2.
L. Paleoc.; N.A.

†*Palaeanodon* Matthew, 1918:621.
L. Paleoc.-E. Eoc.; N.A.

†*Brachianodon* Gunnell & Gingerich, 1993:367.
M. Eoc.; N.A.

†*Metacheiromys* Wortman, 1903:347 (197).
M. Eoc.; N.A.

Family **Manidae** Gray, 1821:305. Pangolins, scaly anteaters.
[= Manina Gray, 1825:343; Manididae Gray, 1865:362; Manidae Turner, 1851:219; MANIDA Kalandadze & Rautian, 1992:100.[1]] [Including †Eurotamanduidae Szalay & Schrenk, 1994:48A.]
M. Eoc., ?L. Eoc., E. Olig.-L. Mioc., Pleist.; Eu. ?L. Eoc., E. Plioc., ?L. Plioc., Pleist., R.; Af.[2] L. Eoc.; N.A. L. Mioc., Pleist., R.; As. Pleist., R.; E. Indies.

[1] Named as a suborder of PHOLIDOTA and excluding †PALAEANODONTA.
[2] Bones of an animal said to belong to an unnamed manid have been collected from the late Eocene Jebel Qatrani Formation of Egypt.

†*Eurotamandua* Storch, 1981:253.
M. Eoc.; Eu.

Subfamily **Maninae** Gray, 1821:305.
[= Manidae Gray, 1821:305; Manina Gray, 1825:343[1]; Maninae Pocock, 1924:723; Pholidotina Gray, 1873.]
M. Eoc., ?L. Eoc., E. Olig.-L. Mioc., Pleist.; Eu. L. Eoc.; N.A. L. Mioc., Pleist., R.; As. Pleist., R.; E. Indies.

[1] As tribe.

†*Eomanis* Storch, 1978:506.
M. Eoc.; Eu.

†*Necromanis* Filhol, 1893:132-134.[1]
[Including †*Leptomanis* Filhol, 1893:134-135; †*Necrodasypus* Filhol, 1893:136-139[2]; †*Teutomanis* Ameghino, 1905:215; †*Galliaetatus* Ameghino, 1905:176.[3]]
?L. Eoc., E. Olig.-L. Mioc.; Eu.

[1] Usually dated as "1894." However, fide Palmer (1904:372), the paper appeared on Dec. 15, 1893.
[2] Possibly = †*Leptomanis*.
[3] Objective synonym of †*Teutomanis*.

†*Patriomanis* Emry, 1970:465.
L. Eoc.; N.A.

Manis Linnaeus, 1758:36. Asiatic pangolins.
[= *Pholidotus* Brisson, 1762:12, 18-20[1]; *Quaggelo* Frisch, 1775:5[2]; *Pangolinus* Rafinesque, 1821:214[3]; *Pangolin* Gray, 1873:193.] [Including *Phatages* Sundevall, 1843:258-261[4]; *Phatagenus* Sundevall, 1843:273[5]; *Paramanis* Pocock, 1924:722.[4]]
L. Mioc., Pleist., R.; As. Pleist., R.; E. Indies. Pleist.; Eu.

[1] Or Storr, 1780. Brisson's work was not constistently binominal. See the International Code of Zoological Nomenclature, Article 11(c)(i), for confusion.
[2] Unavailable.
[3] *Nomen nudum* in Rafinesque, 1815.
[4] Proposed as subgenus.
[5] Objective synonym of *Phatages*.

Subfamily **Smutsiinae** Gray, 1873. African pangolins.
[= Smutsiinae Pocock, 1924:723.] [Including Uromaninae Pocock, 1924:723.]
E. Plioc., ?L. Plioc., Pleist., R.; Af.[1]

[1] Tropical Africa.

Tribe **Smutsiini** Gray, 1873:11.[1]
[= Smutsiana Gray, 1873:11[2]; Smutsiinae Pocock, 1924:723.]
E. Plioc., ?L. Plioc., Pleist., R.; Af.[3]

[1] New suffix.
[2] Proposed as tribe.
[3] Sub-Saharan Africa.

Smutsia Gray, 1865:369-370. African ground pangolin.
E. Plioc., ?L. Plioc., Pleist., R.; Af.[1]

[1] Sub-Saharan Africa.

Tribe **Uromanini** Pocock, 1924:723, **new rank**.
[= Uromaninae Pocock, 1924:723.]
R.; Af.[1]

[1] Tropical Africa.

Phataginus Rafinesque, 1821:214.[1] African tree pangolin.
>   [= *Phatagin* Gray, 1865:363-365; *Triglochinopholis* Fitzinger, 1872:27-37.]
>   R.; Af.[2]

[1] Proposed as subgenus of *Manis*. *Nomen nudum* in Rafinesque, 1815.
[2] Tropical Africa.

Uromanis Pocock, 1924:722. Four-toed pangolin.
>   R.; Af.[1]

[1] Tropical Africa.

Suborder †ERNANODONTA Ding, 1987:88.[1]

[1] Proposed as suborder of order EDENTATA by Ding (Ting).

Family †Ernanodontidae Ding, 1979:57.[1]
>   L. Paleoc., ?L. Eoc.; As.[2] ?E. Eoc.; N.A.

[1] (Ting).
[2] A possible ernanodontid is represented by skeletal remains in the American Museum of Natural History, collected by the Central Asiatic Expeditions from the late Eocene of the People's Republic of China.

†*Asiabradypus* Nessov, 1987:206.
>   L. Paleoc.; As.[1] ?E. Eoc.; N.A.

[1] Kazakhstan.

†*Ernanodon* Ding, 1979:57.[1]
>   L. Paleoc.; As.

[1] (Ting).

Order †CREODONTA Cope, 1875:444.[1]

>   [= †SUBDIDELPHIA Trouessart, 1879:2; †PSEUDOCREODI Matthew, 1909:327; PARACARNIVORA Kretzoi, 1929:1349[2]; †CREOPHAGA Kretzoi, 1945:62[3]; †HYAENODONTIA Romer, 1966:383; †CREODONTIFORMES Kinman, 1994:37.]

[1] Proposed as suborder by Cope. Used as order by McKenna, 1975:41. †CREODONTA was proposed for a more inclusive assemblage than a brigade comprising just oxyaenoids plus hyaenodontoids, but it is a familiar term and is retained here. The more precise term †PSEUDOCREODI, used by Matthew at about infraordinal rank, has fallen into disuse.
[2] In part. Kretzoi's PARACARNIVORA was proposed for †PSEUDOCREODI plus AELUROIDEA.
[3] In part. Order †CREOPHAGA Kretzoi, created only a page after his PARACARNIVORA, also included †PSEUDOCREODI, this time in company with †Deltatheridioidea, †ACREODI (subsuming mesonychids, triisodonts, and †Ptolemaiidae). Although priority is of no official concern above the rank of superfamily, CREOPHAGA Haeckel, 1866:clvii, equivalent to the marsupial superorder DASYUROMORPHIA, long predates Kretzoi's order.

†*Prionogale* Schmidt-Kittler & Heizmann, 1991:6.
>   E. Mioc.; Af.

Subfamily †Koholiinae Crochet, 1988:1797.[1]
>   E. Eoc.; Af.

[1] Systematic position very uncertain.

†*Koholia* Crochet, 1988:1798.
>   E. Eoc.; Af.[1]

[1] N. Africa (Saharan Atlas, Algeria).

Family †Hyaenodontidae Leidy, 1869:38.
>   [= †Hyaenodontoidea Trouessart, 1885:8; †Hyaenodontida Haeckel, 1895:578, 583.]
>   [Including †Proviverridae Schlosser, 1886:293; †Limnocyonidae Gazin, 1946:337.]
>   ?Paleoc., E. Eoc.-L. Olig.; N.A. E. Eoc.-L. Mioc.; As. E. Eoc.-E. Mioc.; Eu. L. Eoc.-E. Olig., E.-M. Mioc., ?L. Mioc.; Af.

†*Parvagula* Lange-Badré, in Godinot, Crochet, Hartenberger, Lange-Badré, Russell & Sigé, 1987:274.
>   E. Eoc.; Eu.

†*Tylodon* Gervais, 1848:50.[1]
>   [= †*Tulodon* Zittel, 1893:599.]
>   Eoc.; Eu.

[1] Validity doubtful and, in any case, a *nomen oblitum*.

Subfamily †**Limnocyoninae** Wortman, 1902:117.
        [= †Limnocyonidae Gazin, 1946:337.]
        ?Paleoc., E.-M. Eoc.; N.A. ?E. Eoc., L. Eoc., Olig.; Eu.[1] M. Eoc.; As.[2]
        [1] A possible limnocyonine has been reported from the early Eocene of France.
        [2] E. Asia.

Tribe †**Limnocyonini** Wortman, 1902:117.
        [= †Limnocyoninae Wortman, 1902:117; †Limnocyonini Van Valen, 1966:111.]
        ?Paleoc., E.-M. Eoc.; N.A. M. Eoc.; As.[1] L. Eoc., Olig.; Eu.
        [1] E. Asia.

    †*Prolimnocyon* Matthew, 1915:67.
        ?Paleoc., E. Eoc.; N.A.

    †*Thinocyon* Marsh, 1872:204.
        [Including †*Entomodon* Marsh, 1872:214.]
        ?M. Eoc.; As.[1] M. Eoc.; N.A.
        [1] E. Asia.

    †*Limnocyon* Marsh, 1872:126-127.
        [= †*Telmatocyon* Marsh, 1899:397.]
        M. Eoc.; N.A.

    †*Oxyaenodon* Wortman, 1899:145.[1]
        M. Eoc.; N.A.
        [1] *Nomen nudum* in Matthew, 1899.

    †*Prolaena* Xu, Yan, Zhou, Han & Zhang, 1979:424.
        M. Eoc.; As.[1]
        [1] E. Asia.

    †*Thereutherium* Filhol, 1876:289.
        L. Eoc., Olig.; Eu.

Tribe †**Machaeroidini** Matthew, 1909:330.
        [= †Machairoidinae Matthew, 1909:330; †Machaeroidinae Denison, 1938:181;
        †Machaeroidini Van Valen, 1966:111.]
        E.-M. Eoc.; N.A.

    †*Machaeroides* Matthew, 1909:461.
        E.-M. Eoc.; N.A.

    †*Apataelurus* Scott, 1937:455.
        M. Eoc.; N.A.

Subfamily †**Hyaenodontinae** Leidy, 1869:38.
        [= †Hyaenodontidae Leidy, 1869:38; †Hyaenodontinae Trouessart, 1885:9.] [Including
        †Proviverridae Schlosser, 1886:293; †Proviverrinae Matthew, 1909:465;
        †Hyaenaelurinae Pilgrim, 1932:166; †Pterodontinae Polly, 1996:315.]
        E. Eoc.-L. Mioc.; As. E. Eoc.-E. Mioc.; Eu. E. Eoc.-L. Olig.; N.A. L. Eoc.-E.
        Olig., E.-M. Mioc., ?L. Mioc.; Af.[1]
        [1] A gigantic hyaenodontine has been reported from the middle or late Miocene Ngorora Fm., Kenya.

Tribe †**Proviverrini** Schlosser, 1886:293.
        [= †Proviverridae Schlosser, 1886:293; †Proviverrida Haeckel, 1895:578, 583;
        †Proviverrinae Matthew, 1909:465; †Proviverrini Van Valen, 1966:111.] [Including
        †Stypolophinae Trouessart, 1885:11.]
        E.-L. Eoc., ?E. Mioc., M. Mioc.; As. E.-L. Eoc., ?E. Olig.; Eu. E.-M. Eoc., ?L.
        Eoc.; N.A. L. Eoc.-E. Olig., E.-M. Mioc.; Af.

    †*Prototomus* Cope, 1874:13.
        E. Eoc., ?M. Eoc.; Eu. E. Eoc.; N.A.

    †*Arfia* Van Valen, 1965:639, 659.
        E. Eoc.; Eu. E. Eoc.; N.A.

†*Sinopa* Leidy, 1871:115.
        [= †*Stypolophus* Cope, 1872:1.]
        M. Eoc.; N.A.

†*Gazinocyon* Polly, 1996:304.
        E. Eoc.; N.A.

†*Tritemnodon* Matthew, 1906:205.
        E.-M. Eoc.; N.A.

†*Acarictis* Gingerich & Deutsch, 1989:332.
        E. Eoc.; N.A.

†*Protoproviverra* Lemoine, 1891:272.[1]
        E.-M. Eoc.; Eu.

[1] †*Protoproviverra* Lemoine, May, 1891 antedates †*Protoproviverra* Ameghino, August, 1891.

†*Galecyon* Gingerich & Deutsch, 1989:362.
        E. Eoc.; N.A.

†*Proviverra* Rütimeyer, 1862:80-86.
        [= †*Prorhyzaena* Rütimeyer, 1891:105-106; †*Leonhardtina* Matthes, 1952:223;
        †*Geiselotherium* Matthes, 1952:225.]
        E.-M. Eoc.; Eu.

†*Prodissopsalis* Matthes, 1952:206.
        [= †*Imperatoria* Matthes, 1952:214.]
        ?E. Eoc., M. Eoc.; Eu.

†*Paratritemnodon* Rao, 1973:1.
        E.-M. Eoc.; As.[1]
        [1] S. Asia.

†*Pyrocyon* Gingerich & Deutsch, 1989:360.
        E. Eoc.; N.A.

†*Propterodon* Martin, 1906:455.[1]
        M. Eoc.; As. ?L. Eoc.; N.A.
        [1] See Van Valen (1966:75) and Dashzeveg (1985) for the confusing history of this genus.

†*Cynohyaenodon* Filhol, 1873:227.
        [= †*Pseudosinopa* Depéret, 1917:172.]
        M.-L. Eoc., ?E. Olig.; Eu.

†*Quercitherium* Filhol, 1880:48.
        [= †*Quercytherium* of authors.]
        M.-L. Eoc.; Eu.

†*Praecodens* Lange-Badré, 1981:25.
        M. Eoc.; Eu.

†*Matthodon* Lange-Badré & Haubold, 1990:610.
        M. Eoc.; Eu.

†*Proviverroides* Bown, 1982:50.
        M. Eoc.; N.A.

†*Paracynohyaenodon* Martin, 1906:424.
        M.-L. Eoc., ?E. Olig.; Eu.

†*Alienetherium* Lange-Badré, 1981:22.
        M. Eoc.; Eu.

†*Allopterodon* Ginsburg, in Ginsburg et al., 1977:313.
        [= †*Hurzelerius* Lange-Badré, 1984:739.]
        M. Eoc.; Eu.

†*Eurotherium* Polly & Lange-Badré, 1993:991.
        M. Eoc.; Eu.

†*Masrasector* Simons & Gingerich, 1974:158.
        L. Eoc.; Af.[1] L. Eoc.; As.[2]
        [1] N. Africa (Egypt).
        [2] S.W. Asia (Oman).

†*Metasinopa* Osborn, 1909:423.
L. Eoc.-E. Olig., E. Mioc.; Af.

†*Anasinopa* R. J. G. Savage, 1965:259.
E. Mioc.; Af. ?E. Mioc.; As.[1]
[1] W. Asia (Israel).

†*Dissopsalis* Pilgrim, 1910:64.
M. Mioc.; Af. M. Mioc.; As.

Tribe †**Hyaenodontini** Leidy, 1869:38.
[= †Hyaenodontidae Leidy, 1869:38; †Hyaenodontinae Trouessart, 1885:9;
†Hyaenodontini Van Valen, 1966:111.] [Including †Hyaenaelurinae Pilgrim,
1932:166; †Hyainailourini Ginsburg, 1980; †Pterodontinae Polly, 1996:315.]
E. Eoc.-L. Mioc.; As. E. Eoc.-E. Mioc.; Eu. M. Eoc.-L. Olig.; N.A. L. Eoc., E.
Mioc.; Af.

†*Oxyaenoides* Matthes, 1967:452-456.
?E. Eoc., M. Eoc.; Eu.

†*Francotherium* Rich, 1971:20.
E. Eoc.; Eu.

†*Pterodon* de Blainville, 1839:23.
[= †*Paroxyaena* R. Martin, 1906:598; †*Schizophagus* Lange-Badré, 1975:675-682.]
[Including †*Metapterodon* Stromer, in Kaiser, 1926:110; †*Isohyaenodon* R. J. G.
Savage, 1965:280.[1]]
M. Eoc.-E. Olig., M.-L. Mioc.; As. M.-L. Eoc.; Eu. L. Eoc., E. Mioc.; Af. L.
Eoc.; N.A.
[1] Proposed as subgenus.

†*Hyaenodon* Laizer & Parieu, 1838:442.[1]
[= or including †*Taxotherium* de Blainville, 1841:55-72; †*Pseudopterodon* Schlosser,
1887:199.] [Including †*Neohyaenodon* Thorpe, 1922:278[2]; †*Protohyaenodon* Stock,
1933:435[3]; †*Megalopterodon* Dashzeveg, 1964:264.]
M. Eoc.-L. Olig.; As. ?M. Eoc., L. Eoc.-L. Olig.; Eu. M. Eoc.-L. Olig.; N.A.
[1] Described as a subgenus of *Didelphis*.
[2] Proposed as genus; made subgenus of †*Hyaenodon* by Mellett, 1977:13.
[3] Subgenus as defined.

†*Hemipsalodon* Cope, 1885:163.
M.-L. Eoc.; N.A.

†*Consobrinus* Lange-Badré, 1979:167.
L. Eoc. and/or Olig.; Eu.

†*Paenoxyaenoides* Lange-Badré, 1979:164.
L. Eoc. and/or Olig.; Eu.

†*Parapterodon* Lange-Badré, 1979:88.
L. Eoc.; Eu.

†*Ischnognathus* Stovall, 1948:85.
M. Eoc.; N.A.

†*Hyainailouros* Biedermann, 1863:20.
[= †*Hyaenailurus* Rütimeyer, 1867:52; †*Hyaenaelurus* Stehlin, 1907:525-550.] [= or
including †*Sivapterodon* Ginsburg, 1980:45.]
E. Mioc.; Af. E.-M. Mioc.; As. E. Mioc.; Eu.

†*Leakitherium* R. J. G. Savage, 1965:276.
E. Mioc.; Af.

†*Megistotherium* R. J. G. Savage, 1973:486.
E. Mioc.; Af. E. Mioc.; As.

Tribe †**Teratodontini** R. J. G. Savage, 1965:246, **new rank**.
[= †Teratodontidae R. J. G. Savage, 1965:246.]
E. Mioc.; Af.

†*Teratodon* R. J. G. Savage, 1965:247.
E. Mioc.; Af.

Tribe †**Apterodontini** Szalay, 1967:2.
L. Eoc.; Af. E. Olig.; Eu.

†*Apterodon* P. Fischer, 1881:288-290.
[Including †*Dasyurodon* Andreae, 1887:125-133.]
L. Eoc.; Af. E. Olig.; Eu.

Family †**Oxyaenidae** Cope, 1877:89.
[= †Oxyaenoidea Osborn, 1910:527.]
L. Paleoc.-M. Eoc.; N.A. E. Eoc.; Eu. M. Eoc.; As.

Subfamily †**Oxyaeninae** Cope, 1877:89.
[= †Oxyaenidae Cope, 1877:89; †Oxyaeninae Trouessart, 1885:15.]
L. Paleoc.-M. Eoc.; N.A. E. Eoc.; Eu. M. Eoc.; As.

†*Oxyaena* Cope, 1874:599.
[Including †*Dipsalidictis* Matthew, 1915:63.]
L. Paleoc.-E. Eoc.; N.A. E. Eoc.; Eu.

†*Dipsalidictides* Denison, 1938:167.
E. Eoc.; N.A.

†*Patriofelis* Leidy, 1870:10.
[= or including †*Limnofelis* Marsh, 1872:202; †*Oreocyon* Marsh, 1872:406.[1]]
[Including †*Protopsalis* Cope, 1880:745; †*Argillotherium* Davies, 1884:438.[2]]
E. Eoc.; Eu. E.-M. Eoc.; N.A.

[1] Not †*Oreocyon* Krumbiegel, 1949, a genus of Canidae.
[2] Gregory Buckley (personal communication, 1987).

†*Sarkastodon* Granger, 1938:1.
M. Eoc.; As.

Subfamily †**Tytthaeninae** Gunnell & Gingerich, 1991:144.
L. Paleoc.; N.A.

†*Tytthaena* Gingerich, 1980:570.
L. Paleoc.; N.A.

Subfamily †**Ambloctoninae** Cope, 1877:89.
[= †Ambloctonidae Cope, 1877:89; †AMPHICREODI Gregory & Hellman, 1939:313.[1]] [Including †Palaeonictidae Osborn, 1892:103; †Palaeonictida Haeckel, 1895:578, 583; †Palaeonictinae Denison, 1938:174.]
L. Paleoc.-E. Eoc.; N.A. E. Eoc.; Eu.

[1] Proposed as infraorder coordinate with †PROCREODI, †ACREODI, †PSEUDOCREODI, and EUCREODI (but excluding "suborder" PINNIPEDIA).

†*Dipsalodon* Jepsen, 1930:524.
L. Paleoc.; N.A.

†*Ambloctonus* Cope, 1875:7.
[= †*Amblyctonus* Cope, 1880:79, 80; †*Amblyctomus* Cope, 1882:360.]
E. Eoc.; N.A.

†*Palaeonictis* de Blainville, 1842:79.
L. Paleoc.-E. Eoc.; N.A. E. Eoc.; Eu.

†*Dormaalodon* Lange-Badré, 1987:829, 831.
E. Eoc.; Eu.

Order **CARNIVORA** Bowdich, 1821:33. Carnivores.
[= Carnivori Vicq d'Azyr, 1792:civ; CARNARIA Haeckel, 1866:clix; CARNIVORAMORPHA Wyss & J. J. Flynn, 1993:37[1]; CARNIVORIFORMES Kinman, 1994:37.] [Including FISSIPEDA Blumenbach, 1791; CYNOFELIFORMIA Ginsburg, 1982:256.[2]]

[1] Not CARNIVORAMORPHA Kalandadze & Rautian, 1992:53.
[2] Paraphyletic.

†*Aelurotherium* Adams, 1896:442.
  M. Eoc.; N.A.

†*Eosictis* Scott, 1945:209.
  M. Eoc.; N.A.

†*Elmensius* Kretzoi, 1929:1323.
  E. Mioc.; Eu.

†*Kelba* R. J. G. Savage, 1965:244.
  E. Mioc.; Af.

†*Vishnucyon* Pilgrim, 1932:26.[1]
  M. Mioc.; As.
  [1] Possibly a creodont.

†*Vinayakia* Pilgrim, 1932:157.
  L. Mioc.; As.

†*Notoamphicyon* Ameghino, 1904:122.
  L. Mioc.; S.A.

Suborder **FELIFORMIA** Kretzoi, 1945:62.[1]
  [= EPIMYCTERI Cope, 1882:473; PARACARNIVORA Kretzoi, 1945:61[2];
  AELUROPODA Gray, 1869:5.[3]] [Including AELUROIDEA Flower, 1869:22[4];
  HYAENOIDEA Kretzoi, 1945:81[5]; †DIDYMICTIDA J. Flynn & Galiano, 1982:24.[6]]
  [1] Proposed as order.
  [2] In part. Kretzoi's PARACARNIVORA also included †PSEUDOCREODI.
  [3] In part. Proposed as a section of CARNIVORA, containing felids, viverrids, and certain mustelids.
  [4] Proposed as "primary group"; subsequently used as infraorder.
  [5] Proposed as suborder.
  [6] Proposed as infraorder.

†*Palaeogale* von Meyer, 1846:474.
  [= †*Bunaelurus* Cope, 1873:8.]
  L. Eoc.-L. Olig., ?E. Mioc.; N.A. E. Olig.; As. E. Olig.-E. Mioc.; Eu.

†*Aeluropsis* Lydekker, 1884:316-317.
  L. Mioc.; As.

†*Mellivorodon* Lydekker, 1884:185-186.
  L. Mioc.; As.

Family †**Viverravidae** Wortman & Matthew, 1899:136.
  [Including †Quercygalidae Kretzoi, 1945:81.]
  E. Paleoc., ?L. Paleoc.; As. E. Paleoc.-M. Eoc.; N.A. E.-L. Eoc.; Eu.

†*Pappictidops* Chiu & Li, 1977:99.[1]
  E. Paleoc.; As.[2]
  [1] (Qiu & Li).
  [2] E. Asia.

Subfamily †**Viverravinae** Wortman & Matthew, 1899:136.
  [= †Viverravidae Wortman & Matthew, 1899:136; †Viverravinae Matthew, 1909:345;
  †Viverravini Kalandadze & Rautian, 1992:62; †Viverravina Kalandadze & Rautian,
  1992:62.]
  E. Paleoc.-M. Eoc.; N.A. E.-L. Eoc.; Eu.

†*Pristinictis* Fox & Youzwyshyn, 1994:384.
  L. Paleoc.; N.A.

†*Simpsonictis* MacIntyre, 1962:3.
  E.-L. Paleoc.; N.A.

†*Viverravus* Marsh, 1872:127.
  [Including †*Triacodon* Marsh, 1871:123.[1]]
  L. Paleoc.-M. Eoc.; N.A. ?E. Eoc., L. Eoc.; Eu.
  [1] Not †*Triacodon* Cope, 1872, a synonym of †*Uintacyon*. †*Triacodon* Marsh, 1871, was rejected by the
  International Commission on Zoological Nomenclature, Opinion 1517, 1988.

†*Quercygale* Kretzoi, 1945:80.
　　　[= †*Humbertia* de Beaumont, 1965:142.]
　　　M.-L. Eoc.; Eu.

†*Simamphicyon* Viret, 1942:95.
　　　M. Eoc., ?L. Eoc.; Eu.

†*Tapocyon* Stock, 1934:423.
　　　M. Eoc.; N.A.

†*Plesiomiacis* Stock, 1935:120.[1]
　　　M. Eoc.; N.A.
　　　[1] Described as a subgenus of †*Viverravus*.

Subfamily †**Ictidopappinae** Van Valen, 1969:123, **new rank**.[1]
　　　[= †Ictidopappini Van Valen, 1969:123.]
　　　E. Paleoc.; N.A.
　　　[1] Tedford.

†*Ictidopappus* Simpson, 1935:237.
　　　E. Paleoc.; N.A.

Subfamily †**Didymictinae** J. Flynn & Galiano, 1982:24, **new rank**.[1]
　　　[= †DIDYMICTIDA J. Flynn & Galiano, 1982:24[2]; †Didymictidae J. Flynn & Galiano, 1982:24.]
　　　E. Paleoc.-M. Eoc.; N.A.   ?L. Paleoc.; As.   E. Eoc.; Eu.
　　　[1] Tedford.
　　　[2] Proposed as infraorder.

†*Didymictis* Cope, 1875:5.
　　　L. Paleoc.-E. Eoc., ?M. Eoc.; N.A.   E. Eoc.; Eu.

†*Protictis* Matthew, 1937:101.[1]
　　　[Including †*Bryanictis* MacIntyre, 1966:176[2]; †*Protictoides* J. Flynn & Galiano, 1982:29[2]; †*Intyrictis* Gingerich & Winkler, 1985:121.]
　　　E.-L. Paleoc., M. Eoc.; N.A.
　　　[1] Proposed as subgenus of †*Didymictis*; raised to generic rank by MacIntyre, 1966:148.
　　　[2] As subgenus.

†*Raphictis* Gingerich & Winkler, 1985:122.
　　　?L. Paleoc.; As.[1]   L. Paleoc.; N.A.
　　　[1] E. Asia.

Family †**Nimravidae** Cope, 1880:835. False sabre-tooths.
　　　[Including †Ailuromachairodontidae Kretzoi, 1929:1340; †Dinailurictidae Kretzoi, 1929:1341.]
　　　?M. Eoc., E.-L. Olig., M.-L. Mioc.; As. L. Eoc.-L. Olig., L. Mioc.; N.A. E. Olig.-L. Mioc.; Eu. E.-M. Mioc.; Af.

Subfamily †**Hoplophoneinae** Kretzoi, 1929:1334.
　　　[= †Hoplophoneini de Beaumont, 1964:841.] [Including †Eusmilini Ginsburg, 1961:152; †Eusmilini de Beaumont, 1964:841.]
　　　?M. Eoc., ?E. Olig.; As. L. Eoc.-L. Olig.; N.A. E.-L. Olig.; Eu.

†*Eusmilus* Gervais, 1876:53-54.
　　　[Including †*Pareusmilus* Kretzoi, 1929:1305; †*Ekgmoiteptecela* Macdonald, 1963:221.]
　　　?M. Eoc.; As. E.-L. Olig.; Eu. L. Olig.; N.A.

†*Hoplophoneus* Cope, 1874:23.
　　　[= †*Hoplophomus* Studder, 1882:153.[1]] [Including †*Dinotomius* Williston, 1895:170.]
　　　L. Eoc.-E. Olig.; N.A. ?E. Olig.; As.
　　　[1] Misprint.

Subfamily †**Nimravinae** Cope, 1880:835.
　　　[= †Nimravidae Cope, 1880:835; †Nimravinae Trouessart, 1885:6, 92; †Nimravini de Beaumont, 1964:840.] [Including †Dinictinae Kretzoi, 1929:1334; †Dinictini de

Beaumont, 1964:840; †Pogonodontinae Kretzoi, 1929:1333; †Dinailurictinae Kretzoi, 1929:1337.]

L. Eoc.-L. Olig.; N.A.   E.-L. Olig.; As.   E.-L. Olig.; Eu.

†*Dinictis* Leidy, 1854:127, 156.
> [= †*Deinictis* Leidy, 1856:91.]   [Including †*Daptophilus* Cope, 1873:2; †*Metadeiniktis* Kretzoi, 1929:1303.]
> L. Eoc.-E. Olig.; N.A.

†*Eofelis* Kretzoi, 1938:103.
> [Including †*Nimraviscus* Kretzoi, 1945:68.]
> E. Olig.; Eu.

†*Nimravus* Cope, 1879:168.
> [= †*Aelurogale* Filhol, 1872:10[1]; †*Archaelurus* Cope, 1879:798; †*Ailurictis* Trouessart, 1886:954; †*Aelurictis* Trouessart, 1899:343.[2]]   [Including †*Prionedes* Jourdan, 1862:132; †*Ictidailurus* Kretzoi, 1945:63, 66; †*Nimravinus* Kretzoi, 1945:66, 68.]
> E.-L. Olig.; As.   E.-L. Olig.; Eu.   E.-L. Olig.; N.A.

> [1] Not *Ailurogale* Fitzinger, 1869, a synonym of *Felis (Ictailurus)*. According to the Principle of Homonymy, *Ailurogale* Fitzinger is is not a homomyn of and therefore does not preoccupy †*Aelurogale* Filhol, which antedates †*Nimravus*. For reasons of nomenclatural stability, however, various authors have retained †*Nimravus*.
> [2] A misspelling or unjustified emendation of †*Ailurictis*.

†*Dinailurictis* Helbing, 1922:568-575.
> [= †*Dinaelurictis* Schlosser, in Zittel, 1923:479.]
> E. Olig.; Eu.

†*Quercylurus* Ginsburg, 1979:41.
> E. Olig.; Eu.

†*Dinaelurus* Eaton, 1922:437.
> L. Olig.; N.A.

†*Pogonodon* Cope, 1880:142-143.
> L. Olig.; N.A.

Subfamily †**Barbourofelinae** C. B. Schultz, M. R. Schultz & L. D. Martin, 1970:2, **new rank**.
> [= †Barbourofelini C. B. Schultz, M. R. Schultz & L. D. Martin, 1970:2.]   [Including †Ailuromachairodontinae Kretzoi, 1929:1336.]
> E.-M. Mioc.; Af.   E.-L. Mioc.; Eu.   M.-L. Mioc.; As.   L. Mioc.; N.A.

†*Syrtosmilus* Ginsburg, 1978:73-74.
> E. Mioc.; Af.[1]
> [1] N. Africa.

†*Prosansanosmilus* Heizmann, Ginsburg & Bulot, 1980:9.
> E. Mioc.; Eu.

†*Vampyrictis* Kurtén, 1978:193.
> M. Mioc.; Af.[1]
> [1] N. Africa.

†*Vishnusmilus* Kretzoi, 1938:110.
> M. Mioc.; As.

†*Sansanosmilus* Kretzoi, 1929:1307.
> [Including †*Albanosmilus* Kretzoi, 1929:1306[1]; †*Ailuromachairodus* Kretzoi, 1929:1319; †*Grivasmilus* de Villalta Comella & Crusafont Pairó, 1952:308.[2]]
> M. Mioc.; As.   M.-L. Mioc.; Eu.

> [1] "†*Sansanosmilus* was proposed at the same time as †*Albanosmilus* but on the next page. Unless other things are equal, insistence on page (*a fortiori*, line) priority is a ridiculous fetish, and I select †*Sansanosmilus* for preservation because its type is better known..." (Simpson 1945:120).
> [2] Objective synonym of †*Albanosmilus*.

†*Barbourofelis* C. B. Schultz, M. R. Schultz & L. D. Martin, 1970:2.
> L. Mioc.; As.   L. Mioc.; N.A.

Family **Felidae** Fischer de Waldheim, 1817:372. Cats.

> [= Felini Fischer de Waldheim, 1817:372; Felidae Gray, 1821:302; Felina Haeckel, 1866:clix[1]; Feloidae Hay, 1930:538; Feloidea Simpson, 1931:277; Felida Gregory & Hellman, 1939:320[2]; Felinoidea Brunet, 1979:173; Euailuroida Kretzoi, 1929:1346.[3]] [Including Lyncina Gray, 1867:276; Lyncidae Schulze, 1900:222; Guepardidae Gray, 1869:39; †Machaerodontidae Woodward, 1898:399; †Megantereontidae Kretzoi, 1929:1337; †Megantereontinae Kretzoi, 1929:1337; †Proaeluridae Kretzoi, 1945:79, 81.]
>
> E. Olig.-R.; Eu. E. Mioc.-R.; Af. M. Mioc.-R.; As. M. Mioc.-R.; N.A. L. Plioc.-R.; S.A. E. Pleist.-R.; E. Indies. L. Pleist.-R.; Cent. A.[4]

[1] Used as a family.
[2] Used as a section for Felinae, Pantherinae, and Acinonychinae.
[3] Proposed as a "tribe" but ranked between superfamily and family. Not based on a generic name.
[4] Also Recent worldwide in domestication.

Subfamily †**Proailurinae** Zittel, 1893:665.

> [= †Proaelurinae Zittel, 1893:665; †Proailurinae Pilgrim, 1931:125[1]; †Proaeluridae Kretzoi, 1945:81; †Proailurini de Beaumont, 1964:840.]
>
> E. Olig.-E. Mioc.; Eu.

[1] Emendation of †Proaelurinae.

†*Proailurus* Filhol, 1879:192.

> [= †*Proaelurus* Forbes, 1881:15.] [Including †*Brachictis* Kretzoi, 1945:79, 81.]
>
> E. Olig.-E. Mioc.; Eu.

Subfamily **Felinae** Fischer de Waldheim, 1817:372.

> [= Felini Fischer de Waldheim, 1817:372; Felina Gray, 1825:339[1]; Felinae Trouessart, 1885:6, 92; Felida Haeckel, 1895:579.] [Including Lyncina Gray, 1867:276; Lyncini Kalandadze & Rautian, 1992:72; Neofelinae Kretzoi, 1929:1338; Neofelina Kalandadze & Rautian, 1992:72; †Pseudaelurinae Kretzoi, 1929:1338; †Pseudaelurini de Beaumont, 1964:840; †Metailurini de Beaumont, 1964:840; †Metailurinae Crusafont Pairó & Aguirre, 1972:221; Profelina Kalandadze & Rautian, 1992:72; Therailurini Kalandadze & Rautian, 1992:73.]
>
> E. Mioc., ?M. Mioc., E. Plioc.-R.; Af. E. Mioc.-R.; Eu. M. Mioc.-E. Pleist., ?M. and/or ?L. Pleist., R.; As. M. Mioc.-R.; N.A. E. and/or M. Pleist., L. Pleist.-R.; E. Indies. E. Pleist.-R.; S.A. L. Pleist.-R.; Cent. A.[2]

[1] As tribe.
[2] Also Recent worldwide in domestication.

†*Pseudaelurus* Gervais, 1850:123.

> [= †*Pseudelurus* Filhol, 1891:73.[1]] [Including †*Parapseudailurus* Kretzoi, 1929:1321, 1322; †*Hyperailurictis* Kretzoi, 1929:1295; †*Afrosmilus* Kretzoi, 1929:1296; †*Styriofelis* Kretzoi, 1929; †*Miopanthera* Kretzoi, 1938:103; †*Sansanailurus* Kretzoi, 1938:107; †*Schizailurus* Viret, 1951:90, 92.[2]]
>
> E. Mioc.; Af. E.-L. Mioc.; Eu. M. Mioc.; As. M.-L. Mioc.; N.A.

[1] Misprint?
[2] Proposed as subgenus. Objective synonym of †*Miopanthera*.

†*Sivaelurus* Pilgrim, 1913:291, 314.

> M. Mioc., E. Plioc.; As.

†*Vishnufelis* Pilgrim, 1932:206.

> M. Mioc.; As.

†*Pikermia* Kretzoi, 1938:109.

> L. Mioc.; Eu.

†*Abelia* Kretzoi, 1938:109.

> L. Mioc.; Eu.

†*Nimravides* Kitts, 1958:368.

> L. Mioc.; N.A.

†*Pratifelis* Hibbard, 1934:240.

> L. Mioc.; N.A.

†*Adelphailurus* Hibbard, 1934:243.
 [Including †*Stenailurus* Crusafont Pairó & Aguirre, 1972:211-225.]
 L. Mioc.; Eu. L. Mioc.; N.A.

†*Metailurus* Zdansky, 1924:123.
 L. Mioc.-E. Plioc.; As. L. Mioc.; Eu.

†*Dinofelis* Zdansky, 1924:137.
 [= †*Therailurus* Piveteau, 1948:115.]
 L. Mioc., L. Plioc.; As. L. Mioc.-E. Pleist.; Eu. E. Plioc.-E. Pleist.; Af. L. Plioc.;
 N.A.

†*Dolichofelis* Kretzoi, 1929.
 Plioc.; Eu.

†*Sivapardus* Bakr, 1969:135.
 Plioc.; As.

Neofelis Gray, 1867:265. Clouded leopard.
 Pleist., R.; E. Indies.[1] Pleist., R.; As.
 [1] Sumatra, Borneo.

†*Jansofelis* Bonifay, 1971:250.
 M. Pleist.; Eu.

Felis Linnaeus, 1758:41. Cats.
 [= *Catus* Frisch, 1775: table 12[1]; *Otailurus* Severtzov, 1858:388, 390[2]; *Mamfelisus*
 Herrera, 1899:17[3]; *Poliailurus* Lonnberg, 1925:2; *Avitofelis* Kretzoi, 1930:1-22.] [=
 or including *Catolynx* Severtzov, 1858:385, 390[4]; †*Proconsuloides* Lonnberge, 1937.]
 [Including *Chaus* Gray, 1843:xx, 44-46[5]; *Eremaelurus* Ognev, 1926:356.]
 ?M. Mioc., E. Plioc., E. Pleist.-R.; Af.[6] L. Mioc., L. Plioc.-E. Pleist., ?M. and/or
 ?L. Pleist., R.; As.[7] L. Mioc.-R.; Eu. E. Plioc.-R.; N.A. E. and/or M. Pleist., L.
 Pleist.-R.; E. Indies. E. Pleist.-R.; S.A. L. Pleist.-R.; Cent. A.[8]
 [1] *Nomen nudum.*
 [2] Proposed as subgenus of *Felis.*
 [3] See the International Code of Zoological Nomenclature, Article 1(b)(8) and Palmer (1904:25-26).
 [4] Proposed as subgenus of *Felis.*
 [5] Sometimes as subgenus.
 [6] Record in ?middle Miocene based on an undescribed new species from Beni Mellal, Morocco; subgenus
  indeterminate.
 [7] *Felis,* subgenus indet., in the late Miocene of Afghanistan and China.
 [8] Also Recent worldwide in domestication.

 (*Felis*) Linnaeus, 1758:41. Old World wild cat, sand cat, domestic cat.
  L. Plioc.-E. Pleist., R.; As. M. Pleist.-R.; Eu. L. Plioc.-R.; Af.[1]
  [1] Also Recent worldwide in domestication.

 (*Lynx*) Kerr, 1792:41, 155-158.[1] Lynx, bobcat, caracal.
  [= *Lynceus* Gray, 1821:302[2]; *Lyncus* Gray, 1825:339; *Lynchus* Jardine, 1834:274.]
  [Including *Pardina* Kaup, 1829:53, 57; *Caracala* Gray, 1843:xx[3]; *Caracal* Gray,
  1843:46[4]; *Urolynchus* Severtzov, 1858:389; *Cervaria* Gray, 1867:276-277[5]; *Eucervaria*
  Palmer, 1903.[6]]
  L. Mioc.-R.; Eu. E. Plioc., E. Pleist.-R.; Af. E. Plioc.-R.; N.A. L. Plioc., Pleist.,
  R.; As.
  [1] Sometimes as genus.
  [2] Not *Lynceus* Müller, 1785, a crustacean.
  [3] *Nomen nudum.*
  [4] Sometimes as subgenus of *Felis* or *Lynx.*
  [5] Proposed as subgenus of *Lyncus.* Not *Cervaria* Walker, 1866, a genus of Lepidoptera.
  [6] New name for *Cervaria* Gray.

 (*Puma*) Jardine, 1834:266-267. Puma, mountain lion, cougar, American panther.
  L. Plioc.; As. L. Plioc.-R.; N.A. ?E. Pleist., M. Pleist.-R.; S.A. L. Pleist.-R.;
  Cent. A.

 (*Prionailurus*) Severtzov, 1858:387. Dwarf tiger cat, fishing cat, rusty-spotted cat.
  [= or including *Viverriceps* Gray, 1867:268.] [Including *Zibethailurus* Severtzov,
  1858:387, 390; *Prionofelis* Kretzoi, 1929; *Mayailurus* Imaizumi, 1967.]

E. and/or M. Pleist., L. Pleist.-R.; E. Indies.[1] R.; As.[2]

[1] Sumatra, Java, Borneo, Philippines.
[2] Including the Ryukyu Islands.

(*Herpailurus*) Severtzov, 1858:390. Eyra, jaguarundi.
E. Pleist.-R.; N.A.[1] Pleist., R.; S.A. R.; Cent. A.

[1] S. North America.

(*Lynchailurus*) Severtzov, 1858:386, 390.[1] Kodkod, grass cat, pampa cat, gato pajero.
[= *Pajeros* Gray, 1867:269-270; *Mungofelis* Antonius, 1933:13.] [= or including *Dendrailurus* Severtzov, 1858:386, 390.[2]] [Including *Pseudolynx* Schwangart, 1941:36.[3]]
E. Pleist., R.; S.A.

[1] Subgenus as described.
[2] Described as subgenus of *Felis*. Probably invalid.
[3] Proposed as subgenus of *Lynchailurus*. Not *Pseudolynx* Girault, 1916, an insect.

(*Oreailurus*) Cabrera, 1940:16.
[= *Colocolo* Pocock, 1941:272; *Montifelis* Schwangart, 1941:37.]
R.; S.A.

(*Leptailurus*) Severtzov, 1858:389. Serval.
[= *Galeopardus* Heuglin & Fitzinger, 1866:557; *Servalina* Greve, 1894:76-77.]
[Including *Serval* Brehm, 1864.]
M. Pleist., R.; Af.

(*Leopardus*) Gray, 1842:260. Ocelot, margay, guina.
[= *Oncoides* Severtzov, 1858:386, 390[1]; *Pardalis* Gray, 1867:270-272.[1]] [= or including *Pardalina* Gray, 1867:266.] [Including *Oncifelis* Severtzov, 1858:386, 390[1]; *Noctifelis* Severtzov, 1858:386, 390[2]; *Margay* Gray, 1867:271-272; *Oncilla* Allen, 1919:358.]
L. Pleist.-R.; N.A.[3] L. Pleist.-R.; S.A. R.; Cent. A.

[1] Proposed as subgenus of *Felis*.
[2] Proposed as subgenus of *Felis*. Not "*Noctifelis*" I. Geoffroy Saint-Hilaire, 1844 (not adopted).
[3] S. North America.

(*Pardofelis*) Severtzov, 1858:387. Marble cat.
R.; E. Indies.[1] R.; As.

[1] Sumatra and Borneo.

(*Otocolobus*) Brandt, 1841:38. Manul.
[= *Trichaelurus* Satunin, 1905:495.]
R.; As.

(*Ictailurus*) Severtzov, 1858:387-388, 390.[1] Flat-headed cat.
[= *Ailurin* Gervais, 1855:86[2]; *Ailurogale* Fitzinger, 1869:249-251; *Aelurina* Gill, 1871:60[3]; *Ailurina* Trouessart, 1885:100[3]; *Plethaelurus* Cope, 1872:475.]
R.; E. Indies.[4] R.; As.[5]

[1] Subgenus as described.
[2] Probably used as a common name rather than as a subgenus of *Felis*, so not given priority here.
[3] Misspelling or unjustified emendation of *Ailurin* Gervais.
[4] Sumatra, Borneo.
[5] S. Asia.

(*Microfelis*) Roberts, 1926:250. Blackfooted cat.
R.; Af.[1]

[1] South Africa.

(*Profelis*) Severtzov, 1858:386. Golden cat, Borneo marble cat.
[Including *Catopuma* Severtzov, 1858:387; *Chrysailurus* Severtzov, 1858:389, 390; *Badiofelis* Pocock, 1932:749.]
R.; Af. R.; E. Indies.[1] R.; As.

[1] Sumatra, Borneo.

Subfamily **Pantherinae** Pocock, 1917:332.
[= Pantherini Kalandadze, 1992:72.] [Including Leonida Haeckel, 1895:579.[1]]

L. Mioc., ?E. Plioc., L. Plioc.-R.; Eu.[2] E. Plioc.-R.; N.A. L. Plioc.-R.; Af. E. Pleist.-R.; E. Indies. ?E. Pleist., M. Pleist.-R.; As. ?E. Pleist., M. Pleist.-R.; S.A. R.; Cent. A.

[1] A division (with Felida) of the felids that should probably be suppressed.
[2] Now extinct.

†*Leontoceryx* Kretzoi, 1938:103.
L. Mioc.; Eu.[1]

[1] E. Europe.

†*Dromopanthera* Kretzoi, 1929.
Plioc.; Eu.

†*Schaubia* Camp, Welles & Green, 1949:330.[1]
[= †*Brachyprosopus* Schaub, 1943:183-188.[2]]
L. Plioc.; Eu.

[1] Not †*Schaubia* Hoffstetter, 1949 (but later), a megathere, synonym of †*Eremotherium*.
[2] Not †*Brachyprosopus* E. C. Olson, 1937, a dicynodont.

†*Viretailurus* Hemmer, 1965.
E. Pleist.; Eu.

Panthera Oken, 1816:1052.[1] Roaring cats.
[= *Pardus* Fitzinger, 1868:459.]
E. Plioc.-R.; N.A. L. Plioc.-R.; Af. E. Pleist.-R.; E. Indies. ?E. Pleist., M. Pleist.-R.; As. E. Pleist.-R.; Eu. ?E. Pleist., M. Pleist.-R.; S.A. R.; Cent. A.

[1] Conserved by the International Commission on Zoological Nomenclature, Opinion 1368, 1985.

(*Panthera*) Oken, 1816:1052. Leopard, panther.
E. Pleist.-R.; Af. E.-L. Pleist.; Eu. L. Pleist.-R.; E. Indies. L. Pleist.-R.; As.

(*Jaguarius*) Severtzov, 1858:386, 390.[1] Jaguar.
[= *Pardotigris* Kretzoi, 1929:1325.]
E. Plioc.-R.; N.A. M. Pleist.-R.; S.A. R.; Cent. A.

[1] Subgenus as described.

(*Leo*) Oken, 1816:1070-76. Lion.
[= *Leo* Frisch, 1775[1]; *Leo* Brehm, 1829:637-638; *Leoninae* Wagner, in Schreber, 1841:460-469[2]; *Leonina* Greve, 1894:60-64.]
L. Plioc.-R.; Af. Pleist., R.; As. Pleist.; S.A. M. Pleist.-R.; Eu.[3] L. Pleist.; N.A.

[1] Invalid. Not published in a consistently binominal work. See the International Code of Zoological Nomenclature, Article 11(c).
[2] Used at subgeneric rank; an apparent but not actual prior synonym of Pantherinae.
[3] Reported in Europe in historical times, but now extinct there.

(*Tigris*) Gray, 1843:40. Tiger.
[= *Tigris* Frisch, 1775:13[1]; *Tigris* Oken, 1816:1066-70[1]; *Tigrinae* Wagner, in Schreber, 1841:469-474[2]; *Tigrina* Greve, 1894:48-55.] [Including †*Feliopsis* Stremme, 1911:86; †*Pachyailurus* Kretzoi, 1929.[1]]
E. Pleist.-R.; E. Indies.[3] ?E. Pleist., M. Pleist.-R.; As.[4]

[1] Invalid.
[2] Used at subgeneric rank; an apparent but not actual prior synonym of Pantherinae.
[3] Sumatra, Java, Bali.
[4] Including Japan in the late Pleistocene.

(*Uncia*) Gray, 1854:394.[1] Snow leopard, irbis, ounce.[2]
Pleist., R.; As.

[1] Sometimes as genus.
[2] "'Ounce', originally the lynx and later, broadly, any medium-sized cat, lynx, panther, etc., has now somehow become attached more or less exclusively to the irbis. 'Ounce' is equivocal in this sense, and irbis is a better name" (Simpson 1945:120).

Subfamily **Acinonychinae** Pocock, 1917:332.
[= Guepardina Gray, 1867:277; Guepardidae Gray, 1869:39; Acinomychini de Beaumont, 1964:840.[1]]
E. Plioc.-E. Pleist., L. Pleist.-R.; Af. L. Plioc.-R.; As. L. Plioc.-M. Pleist.; Eu. L. Plioc.-E. Pleist., L. Pleist.; N.A.

[1] Misspelling.

†*Sivapanthera* Kretzoi, 1929:1324.

[= †*Sivafelis* Pilgrim, 1932:199.[1]]

?E. Plioc.; Af. E. Pleist.; As.

[1] "†*Sivapanthera* was one of the many names that Kretzoi based on the work of others. When Pilgrim returned to the subject, he agreed as to the validity of a separate genus for these cheetah-like fossils, but not as to the scope of the genus. Because of this disagreement he proposed a new name and placed †*Sivapanthera* as a synonym of it. Of course †*Sivafelis* was thus a synonym, not †*Sivapanthera*, and †*Sivafelis* was invalid on the face of it as proposed and has no standing" (Simpson 1945:120).

Acinonyx Brookes, 1828:16, 33. Cheetah, guepard, hunting leopard.

[= *Cynailurus* Wagler, 1830:30[1]; *Cynaelurus* Gloger, 1841:xxix, 63; *Guepardus* Duvernoy, 1834:145[2]; *Guepar* Boitard, 1842:234; *Gueparda* Gray, 1843:xx, 46; *Cynofelis* Lesson, 1842:48; *Paracinonyx* Kretzoi, 1929; *Acinomyx* de Beaumont, 1964:840.[3]] [Including †*Valdarnius* Kretzoi, 1929; †*Abacinonyx* Kretzoi, 1929:11; †*Miracinonyx* Adams, 1979:1155-1158.[4]]

E. Plioc.-E. Pleist., L. Pleist.-R.; Af. L. Plioc.-R.; As. L. Plioc.-M. Pleist.; Eu. L. Plioc.-E. Pleist., L. Pleist.; N.A.

[1] "*Cynaelurus*, Wagner" in Gray, 1869:39.
[2] Proposed as subgenus of *Felis*.
[3] Misspelling.
[4] Proposed as subgenus.

Subfamily †**Machairodontinae** Gill, 1872:59. Sabre-tooths.

[= †Machaerodontinae Gill, 1872:59; †Machairodontoida Kretzoi, 1929:1342[1]; †Machairodontinae Hay, 1930:340[2]; †Macherodontida Gregory & Hellman, 1939:320; †Macherodontinae Gregory & Hellman, 1939:320.[3]] [Including †Epimachairodontinae Kretzoi, 1929:1335; †Smilodontinae Kretzoi, 1929:1336; †Megantereontinae Kretzoi, 1929:1337; †Megantereontidae Kretzoi, 1929:1341; †Paramachaerodontinae Kretzoi, 1938:107; †Pontosmilinae Kretzoi, 1938:109.]

M. Mioc.-M. Pleist.; Af. M. Mioc.-M. Pleist.; As. L. Mioc.-L. Pleist.; Eu. L. Mioc.-L. Pleist.; N.A. L. Plioc.-L. Pleist.; S.A. Pleist.; E. Indies. L. Pleist.; Cent. A.

[1] Described as a "tribe."
[2] Emendation of †Machaerodontinae.
[3] "†Macherodontinae auct." of Gregory & Hellman.

Tribe †**Machairodontini** Gill, 1872:59.

[= †Machairodontini de Beaumont, 1964:840.]

M. Mioc.-E. Pleist.; Af. M. Mioc.-M. Pleist.; As. L. Mioc.-L. Pleist.; Eu. L. Mioc., L. Plioc.-L. Pleist.; N.A. Pleist.; E. Indies.[1]

[1] Java.

†*Paramachaerodus* Pilgrim, 1913:291.

[Including †*Sivasmilus* Kretzoi, 1929:1297; †*Pontosmilus* Kretzoi, 1929:1298; †*Propontosmilus* Kretzoi, 1929:1298; †*Protamphimachairodus* Kretzoi, 1929:1316.]

M.-L. Mioc.; As. L. Mioc.; Eu.

†*Machairodus* Kaup, 1833:24-28.

[= †*Cultridens* Kaup, 1833:24[1]; †*Machaerodus* Agassiz, 1846:219.] [Including †*Heterofelis* Cook, 1922:7[2]; †*Amphimachairodus* Kretzoi, 1929:1316.]

L. Mioc.-E. Pleist.; Af. L. Mioc., Plioc.; As. L. Mioc.-L. Plioc.; Eu. L. Mioc.; N.A.

[1] Error.
[2] Subgenus as defined.

†*Homotherium* Fabrini, 1890:176. Scimitar cat.

[= or including †*Ormenalurus* Jourdan, 1866.[1]] [Including †*Epimachairodus* Kretzoi, 1929:1310; †*Dinobastis* Cope, 1893:896-897.]

L. Mioc.-L. Plioc.; Af. L. Mioc.-L. Pleist.; Eu. E. Plioc.-M. Pleist.; As. L. Plioc.-L. Pleist.; N.A. E. and/or M. Pleist.; E. Indies.[2]

[1] Assignment uncertain. If correct, then this taxon may be a good candidate for suppression.
[2] Java.

†*Miomachairodus* Schmidt-Kittler, 1976:107.

> M. Mioc.; Af.[1] M. Mioc.; As.[2]

[1] N. Africa (Tunisia).
[2] W. Asia (Turkey).

†*Ischyrosmilus* Merriam, 1918:524.

> L. Plioc.; N.A.

†*Hemimachairodus* G. H. R. von Koenigswald, 1974:267.

> Pleist.; E. Indies.[1]

[1] Java.

Tribe †**Smilodontini** Kretzoi, 1929:1336.

> [= †Smildontinae Kretzoi, 1929:1336; †Smilodontini J. L. Kraglievich, 1948:6.]
> [Including †Megantereontinae Kretzoi, 1929:1337; †Megantereontidae Kretzoi, 1929:1341; †Megantereontini Kalandadze & Rautian, 1992:73.]
> E. Plioc.-L. Pleist.; N.A. L. Plioc.-M. Pleist.; Af. L. Plioc.-M. Pleist.; As. L. Plioc.-E. Pleist.; Eu. L. Plioc.-L. Pleist.; S.A. E. and/or M. Pleist.; E. Indies. L. Pleist.; Cent. A.

†*Megantereon* Croizet & Jobert, 1828:200-201.

> [= or including †*Steneodon* Croizet, 1833:86.[1]] [Including †*Toscanius* Kretzoi, 1929:1309[2]; †*Telosmilus* Kretzoi, 1929:1331; †*Toscanosmilus* Kretzoi, 1938:108; †*Promegantereon* Kretzoi, 1938:108.]
> E.-L. Plioc.; N.A. L. Plioc.-M. Pleist.; Af. L. Plioc.-M. Pleist.; As. L. Plioc.-E. Pleist.; Eu. E. and/or M. Pleist.; E. Indies.[3]

[1] In part.
[2] "Kretzoi supposes †*Drepanodon* Bronn, 1853, to be the correct name for this genus, but proposes †*Toscanius* in case this should prove incorrect. Scott and Jepsen (1936) used †*Drepanodon* for †*Hoplophoneus*. The status of †*Drepanodon* is dubious and disputed, owing in part to the rarity of some early publications involved, but it is likely that †*Drepanodon* (which apparently dates from Nesti, 1826, not Bronn, 1853, as Kretzoi has it, or Leidy, 1857, as Scott and Jepsen concluded) is not available for any cat but is a synonym of *Ursus* and can be forgotten. This validates †*Hoplophoneus* and also †*Toscanius*, if the latter be supposed really separable from †*Megantereon*, which I think very improbable" (Simpson 1945:120-121).
[3] Java.

†*Smilodon* Lund, 1842:190-193, 198.

> [Including †*Munifelis* Muñiz, 1845[1]; †*Trucifelis* Leidy, 1868:175[2]; †*Machaerodus* Lydekker, 1884:33[3]; †*Smilodontopsis* Brown, 1908:188; †*Prosmilodon* Rusconi, 1929:5[4]; †*Smilodontidion* J. L. Kraglievich, 1948:6.]
> L. Plioc.-L. Pleist.; N.A. L. Plioc.-L. Pleist.; S.A. L. Pleist.; Cent. A.

[1] Published in a newspaper article.
[2] Proposed as subgenus; given generic rank by Leidy, 1869:366.
[3] Not †*Machaerodus* Agassiz, 1846, a synonym of †*Machairodus*.
[4] Proposed as subgenus.

Family **Viverridae** Gray, 1821:301. Civets, Asiatic palm civets.

> [= Viveridae Gray, 1821:301; Viverrina Gray, 1825:339; Viverridae Bonaparte, 1845:3[1]; Viverrina Haeckel, 1866:clix.[2]] [Including Arctictidae Cope, 1882; Genettidae Rochebrune, 1883; Paradoxuridae Rochebrune, 1883; Eupleridae Chenu, 1852:165; Cynogalidae Gray, 1869; Cryptoproctidae Flower, 1869:37.]
> ?L. Eoc., E. Olig.-E. Plioc., R.; Eu. E. Olig., E.-L. Mioc., ?Plioc., E. Pleist.-R.; As. E. Mioc.-R.; Af. R.; Madagascar. R.; E. Indies. R.; Mediterranean.

[1] Emendation of Viveridae.
[2] Used as a family.

†*Legetetia* Schmidt-Kittler, 1987:97.

> E. Mioc.; Af.

Subfamily †**Stenoplesictinae** Schlosser, 1923:473.

> [= †Stenoplesictini de Beaumont, 1964:840.]
> ?L. Eoc., E.-L. Olig., ?E. Mioc., M. Mioc., ?L. Mioc.; Eu. E. Olig.; As. E. Mioc.; Af.

†*Stenoplesictis* Filhol, 1880:345.

> ?L. Eoc., E. Olig., ?L. Olig.; Eu. E. Olig.; As. E. Mioc.; Af.

†*Palaeoprionodon* Filhol, 1880:1579.
L. Eoc. and/or E. Olig.; Eu. E. Olig.; As.

†*Stenogale* Schlosser, in Roger, 1887:135-136.
?L. Eoc., E.-L. Olig., Mioc.; Eu.

†*Mioprionodon* Schmidt-Kittler, 1987:111.
E. Mioc.; Af.

†*Pseudictis* Schlosser, in Roger, 1887:136.
M. Mioc.; Eu.

Subfamily **Viverrinae** Gray, 1821:301.
[= Viveridae Gray, 1821:301; Viverrina Gray, 1825:339[1]; Viverrinae Gill, 1872:4; Viverrini Winge, 1895:47.] [Including Prionodontina Gray, 1864:507; Prionodontinae Gregory & Hellman, 1939:320; Prionodontini Simpson, 1945:116; Genettina Gray, 1864:49; Genettidae Rochebrune, 1883; Linsanginae Pocock, 1915:350; †Semigenettae Helbing, 1928:241.]
E. Olig., E. Mioc.-E. Plioc., R.; Eu.[2] E.-L. Mioc., ?Plioc., E.-M. Pleist., ?L. Pleist., R.; As.[3] M. Mioc.-R.; Af. R.; E. Indies. R.; Mediterranean.[4]
[1] As tribe.
[2] Viverrinae, *incertae sedis* (as "*Viverra*" *simplicidens*) in the early Oligocene of Europe.
[3] Viverrine genus and sp. indet. in the early Miocene of Israel.
[4] Balearic Islands.
Also Madagascar and the Comoro Islands by introduction.

†*Semigenetta* Helbing, 1927:305.
E.-M. Mioc.; Eu. M. Mioc.; As.

Viverra Linnaeus, 1758:43-44. Oriental civets, makulele, tenggalong.
[= *Vivera* Gray, 1821:301.] [Including *Moschothera* Pocock, 1933:441[1]; †*Viverrictis* de Beaumont, 1973:288[1]; †*Megaviverra* Qiu, 1980:310.[1]]
E. Mioc.-E. Plioc.; Eu. M.-L. Mioc., Plioc. and/or Pleist., R.; As. E.-L. Plioc., Pleist.; Af. R.; E. Indies.
[1] As subgenus.

Genetta G. Cuvier, 1816:116.[1] Genets.
[Including *Pseudogenetta* Dekeyser, 1949:421[2]; *Paragenetta* Kuhn, 1960:155.[2]]
M. Mioc.-L. Plioc., L. Pleist.-R.; Af. R.; As.[3] R.; Mediterranean.[4] R.; Eu.[5]
[1] *Genetta* Oken, 1816:1010-1012 is non-binominal.
[2] As subgenus.
[3] W. Asia.
[4] Balearic Islands.
[5] S. Europe.

†*Galecynus* Owen, 1847:55-60.
M. Mioc.; Eu.

†*Vishnuictis* Pilgrim, 1932:101.
L. Mioc.; As.

Civettictis Pocock, 1915:134. African civet.
L. Plioc.-M. Pleist., R.; Af.

†*Pseudocivetta* G. Petter, 1967:529, 532.
E. Pleist.; Af.

Viverricula Hodgson, 1838:152. Rasse.
E.-M. Pleist., R.; As. R.; E. Indies.[1]
[1] Sumatra, Java, Bali, and Sumbawa.
Also the Philippines, Zanzibar, the Comoro Islands, and Madagascar by introduction.

Poiana Gray, 1864:520-521. African linsang.
R.; Af.

Osbornictis J. A. Allen, 1919:25. Water civet.
R.; Af.

Prionodon Horsfield, 1821. Banded linsang, spotted linsang.
[= *Prionodontidae* Horsfield, 1824:1-4[1]; *Linsang* Muller, 1839:60; *Prionodontes* Lesson, 1842:60; *Linsanga* Lydekker, 1896:20.] [Including *Pardictis* Thomas, 1925:498.[2]]
R.; E. Indies. R.; As.

[1] As "section." Not meant to be a family, but rather a subgenus (of *Felis*). The choice of suffix was particularly unfortunate.
[2] As subgenus.

Subfamily †**Lophocyoninae** Fejfar & Schmidt-Kittler, 1987:16.
[= †Lophocyonini Kalandadze & Rautian, 1992:69.]
E.-M. Mioc.; Eu.

†*Euboictis* Fejfar & Schmidt-Kittler, 1984:58.
E. Mioc.; Eu.[1]

[1] E. Europe (Greece).

†*Sivanasua* Pilgrim, 1931:34.
[= †*Ailuravus* Schlosser, 1916:25[1]; †*Schlossericyon* Crusafont Pairó, 1959:125-149.]
E.-M. Mioc.; Eu.

[1] Not †*Ailuravus* Rütimeyer, 1891, a rodent.

†*Lophocyon* Fejfar & Schmidt-Kittler, 1987:16.
M. Mioc.; Eu.[1]

[1] E. Europe.

Subfamily **Paradoxurinae** Gray, 1864:508, 526.
[= Paradoxurina Gray, 1864:508, 526; Paradoxurinae Gill, 1872:4, 61; Paradoxuridae Rochebrune, 1883; Paradoxurida Gregory & Hellman, 1939:320; Paradoxurini Simpson, 1945:116.] [Including Arctictidina Gray, 1864:57; Arctictidae Cope, 1882:474; Arctogalidiinae Pocock, 1933:969, 977; Arctogalidiini Simpson, 1945:116.]
L. Mioc., M. Pleist.-R.; As.[1] R.; E. Indies.

[1] Paradoxurinae indeterminate in the late Miocene of Asia.

Paguma Gray, 1831:17. Masked palm civet.
M. Pleist.-R.; As. R.; E. Indies.[1]

[1] Sumatra, Borneo.

Macrogalidia Schwarz, 1910:423. Celebes palm civet.
R.; E. Indies.[1]

[1] Sulawesi.

Paradoxurus F. Cuvier, 1821. Palm civets, toddy cats, tupai, tosia.
[= or including *Ictis* Schinz, 1824:110.[1]]
R.; E. Indies.[2] R.; As.

[1] Synonymy and date of publication questionable.
[2] Sumatra, Java, Borneo, Philippines, Lesser Sundas, Mollucas.

Arctogalidia Merriam, 1897:302. Small-toothed palm civet.
[= *Arctogale* Peters, 1863:98.[1]]
R.; E. Indies.[2] R.; As.[3]

[1] Or 1868-75:126; not *Arctogale* Kaup, 1829: 30, a synonym of *Mustela*.
[2] Sumatra, Java, Borneo.
[3] S.E. Asia.

Arctictis Temminck, 1824:xxi. Binturong.
[= *Ictides* F. Cuvier, 1824:252.]
R.; E. Indies.[1]

[1] Sumatra, Java, Borneo, southwestern Philippines.

Subfamily **Hemigalinae** Gray, 1864:508, 524.
[= Hemigalina Gray, 1864:508, 524; Hemigalinae Gill, 1872:4, 62; Hemigalida Gregory & Hellman, 1939:320; Hemigalini Simpson, 1945:116.] [Including Cynogalina Gray, 1865:507; Cynogalidae Gray, 1869; Cynogalinae Gill, 1872:4, 62; Cynogalini Simpson, 1945:116.]
R.; E. Indies. R.; As.

Hemigalus Jourdan, 1837:442-443. Banded palm civet.
> [= *Hemigalea* de Blainville, 1837:595; *Hemigale* Gray, 1864:542; *Hemibatus* Thomas, 1915:613.]
> R.; E. Indies.[1] R.; As.[2]

[1] Sumatra, Borneo.
[2] S.E. Asia.

Diplogale Thomas, 1912:18. Bornean mongoose, Hose's palm civet.
> R.; E. Indies.[1]

[1] Borneo.

Chrotogale Thomas, 1912:17. Owston's palm civet.
> R.; As.

Cynogale Gray, 1837:88. Otter civet, mampalon.
> [= *Lamictis* de Blainville, 1837:595, 596; *Limictis* Blyth, 1840:93[1]; *Potamophilus* Mueller, 1838:142[2]; *Hydrotidasson* Gistel, 1848:x.[3]]
> R.; E. Indies.[4] R.; As.[5]

[1] Misprint.
[2] Preoccupied by *Potamophilus* Germar, 1811:41, a coleopteran, and *Potamophilus* Latreille in Desmarest, 1826:97.
[3] Replacement name for *Potamophilus* Mueller, 1838.
[4] Sumatra, Borneo.
[5] S.E. Asia.

Subfamily **Euplerinae** Chenu, 1852:165.
> [= Eupleridae Chenu, 1850:165[1]; Euplerini Simpson, 1945:116; Euplerinae Wozencraft, in Wilson & Reader, eds., 1993:340.] [Including Fossinae Pocock, 1915:350; Fossini Simpson, 1945:116.]
> R.; Madagascar.

[1] Chenu, 1850-1858:165, fide Palmer (1904:741).

Eupleres Doyere, 1835:45. Small-toothed mongoose, falanouc.
> R.; Madagascar.

Fossa Gray, 1864:518-519.[1] Malagasy civet, fanaloka.
> R.; Madagascar.

[1] "Note that *Cryptoprocta*, and not the genus *Fossa*, is the fossa" (Simpson 1945:116).

Subfamily **Cryptoproctinae** Gray, 1864:508.
> [= Cryptoproctina Gray, 1864:508; Cryptoproctidae Flower, 1869:23, 37; Cryptoproctinae Trouessart, 1885:6, 91; Cryptoproctida Gregory & Hellman, 1939:320; Cryptoproctini de Beaumont, 1964:840.]
> R.; Madagascar.

Cryptoprocta Bennett, 1833:46. Fossa.
> R.; Madagascar.

Family **Herpestidae** Bonaparte, 1845:3. Mongooses.
> [= Herpestina Bonaparte, 1845:3; Herpestidae Gray, 1869:143; Herpestoidei Winge, 1895:46.] [Including Rhinogalidae Gray, 1869; Suricatinae Thomas, 1882:59[1]; Suricatidae Cope, 1882:474[2]; Cynictidae Cope, 1882:474; Mungotoidea Pocock, 1919:515; Mongotidae Pocock, 1919:515.]
> ?L. Olig., E.-M. Mioc., R.; Eu. E. Mioc., L. Mioc.-R.; Af. L. Mioc., Pleist., R.; As. R.; Madagascar. R.; E. Indies.[3]

[1] January, 1882.
[2] November 11, 1882.
[3] Also islands of the Pacific, Indian Ocean, and West Indies by introduction.

Subfamily **Herpestinae** Bonaparte, 1845:3.
> [= Herpestina Bonaparte, 1845:3; Herpestinae Gill, 1872:5, 61; Herpestini Winge, 1895:47.] [Including Rhinogaleacea Gray, 1864:573; Rhinogalina Gray, 1864; Rhinogalidae Gray, 1869:171; Mungosina Gray, 1864:509; Crossarchina Gray, 1864; Cynictidina Gray, 1864:571; Cynictidae Cope, 1882:474; Suricatinae Thomas, 1882:59, ftn.; Suricatidae Cope, 1882:474; Suricatini Simpson, 1945:117.]
> ?L. Olig., E.-M. Mioc., R.; Eu. E. Mioc., L. Mioc.-R.; Af. L. Mioc., Pleist., R.; As. R.; E. Indies.[1]

[1] Also islands of the Pacific, Indian Ocean, and West Indies by introduction.

Herpestes Illiger, 1811:135, 302.[1] Common mongoose, ichneumon, crab-eating mongoose.
[Including *Ichneumon* Frisch, 1775:11[2]; *Urva* Hodgson, 1837:561[3]; *Mesobema* Hodgson, 1841:214, 413[4]; *Calogale* Gray, 1864:560; *Galerella* Gray, 1864:564; *Taeniogale* Gray, 1864:569; *Onychogale* Gray, 1864:570; *Galogale* Wallace, 1876:195[5]; †*Leptoplesictis* Forsyth Major, 1903:534-537[3]; *Xenogale* Allen, 1919:26; *Myonax* Thomas, 1928:408.[6]]
?L. Olig., E.-M. Mioc., R.; Eu. E. Mioc., E. Plioc.-R.; Af. L. Mioc., Pleist., R.; As. R.; E. Indies.[7]

[1] "Pocock believed the prior name *Mungos* to be a synonym of *Herpestes*, so used the former in this sense and made the corresponding changes in family and subfamily names, but Allen maintains that *Mungos* antedates *Ariela*, not *Herpestes*, and acceptance of this permits the maintenance of the more familiar and elegant *Herpestes*" (Simpson 1945:117).
[2] Not *Ichneumon* Linnaeus, 1758, a genus of Hymenoptera.
[3] As subgenus.
[4] Replacement name for *Urva*.
[5] Misprint for *Calogale*.
[6] Objective synonym of *Galerella*.
[7] Sumatra, Java, Borneo, and Philippines.
Also Pacific (Fiji Islands and Hawaii), Indian Ocean (Mauritius and Reunion), West Indies, and Surinam by introduction.

†*Kichechia* R. J. G. Savage, 1965:296.
E. Mioc.; Af.

Atilax F. Cuvier, 1826:liv. Marsh mongoose.
[= *Athylax* de Blaineville, 1837:272.[1]]
L. Mioc., Pleist., R.; Af.
[1] Unjustified emendation.

Helogale Gray, 1861:308. Dwarf mongooses.
[Including *Dologale* Thomas, 1926:183.]
L. Plioc., R.; Af.

Mungos É. Geoffroy Saint-Hilaire & G. Cuvier, 1795:184, 187.[1] Striped mongoose.
[= *Ariela* Gray, 1864:565.]
L. Plioc., Pleist., R.; Af.

[1] "Allen is followed in taking *Herpestes fasciatus* as the type of *Mungos* and Pocock in maintaining '*Ariela*' as distinct from *Herpestes* or *Crossarchus*. Contrary to Pocock's own opinion (also to Thomas and many who have followed Thomas and Pocock), this makes *Herpestes* valid in the usual sense and *Mungos* also valid but in a different sense. If Thomas were followed in making '*Ariela*'=*Crossarchus*, then '*Crossarchus*' would become *Mungos*" (Simpson 1945:117).

Cynictis Ogilby, 1833:48-49. Yellow mongoose, bushy-tailed meerkat.
[Including *Paracynictis* Pocock, 1916:177.]
Pleist., R.; Af.

Ichneumia I. Geoffroy Saint-Hilaire, 1837:251. White-tailed mongoose.
[= *Lasiopus* I. Geoffroy Saint-Hilaire, 1835:37.[1]]
Pleist., R.; Af. R.; As.[2]
[1] Not *Lasiopus* Dejean, 1833, a genus of Coleoptera.
[2] S.W. Asia (Oman).

Suricata Desmarest, 1804:15. Suricate, gray meerkat, slender-tailed meerkat.
[= *Ryzaena* Illiger, 1811:134; *Rysaena* Lesson, 1827:178; *Rizaena* de Blaineville, 1817:339; *Rhyzaena* Wagner, 1841:330; *Surricata* Gray, 1821:302.]
Pleist., R.; Af.

Crossarchus F. Cuvier, in É. Geoffroy Saint-Hilaire & F. Cuvier, 1825:3. Cusimanses.
R.; Af.

Liberiictis Hayman, 1958:448.
R.; Af.[1]
[1] W. Africa.

Rhynchogale Thomas, 1894:139. Meller's mongoose.
[= *Rhinogale* Gray, 1864:509, 573-5.[1]]
R.; Af.

[1] Not *Rhinogale* Gloger, 1841, a synonym of *Melogale* I. Geoffroy Saint-Hilaire, 1831, a mustelid.

Bdeogale Peters, 1850:81-82. Black-legged mongoose.
>> [= *Beleogale* Marshall, 1873:3.[1]] [Including *Galeriscus* Thomas, 1894:522-524.]
>> R.; Af. R.; As.[2]
>> [1] *Lapsus calami?*
>> [2] S.W. Asia (Yemen).

Subfamily **Galidiinae** Gray, 1864:508, 522. Malagasy mongooses.
>> [= Galidiina Gray, 1864:508, 522; Galidiinae Gill, 1872:61.] [Including Galidictinae Mivart, 1882:143; Galidictida Gregory & Hellman, 1939:320.]
>> R.; Madagascar.

Galidictis I. Geoffroy Saint-Hilaire, 1839:33. Malagasy broad-striped mongooses.
>> [= *Galictis* I. Geoffroy Saint-Hilaire, 1837:581[1]; *Musanga* Coues, 1891:3903.]
>> R.; Madagascar.
>> [1] Preoccupied by *Galictis* Bell, 1826, a mustelid.

Mungotictis Pocock, 1915:120. Malagasy narrow-striped mongoose.
>> R.; Madagascar.

Salanoia Gray, 1864:523. Malagasy brown-tailed mongooses.
>> [= *Hemigalidia* Mivart, 1882:143, 188-9.]
>> R.; Madagascar.

Galidia I. Geoffroy Saint-Hilaire, 1837:251-252. Malagasy ring-tailed mongoose.
>> R.; Madagascar.

Family **Hyaenidae** Gray, 1821:302.
>> [= Hyaenadae Gray, 1821:302; Hyenina Gray, 1825:339; Hyaenina Bonaparte, 1850: unpaged chart; Hyaenida Haeckel, 1866:clix[1]; Hyaenidae Gray, 1869:vi, 211.] [Including Protelidae Flower, 1869:37.]
>> E. Mioc.-R.; Af. E. Mioc.-L. Pleist.; Eu. M. Mioc.-R.; As. L. Plioc.-E. Pleist.; N.A. E.-M. Pleist.; E. Indies.
>> [1] Used as a family.

†*Tongxinictis* Werdelin & Solounias, 1991:71.
>> M. Mioc.; As.

Subfamily †**Ictitheriinae** Trouessart, 1897:320.
>> [= †Ictitherinae Trouessart, 1897:320; †Ictitheriinae Dietrich, 1927:368[1]; †Ictitheriini Kalandadze & Rautian, 1992:70.] [Including †Hyaenotheriini Semenov, 1989:91.]
>> E.-L. Mioc.; Af. E. Mioc.-E. Plioc.; Eu. M. Mioc.-E. Plioc.; As.
>> [1] Emendation.

†*Herpestides* de Beaumont, 1967:81.
>> E. Mioc.; Af. E. Mioc.; Eu.

†*Plioviverrops* Kretzoi, 1938:114.
>> [Including †*Jourdanictis* Viret, 1951:73; †*Protoviverrops* de Beaumont & Mein, 1972:384[1]; †*Mesoviverrops* de Beaumont & Mein, 1972:386.[1]]
>> E. Mioc.-E. Plioc.; Eu. L. Mioc.; As.[2]
>> [1] Subgenus as described.
>> [2] W. Asia (Samos).

†*Ictitherium* Wagner, 1848:375.[1]
>> [= †*Galeotherium* Wagner, 1839:163-165.[2]] [Including †*Lepthyaena* Lydekker, 1884:312, 313; †*Sinictitherium* Kretzoi, 1938:114; †*Paraictitherium* Semenov, 1989:69.[3]]
>> M. Mioc.; Af.[4] L. Mioc.-E. Plioc.; As. L. Mioc.-E. Plioc.; Eu.
>> [1] Replacement name for †*Galeotherium* Wagner, 1839.
>> [2] Not †*Galeotherium* Jäger, 1839, a synonym of †*Ursavus*.
>> [3] Subgenus as defined.
>> [4] N. Africa.

†*Thalassictis* Nordmann, in Gervais, 1850:120.[1][Including †*Palhyaena* Gervais, 1859:242[2]; †*Miohyaena* Kretzoi, 1938:114; †*Hyaenictitherium* Kretzoi, 1938:114; †*Hyaenalopex* Kretzoi, 1952:21.]
>> M.-L. Mioc.; As. L. Mioc.; Af.[3] L. Mioc.; Eu.
>> [1] Gervais, 1848-1852.
>> [2] Assignment to †*Thalassictis* is doubtful.
>> [3] N. Africa.

†*Hyaenotherium* Semenov, 1989:94.
L. Mioc., ?E. Plioc.; As. L. Mioc.; Eu.

†*Miohyaenotherium* Semenov, 1989:129.
L. Mioc.; Eu.[1]

[1] E. Europe.

†*Lycyaena* Hensel, 1863:567.
L. Mioc.; As. L. Mioc.; Eu.

†*Tungurictis* Colbert, 1939:67.
?M. Mioc.; Af. M. Mioc.; As. ?M. Mioc.; Eu.

†*Protictitherium* Kretzoi, 1938:113.[1]
M. Mioc.; Af. M. Mioc.; As. M.-L. Mioc.; Eu.

[1] *Nomen nudum* in Kretzoi, in Kadic & Kretzoi, 1930:48.

Subfamily **Hyaeninae** Gray, 1821:302. Hyaenas.
[= Hyaenadae Gray, 1821:302; Hyenina Gray, 1825:339[1]; Hyaeninae Mivart, 1882:143; Hyaenida Gregory & Hellman, 1939:320; Hyaenini Kalandadze & Rautian, 1992:70.] [Including †Chasmaporthetini Kalandadze & Rautian, 1992:71.]
?M. and/or ?L. Mioc., E. Plioc.-R.; Af. L. Mioc.-R.; As. L. Mioc.-L. Pleist.; Eu. L. Plioc.-E. Pleist.; N.A. E.-M. Pleist.; E. Indies.[2]

[1] As tribe.
[2] Java.

†*Palinhyaena* Qiu, Huang & Guo, 1979:200-221.
L. Mioc.; As.

†*Ikelohyaena* Werdelin & Solounias, 1991:71.
E. Plioc.; Af.

Hyaena Brunnich, 1771:32, 42-43. Striped hyaena.
[= *Hyaena* Brisson, 1762:168[1]; *Euhyaena* Falconer, 1868:464[2]; *Hyena* Gray, 1821:302.] [Including †*Pliohyaena* Kretzoi, 1938:116n[3]; †*Pliocrocuta* Kretzoi, 1938:118; †*Anomalopithecus* Arambourg, 1970:16.]
?M. Mioc., E. Plioc.-R.; Af. ?L. Mioc., L. Plioc.-L. Pleist.; Eu. L. Plioc.-R.; As.

[1] Unavailable.
[2] Proposed as subgenus of *Hyaena*.
[3] Proposed as subgenus.

†*Hyaenictis* Gaudry, 1861:723-724.
?L. Mioc.; As. L. Mioc.; Eu. E. Plioc., ?E. Pleist.; Af.

†*Leecyaena* Young & Liu, 1948:273-291.[1]
L. Mioc. and/or E. Plioc.; As.

[1] Not Young & Peiho as sometimes reported.

†*Chasmaporthetes* Hay, 1921:636.
[= †*Ailuraena* Stirton & Christian, 1940:445-448.] [Including †*Lycyaenops* Kretzoi, 1938:115[1]; †*Euryboas* Schaub, 1942:279-286.]
L. Mioc.-E. Pleist.; As. L. Mioc.-E. Pleist.; Eu. E. Plioc., ?L. Plioc. and/or ?E. Pleist.; Af. L. Plioc.-E. Pleist.; N.A.

[1] Assignment to †*Chasmaporthetes* is doubtful.

Pachycrocuta Kretzoi, 1938:118. Brown hyaena.
[Including *Parahyaena* Hendey, 1974:149.[1]]
?L. Mioc., E. Plioc.-R.; Af. L. Plioc.-M. Pleist.; As. L. Plioc.-M. Pleist.; Eu. E.-M. Pleist.; E. Indies.[2]

[1] Proposed as subgenus.
[2] Java.

†*Adcrocuta* Kretzoi, 1938:118.
L. Mioc.; As. L. Mioc.; Eu.

Crocuta Kaup, 1828:1145. Spotted hyaena.
[= *Crocotta* Kaup, 1829:74-78.] [Including †*Eucrocuta* Viret, 1954:51.]

L. Plioc.-R.; Af. L. Plioc., Pleist., R.; As.[1] E. and/or M. Pleist.; E. Indies.[2] M.-L. Pleist.; Eu.

[1] Extinct.
[2] Java.

Subfamily †**Percrocutinae, new.**[1]

M. Mioc.-L. Plioc.; Af. M.-L. Mioc.; As. M.-L. Mioc.; Eu.

[1] Tedford. Formalization of "percrocutoid" hyaenids of Howell and G. Petter, 1985:421-422.

†*Percrocuta* Kretzoi, 1938:117.

[Including †*Capsatherium* Kurtén, 1978:187.]
M. Mioc.-L. Plioc.; Af. M.-L. Mioc.; As. L. Mioc.; Eu.

†*Dinocrocuta* Schmidt-Kittler, 1976:58.[1]

M. Mioc.; Af.[2] M.-L. Mioc.; As.

[1] Proposed as subgenus of †*Percrocuta*.
[2] N. Africa.

†*Belbus* Werdelin & Solounias, 1991:72.

L. Mioc.; As.[1]

[1] Samos Island.

†*Allohyaena* Kretzoi, 1938:116.[1]

[= †*Xenohyaena* Kretzoi, 1938:117.]
M. Mioc.; As.[2] M.-L. Mioc.; Eu.[3]

[1] *Nomen nudum* in Kretzoi, in Kadic & Kretzoi, 1930:48.
[2] W. Asia.
[3] E. Europe.

Subfamily **Protelinae** I. Geoffroy Saint-Hilaire, 1851:xiv.

[= Protelina I. Geoffroy Saint-Hilaire, 1851:xiv; Protelidae Flower, 1869:37; Proteleidae Gray, 1869:213; Protelinae Mivart, 1882:43; Protelida Gregory & Hellman, 1939:320.]
Pleist., R.; Af.

Proteles I. Geoffroy Saint-Hilaire, 1824:355-371. Aardwolf.

[= *Geocyon* Wagler, 1830:30.]
Pleist., R.; Af.

Family **Nandiniidae** Pocock, 1929:898. African palm civets.

[= Nandiniinae Gregory & Hellman, 1939:320; Nandiniini Simpson, 1945:116.]
R.; Af.

Nandinia Gray, 1843:54. African palm civet, tree civet.

R.; Af.

Suborder **CANIFORMIA** Kretzoi, 1943:194.

[= HYPOMYCTERI Cope, 1882:473.]

Family †**Miacidae** Cope, 1880:78.

[= †Miacinae Trouessart, 1885:18; †Miacoidae Teilhard de Chardin, 1915:188; †Miacoidea Simpson, 1931:263, 276; †Miacini Kalandadze & Rautian, 1992:62; †Miacida Haeckel, 1895:578, 583; †EUCREODI Matthew, 1909:327.] [Including †Uintacyonidae Hay, 1902:759.]
L. Paleoc.-L. Eoc.; Eu. L. Paleoc.-L. Eoc.; N.A. E.-L. Eoc.; As.

†*Miacis* Cope, 1872:2.

[Including †*Harpalodon* Marsh, 1872:25 (216?); †*Neovulpavus* Wortman, 1901:445; †*Mimocyon* Peterson, 1919:48; †*Lycarion* Matthew, 1909:343.[1]]
L. Paleoc.-L. Eoc.; Eu. E.-L. Eoc.; N.A. M.-L. Eoc.; As.

[1] Proposed as subgenus.

†*Chailicyon* Chow, 1975:167.

M.-L. Eoc.; As.

†*Vulpavus* Marsh, 1871:124.

[Including †*Phlaodectes* Matthew, 1909:380, 391.[1]]

?E. Eoc.; Eu. E.-M. Eoc.; N.A.

[1] Proposed as subgenus.

†*Uintacyon* Leidy, 1872:277.
[= †*Triacodon* Cope, 1872:469.[1]] [Including †*Miocyon* Matthew, 1909:349, 353.[2]]
L. Paleoc.-M. Eoc.; N.A. ?E. Eoc.; Eu.

[1] Synonymy questionable. Not †*Triacodon* Marsh, 1871, a synonym of †*Viverravus*.
[2] Proposed as subgenus. Raised to generic rank by Zittel, Schlosser & Woodward, 1925 (possibly earlier) and again by H. Bryant, 1992:848.

†*Vassacyon* Matthew, 1909:93.
E. Eoc.; N.A.

†*Oodectes* Wortman, 1901:148-154.
[Including †*Paeneprolimnocyon* Guthrie, 1967:1285.]
E.-M. Eoc.; N.A.

†*Xinyuictis* Zhen, Tung & Chi, 1975:99.
E. Eoc.; As.

†*Ziphacodon* Marsh, 1872:216.[1]
[= †*Xiphacodon* Schlosser, 1890:450.]
M. Eoc.; N.A.

[1] Validity and assignment to the †Miacidae doubtful.

†*Paramiacis* Mathis, 1985:305, 313.
M. Eoc.; Eu.

†*Paroodectes* Springhorn, 1980:171-198.
M. Eoc.; Eu.

†*Prodaphaenus* Wortman & Matthew, 1899:114.[1]
M. Eoc.; N.A.

[1] *Nomen nudum* in Matthew, 1899:20, 49.

†*Palaearctonyx* Matthew, 1909:399.
M. Eoc.; N.A.

Infraorder **CYNOIDEA** Flower, 1869:24.

Family **Canidae** Fischer de Waldheim, 1817:372.
[= Canini Fischer de Waldheim, 1817:372; Canidae Gray, 1821:301; Canina Gray, 1825:339; Canina Haeckel, 1866:clix[1]; Canoidea Simpson, 1931:276.] [Including Megalotidae Gray, 1869:210; Lycaonidae Rochebrune, 1883:133; Vulpidae Rochebrune, 1883:93, 154; Otocyonidae Trouessart, 1885:51.]
M. Eoc.-R.; N.A.[2] L. Mioc.-R.; Eu. L. Mioc., R.; Cent. A. Plioc., E. Pleist.-R.; Af. E. Plioc.-R.; As. L. Plioc., ?E. Pleist., M. Pleist.-R.; S.A. E. and/or M. Pleist., L. Pleist.-R.; E. Indies. L. Pleist.; Mediterranean.[3]

[1] Used as a family.
[2] Including Greenland.
[3] Also Recent worldwide in domestication.

†*Procynodictis* Wortman & Matthew, 1899:121-122.[1]
M. Eoc.; N.A.

[1] *Nomen nudum* in Matthew, 1899:49.

†*Prohesperocyon* Wang, 1994:18.
L. Eoc.; N.A.

Subfamily †**Hesperocyoninae** L. D. Martin, 1989:536-568.
M. Eoc.-M. Mioc.; N.A.

†*Hesperocyon* Scott, 1890:38.
[= †*Pseudocynodictis* Schlosser, 1902:50.]
M. Eoc.-L. Olig.; N.A.

†*Paraenhydrocyon* Wang, 1994:128.
E. Olig.-E. Mioc.; N.A.

†*Caedocyon* Wang, 1994:143.
L. Olig.; N.A.

†*Ectopocynus* Wang, 1994:147.
   E. Olig.-E. Mioc.; N.A.

†*Osbornodon* Wang, 1994:106.
   E. Olig., E.-M. Mioc.; N.A.

†*Mesocyon* Scott, 1890:38.
   [= †*Hypotemnodon* Eyerman, 1894:321.]
   E. Olig.-E. Mioc.; N.A.

†*Cynodesmus* Scott, 1893:659, 660.
   E.-L. Olig.; N.A.

†*Sunkahetanka* Macdonald, 1963:214.
   L. Olig.; N.A.

†*Philotrox* Merriam, 1906:30.
   L. Olig.; N.A.

†*Enhydrocyon* Cope, 1879:56.
   [Including †*Hyaenocyon* Cope, 1879:372.]
   L. Olig.-E. Mioc.; N.A.

Subfamily †**Borophaginae** Simpson, 1945:111.
   [Including †Cynarctinae McGrew, 1937:444.[1]]
   E. Olig.-L. Plioc.; N.A. L. Mioc.; Cent. A.

[1] "Cynarctidae" C. H. Smith, in Jardine, 1842, used for certain viverrids, was not based upon a generic name and has no standing. It cannot preoccupy †Borophaginae or †Cynarctinae McGrew, 1937.

†*Oxetocyon* Green, 1954:218.
   E. Olig.; N.A.

†*Cormocyon* Wang & Tedford, 1992:225.
   L. Olig.-M. Mioc.; N.A.

†*Euoplocyon* Matthew, 1924:103.
   E.-M. Mioc.; N.A.

†*Phlaocyon* Matthew, 1899:54.
   E. Mioc.; N.A.

†*Tomarctus* Cope, 1873:2.
   [= †*Tomarctos* Harksen & Green, 1971.[1]] [Including †*Tephrocyon* Merriam, 1906:6.]
   E.-L. Mioc.; N.A.

[1] Table 1 (a misspelling).

†*Aletocyon* Romer & Sutton, 1927:459-464.
   E. Mioc.; N.A.

†*Bassariscops* Peterson, 1928:96.
   E. Mioc.; N.A.

†*Cynarctoides* McGrew, 1938:324.
   E.-M. Mioc.; N.A.

†*Strobodon* Webb, 1969:43.
   M. Mioc.; N.A.

†*Aelurodon* Leidy, 1858:22.
   [Including †*Prohyaena* Schlosser, 1887:139.]
   M.-L. Mioc.; N.A.

†*Carpocyon* Webb, 1969:275.
   M.-L. Mioc.; N.A.

†*Epicyon* Leidy, 1858:21-22.
   M.-L. Mioc.; N.A.

†*Cynarctus* Matthew, 1902:281-284.
   M.-L. Mioc.; N.A.

†*Osteoborus* Stirton & Vanderhoof, 1933:175.
   L. Mioc.; N.A. L. Mioc.; Cent. A.

†*Borophagus* Cope, 1892:1028.
    [Including †*Hyaenognathus* Merriam, 1903:278; †*Porthocyon* Merriam, 1903:283;
    †*Pliogulo* White, 1941:67[1]; †*Cynogulo* Kretzoi, 1968:164.[2]]
    E.-L. Plioc.; N.A.
[1] Preoccupied by †*Pliogulo* Wosnessensky, 1937, a mustelid.
[2] Replacement for †*Pliogulo* White, 1941.

Subfamily **Caninae** Fischer de Waldheim, 1817:372. Dogs, wolves, foxes.
    [= Canini Fischer de Waldheim, 1817:372; Canina Gray, 1825:339[1]; Caninae Gill,
    1872:63; Lupini Hemprich & Ehrenberg, 1832.] [= or including Lupinae Gray,
    1869:179; Lupida Haeckel, 1895.] [Including Vulpini Hemprich & Ehrenberg, 1832:
    sig. ff; Vulpinae Baird, 1857:121; Vulpidae Rochebrune, 1883:93, 154; Vulpida
    Haeckel, 1895:585; Lycaonina Gray, 1868:495; Lycaonidae Rochebrune, 1883;
    Lycaonini Baryshnikov & Averianov, 1993:171; Megalotina Gray, 1868:523;
    Megalotidae Gray, 1869:210; Otocyonidae Trouessart, 1885; Otocyonini Baryshnikov
    & Averianov, 1993:165; Speothoini Pockock, 1914; Cuoninae Miller, 1924:155;
    Cuonini Baryshnikov & Averianov, 1993:169; Nyctereutini Baryshnikov & Averianov,
    1993:166.]
    E. Mioc.-R.; N.A.[2] L. Mioc.-R.; Eu. Plioc., E. Pleist.-R.; Af. E. Plioc.-R.; As. L.
    Plioc., ?E. Pleist., M. Pleist.-R.; S.A. E. and/or M. Pleist., L. Pleist.-R.; E. Indies.
    L. Pleist.; Mediterranean. R.; Cent. A.[3]
[1] As tribe.
[2] Including Greenland.
[3] Also Recent worldwide in domestication.

†*Leptocyon* Matthew, 1918:189.
    [Including †*Neocynodesmus* Macdonald, 1963:212.]
    E. Mioc., L. Mioc.; N.A.

Tribe **Vulpini** Hemprich & Ehrenberg, 1832: sig. ff.
    [Including Megalotina Gray, 1869:210; Otocyonini Baryshnikov & Averianov,
    1993:165.]
    E. Plioc.-R.; N.A.[1] L. Plioc.-R.; As. L. Plioc.-R.; Eu. E. Pleist.-R.; Af. R.; Cent.
    A. R.; S.A.
[1] Including Greenland.

Vulpes Frisch, 1775.[1] Red fox, kit fox, arctic fox, corsac, fennec.
    [= *Vulpis* Gray, 1821:301; *Mamvulpesus* Herrera, 1899:30.[2]] [Including *Fennecus*
    Desmarest, 1804:18; *Megalotis* Illiger, 1811:131[3]; *Alopex* Kaup, 1829:83, 85;
    Cynalopex C. Hamilton Smith, 1839:222-232; *Leucocyon* Gray, 1868:521;
    †*Xenalopex* Kretzoi, 1954:248.]
    Plioc., E. Pleist.-R.; N.A.[4] L. Plioc.-R.; As.[5] L. Plioc.-R.; Eu. E. Pleist.-R.; Af.
[1] Unpaged table. Conserved under plenary powers by the International Commission on Zoological
  Nomenclature, 1979.
[2] See the International Code of Zoological Nomenclature, Article 1(b)(8), and Palmer (1904:25-26).
[3] Objective synonym of *Fennecus*.
[4] Including Arctic islands of Canada and Greenland.
[5] Including Arctic islands of Siberia.

Urocyon Baird, 1857:138-145. Gray fox.
    E. Plioc.-R.; N.A. R.; Cent. A. R.; S.A.[1]
[1] Northern South America.

†*Prototocyon* Pohle, 1928:18-19.
    [Including †*Sivacyon* Pilgrim, 1932:32.]
    E. Pleist.; Af. E. Pleist.; As.

Otocyon Müller, 1836:1. Bat-eared fox, bakoorjakkal, motlosi.
    [= *Agrodius* H. Smith, 1840:258-261.]
    M. Pleist.-R.; Af.

Tribe **Canini** Fischer de Waldheim, 1817:372.
    [= or including Lupini Hemprich & Ehrenberg, 1832.] [Including Lycaonina Gray,
    1868:495; Lycaonini Baryshnikov & Averianov, 1993:171; Speothoini Pocock, 1914;

Nyctereutini Baryshnikov & Averianov, 1993:166; Cuonini Baryshnikov & Averianov, 1993:169.]

M. Mioc.-E. Plioc., ?L. Plioc., E. Pleist.-R.; N.A. L. Mioc.-R.; Eu. Plioc., E. Pleist.-R.; Af. E. Plioc.-R.; As. L. Plioc., ?E. Pleist., M. Pleist.-R.; S.A. E. and/or M. Pleist., L. Pleist.-R.; E. Indies.[1] L. Pleist.; Mediterranean. R.; Cent. A.[2]

[1] Sumatra and Java.
[2] Also Recent worldwide in domestication.

Pseudalopex Burmeister, 1856:24, 44-54. Culpeo, South American fox.
   [= *Angusticeps* Hilzheimer, 1906:114; *Microcyon* Trouessart, 1906:1186.]
   M. Pleist.-R.; S.A.

†*Dusicyon* C. Hamilton Smith, 1839:248.[1] Falkland Island wolf.
   [= †*Dysicyon* Agassiz, 1846:132; †*Dusocyon* Bourguinat, 1875:24, 29; †*Dasicyon* Trouessart, 1897:299.]
   L. Pleist.-R.; S.A.[2]

[1] Described as subgenus of *Chaon*, a synonym of *Canis*.
[2] Falkland Islands in the Recent.

Lycalopex Burmeister, 1854:95-101.[1] South American field fox.
   [= *Thous* Gray, 1868:514; *Nothocyon* Wortman & Matthew, 1899:124[2]; *Eunothocyon* J. A. Allen, 1905:152.]
   Pleist., R.; S.A.

[1] Described as a subgenus of *Canis*.
[2] Not †*Nothocyon* Matthew, 1899, a borophagine.

Chrysocyon C. Hamilton Smith, 1839:241-247. Guará, aguará-guazú, maned wolf.
   Plioc.; N.A. R.; S.A.

Cerdocyon C. Hamilton Smith, 1839:259-267.[1] Maikong, crab-eating fox, bush fox, zorro de monte.
   [= *Carcinocyon* J. A. Allen, 1905:153.]
   L. Mioc.-E. Plioc.; N.A. L. Pleist.-R.; S.A.

[1] Described as subgenus of *Chaon*, a synonym of *Canis*.

Nyctereutes Temminck, 1838:285.[1] Raccoon dog, hao, tanuki.
   [Including †*Ruscinalopex* Kretzoi, 1938:127; †*Paratanuki* Kretzoi, 1956:188, 262.]
   ?L. Mioc., E. Plioc.-E. Pleist.; Eu.[2] Plioc., E. Pleist.; Af. E. Plioc.-R.; As.

[1] 1838-1839.
[2] Also Recent of E. Europe by introduction.

Atelocynus Cabrera, 1940:14. Small-eared dog, zorro negro.
   R.; S.A.

Speothos Lund, 1839:223-224.[1] Bush dog.
   [= *Cynogale* Lund, 1842:201-203[2]; *Icticyon* Lund, 1843:80; *Abathmodon* Lund, 1843:80[3]; *Cynalicus* Gray, 1846:293; *Cynalius* Gray, 1847:18[4]; *Melictis* Schinz, 1848:176-178; *Cynalycus* Gray, 1869:183.]
   L. Pleist.-R.; S.A.

[1] "The genus was first discovered fossil, and its living, but not generically distinct, representatives were differently named before the relationship was recognized. Many authors continue to use *Icticyon* for the living forms" (Simpson 1945:110).
[2] Not *Cynogale* Gray, 1837, a Sumatran viverrid.
[3] No species mentioned. Not *Abathmodon* Wagner, 1830.
[4] Misprint.

†*Protocyon* Giebel, 1855:851.
   [= †*Palaeocyon* Lund, 1843:79, 79.[1]] [Including †*Palaeospeothos* Spillmann, 1941.[2]]
   L. Plioc., Pleist.; S.A.

[1] Not †*Palaeocyon* de Blainville, 1841, a synonym of †*Arctocyon*.
[2] *Nomen nudum*.

†*Theriodictis* Mercerat, 1891:55-56.
   [= †*Dinocynops* Ameghino, 1898:194.] [Including †*Pleurocyon* Mercerat, 1917:13.[1]]
   M.-L. Pleist.; S.A.

[1] Not †*Pleurocyon* Peterson, 1919, a synonym of †*Simidectes* (†PANTOLESTA).

†*Eucyon* Tedford & Qiu, 1996:27, 36.
    M.-L. Mioc.; N.A. L. Mioc.-E. Plioc.; Eu. E. Plioc.; As.

Canis Linnaeus, 1758:38. Wolf, domestic dog, dingo, coyote, jackals, dire wolf.
    [= *Neocyon* Gray, 1868:506-508; *Mamcanisus* Herrera, 1899:11[1];
    †*Cubacyon* Arredondo & Varona, 1974:1-12; †*Indocyon* Arredondo, 1981:4;
    †*Paracyon* Arredondo, 1981:5.[2]] [= or including *Alopsis* Rafinesque, 1815:59[3];
    Lupus Oken, 1816:1039[4]; *Chaon* C. Hamilton Smith, 1839:129-267.[5]]
    [Including *Thos* Oken, 1816:1037-1039; *Vulpicanis* de Blainville, 1837:279;
    Lyciscus C. Hamilton Smith, 1839:160; *Sacalius* C. Hamilton Smith, 1839:206;
    Oxygous Hodgson, 1841:213; *Lupulus* Gervais, 1855:60[6]; †*Cynotherium* Studiati,
    1857:651[7]; *Simenia* Gray, 1868:494, 506; *Dieba* Gray, 1869:180; †*Macrocyon*
    Ameghino, 1881:306; *Lupulella* Hilzheimer, 1906:363; *Schaeffia* Hilzheimer,
    1906:364; *Alopedon* Hilzheimer, 1906:365; †*Stereocyon* Mercerat, 1917:17;
    †*Aenocyon* Merriam, 1918:532[4]; †*Megacyon* von Koenigswald, 1940:57;
    Oreocyon Krumbiegel, 1949:591[8]; *Dasycyon* Krumbiegel, 1953:102.[9]]
    L. Mioc., L. Plioc.-R.; Eu. L. Mioc.-E. Plioc., E. Pleist.-R.; N.A.
    L. Plioc.-R.; As. E. Pleist.-R.; Af. E. and/or M. Pleist.; E. Indies.[10]
    L. Pleist.; Mediterranean.[11] R.; Cent. A.[12]

[1] See the International Code of Zoological Nomenclature, Article 1(b)(8), and Palmer (1904:25-26).
[2] Not †*Paracyon* Gray, 1827, a synonym of the marsupial †*Thylacinus*.
[3] *Nomen nudum.*
[4] Proposed as subgenus.
[5] Described as a subgenus of *Canis*.
[6] Not *Lupulus* de Blainville, 1843:30-32, a *nomen nudum*.
[7] Not †*Cyotherium* Aymard, 1850, a synonym of †*Cynodictis*, from the French late Eocene.
[8] Not †*Oreocyon* Marsh, 1872, an oxyaenine creodont.
[9] Replacement name for *Oreocyon* Krumbiegel, 1949.
[10] Java.
[11] Sardinia, Sicily.
[12] Also Recent worldwide in domestication.

Cuon Hodgson, 1838:152. Dhole, red dog.
    [= *Chrysaeus* H. Smith, 1839:167; *Primoevus* Hodgson, 1842:39; *Primaevus* Gray,
    1843:xx; *Cyon* Agassiz, 1846:113[1]; *Anurocyon* Heude, 1888:102.] [Including
    †*Mececyon* Stremme, 1911:83; †*Xenocyon* Kretzoi, 1938:93; †*Crassicuon* Kretzoi,
    1941:118; †*Sinicuon* Kretzoi, 1941:118; †*Semicuon* Kretzoi, 1941:119.]
    L. Plioc.-R.; As. E. and/or M. Pleist., L. Pleist.-R.; E. Indies.[2] E.-L. Pleist.; Eu.
    E.-L. Pleist.; N.A.

[1] And most later authors.
[2] Java, Sumatra.

Lycaon Brookes, 1827:151. African hunting dog.
    [= *Cynhyaena* F. Cuvier, 1829:454; *Kynos* Rüppell, 1842:163;
    Hyenoides Boitard, 1842:163-164; *Hyaenoides* Gervais, 1855:53.[1]]
    M. Pleist.-R.; Af.

[1] Misspelling of *Hyenoides*.

Infraorder **ARCTOIDEA** Flower, 1869:15.
    [= ARCTOIFORMIA Ginsburg, 1982:247, 257.[1]] [Including ARCTOMORPHA
    Wolsan, 1993:363[2]; PROCYONIA Kalandadze & Rautian, 1992:62.[3]]

[1] Named as synonym at rank of suborder.
[2] Proposed at an unnamed rank between infraorder and parvorder to include MUSTELIDA and URSIDA,
   but not †Amphicyonidae.
[3] As suborder.

Parvorder **URSIDA** Tedford, 1976:372.[1]
    [= PLANTIGRADES Owen, 1859:45.[2]]

[1] *Ursida* Haeckel, 1895 was at family rank.
[2] *Nomen oblitum.*

†*Adracon* Filhol, 1884:19-21.
    L. Eoc. and/or Olig.; Eu.

Superfamily †**Amphicyonoidea** Haeckel, 1866:clix, **new rank**.
    [= †Amphicyonida Haeckel, 1866:clix; †Amphicyoninae Trouessart, 1885:6.]

Family †**Amphicyonidae** Haeckel, 1866:clix.
> [= †Amphicyonida Haeckel, 1866:clix; †Amphicyonidae Trouessart, 1885:57.]
> M. Eoc.-L. Mioc.; N.A. L. Eoc.-L. Mioc.; Eu. E.-L. Mioc.; Af. E.-L. Mioc.; As.

†*Symplectocyon* Springhorn, 1977:27, 102.
> E. Olig.; Eu.

†*Harpagocyon* Springhorn, 1977:27, 104.
> L. Olig.; Eu.

†*Hubacyon* Kretzoi, 1985:66.
> [Including †*Kanicyon* Kretzoi, 1985:66.[1]]
> L. Mioc.; Eu.
> [1] Proposed as subgenus.

Subfamily †**Daphoeninae** Hough, 1948:576.
> [= †Daphoenidae Hough, 1948:576; †Daphoeninae Hunt, 1974:1031, 1032;
> †Daphoenini Ginsburg, 1977:57, 101.] [Including †Cynodictida Haeckel, 1895:490,
> 578, 584-585[1]; †Haplocyoninae de Bonis, 1966:116.]
> M. Eoc.-E. Mioc.; N.A. L. Eoc.-E. Mioc.; Eu.
> [1] *Nomen oblitum* (or a candidate for suppression). Proposed as a family.

†*Daphoenus* Leidy, 1853:393.[1]
> [= †*Daphaenos* Roger, 1896:44.] [Including †*Proamphicyon* Hatcher, 1902:95-99,
> 105; †*Protemnocyon* Hatcher, 1902:99-104, 105; †*Pericyon* Thorpe, 1922:162, 172;
> †*Amphicynops* Kretzoi, 1941:5.]
> M. Eoc.-E. Mioc.; N.A.
> [1] Apparently originally spelled "†*Daphaenus*", but "ae" and "oe" were often represented by identical
> type in the 1850s. Hay (1902:772) claimed that the original was "†*Daphoenus*", and most authors have
> followed him, as do we for the sake of stability, not accuracy.

†*Cynodictis* Bravard & Pomel, 1850:5.
> [Including †*Cyotherium* Aymard, 1850:113; †*Anictis* Kretzoi, 1945:79, 81.[1]]
> L. Eoc.-E. Olig.; Eu.
> [1] *Nomen nudum?*

†*Daphoenictis* Hunt, 1974:1032.
> L. Eoc.; N.A.

†*Daphoenocyon* Hough, 1948:594.
> L. Eoc.; N.A.

†*Brachyrhynchocyon* Loomis, in Scott & Jepsen, 1936:80.
> [= †*Brachicyon* Loomis, 1931:101.[1]]
> E. Olig.; N.A.
> [1] Not †*Brachycyon* Filhol, 1872, an amphicyonine carnivoran from the Phosphorites of Quercy.

†*Paradaphaenus* Matthew, 1899:62.
> L. Olig.; N.A.

†*Temnocyon* Cope, 1878:6-8.
> [Including †*Haplocyon* Schlosser, 1901:7n; †*Mammacyon* Loomis, 1936:44;
> †*Haplocyonoides* Hürzeler, 1940:224-229; †*Parhaplocyon* de Bonis, 1966:116[1];
> †*Haplocyonopsis* de Bonis, 1973:92.]
> L. Olig.-E. Mioc.; Eu. L. Olig.-E. Mioc.; N.A.
> [1] Proposed as subgenus of †*Haplocyon*.

Subfamily †**Amphicyoninae** Haeckel, 1866:clix.
> [= †Amphicyonida Haeckel, 1866:clix; †Amphicyoninae Trouessart, 1885:6.]
> [Including †Thaumastocyoninae Hürzeler, 1940:230.]
> L. Eoc.-L. Mioc.; Eu. E.-L. Mioc.; Af. E.-L. Mioc.; As. E.-L. Mioc.; N.A.

†*Pseudamphicyon* Schlosser, in Roger, 1887:128.
> L. Eoc.-E. Olig.; Eu.

†*Brachycyon* Filhol, 1872:15-18.
> Olig.; Eu.

†*Amphicyanis* Springhorn, 1977:26, 87.
> Olig.; Eu.

†*Sarcocyon* Ginsburg, 1966:23-44.
    Olig.; Eu.

†*Harpagophagus* de Bonis, 1971:118.
    Olig.; Eu.

†*Cynelos* Jourdan, 1848:14.[1]
    [Including †*Hecubides* R. J. G. Savage, 1965:289.]
    E. Olig.-M. Mioc.; Eu. E. Mioc.; Af. E.-M. Mioc.; N.A.
    [1] 1848-1852.

†*Goupilictis* Ginsburg, 1969:72.
    L. Olig.; Eu.

†*Ysengrinia* Ginsburg, 1965:727.
    L. Olig.; Eu. E. Mioc.; N.A.

†*Amphicyon* Lartet, 1836:219-220.
    [= †*Progenetta* Depéret, 1892:34-35.]
    L. Olig.-L. Mioc.; Eu. E.-L. Mioc.; As. E.-L. Mioc.; N.A.

†*Pseudocyonopsis* Kuss, 1965:1-168.
    L. Olig.; Eu.

†*Pseudocyon* Lartet, 1851:16.
    [Including †*Thaumastocyon* Stehlin & Helbing, 1925:110;
    †*Amphicyonopsis* Viret, 1929:225.]
    L. Olig.-L. Mioc.; Eu. M. Mioc.; N.A.

†*Arctamphicyon* Pilgrim, 1932:24.
    L. Olig.-L. Mioc.; Eu. M.-L. Mioc.; As.

†*Ictiocyon* Crusafont Pairó, Villalta Comella & Truyols Santonja, 1955:1-272.
    E. Mioc.; Eu.

†*Borocyon* Peterson, 1910:263.
    E. Mioc.; N.A.

†*Pseudarctos* Schlosser, 1899:117-121.
    [Including †*Absonodaphoenus* S. J. Olsen, 1958:599; †*Daphaenus* White, 1942:5.[1]]
    E. Mioc.; N.A. M.-L. Mioc.; Eu.
    [1] Variant of †*Daphoenus* for †*D. caroniavorus*.

†*Daphoenodon* Peterson, 1909:620.
    E. Mioc.; N.A.

†*Megamphicyon* Kuss, 1965:1-168.
    E. Mioc.; Eu.

†*Afrocyon* Arambourg, 1961:107-109.
    E. Mioc.; Af.[1]
    [1] N. Africa.

†*Pliocyon* Matthew, 1918:190.
    M. Mioc.; N.A.

†*Ischyrocyon* Matthew, in Matthew & Gidley, 1904:246.
    M.-L. Mioc.; N.A.

†*Hadrocyon* Stock & Furlong, 1926:45.
    M. Mioc.; N.A.

†*Agnotherium* Kaup, 1832:28-30.
    [= †*Agnocyon* Kaup, 1862:16.] [Including †*Tomocyon* Viret, 1929:217.]
    M.-L. Mioc.; Af. M. Mioc.; As. M.-L. Mioc.; Eu.

†*Myacyon* Sudre & Hartenberger, 1991:554.
    L. Mioc.; Af.[1]
    [1] N. Africa.

†*Gobicyon* Colbert, 1939:60.

M. Mioc.; As. M. Mioc.; Eu.

Superfamily **Ursoidea** Fischer de Waldheim, 1817:372.

[= Ursini Fischer de Waldheim, 1817:372; Ursoidea Tedford, 1976:372.[1]]

[1] Redundantly renamed by Wyss & J. J. Flynn, in Szalay, Novacek & McKenna, eds., 1993, vol. 2:38.

Family **Ursidae** Fischer de Waldheim, 1817:372. Bears.

[= Ursini Fischer de Waldheim, 1817:372; Ursinidae Gray, 1821:301; Ursidae Gray, 1825:339; Ursina Haeckel, 1866:clix[1]; Ursida Haeckel, 1895:578, 585.] [Including Ailuropodidae Pocock, 1916:19; †Agriotheriidae Kretzoi, 1929; †Ursavinae Kretzoi, 1945:75, 76; †Ursavinae Hendey, 1980:99.]

E.-L. Mioc., L. Plioc.-R.; As. E. Mioc.-R.; Eu. E. Mioc.-R.; N.A. L. Mioc., M. Pleist.-R.; Af.[2] E. Pleist.-R.; S.A. L. Pleist.-R.; E. Indies. L. Pleist.; Mediterranean. R.; Arctic O. ?R.; Cent. A.

[1] Used as a family.
[2] N. Africa.

†*Ursavus* Schlosser, 1899:103, 108.

[Including †*Galeotherium* Jäger, 1839:71, 200[1]; †*Frickodon* Kretzoi, 1945:76.[2]]

E.-L. Mioc.; As. E.-L. Mioc.; Eu. E.-M. Mioc.; N.A.

[1] Not †*Galeotherium* Wagner, 1839, a synonym of †*Ictitherium*.
[2] Subjective synonymy questionable.

†*Indarctos* Pilgrim, 1913:290.

L. Mioc.; Af.[1] L. Mioc.; As. L. Mioc.; Eu. L. Mioc.; N.A.

[1] N. Africa.

Subfamily **Tremarctinae** Merriam & Stock, 1925:3.

[= Tremarctini Hendey, 1972:119.] [Including †Arctotherinae Ameghino, 1903:159; †Arctodontinae Kretzoi, 1929:1350.]

L. Mioc.-E. Plioc., E.-L. Pleist.; N.A. E. Pleist.-R.; S.A. ?R.; Cent. A.

†*Plionarctos* Frick, 1926:114.

L. Mioc.-E. Plioc.; N.A.

†*Arctodus* Leidy, 1854:90. Short-faced bear.

[Including †*Arctotherium* Bravard, 1860:45-94[1]; †*Arctoidotherium* Lydekker, 1885:157[2]; †*Proarctotherium* Ameghino, 1904:121; †*Dinarctotherium* Barbour, 1916:349-353; †*Tremarctotherium* Kraglievich, 1926:8, 13; †*Megalarctos* L. Kraglievich & C. Ameghino, 1940:584.]

E.-L. Pleist.; N.A. E.-L. Pleist.; S.A.

[1] Or perhaps 1857, as often claimed.
[2] From a manuscript by Bravard, fide Palmer (1904:118).

†*Pararctotherium* Ameghino, 1904:120.

[Including †*Pseudarctotherium* Kraglievich, 1928:25-66.[1]]

L. Pleist.; S.A.

[1] Proposed as subgenus of †*Arctotherium* Bravard.

Tremarctos Gervais, 1855:20-21. Spectacled bear.

Pleist.; N.A. ?R.; Cent. A.[1] R.; S.A.

[1] Possibly Panama.

Subfamily **Ursinae** Fischer de Waldheim, 1817:372.

[= Ursini Fischer de Waldheim, 1817:372; Ursina Gray, 1825:339[1]; Ursinae Burmeister, 1866:144; Arctinae Merriam & Stock, 1925:3.[2]]

E. Plioc.-R.; Eu. E. Plioc.-R.; N.A. L. Plioc.-R.; As. M. Pleist.-R.; Af.[3] L. Pleist.-R.; E. Indies. L. Pleist.; Mediterranean. R.; Arctic O.

[1] As tribe.
[2] Not based upon a generic name.
[3] N. Africa.

Ursus Linnaeus, 1758:47. Brown bear, grizzly bear, black bears, Himalayan bear.

[= *Danis* Gray, 1825:60[1]; *Myrmarctos* Gray, 1864:694; *Ursarctos* Heude, 1898:17; *Melanarctos* Heude, 1898:18; *Mamursus* Herrera, 1899:20[2]; *Mylarctos* Lonnberg,

1923:91.] [Including †*Drepanodon* Nesti, 1826:6; *Euarctos* Gray, 1864:692-694; *Selenarctos* Heude, 1901:2[3]; *Arcticonus* Pocock, 1917:129; *Vetularctos* Merriam, 1918:131; †*Protarctos* Kretzoi, 1945:76; *Ursulus* Kretzoi, 1954:239-265; †*Spelaearctos* Kuzmina, 1982:46.]

E. Plioc.-R.; Eu. E. Plioc.-R.; N.A. L. Plioc.-R.; As. M. Pleist.-R.; Af.[4] L. Pleist.; Mediterranean.[5]

[1] Not *Danis* Fabricius, 1808, a lepidopteran.
[2] See the Internnational Code of Zoological Nomenclature, Article 1(b)(8), and Palmer (1904:25-26).
[3] Sometimes as subgenus.
[4] N. Africa.
[5] Sicily.

Melursus Meyer, 1793:155-160. Sloth bear.
[= *Arceus* Goldfuss, 1809:301-302; *Prochilus* Illiger, 1811:109; *Chondrorhynchus* Fischer de Waldheim, 1814:142-143; *Prochylus* Gray, 1821:301.[1]]
Pleist., R.; As.

[1] *Lapsus calami?*

Helarctos Horsfield, 1825:221. Malayan sun bear.
[= *Helarctus* Gloger, 1841:53.]
Pleist., R.; As. L. Pleist.-R.; E. Indies.[1]

[1] Sumatra, Java and Borneo.

Thalarctos Gray, 1825:62. Polar bear.
[= *Thalassarctos* Gray, 1825:339; *Thalassarctus* Gloger, 1841:xxviii, 54; *Thalassiarchus* Kobelt, 1896:93.]
L. Pleist.-R.; Eu.[1] R.; As.[2] R.; Arctic O. R.; N.A.[3]

[1] N. Europe.
[2] N. Asia.
[3] N. North America including Greenland.

Subfamily **Ailuropodinae** Grevé, 1894:217.
[= Ailuropodae Grevé, 1894:217; Ailuropodidae Pocock, 1916:19; Ailuropodini Hendey, 1972:119.]
L. Mioc., M. Pleist.-R.; As. L. Mioc.; Eu.

†*Agriarctos* Kretzoi, 1942:350.
L. Mioc.; Eu.

†*Ailurarctos* Qiu & Qi, 1989:154.
L. Mioc.; As.

Ailuropoda Milne-Edwards, 1870:342.[1] Giant panda.
[= *Pandarctos* Gervais, 1870:161; *Ailuropus* Milne-Edwards, 1871:92; *Aeluropus* Lydekker, 1891:561.] [Including †*Aelureidopus* Smith Woodward, 1915:425-428.]
M. Pleist.-R.; As.

[1] Not to be confused with AELUROPODA Gray, 1869:5, an unnatural and abandoned higher category (Section) for felids, viverrids, and certain mustelids.

Family †**Hemicyonidae** Frick, 1926:12, **new rank**.[1]
[= †Hemicyoninae Frick, 1926:12.] [Including †Agriotheriidae Kretzoi, 1929; †Agriotheriinae Hendey, 1972:119; †Agriotheriini Hendey, 1972:119; †Agriotheriina Kalandadze & Rautian, 1992:67.]
L. Eoc., L. Olig.-L. Plioc.; As. E. Olig.-E. Plioc.; Eu. E. Mioc., L. Mioc.-E. Plioc.; Af. E. Mioc.-E. Plioc.; N.A.

[1] Tedford.

†*Cephalogale* Jourdan, 1862:126, 129.
[= †*Cephalogalus* Jourdan, 1862:129.]
L. Eoc., L. Olig.-E. Mioc.; As. E. Olig.-E. Mioc.; Eu. E. Mioc.; N.A.

†*Zaragocyon* Ginsburg & Morales, 1995:812.
E. Mioc.; Eu.

†*Hemicyon* Lartet, 1851:16.
[= †*Harpaleocyon* Hürzeler, 1944:131-157.] [Including †*Phoberocyon* Ginsburg, 1955:85-99; †*Plithocyon* Ginsburg, 1955:91-93.]
E. Mioc.; Af. E.-L. Mioc.; Eu. M. Mioc.; As. M.-L. Mioc.; N.A.

†*Agriotherium* Wagner, 1837:335.

>   [= †*Amphiarctos* de Blainville, 1841:96; †*Sivalarctos* de Blainville, 1841:114[1];
>   †*Hyaenarctos* Falconer & Cautley, in Owen, 1845:505.]
>   [Including †*Lydekkerion* Frick, 1926:64, 79.]
>   L. Mioc.-E. Plioc.; Af. L. Mioc.-L. Plioc.; As. L. Mioc.-E. Plioc.; Eu.
>   L. Mioc.-E. Plioc.; N.A.
>   [1] Replacement name for †*Amphiarctos*.

†*Dinocyon* Jourdan, 1861:959.

>   L. Mioc.; Eu.

Superfamily **Phocoidea** Gray, 1821:302.

>   [= PINNIPEDIA Illiger, 1811:138[1]; PINNIGRADA Owen, 1857[2]; PINNIGRADES
>   Owen, 1859:45; AMPHIBIAE Gray, 1821:302[3]; AMPHIBIA Trouessart, 1879:2;
>   PHOCAE Trouessart, 1879:2; Phocadae Gray, 1821:302; Phocidae Gray, 1825:340[4];
>   Phocoidea Smirnov, 1908:1, 36; Phocoidea Wyss & J. J. Flynn, in Szalay, Novacek
>   & McKenna, eds., 1993:38.[5]] [Including Rosmaroidea Gill, 1872:7, 70; Otarioidea
>   Lucas, 1899:1; †Cynodontinae Schlosser, 1911:389; †Amphicynodontinae Simpson,
>   1945:110; †Enaliarctinae Mitchell & Tedford, 1973:218; †Enaliarctidae Repenning
>   & Tedford, 1977:76; PINNIPEDIMORPHA Berta, Ray & Wyss, 1989:61;
>   PINNIPEDIFORMES Berta, 1994:2.]
>   [1] Proposed as order and family.
>   [2] Not seen.
>   [3] Proposed as an order.
>   [4] Emendation.
>   [5] Described as new.

†*Plesiocyon* Schlosser, in Roger, 1887:132.

>   L. Eoc.; Eu.

†*Parictis* Scott, 1893:658-659.

>   [= †*Parietis* Scott, 1893:658-659[1]; †*Parictis* Lydekker, 1894:29.[2]] [Including
>   †*Campylocynodon* Chaffee, 1954:43; †*Subparictis* J. Clark & Guensburg, 1972:11.[3]]
>   L. Eoc.-E. Mioc.; N.A.
>   [1] *Lapsus calami.*
>   [2] Correction of †*Parietis* Scott.
>   [3] Proposed as subgenus.

†*Nothocyon* Matthew, 1899:62.[1]

>   L. Olig.; N.A.
>   [1] Not *Nothocyon* Matthew & Wortman, 1899, a synonym of *Lycalopex*.

†*Adelpharctos* de Bonis, 1971:123.

>   Olig.; Eu.

†*Amphicynodon* Filhol, 1882:32-39.

>   [= †*Cynodon* Aymard, 1848:244.[1]] [Including †*Paracynodon* Schlosser, 1899:115.]
>   E. Olig.; As. Olig.; Eu.
>   [1] Not *Cynodon* Spix, 1829:76, a fish.

†*Drassonax* Galbreath, 1953:77.

>   E. Olig.; N.A.

†*Pachycynodon* Schlosser, 1887:124.

>   E. Olig.; As. Olig.; Eu.

†*Kolponomos* Stirton, 1960:346.

>   E. Mioc.; N.A.

†*Allocyon* Merriam, 1930:230.

>   E. Mioc.; N.A.

†*Enaliarctos* Mitchell & Tedford, 1973:218.

>   E. Mioc.; As.[1] E. Mioc.; N.A.[2]
>   [1] Japan.
>   [2] Pacific North America.

†*Pteronarctos* Barnes, 1989:3.
　　[= †*Pacificotaria* Barnes, 1992:3.]
　　E.-M. Mioc.; N.A.[1]
　[1] Pacific North America.

Family **Otariidae** Gray, 1825:340. Eared seals, fur seals, sea lions.
　　[= Otariina Gray, 1825:340; Otariadae Brookes, 1828:37; Otariidae Gill, 1866:7, 10;
　　Otarioidea Lucas, 1899:1.] [Including Callorhinina Gray, 1869:269;
　　Callorhinae de Muizon, 1978:183.]
　　?Mioc., L. Plioc.-E. Pleist., R.; As.[1] ?Mioc., L. Plioc., ?E. and/or ?M. Pleist.,
　　L. Pleist.-R.; S.A. M. Mioc.-L. Plioc., ?E. Pleist., L. Pleist.-R.; N.A.[2]
　　M. Pleist.-R.; New Zealand. L. Pleist.-R.; Af. L. Pleist.-R.; Aus. R.; Indian O.[3]
　　R.; Atlantic.[4] R.; Antarctica.[5] R.; Pacific.
　　[1] E. Asia.
　　[2] Pacific North America.
　　[3] Islands of the southern Indian Ocean.
　　[4] Islands of the S. Atlantic.
　　[5] Including islands of the Southern Ocean.

†*Pithanotaria* Kellogg, 1925:74.
　　M.-L. Mioc.; N.A.[1]
　[1] Pacific North America.

†*Thalassoleon* Repenning & Tedford, 1977:60.
　　L. Mioc.-E. Plioc.; N.A.[1]
　[1] Pacific North America.

Callorhinus Gray, 1859:359. Northern fur seal.
　　[= *Callirhinus* Gill, 1872:69[1]; *Callotaria* Palmer, 1892:156[2];
　　Callorhynchus Grevé, 1896:322.]
　　?L. Mioc., L. Plioc., L. Pleist.-R.; N.A.[3] R.; As.[4] R.; Pacific.[5]
　　[1] Emendation of *Callorhinus*. Not *Callirhinus* Blanchard, 1850, a genus of Coleoptera,
　　nor *Callirhinus* Girard, 1857, a genus of Reptilia.
　　[2] Replacement name for *Callorhinus*.
　　[3] Pacific North America.
　　[4] E. Asia.
　　[5] N. Pacific (including Bering Sea).

Subfamily **Arctocephalinae** Gray, 1837:583.
　　[= Arctocephalina Gray, 1837:583; Arctocephalida Haeckel, 1895:590;
　　Arctocephalinae von Boetticher, 1934:359; Arctocephalini Mitchell, 1968:1843, 1897.]
　　[Including Gypsophocina Gray, 1874.]
　　?Mioc., L. Plioc., Pleist., R.; S.A. ?E. Pleist., L. Pleist.-R.; N.A.[1]
　　L. Pleist.-R.; Af. L. Pleist.-R.; New Zealand. R.; Indian O.[2] R.; Atlantic.[3]
　　R.; Antarctica.[4] R.; Aus. R.; Pacific.[5]
　　[1] Coastal islands of California and Baja California.
　　[2] Islands of the southern Indian Ocean.
　　[3] Islands of the S. Atlantic.
　　[4] Including islands of the Southern Ocean.
　　[5] Islands of the S. Pacific north to the Galapagos.

Arctocephalus É. Geoffroy Saint-Hilaire & F. Cuvier, 1826:553-554. Southern fur seals.
　　[Including *Arctophoca* Peters, 1866:276[1]; *Gypsophoca* Gray, 1866:236-237.]
　　?Mioc., Pleist., R.; S.A.[2] ?E. Pleist., L. Pleist.-R.; N.A.[3] L. Pleist.-R.; Af.
　　L. Pleist.-R.; New Zealand. R.; Indian O.[4] R.; Atlantic.[5] R.; Antarctica.[6]
　　R.; Aus. R.; Pacific.[7]
　　[1] As subgenus.
　　[2] Argentina in the ?Miocene and Pleistocene.
　　[3] Coastal islands of California and Baja California.
　　[4] Islands of the southern Indian Ocean.
　　[5] Islands of the S. Atlantic.
　　[6] Including islands of the Southern Ocean.
　　[7] Islands of the S. Pacific north to the Galapagos.

†*Hydrarctos* de Muizon, 1978:171.[1]
L. Plioc.; S.A.[2]
[1] Proposed as subgenus of *Arctocephalus*.
[2] Peru.

Subfamily **Otariinae** Gray, 1825:340.
[= Otariina Gray, 1825:340[1]; Otariinae Mitchell, 1968:1897;
Otariini Mitchell, 1968:1897.]
[Including Eumetopiina Gray, 1869:269; Phocarctinae von Boetticher, 1934.]
?Mioc., L. Plioc.-E. Pleist., R.; As.[2] ?E. Pleist., L. Pleist.-R.; N.A.[3]
?E. and/or ?M. Pleist., L. Pleist.-R.; S.A. M. Pleist.-R.; New Zealand.
L. Pleist.-R.; Aus. R.; Pacific.[4]
[1] Proposed as tribe.
[2] N.E. Asia to Japan.
[3] Pacific North America.
[4] N. Pacific south to Galapagos.

Eumetopias Gill, 1866:7, 11. Northern sea lion, Steller sea lion.
[= *Eumetopus* Marschall, 1873:6.]
?Mioc., ?L. Plioc., E. Pleist., R.; As.[1] ?E. Pleist., L. Pleist.-R.; N.A.[2]
Pleist.; S.A. R.; Pacific.[3]
[1] N.E. Asia to Japan.
[2] Pacific North America.
[3] N. Pacific (including Bering Sea).

Otaria Peron, 1816:37, 40. Southern sea lion, South American sea lion.
[= *Otoes* Fischer de Waldheim, 1817:373, 445; *Platyrhynchus* F. Cuvier,
1826:554-555[1]; *Platyrhyncus* F. Cuvier, 1829:465; *Pontoleo* Gloger, 1841:164.[2]]
L. Pleist.-R.; S.A.[3,4]
[1] Not *Platyrhynchus* Desmarest, 1805, a genus of birds.
[2] Replacement name for *Platyrhynchus* F. Cuvier, 1826.
[3] Argentina in the late Pleistocene.
[4] Coasts and islands of southern South America in the Recent.

Zalophus Gill, 1866:7, 11. California sea lion.
L. Plioc.-E. Pleist., R.; As.[1] L. Pleist.-R.; N.A.[2] R.; Pacific.[3]
[1] E. Asia (Japan).
[2] Pacific North America.
[3] N. Pacific south to Galapagos.

Neophoca Gray, 1866:231. Australian sea lion.
M. Pleist.; New Zealand. L. Pleist.-R.; Aus.

Phocarctos Peters, 1866:269.[1] New Zealand sea lion.
L. Pleist.-R.; New Zealand.
[1] Described as subgenus of *Arctocephalus*.

Family **Phocidae** Gray, 1821:302.
[= Phocadae Gray, 1821:302; Phocidae Gray, 1825:340[1]; Phocida Haeckel, 1866:clix;
Phocoidea Smirnov, 1908:1, 14; Rosmaridae Gill, 1866:7, 11[2];
PHOCOMORPHA Berta & Wyss, 1994:41.[3]]
?E. Mioc., M. Mioc., Plioc., E.-M. Pleist., R.; As. E. Mioc.-R.; N.A.
M. Mioc.-R.; Eu. M. Mioc.-E. Plioc., R.; S.A. L. Mioc.-E. Plioc., Pleist., R.; Af.
L. Mioc. and/or E. Plioc., R.; Aus. L. Plioc., M. Pleist., R.; New Zealand.
R.; Indian O. R.; Mediterranean. R.; Atlantic. R.; Arctic O. R.; W. Indies.
R.; Antarctica. R.; Pacific.
[1] Emendation.
[2] Invalid.
[3] No rank given by the authors.

Subfamily **Odobeninae** Allen, 1880:5, 17.
[= Trichecidae Gray, 1821:302[1]; Trichechidae Gray, 1825:340[2]; Trichisina Gray,
1837:582[2]; Trichechina C. L. Bonaparte, 1838:111[2]; Trichechoidea Giebel, 1847:221[3];
Trichecida Haeckel, 1866:clix[3]; Rosmaridae Gill, 1866:7, 11[4]; Odobaenidae Allen,
1880:5, 17; Odobenidae Palmer, 1904:833[5]; Odobaeninae Orlov, 1931:69;
Odobeninae Mitchell, 1968:1843, 1892.]

[Including †Dusignathinae Mitchell, 1968:1897; †Imagotariinae Mitchell, 1968:1895; †Kamtschatarctinae Dubrovo, 1981:970.]
?E. Mioc., M. Mioc., Plioc., E.-M. Pleist.; As.[6] E. Mioc.-E. Plioc., Pleist.; N.A. E. Plioc.-M. Pleist., ?L. Pleist.; Eu. R.; Atlantic.[7] R.; Arctic O. R.; Pacific.[8]

[1] Based on *Trichechus* Linnaeus, 1766, a synonym of *Odobenus*, not on *Trichechus* Linnaeus, 1758, a sirenian. Invalid.
[2] Based on *Trichechus* Linnaeus, 1766.
[3] Based on *Trichechus* Linnaeus, 1766. Proposed as a family, not a superfamily.
[4] Invalid.
[5] Emendation of Odobaenidae Allen, 1880.
[6] E. Asia.
[7] N. Atlantic.
[8] N. Pacific.

†*Oriensarctos* Mitchell, 1968:1881.
        E. Pleist.; As.[1]
    [1] E. Asia (Japan).

Tribe †**Desmatophocini** Hay, 1930:557, **new rank**.[1]
        [= †Desmatophocidae Hay, 1930:557; †Desmatophocinae Mitchell, 1966:4, 39, 40.]
        [Including †Allodesmidae Kellogg, 1931:227.]
        E.-M. Mioc., ?L. Mioc.; N.A.[2] M. Mioc.; As.[3]
    [1] Tedford.
    [2] Pacific North America.
    [3] E. Asia (Japan).

†*Pinnarctidion* Barnes, 1979:16.[1]
        E. Mioc.; N.A.[2]
    [1] Assignment to this subfamily is somewhat questionable.
    [2] Pacific North America.

†*Desmatophoca* Condon, 1906:3.
        E.-M. Mioc., ?L. Mioc.; N.A.[1]
    [1] Pacific North America.

†*Allodesmus* Kellogg, 1922:26.
        [Including †*Atopotarus* Downs, 1956:115-131.]
        M. Mioc.; As.[1] M. Mioc.; N.A.[2]
    [1] E. Asia (Japan).
    [2] Pacific North America.

Tribe **Odobenini** Allen, 1880:5, 17.
        [= Trichecidae Gray, 1821:302[1]; Odobaenidae Allen, 1880:5, 17; Odobenini Deméré, 1994:117; Odobeninae Mitchell, 1968:1843, 1892.] [Including †Imagotariinae Mitchell, 1968:1895; †Kamtschatarctinae Dubrovo, 1981:970.]
        ?E. Mioc., M. Mioc., Plioc., M. Pleist.; As.[2] M. Mioc.-E. Plioc., Pleist.; N.A. E. Plioc.-M. Pleist., ?L. Pleist.; Eu. R.; Atlantic.[3] R.; Arctic O. R.; Pacific.[4]
    [1] Based on *Trichechus* Linnaeus, 1766. Invalid.
    [2] E. Asia.
    [3] N. Atlantic.
    [4] N. Pacific.

†*Kamtschatarctos* Dubrovo, 1981:973.
        M. Mioc.; As.[1]
    [1] E. Asia (Kamchatka).

†*Prototaria* Takeyama & Ozawa, 1984:36.
        E. and/or M. Mioc.; As.[1]
    [1] E. As. (Japan).

†*Neotherium* Kellogg, 1931:296.
        M. Mioc.; N.A.[1]
    [1] Pacific North America.

†*Imagotaria* Mitchell, 1968:1844.
        M.-L. Mioc.; N.A.[1]
    [1] Pacific North America.

†*Pelagiarctos* Barnes, 1988:2.
          M. Mioc.; N.A.[1]
[1] Pacific North America.

Subtribe **Odobenina** Allen, 1880:5, 17, **new rank**.[1] Walruses.
          [= Odobaenidae Allen, 1880:5, 17.]
          L. Mioc.-E. Plioc., Pleist.; N.A. Plioc., M. Pleist.; As.[2] E. Plioc.-M. Pleist., ?L.
          Pleist.; Eu. R.; Atlantic.[3] R.; Arctic O. R.; Pacific.[4]
[1] Tedford.
[2] Tusks similar to those of *Odobenus* have been reported from the Japanese marine Pliocene.
[3] N. Atlantic.
[4] N. Pacific.

†*Aivukus* Repenning & Tedford, 1977:14.
          L. Mioc.; N.A.[1]
[1] Pacific North America.

†*Prorosmarus* Berry & W. K. Gregory, 1906:444, 447.
          [= †*Prorosmerus* Clark & Miller, 1912:167, 168.]
          E. Plioc.; N.A.[1]
[1] Atlantic North America.

†*Alachtherium* Du Bus, 1867:566.
          [= †*Alachterium* Van Beneden, 1871:181.[1]]
          E.-L. Plioc.; Eu.
[1] Misprint.

†*Trichecodon* Lankester, 1865:226-231.
          [= †*Trichechodon* Palmer, 1904:688.]
          L. Plioc.-M. Pleist.; Eu. Pleist.; N.A.[1]
[1] S.E. North America.

Odobenus Brisson, 1762:30.[1] Walrus.
          [= *Trichechus* Linnaeus, 1766:49[2]; *Trichecus* F. Cuvier, 1829:465[3]; *Rosmarus*
          Brunnich, 1772:34, 38-39; *Odobaenus* Fee, in Linnaeus, 1830:59[4]; *Odontobaenus*
          Steenstrup, in Sundevall, 1860:441[5]; *Hodobaenus* Sundevall, 1860:442;
          †*Hemicaulodon* Cope, 1869:190-191.] [Including †*Odobenotherium* Gratiolet,
          1858:620-624.]
          Pleist.; Eu. Pleist.; N.A. M. Pleist.; As.[6] R.; Atlantic.[7] R.; Arctic O. R.; Pacific.[8]
[1] "The older literature almost always used *Trichechus* for the walrus, but this name belongs to the manatee,
     as now generally accepted" (Simpson 1945:122).
[2] Not *Trichechus* Linnaeus, 1758, a sirenian.
[3] Not *Trichecus* Oken, 1816, a sirenian.
[4] Considered by the International Code of Zoological Nomenclature to be an unjustified emendation.
[5] New name for *Odobenus*.
[6] E. Asia (Japan).
[7] N. Atlantic.
[8] N. Pacific.

Subtribe †**Dusignathina** Mitchell, 1968:1897, **new rank**.[1]
          [= †Dusignathinae Mitchell, 1968:1897.]
          L. Mioc.-E. Plioc.; N.A.[2]
[1] Tedford.
[2] Pacific North America.

†*Valenictus* Mitchell, 1961:4.
          L. Mioc.-E. Plioc.; N.A.[1]
[1] Pacific North America.

†*Pliopedia* Kellogg, 1921:213.
          L. Mioc.; N.A.[1]
[1] Pacific North America.

†*Dusignathus* Kellogg, 1927:27.
          L. Mioc.; N.A.[1]
[1] Pacific North America.

†*Pontolis* True, 1905:253.
  [= †*Pontoleon* True, 1905:47.[1]]
  L. Mioc.; N.A.[2]

[1] Not †*Pontoleo* Gloger, 1841, a synonym of *Otaria*. "I question whether †*Pontoleo* preoccupies †*Pontoleon*, but the change to †*Pontolis* has generally been accepted, and the similarity, if not identity, of the two names would lead to confusion, especially as *Pontoleo* is also a name for an otariid (=*Otaria*)" (Simpson 1945:121).
[2] Pacific North America.

†*Gomphotaria* Barnes & Raschke, 1991:2.
  L. Mioc.; N.A.[1]

[1] Pacific North America.

Subfamily **Phocinae** Gray, 1821:302. True seals, hair seals, earless seals.
  [= Phocadae Gray, 1821:302; Phocinae Gill, 1866:5, 8.] [Including Cystophorinae Gill, 1866:6, 9; Stenorhynchinae Gill, 1866:6; Ogmorhininae Turner, 1888:63.]
  M. Mioc.-L. Plioc., ?E. Pleist., L. Pleist.-R.; Eu. M. Mioc., L. Plioc.-R.; N.A. M. Mioc.-E. Plioc., R.; S.A. L. Mioc.-E. Plioc., Pleist., R.; Af. L. Mioc. and/or E. Plioc., R.; Aus.[1] L. Plioc., M. Pleist., R.; New Zealand. E. Pleist., R.; As. R.; Indian O. R.; Mediterranean. R.; Atlantic. R.; Arctic O. R.; W. Indies. R.; Antarctica. R.; Pacific.

[1] "Phocidae" indet. from the late Miocene or early Pliocene of Australia.

†*Prophoca* Van Beneden, 1876:801-802.
  L. Mioc.; Eu.

†*Necromites* Bogachev, 1940:94.
  E. Pleist.; As.

Tribe **Phocini** Gray, 1821:302.
  [= Phocadae Gray, 1821:302; Phocina Gray, 1825:340[1]; Phocini Chapski, 1955:164, 169.] [Including Stemmotopina Gray, 1825:340[1]; Cystophorina Gray, 1837:582; Cystophorini Burns & Fay, 1970:390; Halichoerina Gray, 1869:345[2]; Erignathini Chapski, 1955:165; Histriophocina Chapski, 1955:188.]
  M. Mioc., E.-L. Plioc., L. Pleist.-R.; Eu. M. Mioc., L. Plioc.-R.; N.A. R.; As.[3] R.; Atlantic.[4] R.; Arctic O. R.; Pacific.[5]

[1] As tribe.
[2] Proposed as tribe.
[3] Lake Baikal, Caspian Sea.
[4] N. Atlantic.
[5] N. Pacific.

†*Pontophoca* Kretzoi, 1941:354.
  M. Mioc.; Eu.[1]

[1] E. Europe.

†*Leptophoca* True, 1906:836.
  M. Mioc.; N.A.[1]

[1] Atlantic North America.

†*Praepusa* Kretzoi, 1941:351.
  M. Mioc.; Eu.[1]

[1] E. Europe.

†*Cryptophoca* Koretsky & Ray, 1994:18.
  M. Mioc.; Eu.[1]

[1] E. Europe.

Phoca Linnaeus, 1758:37. Harbor seal, ringed seal, ribbon seal, harp seal.
  [= *Calocephalus* F. Cuvier, 1826:544; *Callocephalus* Heuglin, 1874:56; *Haliphilus* Gray, 1866:466.] [= or including *Ambysus* Rafinesque, 1815:60[1]; *Arctias* Rafinesque, 1815:60[1]; *Halicyon* Gray, 1864:28.] [Including *Pusa* Scopoli, 1777:490[2]; *Pagophilus* Gray, 1844:3[3]; *Pagomys* Gray, 1864:31; *Histriophoca* Gill, 1873:179[2]; *Pagophoca* Trouessart, 1904:287[4]; *Caspiopusa* Dybowski, 1929:414.]

L. Plioc.-R.; N.A. ?L. Pleist., R.; Eu.[5] R.; As.[6] R.; Atlantic.[7] R.; Arctic O. R.; Pacific.[8]

[1] *Nomen nudum.*
[2] As subgenus.
[3] Described as subgenus of *Callocephalus*; used as subgenus of *Phoca*. Not preoccupied by *Pagophila* Kaup, 1829, a genus of birds.
[4] New name for *Pagophilus*.
[5] Including the Sea of Azov in the Pleistocene.
[6] Lake Baikal, Caspian Sea.
[7] N. Atlantic (Including Baltic Sea and North Sea).
[8] N. Pacific (including Bering Sea).

†*Phocanella* Van Beneden, 1876:799.
[= †*Procanella* C. O. Waterhouse, 1902:304.[1]]
M. Mioc., E.-L. Plioc.; Eu.
[1] Misprint.

†*Platyphoca* Van Beneden, 1876:798.
Plioc.; Eu.

†*Gryphoca* Van Beneden, 1876:798-799.
Plioc.; Eu.

Erignathus Gill, 1866:5, 9. Bearded seal.
L. Pleist.; Eu. L. Pleist.; N.A.[1] R.; Atlantic.[2] R.; Arctic O. R.; Pacific.[3]
[1] Atlantic North America.
[2] N. Atlantic.
[3] N. Pacific.

Halichoerus Nilsson, 1820:376. Gray seal.
[= *Halychoerus* Boitard, 1842:198.]
?L. Pleist.; Eu. ?L. Pleist.; N.A.[1] R.; Atlantic.[2] R.; Arctic O.[3]
[1] Atlantic North America.
[2] N. Atlantic.
[3] Kola Peninsula to Novaya Zemlya.

Cystophora Nilsson, 1820:382. Hooded seal.
[= *Stemmatopus* F. Cuvier, 1826:550; *Cystophoca* Shufeldt, 1890:222; *Semmatopis* Gloger, 1841:163; *Stemmatops* Van der Hoeven, 1855:992.]
R.; Atlantic.[1] R.; Arctic O.
[1] N. Atlantic.

Tribe **Monachini** Gray, 1869:345.
[= Monachina Gray, 1869:345[1]; Monachinae Trouessart, 1897:378; Monachini Scheffer, 1958:112.] [Including Stenorhynchina Gray, 1825:340[1]; Stenorhynchinae Gill, 1866:6; Miroungini de Muizon, 1982:199.]
M. Mioc.-L. Plioc., ?E. Pleist., L. Pleist.-R.; Eu.
M. Mioc., Plioc. and/or Pleist., R.; N.A. M. Mioc.-E. Plioc., R.; S.A.
L. Mioc.-E. Plioc., Pleist., R.; Af. L. Plioc., M. Pleist., R.; New Zealand.
R.; Indian O. R.; Mediterranean. R.; Atlantic. R.; W. Indies. R.; Antarctica.
R.; Aus. R.; Pacific.
[1] Proposed as tribe.

Subtribe **Monachina** Gray, 1869:345, **new rank.**[1]
[= Monachini Scheffer, 1958:112.] [Including Miroungini de Muizon, 1982:199.]
M. Mioc.-L. Plioc., ?E. Pleist., L. Pleist.-R.; Eu.
M. Mioc., Plioc. and/or Pleist., R.; N.A.[2] M. Mioc., R.; S.A.
L. Mioc., Pleist., R.; Af. M. Pleist., R.; New Zealand. R.; Indian O.
R.; Mediterranean. R.; Atlantic. R.; W. Indies. R.; Antarctica. R.; Pacific.
[1] Tedford. Proposed as tribe.
[2] A small monachine with affinities to †*Properiptychus* and †*Pristiphoca* is present in the Pliocene Yorktown Fm., North Carolina.

†*Palmidophoca* Ginsburg & Janvier, 1975:78.
M. Mioc.; Eu.

†*Monachopsis* Kretzoi, 1941:353.
>> M. Mioc.; Eu.[1]
> [1] E. Europe.

†*Monotherium* Van Beneden, 1876:800-801.
>> [= †*Monatherium* Lydekker, 1885:206-207.]
>> ?M. Mioc., L. Mioc.; Eu. M. Mioc.; N.A.

†*Pristiphoca* Gervais, 1852:308-309.[1]
>> [Including †*Miophoca* Zapfe, 1937:271-276.]
>> M. Mioc., E. Plioc.; Eu. ?L. Mioc.; Af.[2]
> [1] Proposed as a subgenus of *Phoca*; raised to generic rank by Gervais, 1859:272-273.
> [2] N. Africa (Egypt).

†*Properiptychus* Ameghino, 1897:441.
>> M. Mioc.; S.A.

†*Messiphoca* de Muizon, 1981:183.
>> L. Mioc.; Af.[1]
> [1] N. Africa.

†*Mesotaria* Van Beneden, 1876:796-797.
>> Plioc.; Eu.

†*Callophoca* Van Beneden, 1876:798.
>> Plioc.; Eu.

†*Pliophoca* Tavani, 1942:97-113.
>> L. Plioc.; Eu.

Mirounga Gray, 1827:179. Elephant seals.
>> [= *Macrorhinus* F. Cuvier, 1826:551-553[1]; *Rhinophoca* Wagler, 1830:27.]
>> Pleist., R.; Af.[2] Pleist., R.; N.A.[3] M. Pleist., R.; New Zealand. R.; Indian O. R.; Atlantic.[4] R.; S.A. R.; Antarctica.[5] R.; Pacific.
> [1] Not *Macrorhinus* Latreille, 1825, a coleopteran.
> [2] S. Africa.
> [3] Pacific North America.
> [4] S. Atlantic.
> [5] Including islands of the southern Ocean.

Monachus Fleming, 1822:187. Monk seals.
>> [= *Pelagios* F. Cuvier, 1824:196; *Pelagius* F. Cuvier, 1826; *Pelagocyon* Gloger, 1841:163; *Rigoon* Gistel, 1854[1]; *Heliophoca* Gray, 1854:201; *Mammonachus* Herrera, 1899:13.[2]]
>> ?E. Pleist., L. Pleist.-R.; Eu.[3] R.; Af.[4] R.; Mediterranean. R.; Atlantic.[5] R.; W. Indies.[6] R.; Pacific.[7]
> [1] New name for *Pelagius*.
> [2] See the International Code of Zoological Nomenclature, Article 1(b)(8) and Palmer (1904:25-26).
> [3] Black Sea in the Recent.
> [4] Coastal waters of N.W. Africa.
> [5] Madeira and Canary Islands.
> [6] Now extinct.
> [7] Central Pacific (Hawaii).

Subtribe **Lobodontina** Gray, 1869:345, **new rank**.[1]
>> [= Lobodoninae Kellogg, 1922:89; Lobodontinae Hay, 1930:562; Lobodontini Scheffer, 1958:115.] [Including Stenorhynchina Gray, 1825:340[2]; Ogmorhininae Turner, 1888:63.]
>> L. Mioc.-E. Plioc., R.; S.A. E. Plioc., Pleist., R.; Af.[3] L. Plioc., R.; New Zealand. R.; Indian O.[4] R.; Atlantic.[5] R.; Antarctica.[6] R.; Aus. R.; Pacific.[7]
> [1] Tedford. Proposed as tribe.
> [2] Proposed as tribe.
> [3] Southern Africa.
> [4] Southern Indian Ocean.
> [5] S. Atlantic.
> [6] Including the Southern Ocean.
> [7] S. Pacific.

†*Piscophoca* de Muizon, 1981:20.
>> E. Plioc.; S.A.[1]

[1] Peru.

†*Homiphoca* de Muizon & Hendey, 1980:94.
>> E. Plioc.; Af.[1]

[1] S. Africa.

†*Acrophoca* de Muizon, 1981:80.
>> L. Mioc.-E. Plioc.; S.A.[1]

[1] Peru.

Ommatophoca Gray, 1844:3. Ross seal.
>> [= *Ommatophora* Turner, 1848:88.]
>> L. Plioc., R.; New Zealand. R.; S.A. R.; Antarctica.[1]

[1] Including the Southern Ocean.

Lobodon Gray, 1844:2. Crabeater seal.
>> Pleist.; Af.[1] R.; S.A.[2] R.; Antarctica.[3] R.; New Zealand. R.; Aus.

[1] S. Africa.
[2] Southern Atlantic coast (Argentina).
[3] Including the Southern Ocean.

Hydrurga Gistel, 1848:xi. Leopard seal.
>> [= *Stenorhinchus* F. Cuvier, 1826:549[1]; *Stenorhynchus* Lesson, 1827:199[2];
>> *Stenorhyncus* Cuvier, 1829:463; *Stenorhincus* McMurtrie, 1834:71; *Stenorhynchotes*
>> Turner, 1888:63[3]; *Ogmorhinus* Peters, 1875:393.]
>> R.; Af.[4] R.; Indian O.[5] R.; Atlantic.[6] R.; S.A.[7] R.; Antarctica.[8] R.; New Zealand.
>> R.; Aus.[9] R.; Pacific.[10]

[1] Not *Stenorhynchus* Lamarck, 1819, a crustacean.
[2] Emendation.
[3] Replacement name.
[4] Coasts of southern Africa.
[5] Southern Indian Ocean.
[6] S. Atlantic.
[7] Islands and coasts of southern South America.
[8] Including islands of the Southern Ocean.
[9] S. Australia.
[10] S. Pacific.

Leptonychotes Gill, 1872:70. Weddell seal.
>> [= *Leptonyx* Gray, 1837:582[1]; *Poecilophoca* Lydekker, 1891:605.]
>> R.; S.A.[2] R.; Antarctica. R.; New Zealand. R.; Aus.

[1] Not *Leptonyx* Swainson, 1821, a genus of birds.
[2] Southern coasts.

Parvorder **MUSTELIDA** Tedford, 1976:372.

Family **Mustelidae** Fischer de Waldheim, 1817:372.
>> [= Mustelini Fischer de Waldheim, 1817:372; Mustelladae Gray, 1821:301; Mustelina
>> Gray, 1825:339; Mustelidae Swainson, 1835:321; Mustelina Haeckel, 1866:clix[1];
>> Musteloidea Kretzoi, 1929:1350; Musteloidea Gregory & Hellman, 1939:313;
>> Musteloidea Kretzoi, 1945:61.] [Including Enhydrina Gray, 1825:340; Enhydridae H.
>> Smith, 1842:248; Lutrina Bonaparte, 1838:111; Lutridae De Kay, 1842:xv, 39;
>> Arctogalidae C. H. Smith, in Jardine, 1842:193; Melinidae Gray, 1869:79-142;
>> Galeidae Schulze, 1900:220; †Semantoridae Orlov, 1931:69; †Semantorinae Tedford,
>> 1976:372; †Brachypsalini Webb, 1969:68; †Potamotherini Sokolov, 1973:51[2];
>> †Potamotheriinae Willemsen, 1992:8; Mephitidae Ledje & Arnason, 1996:641, 647.]
>> E.-L. Olig., M. Mioc.-R.; As. E. Olig.-R.; Eu. ?L. Olig., E. Mioc.-R.; N.A. E.
>> Mioc.-R.; Af. ?L. Plioc., E. Pleist., ?M. Pleist., L. Pleist.-R.; Mediterranean. L.
>> Plioc.-R.; S.A. E. Pleist., ?M. Pleist., R.; E. Indies. R.; Cent. A. R.; Pacific.[3]

[1] Used as a family.
[2] Proposed as a tribe of Lutrinae.
[3] Also Recent of New Zealand and islands of the Atlantic by introduction.

†*Luogale* Schmidt-Kittler, 1987:92.
>> E. Mioc.; Af.

†*Bathygale* Wolsan, 1993:377.
    E. Mioc.; Eu.

†*Plesiogale* Pomel, 1847:380.
    E. Olig.-E. Mioc.; Eu.

†*Plesictis* Pomel, 1846:366.
    E. Olig., ?L. Olig.; As. E. Olig.-E. Mioc.; Eu. ?L. Olig.; N.A.

†*Zodiolestes* Riggs, 1942:61.
    E. Mioc.; N.A.

†*Stromeriella* Dehm, 1950:99.
    E.-M. Mioc.; Eu.

†*Franconictis* Wolsan, 1993:375.
    E. Mioc.; Eu.

†*Paragale* G. Petter, 1967:19.[1]
    E. Mioc.; Eu.
    [1] A "paleomustelid".

†*Oligobunis* Cope, 1881:497.[1]
    E. Mioc.; N.A.
    [1] A "paleomustelid".

†*Aelurocyon* Peterson, 1906:68.[1]
    E. Mioc.; N.A.
    [1] A "paleomustelid".

†*Megalictis* Matthew, 1907:195.[1]
    E. Mioc.; N.A.
    [1] A "paleomustelid".

†*Paroligobunis* Peterson, 1910:269.[1]
    E. Mioc.; N.A.
    [1] A "paleomustelid".

†*Amphicticeps* Matthew & Granger, 1924:4.
    E. Olig.; As.

†*Promartes* Riggs, 1942:59.[1]
    E. Mioc.; N.A.
    [1] A "paleomustelid".

†*Brachypsalis* Cope, 1890:951-952.[1]
    [Including †*Brachypsaloides* Webb, 1969:58.]
    M.-L. Mioc.; N.A.
    [1] A "paleomustelid".

†*Sivalictis* Pilgrim, 1932:77.
    M. Mioc.; As.

†*Semantor* Orlov, 1931:67-70.
    L. Mioc.; As.[1]
    [1] W. Siberia.

†*Potamotherium* É. Geoffroy Saint-Hilaire, 1833:80-81.
    [= †*Lutrictis* Pomel, 1847:380; †*Stephanodon* von Meyer, 1847:183; †*Potamophilus* P. Gervais, 1852:2, unnumbered.[1]]
    L. Olig.-L. Mioc.; Eu. E.-M. Mioc.; N.A.
    [1] Preoccupied by *Potamophilus* Germar, 1811, a coleopteran, and by several other homonyms.

Subfamily **Lutrinae** Bonaparte, 1838:111. Otters.
    [= Lutrina Bonaparte, 1838:111[1]; Lutridae deKay, 1842:39; Lutrinae Baird, 1857:3, 183.] [Including Enhydrina Gray, 1825:340[2]; Enhydrinae Gill, 1872:6, 66; Latacina Bonaparte, 1838:213; Lataxinae Burmeister, 1850:13; †Mionictini Ginsburg, 1968:232.]

E.-M. Mioc., ?L. Mioc., E. Plioc.-R.; Af. E. Mioc.-R.; Eu. M. Mioc.-R.; As. M. Mioc.-R.; N.A. E. Pleist., ?M. Pleist., R.; E. Indies. ?E. and/or ?M. Pleist., L. Pleist., ?R.; Mediterranean. E. Pleist.-R.; S.A. R.; Cent. A. R.; Pacific.

[1] As subfamily.
[2] Proposed as tribe of Phocidae.

†*Kenyalutra* Schmidt-Kittler, 1987:94.
    E. Mioc.; Af.

†*Torolutra* G. Petter, Pickford & F. C. Howell, 1991:950.
    ?L. Mioc., E. Plioc., ?L. Plioc.; Af.

†*Mionictis* Matthew, 1924:136.
    M. Mioc.; As.[1] M. Mioc.; N.A.

[1] S.W. Asia (Saudi Arabia).

†*Siamogale* Ginsburg, Ingavat & Tassy, 1983:953.
    M. Mioc.; As.[1]

[1] S.E. Asia (Thailand).

†*Lutravus* Furlong, 1932:95.
    L. Mioc.; N.A.

Tribe **Lutrini** Bonaparte, 1838:111.
    [= Lutrina Bonaparte, 1838:111[1]; Lutrini Sokolov, 1973:51.] [Including Hydrictini Davis, 1978.]
    E. Mioc.-R.; Eu. L. Mioc. and/or E. Plioc., L. Plioc.-R.; Af. E. Plioc.-R.; As. E. Plioc.-R.; N.A. E. Pleist., ?M. Pleist., R.; E. Indies. ?E. and/or ?M. Pleist., L. Pleist., ?R.; Mediterranean. E. Pleist.-R.; S.A. R.; Cent. A.

[1] As subfamily.

†*Paralutra* Roman & Viret, 1934:17.
    E.-L. Mioc.; Eu.

Lutra Brunnich, 1772:34, 42.[1] River otters.
    [= *Lutris* Duméril, 1806:12[2]; *Lutrix* Rafinesque, 1815:59[3]; *Lutronectes* Gray, 1867:180-182; *Mamlutraus* Herrera, 1899:20.[4]] [Including *Lontra* Gray, 1843:118[5]; *Latax* Gray, 1843:119[6]; *Lataxina* Gray, 1843:70[7]; *Lataxia* Gervais, 1855:118[8]; *Barangia* Gray, 1865:123; *Nutria* Gray, 1865:128-9; *Hydrogale* Gray, 1865:131-2[9]; *Hydrictis* Pocock, 1921:543[10]; †*Nesolutra* Bate, 1935; *Basarabictis* Kretzoi, 1942:320.]
    L. Mioc. and/or E. Plioc., L. Plioc.-R.; Af. E. Plioc.-R.; As. E. Plioc.-R.; Eu. E. and/or M. Pleist., R.; E. Indies.[11] Pleist.; Mediterranean.[12] E. Pleist.-R.; N.A. E. Pleist.-R.; S.A. R.; Cent. A.

[1] *Lutra* Brisson, 1762, is not available.
[2] Misprint for *Lutra*.
[3] New name for *Lutra*.
[4] See the International Code of Zoological Nomenclature, Article 1(b)(8), and Palmer (1904: 25-26).
[5] Sometimes as subgenus of *Lutra*; sometimes as separate genus.
[6] Objective synonym of *Lontra*. Not *Latax* Gloger, 1827, a synonym of *Enhydra*.
[7] Objective synonym of *Lontra*.
[8] Replacement name for *Latax* Gray.
[9] Objective synonym of *Hydrictis* Pocock, 1921. Not *Hydrogale* Kaup, 1829, a synonym of the soricid genus *Neomys*, nor *Hydrogale* Pomel, 1848, a synonym of *Sorex*.
[10] Sometimes as subgenus; sometimes as separate genus.
[11] Sumatra, Java, Borneo.
[12] Malta, Sicily.

†*Algarolutra* Malatesta & Willemsen, 1989:285-286.
    L. Pleist.; Mediterranean.[1]

[1] Corsica, Sardinia.

†*Sardolutra* Willemsen, 1992:35.
    L. Pleist. and/or R.; Mediterranean.[1]

[1] Sardinia.

†*Satherium* Gazin, 1934:143.[1]
    E. Plioc.-E. Pleist.; N.A.

[1] Proposed as subgenus of *Lutra*.

Pteronura Gray, 1837:580. Giant otter, flat-tailed otter, saro.
>   [= *Saricovia* Lesson, 1842:72.]
>   L. Pleist.-R.; S.A.

Lutrogale Gray, 1865:127.[1] Simung.
>   [Including †*Isolalutra* Symeonidis & Sondaar, 1975:11-24.]
>   E. Pleist., R.; E. Indies.[2,3] L. Pleist.; Mediterranean.[4] R.; As.

>   [1] "Usually believed to be the barang, rather than the simung, of Java, but Pohle has shown that this is an error" (Simpson 1945:115).
>   [2] Java in the early Pleistocene.
>   [3] Sumatra and ?Borneo in the Recent.
>   [4] Crete.

Tribe **Aonychini** Sokolov, 1973:51.
>   [= Aonyxina Sokolov, 1973:51; Aonychini Davis, 1978.]
>   M. Mioc., M.-L. Pleist.; Eu. L. Mioc. and/or Plioc., R.; As. L. Plioc.-R.; Af.
>   Pleist. and/or R.; Mediterranean. R.; E. Indies.

†*Limnonyx* Crusafont Pairó, 1950:131-142.
>   M. Mioc.; Eu.

†*Cyrnaonyx* Helbing, 1935:337.
>   M.-L. Pleist.; Eu.

Aonyx Lesson, 1827:157. African small-clawed otter, clawless otter.
>   [= *Anahyster* Murray, 1861:157-158.] [Including *Paraonyx* Hinton, 1921:195.[1]]
>   L. Mioc. and/or Plioc.; As. L. Plioc.-R.; Af.

>   [1] Sometimes as subgenus.

Amblonyx Rafinesque, 1832:62.[1] Oriental small-clawed otter.
>   [= *Leptonyx* Lesson, 1842:72[2]; *Micraonyx* J. A. Allen, 1919:24.[3]]
>   R.; E. Indies.[4] R.; As.[5]

>   [1] Described as subgenus of *Lutra*; sometimes considered subgenus of *Aonyx*.
>   [2] Not *Leptonyx* Swainson, 1821, a bird, nor *Leptonyx* Gray, 1837, a synonym of *Leptonychotes*, a phocid seal.
>   [3] Replacement name for *Leptonyx* Lesson.
>   [4] Sumatra, Java, Borneo, S.W. Philippines.
>   [5] S. and S.E. Asia.

†*Megalenhydris* Willemsen & Malatesta, 1987:85.[1]
>   Pleist. and/or R.; Mediterranean.[2]

>   [1] Tentatively assigned to Aonychini.
>   [2] Sardinia.

Tribe **Enhydrini** Gray, 1825:340.
>   [= Enhydrina Gray, 1825:340[1]; Enhydrini Sokolov, 1973:51.] [Including
>   Enhydriodonina Sokolov, 1973:51.[2]]
>   M. Mioc., ?L. Mioc., E.-L. Plioc.; Af. M. Mioc.-L. Plioc., R.; As.[3] L. Mioc., E.
>   Pleist.; Eu. L. Mioc.-E. Plioc., ?L. Plioc., E. Pleist.-R.; N.A.[4] R.; Pacific.[5]

>   [1] Proposed as tribe of Phocidae; as tribe of Mustelidae in Gray, 1865:102.
>   [2] Proposed as subtribe of Aonyxini.
>   [3] Pacific coast of N.E. Asia.
>   [4] Pacific North America.
>   [5] Northern Pacific.

†*Vishnuonyx* Pilgrim, 1932:93.
>   M. Mioc.; Af. M. Mioc.; As.

†*Sivaonyx* Pilgrim, 1931:74.
>   L. Mioc.-E. Plioc.; As. L. Mioc.; Eu.

†*Enhydriodon* Falconer, 1868:331-338.
>   [= †*Amyxodon* Cautley & Falconer, 1835:707.[1]]
>   L. Mioc.-L. Plioc.; As. ?L. Mioc.; Eu. E.-L. Plioc.; Af.

>   [1] *Nomen nudum.*

†*Tyrrhenolutra* Hürzeler, 1987:40.
>   L. Mioc.; Eu.

†*Enhydritherium* Berta & Morgan, 1985:810.
    ?L. Mioc.; Af. L. Mioc.; Eu. L. Mioc.-E. Plioc., ?L. Plioc.; N.A.

†*Paludolutra* Hürzeler, 1987:34.
    L. Mioc.; Eu. ?Plioc.; Af.

Enhydra Flemming, 1822:187.[1] Sea otter.
    [= *Pusa* Oken, 1816:985[2]; *Latax* Gloger, 1827:511; *Enhydrus* Dahl, 1823[3]; *Enydris* J.
    B. Fischer, 1829:228; *Enhydris* Temminck, 1838:285; *Euhydris* Jordan, 1888:339;
    Enhydria Zittel, 1893:652; *Enhydrus* MacLeay, 1925.[3]]
    E. Pleist.; Eu. E. Pleist.-R.; N.A.[4] R.; As.[5] R.; Pacific.[6]

    [1] Not preoccupied by *Enhydris* Latreille, 1802, a reptile, nor by *Enhydrus* Rafinesque, 1815, an incorrect
      spelling of *Enhydris* Latreille, 1802, that is now rejected (International Commission on Zoological
      Nomenclature, Opinion 710).
    [2] Not *Pusa* Scopoli, 1777, a synonym of *Phoca*.
    [3] Suppressed.
    [4] Pacific North America.
    [5] Formerly Pacific coast of N.E. Asia south to Hokkaido, Japan, but now extinct in much of its range.
    [6] Northern Pacific.

Subfamily **Mephitinae** Bonaparte, 1845:1. Skunks.
    [= Mephitina Bonaparte, 1845:1; Mephitinae Gill, 1872:65; Mephitidae Ledje &
    Arnason, 1996:641, 647.] [Including Myadina Gray, 1825:339[1]; Mydaina Gray,
    1864:506; Mydainae Pocock, 1921:834; Mydaini G. Petter, 1971:567.]
    E.-L. Mioc.; Eu. L. Mioc.-E. Plioc.; As. L. Mioc.-R.; N.A. L. Plioc.-R.; S.A. R.;
    E. Indies. R.; Cent. A.

    [1] As tribe.

†*Miomephitis* Dehm, 1950:111.
    E. Mioc.; Eu.

†*Plesiomeles* Viret & Crusafont Pairo, 1955:447.
    [Including †*Grivamephitis* de Beaumont, 1973:93.[1]]
    M.-L. Mioc.; Eu.

    [1] Proposed as subgenus.

†*Nannomephitis* Kretzoi, 1952:8.
    L. Mioc.; Eu.

†*Promephitis* Gaudry, 1861:722.
    L. Mioc.-E. Plioc.; As. L. Mioc.; Eu.

†*Pliogale* Hall, 1930:149.
    L. Mioc.; N.A.

†*Martinogale* Hall, 1930:147.
    L. Mioc.; N.A.

†*Mesomephitis* G. Petter, 1967:100.
    L. Mioc.; Eu.

†*Brachyopsigale* Hibbard, 1954:227.
    E. Plioc.; N.A.

†*Buisnictis* Hibbard, 1950:164.
    E.-L. Plioc.; N.A.

Mydaus F. Cuvier, in Geoffroy Saint-Hilaire & F. Cuvier, 1821:pl. and 2 pp.. Stink badgers,
    teledu.
    [= *Mydaon* Gloger, 1841:57.] [Including *Suillotaxus* Lawrence, 1939:63.]
    R.; E. Indies.[1]

    [1] Sumatra, Java, Borneo, southwest Philippines.

Spilogale Gray, 1865:150. Spotted skunk, Mexican polecat.
    [= †*Hemiacis* Cope, 1869:3.[1]]
    L. Plioc.-R.; N.A. R.; Cent. A.

    [1] *Nomen nudum*.

Mephitis É. Geoffroy Saint-Hilaire & G. Cuvier, 1795:187. Striped skunk.
    [= *Chincha* Lesson, 1842:67; *Mephritis* Gray, 1821:301[1]; *Mephites* Gray, 1847:20;
    Mammephitisus Herrera, 1899:30.[2]] [Including *Leucomitra* Howell, 1901:39.[3]]

L. Plioc.-R.; N.A. R.; Cent. A.

[1] *Lapsus calami?*
[2] See the International Code of Zoological Nomenclature, Article 1(b)(8), and Palmer (1904:25-26).
[3] Proposed as subgenus of *Chincha*.

Conepatus Gray, 1837:581. Hog-nosed skunk.
  [= *Mamconepatus* Herrera, 1899:4.[1]] [= or including *Thiosmus* Lichtenstein, 1838:272.] [Including *Marputius* Gray, 1837:581; *Oryctogale* Merriam, 1902:161-162.[2]]
  L. Plioc.-R.; S.A. M. Pleist.-R.; N.A. R.; Cent. A.

[1] See the International Code of Zoological Nomenclature, Article 1(b)(8), and Palmer (1904:25-26).
[2] As subgenus.

†*Brachyprotoma* Brown, 1908:176.[1] Short-faced skunk.
  E.-L. Pleist.; N.A.

[1] *Nomen nudum* in Hay, 1905:300.

†*Osmotherium* Cope, 1896:385-386.
  [Including †*Pelycictis* Cope, 1896:390-391.]
  Pleist.; N.A.

Subfamily **Mustelinae** Fischer de Waldheim, 1817:372.
  [= Mustelini Fischer de Waldheim, 1817:372; Mustelina Gray, 1825:339[1]; Mustelinae Gill, 1872:5, 64; Musteleae Trouessart, 1904:187-215.[2]] [Including Martina Wagner, 1841:216; Martinae Burmeister, 1850:12; Zorillina Gray, 1865:103[3]; Zorillinae Gill, 1872:6, 65; Putoriinae Schlosser, in Zittel & Schlosser, 1911:392; Tayrinae Pocock, 1921:833; Grisoninae Pocock, 1921:835; Ictonychinae Pocock, 1921:835; Lyncodontinae Pocock, 1921:836; Taxidiinae Pocock, 1920:436; Taxideinae Pocock, 1922:835[4]; Taxidini G. Petter, 1971:567; Taxideini Kalandadze & Rautian, 1992:64; Galictinae Reig, 1956:225.[5]]
  E. Mioc.-R.; Eu. E. Mioc.-R.; N.A. M. Mioc., ?E. Pleist., M. Pleist.-R.; Af. M. Mioc.-R.; As. ?L. Plioc., E. Pleist., R.; Mediterranean. E. Pleist.-R.; S.A. R.; E. Indies. R.; Cent. A.[6]

[1] Proposed as tribe of Felidae.
[2] Used as a tribe.
[3] Proposed as tribe of Mustelidae.
[4] Correction of Taxidiinae.
[5] Objective synonym of Grisoninae.
[6] Also Recent of Atlantic islands (Iceland, the Azores, and possibly Sao Tomé) and New Zealand by introduction.

Martes Pinel, 1792:55. Martens, sable, fisher, pine martens.
  [= *Martes* Frisch, 1775:11[1]; *Foina* Gray, 1865:108.[2]] [Including *Zibellina* Kaup, 1829:31, 34; †*Hydrocyon* Lartet, 1851:17[3]; *Pekania* Gray, 1865:107-108[2]; *Charronia* Gray, 1865:108[4]; *Lamprogale* Ognev, 1928:26[5]; †*Sansanictis* Ginsburg, 1961:107.]
  E. Mioc.-R.; Eu. M. Mioc.; Af.[6] ?M. Mioc., L. Mioc.-R.; As. L. Mioc.-R.; N.A. R.; E. Indies.[7] R.; Mediterranean.

[1] Invalid. Not published in a consistently binominal work.
[2] Described as subgenus of *Martes*.
[3] Not *Hydrocyon* G. Cuvier, 1820, a fish.
[4] Described as subgenus of *Martes*. Not preoccupied by *Charonia* Gistel, 1848, a mollusk.
[5] Replacement name for *Charronia* Gray, 1865.
[6] N. Africa.
[7] Isolates in Sumatra, Java and Borneo.

†*Proputorius* Filhol, 1890:112.
  Mioc.; Eu. L. Mioc.; As.

†*Dinogale* Cook & Macdonald, 1962:564.
  E. Mioc.; N.A.

†*Miomustela* Hall, 1930:151.
  E.-M. Mioc.; N.A.

†*Plionictis* Matthew, 1924:135.
  M.-L. Mioc.; N.A.

†*Anatolictis* Schmidt-Kittler, 1976:23.
>   M. Mioc.; As.[1]
>   [1] W. Asia (Turkey).

†*Sthenictis* Peterson, 1910: Errata et Corrigenda.[1]
>   [= †*Brachygale* Peterson, 1910:277, 278.]
>   M.-L. Mioc.; N.A.
>   [1] Originally printed as †*Brachygale* on pages 277 and 278, but changed to †*Sthenictis* by Peterson in the Errata et Corrigenda.

†*Paramartes* Kretzoi, 1952:14.
>   L. Mioc.; Eu.

†*Marcetia* G. Petter, 1967:103.
>   M. Mioc.; Eu.

†*Circamustela* G. Petter, 1967:106.
>   M. Mioc.; Eu.

†*Cernictis* Hall, 1935:137-138.
>   L. Mioc.; N.A.

†*Pliotaxidea* Hall, 1944:11.
>   L. Mioc.; N.A.

†*Sinictis* Zdansky, 1924:28.
>   L. Mioc.; As. L. Mioc.; Eu.

Mustela Linnaeus, 1758:45-47. Weasels, ermine, stoat, ferret, polecat, minks, kolonok.
>   [= *Mustella* Scopoli, 1777:491, 498; *Mustelina* Bogdanov, 1871:167 (sep.); *Eumustela* Acloque, 1899:62; *Mammustelaus* Herrera, 1899:20.[1]] [Including *Putorius* Frisch, 1775:11[2]; *Putorius* Cuvier, 1817:147[3]; *Arctogale* Kaup, 1829:30; *Ictis* Kaup, 1829:35, 40-41[4]; *Foetorius* Keyserling & Blasius, 1840:68; *Gale* Wagner, 1841:234; *Lutreola* Wagner, 1841:239-242[5]; *Vison* Gray, 1843:64-65[3]; *Gymnopus* Gray, 1865:118-119[6]; *Hydromustela* Bogdanov, 1871:167; *Kolonokus* Satunin, 1911:264; *Plesiogale* Pocock, 1921:805[7]; *Grammogale* Cabrera, 1940:15[3]; *Pocockictis* Kretzoi, 1947:285.[8]]
>   L. Mioc.-R.; Eu.[9] L. Mioc.-R.; N.A.[9] E. Plioc.-R.; As.[9] Pleist., R.; S.A. R.; Af.[10] R.; E. Indies. R.; Cent. A.[11]
>   [1] See the International Code of Zoological Nomenclature, Article 1(b)(8) and Palmer (1904: 25-26).
>   [2] Invalid. Not published in a consistently binominal work. See the International Code of Zoological Nomenclature, Article 11(c).
>   [3] Sometimes as subgenus.
>   [4] Preoccupied by *Ictis* Schinz, 1824(?), a paradoxine genus.
>   [5] Described as subgenus; raised to generic rank by Merriam, 1888:433.
>   [6] *Nomen nudum* in Gray, 1842.
>   [7] Replacement name for *Gymnopus* Gray, 1865, but preoccupied by †*Plesiogale* Pomel, 1846.
>   [8] Replacement name for *Plesiogale* Pocock, 1921.
>   [9] "As with most living genera, pre-Pleistocene records may be viewed with doubt since they almost always use the generic name in a broader sense than among recent mammals and often prove to be erroneous when the species become better known" (Simpson 1945:113).
>   [10] N. Africa.
>   [11] Also Recent of Atlantic islands (Iceland, the Azores, and possibly Sao Tomé) and New Zealand by introduction.

Vormela Blasius, 1884:9. Marbled polecat.
>   [Including †*Pliovormela* Kormos, 1934:132.]
>   Plioc. and/or E. and/or M. Pleist., L. Pleist.-R.; As. Plioc. and/or E. and/or M. Pleist., L. Pleist.-R.; Eu.[1]
>   [1] E. Europe.

†*Trigonictis* Hibbard, 1941:344.
>   E. Plioc.-E. Pleist.; N.A.

†*Canimartes* Cope, 1892:1029.
>   L. Plioc.; N.A.

†*Pannonictis* Kormos, 1931:167-177.
>   [Including †*Xenictis* Kretzoi, 1938:125.]

Plioc. and/or E. Pleist.; As. E. Plioc.-M. Pleist.; Eu. L. Plioc. and/or E. Pleist.;
Mediterranean.[1] L. Plioc.-E. Pleist.; N.A.

[1] Sicily.

†*Tisisthenes* R. A. Martin, 1973:924.
L. Plioc.; N.A.

†*Enhydrictis* Stefani, 1891:222-239.
[= †*Enhydrichtus* Stefani, 1891:222-239; †*Enhydrictis* Forsyth Major, 1902:87.[1]]
L. Plioc.-E. Pleist.; Eu. E. Pleist.; As.[2] E. Pleist.; Mediterranean.

[1] A justified emendation of Stefani's earlier name. See the International Code of Zoological Nomenclature, Article 33(b)(ii).
[2] W. Asia.

†*Baranogale* Kormos, 1934:129-158.
L. Plioc.-E. Pleist.; Eu.

Taxidea Waterhouse, 1839:153. American badger.
[= *Mamtaxideaus* Herrera, 1899:27.[1]]
L. Plioc.-R.; N.A.

[1] See the International Code of Zoological Nomenclature, Article 1(b)(8) and Palmer (1904:25-26).

†*Sminthosinus* Bjork, 1970:26.
L. Plioc.; N.A.

†*Stipanicicia* Reig, 1956:225.
E. Pleist.; S.A.

Lyncodon Gervais, 1844:685. Patagonian weasel.
[= *Lycodon* Gray, 1869:134.]
E. Pleist.-R.; S.A.

Poecilictis Thomas & Hinton, 1920:367. Libyan striped weasel.
?E. Pleist., R.; Af.[1,2]

[1] Angola in the early Pleistocene?
[2] N. Africa in the Recent.

Galictis Bell, 1826:552. Grisons.
[= *Mustela* Bechstein, 1800[1]; *Grison* Oken, 1816; *Grison* Palmer, 1904:300, 831[2]; *Gulo* Desmarest, 1820[3]; *Huro* I. Geoffroy Saint-Hilaire, 1835:37; *Galictes* Bell, 1837:45; *Gallictis* Waterhouse, 1839:21[4]; *Grisonia* Gray, 1865:122; *Mamgalictisus* Herrera, 1899:22.[5]] [Including *Grisonella* Thomas, 1912:46.]
Pleist., R.; N.A.[6] Pleist., R.; S.A. R.; Cent. A.

[1] Not *Mustela* Linnaeus, 1758, another musteline.
[2] *Grison* Oken, 1816, was published in a work that was not consistently binominal. See the International Code of Zoological Nomenclature, Article 11(c), in which the key word is "consistently."
[3] Not *Gulo* Pallas, 1780:25, a gulonine mustelid.
[4] Misprint.
[5] See the International Code of Zoological Nomenclature, Article 1(b)(8) and Palmer (1904:25-26).
[6] Southern North America (Mexico north to San Luis Potosí).

†*Oxyvormela* Rabeder, 1973:675.
E. Pleist.; Eu.

Ictonyx Kaup, 1835:352, 353. Zorille, African polecat, stinkmuishond.
[= *Zorilla* I. Geoffroy Saint-Hilaire, 1826:215-216[1]; *Rhabdogale* Wiegmann, 1838:278-279; *Ozolictis* Gloger, 1841:xxix, 74-75[2]; *Ictomys* Roberts, 1936:228.[3]]
M. Pleist.-R.; Af.

[1] Proposed as a subgenus of "Muffer"; raised to generic rank by F. Cuvier, 1829:449. *Zorilla* was suppressed by the International Commission on Zoological Nomenclature, Opinion 818, 1967.
[2] Replacement name. This name was also mistakenly applied to *Thiosmus* Lichtenstein, 1838, itself a synonym of *Conepatus.*
[3] Misprint.

Poecilogale Thomas, 1883:370-371. White-naped weasel.
R.; Af.

Eira C. H. Smith, in Jardine, 1842:201.[1] Taira.
[= *Galera* Gray, 1843:xx, 67[2]; *Tayra* Palmer, 1904:252, 830[3]; *Eirara* Lund, 1839:232[4]; *Eraria* Gray, 1843:xx.]

R.; N.A.[5] R.; Cent. A. R.; S.A.

[1] The date may be 1839 (Palmer 1904:252).
[2] *Galera* Browne, 1789:475 was non-Linnaean.
[3] *Tayra* Oken, 1816:1001 was non-Linnaean.
[4] A candidate for suppression. Fide Palmer (1904:252), *Eirara* Lund, 1839:245 (in another earlier paper) was a *nomen nudum*.
[5] Southern North America (tropical Mexico north to Sinaloa and Tamaulipas).

Subfamily †**Leptarctinae** Gazin, 1936:207.
E.-L. Mioc.; N.A. M. Mioc.; As. M.-L. Mioc.; Eu.

†*Leptarctus* Leidy, 1856:311.
[Including †*Mephititaxus* White, 1941:92; †*Hypsoparia* Dorr, 1954:179.]
E.-L. Mioc.; N.A. M. Mioc.; As.

†*Craterogale* Gazin, 1936:199.
E. Mioc.; N.A.

†*Trocharion* Forsyth Major, 1903:534-538.
M.-L. Mioc.; Eu.

Subfamily **Mellivorinae** Gray, 1865:103.
[= Mellivorina Gray, 1865:103[1]; Mellivorinae Gill, 1872:6, 64; Mellivorini Webb, 1969:68.] [Including †Peruniinae Orlov, 1947:5; †Ischyrictini Ginsburg, 1977:226.]
L. Olig., M.-L. Mioc., Plioc., R.; As.[2] E.-L. Mioc.; Eu. M. Mioc., E. Plioc.-R.; Af. L. Mioc., L. Plioc.; N.A.

[1] Proposed as tribe of Mustelidae.
[2] An undescribed mellivorine mustelid occurs in the late Oligocene Shand Mbr. of the Hsanda Gol Fm., Mongolia.

†*Dehmictis* Ginsburg & Morales, 1992:112.
E. Mioc.; Eu.

†*Hoplictis* Ginsburg, 1961:100.[1]
E.-M. Mioc.; Eu.

[1] Described as a subgenus of †*Ischyrictis*.

†*Ischyrictis* Helbing, 1930:644.
[Including †*Laphyctis* Viret, 1933:18.]
E.-M. Mioc.; Eu. M. Mioc.; As.

†*Eomellivora* Zdansky, 1924:61.
[Including †*Sivamellivora* Kretzoi, 1942:320.]
M.-L. Mioc.; Eu. L. Mioc.; As. L. Mioc.; N.A.

†*Hadrictis* Pia, 1939:538.
L. Mioc.; Eu.

†*Mellalictis* Ginsburg, 1977:226.
M. Mioc.; Af.[1]

[1] N. Africa.

†*Promellivora* Pilgrim, 1932:65.
L. Mioc.; As.

†*Beckia* L. J. Bryant, 1968:2.
L. Mioc.; N.A.

†*Perunium* Orlov, 1947:5.
[= or including †*Pliogulo* Wosnessensky, 1937.[1]]
L. Mioc.; Eu.[2]

[1] *Nomen oblitum*.
[2] E. Europe.

Mellivora Storr, 1780:34, table A. Ratel, honey badger.
[= *Melivora* Gray, 1847:19; *Ratelus* Gray, 1825; *Ratellus* Gray, 1827:118; *Rattelus* Swainson, 1835:158-160; *Ursitaxus* Hodgson, 1835:522, 564; *Ursotaxus* Blyth, 1840:86; *Melitoryx* Gloger, 1841:xxix, 57[1]; *Lipotus* Sundevall, 1843:199.[2]]
L. Mioc.; Eu. E. Plioc.-R.; Af. Plioc., R.; As.

[1] New name for *Mellivora*.
[2] New name for *Ratelus*.

†*Ferinestrix* Bjork, 1970:19.
L. Plioc.; N.A.

Subfamily **Guloninae** Gray, 1825:339.
[= Gulonina Gray, 1825:339[1]; Guloneae Trouessart, 1904:187-215[2]; Guloninae Miller, 1912:432; Gulonini Webb, 1969:68.]
E. Mioc., L. Mioc.-E. Plioc., M. Pleist.-R.; Eu. M. Mioc.-E. Plioc., M. Pleist.-R.; As. L. Mioc., E. Pleist.-R.; N.A.
[1] Proposed as tribe.
[2] Used as a tribe.

†*Iberictis* Ginsburg & Morales, 1992:116.
E. Mioc.; Eu.

†*Plesiogulo* Zdansky, 1924:38.
M. Mioc.-E. Plioc.; As. L. Mioc.-E. Plioc.; Eu. L. Mioc.; N.A.

Gulo Pallas, 1780:25. Wolverine, glutton, carajou.
[= *Gulo* Frisch, 1775:17.[1]]
E. Pleist.-R.; N.A. M. Pleist.-R.; As. M. Pleist.-R.; Eu.
[1] Invalid. Not published in a consistently binominal work. See the International Code of Zoological Nomenclature, Article 11(c).

Subfamily **Melinae** Bonaparte, 1838:111.
[= Taxina Gray, 1825:339[1]; Melina C. L. Bonaparte, 1838:111; Melinae Burmeister, 1850:13; Melinidae Gray, 1869:120; Melini G. Petter, 1971:567.] [Including Helictidina Gray, 1865:103, 152[2]; Helictidinae Gill, 1872:6, 64; Helictidini G. Petter, 1971:567.]
E. Mioc.-R.; Eu. M. Mioc.-R.; As. R.; E. Indies.
[1] As tribe.
[2] Proposed as tribe of Mustelidae.

†*Trochictis* von Meyer, 1842:584.
E.-M. Mioc.; Eu. M. Mioc.; As.[1]
[1] W. Asia (Turkey).

†*Taxodon* Lartet, 1851:15-16.
[Including †*Melidellavus* Ginsburg, 1961:120.]
M. Mioc.; Eu.

†*Trochotherium* Fraas, 1870:161-164.
L. Mioc.; Eu.

†*Melodon* Zdansky, 1924:55.
[Including †*Parameles* Young & Liu, 1950:74.[1]]
L. Mioc.-E. Plioc., ?M. Pleist.; As.
[1] Not *Perameles* É. Geoffroy Saint-Hilaire, 1804: 150, a bandicoot; nor †*Parameles* Roschchin, 1949, a synonym of †*Arctomeles*. Synonymy tentative.

†*Lartetictis* Ginsbourg & Morales, 1996:664.
M. Mioc.; Eu.

†*Adroverictis* Ginsbourg & Morales, 1996:667.
M. Mioc.; As.[1] L. Mioc.; Eu.
[1] W. Asia (Turkey).

†*Parataxidea* Zdansky, 1924:47.
L. Mioc.; As. L. Mioc.; Eu.

†*Promeles* Zittel, 1893:650-651.
L. Mioc.; Eu.

†*Palaeomeles* de Villalta Comella & Crusafont Pairó, 1943:182.
L. Mioc.; Eu.

†*Sabadellictis* G. Petter, 1963:1-44.
L. Mioc.; Eu.

†*Polgardia* Kretzoi, 1952:12.
L. Mioc.; Eu.

†*Arctomeles* Stach, 1951:129-157.

[= or including †*Parameles* Roschchin, 1949:97-106.[1]]
E. Plioc.; Eu.[2]

[1] Not *Perameles* É. Geoffroy Saint-Hilaire, 1804, a bandicoot; nor †*Parameles* Young & Liu, 1950:74, a probable synonym of †*Melodon.*
[2] E. Europe.

Meles Boddaert, 1785:45. Old World badger.

[= *Meles* Brisson, 1762:13[1]; *Taxus* É. Geoffroy Saint-Hilaire & G. Cuvier, 1795:187; *Melesium* Rafinesque, 1815:59[2]; *Eumeles* Gray, 1865:140; *Meledes* Kastschenko, 1925:21.[3]]
E. Plioc.-R.; As. L. Plioc.-R.; Eu.

[1] Not proposed in a consistently binominal work. However, see the International Code of Zoological Nomenclature, Article 11(c)(i) for confusion.
[2] New name for *Taxus.*
[3] Proposed as subgenus.

Arctonyx F. Cuvier, 1825:3. Hog badger.

[= *Syarchus* Gloger, 1841:xxviii, 55; *Synarchus* Gray, 1865:137; *Trichomanis* Hubrecht, 1891:241-242.]
E. Pleist.-R.; As. R.; E. Indies.[1]

[1] Sumatra.

Melogale I. Geoffroy Saint-Hilaire, 1831:137. Ferret badgers.

[= *Rhinogale* Gloger, 1841:xxix, 75.] [Including *Helictis* Gray, 1831:94[1]; *Nesictis* Thomas, 1922:194.]
R.; E. Indies.[2] R.; As.

[1] Published later than *Melogale.*
[2] Borneo, Java.

Family **Procyonidae** Gray, 1825:339.

[= Procyonina Gray, 1825:339; Procyonidae Bonaparte, 1850: unpaged chart.; Procyonoidea Tedford, 1976:372.] [Including Cercoleptidae Bonaparte, 1838:111; Cercoleptina Gray, 1864:706; Nasuidae Gray, 1869:214 or 238; Bassariscidae Gray, 1869:246; Ailuridae Flower, 1869:15; Bassaricyonidae Coues, 1887:516; †Amphictidae Winge, 1895:46, 51; †Broilianini Dehm, 1950:109; †Broilianinae de Beaumont, 1964:835.]
L. Eoc., E. Mioc.-R.; N.A. E. Olig.-L. Plioc.; Eu. L. Mioc., R.; As. L. Mioc.-L. Plioc., E. and/or M. Pleist., ?L. Pleist., R.; S.A. R.; Cent. A. R.; W. Indies.[1]

[1] Probably by introduction.

†*Bavarictis* Mödden, 1981:125-147.

L. Olig.; Eu.

†*Amphictis* Pomel, 1853:63-64.

[= †*Ichneugale* Jourdan, 1852[1]; †*Ichneugales* Jourdan, 1861:1012.[1]]
E. Olig.-E. Mioc.; Eu. E. Mioc.; N.A.

[1] *Nomen nudum.*

†*Broiliana* Dehm, 1950:80.

E. Mioc.; Eu.

†*Angustictis* Wolsan, 1993:372.

E. Mioc.; Eu.

Subfamily **Bassariscinae** Gray, 1869:246.

[= Bassaridae Gray, 1869:214; Bassariscidae Gray, 1869:246; Bassariscinae Pocock, 1921:421.] [Including Cercoleptidina C. L. Bonaparte, 1838:7[1]; Cercoleptina C. L. Bonaparte, 1850: unpaged chart.; Potidae Degland, 1854[2]; Potosinae Trouessart, 1904:183; Bassaricyoninae Pocock, 1921:422.]
L. Eoc., M. Mioc.-R.; N.A. E. Olig.; Eu. R.; Cent. A. R.; S.A.

[1] As subfamily.
[2] See the International Code of Zoological Nomenclature, Article 23(b).

†*Pseudobassaris* Pohle, 1917:408.

[Including †*Mustelavus* J. Clark, in Scott & Jepsen, 1936:107; †*Mustelictis* Lange, 1969:2870-2872.]
L. Eoc.; N.A. E. Olig.; Eu.

Bassariscus Coues, 1887:516. Ring-tailed cats, cacomistles.

[= *Bassaris* Lichtenstein, 1831:512-513[1]; *Mambassarisus* Herrera, 1899:26.[2]]
[Including *Wagneria* Jentink, 1886:127-9[3]; *Jentinkia* Trouessart, 1904:184;
†*Probassariscus* Merriam, 1911:205-207; †*Rhapsodus* Linares, 1981:114.[4]]
M. Mioc.-R.; N.A.

[1] Not *Bassaris* Hübner, 1816-1821, a genus of Lepidoptera.
[2] See the International Code of Zoological Nomenclature, Article 1(b)(8) and Palmer (1904:25-26).
[3] Objective synonym of *Jentinkia*. Not *Wagneria* Robineau-Desvoidy, 1830, a genus of Diptera.
[4] Objective synonym of †*Probassariscus*.

Bassaricyon J. A. Allen, 1876:20-23. Olingos.
R.; Cent. A. R.; S.A.

Potos É. Geoffroy Saint-Hilaire & G. Cuvier, 1795:187. Kinkajou.

[= *Kinkajou* Lacépède, 1799:7; *Kinkaschu* G. Fischer de Waldheim, 1813:14;
Cercoleptes Illiger, 1811:127; *Mamcercolepteus* Herrera, 1899:19.[1]]
R.; N.A.[2] R.; Cent. A. R.; S.A.

[1] See the International Code of Zoological Nomenclature, Article 1(b)(8) and Palmer (1904:25-26).
[2] Southern North America (tropical Mexico north to Guerrero and S. Tamaulipas).

Subfamily **Procyoninae** Gray, 1825:339.

[= Procyonina Gray, 1825:339[1]; Procyoninae Gill, 1872:6, 67.] [Including Nasuidae
Gray, 1869:214 or 238; Nasuinae Gill, 1872:6, 67.]
E. Mioc.-R.; N.A. L. Mioc.-L. Plioc., E. and/or M. Pleist., ?L. Pleist., R.; S.A. R.;
Cent. A. R.; W. Indies.[2]

[1] Proposed as tribe.
[2] Probably by introduction.

†*Edaphocyon* J. A. Wilson, 1960:988.
E.-M. Mioc.; N.A.

†*Arctonasua* Baskin, 1982:72.
M.-L. Mioc.; N.A.

†*Protoprocyon* Linares, 1981:113-121.
[= †*Lichnocyon* Baskin, 1982:82.]
L. Mioc.; N.A.

†*Paranasua* Baskin, 1982:84.
L. Mioc.; N.A.

†*Cyonasua* Ameghino, 1885:19-22.
[Including †*Amphinasua* Moreno & Mercerat, 1891:235; †*Pachynasua* Ameghino,
1904:119.]
L. Mioc.-L. Plioc.; S.A.

Nasua Storr, 1780:35, table A. Coatis, coatimundis.

[= *Coati* Lacépède, 1799:7[1]; *Nasica* South, 1845:343[2]; *Mamnasuaus* Herrera,
1899:26.[3]]
E. Plioc., R.; N.A.[4] Pleist., R.; S.A. R.; Cent. A.

[1] *Coati* Frisch, 1775 is not available.
[2] Misprint.
[3] See the International Code of Zoological Nomenclature, Article 1(b)(8) and Palmer (1904:25-26).
[4] Southwestern North America.

†*Chapalmalania* Ameghino, 1908:424.
E.-L. Plioc.; S.A.

Procyon Storr, 1780:35, table A. Raccoons.

[= *Lotor* G. Cuvier & É. Geoffroy Saint-Hilaire, 1795:187; *Campsiurus* Link,
1795:52, 87; *Mamprocyonus* Herrera, 1899:18.[1]] [Including *Euprocyon* Gray,
1864:705-706; †*Mixophagus* Cope, 1869:3.]
L. Plioc.-R.; N.A. Pleist., R.; S.A. R.; Cent. A. R.; W. Indies.[2]

[1] See the International Code of Zoological Nomenclature, Article 1(b)(8) and Palmer (1904:25-26).
[2] Probably by introduction.

†*Brachynasua* Kraglievich & C. Ameghino, 1925:181-191.
E. and/or M. Pleist.; S.A.

Nasuella Hollister, 1915:118, 148. Mountain coati, little coatimundi.
R.; S.A.

Subfamily †**Simocyoninae** Dawkins, 1868:1.
[= †Simocyonidae Dawkins, 1868:1; †Simocyoninae Zittel, 1893:632.]
M.-L. Mioc.; Eu. M.-L. Mioc.; N.A. L. Mioc.; As.

†*Alopecocyon* Camp & VanderHoof, 1940:320.
[= †*Alopecodon* Viret, 1933:9[1]; †*Viretius* Kretzoi, 1947:286.]
M.-L. Mioc.; Eu.

[1] Not †*Alopecodon* Broom, 1908, a reptile.

†*Actiocyon* Stock, 1947:85.
M. Mioc.; N.A.

†*Simocyon* Wagner, 1858:366.[1]
[= †*Pseudocyon* Wagner, 1857:123.[2]] [Including †*Metarctos* Gaudry, 1860:926;
†*Amphalopex* Kaup, 1861:15; †*Pliocyon* Thorpe, 1921:477[3]; †*Araeocyon* Thorpe,
1922:97; †*Protursus* Crusafont Pairó & Kurtén, 1976:22.[4]]
L. Mioc.; As. L. Mioc.; Eu. L. Mioc.; N.A.

[1] Replacement name.
[2] Not †*Pseudocyon* Lartet, 1851:16, an amphicyonine.
[3] Not †*Pliocyon* Matthew, 1918, an amphicyonine.
[4] *Nomen nudum* in Crusafont Pairó & Kurtén, 1971:155; 1973:59.

Subfamily **Ailurinae** Gray, 1843:xxi.
[= Ailurina Gray, 1843:xxi; Ailuridae Flower, 1869:15; Ailuridae Gray, 1869:214;
Ailurinae Trouessart, 1885:25.]
E.-L. Plioc.; Eu. E. Plioc.; N.A. R.; As.

†*Parailurus* Schlosser, 1899:9-19.
E.-L. Plioc.; Eu. E. Plioc.; N.A.

Ailurus F. Cuvier, in É. Geoffroy Saint-Hilaire & F. Cuvier, 1825: pl. and 3 pp.. Lesser panda,
wah, red panda.
[= *Arctaelurus* Gloger, 1841:xxviii, 55; *Aelurus* Agassiz, 1846:9[1]; *Aelurus* Flower,
1870:752-769.]
R.; As.

[1] Not *Aelurus* Klug, 1842: 42.

Grandorder **LIPOTYPHLA** Haeckel, 1866:clx.[1] Insectivores.
[= SUBTERRANEA Illiger, 1811:123; INSECTIVORES G. Cuvier, 1817;
INSECTIVORA Bowdich, 1821:24, 31[2]; GRADIENTIA Gill, 1871:7;
LIPOTYPHLIFORMES Kinman, 1994:37.] [Including EUINSECTIVORA Heim de
Balsac & Bourlière, 1954:1656[3]; ERINACEOTA Van Valen, 1967:261.[4]]

[1] Proposed as suborder. Used as an order by Gregory, 1910, and as a grandorder by McKenna, 1975:41.
[2] Not to be confused with INSECTIVORAE J. E. Gray, 1821:299, which was used for what later became
MICROCHIROPTERA.
[3] Heim de Balsac & Bourlière divided lipotyphlans into suborders CHRYSOCHLOROIDA and
EUINSECTIVORA.
[4] Proposed as suborder to include Erinaceoidea and Soricoidea.

†*Ndamathaia* Jacobs, Anyonge & Barry, 1987:10.
E. Mioc.; Af.

Family †**Adapisoriculidae** Van Valen, 1967:271.
[= †Adapisoriculinae Van Valen, 1967:271; †Adapisoriculidae Gheerbrant & D. E.
Russell, 1989:3.]
E. Paleoc.-E. Eoc.; Eu. L. Paleoc.-E. Eoc.; Af.

†*Adapisoriculus* Lemoine, 1885:205, 212-213.
[= †*Nycticonodon* Quinet, 1964:4.[1]]
L. Paleoc.-E. Eoc.; Eu.

[1] Invalid junior objective synonym and also invalid under the International Code of Zoological
Nomenclature, Article 13(b).

†*Afrodon* Gheerbrant, 1988:1303.
L. Paleoc.-E. Eoc.; Af. L. Paleoc.; Eu.

†*Bustylus* Gheerbrant & D. E. Russell, 1991:468.
E.-L. Paleoc.; Eu.

Order **CHRYSOCHLORIDEA** Broom, 1915:353.[1]
[= CHRYSOCHLOROIDA Heim de Balsac & Bourlière, 1954:1653[2];
CRYSOCHLOROIDA Heim de Balsac & Borliere, 1955:1653; CHRYSOCHLORIDA
Butler, 1972:261[3]; CHRYSOCHLOROMORPHA MacPhee & Novacek, 1993:13,
26.[3]]

[1] Considered to be of ordinal rank by Broom, who attributed the name to Dobson, but Dobson's use of the
name, in 1882, was at the rank of superfamily.
[2] Proposed as a suborder. All other lipotyphlans were placed in a suborder EUINSECTIVORA.
[3] Proposed as a suborder of order LIPOTYPHLA.

Family **Chrysochloridae** Gray, 1825:335. Golden moles.
[= Chrysochlorina Gray, 1825:335; Chrysochloridae Mivart, 1868:150;
Chrysochloridea Dobson, 1882:2[1]; Chrysochloridoidea Gill, 1872:19; Chrysochloroidea
Gregory, 1910:267.]
E. Mioc., L. Plioc., Pleist., R.; Af.

[1] Used as superfamily.

Subfamily †**Prochrysochlorinae** Butler, 1984:170.
E. Mioc.; Af.

†*Prochrysochloris* Butler & Hopwood, 1957:11.
E. Mioc.; Af.

Subfamily **Chrysochlorinae** Gray, 1825:335. Golden moles, kruipmollen.
[= Chrysochlorina Gray, 1825:335; Chrysochloridini Winge, 1917:124.] [Including
Eremitalpinae Simonetta, 1957; Amblysominae Simonetta, 1957.]
L. Plioc., Pleist., R.; Af.

Chrysochloris Lacépède, 1799:7. Cape golden moles, kaapse kruipmollen.
[= *Chrysoris* Rafinesque, 1815:59[1]; *Chrysochlora* de Blainville, 1840:111[2]; *Aspalax*
Wagler, 1830:14[3]; *Engyscopus* Gistel, 1848:viii.[4]] [= or including *Ducantalpa* Boitard,
1842:118.]
L. Plioc., Pleist., R.; Af.

[1] New name for *Chrysochloris*.
[2] Replacement name. Not *Chrysochlora* Latreille, 1825?, a dipteran.
[3] Replacement name. Not *Aspalax* Desmarest, 1804, a synonym of the rodent *Spalax*.
[4] Replacement name.

†*Proamblysomus* Broom, 1941:215-216.
L. Plioc. and/or Pleist.; Af.[1]

[1] S. Africa.

Chrysospalax Gill, 1884:137. Giant golden moles, reuse kruipmollen.
[Including *Bematiscus* Cope, 1892:127.]
L. Plioc., Pleist., R.; Af.[1]

[1] S. Africa.

Amblysomus Pomel, 1848:247.[1] African golden moles, kruipmollen.
[Including *Calcochloris* Mivart, 1867:282; *Chalcochloris* Palmer, 1904:151[2];
Chrysotricha Broom, 1907:303[3]; *Neamblysomus* Roberts, 1924:64; *Huetia* Forcart,
1942:2.[4]]
L. Plioc., Pleist., R.; Af.

[1] *Amblysomus* was supposed by Palmer (1904:92) to be preoccupied by *Amblysoma* Westwood, 1841, a
genus of Hymenoptera.
[2] Correction of *Calcochloris*.
[3] Objective synonym of *Calcochloris*.
[4] Described as subgenus of *Chrysochloris*.

Chlorotalpa Roberts, 1924:64.
[Including *Carpitalpa* Lundholm, 1955:288[1]; *Kilimatalpa* Lundholm, 1955:288.[1]]
L. Plioc., Pleist., R.; Af.

[1] Described as subgenus.

Cryptochloris Shortridge & Carter, 1938:284. DeWinton's golden moles, DeWintonse kruipmollen.
R.; Af.[1]
[1] S. Africa.

Eremitalpa Roberts, 1924:63. Grant's desert golden mole, woestyn kruipmollen.
R.; Af.[1]
[1] S. Africa.

Order **ERINACEOMORPHA** Gregory, 1910:464.[1]
[= ERINACEI Peters, 1864:20; ACULEATA Haeckel, 1866:cxlix[2]; ERINACEOMORPHA Saban, 1954:429.]
[1] Proposed as section of suborder LIPOTYPHLA of order INSECTIVORA; raised to suborder by Saban, 1954:429; raised to ordinal rank by McKenna, 1975:41.
[2] *Nomen oblitum.* Not ACULEATA É. Geoffroy Saint-Hilaire, 1795, a synonym of Tachyglossidae, nor ACULEATA Illiger, 1811:89, a subset of caviomorph rodents.

†*Adunator* D. E. Russell, 1964:47.
[Including †*Mckennatherium* Van Valen, 1965:435; †*Mackennatherium* Van Valen, 1965:435[1]; †*Diacocherus* Gingerich, 1983:238.]
E.-L. Paleoc.; N.A. L. Paleoc.; Eu.
[1] Our emendation. See the International Code of Zoological Nomenclature, Appendix D, Recommendation 21(a).

†*Talpavoides* Bown & Schankler, 1982:50.
E. Eoc.; N.A.

†*Talpavus* Marsh, 1872:128.
?E. Eoc.; Eu. E.-M. Eoc.; N.A.

†*Seia* D. E. Russell & Gingerich, 1981:278.
E. and/or M. Eoc.; As.[1]
[1] S. Asia (Pakistan).

Subfamily †**Litocherinae** Gingerich, 1983:228.
L. Paleoc.; N.A.

†*Litocherus* Gingerich, 1983:232.
L. Paleoc.; N.A.

Subfamily †**Diacodontinae** Trouessart, 1879:223, 235.
E. Eoc.; N.A.

†*Diacodon* Cope, 1875:11-12.
E. Eoc.; N.A.

Family †**Sespedectidae** Novacek, 1985:3, **new rank.**
[= †Sespedectinae Novacek, 1985:3.] [Including †Scenopagidae Butler, 1988:125.]
?L. Paleoc., E. Eoc.-E. Olig.; N.A. ?E. Eoc.; Eu.

Subfamily †**Scenopaginae** Novacek, 1985:1, 23.[1]
[= †Scenopagidae Butler, 1988:125.]
?L. Paleoc., E. Eoc.-E. Olig.; N.A. ?E. Eoc.; Eu.
[1] Novacek did not clearly state that the subfamily was new, but we have found no earlier usage.

†*Scenopagus* McKenna & Simpson, 1959:1.
?L. Paleoc., E.-M. Eoc.; N.A. ?E. Eoc.; Eu.

†*Ankylodon* Patterson & McGrew, 1937:269.
M. Eoc.-E. Olig.; N.A.

Subfamily †**Sespedectinae** Novacek, 1985:3.
M. Eoc.; N.A.

†*Proterixoides* Stock, 1935:214.
M. Eoc.; N.A.

†*Crypholestes* Novacek, 1980:1135.
[= †*Cryptolestes* Novacek, 1976:18.[1]]

M. Eoc.; N.A.

[1] Not †*Cryptolestes* Ganglbaur, 1899, a coleopteran, nor *Cryptolestes* Tate, 1934, a caenolestid marsupial.

†*Sespedectes* Stock, 1935:218.
M. Eoc.; N.A.

Family †**Amphilemuridae** Hill, 1953:491.
[= †Amphilemuriden Heller, 1935:296[1]; †Amphilemuridae Hill, 1953:491.] [Including †Dormaaliidae Quinet, 1964:14.]
E.-L. Eoc.; Eu. E. Eoc., ?M. Eoc.; N.A. ?E. Olig.; As.[2]

[1] Informally as "neue Familie der Amphilemuriden." See the International Code of Zoological Nomenclature, Article 11(f)(iii), which legitimizes only certain non-Latinized names published prior to 1900.
[2] A possible amphilemurid has been reported from the Caijiachong Formation, early Oligocene of Yunnan, PRC.

Subfamily †**Placentidentinae** D. E. Russell, Louis & D. E. Savage, 1973:42.
[= †Placentidentine D. E. Russell, Louis & D. E. Savage, 1973:42[1]; †Placentidentidae Benton, 1987:31.[2]]
E. Eoc.; Eu.

[1] Proposed as subfamily.
[2] Authorship probably different and earlier, but not yet certain. Benton's listing was taken from an unspecified earlier paper. Also listed at family rank by Carroll, 1988.

†*Placentidens* D. E. Russell, Louis & D. E. Savage, 1973:42.
E. Eoc.; Eu.

Subfamily †**Amphilemurinae** Hill, 1953:491, **new rank**.
[= †Amphilemuriden Heller, 1935:296[1]; †Amphilemuridae Hill, 1953:491; †Amphilemurini Novacek, 1985:23.] [Including †Dormaaliidae Quinet, 1964:14; †Dormaalinae D. E. Russell, Louis & D. E. Savage, 1975:136.]
E.-L. Eoc.; Eu. E. Eoc., ?M. Eoc.; N.A.

[1] As "Familie der Amphilemuriden."

†*Macrocranion* Weitzel, 1949:15.
[= †*Aculeodens* Weitzel, 1949:19.] [Including †*Messelina* Tobien, 1962:36; †*Dormaalius* Quinet, 1964:12.]
E.-M. Eoc.; Eu. E. Eoc., ?M. Eoc.; N.A.

†*Amphilemur* Heller, 1935:296.
[Including †*Alsaticopithecus* Hürzeler, 1948:343.]
M. Eoc.; Eu.

†*Pholidocercus* von Koenigswald & Storch, 1983:447, 454.
M. Eoc.; Eu.

†*Gesneropithex* Hürzeler, 1946:355.
M.-L. Eoc.; Eu.

Family †**Adapisoricidae** Schlosser, 1887:91, 138.
[= †Adapisoricida Haeckel, 1895:578, 582.]
L. Paleoc.-E. Eoc.; Eu.

†*Adapisorex* Lemoine, 1883:76.
L. Paleoc.-E. Eoc.; Eu.

†*Neomatronella* D. E. Russell, Louis & D. E. Savage, 1975:177.
[= †*Matronella* D. E. Russell, Louis & D. E. Savage, 1975:136.[1]]
E. Eoc.; Eu.

[1] Preoccupied by an invertebrate generic name.

Family †**Creotarsidae** Hay, 1930:487.
[= †Creotarsinae Van Valen, 1967:261; †Creotarsiinae Kalandadze & Rautian, 1992:58.[1]]
E. Eoc.; N.A.

[1] Invalid emendation or *lapsus calami*.

†*Creotarsus* Matthew, 1918:611.
E. Eoc.; N.A.

Superfamily **Erinaceoidea** Fischer de Waldheim, 1817:372.
[= Erinacini Fischer de Walheim, 1817:372; Erinaceoidea Gill, 1872:18.]

†*Leipsanolestes* Simpson, 1928:6.
L. Paleoc.; N.A.

†*Eolestes* Bown & Schankler, 1982:52.
E. Eoc.; N.A.

†*Dartonius* Novacek, Bown & Schankler, 1985:15.
E. Eoc.; N.A.

Family **Erinaceidae** Fischer de Waldheim, 1817:372. Hedgehogs, moon rats.
[= Erinacini Fischer de Waldheim, 1817:372; Erinacidae Gray, 1821:300; Erinaceidae Bonaparte, 1838:111; Erinacei Peters, 1864; Erinaceidea Haeckel, 1866:clx.] [Including Hylomidae J. Anderson, 1879:138; †Parasoricidae Schlosser, 1887:91; Echinosoricidae Shaw & Wong, 1959; †Changlelestidae Tong & Wang, 1993:19-32.[1]]
?E. Paleoc., L. Paleoc., M. Eoc., E. Olig.-L. Mioc.; N.A. E. Eoc.-E. Plioc., E. Pleist.-R.; As. L. Eoc.-E. Plioc., L. Plioc. and/or Pleist., R.; Eu. Olig., ?Plioc., R.; Mediterranean.[2] E.-L. Mioc., ?E. Plioc., L. Plioc., Pleist., R.; Af. R.; E. Indies.[3]
[1] (Tung & Wang).
[2] Erinaceidae indet. in the ?Pliocene of Sardinia.
[3] Also New Zealand and the Atlantic (Canary Islands) by introduction.

†*Litolestes* Jepsen, 1930:513.
?E. Paleoc., L. Paleoc.; N.A.

†*Cedrocherus* Gingerich, 1983:237.
L. Paleoc.; N.A.

†*Scaptogale* Trouessart, 1897:207.
[= †*Echinogale* Pomel, 1848:163, 251.[1]]
Mioc.; Eu.
[1] Not *Echinogale* Wagner, 1841, a genus of tenrecs.

Subfamily †**Tupaiodontinae** Butler, 1988:124.
[Including †Changlelestidae Tong & Wang, 1993:20.[1]]
E. Eoc.-E. Olig., ?L. Olig.; As. M. Eoc.; N.A.
[1] (Tung & Wang).

†*Entomolestes* Matthew, 1909:541.
M. Eoc.; N.A.

†*Tupaiodon* Matthew & Granger, 1924:1.
M. Eoc., Olig.; As.

†*Changlelestes* Tong & Wang, 1993:20.[1]
E. Eoc.; As.
[1] (Tung & Wang).

†*Ictopidium* Zdansky, 1930:7.
L. Eoc.-E. Olig.; As.

Subfamily **Galericinae** Pomel, 1848:149. Moon rats, gymnures.
[= Galerices Pomel, 1848:249; Galericini Butler, 1948:488.] [Including Gymnurinae Gill, 1872:19; Hylomidae J. Anderson, 1879:138; Hylomyinae Frost, Wozencraft & Hoffmann, 1991:23[1]; Parasoricidae Schlosser, 1887:91; Echinosoricinae Cabrera, 1925:57, 58; Echinosoricidae Shaw & Wong, 1959.]
M.-L. Eoc., Olig., E. Mioc.-E. Plioc., R.; As.[2] L. Eoc.-E. Plioc.; Eu. Olig.; Mediterranean. E. Olig.-L. Mioc.; N.A. E.-L. Mioc., ?E. Plioc.; Af. R.; E. Indies.
[1] Frost, Wozencraft & Hoffmann rejected emendation of Pomel's *Galerices* because it was not "generally accepted" before Butler's use in 1948. *Galerices* and its emendation certainly have been accepted by many workers since 1948, however, although primarily by paleontologists familiar with both Recent and extinct mammals.
[2] Unnamed galericine hedgehogs have been reported from the late Eocene and early and "middle" Oligocene of Kazakhstan.

Tribe **Galericini** Pomel, 1848:249.
> [= Galerices Pomel, 1848:249; Galericini Butler, 1948:488.] [Including Parasoricidae Schlosser, 1887:91; Gymnurini Winge, 1917:124; Echinosoricini Kalandadze & Rautian, 1992:59.]
> M. Eoc., E. Mioc.-E. Plioc., R.; As.[1] Olig.; Mediterranean. E. Olig.-E. Plioc.; Eu. L. Olig.-L. Mioc.; N.A. E.-L. Mioc., ?E. Plioc.; Af. R.; E. Indies.
>
> [1] S.E. Asia in the Recent.

†*Eochenus* Wang & Li, 1990:168.
> M. Eoc.; As.

†*Galerix* Pomel, 1848:164.
> [Including †*Parasorex* Meyer, 1865:844-845.]
> E. Olig.-E. Plioc.; Eu. E.-M. Mioc., ?L. Mioc.; Af. E.-L. Mioc.; As.

†*Tetracus* Aymard, 1850:81, 82, 105.
> [= or including †*Camphotherium* Filhol, 1884:62, 63; †*Gomphotherium* Filhol, 1884[1]; †*Comphotherium* Filhol, 1884:11-13.]
> Olig.; Mediterranean.[2] Olig.; Eu.
>
> [1] Not †*Gomphotherium* Burmeister, 1837, a proboscidean.
> [2] Balearic Islands.

†*Ocajila* Macdonald, 1963:166.
> L. Olig.-E. Mioc.; N.A.

†*Pseudogalerix* Gaillard, 1928:110.[1]
> E.-L. Mioc.; Eu.
>
> [1] Or, if a *nomen nudum* in 1928, then Gaillard, 1929:46.

†*Lantanotherium* Filhol, 1888:25.
> [= †*Lanthanotherium* Filhol, 1891:23.[1]] [Including †*Rubitherium* Crusafont Pairó & Villalta, 1955:11-172.[2]]
> Mioc.; Af. ?E. and/or ?M. Mioc., L. Mioc.; As. M.-L. Mioc.; Eu. M.-L. Mioc.; N.A.
>
> [1] Unjustified emendation [see the International Code of Zoological Nomenclature, Article 33(b)(iii)].
> [2] Proposed as subgenus.

Echinosorex de Blainville, 1838:742.[1] Bulau, moon rat, gymnure.
> [= *Gymnura* Lesson, 1827:171.[2]]
> ?M. Mioc., R.; As.[3,4] R.; E. Indies.[5]
>
> [1] Originally *Echino-Sorex*, a subgenus of *Sorex*.
> [2] Not *Gymnura* Van Hasselt, 1823, a ray, nor *Gymnura* Kuhl, 1824.
> [3] S. Asia in the middle Miocene.
> [4] S.E. Asia in the Recent.
> [5] Sumatra and Borneo.

†*Schizogalerix* Engesser, 1980:63.
> M. Mioc.-E. Plioc.; As.[1] L. Mioc., ?E. Plioc.; Af.[2] L. Mioc.; Eu.
>
> [1] W. Asia (Turkey).
> [2] N. Africa.

†*Deinogalerix* Freudenthal, 1972:3.
> L. Mioc.; Eu.

Hylomys Müller, 1839:50, 60. Lesser gymnures.
> [= *Hyllomys* Pomel, 1848:251.] [Including *Neotetracus* Trouessart, 1909:389; *Neohylomys* Shaw & Wong, 1959:422.]
> L. Mioc., R.; As.[1] R.; E. Indies.[2]
>
> [1] S.E. Asia.
> [2] Sumatra, Java, Borneo.

Podogymnura Mearns, 1905:436. Philippine gymnures.
> R.; E. Indies.[1]
> [1] Philippines.

Tribe †**Neurogymnurini** Butler, 1948:488.
> [= †Cayluxotheriini Winge, 1917:123.[1]]
> L. Eoc.-L. Olig.; Eu.
> [1] Winge placed this *nomen oblitum* in quotation marks.

†*Neurogymnurus* Filhol, 1877:52.
[= †*Cayluxotherium* Filhol, 1880:1579; †*Necrogymnurus* Lydekker, 1891:621[1];
†*Caluxotherium* Waterhouse, 1902:58.]
L. Eoc.-L. Olig.; Eu.
[1] And of most authors prior to Simpson, 1945.

Tribe †**Protericini** Butler, 1948:488.
E. Olig.; N.A.

†*Proterix* Matthew, 1903:228.
E. Olig.; N.A.

Subfamily **Erinaceinae** Fischer de Waldheim, 1817:372. Hedgehogs.
[= Erinacini Fischer de Waldheim, 1817:372; Erinaceina C. L. Bonaparte, 1837:9[1];
Erinaceinae Gill, 1872:18.]
E.-L. Olig., ?E. Mioc., M. Mioc., ?L. Mioc., E. Pleist.-R.; As. ?E. Olig., L. Olig.-
L. Mioc., L. Plioc. and/or Pleist., R.; Eu. E.-M. Mioc., L. Plioc., Pleist., R.; Af.
E.-L. Mioc.; N.A. R.; Mediterranean.[2]
[1] As subfamily.
[2] Also the Atlantic (Canary Islands) and New Zealand by introduction.

Tribe †**Amphechinini** Gureev, 1979:147.
[= †Amphechina Kalandadze & Rautian, 1992:59.]
E.-L. Olig., ?E. Mioc., M. Mioc.; As. ?E. Olig., L. Olig.-L. Mioc., L. Plioc. and/or
E. Pleist.; Eu. E.-M. Mioc.; Af. E.-M. Mioc.; N.A.

†*Amphechinus* Aymard, 1850:109-110.
[Including †*Palaeoerinaceus* Filhol, 1879:12; †*Postpalerinaceus* Crusafont Pairó &
Villalta, 1948:320-333[1]; †*Protechinus* Lavocat, 1961:68.]
?E. Olig., L. Olig.-L. Mioc., L. Plioc. and/or E. Pleist.; Eu. E.-M. Mioc.; Af. ?E.
Mioc., M. Mioc.; As. E.-M. Mioc.; N.A.
[1] Proposed as subgenus.

†*Palaeoscaptor* Matthew & Granger, 1924:2.
E.-L. Olig., ?E. Mioc.; As.

†*Parvericius* Koerner, 1940:837-862.
L. Olig.; As. E.-M. Mioc.; N.A.

†*Dimylechinus* Hürzeler, 1944:460.
E. Mioc.; Eu.

Tribe **Erinaceini** Fischer de Waldheim, 1817:372.
[= Erinacini Fischer de Waldheim, 1817:372; Erinacina Gray, 1825:339[1]; Erinaceini
Butler, 1948:488.]
E. Mioc., L. Plioc., Pleist., R.; Af. E.-L. Mioc.; N.A. M. Mioc., ?L. Mioc., E.
Pleist.-R.; As. M. Mioc., ?L. Mioc., Pleist., R.; Eu. R.; Mediterranean.[2]
[1] As tribe.
[2] Also the Atlantic (Canary Islands) and New Zealand by introduction.

†*Gymnurechinus* Butler, 1956:3.
E. Mioc.; Af.

†*Stenoechinus* Rich & Rasmussen, 1973:7.
E. Mioc.; N.A.

†*Untermannerix* Rich, 1981:15.
M.-L. Mioc.; N.A.

Erinaceus Linnaeus, 1758:52. Hedgehog.
[= or including *Setiger* É. Geoffroy Saint-Hilaire, 1803:70-72[1]; *Peroechinus* Fitzinger,
1866:565; *Herinaceus* Mina-Palumbo, 1868:37.] [Including *Atelerix* Pomel,
1840:251; *Aethechinus* Thomas, 1918:194; *Mesechinus* Ognev, 1951:8.[2]]
?L. Mioc., E. Pleist.-R.; As. ?L. Mioc., Pleist., R.; Eu. L. Plioc., Pleist., R.; Af.
R.; Mediterranean.[3]
[1] Not *Setiger* G. Cuvier, 1800, a synonym of the tenrec genus *Hemicentetes*.
[2] Proposed as subgenus.

[3] Balearic Islands, Corsica, Sardinia, Sicily, Malta, Crete.
Also the Atlantic (Canary Islands) and New Zealand by introduction.

Hemiechinus Fitzinger, 1866:565.

> [= *Ericius* Sundevall, 1842:223[1]; *Erinaceolus* Ognev, 1928:168.] [Including
> *Paraechinus* Trouessart, 1879:242[2]; *Macroechinus* Satunin, 1907:189; †*Mioechinus*
> Butler, 1948:479.]
> M. Mioc., L. Pleist.-R.; As. M. Mioc.; Eu. R.; Af.[3]

[1] Not *Ericius* Tilesius von Tilenau, 1813, a genus of fishes. Proposed as subgenus of *Erinaceus*.
[2] Described as a subgenus of *Erinaceus*.
[3] N. Africa.

Subfamily †**Brachyericinae** Butler, 1948:488.

> [= †Brachyericini Butler, 1948:488; †Brachyericinae McKenna & Holton, 1967:2;
> †Brachyericina Kalandadze & Rautian, 1992:59.]
> E.-L. Olig., ?E. Mioc.; As. E.-L. Mioc.; N.A.

†*Exallerix* McKenna & Holton, 1967:3.

> E. Olig.; As.

†*Metexallerix* Qiu & Gu, 1988:199.

> L. Olig., ?E. Mioc.; As.

†*Brachyerix* Matthew, 1933:2.

> E.-L. Mioc.; N.A.

†*Metechinus* Matthew, 1929:95.

> L. Mioc.; N.A.

Superfamily **Talpoidea** Fischer de Waldheim, 1817:372.

> [= Talpini Fischer de Waldheim, 1817:372; TALPINA Peters, 1864:20; Talpoidea
> Novacek, 1975:10.]

Family †**Proscalopidae** K. M. Reed, 1961:473.

> [= †Proscalopinae K. M. Reed, 1961:473; †Proscalopidae C. A. Reed & Turnbull,
> 1965:164, 165; †Arctoryctinae C. A. Reed & Turnbull, 1965:164, 165.]
> ?E. Olig.; As. E. Olig.-M. Mioc.; N.A.

†*Oligoscalops* K. M. Reed, 1961:474.

> ?E. Olig.; As. E. Olig.; N.A.

†*Proscalops* Matthew, 1901:370, 375-376.

> [= †*Arctoryctes* Matthew, 1907:172.]
> E. Olig.-M. Mioc.; N.A.

†*Mesoscalops* K. M. Reed, 1960:2.

> E.-M. Mioc.; N.A.

Family **Talpidae** Fischer de Waldheim, 1817:372. Moles.

> [= Talpini Fischer de Waldheim, 1817:372; Talpidae Gray, 1825:335; Talpida Haeckel,
> 1866:clx.] [= or including Orycteri de Blainville, 1834.] [Including Mygaladae Gray,
> 1821:300; Myogalidae A. Milne-Edwards, 1868:267, 272[1]; Scalopidae Cope,
> 1889:876.]
> ?M. Eoc., L. Eoc.-R.; Eu.[2] Olig., E. Mioc.-R.; As. Olig., E. Mioc., ?E. Plioc., L.
> Plioc. and/or E. Pleist., ?L. Pleist., R.; Mediterranean.[3,4] L. Olig.-R.; N.A.

[1] A. Milne-Edwards, 1868-1874.
[2] Talpidae indet. possibly in the middle Eocene of Europe.
[3] Talpid indet. reported from the Pliocene of Capo Mannu, Sardinia.
[4] Sardinia and the Balearic Islands.

†*Eotalpa* Sigé, Crochet & Insole, 1977:144.

> L. Eoc.; Eu.

†*Galeospalax* Pomel, 1848:161.

> L. Olig.; Eu.

†*Quadrodens* Macdonald, 1970:21.

> [= †*Palaeoscalopus* Macdonald, 1940:23.]
> L. Olig.; N.A.

†*Nuragha* de Bruijn & Rümke, 1974:64.
    E. Mioc.; Mediterranean.[1]
    [1] Sardinia.

†*Desmanodon* Engesser, 1980:116.
    M. Mioc.; As.[1] M. Mioc.; Eu.
    [1] W. Asia (Turkey).

†*Hyporyssus* Pomel, 1848:161.
    L. Mioc.; Eu.

Subfamily **Uropsilinae** Dobson, 1883:128.
    [= Uropsili Dobson, 1883:128; Uropsilinae Thomas, 1912; Uropsilini Winge, 1917:123.]
    Olig., R.; As.[1] E.-L. Mioc.; N.A.
    [1] Unnamed uropsiline genera have been reported from the Oligocene Aksyrskaya Svita and Buranskaya Svita, Kazakhstan.

†*Mystipterus* Hall, 1930:319.
    [Including †*Mydecodon* Wilson, 1960:38.]
    E.-L. Mioc.; N.A.

Uropsilus A. Milne-Edwards, in David, 1871:92. Shrew mole.
    [Including *Nasillus* Thomas, 1911:49; *Rhynchonax* Thomas, 1911:49.]
    R.; As.

Subfamily **Talpinae** Fischer de Waldheim, 1817:372.
    [= Talpini Fischer de Waldheim, 1817:372; Talpina Gray, 1825:339; Talpina Pomel, 1848:246; Talpinae Murray, 1866:319.]
    L. Eoc.-R.; Eu. Olig., E. Mioc.-R.; As. ?Olig., L. Plioc. and/or E. Pleist., ?L. Pleist., R.; Mediterranean.[1] L. Olig.-R.; N.A.
    [1] Sardinia and the Balearic Islands.

†*Teutonotalpa* Hutchison, 1974:233.
    E. Mioc.; Eu.

†*Achlyoscapter* Hutchison, 1968:96.
    M. Mioc.-E. Plioc.; N.A.

Tribe **Scaptonychini** Van Valen, 1967:263.
    [= Scaptonychina Kalandadze & Rautian, 1992:55.]
    L. Eoc.-E. Mioc., ?M. Mioc., ?L. Plioc. and/or ?E. Pleist.; Eu. ?Olig.; Mediterranean.[1] M. Mioc., R.; As.[2,3]
    [1] Balearic Islands.
    [2] W. Asia (Turkey) in the middle Miocene.
    [3] E. Asia in the Recent.

†*Myxomygale* Filhol, 1890:177.
    L. Eoc.-E. Mioc.; Eu. ?Olig.; Mediterranean.[1] M. Mioc.; As.[2]
    [1] Balearic Islands.
    [2] W. Asia (Turkey).

†*Geotrypus* Pomel, 1848:159.
    [= or including †*Protalpa* Filhol, 1877:52.]
    L. Eoc.-L. Olig., E. and/or M. Mioc., ?L. Plioc. and/or ?E. Pleist.; Eu.

Scaptonyx Milne-Edwards, 1871:92.
    ?M. Mioc., ?L. Plioc. and/or ?E. Pleist.; Eu. R.; As.[1]
    [1] E. Asia.

Tribe **Talpini** Fischer de Waldheim, 1817:372.
    [= Talpina Gray, 1825:339[1]; Talpae Dobson, 1883:128.]
    ?Olig., L. Plioc. and/or E. Pleist., ?L. Pleist., R.; Mediterranean.[2] E. Mioc.-R.; Eu. E. Plioc., Pleist., R.; As.
    [1] As tribe.
    [2] Mallorca, Sardinia.

Talpa Linnaeus, 1758:52-53. Common Old World mole.

> [Including *Mogera* Pomel, 1848:246[1]; *Heterotalpa* Peters, 1863:86[2]; *Talpops* Gervais, 1868:92[2]; *Parascaptor* Gill, 1875:110; *Asioscalops* Stroganov, 1941:271[2]; *Asioscaptor* Schwartz, 1948:36.[3]]
> ?Olig., L. Plioc. and/or E. Pleist., ?L. Pleist., R.; Mediterranean.[4] E. Mioc.-R.; Eu. Pleist., R.; As.[5]

> [1] Proposed as subgenus; sometimes as separate genus.
> [2] Proposed as subgenus.
> [3] New name for *Asioscalops*.
> [4] Mallorca, Sardinia.
> [5] Including Japan in the late Pleistocene.

Scaptochirus A. Milne-Edwards, 1867:375.

> [= *Chiroscaptor* Heude, 1898:36-40.]
> E. Plioc., Pleist., R.; As.

Euroscaptor Miller, 1940:443.

> [= *Eoscalops* Stroganov, 1941:270.]
> R.; As.[1]

> [1] E. Asia (Japan).

Nesoscaptor Abe, Shiraishi & Arai, 1991:48.

> R.; As.[1]

> [1] West coast of Uotsuri-jima, Ryukyu Islands, Japan.

Tribe **Scalopini** Trouessart, 1879:49. American moles.

> [= Scalopeae Trouessart, 1879:49; Scalopes Dobson, 1883:128; Scalopidae Cope, 1889:876; Scalopinae Thomas, 1912:397; Scalopini Van Valen, 1967:264.]
> L. Olig.-L. Mioc., ?L. Plioc. and/or ?E. Pleist.; Eu. L. Olig.-R.; N.A. E. Mioc.-R.; As.

†*Hugueneya* van den Hoek Ostende, 1989:15.

> L. Olig.-E. Mioc.; Eu.

†*Yunoscaptor* Storch & Qiu, 1991:615.

> L. Mioc.; As.

†*Yanshuella* Storch & Qiu, 1983:111.

> M. Mioc., L. Mioc. and/or E. Plioc., E. Pleist.; As. L. Mioc. and/or E. Plioc.; N.A.

Subtribe **Parascalopina** Hutchison, 1968:8.

> L. Olig.-L. Mioc., M. Pleist.-R.; N.A. E. Mioc., E. Plioc.-R.; As. E.-L. Mioc., ?L. Plioc. and/or ?E. Pleist.; Eu.

†*Domninoides* Green, 1956:152.

> L. Olig., M.-L. Mioc.; N.A. L. Mioc.; Eu.

†*Scalopoides* R. W. Wilson, 1960:42.

> E.-L. Mioc.; N.A. ?M. Mioc.; Eu.

†*Proscapanus* Gaillard, 1899:23.

> [= †*Alloscapanus* Baudelot, 1968:125.]
> E. Mioc.; As. E.-M. Mioc.; Eu.

Scapanulus Thomas, 1912:396.

> E. Plioc.-R.; As. ?L. Plioc. and/or ?E. Pleist.; Eu.

Parascalops True, 1894:242. Hairy-tailed mole.

> [= *Perascalops* Beddard, 1902:518.]
> M. Pleist.-R.; N.A.

Subtribe **Scalopina** Trouessart, 1879:49.

> [= Scalopeae Trouessart, 1879:49.] [Including Scapanei Winge, 1917:128-129.]
> M. Mioc.-R.; N.A.

†*Scapanoscapter* Hutchison, 1968:83.

> M. Mioc.; N.A.

Scapanus Pomel, 1848:247. Western American mole.

> [= *Scapaneus* Winge, 1917:123.] [Including †*Xeroscapheus* Hutchison, 1968:87.[1]]

L. Mioc.-R.; N.A.

[1] Proposed as subgenus.

Scalopus É. Geoffroy Saint-Hilaire, 1803:77-78. American mole.
[= *Scalpos* Brooks, 1910:28.[1]] [= or including *Scalops* Illiger, 1811:126; *Talpasorex*
Lesson, 1827:124-125[2]; †*Hesperoscalops* Hibbard, 1941:337.]
L. Mioc.-R.; N.A.

[1] Misprint.
[2] Not *Talpasorex* Schinz, 1821:191-192, a synonym of *Condylura*.

Tribe **Urotrichini** Dobson, 1883:128.
[= Urotrichi Dobson, 1883:128; Urotrichina Kalandadze & Rautian, 1992:55.]
Olig., M. Mioc., L. Mioc. and/or E. Plioc., L. Pleist.-R.; As.[1] L. Olig.-L. Mioc.,
?L. Plioc. and/or ?E. Pleist.; Eu. ?L. Mioc., R.; N.A.

[1] Cf. Urotrichini reported from the Oligocene Buranskaya Svita, Zaisan basin, Kazakhstan.

†*Paratalpa* Lavocat, 1951:30.
L. Olig.-L. Mioc.; Eu.

†*Palurotrichus* Ziegler, 1985:146.
E. Mioc.; Eu.

Urotrichus Temminck, 1838:285-286.[1]
[Including *Dymecodon* True, 1886:97-98; *Dimecodon* Coues, 1889:1621.]
?M. Mioc.; Eu. L. Pleist.-R.; As.[2]

[1] 1838-1839 (?1841).
[2] E. Asia (Japan).

Neurotrichus Günther, 1880:441. "American" shrew mole.
?L. Mioc., R.; N.A. ?L. Plioc. and/or ?E. Pleist.; Eu.

†*Quyania* Storch & Qiu, 1983:91.
M. Mioc., L. Mioc. and/or E. Plioc.; As.

Tribe **Condylurini** Trouessart, 1879:51.
[= Condylureae Trouessart, 1879:51; Condylurae Dobson, 1883:128; Condylurinae
Thomas, 1912:397.]
?L. Mioc. and/or ?E. Plioc.; As. L. Mioc., M. Pleist.-R.; N.A. ?E. Plioc., L. Plioc.;
Eu.

Condylura Illiger, 1811:125. Star-nosed mole.
[= *Talpasorex* Schinz, 1821:191-192; *Astromycter* Harris, 1825:400; *Rhinaster* Wagler,
1830:14; *Astromyctes* Gray, 1843:xxi, 76; *Astromydes* Blyth, 1863:87.]
?L. Mioc. and/or ?E. Plioc.; As. L. Mioc., M. Pleist.-R.; N.A. ?E. Plioc., L. Plioc.;
Eu.[1]

[1] E. Europe.

Subfamily **Desmaninae** Mivart, 1871. Desmans.
[= Myaladae Gray, 1821:300; Myogalina C. L. Bonaparte, 1837:9[1]; Mygalina Pomel,
1848:247; Myogalidae A. Milne-Edwards, 1868:267, 272[2]; Myogalini Winge,
1917:123; Desmaninae Thomas, 1912; Desmanini Kalandadze & Rautian, 1992:55.]
Olig., E. Mioc.-E. Plioc.; As.[3,4] Olig.; Mediterranean.[5] E. Olig.-R.; Eu. L. Mioc.-
E. Plioc.; N.A.

[1] As subfamily.
[2] 1868-1874.
[3] An unnamed genus of Desmaninae has been reported from the "M." Oligocene of the Buranskaya Svita,
Kazakhstan.
[4] Also Recent of W. Asia by introduction.
[5] An indeterminate desmanine has been reported from the Oligocene of Sineu, Mallorca.

†*Mygatalpa* Schreuder, 1940:321, 324.
E. Olig.-E. Mioc.; Eu.

†*Asthenoscapter* Hutchison, 1974:215.
[= †*Astenoscapter* Bendukidze, 1993:19.[1]]
E. Mioc.; As. E.-M. Mioc.; Eu.

[1] Misspelling.

†*Desmanella* Engesser, 1972:115.

    L. Olig.-E. Mioc., L. Mioc.-E. Plioc.; Eu. M.-L. Mioc.; As.

†*Mygalea* Schreuder, 1940:318.

    [= †*Migalea* Bendukidze, 1993:11.[1]]

    E.-M. Mioc.; As. E.-L. Mioc.; Eu.

[1] Misspelling.

†*Lemoynea* Bown, 1980:112.

    L. Mioc.-E. Plioc.; N.A.

†*Archaeodesmana* Topachevsky & Pashkov, 1983:43.[1]

    [= †*Dibolia* Rümke, 1985:115[2]; †*Ruemkelia* Rzebik-Kowalska & Pawlowski, 1994:75.[3]]

    M. Mioc., E. Plioc.; As. L. Mioc.-L. Plioc.; Eu.

[1] Described as subgenus of *Desmana*.
[2] Not *Dibolia* Latreille, 1829: 155, a genus of chrysomelid beetles.
[3] Replacement name for †*Dibolia* Rümke.

Desmana Güldenstaedt, 1777:108. Desman, water mole.

    [= *Desman* Lacépède, 1799:7; *Mygale* G. Cuvier, 1800: table 1[1]; *Desmanus* Rafinesque, 1815:59; *Myale* Gray, 1821:300[2]; *Myogalea* J. B. Fischer, 1829:250-251[3]; *Caprios* Wagler, 1830:14[3]; *Myogale* Brandt, 1836:176, 182.] [Including †*Palaeospalax* Owen, 1846:25-27; †*Praedesmana* Topachevsky & Pashkov, 1983:41[4]; †*Pliodesmana* Topachevsky & Pashkov, 1983:42[4]; †*Galemodesmana* Topachevsky & Pashkov, 1983:43.[4]]

    ?E. Plioc.; As.[5] E. Plioc.-R.; Eu.[6]

[1] Not *Mygale* Latreille, 1802 (or earlier), a genus of spiders that conceivably could have priority.
[2] *Lapsus calami*?
[3] New name for *Mygale* Cuvier.
[4] Subgenus as defined.
[5] China in the early Pliocene? Also Recent of W. Asia by introduction.
[6] E. Europe in the Recent.

Galemys Kaup, 1829:118, 119. Spanish desman, almizclera.

    [= *Mygalina* I. Geoffroy Saint-Hilaire, in Gervais, 1835:45; *Galomys* Agassiz, 1846:159.]

    E. Plioc.-R.; Eu.

†*Mygalinia* Schreuder, 1940:316.

    [= †*Migalinia* Bendukidze, 1993:15.[1]]

    M. Mioc.; As. L. Mioc.; Eu.

[1] Misspelling.

†*Desmagale* Kretzoi, 1954:239-265.

    Pleist.; Eu.

Subfamily †**Gaillardiinae** Hutchison, 1968:8.

    [= †Gaillardinae Hutchison, 1968:8.]

    E.-L. Mioc.; N.A.

†*Gaillardia* Matthew, 1932:5.

    [= †*Hydroscapheus* Shotwell, 1956:724.]

    E.-L. Mioc.; N.A.

Family †**Dimylidae** Schlosser, 1887:103.

    [= †Dimylinae Gaillard, 1899:31; †Dimylini Winge, 1917:124.] [Including †Plesiodimylinae Hürzeler, 1944:40; †Cordylodinae Wegner, 1913:221; †Cordylodontinae Müller, 1967:53-77.]

    E. Olig.-L. Mioc.; Eu. M. Mioc.; As.[1]

[1] W. Asia.

†*Exoedaenodus* Hürzeler, 1944:31.

    E.-L. Olig.; Eu.

†*Dimylus* von Meyer, 1846:473.

    L. Olig.-E. Mioc.; Eu.

†*Dimyloides* Hürzeler, 1944:22.
[Including †*Pseudocordylodon* Hürzeler, 1944:26.]
L. Olig.-E. Mioc.; Eu.

†*Plesiodimylus* Gaillard, 1897:1248-1250.
E.-L. Mioc.; Eu. M. Mioc.; As.[1]
[1] W. Asia (Turkey).

†*Chainodus* Ziegler, 1990:48.
E. Mioc., ?M. Mioc.; Eu.

†*Cordylodon* von Meyer, 1859:174.
E. Mioc.; Eu.

†*Metacordylodon* Schlosser, in Zittel, 1911:368.
M. Mioc.; Eu.

†*Turkodimylus* van den Hoek Ostende, 1995:25.
E. Mioc.; As.[1]
[1] W. Asia (Anatolia).

Order **SORICOMORPHA** Gregory, 1910:465.[1]
[Including SOLENODONTA Kalandadze & Rautian, 1992:54[2]; SORICOTA
Kalandadze & Rautian, 1992:55.[3]]
[1] Proposed as section (roughly infraordinal rank) by Gregory. Used as a suborder by Saban, 1954:428; used as order by McKenna, 1975:41.
[2] In part. Named as an infraorder.
[3] Named as an infraorder.

†*Paranyctoides* Fox, 1979:119.
L. Cret.; As. L. Cret.; N.A.

†*Geomana* Brunner, 1957:367.
Pleist.; Eu.

Family †**Otlestidae** Nessov, 1985:15.
[= †Otlestinae Nessov, 1985:15; †Otlestidae Kielan-Jaworowska & Dashzeveg, 1989:348.]
L. Cret.; As.

†*Otlestes* Nessov, 1985:15.
L. Cret.; As.

Family †**Geolabididae** McKenna, 1960:134.[1]
[= †Geolabidinae McKenna, 1960:134; †Geolabididae Butler, 1972:261.] [Including †Centetodontinae Trouessart, 1879:5, 60[2]; †Metacodontidae Butler, 1948:491; †Metacodontinae Saban, 1954:429.]
L. Cret., E. Eoc.-E. Mioc.; N.A.
[1] See the International Code of Zoological Nomenclature, Article 40.
[2] *Nomen oblitum.*

†*Batodon* Marsh, 1892:258.
L. Cret.; N.A.

†*Centetodon* Marsh, 1872:209.
[= †*Hypacodon* McKenna, 1960:148.] [Including †*Embassis* Cope, 1873:4; †*Geolabis* Cope, 1884:807; †*Protictops* Peterson, 1934:374; †*Metacodon* J. Clark, in Scott & Jepsen, 1936:22.]
E. Eoc.-E. Mioc.; N.A.

†*Marsholestes* McKenna & Haase, 1992:256.
[= †*Myolestes* Matthew, 1909:541.[1]]
M. Eoc.; N.A.
[1] Not *Myolestes* Brethes, 1905: 338, a dipteran.

†*Batodonoides* Novacek, 1976:26.
M. Eoc.; N.A.

Superfamily **Soricoidea** Fischer de Waldheim, 1817:372.
> [= Soricini Fischer de Waldheim, 1817:372; Soricina Gray, 1825:339; SORICES Peters, 1864:20; Soricoidea Gill, 1872:18.] [Including †Nesophontoidea Saban, 1954:428.]

Family †**Nesophontidae** Anthony, 1916:728.
> [= †Nesophontini Winge, 1917:123; †Nesophontoidea Saban, 1954:428.]
> Pleist., R.; W. Indies.

> †*Nesophontes* Anthony, 1916:725. Extinct West Indian shrews.
>> Pleist., R.; W. Indies.[1]
>> [1] Puerto Rice, Hispaniola, Cuba, Cayman Brac.

Family †**Micropternodontidae** Stirton & Rensberger, 1964:59.
> [= †Micropternodidae Stirton & Rensberger, 1964:59; †Micropternodontidae Van Valen, 1965:639; †Micropternodontinae Van Valen, 1966:110; †Micropternodontinae Robinson, 1968:129, 134.]
> E. Paleoc.-M. Eoc.; As. ?E. Eoc.; Eu. ?M. Eoc., L. Eoc.-L. Olig.; N.A.[1]
> [1] An unnamed micropternodontid taxon has been reported from the middle Eocene Mission Valley Fm., California.

> †*Carnilestes* Wang & Zhai, 1995:132.
>> E. Paleoc.; As.

> †*Prosarcodon* McKenna, Xue & Zhou, 1984:4.
>> E. Paleoc.; As.[1]
>> [1] E. Asia.

> †*Jarveia* Nessov, 1987:209.
>> L. Paleoc.; As.[1]
>> [1] Kazakhstan.

> †*Hyracolestes* Matthew & Granger, 1925:10.
>> L. Paleoc.; As.[1] ?E. Eoc.; Eu.[2]
>> [1] E. Asia.
>> [2] Two specimens from the French early Eocene have been described as cf. †*Hyracolestes*.

> †*Sarcodon* Matthew & Granger, 1925:11.
>> [= †*Opisthopsalis* Matthew, Granger & Simpson, 1929:8.]
>> L. Paleoc.-E. Eoc.; As.

> †*Sinosinopa* Qi, 1987:15.[1]
>> M. Eoc.; As.[2]
>> [1] Technically a *nomen nudum* in Qi, 1979.
>> [2] E. Asia.

> †*Micropternodus* Matthew, 1903:204.
>> [= †*Kentrogomphios* White, 1954:404; †*Cryptoryctes* C. A. Reed, 1954:103.]
>> L. Eoc.-L. Olig.; N.A.

> †*Clinopternodus* J. Clark, 1937:308.
>> [= †*Clinodon* J. Clark, in Scott & Jepsen, 1936:13.[1]]
>> L. Eoc.; N.A.
>> [1] Not *Clinodon* Regan, 1920: 45, a genus of fishes.

Family †**Apternodontidae** Matthew, 1910:35.
> [= †Apternodontinae Matthew, 1910:35[1]; †Apternodontidae Matthew, in Osborn, 1910:519.[2]]
> L. Paleoc.-E. Olig.; N.A.[3] M. Eoc.; As.[4]
> [1] March.
> [2] October.
> [3] Late Paleocene record based on skull from Silver Coulee in the Princeton Collection, now at Yale University.
> [4] Asian record based on an unpublished specimen in Moscow from Khaichin Ula, Mongolia.

> †*Parapternodus* Bown & Schankler, 1982:67.
>> E. Eoc.; N.A.

†*Oligoryctes* Hough, 1956:538.
   [= †*Oligorcytes* Hough, 1956:538.[1]]
   M.-L. Eoc.; N.A.
   [1] One among the many misprints in that paper.

†*Apternodus* Matthew, 1903:202.
   M. Eoc.-E. Olig.; N.A.

Family **Solenodontidae** Gill, 1872:19.
   [= Solenodontinae Gill, 1872:19; Solenodontidae Dobson, 1882:87; Solenodontini Winge, 1917:124.]
   Pleist., R.; W. Indies.

Solenodon Brandt, 1833:459. Alamiqui.
   [Including *Atopogale* Cabrera, 1925:177; †*Antillogale* Patterson, 1962:2.]
   Pleist., R.; W. Indies.[1]
   [1] Cuba, Hispaniola.

Family †**Plesiosoricidae** Winge, 1917:124.
   [= †Plesiosoricini Winge, 1917:124[1]; †Plesiosoricidae Romer, 1966:381; †Plesiosoricidae Van Valen, 1967:264.] [Including †Butselidae Quinet & Misonne, 1965:6; †Butseliidae Quinet & Misonne, 1965:6.[2]]
   ?E. Eoc., M. Eoc., Olig., M. Mioc.; As. E. Olig.-L. Mioc.; Eu. E.-L. Mioc., ?E. Plioc.; N.A.
   [1] Winge attached quotation marks.
   [2] Our emendation of †Butselidae.

†*Pakilestes* D. E. Russell & Gingerich, 1981:282.
   E. and/or M. Eoc.; As.[1]
   [1] S. Asia (Pakistan).

†*Ernosorex* Wang & Li, 1990:176.
   M. Eoc.; As.

†*Butselia* Quinet & Misonne, 1965:6.
   E. Olig.; Eu.

†*Pseudoneurogymnurus* Gureev, 1979:123.
   Olig.; As.

†*Plesiosorex* Pomel, 1848:162.
   [= †*Theridosorex* Jourdan, 1859:55.[1]] [= or including †*Hibbarderix* Martin & Green, 1984:30.]
   L. Olig.-L. Mioc.; Eu. E.-M. Mioc.; N.A. M. Mioc.; As.[2]
   [1] Proposed as a synonym.
   [2] W. Asia (Anatolia).

†*Meterix* Hall, 1929:230.
   L. Mioc., ?E. Plioc.; N.A.

Family †**Nyctitheriidae** Simpson, 1928:2.
   [Including †Ceutholestidae Rose & Gingerich, 1987:18.]
   ?E. Paleoc., L. Paleoc.-M. Eoc.; N.A. L. Paleoc.-E. Eoc.; As. L. Paleoc.-E. Olig.; Eu. Olig.; Mediterranean.

†*Voltaia* Nessov, 1987:207.
   L. Paleoc.; As.[1]
   [1] Kazakhstan.

†*Plagioctenodon* Bown, 1979:62.
   ?L. Paleoc., E. Eoc.; N.A.

†*Ceutholestes* Rose & Gingerich, 1987:18.
   L. Paleoc.; N.A.

†*Limaconyssus* Gingerich, 1987:304.
   L. Paleoc.; N.A.

†*Wyonycteris* Gingerich, 1987:306.
   L. Paleoc.; Eu. L. Paleoc.; N.A.

Subfamily †**Nyctitheriinae** Simpson, 1928:2.
[= †Nyctitheriidae Simpson, 1928:2; †Nyctitheriinae Van Valen, 1967:262.]
[Including †Saturniinae Gureev, 1971:1-254.]
?E. Paleoc., L. Paleoc.-M. Eoc.; N.A. L. Paleoc.-E. Olig.; Eu. E. Eoc.; As. Olig.;
Mediterranean.

†*Leptacodon* Matthew & Granger, 1921:2.
?E. Paleoc., L. Paleoc.-E. Eoc.; N.A. E. Eoc.; Eu.

†*Remiculus* D. E. Russell, 1964:72.
L. Paleoc.; Eu.

†*Saturninia* Stehlin, 1940:301.
[= or including †*Cryptotopos* Crochet, 1974:127.]
E. Eoc.-E. Olig.; Eu. Olig.; Mediterranean.[1]
[1] Balearic Islands.

†*Pontifactor* West, 1974:983.
?L. Paleoc., E.-M. Eoc.; N.A. ?E. Eoc.; Eu.

†*Oedolius* D. E. Russell & Dashzeveg, 1986:275.
E. Eoc.; As.

†*Bumbanius* D. E. Russell & Dashzeveg, 1986:269.
E. Eoc.; As.

†*Nyctitherium* Marsh, 1872:8.
[Including †*Nyctilestes* Marsh, 1872:24.]
E.-M. Eoc.; N.A.

†*Scraeva* Cray, 1973:37.
[= †*Spalacodon* Charlesworth, in S. Wood, 1844:350[1]; †*Arvaldus* Cray, 1973:45.]
M.-L. Eoc.; Eu.
[1] *Nomen nudum* and *nomen oblitum.*

Subfamily †**Amphidozotheriinae** Sigé, 1976:79.
E. Eoc.; N.A. L. Eoc.-E. Olig.; Eu.

†*Plagioctenoides* Bown, 1979:65.
E. Eoc.; N.A.

†*Amphidozotherium* Filhol, 1876:48-49.
L. Eoc.; Eu.

†*Paradoxonycteris* Revilliod, 1922:150.[1]
L. Eoc.; Eu.
[1] *Nomen nudum* in Revilliod, 1919:93, 95.

†*Darbonetus* Crochet, 1974.
E. Olig.; Eu.

Family **Soricidae** Fischer de Waldheim, 1817:372. Shrews.
[= Soricini Fischer de Waldheim, 1817:372; Soricidae Gray, 1821:300; Soricina Gray,
1825:339; Soricida Haeckel, 1866:clx.]
M. Eoc.-R.; N.A. L. Eoc.-R.; Eu. E. Olig., E. Mioc.-R.; As. Olig., E. Plioc.-R.;
Mediterranean. ?E. and/or ?M. Mioc., L. Mioc., ?E. Plioc., L. Plioc.-R.; Af.[1] Pleist.,
R.; S.A.[2] L. Pleist.-R.; Atlantic. R.; Madagascar. R.; Indian O. R.; E. Indies. R.;
Cent. A.[3]
[1] An indet. soricid has been reported from the late Miocene Otavi breccias, Namibia.
[2] Northwestern South America.
[3] Also many Pacific islands and New Guinea by introduction.

†*Mysarachne* Pomel, 1848:162, 247.[1]
L. Olig.; Eu.
[1] This genus is probably best forgotten, but was discussed briefly in 1951 by Lavocat, who believed the name
to date from 1853.

†*Planisorex* Hibbard, in Skinner & Hibbard, 1972:78.
L. Plioc., Pleist.; N.A.

Subfamily †**Heterosoricinae** Viret & Zapfe, 1951:419.
> [= †Heterosoricidae Reumer, 1987:189-192.]
> M. Eoc.-M. Mioc.; N.A. L. Eoc.-L. Mioc.; Eu. E. Olig., E. Mioc.-E. Plioc.; As. Olig.; Mediterranean.

†*Domnina* Cope, 1873:1.
> [= †*Protosorex* Scott, 1894:446; †*Miothen* Cope, 1873:5.]
> M. Eoc.-E. Mioc.; N.A.

†*Quercysorex* Engesser, 1975:655.
> L. Eoc.-E. Olig.; Eu.

†*Trimylus* Roger, 1885:106, 107.
> [= †*Heterosorex* Gaillard, 1915:83-98.]
> E. Olig.-M. Mioc.; N.A. E.-L. Mioc.; Eu. L. Mioc.; As.

†*Dinosorex* Engesser, 1972:80.
> Olig.; Mediterranean.[1] L. Olig.-M. Mioc.; Eu. E.-M. Mioc., E. Plioc.; As.
> [1] Balearic Islands.

†*Mongolosorex* Qiu, 1996:29.
> M. Mioc.; As.

†*Gobisorex* Sulimski, 1970:66.
> [= †*Gobiosorex* Bendukidze, 1993:22.[1]]
> E. Olig., E. Mioc.; As.
> [1] Misspelling.

†*Pseudotrimylus* Gureev, 1971:1-254.
> E. Mioc.; N.A.

†*Wilsonosorex* J. E. Martin, 1978:637.[1]
> E. Mioc.; N.A.
> [1] †*Wilsonosorex conulatus* J. E. Martin, 1978:637 is selected here as type species.

†*Ingentisorex* Hutchison, 1966:9.
> M. Mioc.; N.A.

†*Paradomnina* Hutchison, 1966:5.
> M. Mioc.; N.A.

Subfamily **Soricinae** Fischer de Waldheim, 1817:372.
> [= Soricini Fischer de Waldheim, 1817:372; Soricina Gray, 1825:339; Sorexineae Lesson, 1842:87; Soricinae Murray, 1866:231; Blarinae Stirton, 1930:219.[1]]
> [Including Nectogalinae J. Anderson, 1879:149; †Amblycoptinae Kormos, 1926:391; †Crocidosoricinae Reumer, 1987:189-192.]
> E. Olig.-R.; Eu. E. Mioc., L. Mioc.-R.; N.A. M. Mioc.-R.; As. E. Plioc.-R.; Mediterranean. ?L. Plioc.; Af. Pleist., R.; S.A.[2] R.; Cent. A.
> [1] *Lapsus.* Elsewhere in the same paper Stirton recognized only a "*Blarina* group" and stated that he did not intend to give it subfamily rank.
> [2] Northwestern South America.

†*Hemisorex* Baudelot, 1967:292.[1]
> ?E. Mioc., M. Mioc.; Eu.
> [1] Possibly referable to Nectogalini.

†*Deinsdorfia* Heller, 1963:5-7.[1]
> E. Plioc., Pleist.; Eu.
> [1] Possibly referable to Nectogalini.

†*Zelceina* Sulimski, 1962:441-502.[1]
> L. Mioc.; As. E.-L. Plioc.; Eu.[2]
> [1] Possibly referable to Soricini.
> [2] E. Europe.

†*Peisorex* Kowalski & Li, 1963:138.[1]
> Pleist.; As.[2]
> [1] Possibly referable to Blarinini.
> [2] E. Asia.

Tribe **Soricini** Fischer de Waldheim, 1817:372.
>>> [= Soricina Gray, 1825:339.[1]] [Including †Oligosoricini Gureev, 1971:1-254; †Crocidosoricinae Reumer, 1987:190.]
>>> E. Olig.-R.; Eu. E. Mioc., L. Mioc.-R.; N.A. M. Mioc.-R.; As. R.; Cent. A.[2]
>>> [1] Used as tribe.
>>> [2] Southern Mexico and Guatemala.

†*Srinitium* Hugueney, 1976:981.
>>> E. Olig.; Eu.

†*Crocidosorex* Lavocat, 1951:23.
>>> [Including †*Oligosorex* Kretzoi, 1959:247-248; †*Kretzoia* Gibert, 1974.[1]]
>>> L. Olig.-E. Mioc.; Eu.
>>> [1] Objective synonym of †*Oligosorex* (if technically published).

†*Ulmensia* Ziegler, 1989:13.
>>> L. Olig.; Eu.

†*Carposorex* Crochet, 1975:646.
>>> ?L. Olig., E. Mioc.; Eu.

†*Lartetium* Ziegler, 1989:48.
>>> E.-M. Mioc.; Eu.

†*Clapasorex* Crochet, 1975:639.
>>> E. Mioc.; Eu.

†*Florinia* Ziegler, 1989:45.
>>> E. Mioc.; Eu.

†*Antesorex* Repenning, 1967:30.
>>> E. Mioc.; N.A.

†*Anchiblarinella* Hibbard & Jammot, 1971:380.
>>> L. Mioc.; N.A.

Blarinella Thomas, 1911:166.
>>> M. Mioc.-E. Plioc., M. Pleist., R.; As. L. Mioc.-L. Plioc.; Eu.[1]
>>> [1] E. Europe.

†*Alloblairinella* Storch, 1995:228.
>>> L. Mioc.; As. E. Plioc.; Eu.

†*Alluvisorex* Hutchison, 1966:18.
>>> [Including †*Parydrosorex* R. L. Wilson, 1968:106.]
>>> L. Mioc., ?E. Plioc.; N.A.

Sorex Linnaeus, 1758:53. Common shrew.
>>> [= *Musaraneus* Brisson, 1762:126; *Oxyrhin* Kaup, 1829:120; *Amphisorex* Duvernoy, 1835:23[1]; *Corsira* Gray, 1838:123; *Soricidus* Altobello, 1927:6.] [Including *Otisorex* De Kay, 1842:22[2]; *Hydrogale* Pomel, 1848:248[3]; *Neosorex* Baird, 1857:xxxii, 11; *Atophyrax* Merriam, 1884:217-222; *Homalurus* Schulze, 1890:28[2]; *Otiosorex* Ognev, 1928:174; *Eurosorex* Stroganov, 1953:22[4]; *Asorex* Mezzherin, 1965:167; *Ognevia* Dolgov & Heptner, 1967:1422; *Kratochvilia* Vorontsov & Kral, 1986:49; *Fredgia* Vorontsov & Kral, 1986:49; *Yudinia* Vorontsov & Kral, 1986:49; *Dolgovia* Vorontsov & Kral, 1986:49.]
>>> L. Mioc.-R.; As. L. Mioc.-R.; N.A. E. Plioc.-R.; Eu. R.; Cent. A.[5]
>>> [1] Described as subgenus of *Sorex*.
>>> [2] Sometimes as subgenus.
>>> [3] Not *Hydrogale* Kaup, 1829, a synonym of *Neomys*.
>>> [4] Proposed as subgenus.
>>> [5] Southern Mexico and Guatemala.

†*Petenyia* Kormos, 1934:301.[1]
>>> [= or including †*Parapetenyia* Jamot, 1977:172[2]; †*Wezea* Jamot, 1977:178.[3]]
>>> L. Mioc.-M. Pleist.; Eu. L. Plioc.; As.
>>> [1] *Nomen nudum* in Kormos, 1930:57.
>>> [2] Thesis.
>>> [3] Described as a subgenus of †*Parapetenyia*.

†*Cokia* Storch, 1995:232.
    L. Mioc.; As. E. Plioc.; Eu.

†*Paenepetenyia* Storch, 1995:230.
    L. Mioc.; As.

†*Dimylosorex* Rabeder, 1972:636.
    Pleist.; Eu.

†*Drepanosorex* Kretzoi, 1941:109.[1]
    L. Plioc.-M. Pleist.; Eu.
    [1] Proposed as possible subgenus of *Sorex*. Used at generic rank by Repenning, 1967.

Microsorex Baird, in Coues, 1877:643, 646. Pygmy shrew.
    M. Pleist.-R.; N.A.

Tribe **Nectogalini** Anderson, 1879:149, **new rank**.
    [= Nectogalinae J. Anderson, 1879:149.] [Including Hydrosoridae Anonymous,
    1838:427[1]; Crossopinae A. Milne-Edwards, 1872:257; Anourosoricinae J. Anderson,
    1879:159; Anourosoricini Storch & Zazhigin, 1996:257, 259; Soriculi Winge,
    1917:141; †Amblycoptinae Kormos, 1926:352-370; Neomyini Repenning, 1967:45;
    †Beremendina Gureev, 1971:1-254; Soriculina Kretzoi, 1965; †Amblycoptini Rzebik-
    Kowalska, 1994:139.]
    L. Mioc.-R.; As. L. Mioc.-R.; Eu. L. Mioc.-R.; N.A. E. Plioc.-R.; Mediterranean.
    ?L. Plioc.; Af.
    [1] Jardine?

Notiosorex Baird, 1877:643. Crawford shrew.
    [Including *Megasorex* Hibbard, 1950:129.]
    L. Mioc.-R.; N.A.

†*Allopachyura* Kormos, 1934:296-321.
    [Including †*Petenyiella* Kretzoi, 1956:260.[1]]
    ?L. Mioc., E.-L. Plioc., Pleist.; Eu.[2]
    [1] In 1967 Repenning continued the use of †*Petenyiella* with comments to the effect that Kormos' name
    would appear to have no validity because it was conditional. Article 15 of the International Code of
    Zoological Nomenclature appears to be applicable, however.
    [2] E. Europe.

†*Amblycoptus* Kormos, 1926:352-370.
    L. Mioc.; Eu.[1]
    [1] E. Europe.

†*Crusafontina* Gibert, 1975:117.
    [Including †*Anouroneomys* Hutchison & Bown, in Bown, 1980:105.]
    L. Mioc.; Eu. L. Mioc.-E. Plioc.; N.A.

Anourosorex Milne-Edwards, 1870:341. Short-tailed shrew, mole shrew.
    [= *Anaurosaurex* Günther, 1871:9; *Anurosorex* Anderson, 1875:282.]
    L. Mioc., Plioc., E. Pleist.-R.; As.[1] L. Mioc.; Eu.
    [1] E. Asia.

†*Hesperosorex* Hibbard, 1957:328.
    L. Mioc., ?E. Plioc.; N.A.

Soriculus Blyth, 1854:733. Oriental shrew.
    [Including *Chodsigoa* Kastschenko, 1907:251; *Episoriculus* Ellerman & Morrison-
    Scott, 1951:56[1]; †*Asoriculus* Kretzoi, 1959:238.[2]]
    L. Mioc.-M. Pleist.; Eu. E.-L. Plioc., M. Pleist., R.; As.[3] ?L. Plioc.; Af. Pleist.;
    Mediterranean.[4]
    [1] Proposed as a subgenus of *Soriculus*. Elevated to generic rank by Repenning, 1967:47.
    [2] Proposed as a subgenus of *Soriculus*. Included in †*Episoriculus* by Repenning, 1967.
    [3] Including Rhodes in the early Pliocene.
    [4] Sardinia.

Neomys Kaup, 1829:117. Old World water shrew.
    [= *Leucorrhynchus* Kaup, 1829:118; *Hydrogale* Kaup, 1829:123; *Crossopus* Wagler,
    1832:275; *Hydrosorex* Duvernoy, 1835:19; *Pinalia* Gray, 1838:126[1]; *Myosictis*
    Pomel, 1854:14.]

Plioc., E. Pleist.-R.; Eu. R.; As.

[1] MS synonym of *Crossopus*.

†*Anourosoricodon* Topachevsky, 1966:91.
Plioc.; Eu.[1]

[1] E. Europe.

†*Paranourosorex* Rzebik-Kowalska, 1975:169.
L. Mioc.; As. Plioc.; Eu.[1]

[1] E. Europe.

†*Nesiotites* Bate, 1945:741.[1]
E. Plioc.-R.; Mediterranean.[2]

[1] The paper bears the date 1944 but was published in January 1945.
[2] Balearic Islands, Sardinia, and Corsica.

†*Beckiasorex* Dalquest, 1972:571.
L. Plioc.; N.A.

†*Neomysorex* Rzebik-Kowalska, 1981:228.
L. Plioc.; Eu.

†*Beremendia* Kormos, 1930:57.[1]
L. Plioc.; As. L. Plioc.-M. Pleist.; Eu.

[1] Later considered a *nomen nudum* by Kormos but based on a previously described species and named prior to 1931 [see the International Code of Zoological Nomenclature, Article 12 (a)]. The first adequate description is arguably by Kormos, 1934:299 and he gave the name that date.

†*Nectogalinia* Gureev, 1979:458.
L. Plioc.-M. Pleist.; As.

†*Macroneomys* Fejfar, 1966:680-691.
E. Plioc., Pleist.; Eu.

Nectogale Milne-Edwards, 1870:341. Tibetan water shrew, web-footed shrew.
R.; As.[1]

[1] E. Asia.

Chimarrogale Anderson, 1877:262-263. Asiatic water shrew.
[Including *Crossogale* Thomas, 1921:243.]
R.; As.[1]

[1] E. Asia.

Tribe **Blarinini** Stirton, 1930:219.
[= Blarinae Stirton, 1930:219[1]; Blarinini Repenning, 1967:37.]
?L. Mioc., E.-L. Plioc.; Eu. L. Mioc.-R.; N.A. E. Plioc., Pleist.; As.[2,3] Pleist., R.; S.A.[4] R.; Cent. A.

[1] Judged from comments elsewhere in the same paper, a *lapsus*.
[2] W. Asia (Rhodes) in the early Pliocene.
[3] E. Asia (Japan) in the Pleistocene.
[4] Northwestern South America.

†*Tregosorex* Hibbard & Jammot, 1971:377.
L. Mioc.; N.A.

†*Adeloblarina* Repenning, 1967:37.
L. Mioc.; N.A.

†*Paracryptotis* Hibbard, 1950:121-122.
?L. Mioc.; Eu. L. Mioc.-L. Plioc.; N.A.

Cryptotis Pomel, 1848:249.[1] Small-eared shrews, least shrew.
[= *Soriciscus* Coues, 1877:649.] [Including *Xenosorex* Schaldach, 1966:289.]
L. Mioc.-R.; N.A. Pleist., R.; S.A.[2] R.; Cent. A.

[1] "The only insectivore of South America and an obviously recently immigrant North American marginal form, not a true Neotropical animal" (Simpson 1945:51).
[2] Northwestern South America.

†*Mafia* Reumer, 1984:78.
E.-L. Plioc.; Eu.[1]

[1] E. Europe.

†*Sulimskia* Reumer, 1984:86.
  E. Plioc.; As.[1] E.-L. Plioc.; Eu.[2]

  [1] W. Asia (Rhodes).
  [2] E. Europe.

†*Blarinoides* Sulimski, 1959:144.
  E.-L. Plioc.; Eu.

Blarina Gray, 1838:124. Short-tailed shrew.
  [= *Brachysorex* Duvernoy, 1842:37-41[1]; *Mamblarinaus* Herrera, 1899:20.[2]]
  L. Plioc.-R.; N.A.

  [1] Described as a subgenus of *Sorex*.
  [2] See the International Code of Zoological Nomenclature, Article 1(b)(8) and Palmer (1904:25-26).

†*Shikamainosorex* Hasegawa, 1957:65.
  Pleist.; As.[1]

  [1] E. Asia (Japan).

Subfamily **Crocidurinae** Milne-Edwards, 1868:256.[1]
  [= Crocidurini Pavlinov & Rossolimo, 1987:25.[2]] [Including Scutisoricinae Allen, 1917:781; †Myosoricina Kretzoi, 1965.[3]]
  ?E. and/or ?M. Mioc., L. Mioc. and/or E. Plioc., L. Plioc.-R.; Af.[4] ?E. Mioc., M. Mioc.-R.; As. M. Mioc., ?L. Mioc., E. Pleist.-R.; Eu. E. Pleist.-R.; Mediterranean. L. Pleist.-R.; Atlantic. R.; Madagascar. R.; Indian O. R.; E. Indies.[5]

  [1] 1868-1874.
  [2] Possibly preoccupied by Crocidurini Gureev, 1971, published in a paper we have not seen.
  [3] Proposed as subtribe.
  [4] Crocidurinae indet. in the late Miocene or early Pliocene at Sahabi, Libya.
  [5] Also New Guinea and many islands of the Pacific by introduction.

Crocidura Wagler, 1832:275. Musk shrews, white-toothed shrews.
  [= *Leucodon* Fatio, 1869:132.] [Including *Paurodus* Schulze, 1897:90[1]; *Heliosorex* Heller, 1910:6[2]; *Praesorex* Thomas, 1913:320[2]; *Afrosorex* Hutterer, 1986:26.[1]]
  ?E. Mioc., L. Plioc.-R.; Af. Mioc., E. Plioc.-R.; As.[3] ?M. and/or ?L. Mioc., E. Pleist.-R.; Eu. E. Pleist.-R.; Mediterranean.[4] L. Pleist.-R.; Atlantic.[5] R.; Indian O.[6] R.; E. Indies.[7]

  [1] Described as subgenus.
  [2] Sometimes as subgenus.
  [3] Including the Andaman and Nicobar Islands, Taiwan, and Japan.
  [4] Including Sardinia, Sicily, and Malta in the Pleistocene and Recent, and the Balearic Islands, Corsica, Crete, and Cyprus in the Recent.
  [5] Canary Islands.
  [6] Christmas Island.
  [7] Sumatra and Borneo to the Philippines, Timor, and the Moluccas.

†*Soricella* Doben-Florin, 1964:48.
  M. Mioc.; Eu.

†*Miosorex* Kretzoi, 1959:249.
  ?M. Mioc.; Af. M. Mioc.; Eu.

†*Similisorex* Stogov & Savinov, 1965:92.
  M.-L. Mioc.; As.

Myosorex Gray, 1838:124. Mouse shrews.
  [Including *Surdisorex* Thomas, 1906:223; *Congosorex* Heim de Balsac & Lamotte, 1956:167.[1]]
  L. Plioc.-R.; Af.

  [1] Subgenus as described.

Suncus Ehrenberg, in Hemprich & Ehrenberg, 1832.[1] Thick-tailed shrew, house shrew, dwarf shrews.
  [Including *Pachyura* de Selys-Longchamps, 1839:32[2]; *Paradoxodon* Wagner, 1855:805[3]; *Plerodus* Schulze, 1897:90[4]; *Podihik* Deraniyagala, 1958:5.]
  L. Plioc.-R.; Af. L. Pleist.-R.; As. R.; Madagascar.[5] R.; E. Indies. R.; Mediterranean.[6] R.; Eu.[7]

  [1] Unpaged. The smallest living mammal belongs to this genus.
  [2] Not *Pachyurus* Agassiz, 1829, a genus of fishes.

[3] Described as a subgenus of *Sorex*.
[4] Described as subgenus of *Crocidura*.
[5] Including the Comoro Islands.
[6] Sardinia, Crete (by introduction?), Cyprus.
[7] Also islands of the Indian Ocean (Mauritius, Réunion, and the Maldives), New Guinea, and many Pacific islands by introduction.

Sylvisorex Thomas, 1904:12. Forest shrew.
    E. Pleist., R.; Af.

Diplomesodon Brandt, 1853:299.
        L. Plioc., Pleist.; Af.[1] R.; As.
[1] South Africa.

Feroculus Kelaart, 1852:31.
        R.; As.[1]
[1] S. Asia (Sri Lanka).

Scutisorex Thomas, 1913:321. Armored shrew.
        R.; Af.

Ruwenzorisorex Hutterer, 1986:260.
        R.; Af.

Solisorex Thomas, 1924:93.
        R.; As.[1]
[1] S. Asia (Sri Lanka).

Paracrocidura Heim de Balsac, 1956:137.
        R.; Af.

Subfamily †**Limnoecinae** Repenning, 1967:24.
        E.-L. Mioc., Plioc.; N.A. L. Mioc.; Eu.

†*Angustidens* Repenning, 1967:25.
        E. Mioc.; N.A. L. Mioc.; Eu.

†*Limnoecus* Stirton, 1930:218.
        [Including †*Stirtonia* Gureev, 1979:372.[1]]
        E.-L. Mioc., Plioc.; N.A. L. Mioc.; Eu.
[1] Preoccupied by †*Stirtonia* Hershkovitz, 1970:6, an alouattine primate.

Subfamily †**Allosoricinae** Fejfar, 1966:221-248.
        [= †Allosoricini Rzebik-Kowalska, 1990:320.]
        E. Mioc.-L. Plioc.; Eu. L. Mioc., L. Plioc.; As.

†*Paenelimnoecus* Baudelot, 1972:100.
        E. Mioc.-L. Plioc.; Eu. L. Mioc., L. Plioc.; As.

†*Allosorex* Fejfar, 1966:221-248.
        ?L. Mioc., E. Plioc.; Eu.

Superfamily **Tenrecoidea** Gray, 1821:301.[1]
        [= Centetoidea Gill, 1872:19; Tenrecoidea Simpson, 1931:268; Zalambdodonta Gill, 1884:12[2]; TENRECOMORPHA Butler, 1972:261.]
[1] Proposed as suborder of LIPOTYPHLA, coordinate with ERINACEOMORPHA, SORICOMORPHA, and CHRYSOCHLORIDA.
[2] In part. Solenodontids were included as well.

Family **Tenrecidae** Gray, 1821:301. Tenrecs.
        [= Tenrecina Gray, 1825:339; Centetida Haeckel, 1866:clx, in part; Centetidae Murray, 1866:xiv, 232.] [Including Oryzorictidae Gill, 1882:120.]
        E. Mioc., M. and/or L. Mioc., R.; Af.[1] ?Pleist., R.; Madagascar.
[1] A "tenrecoid" has been reported from the middle-late Miocene Otavi breccias, Namibia.

Subfamily **Geogalinae** Trouessart, 1879:57.
        E. Mioc.; Af. ?Pleist., R.; Madagascar.

†*Parageogale* Butler, 1984:168.
        [= †*Butleriella* Poduschka & Poduschka, 1985:139.]
        E. Mioc.; Af.

Geogale Milne-Edwards & A. Grandidier, 1872:1.
[Including *Cryptogale* G. Grandidier, 1928:64.]
?Pleist., R.; Madagascar.

Subfamily †**Protenrecinae** Butler, 1969:26.
E. Mioc.; Af.

†*Protenrec* Butler & Hopwood, 1957:21.
E. Mioc.; Af.

†*Erythrozootes* Butler & Hopwood, 1957:24.
E. Mioc.; Af.

Subfamily **Potamogalinae** Allman, 1865:467. African water shrews.
[= Potamogalidae Allman, 1865:467; Potamogalini Winge, 1917:124; Potamogalinae Heim de Balsac & Bourliere, 1955:1665; Mystomyidae Cope, 1883:83; Mythomyidae Cope, 1884:261.]
R.; Af.

Potamogale Du Chaillu, 1860:363. Jes, giant African water shrew, otter shrew.
[= *Mystomys* Gray, 1861:61; *Mythomys* Gray, 1861:275; *Bayonia* Barboza du Bocage, 1865:402-404.]
R.; Af.

Micropotamogale Heim de Balsac, 1954:102. Small African water shrews, beamos.
[Including *Mesopotamogale* Heim de Balsac, 1956:2257[1]; *Kivugale* Kretzoi, 1961:139.[2]]
R.; Af.[3]

[1] As subgenus.
[2] Objective synonym of *Mesopotamogale*.
[3] W. Africa.

Subfamily **Tenrecinae** Gray, 1821:301.
[= Tenrecidae Gray, 1821:301; Tenrecina Gray, 1825:339[1]; Tenrecinae Cabrera, 1925:183, 191; Centetina C. L. Bonaparte, 1837:9[2]; Centetinae Murray, 1866:319; Centetini Winge, 1917:124.] [Including Echinogalinae Murray, 1866:319.]
R.; Madagascar.

[1] As tribe.
[2] As subfamily.

Echinops Martin, 1838:17-19. Small Malagasy hedgehog.
[= *Echinogale* Wagner, 1841:29-30.]
R.; Madagascar.

Tenrec Lacépède, 1799:156. Tenrec.
[= *Setifer* Tiedemann, 1808:384[1]; *Centetes* Illiger, 1811:124; *Tanrecus* de Blainville, 1838:742.]
R.; Madagascar.
[1] Not *Setifer* Froriep, 1806, another tenrec.

Hemicentetes Mivart, 1871:58-65, 72.[1] Streaked tenrecs.
[= *Setiger* G. Cuvier, 1800: table 1; *Eteocles* Gray, 1821:301; *Echinodes* Pomel, 1848:251[2]; *Echinodes* Trouessart, 1879:274[3]; *Ericius* Giebel, 1871:57[4]; *Ericus* Bergroth, 1902:129.]
R.; Madagascar.
[1] We follow Van Valen in conserving *Hemicentetes*. Technically, the matter should be submitted to the International Commission on Zoological Nomenclature.
[2] *Nomen nudum.*
[3] Preoccupied by *Echinodes* Le Conte, 1869, a beetle.
[4] Preoccupied by *Ericius* Tilesius, 1813, a fish.

Setifer Froriep, 1806:15. Hedgehog tenrec, large Malagasy hedgehog.
[= *Ericulus* I. Geoffroy Saint-Hilaire, 1837:60; *Tendrac* de Blainville, 1838:742; *Hericulus* Gloger, 1841:78; *Dasogale* G. Grandidier, 1930:85.[1]]
R.; Madagascar.
[1] Published in 1930, not 1928 as stated in the original paper.

Subfamily **Oryzorictinae** Trouessart, 1879:57.
> [= Oryzorycteae Trouessart, 1879:57; Oryzorictinae Dobson, 1882:2, 71.]
> R.; Madagascar.

Limnogale Forsyth Major, 1896:318-320. Voalavondrano, web-footed tenrec.
> R.; Madagascar.

Microgale Thomas, 1882:319. Long-tailed tenrecs, shrew-like tenrecs.
> [Including *Nesogale* Thomas, 1918:303; *Leptogale* Thomas, 1918:306; *Paramicrogale* G. Grandidier & Petit, 1931:127.]
> R.; Madagascar.

Oryzorictes A. Grandidier, 1870:50. Rice tenrecs.
> [= *Oryzoryctes* Trouessart, 1879:57.[1]] [Including *Nesoryctes* Thomas, 1918:307.]
> R.; Madagascar.
> [1] Invalid emendation.

Grandorder **ARCHONTA** Gregory, 1910:322.[1]
> [Including VOLITANTIA Illiger, 1811:116[2]; PRIMATOMORPHA Beard, in Szalay, Novacek & McKenna, 1993:129, 145.[3]]
> [1] Proposed as superorder by Gregory. Used as a grandorder by McKenna, 1975:41. The content of this taxon is modified from Gregory's original definition by the omission of the elephant shrews (MACROSCELIDEA).
> [2] Illiger's VOLITANTIA contained extant dermopterans and bats.
> [3] Proposed as a mirorder, for DERMOPTERA and PRIMATES, plus (as taxa *incertae sedis*) †Purgatoriidae, †Palaechthonidae, †Microsyopidae, †Picrodontidae, and †*Berruvius*. Antedated by PRIMATOMORPHA Kalandadze & Rautian, 1992:78, named as a cohort, but containing both PRIMATES and RODENTIA. The International Code of Zoological Nomenclature does not apply in matters above family-group rank, however.

Order **CHIROPTERA** Blumenbach, 1779:58, 74. Bats.
> [= DERMAPTERA Aristotle, 330 B.C.[1]; CHEIROPTERA Gray, 1821:299[2]; CHEIROPTERA Flemming, 1822:xxxii; PTETICA Ameghino, 1889[3]; CHIROPTERIFORMES Kinman, 1994:37.]
> [1] Non-Linnean but noted here for the sake of interest. Date approximate.
> [2] Proposed as a class.
> [3] Not seen, but see Ameghino's Obras Completas 6:586. PTETICA included not only CHIROPTERA but a hypothetical "PROCHIROPTERA."

Suborder **MEGACHIROPTERA** Dobson, 1875:346.
> [= FRUCTIVORAE Gray, 1821:299[1]; FRUGIVORA Giebel, 1855:991; Pterocynes Haeckel, 1866:clx.[2]]
> [1] Named at ordinal rank. The term has long actual priority over MEGACHIROPTERA, but the International Code of Zoological Nomenclature does not govern above family-group ranks.
> [2] *Nomen oblitum.*

Family **Pteropodidae** Gray, 1821:299. Old World fruit bats, flying foxes.
> [= Pteropidae Gray, 1821:299; Pteropina Gray, 1825:338; Pteropodidae Bonaparte, 1838:111; Pteropodida Haeckel, 1866:clx.] [Including Cephalotidae Gray, 1821:299; Hypodermida Haeckel, 1866:clx.]
> L. Eoc., L. Mioc., R.; As.[1,2] E. Olig.-M. Mioc.; Eu.[3] E. Mioc., L. Plioc., ?E. Pleist., R.; Af.[4] Pleist., R.; Madagascar. L. Pleist.-R.; E. Indies. L. Pleist.-R.; New Guinea. R.; Indian O. R.; Mediterranean. R.; Aus. R.; Pacific.[5]
> [1] Pteropodid genus indet. in the late Eocene of Thailand.
> [2] Pteropodid genus indet. in the late Miocene of Lufeng, Yunnan, China.
> [3] Indet. pteropodids reported from the late Oligocene-middle Miocene of France.
> [4] Pteropodid genus indet. in the late Pliocene or early Pleistocene of Koobi Fora.
> [5] W. Pacific.

Subfamily †**Archaeopteropodinae** Simpson, 1945:54.
> E. Olig.; Eu.

†*Archaeopteropus* Meschinelli, 1903:1344.
> E. Olig.; Eu.[1]
> [1] Italy.

Subfamily †**Propottininae** Butler, 1984:175.
> E. Mioc.; Af.

†*Propotto* Simpson, 1967:50.
E. Mioc.; Af.

Subfamily **Pteropodinae** Gray, 1821:299.
[= Pteropidae Gray, 1821:299; Pteropodina C. L. Bonaparte, 1837:8[1]; Pteropodinae Flower & Lydekker, 1891:650.] [Including Harpyionycterinae Miller, 1907:77; Nyctymeninae Miller, 1907:75.]
L. Plioc., R.; Af. Pleist., R.; Madagascar. L. Pleist.-R.; E. Indies. L. Pleist.-R.; New Guinea. R.; Indian O. R.; As. R.; Mediterranean. R.; Aus. R.; Pacific.[2]
[1] As subfamily.
[2] W. Pacific.

Tribe **Pteropodini** Gray, 1821:299.
[= Pteropidae Gray, 1821:299; Pteropina Gray, 1825:338[1]; Pteropodini Koopman & Jones, 1970:23.]
L. Plioc., R.; Af. Pleist., R.; Madagascar. L. Pleist.-R.; E. Indies. L. Pleist.-R.; New Guinea. R.; Indian O. R.; As. R.; Mediterranean. R.; Aus. R.; Pacific.[2]
[1] As tribe.
[2] W. Pacific.

Subtribe **Rousettina** Koopman & Jones, 1970:23.
L. Plioc., R.; Af. L. Pleist.-R.; E. Indies. R.; Madagascar. R.; As. R.; Mediterranean. R.; New Guinea. R.; Pacific.[1]
[1] W. Pacific.

Eidolon Rafinesque, 1815:54. Yellow-haired fruit bats, straw-colored fruit bats.
[= *Pterocyon* Peters, 1861:423; *Leiponyx* Jentink, 1881:59-61; *Liponyx* Forbes, 1882:13.[1]]
L. Plioc., R.; Af. R.; Madagascar. R.; As.[2]
[1] Preoccupied by *Liponyx* Viellot, 1816, a genus of birds.
[2] S.W. Asia.

Rousettus Gray, 1821:299. Dog bats, rousette bats.
[= *Cercopteropus* Burnett, 1829:269; *Eleutherura* Gray, 1844:29[1]; *Cynonycteris* Peters, 1852:25.] [Including *Xantharpyia* Gray, 1843:xix, 37-38; *Senonycteris* Gray, 1870:115-116; *Lissonycteris* Andersen, 1912:23[2]; *Stenonycteris* Andersen, 1912:23.[2]]
L. Pleist.-R.; E. Indies. R.; Af. R.; Madagascar. R.; As. R.; Mediterranean.[3] R.; New Guinea. R.; Pacific.[4]
[1] *Nomen nudum.*
[2] Subgenus as described.
[3] Cyprus.
[4] W. Pacific east to the Solomon Islands.

Boneia Jentink, 1879:117.
R.; E. Indies.[1]
[1] Sulawesi.

Myonycteris Matschie, 1899:61, 63.[1] Little-collared fruit bats.
[Including *Phygetis* Andersen, 1912:579.[2]]
R.; Af.[3]
[1] Described as subgenus of *Xantharpyia.*
[2] Subgenus.
[3] Tropical Africa, Sao Tomé Island.

Subtribe **Pteropodina** Gray, 1821:299.
[= Pteropidae Gray, 1821:299; Pteropina Gray, 1825:338[1]; Pteropodina Koopman & Jones, 1970:23.]
Pleist., R.; Madagascar. R.; Indian O. R.; E. Indies. R.; As.[2] R.; Aus. R.; New Guinea. R.; Pacific.[3]
[1] As tribe.
[2] S. and S.E. Asia.
[3] W. Pacific east to Samoa and the Cook Islands.

Pteropus Erxleben, 1777:130.[1] Fruit bats, flying foxes.
> [= *Spectrum* Lacépède, 1799:15.[2]] [Including *Eunycteris* Gray, 1866:64[3]; *Pselaphon* Gray, 1870:110[4]; *Sericonycteris* Matschie, 1899:7, 30[5]; *Desmalopex* Miller, 1907:60.]
> Pleist., R.; Madagascar.[6] R.; Indian O.[7] R.; E. Indies.[8] R.; As.[9] R.; Aus. R.; New Guinea. R.; Pacific.[10]

[1] Often attributed to Brisson, 1762:153-155, but his names were published in a work that was not consistently binominal.
[2] Not *Spectrum* Scopoli, 1777, a genus of lepidopterans. Revived by Matschie, 1899, as subgenus of *Pteropus*, but with a different type species.
[3] Revived by Matschie, 1899, as subgenus of *Pteropus*, but with a different type species.
[4] Not *Pselaphus* Herbst, 1792, a genus of Coleoptera.
[5] Proposed as subgenus of *Pteropus*.
[6] Madagascar includes the Comoro Islands.
[7] Seychelles, Mascarenes, Maldives, and Christmas Island.
[8] Greater Sundas, Lesser Sundas, Moluccas, and Philippines including Palawan Island.
[9] S. and S.E. Asia (including the Andaman and Nicobar Islands, Taiwan, and S. Japan).
[10] Carolines and Bismarcks east to Somoa and the Cook Islands.

Acerodon Jourdan, 1837:156.
> R.; E. Indies.[1]

[1] Lesser Sundas, Sulawesi, Philippines.

Pteralopex Thomas, 1888:155.
> R.; Pacific.[1]

[1] Solomon and Fiji Islands.

Styloctenium Matschie, 1899:33. Stripe-faced fruit bat.
> R.; E. Indies.[1]

[1] Sulawesi and Togian Islands.

Neopteryx Hayman, 1946:569. Small-toothed fruit bat.
> R.; E. Indies.[1]

[1] Sulawesi.

Subtribe **Dobsoniina** Koopman & Jones, 1970:23.
> [= Hypodermida Haeckel, 1866:clx.[1]]
> L. Pleist.-R.; New Guinea. R.; E. Indies.[2] R.; Aus. R.; Pacific.[3]

[1] Used as a family. This name should be suppressed.
[2] Lesser Sundas, Moluccas, Sulawesi, Philippines.
[3] W. Pacific.

Dobsonia Palmer, 1898:114. Bare-backed fruit bats, spinal-winged fruit bats.
> [= *Hypoderma* I. Geoffroy Saint-Hilaire, 1828:706-708.[1]]
> R.; E. Indies.[2] R.; Aus. R.; New Guinea. R.; Pacific.[3]

[1] Not *Hypoderma* Latreille, 1825, a dipteran.
[2] Lesser Sundas, Moluccas, Sulawesi, Philippines.
[3] W. Pacific east to Solomon Islands.

Aproteles Menzies, 1977:330.[1]
> L. Pleist.-R.; New Guinea.

[1] First discovered as fossils.

Tribe **Harpyionycterini** Miller, 1907:77.
> [= Harpyionycterinae Miller, 1907:77; Harpyionycterini Koopman & Jones, 1970:23.]
> R.; E. Indies.[1]

[1] Sulawesi, Philippines.

Harpyionycteris Thomas, 1896:243. Harpy fruit bats.
> R.; E. Indies.[1]

[1] Sulawesi, Philippines.

Tribe **Epomophorini** Gray, 1866:65.
> [= Epomophorina Gray, 1866:65; Epomophorini Koopman & Jones, 1970:23.]
> R.; Af. R.; As.[1]

[1] S.W. Asia (Saudi Arabia).

Plerotes Andersen, 1910:97.
> R.; Af.[1]

[1] Tropical Africa.

Hypsignathus H. Allen, 1861:156. Hammer-headed fruit bats, big-lipped bats.
[= *Sphyrocephalus* Murray, 1862:8-11[1]; *Zygaenocephalus* Murray, 1862: plate 1; *Spyrocephalus* Dobson, 1878:6.[2]]
R.; Af.[3]

[1] Not *Sphyrocephala* Westwood, 1848, a dipteran; nor *Sphyrocephalus* Schmarda, 1859, a "worm."
[2] Misprint.
[3] Tropical Africa.

Epomops Gray, 1866:65. Epauleted bats.
R.; Af.[1]

[1] Tropical Africa.

Epomophorus Bennett, 1836:149. Epauleted fruit bats.
R.; Af. R.; As.[1]

[1] S.W. Asia (Saudi Arabia).

Micropteropus Matschie, 1899:36, 57.[1] Dwarf epauleted bats.
R.; Af.[2]

[1] Described as subgenus of *Epomophorus*.
[2] Tropical Africa.

Nanonycteris Matschie, 1899:36, 58.[1] Little flying cows.
R.; Af.[2]

[1] Described as subgenus of *Epomophorus*.
[2] W. Africa.

Scotonycteris Matschie, 1894:200.
R.; Af.[1]

[1] W. Africa.

Casinycteris Thomas, 1910:111. Short-palate fruit bat.
R.; Af.[1]

[1] Cameroon to E. Zaire.

Tribe **Cynopterini** Gray, 1866:64.
[= Cynopterina Gray, 1866:64; Cynopterini Koopman & Jones, 1970:23.]
R.; E. Indies. R.; As. R.; Aus. R.; New Guinea. R.; Pacific.[1]

[1] W. Pacific.

Subtribe **Cynopterina** Gray, 1866:64.[1]
[= Cynopterina Koopman & Jones, 1970:23.[2]]
R.; E. Indies. R.; As.

[1] Proposed as division of Pteropodidae.
[2] As a subtribe.

Cynopterus F. Cuvier, 1825:248.[1] Dog-faced fruit bats, small fox bats.
[Including *Pachysoma* I. Geoffroy Saint-Hilaire, 1828:703-5[2]; *Niadius* Miller, 1906:83.]
R.; E. Indies.[3] R.; As.[4]

[1] Date of publication possibly 1824.
[2] Not *Pachysoma* MacLeay, 1821, a genus of Coleoptera.
[3] The Mentawais, Sumatra, Java, Borneo, Sulawesi, Lesser Sundas Islands, and Philippines.
[4] S. and S.E. Asia, including the Andaman and Nicobar Islands.

Ptenochirus Peters, 1861:707.[1]
R.; E. Indies.[2]

[1] Described as subgenus of *Pachysoma*, an antedated synonym of *Cynopterus*.
[2] Philippines, including Palawan Island group.

Megaerops Peters, 1865:256.[1] Tailless fruit bats.
[= *Megaera* Temminck, 1841:274.[2]]
R.; E. Indies.[3] R.; As.[4]

[1] Possibly 1863.
[2] 1835-1841. Not *Megaera* Wagler, 1830, a reptile; nor *Megaera* Robineau-Desvoidy, 1830, a dipteran.
[3] Sumatra, Java, Borneo, Philippines.
[4] S.E. Asia.

Dyacopterus Andersen, 1912:651.
> R.; E. Indies.[1] R.; As.[2]

[1] Sumatra, Borneo, Philippines.
[2] S.E. Asia (Malaysia).

Balionycteris Matschie, 1899:80. Spotted-winged fruit bats.
> R.; E. Indies.[1] R.; As.[2]

[1] Borneo.
[2] S.E. Asia (Malay Peninsula).

Chironax Andersen, 1912:658. Black-capped fruit bats.
> R.; E. Indies.[1] R.; As.[2]

[1] Sumatra, Nias, Java, Borneo, and Sulawesi.
[2] S.E. Asia (Malay Peninsula).

Thoopterus Matschie, 1899:73, 77.[1] Swift fruit bat.
> R.; E. Indies.[2]

[1] Described as subgenus of *Cynopterus*.
[2] Sulawesi, Moluccas.

Sphaerias Miller, 1906:83. Blanford's fruit bat.
> R.; As.[1]

[1] N. India and Tibet to S.E. Asia.

Aethalops Thomas, 1923:178. Pygmy fruit bats.
> [= *Aethalodes* Thomas, 1923:251.[1]]
> R.; E. Indies.[2] R.; As.[3]

[1] Not *Aethalodes* Gahan, 1888, a beetle.
[2] Borneo, Sumatra, Java, Lombok.
[3] S.E. Asia (Malaya).

Penthetor Andersen, 1912:665. Lucas' short-nosed fruit bats.
> R.; E. Indies.[1] R.; As.[2]

[1] Borneo.
[2] S.E. Asia (Malaysia).

Latidens Thonglongya, 1972:151.
> R.; As.[1]

[1] S. Asia (India).

Alionycteris Kock, 1969:319.
> R.; E. Indies.[1]

[1] Philippines (Mindanao).

Otopteropus Kock, 1969:329. Small-eared fruit bat.
> R.; E. Indies.[1]

[1] Philippines (Luzon).

Haplonycteris Lawrence, 1939:31.
> R.; E. Indies.[1]

[1] Philippines, including Palawan Island group.

Subtribe **Nyctimenina** Miller, 1907:75.
> [= Cephalotidae Gray, 1821:299; Harpyidae C. Hamilton Smith, in Jardine, 1842:115; Harpiana Gray, 1866:64; Nyctymeninae Miller, 1907:75; Nyctimenina Koopman & Jones, 1970:23.]
> R.; E. Indies.[1] R.; Aus. R.; New Guinea. R.; Pacific.[2]

[1] Philippines, Sulawesi, Timor, Moluccas.
[2] W. Pacific.

Paranyctimene Tate, 1942:1. Lesser tube-nosed fruit bats.
> R.; New Guinea.

Nyctimene Borkhausen, 1797:86. Tube-nosed fruit bats.
> [= *Nyctymene* Bechstein, 1800:615, 736; *Cephalotes* É. Geoffroy Saint-Hilaire, 1810:104-106; *Harpyia* Illiger, 1811:118[1]; *Gelasinus* Temminck, 1837:100.]
> [Including *Uronycteris* Gray, 1862:262[2]; *Bdelygma* Matschie, 1899:82, 84.]
> R.; E. Indies.[3] R.; Aus.[4] R.; New Guinea. R.; Pacific.[5]

[1] Not *Harpyia* Ochsenheimer, 1810, a genus of Lepidoptera.

[2] Described as subgenus of *Cynopterus*.
[3] Philippines, Sulawesi, Timor, Moluccas.
[4] N. E. Australia.
[5] W. Pacific east to Bismarcks, Solomon Islands, and Santa Cruz Islands.

Subfamily **Macroglossinae** Gray, 1866:64.
[= Macroglossina Gray, 1866:64; Macroglossinae Trouessart, 1897:89; Carponycterinae Flower & Lydekker, 1891:659; Kiodontinae Palmer, 1898:111.]
?L. Pleist., R.; E. Indies. R.; Af.[1] R.; As.[2] R.; Aus. R.; New Guinea. R.; Pacific.[3]

[1] Tropical Africa.
[2] S. and S.E. Asia.
[3] W. Pacific.

Tribe **Macroglossini** Gray, 1866:64.
[= Macroglossini Koopman & Jones, 1970:23.]
?L. Pleist., R.; E. Indies. R.; Af.[1] R.; As.[2] R.; Aus. R.; New Guinea. R.; Pacific.[3]

[1] Tropical Africa.
[2] S. and S.E. Asia.
[3] W. Pacific east to Bismarcks and Solomon Islands.

Eonycteris Dobson, 1873:204. Dawn bats.
[= *Callinycteris* Jentink, 1889:209-212.]
?L. Pleist., R.; E. Indies.[1] R.; As.[2]

[1] Philippines, Borneo, Sulawesi, Sumatra, Java, Lesser Sundas, and Moluccas.
[2] India and S.E. Asia, including the Andaman Islands.

Megaloglossus Pagenstecher, 1885:245.[1] African long-tongued fruit bat.
[= *Trygenycteris* Lydekker, 1891:655.[2]]
R.; Af.[3]

[1] Not preoccupied by *Megaglossa* Rondani, 1865, a genus of Diptera.
[2] New name for *Megaloglossus* Pagenstecher, 1885, thought to be preoccupied.
[3] Tropical Africa.

Macroglossus Schinz, 1824:71.[1] Long-tongued fruit bats.
[= *Macroglossa* Lesson, 1827:115; *Kiodotus* Blyth, 1840:69[2]; *Rhynchocyon* Gistel, 1848:ix[3]; *Carponycteris* Lydekker, 1891:654[2]; *Odontonycteris* Jentink, 1902:140.]
R.; E. Indies.[4] R.; As.[5] R.; Aus.[6] R.; New Guinea. R.; Pacific.[7]

[1] Not *Macroglossum* Scopoli, 1777, a genus of lepidopterans. This was formerly thought to constitute preoccupation and for this reason *Kiodotus*, *Rhynchocyon*, and *Carponycteris* were successively proposed to replace *Macroglossus*, but according to the International Code of Zoological Nomenclature, Art. 56(b), *Macroglossus* is valid. The same rule assures the validity of *Megaloglossus* in place of the often used replacement, *Trygenycteris*.
[2] New name for *Macroglossus* Schinz, 1824, thought to be preoccupied.
[3] New name for *Macroglossus* Schinz, 1824, thought to be preoccupied. Not *Rhynchocyon* Peters, 1847, a macroscelidean.
[4] Philippines, Borneo, Sulawesi, Moluccas, Lesser Sundas, Java, Mentawai Islands, Nias.
[5] S.E. Asia.
[6] N. Australia.
[7] W. Pacific east to Solomon Islands.

Syconycteris Matschie, 1899:95, 98.[1] Blossom fruit bats.
R.; E. Indies.[2] R.; Aus.[3] R.; New Guinea. R.; Pacific.[4]

[1] Described as subgenus of *Macroglossus*.
[2] Moluccas.
[3] E. Australia.
[4] Bismarcks.

Tribe **Notopterini** Koopman & Jones, 1970:23.
R.; Pacific.[1]

[1] W. Pacific.

Melonycteris Dobson, 1877:119.
[= *Cheiropteruges* Ramsay, 1877:17-19.[1]] [Including *Nesonycteris* Thomas, 1887:147.[2]]
R.; Pacific.[3]

[1] Proposed as subgenus of *Pteropus*.
[2] Subgenus.
[3] Bismarcks and Solomon Islands.

Notopteris Gray, 1859:36. Long-tailed fruit bat.
R.; Pacific.[1]
[1] W. Pacific.

Suborder **MICROCHIROPTERA** Dobson, 1875:346.
[= INSECTIVORAE Gray, 1821:299[1]; Nycterides Haeckel, 1866:clx[2];
ENTOMOPHAGA Murray, 1866:xiv[3]; ANIMALIVORA Gill, 1872:16.] [= or
including Gymnorhina Giebel, 1855:926.] [Including PHYLLOSTOMATIA Van
Valen, 1979:109[4]; VESPERTILIONIA Van Valen, 1979:109[4]; †EOCHIROPTERA
Van Valen, 1979:108-109.[5]]
[1] Not to be confused with INSECTIVORA of other authors. Proposed as an order of bats, for the same content as that of MICROCHIROPTERA Dobson.
[2] *Nomen oblitum.*
[3] Not ENTOMOPHAGA Owen, 1859, used for didelphoid marsupials.
[4] Proposed as infraorder.
[5] Proposed as suborder.

†*Ageina* D. E. Russell, Louis & D. E. Savage, 1973:35.
E. Eoc.; Eu. E. Eoc.; N.A.

†*Australonycteris* Hand, Novacek, Godthelp & Archer, 1994:376.
E. Eoc.; Aus.

†*Provampyrus* Schlosser, 1911:72.
[= †*Vampyravus* Schlosser, 1910:507.[1]]
L. Eoc.; Af.[2]
[1] Probably *nomen nudum* but possibly a senior synonym of †*Provampyrus* and †*Philisis.*
[2] N. Africa.

†*Chadronycteris* Ostrander, 1983:131.
L. Eoc.; N.A.

†*Alastorius* Strand, 1928:59.
[= †*Alastor* Weithofer, 1887:285.[1]]
E. Olig.; Eu.
[1] Not †*Alastor* Boisduval, 1869, a lepidopteran.

Family †**Archaeonycteridae** Revilliod, 1917:190.
[= †Archaeonycteridae Revilliod, 1917:190; †Archaeonycteridinae D. E. Russell &
Sige, 1970:171; †Archaeonycterididae Habersetzer & Storch, 1987:117.] [Including
†Icaronycteridae Jepsen, 1966:1334; †Icaronycterididae Habersetzer & Storch,
1987:125.]
?L. Paleoc., E.-M. Eoc.; N.A. E.-M. Eoc.; Eu.

†*Archaeonycteris* Revilliod, 1917.
E.-M. Eoc.; Eu.

†*Icaronycteris* Jepsen, 1966:1333.
?L. Paleoc., E.-M. Eoc.; N.A. ?E. Eoc.; Eu.

Family †**Palaeochiropterygidae** Revilliod, 1917:190.
[= †Palaeochiropterygoidea D. E. Russell & Sige, 1970:170; †Palaeochiropteryginae
D. E. Russell & Sige, 1970:171; †Palaeochiropterigidae Habersetzer & Storch,
1992:465.]
E.-M. Eoc.; Eu. ?E. Eoc.; Aus.[1]
[1] Palaeochiropterygoid-like bats have been reported from the Tngamarra l.f., early Eocene of Australia.

†*Palaeochiropteryx* Revilliod, 1917.
E.-M. Eoc.; Eu.

†*Matthesia* Sigé & D. E. Russell, 1980:114.
M. Eoc.; Eu.

†*Cecilionycteris* Heller, 1935:311.
M. Eoc.; Eu.

Family †**Hassianycterididae** Habersetzer & Storch, 1987:129.
E.-M. Eoc.; Eu.

†*Hassianycteris* Smith & Storch, 1981:153.
E.-M. Eoc.; Eu.

Family **Emballonuridae** Gervais, in de Castelnau, 1855:62*n*. Sac-winged bats, sheath-tailed bats.
[= Emballonurina Gervais, in de Castelnau, 1855:62*n*; Emballonuridae Dobson, 1875:349; Emballonuroidea Weber, 1928:xiv, 154.]
M. Eoc.-E. Olig., ?E. Mioc.; Eu. E.-M. Mioc., L. Plioc., R.; Af. ?Pleist., R.; Madagascar. Pleist., R.; S.A. R.; Indian O.[1] R.; E. Indies. R.; As. R.; N.A.[2] R.; Cent. A. R.; Aus. R.; New Guinea. R.; Pacific.[3]
[1] Mascarene and Seychelle Islands.
[2] Southern North America (Mexico).
[3] W. Pacific.

Subfamily **Taphozoinae** Jerdon, 1867.
M. Eoc.-E. Olig., ?E. Mioc.; Eu. E.-M. Mioc., L. Plioc., R.; Af. R.; Madagascar. R.; Indian O.[1] R.; E. Indies. R.; As. R.; Aus. R.; New Guinea. R.; Pacific.[2]
[1] Mascarene Islands.
[2] Solomon Islands.

Tribe †**Vespertiliavini** Robbins & Sarich, 1988:10.
M. Eoc.-E. Olig.; Eu.

†*Vespertiliavus* Schlosser, 1887:70-75.
M. Eoc.-E. Olig.; Eu.

Tribe **Taphozoini** Jerdon, 1874:30.
[= Taphozoinae Jerdon, 1874:30; Taphozoini Robbins & Sarich, 1988:10.]
E.-M. Mioc., L. Plioc., R.; Af. ?E. Mioc.; Eu. R.; Madagascar. R.; Indian O.[1] R.; E. Indies. R.; As. R.; Aus. R.; New Guinea. R.; Pacific.[2]
[1] Mascarene Islands.
[2] Solomon Islands.

Taphozous É. Geoffroy Saint-Hilaire, 1818:113. Tomb bats.
[Including *Liponycteris* Thomas, 1922:267.[1]]
E.-M. Mioc., L. Plioc., R.; Af. ?E. Mioc.; Eu. R.; Madagascar. R.; Indian O.[2] R.; E. Indies. R.; As.[3] R.; Aus. ?R.; New Guinea.
[1] Subgenus.
[2] Mascarene Islands (Mauritius, Reunion).
[3] S.W., S. and S.E. Asia, including the Andaman Islands.

Saccolaimus Temminck, 1838:14.
[= *Taphonycteris* Dobson, 1875:548, 555-556.]
R.; Af. R.; E. Indies. R.; As.[1] R.; Aus. R.; New Guinea. R.; Pacific.[2]
[1] S. and S.E. Asia, including the Nicobars.
[2] Solomon Islands.

Subfamily **Emballonurinae** Gervais, 1855:62*n*.
[= Emballonurina Gervais, 1855:62*n*; Emballonurina Gray, 1866:92; Emballonurinae Flower & Lydekker, 1891:666.] [Including Diclidurinae Miller, 1907:94.]
L. Plioc., R.; Af. ?Pleist., R.; Madagascar. Pleist., R.; S.A. R.; Indian O.[1] R.; E. Indies. R.; As.[2] R.; N.A.[3] R.; Cent. A. R.; Aus. R.; New Guinea. R.; Pacific.[4]
[1] Seychelle Islands.
[2] S.E. and S.W. Asia.
[3] Southern North America (Mexico).
[4] W. Pacific.

Tribe **Emballonurini** Gervais, 1855:62*n*.
[= Emballonurina Gervais, 1855:62*n*; Emballonurini Robbins & Sarich, 1988:10.]
L. Plioc., R.; Af. ?Pleist., R.; Madagascar. R.; Indian O.[1] R.; E. Indies. R.; As.[2] R.; Aus. R.; New Guinea. R.; Pacific.[3]
[1] Seychelle Islands.
[2] S.E. and S.W. Asia.
[3] W. Pacific.

Mosia Gray, 1843:117.
> R.; E. Indies.[1] R.; New Guinea. R.; Pacific.[2]

[1] Sulawesi, Moluccas.
[2] Solomon Islands.

Emballonura Temminck, 1838:22. Old World sheath-tailed bats.
> ?Pleist., R.; Madagascar. R.; E. Indies.[1] R.; As.[2] R.; Aus. R.; New Guinea. R.; Pacific.[3]

[1] Sumatra, Java, Borneo, Sulawesi, Moluccas, Mentawais, Philippines.
[2] S.E. Asia.
[3] Western Pacific.

Coleura Peters, 1867:479. African sheath-tailed bats.
> L. Plioc., R.; Af.[1] R.; Indian O.[2] R.; As.[3]

[1] Tropical Africa.
[2] Seychelle Islands.
[3] S.W. Asia.

Tribe **Diclidurini** Gray, 1866:92.
> [= Diclidurina Gray, 1866:92; Diclidurinae Miller, 1907:94; Diclidurini Robbins & Sarich, 1988:10.]
> Pleist., R.; S.A. R.; N.A.[1] R.; Cent. A.

[1] Southern North America (Mexico).

Rhynchonycteris Peters, 1867:477.[1] Proboscis bats.
> [= *Proboscidea* Spix, 1823:61[2]; *Rhynchiscus* Miller, 1907:65.[3]]
> R.; Cent. A.[4] R.; S.A.

[1] Not preoccupied by *Rhinchonycteris* Tschudi, 1844-1846:71, a manuscript name suppressed in favor of *Choeronycteris*.
[2] Not *Proboscidea* Brugière, 1791, a nematode.
[3] New name for *Rhynchonycteris*, thought to be preoccupied.
[4] Includes Mexico north to Veracruz.

Centronycteris Gray, 1838:499. Shaggy-haired bat.
> R.; Cent. A.[1] R.; S.A.

[1] Including Mexico north to Veracruz.

Balantiopteryx Peters, 1867:476. Least sac-winged bats.
> R.; N.A.[1] R.; Cent. A. R.; S.A.

[1] Southern North America (Mexico north to Sonora and Baja California).

Saccopteryx Illiger, 1811:121. Two-lined bats, pouch-winged bats.
> [Including *Urocryptus* Temminck, 1838:31-34.[1]]
> R.; N.A.[2] R.; Cent. A. R.; S.A.

[1] 1838-1839.
[2] Southern North America (Mexico north to Jalisco).

Cormura Peters, 1867:475.
> [= *Myropteryx* Miller, 1906:59.]
> R.; Cent. A.[1] R.; S.A.[2]

[1] Nicaragua southward.
[2] Northern South America to Peru and Brazil.

Peropteryx Peters, 1867:472. Doglike bats.
> [Including *Peronymus* Peters, 1868:145.[1]]
> Pleist., R.; S.A.[2] R.; N.A.[3] R.; Cent. A.

[1] Subgenus.
[2] Including Grenada, Trindad and Tobago.
[3] Southern Mexico north to Guerrero.

Cyttarops Thomas, 1913:134. Short-eared bat.
> R.; Cent. A.[1] R.; S.A.[2]

[1] Nicaragua southward.
[2] Northern South America to Brazil.

Diclidurus Wied-Neuwied, 1820:1629.[1] Ghost bats.
> [Including *Depanycteris* Thomas, 1920:271[2]; *Drepanycteris* Simpson, 1945:56.[3]]
> R.; N.A.[4] R.; Cent. A. R.; S.A.

[1] Prince Maximilian zu Wied-Neuwied.

[2] Subgenus.
[3] Misspelling.
[4] Southern North America (Mexico north to Nayarit).

Infraorder **YINOCHIROPTERA** Koopman, 1984:26.[1]

[1] Possibly not published until 1985.

Superfamily **Rhinopomatoidea** Bonaparte, 1838:112.
[= Rhinopomina Bonaparte, 1838:112; Rhinopomatoidea Van Valen, 1979:105, 109.]

Family **Rhinopomatidae** Bonaparte, 1838:112.
[= Rhinopomina Bonaparte, 1838:112; Rhinopomatidae Dobson, 1872:221; Rhinopomata Dobson, 1878:353; Rhinopomatini Winge, 1892:24.]
R.; Af. R.; E. Indies.[1] R.; As.

[1] Sumatra.

Rhinopoma É. Geoffroy Saint-Hilaire, 1813:113. Mouse-tailed bats, long-tailed bats.
[= *Rhynopoma* Bowdich, 1821:30; *Rhinopomus* Gervais, 1854:202.]
R.; Af. R.; E. Indies.[1] R.; As.

[1] Sumatra.

Family **Craseonycteridae** Hill, 1974:303.
R.; As.[1]

[1] S.E. Asia.

Craseonycteris Hill, 1974:304. Bumblebee bat.
R.; As.[1]

[1] S.E. Asia (Thailand).

Superfamily **Rhinolophoidea** Gray, 1825:338.
[= Rhinolophina Gray, 1825:338; Rhinolophoidea Weber, 1928:152; Rhinolophoidea Bell, 1836.]

Family **Megadermatidae** H. Allen, 1864:1. False vampire bats.
[= Megadermidae Gill, 1872:17; Megadermata Peters, 1865:256.]
L. Eoc., L. Olig.-E. Plioc.; Eu. E.-M. Mioc., L. Plioc.-E. Pleist., R.; Af.[1,2] M. Mioc., E. Plioc., L. Pleist.-R.; Aus. Pleist., R.; As. R.; E. Indies.

[1] "Megadermatidae" indet. in the early Miocene of Rusinga, E. Africa.
[2] Megadermatidae genus nov. in the late Pliocene of Makapansgat, S. Africa.

†*Necromantis* Weithofer, 1887:286.[1]
[= †*Necromanter* Lydekker, 1888:31; †*Necronycteris* Palmer, 1903:873.]
L. Eoc.; Eu.

[1] Not *Necromantes* Gistel, 1848, a mollusk.

Lavia Gray, 1838:490. Yellow-winged bat.
[= *Livia* Agassiz, 1846:6.[1]]
R.; Af.[2]

[1] Misprint.
[2] Tropical Africa.

Cardioderma Peters, 1873:488.[1] Heart-nosed bat.
E. Pleist., R.; Af.[2]

[1] Described as subgenus of *Megaderma*.
[2] E. Africa.

Megaderma É. Geoffroy Saint-Hilaire, 1810:197. Asiatic false vampire bats.
[= *Spasma* Gray, 1866:83.[1]] [Including *Eucheira* Hodgson, 1847:891[2]; *Lyroderma* Peters, 1872:195[3]; †*Miomegaderma* Gaillard, 1929:53; †*Afropterus* Lavocat, 1961:83.]
L. Olig.-E. Plioc.; Eu. M. Mioc., E. Pleist.; Af.[4] E. Plioc.; Aus. Pleist., R.; As.[5,6] R.; E. Indies.[7]

[1] Described as subgenus.
[2] Objective synonym of *Lyroderma*. Not *Eucheira* Westwood, 1836.
[3] Subgenus as described.
[4] N. Africa.
[5] S.W. Asia in the Pleistocene.
[6] S. and S.E. Asia including Andaman Islands in the Recent.
[7] Mentawais, Sumatra, Java, Borneo, Sulawesi, Philippines, and Moluccas.

Macroderma Miller, 1906:84. Australian false vampire bats, Australian ghost bat.
M. Mioc., E. Plioc., L. Pleist.-R.; Aus.

Family **Nycteridae** Van der Hoeven, 1855:1028.
[= Nycterina Van der Hoeven, 1855:1028; Nycteridae Dobson, 1875:347.]
?L. Olig.; Eu.[1] E. Mioc., L. Plioc., ?E. Pleist., R.; Af.[2] R.; Madagascar. R.; E. Indies. R.; As.

[1] Teeth of an indet. ?nycterid have been reported from the late Oligocene of Verneuil, France.
[2] A nycterid humerus has been reported from the early Miocene of Chamtwara, Kenya.

Nycteris É. Geoffroy Saint-Hilaire & G. Cuvier, 1795:186.[1] Hispid bats, slit-faced bats, hollow-faced bats.
[Including *Petalia* Gray, 1838.[2]]
L. Plioc., ?E. Pleist., R.; Af. R.; Madagascar. R.; E. Indies.[3] R.; As.[4]

[1] As Simpson (1945:56) noted, *Nycteris* É. Geoffroy Saint-Hilaire & G. Cuvier, 1795, was a *nomen nudum*. According to the International Code of Zoological Nomenclature, then, the correct name of this genus would be *Petalia* Gray, 1838; *Nycteris* would date from Borkhausen, 1797, and would not be this genus but the same as *Lasiurus* Gray, 1831. In 1929, however, the International Commission on Zoological Nomenclature (Opinion 111) suspended the Rules in this case and made *Nycteris* É. Geoffroy Saint-Hilaire & G. Cuvier, 1795, the valid name.
[2] Proposed as subgenus.
[3] Borneo, Sumatra, Java, Bali, and Kangean Island.
[4] S.W. and S.E. Asia.

Family **Rhinolophidae** Gray, 1825:242.
[= Rhinolophina Gray, 1825:242; Rhinolophidae Bell, 1836:599.] [Including Phyllorrhina Koch, 1860[1]; Phyllorhinidae Rochebrune, 1883[1]; Hipposideridae Miller, 1907:109.]
?E. Eoc., E.-L. Mioc., L. Plioc.-R.; Af.[2] M. Eoc.-R.; Eu. M.-L. Mioc., E. Pleist.-R.; As. M. Mioc., E. Plioc., L. Pleist.-R.; Aus. ?Pleist., R.; Madagascar. Pleist., R.; Mediterranean. L. Pleist.-R.; E. Indies. R.; New Guinea. R.; Pacific.[3]

[1] Unavailable.
[2] A single tooth from the early Eocene of Chambi, Tunisia, has been identified as an indet. rhinolophoid.
[3] W. Pacific.

†*Vaylatsia* Sigé, 1990:1131.
E. Olig.; Eu.

Subfamily **Rhinolophinae** Gray, 1825:242.
[= Rhinolophina Gray, 1825:242; Rhinolophinae Dobson, 1875:106.]
M. Eoc.-R.; Eu. M.-L. Mioc., L. Plioc.-R.; Af. E. Pleist.-R.; As. Pleist., R.; Mediterranean. L. Pleist.-R.; E. Indies. L. Pleist.-R.; Aus. R.; New Guinea. R.; Pacific.[1]

[1] W. Pacific east to Bismarcks.

Rhinolophus Lacépède, 1799:15. Horseshoe bats.
[= *Rhinocrepis* Gervais, 1836:617.[1]] [Including *Phyllorhina* Leach, 1816:5[2]; *Aquias* Gray, 1847:15-16; *Phyllotis* Gray, 1866:81[3]; *Coelophyllus* Peters, 1866:427; *Euryalus* Matschie, 1901:225; *Rhinophyllotis* Troughton, 1941:342[4]; *Rhinomegalophus* Bourret, 1951:607.]
M. Eoc.-R.; Eu. M.-L. Mioc., L. Plioc.-R.; Af. E. Pleist.-R.; As. Pleist., R.; Mediterranean.[5,6] L. Pleist.-R.; E. Indies. L. Pleist.-R.; Aus. R.; New Guinea. R.; Pacific.[7]

[1] Gervais cited Cuvier & Geoffroy Saint-Hilaire, 1795, as the authors.
[2] *Nomen nudum*. Not *Phyllorhina* Bonaparte, 1831, a *nomen nudum*, nor *Phyllorrhina* Bonaparte 1837, a synonym of *Hipposideros*.
[3] Not *Phyllotis* Waterhouse, 1837, a murid rodent.
[4] *Nomen nudum* in Iredale & Troughton, 1934:92.
[5] Including Menorca in the Pleistocene.
[6] Including Sardinia in the subrecent.
[7] W. Pacific east to Bismarcks.

†*Palaeonycteris* Pomel, 1854:9-10.
E. Mioc.; Eu.

Subfamily **Rhinonycterinae** Gray, 1866:81, **new rank**.[1] Old World leaf-nosed bats.

> [= Rhinonycterina Gray, 1866:81.] [Including Phyllorrhina C. Koch, 1860:26, 34[2]; Phyllorhininae Dobson, 1875:347; Hipposiderinae Lydekker, in Flower & Lydekker, 1891:657.]
>
> M. Eoc.-E. Plioc.; Eu. E.-L. Mioc., L. Plioc., R.; Af. M.-L. Mioc., M. Pleist.-R.; As.[3,4] M. Mioc., E. Plioc., R.; Aus.[5] ?Pleist., R.; Madagascar. L. Pleist.-R.; E. Indies. R.; New Guinea. R.; Pacific.[6]

[1] Possibly there will be those who wish to petition for the suppression of Rhinonycterinae in favor of the more familiar term Hipposiderinae and other family-group names based on *Hipposideros*.
[2] Unavailable because of various confusions. See Palmer (1904:535).
[3] Indet. hipposiderid in the middle Miocene of Thailand.
[4] "Hipposiderid" genus and species indet. in the late Miocene of S. China.
[5] "Hipposiderids" reported from the early Pliocene of Queensland.
[6] W. Pacific.

Tribe **Rhinonycterini** Gray, 1866:81.

> [= Rhinonycterina Gray, 1866:81.[1]] [Including Hipposiderinae Lydekker, in Flower & Lydekker, 1891:657; Hipposiderini Koopman & Jones, 1970:25.]
>
> M. Eoc.-E. Plioc.; Eu. E.-L. Mioc., L. Plioc., R.; Af. M. Mioc., R.; Aus. ?Pleist., R.; Madagascar. M. Pleist.-R.; As. L. Pleist.-R.; E. Indies. R.; New Guinea. R.; Pacific.[2]

[1] Rank not stated in the original description but Gray's names with the ending "ina" were usually called tribes.
[2] W. Pacific.

Subtribe **Hipposiderina** Lydekker, in Flower & Lydekker, 1891:657.

> [= Hipposiderinae Lydekker, in Flower & Lydekker, 1891:657; Hipposiderina Koopman, 1994:60.]
>
> M. Eoc.-E. Plioc.; Eu. E.-L. Mioc., L. Plioc., R.; Af. ?Pleist., R.; Madagascar. M. Pleist.-R.; As. L. Pleist.-R.; E. Indies. R.; Aus. R.; New Guinea. R.; Pacific.[1]

[1] W. Pacific to Solomon Islands.

Hipposideros Gray, 1831:37.

> [Including *Phyllorrhina* Bonaparte, 1837: unpaged[1]; *Macronycteris* Gray, 1866:82; *Gloionycteris* Gray, 1866:82; *Chrysonycteris* Gray, 1866:82; *Rhinophylla* Gray, 1866:82[2]; *Speorifera* Gray, 1866:82; *Doryrhina* Peters, 1871:314[3]; *Sideroderma* Peters, 1871:324-325[3]; *Ptychorhina* Peters, 1871:325-326[3]; *Cyclorhina* Peters, 1871:326-327[4]; *Thyreorhina* Peters, 1871:327-328[3]; *Syndesmotis* Peters, 1871:329-330[5]; *Syndesmotus* Waterhouse, 1902:362[6]; †*Pseudorhinolophus* Schlosser, 1887:55.[5]]
>
> M. Eoc.-E. Plioc.; Eu. E.-L. Mioc., L. Plioc., R.; Af. ?Pleist., R.; Madagascar. M. Pleist.-R.; As.[7] L. Pleist.-R.; E. Indies. R.; Aus. R.; New Guinea. R.; Pacific.[8]

[1] Described as subgenus of *Rhinolophus*. Not *Phyllorhina* Leach, 1816, *nomen nudum*, a synonym of *Rhinolophus*. *Phyllorhina* Bonaparte, 1831:16, was a *nomen nudum*.
[2] Not *Rhinophylla* Peters, 1865, a member of the Carolliinae.
[3] Proposed as subgenus of "*Phyllorhina*" (= *Phyllorrhina*).
[4] Proposed as a section of a subgenus of "*Phyllorhina*" (= *Phyllorrhina*).
[5] As subgenus.
[6] Objective synonym of *Syndesmotis*.
[7] S.W. Asia (Yemen), S. Asia, S.E. Asia.
[8] W. Pacific east to the New Hebrides.

Anthops Thomas, 1888:156. Flower-faced bat.

> R.; Pacific.[1]

[1] Solomon Islands.

Aselliscus Tate, 1941:2. Tate's trident-nosed bats.

> R.; E. Indies.[1] R.; As.[2] R.; New Guinea. R.; Pacific.[3]

[1] Moluccas.
[2] S.E. Asia.
[3] W. Pacific.

Asellia Gray, 1838:493. Trident leaf-nosed bat.

> E.-L. Mioc.; Eu. R.; Af. R.; As.[1]

[1] S.W. Asia and islands of the Red Sea to Pakistan.

Subtribe **Rhinonycterina** Gray, 1866:81.[1]
　　　　　[= Rhinonycterina Koopman, 1994:68.[2]]
　　　　　L. Olig.-M. Mioc.; Eu. M. Mioc., R.; Aus. ?Pleist., R.; Madagascar. R.; Af. R.;
　　　　　As.[3]
　　[1] Rank not stated in the original description but Gray's names with the ending "ina" were usually called
　　　tribes.
　　[2] As subtribe.
　　[3] S.W. Asia.

†*Brachipposideros* Sigé, 1968:83.[1]
　　　　　L. Olig.-M. Mioc.; Eu. M. Mioc.; Aus.
　　[1] Described as subgenus of *Hipposideros.*

Rhinonycteris Gray, 1866:81.[1] Golden horseshoe bat, orange leaf-nosed bat.
　　　　　[= *Rhinonicteris* Gray, 1847:16.[2]]
　　　　　R.; Aus.
　　[1] Corrected spelling.
　　[2] Error in Latinization.

Cloeotis Thomas, 1901:28. African trident-nosed bat.
　　　　　R.; Af.[1]
　　[1] S.E. Africa.

Triaenops Dobson, 1871:455. Triple nose-leaf bats.
　　　　　?Pleist., R.; Madagascar. R.; Af.[1] R.; As.[2]
　　[1] Tropical Africa and possibly Egypt.
　　[2] S.W. Asia.

Tribe †**Palaeophyllophorini** Revilliod, 1917:41.[1]
　　　　　[= †Palaeophyllophorinae Revilliod, 1917:41[1]; †Palaeophyllophorini Koopman &
　　　　　Jones, 1970:25.]
　　　　　M. Eoc.-M. Mioc.; Eu.
　　[1] 1917-1922.

†*Palaeophyllophora* Revilliod, 1917:36.
　　　　　M. Eoc.-E. Mioc.; Eu.

†*Paraphyllophora* Revilliod, 1922:160.
　　　　　L. Eoc. and/or E. Olig., M. Mioc.; Eu.

Tribe **Coelopini** Tate, 1941:11.
　　　　　[= Coelopinae Tate, 1941:11; Coelopsini Koopman & Jones, 1970:25.]
　　　　　?M. and/or ?L. Mioc.; Af. R.; E. Indies. R.; As.[1]
　　[1] S.E. Asia.

Paracoelops Dorst, 1947:436.
　　　　　R.; As.[1]
　　[1] S.E. Asia (Vietnam).

Coelops Blyth, 1848:251. Tailless leaf-nosed bats.
　　　　　[Including *Chilophylla* Miller, 1910:395.]
　　　　　?M. and/or ?L. Mioc.; Af. R.; E. Indies.[1] R.; As.[2]
　　[1] Java, Bali, Borneo, Philippines.
　　[2] S.E. Asia.

Infraorder **YANGOCHIROPTERA** Koopman, 1984:26.[1]
　　[1] Possibly not published until 1985.

Family **Mystacinidae** Dobson, 1875:349.
　　　　　[= Mystacinae Dobson, 1875:349; Mystacopidae Miller, 1907:239; Mystacinidae
　　　　　Simpson, 1945:60.]
　　　　　R.; New Zealand.

Mystacina Gray, 1843:296.[1] New Zealand short-tailed bat.
　　　　　[= *Mystacops* Lydekker, 1891:671.]
　　　　　R.; New Zealand.
　　[1] Not *Mystacinus* Boie, 1822, a bird.

Superfamily **Noctilionoidea** Gray, 1821:299.[1]
[= Noctilionidae Gray, 1821:299; Noctilionoidea Van Valen, 1979:105, 109.]
[1] This has priority over Phyllostomatoidea Gray, 1825. See the International Code of Zoological Nomenclature, Article 23(d).

Family **Noctilionidae** Gray, 1821:299.
[= Noctilionina Gray, 1825:339.[1]]
L. Mioc., R.; S.A. ?L. Pleist., R.; W. Indies. R.; N.A.[2] R.; Cent. A.
[1] As tribe.
[2] Southern North America (Mexico).

Noctilio Linnaeus, 1766:88. Hare-lipped bats, bulldog bats, fisherman bats.
[Including *Dirias* Miller, 1906:84.[1]]
L. Mioc., R.; S.A. ?L. Pleist., R.; W. Indies. R.; N.A.[2] R.; Cent. A.
[1] Subgenus.
[2] Southern North America (Mexico north to Sinaloa).

Family **Mormoopidae** de Saussure, 1860:286. Spectacled bats.
[= Mormopsins de Saussure, 1860:286; Mormopida Koch, 1862:358[1]; Mormopsina Gray, 1866:93; Mormopidae Gill, 1872:16; Mormoopinae Rehn, 1901:297; Mormoopidae Smith, 1972:4, 15, 41.] [Including Phyllodiana Gray, 1866:93; Lobostominae Dobson, 1875:350; Chilonycteriinae Flower & Lydekker, 1891:672.]
?E. Pleist., R.; Cent. A. L. Pleist.-R.; N.A.[2] L. Pleist.-R.; W. Indies. R.; S.A.
[1] 1862-1863.
[2] Southern North America.

Pteronotus Gray, 1838:500.[1] Naked-backed bats, leaf-lipped bats, mustached bats.
[= *Dermonotus* Gill, 1901:177.] [Including *Chilonycteris* Gray, 1839:4-5[2]; *Lobostoma* Gundlach, 1840:356-358[3]; *Phyllodia* Gray, 1843:50.[2]]
?E. Pleist., R.; Cent. A. L. Pleist.-R.; W. Indies. R.; N.A.[4] R.; S.A.
[1] Not *Pteronotus* Rafinesque, 1815, a *nomen nudum*.
[2] Subgenus.
[3] Objective synonym of *Chilonycteris*.
[4] Southern North America (Mexico north to S. Tamaulipas, S. Sonora, and S. Baja California).

Mormoops Leach, 1821:76-78. Cinnamon bats, leaf-chinned bats, ghost-faced bats.
[= *Aello* Leach, 1821:69-71.]
L. Pleist.-R.; N.A.[1] L. Pleist.-R.; W. Indies. R.; Cent. A. R.; S.A.
[1] Southern North America.

Family **Phyllostomidae** Gray, 1825:242. New World leaf-nosed bats.
[= Phyllostomina Gray, 1825:242; Phyllostomidae Waterhouse, 1839:1; Phyllostomatidae Coues & Yarrow, 1875:80.] [= or including Istiophora Wagner, 1855:ix-xxvi.[1]] [Including Vampiridae C. L. Bonaparte, 1837:8; Desmodina C. L. Bonaparte, 1845:5; Desmodidae I. Geoffroy Saint-Hilaire, in Chenu, 1850:102[2]; Desmodontidae Gill, 1884:175.]
?E. and/or ?M. Mioc., E. Pleist.-R.; N.A.[3,4] M. Mioc., Pleist., R.; S.A. ?E. and/or ?M. Pleist., L. Pleist.-R.; W. Indies. R.; Cent. A.
[1] Not based upon a generic name. Used as family. Also used as a family by Giebel, 1855:967.
[2] 1850-1858.
[3] Possible phyllostomids of uncertain generic allocation in the early and middle Miocene of California.
[4] Southern North America in the Pleistocene and Recent.

Subfamily **Phyllostominae** Gray, 1825:242.
[= Phyllostomina Gray, 1825:242; Phyllostomatinae Flower & Lydekker, 1891:673; Phyllostominae Miller, 1907:122; Philostomini Baker, Hood & Honeycutt, 1989:232.] [Including Vampyrina C. L. Bonaparte, 1837:8[1]; Vampyrina Gervais, 1855:44; Vampyrinae Baker, Hood & Honeycutt, 1989:232; Lonchorhinina Gray, 1866:113[2]; Macrophyllina Gray, 1866:113[2]; Trachyopina Gray, 1866:115.[2]]
M. Mioc., Pleist., R.; S.A. L. Pleist.-R.; N.A.[3] L. Pleist.-R.; W. Indies. R.; Cent. A.
[1] Proposed as subfamily; used as tribe by Gray, 1866:113.
[2] Proposed as tribe.
[3] Southern North America.

Micronycteris Gray, 1866:113. Large-eared bats, little big-eared bats.
  [Including *Schizostoma* Gervais, 1855:49[1]; *Schizastoma* Gray, 1862:38[2];
  Glyphonycteris Thomas, 1896:301-303[3]; *Xenoctenes* Miller, 1907:124[3];
  Lampronycteris Sanborn, 1949:216[3]; *Neonycteris* Sanborn, 1949:216[3]; *Trinycteris*
  Sanborn, 1949:216[3]; *Barticonycteris* Hill, 1964:557.[3]]
  Pleist., R.; S.A.[4] R.; N.A.[5] R.; Cent. A. R.; W. Indies.

[1] Not *Schizostoma* Bronn, 1835, a mollusk.
[2] A candidate for suppression.
[3] Subgenus.
[4] Including Grenada.
[5] Southern North America (Mexico north to Tamaulipas and Nayarit).

Macrotus Gray, 1843:21.[1] Big-eared bats.
  [= *Otopterus* Lydekker, 1891:673.]
  L. Pleist.-R.; N.A.[2] L. Pleist.-R.; W. Indies. R.; Cent. A.

[1] Not preoccupied by *Macrotis* Reid, 1837, a peramelid marsupial, nor *Macrotis* Dejean, 1833, a genus of
 Coleoptera; not *Macrotus* Leach, 1816, *nomen nudum* (possibly a synonym of the bat *Plecotus*).
[2] Southern North America (north to southern California and Nevada).

Lonchorhina Tomes, 1863:81. Sword-nosed bat.
  R.; Cent. A.[1] ?R.; W. Indies. R.; S.A.[2]

[1] Including southern Mexico north to Oaxaca.
[2] Tropical South America including Trinidad.

Macrophyllum Gray, 1838:489.[1] Long-legged bat.
  [= *Dolichophyllum* Lydekker, 1891:673.]
  R.; Cent. A. R.; S.A.

[1] Not preoccupied by *Macrophylla* Hope, 1837, a beetle.

Tonatia Gray, in Griffith, 1827:71. Round-eared bats.
  [= *Tylostoma* Gervais, 1855:44-45, 49[1]; *Anthorhina* Lydekker, in Flower & Lydekker,
  1891:674[2]; *Anthorina* Palmer, 1904:108.[3]] [Including *Lophostoma* d'Orbigny, in
  Gray, 1838:489.]
  Pleist., R.; S.A. ?L. Pleist., R.; W. Indies.[4] R.; Cent. A.[5]

[1] Not *Tylostoma* Sharpe, 1849, a genus of gastropods.
[2] Replacement name for *Tylostoma* Gervais.
[3] Misspelling of *Anthorhina*.
[4] Jamaica (now extinct).
[5] Including Southern Mexico.

Mimon Gray, 1847:14.
  Pleist., R.; S.A. R.; Cent. A.[1]

[1] Including Mexico north to Veracruz.

Phyllostomus Lacépède, 1799:16. Spear-nosed bats, javelin bat.
  [= *Phyllostoma* G. Cuvier, 1800: Table I.]
  Pleist., R.; S.A. R.; Cent. A.[1]

[1] Including Mexico north to Veracruz.

Phylloderma Peters, 1865:512.[1]
  [= *Guandira* Gray, 1866:114.[2]]
  R.; Cent. A. R.; S.A.

[1] Proposed as subgenus; raised to generic rank by Dobson, 1878:482.
[2] *Nomen nudum* in Gray, 1843: xviii, 194.

Trachops Gray, 1847:14. Fringe-lipped bat.
  [= *Trachyops* Peters, 1878:481-482.]
  R.; Cent. A.[1] R.; S.A.

[1] Including Mexico north to Oaxaca.

†*Notonycteris* D. E. Savage, 1951:357.
  M. Mioc.; S.A.[1]

[1] Colombia.

Chrotopterus Peters, 1865:505. Peters' false vampire bat.
  Pleist., R.; S.A. R.; Cent. A.[1]

[1] Including Mexico north to Veracruz and Oaxaca.

Vampyrum Rafinesque, 1815:54. Linnaeus' false vampire bat.
> [= *Vampyrus* Leach, 1821:79-80.]
> R.; Cent. A.[1] ?R.; W. Indies. R.; S.A.
> [1] Including Mexico north to Veracruz.

Subfamily **Glossophaginae** Bonaparte, 1845:5.
> [= Glossophagina Bonaparte, 1845:5; Glossophaginae Gill, 1872:17; Glossophagini Baker, Hood & Honeycutt, 1989:232.]
> ?E. and/or ?M. Pleist., L. Pleist.-R.; W. Indies. Pleist., R.; S.A. L. Pleist.-R.; N.A. R.; Cent. A.

Tribe **Brachyphyllini** Gray, 1866:115.
> [= Brachyphyllina Gray, 1866:115[1]; Brachyphyllini Koopman & Jones, 1970:25; Brachyphyllinae Baker, 1979:113, 122.]
> L. Pleist.-R.; W. Indies.
> [1] Proposed as tribe.

Brachyphylla Gray, 1834:122. Cavern leaf-nosed bat.
> L. Pleist.-R.; W. Indies.

Tribe **Phyllonycterini** Miller, 1907:171, **new rank.**[1] Flower bats.
> [= Phyllonycterinae Miller, 1917:171.]
> Pleist., R.; W. Indies.
> [1] Koopman.

Erophylla Miller, 1906:84. Brown flower bat.
> ?L. Pleist., R.; W. Indies.[1]
> [1] Greater Antilles, Bahamas, Cayman Islands.

Phyllonycteris Gundlach, 1860:817-819. Pallid flower bats.
> [Including *Reithronycteris* Miller, 1898:334[1]; *Rhithronycteris* Elliot, 1904:687.[2]]
> Pleist., R.; W. Indies.
> [1] Subgenus.
> [2] Variant spelling.

Tribe **Glossophagini** Bonaparte, 1845:5. Long-nosed bats.
> [= Glossophagina Bonaparte, 1845:5; Glossophagini Baker, Hood & Honeycutt, 1989:232.]
> Pleist., R.; S.A. L. Pleist.-R.; N.A.[1] L. Pleist.-R.; W. Indies. R.; Cent. A.
> [1] Southern North America.

Glossophaga É. Geoffroy Saint-Hilaire, 1818:418. Long-tongued bats.
> Pleist., R.; S.A. R.; N.A.[1] R.; Cent. A. R.; W. Indies.
> [1] Southern North America (Mexico north to Sonora and Tamaulipas).

Monophyllus Leach, 1821:75. West Indian long-tongued bats.
> [= *Monophylla* Lydekker, in Flower & Lydekker, 1891:674.]
> L. Pleist.-R.; W. Indies.

Lichonycteris Thomas, 1895:55. Brown long-nosed bat.
> R.; Cent. A. R.; S.A.

Leptonycteris Lydekker, in Flower & Lydekker, 1891:674. Saussure's long-nosed bat.
> [= *Ischnoglossa* de Saussure, 1860:491-493.[1]]
> L. Pleist.-R.; N.A.[2] R.; Cent. A. R.; S.A.[3]
> [1] Not *Ischnoglossa* Kraatz, 1856, a genus of Coleoptera.
> [2] Southern North America north to Arizona, New Mexico and Texas.
> [3] Northern South America including the Netherlands Antilles.

Anoura Gray, 1838:490. Tailless bats.
> [Including *Lonchoglossa* Peters, 1868:364.[1]]
> Pleist., R.; S.A.[2] R.; N.A.[3] R.; Cent. A.
> [1] Sometimes as subgenus.
> [2] Including Grenada.
> [3] Southern North America (Mexico north to Sinaloa and Tamaulipas).

Hylonycteris Thomas, 1903:286. Underwood's long-tongued bat.
R.; N.A.[1] R.; Cent. A.
[1] Southern North America (Mexico north to Nayarit and Veracruz).

Scleronycteris Thomas, 1912:404.
R.; S.A.

Choeroniscus Thomas, 1928:123. Godman's long-nosed bat.
R.; N.A.[1] R.; Cent. A. R.; S.A.
[1] Southern North America (Mexico north to Sinaloa).

Choeronycteris Lichtenstein, in Tschudi, 1844:70.[1] Hog-nosed bat.
R.; N.A.[2] R.; Cent. A.
[1] Proposed as subgenus of *Glossophaga*.
[2] Southern North America.

Musonycteris Schaldach & McLaughlin, 1960:2. Colima long-nosed bat, trumpet-nosed bat.
R.; N.A.[1]
[1] Southern North America (Mexico).

Tribe **Lonchophyllini** Griffiths, 1982:33, 43, **new rank.**[1]
[= Lonchophyllinae Griffiths, 1982:33, 43.]
R.; Cent. A. R.; S.A.
[1] Koopman.

Lionycteris Thomas, 1913:270. Little long-tongued bat.
R.; S.A.[1]
[1] South America north to Panama.

Lonchophylla Thomas, 1903:458. Thomas' long-tongued bat.
R.; Cent. A. R.; S.A.

Platalina Thomas, 1928:120.
R.; S.A.[1]
[1] Peru.

Subfamily **Stenodermatinae** Gervais, in de Castelnau, 1855:32n.
[= Stenodermina Gervais, in de Castelnau, 1855:32n; Stenoderminae Gill, 1872:17; Stenodermatini Baker, Hood & Honeycutt, 1989:232.] [Including Centurionina Gray, 1866:118; Centurioninae Rehn, 1901:296; Sturnirinae Miller, 1907:38.]
Pleist., R.; S.A. ?L. Pleist., R.; W. Indies. R.; N.A.[1] R.; Cent. A.
[1] Southern N.A.

Tribe **Carolliini** Miller, 1924:53, **new rank.**
[= Carolliinae Miller, 1924:53.] [Including Hemiderminae Miller, 1907:144.]
Pleist., R.; S.A. R.; N.A.[1] R.; Cent. A. ?R.; W. Indies.
[1] Southern North America (Mexico).

Carollia Gray, 1838:488.[1] Short-tailed bats.
[Including *Hemiderma* Gervais, 1855:43.]
Pleist., R.; S.A. R.; N.A.[2] R.; Cent. A. ?R.; W. Indies.
[1] Not preoccupied by *Carolia* Cantraine, 1837, a genus of mollusks.
[2] Southern North America (Mexico north to Colima and San Luis Potosí).

Rhinophylla Peters, 1865:355.
R.; S.A.[1]
[1] Tropical South America.

Tribe **Stenodermatini** Gervais, in de Castelnau, 1855:32n.
[= Stenodermina Gervais, in de Castelnau, 1855:32n; Stenodermini Koopman & Jones, 1970:25; Stenodermatini Baker, Hood & Honeycutt, 1989:232.]
Pleist., R.; S.A. ?L. Pleist., R.; W. Indies. R.; N.A.[1] R.; Cent. A.
[1] Southern North America.

Subtribe **Sturnirina** Miller, 1907:33, **new rank.**
[= Sturnirinae Miller, 1907:38; Sturnirini Koopman & Jones, 1970:25.]
Pleist., R.; S.A. R.; N.A.[1] R.; Cent. A. R.; W. Indies.
[1] Southern North America (Mexico).

Sturnira Gray, 1842:257. Yellow-shouldered bats, American epauleted bats.
> [= *Nyctiplanus* Gray, 1849:58.] [Including *Corvira* Thomas, 1915:310[1]; *Sturnirops* Goodwin, 1938:1.]
> Pleist., R.; S.A. R.; N.A.[2] R.; Cent. A. R.; W. Indies.
> [1] Subgenus.
> [2] Southern North America (Mexico north to Sonora and Tamaulipas).

Subtribe **Stenodermatina** Gervais, in de Castelnau, 1855:32*n*, **new rank**.
> [= Stenodermina Gervais, in de Castelnau, 1855:32*n*; Stenodermatini Baker, Hood & Honeycutt, 1989:232.] [Including Centurionina Gray, 1866:118.[1]]
> Pleist., R.; S.A. ?L. Pleist., R.; W. Indies. R.; N.A.[2] R.; Cent. A.
> [1] Proposed as tribe.
> [2] Southern North America.

Uroderma Peters, 1865:588. Tent-building bats.
> R.; N.A.[1] R.; Cent. A. R.; S.A.
> [1] Southern North America (Mexico north to Michoacan).

Platyrrhinus de Saussure, 1860:429.[1] White-striped bats, broad-nosed bats.
> [= *Artibaeus* Gervais, 1856[2]; *Vampyrops* Peters, 1865:356.]
> Pleist., R.; S.A. R.; Cent. A.[3]
> [1] Not *Platyrrhinus* Fabricius, 1801, an emendation of *Platyrhinus* Clairville, 1798, a beetle.
> [2] Not *Artibeus* Leach, of which it is an unjustified emendation.
> [3] Including Mexico north to Oaxaca.

Vampyrodes Thomas, 1900:270.[1] Great striped-faced bat.
> R.; Cent. A.[2] R.; S.A.
> [1] Described as subgenus of *Vampyrops*.
> [2] Including Mexico north to Oaxaca.

Vampyressa Thomas, 1900:270.[1] Yellow-eared bats.
> [Including *Vampyriscus* Thomas, 1900:270[2]; *Metavampyressa* Peterson, 1968.[3]]
> R.; Cent. A.[4] R.; S.A.
> [1] Described as subgenus of *Vampyrops*.
> [2] Described as subgenus of *Vampyrops*; used as subgenus of *Vampyressa* by Koopman, in Wilson & Reeder, eds., 1993:193.
> [3] Subgenus.
> [4] Including Mexico north to Oaxaca.

Mesophylla Thomas, 1901:143.
> R.; Cent. A. R.; S.A.

Ectophylla H. Allen, 1892:441. White bat.
> R.; Cent. A. R.; S.A.[1]
> [1] W. Colombia.

Chiroderma Peters, 1860:747-748. Big-eyed bats, white-lined bats.
> Pleist., R.; S.A. R.; N.A.[1] R.; Cent. A. R.; W. Indies.[2]
> [1] Southern North America (Mexico north to Chihuahua).
> [2] Lesser Antilles.

Artibeus Leach, 1821:75. Fruit-eating bats, neotropical fruit bats, little fruit-eating bat.
> [Including *Dermanura* Gervais, 1856:36[1]; *Enchisthenes* Andersen, 1906:420[2]; *Koopmania* Owen, 1991:21.[3]]
> Pleist., R.; S.A. ?L. Pleist., R.; W. Indies. R.; N.A.[4] R.; Cent. A.
> [1] Subgenus.
> [2] Subgenus as defined. Raised to generic rank by R. D. Owen, 1987.
> [3] Proposed as genus; used as subgenus by Koopman, in Wilson & Reeder, eds., 1993:187.
> [4] Southern North America (north to Arizona).

Ardops Miller, 1906:84. Tree bat.
> R.; W. Indies.[1]
> [1] Lesser Antilles.

Phyllops Peters, 1865:356. Falcate-winged bats.
> ?L. Pleist., R.; W. Indies.[1]
> [1] Greater Antilles (Cuba, Hispaniola).

Ariteus Gray, 1838:491. Jamaican fig-eating bat.
　　?L. Pleist., R.; W. Indies.[1]
　　[1] Greater Antilles (Jamaica).

Stenoderma É. Geoffroy Saint-Hilaire, 1813:114. Red fruit bat.
　　?L. Pleist., R.; W. Indies.[1]
　　[1] Puerto Rican Bank.

Pygoderma Peters, 1863:83.[1] Ipanema bat.
　　R.; S.A.
　　[1] Described as subgenus of *Stenoderma*; raised to generic rank by Peters, 1865:357.

Ametrida Gray, 1847:15.
　　R.; S.A.

Sphaeronycteris Peters, 1882:988.
　　R.; S.A.

Centurio Gray, 1842:259. Centurion bat, wrinkle-faced bat, lattice-winged bat.
　　R.; N.A.[1] R.; Cent. A. R.; S.A.[2]
　　[1] Southern North America (Mexico north to Sinaloa and Tamaulipas).
　　[2] Northern South America.

Subfamily **Desmodontinae** Bonaparte, 1845:5. Vampire bats.
　　[= Desmodina Bonaparte, 1845:5; Desmodidae I. Geoffroy Saint-Hilaire, in Chenu, 1850:102[1]; Desmodontidae Gill, 1884:175; Desmodoidea Gill, 1872:302.]
　　E. Pleist.-R.; N.A.[2] Pleist., R.; S.A. ?L. Pleist., R.; W. Indies.[3] R.; Cent. A.
　　[1] 1850-1858.
　　[2] Southern North America.
　　[3] Cuba (now extinct).

Diphylla Spix, 1823:68. Hairy-legged vampire bat.
　　[= *Diphydia* Gray, 1829:29.]
　　R.; N.A.[1] R.; Cent. A. R.; S.A.
　　[1] Southern North America (north to S. Texas).

Diaemus Miller, 1906:84.
　　R.; N.A.[1] R.; Cent. A. R.; S.A.
　　[1] Southern North America (Mexico north to Tamaulipas).

Desmodus Wied-Neuwied, 1824:231.[1] Common vampire bat, white-winged vampire bat.
　　E. Pleist.-R.; N.A.[2,3] Pleist., R.; S.A. ?L. Pleist., R.; W. Indies.[4] R.; Cent. A.
　　[1] Prince Maximilian zu Wied-Neuwied.
　　[2] Florida in the Pleistocene.
　　[3] Mexico north to Sonora and Tamaulipas in the Recent.
　　[4] Cuba in the ?late Pleistocene and subrecent. Now extinct.

Superfamily **Vespertilionoidea** Gray, 1821:299.
　　[= Vespertilionidae Gray, 1821:299; Vespertilionina Gray, 1825:339; Vespertilionoidea Gill, 1872:302.]

Family †**Philisidae** Sigé, 1985:179.
　　E. Eoc., L. Eoc.; Af.

†*Dizzya* Sigé, 1991:358.
　　E. Eoc.; Af.

†*Philisis* Sigé, 1985:179.[1]
　　L. Eoc.; Af.
　　[1] May be a junior synonym of †*Provampyrus*.

Family **Molossidae** Gervais, in de Castelnau, 1855:53*n*.
　　[= Molossidae Gill, 1872:17.]
　　M. Eoc., L. Plioc., M. Pleist.-R.; N.A. L. Eoc.-E. Olig., E.-L. Mioc., R.; Eu. L. Olig., M.-L. Mioc., Pleist., R.; S.A.[1] E.-L. Mioc., E. Pleist., R.; Af. E. and/or M. Mioc., L. Pleist.-R.; Aus. M. Mioc., L. Pleist.-R.; As. ?Pleist., R.; Madagascar. L. Pleist.-R.; E. Indies. L. Pleist.-R.; W. Indies. R.; Indian O.[2] R.; Mediterranean. R.; Atlantic.[3] R.; Cent. A. R.; New Guinea. R.; Pacific.[4]
　　[1] Molossidae, genus indeterminate, in the late Miocene of Rio Acre, Brazil.

[2] Mauritius and the Cocos Keeling Islands.
[3] Madeira and the Canary Islands.
[4] W. Pacific.

Subfamily **Tomopeatinae** Miller, 1907:237.
R.; S.A.[1]

[1] Peru.

Tomopeas Miller, 1900:570. Peruvian crevice-dwelling bat.
R.; S.A.[1]

[1] Peru.

Subfamily **Molossinae** Gervais, in de Castelnau, 1855:53n. Mastiff bats.
[= Molossina Gervais, 1855:53n; Molossinae Legendre, 1984:399, 425.] [Including Cheiromelinae Legendre, 1984:399, 425; Tadaridinae Legendre, 1984:399, 426.]
M. Eoc., L. Plioc., M. Pleist.-R.; N.A. L. Eoc.-E. Olig., E.-L. Mioc., R.; Eu. L. Olig., M. Mioc., Pleist., R.; S.A. E.-L. Mioc., E. Pleist., R.; Af.[1,2] E. and/or M. Mioc., L. Pleist.-R.; Aus. M. Mioc., L. Pleist.-R.; As.[3] ?Pleist., R.; Madagascar. L. Pleist.-R.; E. Indies. L. Pleist.-R.; W. Indies. R.; Indian O.[4] R.; Mediterranean. R.; Atlantic.[5] R.; Cent. A. R.; New Guinea. R.; Pacific.[6]

[1] Molossidae [s.s.] indet. in the early Miocene of Napak, Uganda.
[2] Molossidae [s.s.], genus indet., in the late Miocene Otavi breccias, Namibia.
[3] Molossidae [s.s] indet. reported from the middle Miocene Mae Long locality, N. Thailand.
[4] Mascarene Islands, Seychelles, and Cocos Keeling Islands.
[5] Madeira and the Canary Islands.
[6] Melanesia.

†*Wallia* Storer, 1984:40.
M. Eoc.; N.A.

†*Cuvierimops* Legendre & Sigé, 1983:347-361.
L. Eoc.-E. Olig.; Eu.

†*Petramops* Hand, 1990:176.
E. and/or M. Mioc.; Aus.

Mormopterus Peters, 1865:468.[1] Goblin bats, flat-headed bats.
[Including *Platymops* Thomas, 1906:499[2]; *Sauromys* Roberts, 1917:5[3]; *Micronomus* Troughton, 1943:361[4]; †*Hydromops* Legendre, 1984:719[5]; †*Neomops* Legendre, 1984:719.[5]]
L. Olig., R.; S.A. E.-M. Mioc.; Eu. ?Pleist., R.; Madagascar. R.; Af. R.; Indian O.[6] R.; E. Indies.[7] R.; W. Indies.[8] R.; Aus. R.; New Guinea.

[1] Described as subgenus of *Nyctinomus* Geoffroy Saint-Hilaire.
[2] Subgenus.
[3] Described as subgenus of *Platymops*; used as subgenus of *Mormopterus*.
[4] Sometimes as subgenus. *Nomen nudum* in Iredale & Troughton, 1934:iii, 100.
[5] Described as subgenus of *Mormopterus*.
[6] Mauritius, Reunion.
[7] Sumatra, Moluccas.
[8] Cuba.

Molossops Peters, 1865:575.[1] Dog-faced bats, South American flat-headed bat.
[Including *Cynomops* Thomas, 1920:189[2]; *Neoplatymops* Peterson, 1965[2]; *Cabreramops* Ibanez, 1980:105.[2]]
R.; N.A.[3] R.; Cent. A. R.; S.A.

[1] Described as subgenus of *Molossus*.
[2] Subgenus.
[3] Southern North America (Mexico north to Nayarit).

Myopterus É. Geoffroy Saint-Hilaire, 1813:113.
[Including *Eomops* Thomas, 1905:572.]
R.; Af.[1]

[1] Tropical Africa.

Cheiromeles Horsfield, 1824. Naked bats, hairless bats.
L. Pleist.-R.; E. Indies.[1] R.; As.[2]

[1] Sumatra, Java, Borneo, Sulawesi, Philippines.
[2] S.E. Asia.

Tadarida Rafinesque, 1814:55. Free-tailed bats.
    [= *Dinops* Savi, 1825:230; *Mammnyctinomus* Herrera, 1899:20.[1]] [Including
    Nyctinomus É. Geoffroy Saint-Hilaire, 1818[2]; *Austronomus* Troughton, 1941:360;
    Rhizomops Legendre, 1984:399, 427.]
    L. Eoc.-E. Olig., E.-L. Mioc., R.; Eu. M. Mioc., E. Pleist., R.; Af. L. Plioc., M.
    Pleist.-R.; N.A. L. Pleist.-R.; As.[3] L. Pleist.-R.; W. Indies. L. Pleist.-R.; Aus. R.;
    Madagascar. R.; Mediterranean.[4] R.; Atlantic.[5] R.; Cent. A. R.; S.A. R.; New
    Guinea.
[1] See the International Code of Zoological Nomenclature, Article 1(b)(8), and Palmer (1904:25-26).
[2] Palmer (1904:466) gives the date of *Nyctinomus* as 1813. If this is true then *Nyctinomus* should be
  suppressed.
[3] S.W. Asia in the late Pleistocene.
[4] Sicily.
[5] Madeira and the Canary Islands.

Chaerephon Dobson, 1874:144.[1]
    [Including *Lophomops* Allen, 1917:460.[2]]
    L. Pleist.-R.; E. Indies.[3] R.; Af. R.; Madagascar. R.; Indian O.[4] R.; As. R.; Aus.
    R.; New Guinea. R.; Pacific.[5]
[1] Described as subgenus of *Nyctinomus*.
[2] As subgenus of *Chaerephon*.
[3] Sumatra, Java and Borneo to the Philippines and Lesser Sunda Islands.
[4] Cocos Keeling Islands and Seychelles.
[5] Melanesia.

Mops Lesson, 1842:18.
    [Including *Xiphonycteris* Dollman, 1911:210[1]; *Allomops* Allen, 1917:470;
    Philippinopterus Taylor, 1934:310; †*Meganycteris* Rachl, 1983:1-284.[2]]
    M. Mioc.; Eu. R.; Af. R.; Madagascar. R.; E. Indies.[3] R.; As.[4]
[1] Subgenus.
[2] Published PhD thesis.
[3] Sumatra, Borneo, Sulawesi, Philippines.
[4] S.W. Asia (Yemen) and S.E. Asia (Malaya).

Otomops Thomas, 1913:90. Big-eared free-tailed bats.
    R.; Af. R.; Madagascar. R.; E. Indies.[1] R.; As.[2] R.; New Guinea.
[1] Java and Alor.
[2] S. Asia (India).

Nyctinomops Miller, 1902:393.
    R.; N.A. R.; Cent. A. R.; W. Indies.[1] R.; S.A.
[1] Greater Antilles.

Eumops Miller, 1906:85. Bonneted bats.
    [Including †*Molossides* G. M. Allen, 1932:257.]
    L. Plioc., L. Pleist.-R.; N.A. Pleist., R.; S.A. R.; Cent. A. R.; W. Indies.[1]
[1] Greater Antilles.

†*Kiotomops* Takai, Setoguchi, Villarroel, Cadena & Shigehara, 1991:2.
    M. Mioc.; S.A.[1]
[1] Colombia.

Promops Gervais, in de Castelnau, 1855:58. Dome-palate mastiff bats.
    Pleist., R.; S.A. R.; N.A.[1] R.; Cent. A.
[1] Southern North America (Mexico north to Jalisco).

Molossus É. Geoffroy Saint-Hilaire, 1805:151. Velvety free-tailed bats, velvety mastiff bats.
    [= *Dysopes* Illiger, 1811:122.]
    ?L. Pleist., R.; W. Indies. ?L. Pleist., R.; S.A. R.; N.A.[1] R.; Cent. A.
[1] Southern North America (Mexico north to Sinaloa and Tamaulipas).

Family **Natalidae** Gray, 1866:90.
    [= Natalinia Gray, 1866:90; Natalidae Miller, 1899:245.]
    E. Eoc., R.; N.A.[1,2] Pleist., R.; W. Indies. Pleist., R.; S.A. R.; Cent. A.
[1] Wyoming in the early Eocene.
[2] Southern North America (Mexico) in the Recent.

†*Honrovits* Beard, Sigé & Krishtalka, 1992:736.
    E. Eoc.; N.A.[1]
    [1] Wyoming.

Natalus Gray, 1838:496. Funnel-eared bats, tall-crowned bats, graceful bats.
    [= or including *Spectrellum* Gervais, 1855:51.] [Including *Nyctiellus* Gervais,
    1855:84[1]; *Chilonatalus* Miller, 1898:326-328[2]; *Phodotes* Miller, 1906:85.]
    Pleist., R.; W. Indies. Pleist., R.; S.A. R.; N.A.[3] R.; Cent. A.
    [1] Subgenus.
    [2] Subgenus as described.
    [3] Southern North America (Mexico north to Baja California, Sonora, and Nuevo Leon).

Family **Furipteridae** Gray, 1866:91. Smoky bats, thumbless bats.
    [= Furipterina Gray, 1866:91; Furipteridae Miller, 1907:84, 186.]
    R.; Cent. A. R.; S.A.

Furipterus Bonaparte, 1837:3. Fury.
    [= *Furia* F. Cuvier, 1828.[1]]
    R.; Cent. A. R.; S.A.[2]
    [1] Not "*Furia* Linnaeus, 1758."
    [2] Tropical South America.

Amorphochilus Peters, 1877:185.
    R.; S.A.[1]
    [1] Western South America.

Family **Thyropteridae** Miller, 1907:84, 186.
    R.; Cent. A. R.; S.A.

Thyroptera Spix, 1823:61. Disk-winged bats, tri-color bats, New World sucker-footed bats.
    R.; Cent. A.[1] R.; S.A.
    [1] Including Mexico north to Veracruz.

Family **Myzopodidae** Thomas, 1904:5.
    E. Pleist.; Af.[1] R.; Madagascar.
    [1] E. Africa.

Myzopoda Milne-Edwards & A. Grandidier, 1878:220. Golden bat, Old World sucker-footed bat.
    E. Pleist.; Af.[1] R.; Madagascar.
    [1] E. Africa.

Family **Vespertilionidae** Gray, 1821:299. Common bats.
    [= Vespertilionina Gray, 1825:339.]
    M. Eoc.-R.; Eu. Olig., M.-L. Mioc., ?E. Plioc., E. Pleist.-R.; As.[1] E. Olig., E.
    Mioc.-R.; N.A. M. Mioc., ?L. Mioc., E. Plioc.-R.; Af. E. Plioc., L. Pleist.-R.;
    Aus.[2] E.-M. Pleist., R.; Mediterranean. E. and/or M. Pleist., L. Pleist.-R.; W.
    Indies. Pleist., R.; S.A. L. Pleist.-R.; E. Indies. R.; Madagascar. R.; Indian O. R.;
    Atlantic. R.; Cent. A. R.; New Zealand. R.; New Guinea. R.; Pacific.
    [1] ?Vespertilionoidea in the early Oligocene of Yunnan; Vespertilionidae new genus in the Buranian suite,
        Oligocene of Kazakhstan.
    [2] Vespertilionids reported from the early Pliocene of Queensland.

†*Shanwangia* Yang, 1977:77.
    M. Mioc.; As.[1]
    [1] E. Asia (China).

†*Potamonycteris* Czaplewski, 1991:716.
    M. Mioc.; N.A.

†*Plionycteris* Lindsay & Jacobs, 1985:9.
    E. Plioc.; N.A.

Subfamily **Vespertilioninae** Gray, 1821:299.
    [= Vespertilionidae Gray, 1821:299; Vespertilionina Gray, 1825:339; Vespertilioninae
    Miller, 1897:41, 54.] [Including Plecotinae Miller, 1897:46; Antrozoinae Miller,
    1897:41; Nyctophilinae Miller, 1907:234; Leuconoides Menu, 1987:82.[1]]
    M. Eoc.-R.; Eu. E. Olig., E. Mioc.-R.; N.A. M. Mioc., ?L. Mioc., E. Plioc.-R.; Af.
    ?M. Mioc., L. Mioc., ?E. Plioc., E. Pleist.-R.; As. E.-M. Pleist., R.; Mediterranean.

?E. and/or ?M. Pleist., L. Pleist.-R.; W. Indies. Pleist., R.; S.A. L. Pleist.-R.; Aus. R.; Madagascar. R.; Indian O. R.; E. Indies. R.; Atlantic. R.; Cent. A. R.; New Zealand. R.; New Guinea. R.; Pacific.

[1] As emended; published as "Les leuconöides." This idiosyncratic term apparently meets the requirements of the International Code of Zoological Nomenclature, Article 11.

†*Pleistomyotis* Kuramoto & Yoon, in Yoon, Kuramoto & Uchida, 1983:17.

M. Pleist.; As.[1]

[1] E. Asia (Japan).

Tribe **Myotini** Tate, 1942:229.

[= Leuconoformes Menu, 1987:82, 133.[1]]

M. Eoc.-R.; Eu. E. Olig., E. Mioc.-R.; N.A. M. Mioc., ?L. Mioc., L. Plioc.-R.; Af. ?M. Mioc., L. Mioc., E. Pleist.-R.; As. Pleist., R.; S.A. R.; Madagascar. R.; E. Indies. R.; Mediterranean. R.; Atlantic. R.; Cent. A. R.; W. Indies. R.; Aus. R.; New Guinea. R.; Pacific.

[1] Also as "les leuconoformes." The term apparently meets the requirements of the International Code of Zoological Nomenclature, Article 11.

†*Stehlinia* Revilliod, 1919:94.

[= †*Nycterobius* Revilliod, 1922:133[1]; †*Revilliodia* Simpson, 1945:59.] [Including †*Paleunycteris* Revilliod, 1922:144.[2]]

M. Eoc.-L. Olig.; Eu.

[1] Not †*Nycterobius* MacLeay, 1832.
[2] *Nomen nudum* in Revilliod, 1919:95.

Myotis Kaup, 1829:106. Little brown bats, mouse-eared bats.

[Including *Nystactes* Kaup, 1829:108[1]; *Leuconoe* Boie, 1830:256[2]; *Capaccinius* Bonaparte, 1841:iv; *Selysius* Bonaparte, 1841:3[3]; *Brachyotus* Kolenati, 1856:131[4]; *Isotus* Kolenati, 1856:131[5]; *Exochurus* Fitzinger, 1870:75; *Comastes* Fitzinger, 1870:565; *Euvespertilio* Acloque, 1899:38; *Pizonyx* Miller, 1906:85[5]; *Chrysopteron* Jentink, 1910:74[5]; *Cistugo* Thomas, 1912:205[3]; *Dichromyotis* Bianchi, 1917:lxviii[6]; *Megapipistrellus* Bianchi, 1917:lxxvii[7]; *Rickettia* Bianchi, 1917:lxxviii[5]; *Paramyotis* Bianchi, 1917:lxxix[5]; *Anamygdon* Troughton, 1929:87; *Hesperomyotis* Cabrera, 1958:103.]

E. Olig.-R.; Eu. M. Mioc., ?L. Mioc., L. Plioc.-R.; Af. ?M. Mioc., L. Mioc., E. Pleist.-R.; As. ?M. Mioc., L. Mioc.-R.; N.A. Pleist., R.; S.A. R.; Madagascar. R.; E. Indies. R.; Mediterranean. R.; Atlantic.[8] R.; Cent. A. R.; W. Indies. R.; Aus. R.; New Guinea. R.; Pacific.

[1] Objective synonym of *Paramyotis*. Not *Nystactes* Gloger, 1827, a bird.
[2] Usually as subgenus.
[3] Subgenus.
[4] Described as a subgenus of *Vespertilio*. Not *Brachyotus* Gould, 1837, a bird.
[5] Sometimes as subgenus.
[6] Proposed as subgenus of *Myotis*.
[7] Proposed as subgenus of *Pipistrellus*.
[8] Azores.

†*Oligomyotis* Galbreath, 1962:448.

E. Olig.; N.A.

†*Suaptenos* Lawrence, 1943:364.

E. Mioc.; N.A.

†*Miomyotis* Lawrence, 1943:365.

E. Mioc.; N.A.

Lasionycteris Peters, 1866:648. Silver-haired bat.

E. Plioc., L. Pleist.-R.; N.A. R.; Atlantic.[1]

[1] Bermuda.

Tribe **Plecotini** Gray, 1866:90.

[= Plecotina Gray, 1866:90; Plecotinae Miller, 1897:46; Plecotini Tate, 1942:229; Plecotiformes Menu, 1987:82, 90, 133.[1]]

L. Mioc., R.; As. L. Mioc.-R.; Eu. M. Pleist.-R.; N.A. R.; Af. R.; Mediterranean. R.; Atlantic.[2] R.; Cent. A.[3]

[1] Also as "les plécotiformes." Insofar as we can determine, this confusing and idiosyncratic name meets the requirements of the International Code of Zoological Nomenclature, Article 11.
[2] Canary Islands and Cape Verde Islands.
[3] Yucatan Peninsula.

Barbastella Gray, 1821:300. Barbastelles.

[= *Barbastellus* Gray, 1825:243; *Synotus* Keyserling & Blasius, 1839:305.]

E. Pleist.-R.; Eu. R.; Af. R.; As. R.; Mediterranean. R.; Atlantic.[1]

[1] Canary Islands.

Euderma H. Allen, 1892:467. Spotted bat.

R.; N.A.

Idionycteris Anthony, 1923:1.

R.; N.A.

Plecotus É. Geoffroy Saint-Hilaire, 1813:112. Long-eared bats, lump-nosed bats.

[= *Plecautus* F. Cuvier, 1829:415.] [= or including *Macrotus* Leach, 1816:5.[1]] [Including *Corynorhinus* H. Allen, 1865:173-4[2]; *Corinorhinus* Trouessart, 1897:105[3]; †*Paraplecotus* Rabeder, 1974:167, 171.[4]]

L. Mioc., R.; As. L. Mioc.-R.; Eu. M. Pleist.-R.; N.A. R.; Af. R.; Mediterranean. R.; Atlantic.[5] R.; Cent. A.[6]

[1] Probably a *nomen nudum*.
[2] Subgenus.
[3] Misprint.
[4] Described as subgenus of *Plecotus*.
[5] Canary Islands and Cape Verde Islands.
[6] Yucatan Peninsula.

Tribe **Vespertilionini** Gray, 1821:299.[1]

[= Vespertilionidae Gray, 1821:299; Vespertilionina Gray, 1825:339[2]; Vespertilionini Koopman & Jones, 1970:27.] [Including Pipistrellini Tate, 1942:232; Nyctaloides Menu, 1987:94[3]; Nyctaliformes Menu, 1987:95, 133[4]; Eptesiformes Menu, 1987:104, 134[5]; Neoeptesiformes Menu, 1987:95, 123, 134.[6]]

E.-M. Mioc., E. Plioc.-R.; Eu. ?E. Mioc., M. Mioc.-R.; N.A. L. Mioc., ?E. Plioc., ?E. Pleist., M. Pleist., ?L. Pleist., R.; As. E. Plioc.-E. Pleist., R.; Af. E.-M. Pleist., R.; Mediterranean.[7] Pleist., R.; W. Indies. Pleist., R.; S.A. L. Pleist.-R.; Aus. R.; Madagascar. R.; Indian O. R.; E. Indies. R.; Atlantic.[8] R.; Cent. A. R.; New Zealand. R.; New Guinea. R.; Pacific.[9]

[1] All eleven species placed in "collective group" *Attalepharca* Menu, 1987:126 belong to this tribe. As set forth in Articles 1(d), 23(g), and 67(m) of the International Code of Zoological Nomenclature, a collective group, although having the rank of a genus, has no type species and needs no diagnosis. Collective groups were apparently set up to accomodate fragmentary fossils or larval helminths that could not be placed in known genera. All the species Menu referred to *Attalepharca* are living species known by complete adult material, which previously were placed in three genera. Several are well-known, and one (*Eptesicus nilssoni*) is even the type of a genus group name (*Amblyotus* Kolenati, 1858). We are convinced that whatever utility a collective group may have, the concept was not intended to cover well-known living species. In this case, the provisions of the Code that sanction this obscure category were misused. Furthermore, the action was unnecessary because the same purpose could have been served simply by listing the species as *incertae sedis*.
[2] As tribe.
[3] As emended; published as "les nyctaloides." This idiosyncratic term apparently meets the requirements of the International Code of Zoological Nomenclature, Article 11.
[4] Also as "les nyctaliformes." This idiosyncratic term apparently meets the requirements of the International Code of Zoological Nomenclature, Article 11.
[5] Apparently this dubious term meets the requirements of the the International Code of Zoological Nomenclature, Article 11. However, Menu also used "les eptésiformes," which is in French, not Latin.
[6] Also as "les néoeptésiformes." This idiosyncratic term apparently meets the requirements of the International Code of Zoological Nomenclature, Article 11.
[7] Malta.
[8] Azores, Madeira, Canary Islands, and Cape Verde Islands.
[9] W. Pacific.

Eudiscopus Conisbee, 1953:30.[1] Disk-footed bat.

[= *Discopus* Osgood, 1932:236.[2]]

R.; As.[3]

[1] Replacement name for *Discopus* Osgood, 1932.
[2] Not *Discopus* Thomson, 1864, a beetle, nor *Discopus* Zelinka, 1888, a rotifer.
[3] S.E. Asia.

Pipistrellus Kaup, 1829:98. Pipistrelles.

[= or including *Vesperugo* Keyserling & Blasius, 1839:312-318[1]; *Nannugo* Kolenati, 1856:131[2]; *Euvesperugo* Acloque, 1899:35.] [Including *Romicia* Gray, 1838:495; *Hypsugo* Kolenati, 1856:131[3]; *Alobus* Peters, 1868:707[4]; *Scotozous* Dobson, 1875:372-373[5]; *Eptesicops* Roberts, 1926:245; *Neoromicia* Roberts, 1926:245[5]; *Vespadelus* Troughton, 1943:348[6]; *Registrellus* Troughton, 1943:349; *Falsistrellus* Troughton, 1943:349[5]; *Vansonia* Roberts, 1946:304[5]; *Perimyotis* Menu, 1984:409, 415[5]; *Nycterikaupius* Menu, 1987:108[7]; *Arielulus* Hill & Harrison, 1987:250.[8]]
E.-M. Mioc., E. Pleist.-R.; Eu. L. Mioc., M. Pleist., R.; As. ?E. Pleist., R.; Af. M. Pleist., R.; Mediterranean.[9] M. Pleist.-R.; N.A. L. Pleist.-R.; Aus. R.; Madagascar. R.; Indian O.[10] R.; E. Indies.[11] R.; Atlantic.[12] R.; Cent. A. R.; New Guinea. R.; Pacific.[13]

[1] In part.
[2] Described as subgenus of *Vesperugo*.
[3] Described as subgenus of *Vesperugo*; used as subgenus of *Pipistrellus*. Used as a genus by Menu, 1987:123.
[4] Described as subgenus of *Vespertilio*. Not *Alobus* Le Conte, 1856, a genus of Coleoptera.
[5] Subgenus.
[6] Subgenus. Published as *nomen nudum* in Iredale & Troughton, 1934:95.
[7] Proposed as genus.
[8] Subgenus as described.
[9] Malta.
[10] Cocos Keeling Islands and Christmas Island.
[11] Mentawais and Sumatra to the Philippines, Moluccas, and Lesser Sundas.
[12] Cape Verde Islands, Canary Islands, and Madeira.
[13] W. Pacific.

Nyctalus Bowdich, 1825:36. Noctules.

[Including *Atalapha* Rafinesque, 1814:12[1]; *Pterygistes* Kaup, 1829:100; *Noctulinia* Gray, 1842:258; *Panugo* Kolenati, 1856:131.]
E. Pleist.-R.; Eu. R.; Af.[2] R.; As. R.; Mediterranean. R.; Atlantic.[3]

[1] Usually regarded as a *nomen dubium*.
[2] N. Africa.
[3] Azores and Madeira.

Ia Thomas, 1902:163.

[= *Parascotomanes* Bourret, 1942:23.[1]]
R.; As.[2]

[1] Described as subgenus of *Scotomanes*.
[2] S.E. Asia.

Glischropus Dobson, 1875:472. Thick-thumbed bats.

R.; E. Indies.[1] R.; As.[2]

[1] Sumatra, Java, Borneo, Moluccas, Philippines.
[2] S.E. Asia.

†*Paleptesicus* Zapfe, 1970:93.

[= †*Pareptesicus* Zapfe, 1950:58.[1]]
M. Mioc.; Eu.

[1] Not *Pareptesicus* Bianchi, 1917, a synonym of *Eptesicus*.

†*Ancenycteris* Sutton & Genoways, 1974:1.

M. Mioc.; N.A.

Eptesicus Rafinesque, 1820:2. Big brown bats, serotine bats.

[Including *Cnephaeus* Kaup, 1829:103; *Noctula* Bonaparte, 1837[1]; *Cateorus* Kolenati, 1856:131[2]; *Amblyotus* Kolenati, 1858:252[3]; *Pachyomus* Gray, 1866:90; *Nyctiptenus* Fitzinger, 1870:424; *Rhinopterus* Miller, 1906:85[4]; *Scabrifer* G. M. Allen, 1908:46[5]; *Pareptesicus* Bianchi, 1917:lxxvi[6]; *Rhyneptesicus* Bianchi, 1917:lxxvi; †*Miostrellus* Rachl, 1983:1-284.[7]]
?E. Mioc., L. Mioc.-R.; N.A. M. Mioc., E. Plioc.-R.; Eu. L. Mioc., ?E. Plioc., Pleist., R.; As. E. Plioc.-E. Pleist., R.; Af. E. Pleist., R.; Mediterranean.[8] Pleist., R.; W. Indies. Pleist., R.; S.A. R.; Cent. A. R.; Aus.

[1] Proposed as subgenus of *Pipistrellus*.
[2] Proposed as subgenus of *Vesperus*.
[3] Described as subgenus of *Exochura*; sometimes used as subgenus of *Eptesicus*.
[4] Subgenus.

[5] Replacement name for *Rhinopterus*, believed to be preoccupied by *Rhinoptera* Hasselt, 1824, a genus of fishes.
[6] Proposed as subgenus.
[7] Published PhD thesis.
[8] Malta.

Vespertilio Linnaeus, 1758:31. Frosted bats, particolored bats, tsuitate-komori, screen bat.
[= *Marsipolaemus* Peters, 1872:260.[1]] [= or including *Vesperus* Keyserling & Blasius, 1839:313[2]; *Meteorus* Kolenati, 1856:40[3]; *Aristippe* Kolenati, 1863:40[4]; *Adelonycteris* H. Allen, 1892:466.[5]] [Including *Tuitatus* Kishida & Mori, 1931:379.[6]]
?E. Mioc., M. Mioc., L. Plioc.-R.; Eu. R.; As.

[1] Proposed as subgenus of *Vesperus*.
[2] In part. Proposed as subgenus of *Vesperugo*. Not *Vesperus* Latreille, 1829, a genus of Coleoptera.
[3] In part. Proposed as subgenus of *Vesperus*. Not *Meteorus* Haliday, 1835, a genus of Hymenoptera.
[4] In part.
[5] In part. New name for *Vesperus*.
[6] A *nomen nudum*, fide Kuroda (1938: 101), but based upon *Vespertilio aurijunctus* Mori.

Laephotis Thomas, 1901:460.
R.; Af.

Histiotus Gervais, in de Castelnau, 1855:77. Big-eared brown bats.
[Including †*Simonycteris* Stirton, 1931.]
L. Plioc. and/or E. Pleist.; N.A. Pleist., R.; S.A.

Philetor Thomas, 1902:220.
R.; E. Indies.[1] R.; As. R.; New Guinea. R.; Pacific.[2]

[1] Sumatra, Java, Borneo, Philippines, Sulawesi.
[2] W. Pacific (Bismarcks).

Tylonycteris Peters, 1872:703. Club-footed bats, bamboo bats.
R.; E. Indies.[1] R.; As.[2]

[1] Sumatra, Java, Borneo, Sulawesi, Philippines, and Lesser Sundas.
[2] S. and S.E. Asia, including the Andaman Islands.

Mimetillus Thomas, 1904:12. Narrow-winged bat.
R.; Af.[1]

[1] Tropical Africa.

Hesperoptenus Peters, 1868:626.[1]
[Including *Milithronycteris* Hill, 1976:14.[2]]
M. Pleist., R.; As.[3,4] R.; E. Indies.[5]

[1] Described as subgenus of *Vesperus*.
[2] Subgenus.
[3] China in the middle Pleistocene.
[4] S. and S.E. Asia, including the Andaman Islands, in the Recent.
[5] Borneo, Sulawesi.

Chalinolobus Peters, 1866:679. Groove-lipped bats, lobe-lipped bats, butterfly bats.
[Including *Glauconycteris* Dobson, 1875:383.[1]]
L. Pleist.-R.; Aus. R.; Af. R.; New Zealand. R.; New Guinea. R.; Pacific.[2]

[1] Subgenus as described.
[2] W. Pacific.

Tribe **Nycticeini** Gervais, 1855:71.
[= Nycticeina Gervais, 1855:71; Nycticejinae Gill, 1872:17; Nycticeini Tate, 1942:280.] [Including Scotophili Murray, 1966:238; Scotophilini Hill & Harrison, 1987:278.]
?M. Mioc.; Eu. L. Mioc., R.; As.[1] L. Plioc. and/or E. Pleist., R.; Af. L. Pleist.-R.; N.A. R.; Madagascar. R.; Indian O.[2] R.; E. Indies. R.; Cent. A. R.; W. Indies.[3] R.; S.A.[4] R.; Aus. R.; New Guinea.

[1] Samos in the late Miocene.
[2] Reunion Island.
[3] Cuba.
[4] Tropical South America.

Nycticeius Rafinesque, 1819:417. Evening bat, broad-nosed bats.
[= *Nycticejus* Temminck, 1827:xviii.] [Including *Scoteanax* Troughton, 1943:353[1]; *Scotorepens* Troughton, 1943:354[1]; *Nycticeinops* Hill & Harrison, 1987:254.[1]]

?E. Pleist., R.; Af. L. Pleist.-R.; N.A. R.; As.[2] R.; W. Indies.[3] R.; Aus. R.; New Guinea.

[1] Subgenus.
[2] S.W. Asia (Arabia).
[3] Cuba.

Rhogeessa H. Allen, 1866:285. Little yellow bats.
  [Including *Baeodon* Miller, 1906:85.[1]]
  R.; N.A.[2] R.; Cent. A. R.; S.A.[3]

[1] Subgenus.
[2] Southern North America (Mexico north to Sonora and Tamaulipas).
[3] Tropical South America.

Scotoecus Thomas, 1901:263.
  R.; Af. R.; As.[1]

[1] S. Asia.

Scotomanes Dobson, 1875:371. Harlequin bats.
  [Including *Scoteinus* Dobson, 1875:371.]
  R.; As.[1]

[1] S. and S.E. Asia.

Scotophilus Leach, 1821:69, 71.[1] House bats.
  [= *Pachyotus* Gray, 1831:38.]
  ?M. Mioc.; Eu. L. Plioc. and/or E. Pleist., R.; Af. R.; Madagascar. R.; Indian O.[2]
  R.; E. Indies.[3] R.; As.

[1] Not *Scotophila* Hübner, 1816, a lepidopteran.
[2] Reunion Island.
[3] E. Indies east to Aru Islands.

†*Samonycteris* Revilliod, 1922:139.
  L. Mioc.; As.[1]

[1] W. Asia (Samos).

Otonycteris Peters, 1859:223. Desert long-eared bat.
  R.; Af. R.; As.

Tribe **Lasiurini** Tate, 1942:290.
  E. Plioc.-R.; N.A. Pleist., R.; S.A. R.; Atlantic.[1] R.; Cent. A. R.; W. Indies. R.; Pacific.[2]

[1] N. Atlantic.
[2] Galapagos and Hawaii.

Lasiurus Gray, 1831:38. Hoary bats, yellow bats, red bats.
  [= *Nycteris* Borkhausen, 1797[1]; *Atalapha* Peters, 1870:907.[2]] [Including *Dasypterus* Peters, 1871:912.[3]]
  E. Plioc.-R.; N.A. Pleist., R.; S.A. R.; Atlantic.[4] R.; Cent. A. R.; W. Indies. R.; Pacific.[5]

[1] Not *Nycteris* É. Geoffroy Saint-Hilaire & G. Cuvier, 1795, a nycterid bat.
[2] Not *Atalapha* Rafinesque, 1814, a possible synonym of *Nyctalus* but a probable *nomen dubium*.
[3] Subgenus.
[4] N. Atlantic (Bermuda, Iceland, Orkney Islands).
[5] Galapagos and Hawaii.

Tribe **Antrozoini** Miller, 1897:41.
  [= Antrozoinae Miller, 1897:41; Antrozoini Koopman & Jones, 1970:27.]
  L. Mioc.-L. Plioc., L. Pleist.-R.; N.A. L. Pleist.-R.; W. Indies.[1] R.; Cent. A.

[1] Cuba.

Bauerus Van Gelder, 1959:1.[1]
  R.; N.A.[2] R.; Cent. A.

[1] Described as subgenus of *Antrozous*.
[2] Mexico (Nayarit and Veracruz to Chiapas).

Antrozous H. Allen, 1862:248. Pallid bats, desert bats.
  L. Mioc.-L. Plioc., L. Pleist.-R.; N.A. L. Pleist.-R.; W. Indies.[1]

[1] Cuba.

†*Anzanycteris* White, 1969:276.
L. Plioc.; N.A.

Tribe **Nyctophilini** Peters, 1865:524.
[= Nyctophili Peters, 1865:524; Nyctophilinae Miller, 1907:234;
Nyctophilini Koopman & Jones, 1970:27.]
L. Pleist.-R.; Aus. R.; E. Indies.[1] R.; New Guinea.
[1] Lesser Sundas.

Nyctophilus Leach, 1822:78.
[Including *Lamingtona* McKean & Calaby, 1968:372-375.]
L. Pleist.-R.; Aus. R.; E. Indies.[1] R.; New Guinea.
[1] Lesser Sundas (Lembata, ?Timor).

Pharotis Thomas, 1914:381.
R.; New Guinea.

Subfamily **Miniopterinae** Dobson, 1875:349.
[= Miniopteri Dobson, 1875:349; Miniopterinae Miller, 1907:227;
Miniopteridae Mein & Tupinier, 1977:207, 209.]
E. Mioc.-R.; Eu. L. Plioc.-E. Pleist., R.; Af. M. Pleist.-R.; As.
L. Pleist.-R.; E. Indies. L. Pleist.-R.; Aus. R.; Madagascar.
R.; Mediterranean. R.; New Guinea. R.; Pacific.[1]
[1] W. Pacific.

Miniopterus C. L. Bonaparte, 1837. Long-winged bats, long-fingered bats, bent-winged bats.
[= *Miniopteris* Gray, 1866:91; *Minneopterus* Lampe, 1900:12.]
E. Mioc.-R.; Eu. L. Plioc.-E. Pleist., R.; Af. M. Pleist.-R.; As.
L. Pleist.-R.; E. Indies.[1] L. Pleist.-R.; Aus. R.; Madagascar.
R.; Mediterranean. R.; New Guinea. R.; Pacific.[2]
[1] Sumatra to Philippines, Moluccas, and Lesser Sundas.
[2] W. Pacific.

Subfamily **Murininae** Miller, 1907:229.
M. Pleist., R.; As. R.; E. Indies. R.; Aus. R.; New Guinea.

Murina Gray, 1842:258. Tube-nosed insectivorous bats.
[Including *Harpiola* Thomas, 1915:309.[1]]
M. Pleist., R.; As. R.; E. Indies. R.; Aus. R.; New Guinea.
[1] Subgenus as described.

Harpiocephalus Gray, 1842:259. Hairy-winged tube-nosed bat.
[= *Harpyiocephalus* Gray, 1866:90.]
R.; E. Indies.[1] R.; As.[2]
[1] Sumatra, Java, Borneo, Lesser Sunda Islands, and Moluccas.
[2] S. and S.E. Asia.

Subfamily **Kerivoulinae** Miller, 1907:232.
E. Mioc., R.; Af. R.; E. Indies. R.; As. R.; Aus. R.; New Guinea.
R.; Pacific.[1]
[1] W. Pacific.

†*Chamtwaria* Butler, 1984:187.
E. Mioc.; Af.

Kerivoula Gray, 1842:258. Painted bats, woolly bats, plantain bat, kehelvoulha.
[= *Kirivoula* Gervais, 1849:213; *Cerivoula* Blanford, 1891:338-341.]
[Including *Phoniscus* Miller, 1905:229.[1]]
R.; Af. R.; E. Indies.[2] R.; As. R.; Aus. R.; New Guinea. R.; Pacific.[3]
[1] Subgenus.
[2] Sumatra, Java, Borneo, Sulawesi, Lesser Sundas, Moluccas, Philippines, Mentawais.
[3] W. Pacific (Bismarcks).

Order **PRIMATES** Linnaeus, 1758:20.[1] Primates.

[= POLLICATA Illiger, 1811:66[2]; SCANDENTIA Newman, 1843:34; ACMATHERIA Haeckel, 1895:597; PRIMATOMORPHA Kalandadze & Rautian, 1992:78[3]; PRIMATIFORMES Kinman, 1994:37.] [Including ERECTA Illiger, 1811:65[4]; †PLESIADAPOIDEA Romer, 1966:215, 382[5]; †PLESIADAPIFORMES Simons, 1972:284[6]; †PAROMOMYIFORMES Szalay, 1973:74[7]; †MICROSYOPIDA Krishtalka & Schwartz, 1978: fig. 2; †PENEPRIMATES Hoffstetter, 1978:185[8]; PRAESIMII Gingerich, 1984:60; PROPRIMATES Gingerich, 1989:23[9]; †MICROMOMYIFORMES Beard, in Szalay, Novacek & McKenna, 1993:129, 145.[10]]

[1] Order PRIMATES: Gray, 1821:297 was restricted to just one species, *Homo sapiens*; Gray put the remaining members of today's PRIMATES into class QUADRUMANES, a modification of "family" Quadrumana Illiger, 1812:67.

[2] In part. Illiger placed various non-human primates and also his "family" Marsupialia (except kangaroos) in this thoroughly unnatural assemblage.

[3] In part. RODENTIA was also included. Antedates PRIMATOMORPHA Beard, but this has no legal force because the International Code of Zoological Nomenclature does not govern above the family-group of taxa.

[4] Proposed as an order for *Homo sapiens*; used as a family by Haeckel although not based on a generic name.

[5] Proposed as a suborder (p. 215); also listed (p. 382) as a questionable suborder to include †Plesiadapidae, †Carpolestidae, and †Paromomyidae.

[6] Proposed by Simons as an infraorder. Simons credited the name to "Simons and Tattersall, 1972", a work that never appeared in print.

[7] Proposed as suborder; later called a semiorder by Szalay, Rosenberger & Dagosto, 1987:75, 79.

[8] Used in quotation marks by Hoffstetter, but without them by Gingerich, 1989:23, 1990:821.

[9] New name for PRAESIMII. Proposed as suborder.

[10] Proposed as a suborder of order DERMOPTERA.

†*Hallelemur* Schwartz, Tattersall & Haubold, 1983:911.
M. Eoc.; Eu.

†*Arsinoea* Simons, 1992:10744.
L. Eoc.; Af.[1]
[1] N. Africa (Egypt).

Family †**Purgatoriidae** Van Valen & Sloan, 1965:743.
[= †Purgatoriinae Van Valen & Sloan, 1965:743; †Purgatoriidae Gunnell, 1989:11.]
E. Paleoc.; N.A.

†*Purgatorius* Van Valen & Sloan, 1965:743.
E. Paleoc.; N.A.

Family †**Microsyopidae** Osborn, 1892:101.
[= †Microsyopsidae Osborn, 1892:101; †Microsyopoidea Van Valen, 1969:295.]
L. Paleoc.-E. Eoc.; Eu. L. Paleoc.-M. Eoc.; N.A.

Subfamily †**Uintasoricinae** Szalay, 1969:3.
[Including †Navajoviinae Gunnell, 1989:79.]
L. Paleoc.-E. Eoc.; Eu. L. Paleoc.-M. Eoc.; N.A.

†*Avenius* D. Russell, Phélizon & Louis, 1992:244.[1]
E. Eoc.; Eu.
[1] Assignment to †Uintasoricinae is questionable.

Tribe †**Uintasoricini** Szalay, 1969:3, **new rank**.
[= †Uintasoricinae Szalay, 1969:3.]
E.-M. Eoc.; N.A.

†*Niptomomys* McKenna, 1960:63.
L. Paleoc.-E. Eoc.; N.A.

†*Uintasorex* Matthew, 1909:545.
E.-M. Eoc.; N.A.

Tribe †**Navajoviini** Szalay & Delson, 1979:65.
[= †Navajoviinae Gunnell, 1989:79.]
L. Paleoc.; Eu. L. Paleoc., ?E. Eoc.; N.A.

†*Navajovius* Matthew & Granger, 1921:5.
L. Paleoc., ?E. Eoc.; N.A.

†*Berruvius* D. E. Russell, 1964:124.
L. Paleoc.; Eu.

Subfamily †**Microsyopinae** Osborn, 1892:101.
[= †Microsyopsidae Osborn, 1892:101; †Microsyopinae Matthew, 1915:467.]
L. Paleoc.-M. Eoc.; N.A.

†*Arctodontomys* Gunnell, 1985:52.
L. Paleoc.-E. Eoc.; N.A.

†*Microsyops* Leidy, 1872:20.
[= †*Palaeacodon* Leidy, 1872:21; †*Bathrodon* Marsh, 1872:211; †*Mesacodon*
Marsh, 1872:212.] [Including †*Cynodontomys* Cope, 1882:151-152.]
E.-M. Eoc.; N.A.

†*Megadelphus* Gunnell, 1989:110.
E. and/or M. Eoc.; N.A.

†*Craseops* Stock, 1934:349.
M. Eoc.; N.A.

Family †**Micromomyidae** Szalay, 1974:243.
[= †Micromomyini Szalay, 1974:243; †Micromomyinae Gunnell, 1989:82;
†Micromomyidae Beard, 1990:818; †MICROMOMYIFORMES Beard, in
Szalay, Novacek & McKenna, 1993:129, 145.[1]]
L. Paleoc.-E. Eoc.; N.A.
[1] Proposed as a suborder of DERMOPTERA.

Subfamily †**Micromomyinae** Szalay, 1974:243, **new rank**.
[= †Micromomyini Szalay, 1974:243.]
L. Paleoc.-E. Eoc.; N.A.

†*Micromomys* Szalay, 1973:76.
L. Paleoc.-E. Eoc.; N.A.

†*Myrmekomomys* Robinson, 1994:86.
E. Eoc.; N.A.

Subfamily †**Tinimomyinae** Beard & Houde, 1989:389, **new rank**.
[= †Tinimomyina Beard & Houde, 1989:389.]
L. Paleoc.-E. Eoc.; N.A.

†*Tinimomys* Szalay, 1974:244.
L. Paleoc.-E. Eoc.; N.A.

†*Chalicomomys* Beard & Houde, 1989:389.
E. Eoc.; N.A.

Family †**Picromomyidae** Rose & Bown, 1996:306.
E.-M. Eoc.; N.A.

†*Picromomys* Rose & Bown, 1996:307.
E. Eoc.; N.A.

†*Alveojunctus* Bown, 1982:A47.
M. Eoc.; N.A.

Family †**Plesiadapidae** Trouessart, 1897:75.
[= †Plesiadapoidea Bown & Rose, 1976:109[1]; †Plesiadapoidea Gingerich, 1976:95.[2]]
[= or including †Tillarvernidae Piton, 1940:290.[3]]
[Including †Platychoeropidae Lydekker, 1887:3.[4]]
E. Paleoc.-E. Eoc.; N.A. L. Paleoc.-E. Eoc.; Eu.
[1] Used at the level of superfamily for †Plesiadapidae, †Carpolestidae, and †Saxonellidae.
[2] June 1, 1976. Used at superfamilial rank to include †Plesiadapidae, †Carpolestidae,
†Paromomyidae [including †Saxonellinae], and †Picrodontidae.
[3] Not based on a generic name.
[4] *Nomen oblitum.*

†*Pandemonium* Van Valen, 1994:5.
        E. Paleoc.; N.A.

Subfamily †**Saxonellinae** D. E. Russell, 1964:128.
        L. Paleoc.; Eu. L. Paleoc.; N.A.

†*Saxonella* D. E. Russell, 1964:128.
        L. Paleoc.; Eu. L. Paleoc.; N.A.

Subfamily †**Plesiadapinae** Trouessart, 1897:75.
        E. Paleoc.-E. Eoc.; N.A. L. Paleoc.-E. Eoc.; Eu.

†*Pronothodectes* Gidley, 1923:12.
        E.-L. Paleoc.; N.A.

†*Chiromyoides* Stehlin, 1916:1489.
        L. Paleoc.; Eu. L. Paleoc.; N.A.

†*Nannodectes* Gingerich, 1975:138.
        L. Paleoc.; N.A.

†*Plesiadapis* Gervais, 1877:76.
        [= †*Tricuspidens* Lemoine, 1887:192.] [Including †*Nothodectes* Matthew, 1915:482;
        †*Menatotherium* Piton, 1940:290; †*Ancepsoides* D. E. Russell, 1964:115.[1]]
        L. Paleoc.-E. Eoc.; Eu. L. Paleoc.-E. Eoc.; N.A.
        [1] Proposed as subgenus.

†*Platychoerops* Charlesworth, 1855:80.[1]
        [= †*Miolophus* Owen, 1865:339-341; †*Subunicuspidens* Lemoine, 1887:193.]
        L. Paleoc.-E. Eoc.; Eu.
        [1] The original spelling seems to have been †*Platychaerops* as several authors have noted, but the ligation
        "ae" was used for both "ae" and "oe" in other places in the same journal.

Family †**Palaechthonidae** Szalay, 1969:315.
        [= †Palaechthonini Szalay, 1969:315; †Palaechthonidae Gunnell, 1986.[1]]
        E.-L. Paleoc.; N.A.
        [1] Republished by Gunnell, 1989:14.

Subfamily †**Palaechthoninae** Szalay, 1969:315.
        [= †Palaechthonini Szalay, 1969:315; †Palaechthoninae Gunnell, 1989:15.]
        E. Paleoc.; N.A.

†*Palaechthon* Gidley, 1923:6.
        E. Paleoc.; N.A.

†*Premnoides* Gunnell, 1989:16.
        E. Paleoc.; N.A.

†*Anasazia* Van Valen, 1994:20.
        E. Paleoc.; N.A.

†*Palenochtha* Simpson, 1935:231.
        E. Paleoc.; N.A.

Subfamily †**Plesiolestinae** Gunnell, 1989:21.
        E.-L. Paleoc.; N.A.

†*Plesiolestes* Jepsen, 1930:505.
        [Including †*Torrejonia* Gazin, 1968:629; †*Talpohenach* Kay & Cartmill, 1977:45.]
        E.-L. Paleoc.; N.A.

Family †**Picrodontidae** Simpson, 1937:134.
        [= †Picrodontinae Schwartz & Krishtalka, 1977:68; †Picrodontini Schwartz &
        Krishtalka, 1977:68.]
        E.-L. Paleoc.; N.A.

†*Draconodus* Tomida, 1982:38.
        E. Paleoc.; N.A.

†*Picrodus* Douglass, 1908:17.
[= †*Megopterna* Douglass, 1908:18.]
E.-L. Paleoc.; N.A.

†*Zanycteris* Matthew, 1917:569.
L. Paleoc.; N.A.

Suborder **DERMOPTERA** Illiger, 1811:116.[1]
[= PTEROPHORAE Gray, 1821:300[2]; PTENOPLEURA Van der Hoeven, 1858;
GALEOPITHECI Peters, 1864:20; Galeopithecida Haeckel, 1895:593;
GALEOPITHECIA Cabrera, 1925:202; Galeopithecoidea Wilder, 1926:12, 15;
EUDERMOPTERA Beard, in Szalay, Novacek & McKenna, 1993:129, 145[3];
DERMOPTERIFORMES Kinman, 1994:37.] [Including †PROGLIRES Osborn,
1902:203.]

[1] Proposed as a family. Used as a suborder by Dobson, 1883:2; and as an order by Simpson, 1945:53.
[2] Named as an order.
[3] Proposed as an infraorder.

Subfamily †**Thylacaelurinae** Van Valen, 1967:261.
?L. Paleoc., M.-L. Eoc.; N.A.[1]

[1] An unnamed taxon similar to †*Thylacaelurus* has been reported from the Cochrane 2 locality in the early Tiffanian of Canada.

†*Thylacaelurus* L. S. Russell, 1954:96.
M.-L. Eoc.; N.A.

Family †**Paromomyidae** Simpson, 1940:198.
[= †Paromomyinae Simpson, 1940:198; †Paromomyidae Simons, 1961:65;
†Paromomyoidea Schwartz, 1986:28.] [Including †Phenacolemuridae Simpson,
1955:419.]
E. Paleoc.-M. Eoc.; N.A. E.-M. Eoc.; Eu.

Subfamily †**Paromomyinae** Simpson, 1940:198.
[= †Paromomyinae Gunnell, 1989:13.]
E. Paleoc., ?L. Paleoc.; N.A.

†*Paromomys* Gidley, 1923:3.
E. Paleoc., ?L. Paleoc.; N.A.

Subfamily †**Phenacolemurinae** Simpson, 1955:419.
[= †Phenacolemuridae Simpson, 1955:419; †Phenacolemurinae Simons, 1963:81.]
L. Paleoc.-M. Eoc.; N.A. E.-M. Eoc.; Eu.

Tribe †**Simpsonlemurini** Robinson & Ivy, 1994:100.
?E. Eoc.; Eu.[1] E.-M. Eoc.; N.A.

[1] Two species from the early Eocene of Europe assigned to genera of †Phenacolemurini may belong with the †Simpsonlemurini.

†*Simpsonlemur* Robinson & Ivy, 1994:100.
E. Eoc.; N.A.

†*Elwynella* Rose & Bown, 1982:67.
M. Eoc.; N.A.

Tribe †**Phenacolemurini** Simpson, 1955:419.
[= †Phenacolemuridae Simpson, 1955:419; †Phenacolemurini Schwartz & Krishtalka,
1977:68; †Phenacolemurini Robinson & Ivy, 1994:101.]
L. Paleoc.-M. Eoc.; N.A. E.-M. Eoc.; Eu.

†*Ignacius* Matthew & Granger, 1921:5.
L. Paleoc.-M. Eoc.; N.A.

†*Arcius* Godinot, 1984:85.
E.-M. Eoc.; Eu.

†*Pulverflumen* Robinson & Ivy, 1994:104.
    E. Eoc.; N.A.

†*Dillerlemur* Robinson & Ivy, 1994:103.
    L. Paleoc.-E. Eoc.; N.A.

†*Phenacolemur* Matthew, 1915:479.
    L. Paleoc.-M. Eoc.; N.A.

Family †**Plagiomenidae** Matthew, 1918:598.
    E. Paleoc.-M. Eoc., ?L. Eoc., L. Olig.; N.A.

†*Elpidophorus* Simpson, 1927:5.
    E.-L. Paleoc.; N.A.

†*Eudaemonema* Simpson, 1935:231.
    E.-L. Paleoc.; N.A.

Subfamily †**Plagiomeninae** Matthew, 1918:598.
    [= †Plagiomenidae Matthew, 1918:598; †Plagiomeninae D. E. Russell, Louis & D. E.
    Savage, 1973:50.]
    L. Paleoc.-E. Eoc.; N.A.

Tribe †**Worlandiini** Bown & Rose, 1979:97.
    [= †Worlandiinae Bown & Rose, 1979:97; †Worlandiini McKenna, 1990:212, 231.]
    L. Paleoc.-E. Eoc.; N.A.

†*Planetetherium* Simpson, 1928:11.
    L. Paleoc.; N.A.

†*Worlandia* Bown & Rose, 1979:97.
    L. Paleoc.-E. Eoc.; N.A.

Tribe †**Plagiomenini** Matthew, 1918:598, **new rank**.
    [= †Plagiomenidae Matthew, 1918:598.]
    L. Paleoc.-E. Eoc.; N.A.

†*Plagiomene* Matthew, 1918:601.
    L. Paleoc.-E. Eoc.; N.A.

†*Ellesmene* Dawson, McKenna, Beard & Hutchison, 1993:180.
    E. Eoc.; N.A.

Subfamily †**Ekgmowechashalinae** Szalay, 1976:349.
    M. Eoc., ?L. Eoc., L. Olig.; N.A.

Tribe †**Tarkadectini** Szalay & Lucas, 1996:32, **new rank**.
    [= †Tarkadectinae Szalay & Lucas, 1996:32.]
    M. Eoc., ?L. Eoc.; N.A.

†*Tarkadectes* McKenna, 1990:224.
    M. and/or L. Eoc.; N.A.

†*Tarka* McKenna, 1990:214.
    M. Eoc.; N.A.

Tribe †**Ekgmowechashalini** Szalay, 1976:349, **new rank**.
    [= †Ekgmowechashalinae Szalay, 1976:349.]
    L. Olig.; N.A.

†*Ekgmowechashala* Macdonald, 1963:171.
    L. Olig.; N.A.

Family †**Mixodectidae** Cope, 1883:80.
    [= †Oldobotidae Schlosser, 1907:222; †Mixodectoidea Simpson, 1945:53;
    †MIXODECTOMORPHA Saban, 1954:429.[1]]
    E. Paleoc.; N.A.

[1] In part. Proposed as suborder of INSECTIVORA for †Mixodectidae and †Apatemyidae.

†*Mixodectes* Cope, 1883:559.
> [= †*Olbodotes* Osborn, 1902:205.] [Including †*Indrodon* Cope, 1883:318.]
> E. Paleoc.; N.A.

†*Dracontolestes* Gazin, 1941:13.
> E. Paleoc.; N.A.

Family **Galeopithecidae** Gray, 1821:300.
> [= Pleuropteridae Burnett, 1829:268, 269; Galeopithecina Bonaparte, 1837:7[1];
> Galeopithecoidea D. E. Russell, Louis & D. E. Savage, 1973:42; Ptenopleura Haeckel,
> 1866:clix[2]; Galeopteridae Thomas, 1908:254; Cynocephalidae Simpson, 1945:54.[3]]
> L. Eoc., R.; As.[4] R.; E. Indies.

[1] As subfamily.
[2] Used as a family.
[3] Not Cynocephalidae Ameghino, 1889:893, which was based on *Cynocephalus* G. Cuvier & É. Geoffroy Saint-Hilaire, 1795, a genus of baboons now called *Papio*, not on *Cynocephalus* Boddaert, 1768. See the International Code of Zoological Nomenclature, Articles 11(f)(i)(1), 36, and 39.
[4] S.E. Asia.

†*Dermotherium* Ducrocq, Buffetaut, Buffetaut-Tong, Jaeger, Jongkanjanasoontorn & Suteethorn, 1992:373.
> L. Eoc.; As.[1]

[1] S.E. Asia (Thailand).

Cynocephalus Boddaert, 1768:8.[1] Gliding lemurs, flying lemurs, colugos, cobegos, kaguan, kubuk.
> [= *Galeopithecus* Pallas, 1783:208[2]; *Galeopus* Rafinesque, 1815:54[3]; *Pleuropterus*
> Burnett, 1829:268, 269; *Dermopterus* Burnett, 1829:268; *Colugo* Gray, 1870:98[4];
> *Galeopterus* Thomas, 1908:254.]
> R.; E. Indies.[5] R.; As.[6]

[1] Not *Cynocephalus* G. Cuvier & É. Geoffroy Saint-Hilaire, 1795, a genus of baboons.
[2] See International Commission of Zoological Nomenclature, Opinion 1077, 1977.
[3] New name for *Galeopithecus*.
[4] Described as subgenus of *Galeopithecus*.
[5] Java, Borneo, Philippines.
[6] S.E. Asia.

Suborder **EUPRIMATES** Hoffstetter, 1978:185, **new rank**.[1]

[1] Enclosed in quotation marks by Hoffstetter, but these have long been abandoned by enthusiasts for the name. Proposed as "pénordre = (quasi-ordre)" between the ranks of order and suborder. Used as semiorder by Tattersall, Delson & Van Couvering, 1988:xxxi.

Infraorder **STREPSIRRHINI** É. Geoffroy Saint-Hilaire, 1812:156, **new rank**.[1]
> [= Prosimii Illiger, 1811:72[2]; PROSIMIAE Haeckel, 1895:600, 602; STREPSIRHINI
> McKenna, 1975:41[3]; HETERONYCHAE Gray, 1821:298[4]; LEMUROIDEA Mivart,
> 1864[5]; Brachytarsi Haeckel, 1866:clix[6]; HEMIPITHECI Haeckel, 1895:600, 602.]

[1] Proposed as a family.
[2] Named by Illiger as a family but not based on a genus-group name. Used at higher ranks by others later. The original members of 'family" Prosimii were *Lichanotus*, *Lemur*, and *Stenops*, which are all members of É. Geoffroy Saint-Hilaires's STREPSIRRHINI named a year later. We take the stance that in its original meaning Prosimii does not preoccupy STREPSIRRHINI.
[3] Used as a suborder. Used as a semiorder by Goodman, 1975:228, 244.
[4] Described as an order of Gray's class QUADRUMANES.
[5] Proposed as a suborder.
[6] Used as a family.

†*Lushius* Chow, 1961:1.
> M. Eoc.; As.[1]

[1] E. Asia.

Family †**Plesiopithecidae** Simons & Rasmussen, 1994.
> L. Eoc.; Af.

†*Plesiopithecus* Simons, 1992:10744.
> L. Eoc.; Af.

Superfamily **Daubentonioidea** Gray, 1863:151, **new rank**.
> [= Daubentoniadae Gray, 1863:151; Chiromyoidea Pocock, 1918;
> CHIROMYIFORMES Anthony & Coupin, 1931; DAUBENTONIIFORMES Vallois,
> 1955.]

Family **Daubentoniidae** Gray, 1863:151.
> [= Leptodactyla Illiger, 1811:75[1]; Cheiromydae Gray, 1821:309[2]; Cheiromina Gray,
> 1825:338[3]; Chiromydidae C. L. Bonaparte, 1837:9; Chiromydina C. L. Bonaparte,
> 1837:9[4]; Chiromidae C. L. Bonaparte, 1838:111; Cheiromydidae C. L. Bonaparte,
> 1841:256; Cheiromydina C. L. Bonaparte, 1841:256[4]; Chiromyidae C. L. Bonaparte,
> 1850; Chiromyini Giebel, 1855:663; Cheiromyini Huxley, 1872:390[1]; Chiromysidae
> Ameghino, 1889:893; Chiromyida Haeckel, 1895:600, 606; Gliridae Ogilby, 1837[5];
> Daubentoniadae Gray, 1863:151; Daubentoniidae Gray, 1870:2, 96.]
> R.; Madagascar.

[1] Used as a family.
[2] See International Code of Zoological Nomenclature, 1961, Article 40(a).
[3] As tribe.
[4] As subfamily.
[5] Based on *Cheiromys* and therefore invalid.

Daubentonia É. Geoffroy Saint-Hilaire, 1795:195. Aye-aye.
> [= *Cheiromys* G. Cuvier, 1800: plate I; *Cheyromis* É. Geoffroy Saint-Hilaire,
> 1803:181; *Chiromys* Illiger, 1811:75.]
> R.; Madagascar.

Superfamily **Lemuroidea** Gray, 1821:298.
> [= Lemuridae Gray, 1821:298; Lemuroidea Gill, 1872:2, 54.]

Family †**Adapidae** Trouessart, 1879:225.
> [= †Adapoidea Kalandadze & Rautian, 1992:79; †EOSIMIA Ameghino, 1889[1];
> †ADAPIFORMES Szalay & Delson, 1979:105[2]; †ADAPIFORMES Hoffstetter,
> 1977:327-346.[2]] [Including †Limnotheridae Marsh, 1872:205; †Limnotheriidae Marsh,
> 1876:205; †Notharctidae Trouessart, 1879:230; †Sivaladapidae Tattersall, Delson &
> Van Couvering, 1988:xxv.]
> E.-L. Eoc.; Af. E.-L. Eoc., ?E. Olig.; Eu. E.-M. Eoc.; N.A. M. Eoc.-E. Olig., M.-
> L. Mioc.; As.

[1] Not seen. See Ameghino's Obras Completas 6:182.
[2] Proposed as infraorder.

†*Djebelemur* Hartenberger & Marandat, 1992:9.
> E. Eoc.; Af.[1]

[1] N. Africa (Tunisia).

†*Shizarodon* Gheerbrant, Thomas, Roger, Sen & Al-Sulaimani, 1993:182.
> E. Olig.; As.[1]

[1] S.W. Asia (Oman, Arabian Peninsula).

†*Omanodon* Gheerbrant, Thomas, Roger, Sen & Al-Sulaimani, 1993:145.
> E. Olig.; As.[1]

[1] S.W. Asia (Arabian Peninsula).

Subfamily †**Adapinae** Trouessart, 1879:225.
> [= †Adapidae Trouessart, 1879:225; †Adapinae Trouessart, 1879:226.] [Including
> †Caenopithecina Szalay & Delson, 1979:143; †Sivaladapidae Tattersall, Delson & Van
> Couvering, 1988:xxv; †Hoanghoniinae Gingerich, Holroyd & Ciochon, in Fleagle &
> Kay, eds., 1994:166.]
> E.-L. Eoc., ?E. Olig.; Eu. M.-L. Eoc., M.-L. Mioc.; As. M. Eoc.; N.A. L. Eoc.;
> Af.

†*Donrussellia* Szalay, 1976:355.
> [Including †*Palettia* Godinot, 1992:239.[1]]
> E. Eoc.; Eu.

[1] Subgenus as defined.

†*Agerinia* Crusafont Pairó & Golpe-Posse, 1973.
> [= †*Agerina* Crusafont Pairó, 1967:618.[1]]

E.-M. Eoc.; Eu. ?M. Eoc.; As.[2]

[1] Not *Agerina* Leach, 1814; nor †*Agerina* Tjernvik, 1956:197, a trilobite.
[2] S. Asia.

†*Caenopithecus* Rütimeyer, 1862:88.
    M. Eoc.; Eu.

†*Panobius* D. E. Russell & Gingerich, 1987:209.
    M. Eoc.; As.[1]
[1] S. Asia (Pakistan).

Tribe †**Protoadapini** Szalay & Delson, 1979:120.
    [= †Protoadapinae Schwartz, 1986:28.] [Including †Microadapini Szalay & Delson, 1979:134.]
    E.-L. Eoc.; Eu. M. Eoc.; As. M. Eoc.; N.A.

†*Protoadapis* Lemoine, 1878:101.
    [= or including †*Protadapis* Stehlin, 1912.]
    E. Eoc.; Eu.

†*Pronycticebus* G. Grandidier, 1904:9.
    E.-L. Eoc.; Eu.

†*Europolemur* Weigelt, 1933:123.
    [= or including †*Megatarsius* Weigelt, 1933:141.] [Including †*Alsatia* Tattersall & Schwartz, 1983:232.]
    ?M. Eoc.; As.[1] M. Eoc.; Eu.
[1] Coastal southeastern China.

†*Barnesia* Thalmann, 1994:60.
    M. Eoc.; Eu.

†*Microadapis* Szalay, 1974:127.
    M. Eoc.; Eu.

†*Mahgarita* J. A. Wilson & Szalay, 1977:643.
    [= †*Margarita* Wilson & Szalay, 1976:298.[1]]
    M. Eoc.; N.A.
[1] Not †*Margarita* Leach, 1814: 107, a mollusk.

Tribe †**Adapini** Trouessart, 1879:223, 225.
    [= †Adapidae Trouessart, 1879:223, 225; †Adapini Szalay & Delson, 1979:134; †Adapina Szalay & Delson, 1979:136.]
    M. Eoc.; As. M.-L. Eoc., ?E. Olig.; Eu.

†*Adapis* G. Cuvier, 1821.[1]
    [= or including †*Aphelotherium* Gervais, 1848: pl. 34[2]; †*Palaeolemur* Delfortrie, 1873:90-93[3]; †*Paleolemur* Delfortrie, 1873:64.[4]] [Including †*Leptadapis* Gervais, 1852:35-36[5]; †*Arisella* Crusafont Pairó, 1966:16[6]; †*Paradapis* Tattersall & Schwartz, 1983:235.]
    M.-L. Eoc., ?E. Olig.; Eu.[7]
[1] Possibly 1822.
[2] 1848-52, pl. 34.
[3] Actes Soc. Linn. Bordeaux 29.
[4] C.R. Paris 77(1).
[5] Sometimes as subgenus; sometimes as separate genus.
[6] *Nomen nudum*; sometimes doubtfully placed in †*Leptadapis*.
[7] A possible †*Leptadapis* molar has been reported from the early Oligocene of Bouldnor Cliff, England.

†*Simonsia* Schwartz & Tattersall, 1982:182.
    M. Eoc.; Eu.

†*Adapoides* Beard, Qi, Dawson, Wang & Li, 1994:605.
    M. Eoc.; As.[1]
[1] Coastal southeastern China.

†*Cryptadapis* Godinot, 1984:1291.
    L. Eoc.; Eu.

Tribe †**Anchomomyini** Szalay & Delson, 1979:130.
      M.-L. Eoc.; Eu. L. Eoc.; Af.

†*Wadilemur* Simons, 1997:180.
      L. Eoc.; Af.

†*Anchomomys* Stehlin, 1916:1406.
      [Including †*Laurasia* Schwartz & Tattersall, 1983:350.]
      M. Eoc.; Eu. L. Eoc.; Af.

†*Buxella* Godinot, 1988:391.
      M. Eoc.; Eu.

†*Huerzeleris* Szalay, 1974:125.
      L. Eoc.; Eu.

Tribe †**Sivaladapini** Thomas & Verma, 1979:833, **new rank**.
      [= †Sivaladapinae Thomas & Verma, 1979:833[1]; †Sivaladapidae Tattersall, Delson &
      Van Couvering, 1988:xxv.] [Including †Indralorisini Szalay & Delson, 1979:143[2];
      †Hoanghoniinae Gingerich, Holroyd & Ciochon, in Fleagle & Kay, eds., 1994:166.]
      M.-L. Eoc., M.-L. Mioc.; As.
      [1] Nov. 12, 1979.
      [2] Dec. 31, 1979.

†*Hoanghonius* Zdansky, 1930:75.
      [Including †*Rencunius* Gingerich, Holroyd & Ciochon, 1994:166.]
      M.-L. Eoc.; As.[1]
      [1] E. Asia.

†*Sivaladapis* Gingerich & Sahni, 1979:415.
      [Including †*Indoadapis* Chopra & Vasishat, 1980:511.]
      M.-L. Mioc.; As.[1]
      [1] S. Asia.

†*Indraloris* Lewis, 1933:135.
      L. Mioc.; As.[1]
      [1] S. Asia.

†*Sinoadapis* R. Wu & Pan, 1985:2.
      L. Mioc.; As.[1]
      [1] China.

Subfamily †**Notharctinae** Trouessart, 1879:230.
      [= †Limnotheridae Marsh, 1875:239[1]; †Limnotheriidae Hay, 1902:788; †Notharctidae
      Trouessart, 1879:230.]
      E.-M. Eoc.; Eu. E.-M. Eoc.; N.A.
      [1] See the International Code of Zoological Nomenclature, Article 40.

†*Cantius* Simons, 1962:5.
      E. Eoc., ?M. Eoc.; Eu. E. Eoc., ?M. Eoc.; N.A.

Tribe †**Copelemurini** Beard, 1988:450.
      [Including †Smilodectini Schwartz, 1986:29.]
      E.-M. Eoc.; N.A. M. Eoc.; Eu.

†*Copelemur* Gingerich & Simons, 1977:266.
      E. Eoc.; N.A.

†*Smilodectes* Wortman, 1903:212.
      [= †*Aphanolemur* Granger & Gregory, 1917:856.]
      E.-M. Eoc.; N.A. M. Eoc.; Eu.

Tribe †**Notharctini** Trouessart, 1879:230.
      [= †Limnotheridae Marsh, 1875:239; †Notharctinae Trouessart, 1879:230; †Notharctini
      Schwartz, 1986:28; †Notharctini Beard, 1988:440.] [Including †Pelycodontini
      Schwartz, 1986:28.]
      E.-M. Eoc.; N.A.

†*Pelycodus* Cope, 1875:13.
E. Eoc.; N.A.

†*Notharctus* Leidy, 1870:113-114.
[= or including †*Limnotherium* Marsh, 1871:43-44; †*Hipposyus* Leidy, 1872:37; †*Tomitherium* Cope, 1872:2-3; †*Thinolestes* Marsh, 1872:205-206; †*Telmalestes* Marsh, 1872:206; †*Telmatolestes* Marsh, 1872:405; †*Prosinopa* Trouessart, 1897:68.]
E.-M. Eoc.; N.A.

†*Hesperolemur* Gunnell, 1995:449.
M. Eoc.; N.A.

Subfamily †**Azibiinae** Gingerich, 1976:95.
M. Eoc.; Af.

†*Azibius* Sudre, 1975:1539.
M. Eoc.; Af.

Subfamily †**Cercamoniinae** Gingerich, 1975:164.
[= †Cercamoninae Gingerich, 1975:164.]
L. Eoc.; Af. L. Eoc. and/or E. Olig.; Eu.

†*Cercamonius* Gingerich, 1975:164.
L. Eoc. and/or E. Olig.; Eu.

†*Aframonius* Simons, Rasmussen & Gingerich, 1995:578.
L. Eoc.; Af.

Family **Lemuridae** Gray, 1821:298. Lemurs.
[= Lemurum Fischer de Waldheim, 1817:402; Limuridae Flemming, 1822:xxxii; Lemurini Huxley, 1872:390.[1]] [Including †Megaladapidae Forsyth Major, 1893:178.]
R.; Madagascar.[2]
[1] Used as a family, the sister-group of the Cheiromyini.
[2] Including Comoro Islands.

Subfamily **Lepilemurinae** Gray, 1870:132.
[= Lepilemurina Gray, 1870:132; Lepilemurinae Rumpler & Rakotosamimanana, 1971.[1]] [Including †Megaladapidae Forsyth Major, 1893:178; †Megaladapinae Simpson, 1945:62.[2]]
R.; Madagascar.
[1] See the International Code of Zoological Nomenclature, Article 36.
[2] Incorrectly attributed to Flower & Lydekker, 1891:6 by Simpson, but Simpson may not have been the first to use this name as a subfamily.

Lepilemur I. Geoffroy Saint-Hilaire, 1851:341. Sportive lemurs, weasel lemurs, nattocks.
[= *Lepidilemur* Giebel, 1855:1018-1019[1]; *Galeocebus* Wagner, 1855:xii, 147[2]; *Lepidolemur* Peters, 1874:690.] [Including *Mixocebus* Peters, 1874:690-693.]
R.; Madagascar.
[1] Emendation of *Lepilemur*.
[2] Replacement name for *Lepilemur*.

†*Megaladapis* Forsyth Major, 1893:176-179.
[Including †*Peloriadapis* Grandidier, 1899:276[1]; †*Megalindris* Standing, 1908.]
R.; Madagascar.
[1] Subgenus.

Subfamily **Lemurinae** Gray, 1821:298.
[= Lemuridae Gray, 1821:298; Lemurina Gray, 1825:338[1]; Lemurinae Mivart, 1864:638.]
R.; Madagascar.[2]
[1] As tribe.
[2] Including Comoro Islands.

Hapalemur I. Geoffroy Saint-Hilaire, 1851:341. Gentle lemurs.
[= *Hapalolemur* Giebel, 1855:1018.] [Including *Prolemur* Gray, 1870:828-831[1]; *Prohapalemur* Lamberton, 1936.]
R.; Madagascar.
[1] Subgenus as defined.

Lemur Linnaeus, 1758:29-30. Ring-tailed lemur.

> [= *Prosimia* Brisson, 1762:13, 156-158[1]; *Prosimia* Scopoli, 1772:71; *Procebus* Storr, 1780:32-33; *Catta* Fink, 1806:7-8; *Maki* Muirhead, 1819:405; *Mococo* Trouessart, 1878:163; *Odorlemur* Bolwig, 1961:207.]
>
> R.; Madagascar.[2]

[1] Invalid. Not published in a consistently binominal work. However, see the International Code of Zoological Nomenclature, Article 11(c)(i) for confusion.
[2] Including Comoro Islands.

Varecia Gray, 1863:135. Ruffed lemur.

> [Including †*Pachylemur* Lamberton, 1948.]
>
> R.; Madagascar.

Eulemur Simons & Rumpler, 1988:550.[1] Mongoose lemur, black lemur.

> [= *Prosimia* Boddaert, 1785:43, 65.[2]] [Including *Petterus* Groves & Eaglen, 1988:533.[3]]
>
> R.; Madagascar.

[1] Dated Sept. 15, 1988, but legally deposited August 1988.
[2] Preoccupied by *Prosimia* Scopoli, 1772, a synonym of *Lemur*.
[3] Sept. 9, 1988.

Superfamily **Loroidea** Gray, 1821:298.

> [= Loridae Gray, 1821:298; Lorisoidea Szalay, 1975.]

†*Chasselasia* Schwartz & Tattersall, 1982:184.

> M. Eoc.; Eu.

†*Fendantia* Schwartz & Tattersall, 1983:349.[1]

> M. Eoc.; Eu.

[1] Possibly related to the galagines.

Family **Loridae** Gray, 1821:298.[1]

> [= LORISIFORMES Gregory, 1915:435[2]; Lorisidae Gregory, 1915:436.] [Including Nycticebidae Nicholson, 1870:553; Perodicticinidae Gray, in Rochebrune, 1883:39, 151.]
>
> E. Olig., E.-L. Mioc., L. Plioc., Pleist., R.; Af.[3] L. Mioc., R.; As. R.; E. Indies.

[1] Spelling follows Article 29(b)(ii) of the International Code of Zoological Nomenclature.
[2] Used as infraorder.
[3] Loridae, genus indet., from the early Oligocene of Egypt.

Pseudopotto Schwartz, 1996:8.

> R.; Af.

Subfamily **Galagoninae** Gray, 1825:338.[1] Galagos.

> [= Galagonina Gray, 1825:338[2]; Galagininae Mivart, 1864:637; Galaginidae Alston, 1878:10; Galaginae Simpson, 1945:63; Galagidae Hill, 1953:195; Galagonidae Jenkins, 1987:1, 85.]
>
> E.-L. Mioc., L. Plioc., Pleist., R.; Af.[3]

[1] Spelling follows Article 29(b)(ii) of the International Code of Zoological Nomenclature.
[2] Proposed as tribe.
[3] Indet. "Galagidae" have been reported from the late Miocene Otavi breccias, Namibia.

†*Komba* Simpson, 1967:49.

> E.-M. Mioc.; Af.

†*Progalago* MacInnes, 1943:145.

> [Including †*Mioeuoticus* L. S. B. Leakey, in W. W. Bishop, 1962:6-9.]
>
> E. Mioc.; Af.

Galago É. Geoffroy Saint-Hilaire, 1796:49. Bush babies, galagos, ojams.

> [= *Gallacho* Wiegmann, 1838:394[1]; *Otolicnus* Illiger, 1811:74; *Otolincus* McMurtrie, 1831:74[1]; *Otoleneus* McMurtrie, 1834:50[1]; *Otolichnus* Boitard, 1842:91[1]; *Otaclinus* London Encyclopaedia (anon.), 1845:736[1]; *Stolicnus* ("Fleming") Gray, 1870:91.[1]] [Including *Galagoides* A. Smith, 1833:32; *Hemigalago* Dahlbom, 1857:224-230[2]; *Otolemur* Coquerel, 1859:458; *Callotus* Gray, 1863:145; *Otogale* Gray, 1863:139-140; *Sciurocheirus* Gray, 1872:857-858.]

L. Plioc., Pleist., R.; Af.[3]

[1] Emendation or misprint.
[2] Objective synonym of *Galagoides*.
[3] Southern and tropical Africa.

Euoticus Gray, 1863:140.[1] Needle-clawed galagos.
R.; Af.[2]

[1] Proposed as subgenus of *Otogale* Gray, 1863, a synonym of *Galago*.
[2] W. Africa.

Subfamily **Lorinae** Gray, 1821:298. Lorises, pottos.
[= Loridae Gray, 1821:298; Loridina Gray, 1825:338; Lorisinae Flower & Lydekker, 1891:691.] [Including Nycticebinae Mivart, 1864:637; Nycticebinae Schwartz, 1986:31; Perodicticinae Trouessart, 1879:38.]
L. Mioc., R.; As. R.; Af. R.; E. Indies.

Tribe **Nycticebini** Mivart, 1864:637, **new rank**.
[= Nycticebinae Mivart, 1864:637; Nycticebidae Nicholson, 1870:553.] [Including Perodicticina Gray, 1863:132, 150.]
L. Mioc., R.; As. R.; Af. R.; E. Indies.

†*Nycticeboides* Jacobs, 1981:586.
L. Mioc.; As.[1]

[1] S. Asia.

Nycticebus É. Geoffroy Saint-Hilaire, 1812:163. Slow loris, gray loris.
R.; E. Indies. R.; As.[1]

[1] S.E. Asia.

Perodicticus Bennett, 1831:109. Potto.
R.; Af.[1]

[1] Tropical Africa.

Tribe **Lorini** Gray, 1821:298.[1]
[= Loridae Gray, 1821:298; Loridina Gray, 1825:338.[2]]
R.; Af. R.; As.

[1] New suffix.
[2] As tribe.

Arctocebus Gray, 1863:150. Angwantibo, Calabar potto.
R.; Af.[1]

[1] W. Africa.

Loris É. Geoffroy Saint-Hilaire, 1796:48. Slender loris.
[= *Tardigradus* Boddaert, 1784:43, 67[1]; *Lori* Lacépède, 1799:5; *Stenops* Illiger, 1811:73; *Loridium* Rafinesque, 1815:54[2]; *Bradylemur* de Blainville, 1839:12-13.]
R.; As.[3]

[1] Not to be confused with "*Tardigradus*" Brisson, 1762 (= *Bradypus*), published in a work that was not consistently binominal.
[2] New name for *Loris*.
[3] S. Asia (southern India and Sri Lanka).

Family **Cheirogaleidae** Gray, 1872:858.[1]
[= Microcebina Gray, 1870:131; Cheirogaleina Gray, 1872:858; Cheirogaleinae Gregory, 1915:434; Cheirogaleidae Rumpler, 1974:867.]
R.; Madagascar.

[1] Historically, the correct name of this taxon should be "Microcebidae", but stability would be better served by retaining "Cheirogalidae." Microcebina Gray, 1870, should in that case be formally suppressed.

Subfamily **Cheirogaleinae** Gray, 1872:858.[1] Dwarf lemurs.
[= Cheirogaleina Gray, 1872:858; Cheirogaleinae Gregory, 1915:434; Cheirogaleini Tattersall & Schwartz, 1974:188; Cheirogaleidae Rumpler, 1974:865.] [Including Microcebina Gray, 1870:131; Microcebinae Schwartz, 1986:29.]
R.; Madagascar.

[1] See note for Cheirogalidae.

Allocebus Petter-Rousseaux & Petter, 1967:574. Hairy-eared dwarf lemur.
R.; Madagascar.

Microcebus É. Geoffroy Saint-Hilaire, 1834:24.[1] Mouse lemurs.
[= *Murilemur* Gray, 1870:87, 135.] [Including *Scartes* Swainson, 1835:352; *Myscebus* Lesson, 1840:270, 214-216; *Mirza* Gray, 1870:131, 135-136; *Azema* Gray, 1870:132, 134.]
R.; Madagascar.
[1] Not 1828. See Palmer (1904:420).

Cheirogaleus É. Geoffroy Saint-Hilaire, 1812:172. Dwarf lemurs.
[= *Cebugale* Lesson, 1840:207, 213-214; *Chirogale* Gloger, 1841.[1]] [Including *Opolemur* Gray, 1872:853-855; *Altililemur* Elliot, 1913:xxxi, 111; *Altilemur* Weber, 1928:736.[2]]
R.; Madagascar.
[1] And many later authors.
[2] Misspelling.

Subfamily **Phanerinae** Rumpler & Rakotosamimanana, 1971.
[= Phanerini Rumpler & Rakotosamimanana, 1971; Phanerinae Rumpler & Albignac, 1972[1]; Phanerini Tattersall & Schwartz, 1974:188.]
R.; Madagascar.
[1] See the International Code of Zoological Nomenclature, Article 36.

Phaner Gray, 1870:132, 135. Fork-marked lemur.
R.; Madagascar.

Superfamily **Indroidea** Burnett, 1828:306.
[= Indridae Burnett, 1828:306; INDRIIFORMES Tattersall & Schwartz, 1974:188[1]; Indrioidea Tattersall & Schwartz, 1975:311.]
[1] Proposed as infraorder.

Family †**Archaeolemuridae** Forsyth Major, 1896:436.[1]
[= †Nesopithecidae Forsyth Major, 1896:436; †Archaeolemurinae Standing, 1908:72.] [Including †Hadropithecinae Abel, 1931:222.]
R.; Madagascar.
[1] See the International Code of Zoological Zomenclature, Articles 23(d) and 40.

†*Archaeolemur* Filhol, 1895:13.
[= †*Nesopithecus* Forsyth Major, 1896:436; †*Globilemur* Forsyth Major, 1897:46-47.] [Including †*Bradylemur* Grandidier, 1899:346-348.[1]]
R.; Madagascar.
[1] Not *Bradylemur* de Blainville, 1839, a synonym of *Loris*.

†*Hadropithecus* Lorenz von Liburnau, 1899:256.
R.; Madagascar.

Family †**Palaeopropithecidae** Tattersall, 1973.[1] Sloth lemurs.
[= †Palaeopropithecinae Tattersall, 1973; †Palaeopropithecidae Tattersall, Delson & Van Couvering, 1988:xxv.]
R.; Madagascar.
[1] For retention of this family-group name, see the International Code of Zoological Nomenclature, Article 40.

†*Babakotia* Godfrey, Simons, Chatrath & Rakotosamimanana, 1990:83.
R.; Madagascar.

†*Archaeoindris* Standing, 1908:9.
[= †*Lemuridotherium* Standing, 1910:62.]
R.; Madagascar.

†*Palaeopropithecus* Grandidier, 1899:345.[1]
[= †*Thaumastolemur* Filhol, 1895:13[2]; †*Bradytherium* Grandidier, 1901:54-56.]
R.; Madagascar.
[1] Conserved under the plenary powers by the International Commission on Zoological Nomenclature, Opinion 1737, 1993.

[2] Suppressed under the plenary powers by the International Commission of Zoological Nomenclature, Opinion 1737, 1993.

Family **Indridae** Burnett, 1828:306.[1]

[= Lichanotina Gray, 1825:338[2]; Indrisina I. Geoffroy Saint-Hilaire, 1851:67; Indrisidae Alston, 1878:10; Indrisinae Mivart, 1864:637; Indriidae Hill, 1953:516; Indriinae Hill, 1953:555.] [Including Propitheci Winge, 1895; Propithecinae Trouessart, 1897:55.[3]]

R.; Madagascar.

[1] Spelling follows the International Code of Zoological Nomenclature, Article 29(b)(ii).
[2] As tribe.
[3] Credited to Winge.

Indri É. Geoffroy Saint-Hilaire, 1796:46. Indri,[1] babakoto.

[= *Indris* G. Cuvier, 1800: table I[2]; *Lichanotus* Illiger, 1811:72; *Indrium* Rafinesque, 1815:54.[3]]

R.; Madagascar.

[1] The indri's European "discoverer," Sonnerat (c. 1780), dutifully recorded the Malagasy word for "lo, behold," exclaimed by his guide on sighting the animal. The correct Malagasy name is "babakoto."
[2] And many later authors.
[3] New name for *Indri*.

†*Mesopropithecus* Standing, 1905:95-100.

[Including †*Neopropithecus* Lamberton, 1936:370-373.]

R.; Madagascar.

Propithecus Bennett, 1832:20-22. Sifakas.

R.; Madagascar.

Avahi Jourdan, 1834:231.[1] Avahi, woolly lemur.

[= *Microrhynchus* Jourdan, 1834[2]; *Semnocebus* Lesson, 1840:207, 209-212; *Iropocus* Gloger, 1841:43-44.]

R.; Madagascar.

[1] Sometimes confused with *Lichanotus* Illiger, 1811, a synonym of *Indri* É. Geoffroy Saint-Hilaire.
[2] Not *Microrhynchus* Megerle, 1823, a genus of Coleoptera.

Infraorder **HAPLORHINI** Pocock, 1918:19-53, **new rank**.[1]

[1] Used as a suborder by McKenna, 1975:41; as semiorder by Goodman, 1975:228, 244; as a hyporder by Delson, 1977:452.

Parvorder **TARSIIFORMES** Gregory, 1915:437, **new rank**.[1]

[= TARSII Rothchild, 1965[2]; TARSII deBlase, in Parker, ed., 1982:1032.] [Including †OMOMYIFORMES Schmid, 1982.]

[1] Proposed as a series (between the ranks of suborder and family). Used as an infraorder by Simpson, 1945:63; as hyporder by Delson, 1977:452; as a semisuborder by Szalay, Rosenberger & Dagosto, 1987:79. An undescribed "?Primate tarsiiforme" has been reported from the early Eocene of Tunisia.
[2] Possibly 1961.

Superfamily †**Carpolestoidea** Simpson, 1935:9, **new rank**.

[= †Carpolestidae Simpson, 1935:9.]

Family †**Carpolestidae** Simpson, 1935:9.

E. Paleoc.-E. Eoc.; N.A. E. Eoc.; As.

Subfamily †**Carpolestinae** Simpson, 1935:9.

[= †Carpolestidae Simpson, 1935:9; †Carpolestinae Beard & Wang, 1995:14.]

E. Paleoc.-E. Eoc.; N.A. E. Eoc.; As.

†*Elphidotarsius* Gidley, 1923:10.

E.-L. Paleoc.; N.A.

†*Carpodaptes* Matthew & Granger, 1921:6.

?E. Paleoc., L. Paleoc.; N.A.

†*Carpolestes* Simpson, 1928:7.

[= †*Litotherium* Simpson, 1929.]

L. Paleoc.-E. Eoc.; N.A.

†*Carpocristes* Beard & Wang, 1995:14.

L. Paleoc.; N.A. E. Eoc.; As.

Subfamily †**Cronolestinae** Beard & J. Wang, 1995:2.
  E. Eoc.; As.

 †*Chronolestes* Beard & J. Wang, 1995:3.
  E. Eoc.; As.

Superfamily **Tarsioidea** Gray, 1825:338.
  [= Tarsina Gray, 1825:338; Tarsioidea McKenna, 1967:607.[1]]
  [1] Used as superfamily.

Family †**Omomyidae** Trouessart, 1879:225.[1]
  [= †Omomynae Trouessart, 1879:225; †Omomyinae Wortman, 1904:29;
  †Omomyidae Gazin, 1958:47; †Omomyoidea McKenna, 1967:605.]
  [Including †Anaptomorphidae Cope, 1883:80; †Anaptomorpha Haeckel,
  1895:600, 604[2]; †Tetoniidae Abel, 1931:199; †Teilhardidae Quinet, 1964:12.]
  E. Paleoc., E.-M. Eoc.; As. L. Paleoc., L. Eoc.; Af.[3] E.-M. Eoc.; Eu.
  E.-L. Eoc.; N.A.
  [1] See the International Code of Zoological Nomenclature, Article 33.
  [2] Used as a family by Haeckel, synonymous with †Necrolemures.
  [3] Omomyidae indet. reported from the Jebel Quatrani Fm., locality E, late Eocene of Egypt.

 †*Steinius* Bown & Rose, 1984:98.
  E. Eoc.; N.A.

 †*Periconodon* Stehlin, 1916.
  M. Eoc.; Eu.

 †*Asiomomys* Wang & Li, 1990:179.
  M. Eoc.; As.

 †*Kohatius* D. E. Russell & Gingerich, 1980:621.[1]
  E. and/or M. Eoc.; As.[2]
  [1] Assignment to †Omomyidae is questionable.
  [2] S. Asia (Pakistan).

Subfamily †**Decoredontinae** Szalay & Li, 1987:388.[1]
  [= †Decoredontidae Tattersall, Delson & Van Couvering, 1988:xxxii.]
  E. Paleoc.; As.[2]
  [1] Although actually published in 1987, this paper was dated 1986.
  [2] E. Asia.

 †*Decoredon* Xu, 1977:119.
  E. Paleoc.; As.[1]
  [1] E. Asia.

Subfamily †**Omomyinae** Trouessart, 1879:225.
  [= †Omomynae Trouessart, 1879:225; †Omomyinae Wortman, 1904:29;
  †Omomyidae Gazin, 1958:47.]
  L. Paleoc.; Af. E.-L. Eoc.; N.A. M. Eoc.; As.

 †*Altiatlasius* Sigé, Jaeger, Sudre & Vianey-Liaud, 1990:49.
  L. Paleoc.; Af.[1]
  [1] N. Africa (Egypt).

Tribe †**Omomyini** Trouessart, 1879:225.
  [= †Omomynae Trouessart, 1879:225; †Omomyini Szalay, 1976:256;
  †Omomyina Szalay, 1976:257.]
  E.-M. Eoc.; N.A.

 †*Omomys* Leidy, 1869:63.
  [= †*Euryacodon* Marsh, 1872:223.]
  E.-M. Eoc.; N.A.

 †*Chumashius* Stock, 1933:954.
  M. Eoc.; N.A.

Tribe †**Washakiini** Szalay, 1976:285.
  [Including †Macrotarsiini Krishtalka & Schwartz, 1978:518, fig. 1.]
  E.-L. Eoc.; N.A. M. Eoc.; As.

†*Loveina* Simpson, 1940:188.
> E. Eoc.; N.A.

†*Washakius* Leidy, 1873:123.
> [Including †*Yumanius* Stock, 1938:288.]
> M. Eoc.; N.A.

†*Shoshonius* Granger, 1910:249.
> E. Eoc.; N.A.

†*Dyseolemur* Stock, 1934:150.
> M. Eoc.; N.A.

†*Macrotarsius* J. Clark, 1941:562.[1]
> M. Eoc.; As.[2] M.-L. Eoc.; N.A.
>
> [1] Not preoccupied by *Macrotarsus* Link, 1795, a synonym of *Tarsius*.
> [2] Coastal southeastern China.

†*Hemiacodon* Marsh, 1872:212.
> E.-M. Eoc.; N.A.

†*Yaquius* Mason, 1990:2.
> M. Eoc.; N.A.

Tribe †**Mytoniini** Robinson, 1968:318, **new rank**.[1]
> [= †Mytoniinae Robinson, 1968:318; †Ourayiini Gunnell, 1995:168.]
> E.-M. Eoc.; N.A.
>
> [1] See the International Code of Zoological Nomenclature, Articles 23, 36, and 40.

Subtribe †**Mytoniina** Robinson, 1968:318.
> [= †Mytoniinae Robinson, 1968:318; †Mytoniina Szalay, 1976:274.]
> M. Eoc.; N.A.

†*Wyomomys* Gunnell, 1995:172.
> M. Eoc.; N.A.

†*Ageitodendron* Gunnell, 1995:175.
> M. Eoc.; N.A.

†*Ourayia* Gazin, 1958:70.
> [= †*Mytonius* Robinson, 1968:321.]
> M. Eoc.; N.A.

Subtribe †**Utahiina** Szalay, 1976:340.
> [= †Utahiini Szalay, 1976:340; †Utahiina Gunnell, 1995:168.]
> E.-M. Eoc.; N.A.

†*Utahia* Gazin, 1958:66.
> E.-M. Eoc.; N.A.

†*Stockia* Gazin, 1958:68.
> M. Eoc.; N.A.

†*Chipetaia* Rasmussen, 1996:76.
> M. Eoc.; N.A.

Tribe †**Uintaniini** Szalay, 1976:231.
> E.-M. Eoc.; N.A.

†*Uintanius* Matthew, 1915:455.
> [Including †*Huerfanius* Robinson, 1966:35; †*Jemezius* Beard, 1987:458.]
> E.-M. Eoc.; N.A.

Tribe †**Rooneyiini** Szalay, 1976:347.
> L. Eoc.; N.A.

†*Rooneyia* J. A. Wilson, 1966:228.
> L. Eoc.; N.A.

Subfamily †**Anaptomorphinae** Cope, 1883:80.
        [= †Anaptomorphidae Cope, 1883:80; †Anaptomorphinae Simpson, 1940:198.]
        [Including †Tetoniidae Abel, 1931:199; †Teilhardinidae Quinet, 1964:12; †Altaniinae
        Van Valen, 1994:41.]
        E. Eoc.; As.[1] E. Eoc.; Eu. E.-M. Eoc.; N.A.
        [1] E. Asia.

Tribe †**Anaptomorphini** Cope, 1883:80.
        [= †Anaptomorphidae Cope, 1883:80; †Anaptomorphini Szalay, 1976:173.] [Including
        †Altaniinae Van Valen, 1994:41.]
        E. Eoc.; As.[1] E. Eoc.; Eu. E.-M. Eoc.; N.A.
        [1] E. Asia.

Subtribe †**Teilhardinina** Quinet, 1964:12.
        [= †Teilhardinidae Quinet, 1964:12; †Teilhardinina Szalay, 1976:173.]
        E. Eoc.; Eu. E. Eoc.; N.A.

†*Teilhardina* Simpson, 1940:190.
        [= †*Protomomys* Teilhard de Chardin, 1927.[1]]
        E. Eoc.; Eu.
        [1] Suppressed under plenary powers for purposes of the law of priority by the International Commission on
        Zoological Nomenclature, Opinion 947, 1971.

†*Chlororhysis* Gazin, 1958:27.
        E. Eoc.; N.A.

Subtribe †**Tetoniina** Abel, 1931:199.
        [= †Tetoniidae Abel, 1931:199; †Tetoniina Szalay, 1976:196; †Tetoniinae Schwartz,
        1986:29.] [Including †Altaniinae Van Valen, 1994:41.]
        E. Eoc.; As.[1] E.-M. Eoc.; N.A.
        [1] E. Asia.

†*Arapahovius* D. E. Savage & Waters, 1978:3.
        E. Eoc.; N.A.

†*Altanius* Dashzeveg & McKenna, 1977:126.[1]
        E. Eoc.; As.[2]
        [1] In 1984 Rose and Krause suggested †*Altanius* was a member of the paraphyletic †Plesiadapoidea related to
        carpolestids, but their evidence is better interpreted to mean that carpolestids are tarsioid primates, as
        others have long since concluded.
        [2] E. Asia.

†*Tetonius* Matthew, 1915:459.
        [Including †*Paratetonius* Seton, 1940:39; †*Anemorhysis* Gazin, 1958:25; †*Uintalacus*
        Gazin, 1958:29; †*Tetonoides* Gazin, 1962:34; †*Pseudotetonius* Bown, 1974:20;
        †*Mckennamorphus* Szalay, 1976:249; †*Mackennamorphus* Szalay, 1976:249.[1]]
        E. Eoc.; N.A.
        [1] Our emendation, following the International Code of Zoological Nomenclature.

†*Absarokius* Matthew, 1915:463.
        [= or including †*Aycrossia* Bown, 1979:50; †*Strigorhysis* Bown, 1979:60.]
        E.-M. Eoc.; N.A.

†*Tatmanius* Bown & Rose, 1991:467.
        E. Eoc.; N.A.

Subtribe †**Anaptomorphina** Cope, 1883:80.
        [= †Anaptomorphidae Cope, 1883:80; †Anaptomorphina Szalay, 1976:181.]
        M. Eoc.; N.A.

†*Anaptomorphus* Cope, 1872:554.
        [= or including †*Gazinius* Bown, 1979:67.]
        M. Eoc.; N.A.

Tribe †**Trogolemurini** Szalay, 1976:255.
        M. Eoc.; N.A.

†*Trogolemur* Matthew, 1909:546.
   M. Eoc.; N.A.

†*Sphacorhysis* Gunnell, 1995:157.
   M. Eoc.; N.A.

Family †**Microchoeridae** Lydekker, 1887:303.
   [Including †Necrolemures Haeckel, 1895:600, 604; †Necrolemurinae Simpson, 1940:198; †Pseudolorisinae Simpson, 1940:198.]
   E.-L. Eoc.; Eu.

†*Pseudoloris* Stehlin, 1916:1397.
   [Including †*Pivetonia* Crusafont Pairó, 1967:16.]
   M.-L. Eoc.; Eu.

†*Paraloris* Fahlbusch, 1995:19.
   L. Eoc.; Eu.

†*Necrolemur* Filhol, 1873:1111-1112.
   M.-L. Eoc.; Eu.

†*Nannopithex* Stehlin, 1916:1392.
   E.-M. Eoc.; Eu.

†*Microchoerus* S. Wood, 1844:350.[1]
   M.-L. Eoc.; Eu.
   [1] This paper was actually authored by Edward Charlesworth, based on S. Wood's manuscript. In 1846 S. Wood republished the name, spelling it †*"Microchaerus"* on page 5 and †*Microchoerus* on plate 2.

Family †**Afrotarsiidae** Ginsburg & Mein, 1987:1215.
   E. Olig.; Af.[1] ?E. Olig.; As.
   [1] N. Africa.

†*Afrotarsius* Simons & Bown, 1985:476.
   E. Olig.; Af.[1] ?E. Olig.; As.[2]
   [1] N. Africa (Egypt).
   [2] Possibly in Arabian Peninsula.

Family **Tarsiidae** Gray, 1825:338. Tarsiers.
   [= Macrotarsi Illiger, 1811:73[1]; Tarsiorum Fischer de Waldheim, 1817:402; Tarsina Gray, 1825:338[2]; Tarsidae Burnett, 1828:306; Tarsiina Bonaparte, 1850: unpaged chart; Tarsiidae Gill, 1872:3.]
   M. Eoc., M. Mioc.; As. R.; E. Indies.
   [1] Used as a family.
   [2] As tribe.

Tarsius Storr, 1780:33, table A. Tarsier, spectral tarsier, umus, mal, malmag, maomog.
   [= *Tarsier* Lacépède, 1788[1]; *Macrotarsus* Link, 1795:51, 65-66; *Prosimia* Shaw, 1800.[2]] [Including *Rabienus* Gray, 1821:299; *Rubienus* Gray, 1870:96; *Cephalopachus* Swainson, 1835:352; *Hypsicebus* Lesson, 1840:207.]
   ?M. Eoc.; M. Mioc.; As.[3,4] R.; E. Indies.[5]
   [1] 1788-1789.
   [2] Not "*Prosimia* Brisson, 1762" nor *Prosimia* Scopoli, 1772, synonyms of *Lemur*; nor *Prosimia* Boddaert, 1785, a synonym of *Petterus*.
   [3] S.E. China in the middle Eocene.
   [4] S.E. Asia (Thailand) in the middle Miocene.
   [5] Sumatra, Borneo, Sulawesi, Savu, Philippines.

Parvorder **ANTHROPOIDEA** Mivart, 1864:635, **new rank**.[1]
   [= PLATYONYCHAE Gray, 1821:297[2]; SIMIAE Haeckel, 1866:clx[3]; SIMIOIDEA Ameghino, 1909:2074[4]; PITHECOIDEA Pocock, 1918:51.[5]] [Including †PARACATARRINI Delson, 1977:452[6]; †EOSIMIAE Van Valen, 1994:40.[7]]
   [1] Proposed as suborder. Used at ordinal rank by Ameghino and as a semisuborder by Szalay, Rosenberger & Dagosto, 1987:79. Rank of hyporder suggested (as synonym of Simiiformes) by Delson, 1977:452.
   [2] Included some but not all of both platyrrhines and catarrhines.
   [3] Used as an order for monkeys, apes, and humans. Used as a hyporder by Van Valen, 1994:41.
   [4] Used at ordinal rank by Ameghino.
   [5] Used as a suborder subsuming PLATYRHINI and CATARRHINI, but not TARSIIFORMES.
   [6] Proposed as parvorder to which †Parapithecidae was assigned.
   [7] Proposed as a hyporder.

†*Biretia* de Bonis, Jaeger, Coiffait & Coiffait, 1988:929.
   L. Eoc.; Af.[1]
   [1] N. Africa (Algeria).

†*Amphipithecus* Colbert, 1937:1.
   M. Eoc.; As.[1]
   [1] S.E. Asia (Burma).

†*Pondaungia* Pilgrim, 1927:12.
   M. Eoc.; As.[1]
   [1] S.E. Asia (Burma).

Family †**Eosimiidae** Beard, Qi, Dawson, Wang & Li, 1994:607.
   M. Eoc.; As.

†*Eosimias* Beard, Qi, Dawson, Wang & Li, 1994:608.
   M. Eoc.; As.[1]
   [1] Coastal southeastern China.

Family †**Parapithecidae** Schlosser, 1911:58.
   [= †Parapithecoidea de Bonis, Jaeger, Coiffait & Coiffait, 1988:929.]
   [Including †Oreopithecidae Schwalbe, 1915:149-254.]
   L. Eoc.-E. Olig., E.-M. Mioc.; Af. L. Mioc.; Mediterranean.[1] L. Mioc.; Eu.
   [1] Sardinia.

Subfamily †**Parapithecinae** Schlosser, 1911:58.
   L. Eoc.-E. Olig.; Af.[1]
   [1] N. Africa (Egypt).

†*Apidium* Osborn, 1908:271.
   E. Olig.; Af.[1]
   [1] N. Africa (Egypt).

†*Serapia* Simons, 1992:10743.
   L. Eoc.; Af.[1]
   [1] N. Africa (Egypt).

†*Qatrania* Simons & Kay, 1983:624.
   L. Eoc.-E. Olig.; Af.[1]
   [1] N. Africa (Egypt).

†*Parapithecus* Schlosser, 1911:58.
   [Including †*Simonsius* Gingerich, 1978:88.]
   E. Olig.; Af.[1]
   [1] N. Africa (Egypt).

Subfamily †**Oreopithecinae** Schwalbe, 1915:149-254, **new rank**.
   [= †Oreopithecidae Schwalbe, 1915:149-254.]
   E.-M. Mioc.; Af. L. Mioc.; Mediterranean.[1] L. Mioc.; Eu.
   [1] Sardinia.

†*Nyanzapithecus* Harrison, 1986:266.
   E.-M. Mioc.; Af.

†*Oreopithecus* Gervais, 1872:1217-1223.
   L. Mioc.; Mediterranean.[1] L. Mioc.; Eu.[2]
   [1] Sardinia.
   [2] Italy (Tuscany).

Superfamily **Cercopithecoidea** Gray, 1821:297.
   [= Cercopithecidae Gray, 1821:297; Cercopithecoidea Gregory & Hellman, 1923:14; CATARRHINI É. Geoffroy Saint-Hilaire, 1812:86[1]; Simiae catarrhinae Giebel, 1855:1053; CATARHINAE Haeckel, 1895:600, 610[2]; HEOPITHECI Haeckel, 1866:cliv; EOPITHECA Haeckel, 1895:600, 610.]
   [Including †EUCATARRHINI Delson, 1977:452[3]; †EOCATARRHINI Ginsburg & Mein, 1980:73[3]; Papionoidea Kretzoi, 1962:376; †Pliopithecoidea Harrison, 1987:74; †Proconsuloidea Harrison, 1987:74.]

[1] Proposed as group of family Singes. Used as suborder by Simpson, 1945:181; as infraorder by Delson, 1975:170 and McKenna, 1975:41.
[2] Used as a suborder.
[3] Proposed as parvorder.

†*Siamopithecus* Chaimanee, Suteethorn, Jaeger & Ducrocq, 1997:429.
  L. Eoc.; As.

†*Wailekia* Ducrocq, Jaeger, Chaimanee & Suteethorn, 1995:478.
  L. Eoc.; As.

†*Dionysopithecus* Li, 1978:188.
  ?E. Mioc., M. Mioc.; As.

†*Afropithecus* R. E. Leakey & M. G. Leakey, 1986:143.
  [Including †*Heliopithecus* Andrews & Martin, 1987:384.]
  E. Mioc.; Af. E. Mioc.; As.

†*Turkanapithecus* R. E. Leakey & M. G. Leakey, 1986:146.
  E. Mioc.; Af.

†*Otavipithecus* Conroy, Pickford, Senut, Van Couvering & Mein, 1992:144.
  M. Mioc.; Af.

Family †**Pliopithecidae** Zapfe, 1960:261.
  [= †Pliopithecinae Zapfe, 1960:261; †Pliopithecoidea Harrison, 1987:74.]
  E. and/or M. Eoc., L. Eoc.-E. Olig., E.-M. Mioc.; Af. L. Eoc. and/or E. Olig., M.-L. Mioc.; As. M.-L. Mioc.; Eu.

†*Algeripithecus* Godinot & Mahboubi, 1992:324.
  E. and/or M. Eoc.; Af.[1]
 [1] N. Africa (Algeria).

†*Tabelia* Godinot & Mahboubi, 1994:360.
  ?E. Eoc.; Af.

Subfamily †**Propliopithecinae** Straus, 1961:761.
  [= †Propliopithecidae Straus, 1961:761; †Propliopithecinae Delson & Andrews, 1975; †Propliopithecoidea Harrison, 1987:74.]
  L. Eoc. and/or E. Olig.; As. E. Olig.; Af.[1]
 [1] N. Africa.

†*Propliopithecus* Schlosser, 1911:52.
  [Including †*Moeripithecus* Schlosser, 1911:64; †*Aeolopithecus* Simons, 1965:36.]
  L. Eoc. and/or E. Olig.; As.[1] E. Olig.; Af.[2]
 [1] S.W. Asia (Oman).
 [2] N. Africa (Egypt).

†*Aegyptopithecus* Simons, 1965:135.
  L. Eoc. and/or E. Olig.; As. E. Olig.; Af.[1]
 [1] N. Af. (Egypt).

Subfamily †**Pliopithecinae** Zapfe, 1960:261.
  [Including †Crouzeliinae Ginsburg & Mein, 1980:75; †Crouzeliinae Ginsburg & Mein, 1980:75.[1]]
  E.-M. Mioc.; Af. M.-L. Mioc.; As. M.-L. Mioc.; Eu.
 [1] Our emendation.

†*Laccopithecus* R. Wu & Pan, 1984:185.
  L. Mioc.; As.[1]
 [1] China.

†*Dendropithecus* P. Andrews & Simons, 1977:162.
  E. Mioc.; Af. M. Mioc.; As.

†*Simiolus* R. E. Leakey & M. G. Leakey, 1987:369.
  E.-M. Mioc.; Af.

†*Pliopithecus* Gervais, 1849:5.
  [= †*Protopithecus* Lartet, 1851:11-12.[1]] [Including †*Epipliopithecus* Zapfe & Hürzeler, 1957:113-123[2]; †*Plesiopliopithecus* Zapfe, 1961:147[2]; †*Anapithecus* Kretzoi,

1975:579[2]; †*Crouzelia* Ginsburg, 1975:877, 882; †*Krishnapithecus* Ginsburg & Mein, 1980:57, 77.[3]]
M.-L. Mioc.; As. M.-L. Mioc.; Eu.

[1] Not †*Protopithecus* Lund, 1838:14, an alouattine primate.
[2] Proposed as subgenus.
[3] Assignment to †*Pliopithecus* is dubious.

Subfamily †**Oligopithecinae** Simons, 1989:9956.
L. Eoc.; Af.[1] ?L. Eoc.; As.

[1] N. Africa.

†*Proteopithecus* Simons, 1989:9957.
L. Eoc.; Af.[1]

[1] N. Africa (Egypt).

†*Catopithecus* Simons, 1989:9957.
L. Eoc.; Af.[1]

[1] N. Africa (Egypt).

†*Oligopithecus* Simons, 1962:2.
L. Eoc.; Af.[1] ?L. Eoc.; As.[2]

[1] N. Africa (Egypt).
[2] S.W. Asia (Oman).

Family **Cercopithecidae** Gray, 1821:297. Old World monkeys.
[= Cercopithecoidea Gregory & Hellman, 1923:14.] [Including Cynocephalina Gray, 1825:338; Cynocephalidae Ameghino, 1899:893; Papionidae Burnett, 1828:306; Macacidae Owen, 1843:55; Semnopithecidae Owen, 1843:55.]
E. Mioc.-R.; Af. ?E. and/or ?M. Mioc., L. Mioc., ?E. Plioc., L. Plioc.-R.; As.[1] L. Mioc.-R.; Eu.[2] L. Plioc.; Mediterranean.[3] ?M. Pleist., L. Pleist.-R.; E. Indies. R.; Indian O.[4]

[1] A questionable cercopithecid has been reported from the early or middle Miocene of Israel, but the specimen may not belong to a primate.
[2] Gibraltar in the Recent (probably by reintroduction).
[3] Sardinia.
[4] Mauritius (probably by introduction).

Subfamily †**Victoriapithecinae** von Koenigswald, 1969:39-52.
E.-M. Mioc.; Af.

†*Prohylobates* Fourtau, 1918:195-196.
E.-M. Mioc.; Af.

†*Victoriapithecus* G. H. R. von Koenigswald, 1969:41.
?E. Mioc., M. Mioc.; Af.

Subfamily **Colobinae** Blyth, 1875:9.
[= Colobidae Blyth, 1875:9; Colobinae Elliot, 1913:xxv, xliii.] [= or including Anasca Haeckel, 1866:clx.[1]] [Including Presbytina Gray, 1825:338[2]; Nasalinae Groves, 1989.]
L. Mioc.-M. Pleist., ?L. Pleist., R.; Af. L. Mioc., L. Plioc., Pleist., R.; As. L. Mioc.-L. Plioc.; Eu. R.; E. Indies.

[1] Proposed as a family; not based upon a generic name.
[2] Suppressed in favor of Colobinae when *Colobus* and *Presbytis* are both considered to belong to that same subfamily.

†*Dolichopithecus* Depéret, 1889:983.
[Including †*Adelopithecus* Gremyatskii, 1960.]
?L. Mioc., E.-L. Plioc.; Eu.

†*Mesopithecus* Wagner, 1839:310.
[Including †*Anthropodus* de Lapouge, 1894:202-208.]
L. Mioc.; As.[1] L. Mioc.-L. Plioc.; Eu.

[1] S.W. Asia.

†*Microcolobus* Benefit & Pickford, 1986:446.
L. Mioc.; Af.

Tribe **Colobini** Blyth, 1875:9, **new rank**.
>> [= Colobidae Blyth, 1875:9; Colobinae Elliot, 1913:xxv, xliii;
>> Colobina Delson, 1975:171.]
>> ?L. Mioc., E. Plioc.-M. Pleist., ?L. Pleist., R.; Af.

Colobus Illiger, 1811:69. Guerezas, colobs.
>> [= *Colobolus* Gray, 1821.[1]] [= or including *Stachycolobus* Rochebrune, 1887:96,
>> 114-116.] [Including *Guereza* Gray, 1870:5, 19; *Procolobus* Rochebrune, 1887:97;
>> *Piliocolobus* Rochebrune, 1887:105; *Tropicolobus* Rochebrune, 1887:96, 102-104[2];
>> †*Cercopithecoides* Mollett, 1947:298; *Lophocolobus* Pousargues, 1895:98-101.]
>> ?L. Mioc., Plioc., E.-M. Pleist., R.; Af.
>> [1] *Lapsus calami.*
>> [2] Equal or including *Colobus* (*Piliocolobus*).

†*Cercopithecoides* Mollett, 1947:298.
>> L. Plioc., Pleist.; Af.

†*Libypithecus* Stromer, 1913:350-372.
>> L. Mioc. and/or E. Plioc.; Af.

†*Paracolobus* R. E. F. Leakey, 1969:54.
>> E.-L. Plioc.; Af.[1]
>> [1] E. Africa.

†*Rhinocolobus* M. G. Leakey, 1982:154.
>> L. Plioc.; Af.[1]
>> [1] E. Africa.

Tribe **Presbytini** Gray, 1825:338.[1]
>> [= Presbytina Gray, 1825:338.[2]]
>> [Including Semnopithecina Delson, 1977:454; Nasalinae Groves, 1989.]
>> L. Mioc., L. Plioc., Pleist., R.; As. R.; E. Indies.
>> [1] New suffix.
>> [2] Proposed as tribe; used as subtribe by Delson, 1975:171 and by Strasser & Delson, 1987:96.

Presbytis Escholtz, 1821:196. Langurs, hanuman.
>> [Including *Semnopithecus* Desmarest, 1822:532; *Trachypithecus* Reichenbach,
>> 1862:89; *Kasi* Reichenbach, 1862:101; *Lophopithecus* Trouessart, 1878:187;
>> *Presbypithecus* Trouessart, 1879:56; *Corypithecus* Trouessart, 1879:53.]
>> L. Mioc., L. Plioc., Pleist., R.; As. R.; E. Indies.

†*Parapresbytis* Kalmykov & Mashchenko, 1992:136.
>> L. Plioc.; As.

Pygathrix É. Geoffroy Saint-Hilaire, 1812:90. Douc langur, variegated langur,
>> snub-nosed langurs, retrousee-nosed langurs, golden monkey.
>> [Including *Rhinopithecus* Milne-Edwards, 1872:233;
>> *Presbytiscus* Pocock, 1924:17; †*Megamacaca* Hu & Qi, 1978:1-64.]
>> Pleist., R.; As.[1]
>> [1] S.E. Asia.

Nasalis É. Geoffroy Saint-Hilaire, 1812:90. Proboscis monkey, Pagi Island langur.
>> [= *Hanno* Gray, 1821:297; *Rhinolazon* Gloger, 1841;
>> *Rhynchopithecus* Dahlbom, 1857:91.] [Including *Simias* Miller, 1903:66.]
>> R.; E. Indies.[1]
>> [1] Borneo and Mentawai Islands (west of Sumatra).

Subfamily **Cercopithecinae** Gray, 1821:297.
>> [= Cercopithecidae Gray, 1821:297; Cercopithecinae Blanford, 1888:10;
>> Ascoparea Haeckel, 1866:clx.[1]] [Including Cynopithecina I. Geoffroy Saint-Hilaire,
>> 1843:495; Cynopithecinae Mivart, 1865:547; Cynopithecidae Gill, 1872:2, 52.]
>> L. Mioc.-R.; Af. L. Mioc.-R.; Eu.[2] Plioc., E. Pleist.-R.; As.
>> L. Plioc.; Mediterranean.[3] ?M. Pleist., L. Pleist.-R.; E. Indies. R.; Indian O.[4]
>> [1] Not based upon a generic name. Used as a family.
>> [2] Gibraltar in the Recent (probably by reintroduction).
>> [3] Sardinia.
>> [4] Mauritius (probably by introduction).

Tribe **Papionini** Burnett, 1828.

> [= Papionidae Burnett, 1828:306; Papionini Kuhn, 1967.] [Including Cercocebini Jolly, 1966:430; Theropithecini Jolly, 1966:431.]
>
> L. Mioc.-R.; Af. L. Mioc.-R.; Eu.[1] Plioc., E. Pleist.-R.; As. L. Plioc.; Mediterranean.[2] ?M. Pleist., L. Pleist.-R.; E. Indies. R.; Indian O.[3]
>
> [1] Gibraltar in the Recent (probably by reintroduction).
> [2] Sardinia.
> [3] Mauritius (probably by introduction).

Subtribe **Macacina** Owen, 1843:55.

> [= Macacidae Owen, 1843:55; Macacina Delson, 1975.] [Including Cynopithecina I. Geoffroy Saint-Hilaire, 1843:495.]
>
> L. Mioc.-R.; Af.[1] L. Mioc.-R.; Eu.[2] Plioc., E. Pleist.-R.; As. L. Plioc.; Mediterranean.[3] ?M. Pleist., L. Pleist.-R.; E. Indies. R.; Indian O.[4]
>
> [1] N. Africa.
> [2] Gibraltar in the Recent (probably by reintroduction).
> [3] Sardinia.
> [4] Mauritius (probably by introduction).

Macaca Lacépède, 1799:4. Macaques, rhesus monkey, barbary ape, saru.

> [= *Macaco* Oken, 1817:1185; *Macacus* Desmarest, 1820:63; *Magotus* Ritgen, 1824:33; *Inuus* É. Geoffroy Saint-Hilaire, 1812:100; *Pithes* Burnett, 1828:307; *Salmacis* Gloger, 1841:xxvii, 35-36.] [Including *Pithecus* É. Geoffroy Saint-Hilaire & G. Cuvier, 1795:462[1]; *Silenus* Goldfuss, 1820:479; *Magus* Lesson, 1827:43; *Cynopithecus* I. Geoffroy Saint-Hilaire, 1835:16[2]; *Maimon* Wagner, 1839:141[2]; *Rhesus* Lesson, 1840:95; *Pithex* Hodgson, 1841:1212; *Lyssodes* Gistel, 1848:9; *Vetulus* Reichenbach, 1862:125[2]; *Zati* Reichenbach, 1862:130; *Cynamolgus* Reichenbach, 1862:139; *Gymnopyga* Gray, 1866:203[3]; †*Aulaxinuus* Cocchi, 1872:68; †*Szechuanopithecus* Young & Liu, 1950:52; *Cynomacaca* Khajuria, 1953:303.[3]]
>
> L. Mioc.-R.; Af.[4] L. Mioc.-R.; Eu.[5] Plioc., E. Pleist.-R.; As.[6] L. Plioc.; Mediterranean.[7] ?M. Pleist., L. Pleist.-R.; E. Indies.[8] R.; Indian O.[9]
>
> [1] Suppressed by the International Commission on Zoological Nomenclature, Opinion 114, 1929.
> [2] Objective synonym of *Silenus*.
> [3] Objective synonym of *Magus*.
> [4] N. Africa.
> [5] Gibraltar in the Recent (probably by reintroduction).
> [6] Including Japan in the middle? and late Pleistocene and Recent.
> [7] Sardinia.
> [8] Including the Philippines and Sulawesi in the Recent.
> [9] Mauritius (probably by introduction).

†*Paradolichopithecus* Necrasov, Samson & Radulesco, 1961:415.

> L. Plioc., Pleist.; Eu.

†*Procynocephalus* Schlosser, 1924:8.

> L. Plioc. and/or E. Pleist.; As.

Subtribe **Theropithecina** Jolly, 1966:431.

> [= Theropithecini Jolly, 1966:431; Theropithecina Szalay & Delson, 1979:373.]
> E. Plioc.-R.; Af. E. and/or M. Pleist.; As. E. Pleist.; Eu.

Theropithecus I. Geoffroy Saint-Hilaire, 1843:576. Gelada baboon.

> [= *Gelada* Gray, 1843:xvii, 9.[1]] [Including †*Simopithecus* Andrews, 1916:417[2]; †*Brachygnathopithecus* Kitching, 1952:17; †*Omopithecus* Delson, 1993:182.[3]]
> E. Plioc.-R.; Af. E. and/or M. Pleist.; As.[4] E. Pleist.; Eu.
>
> [1] Possibly antedated. See Palmer (1904:292).
> [2] Proposed as genus; reduced to subgenus of *Theropithecus* by Jolly, 1972:95.
> [3] Described as subgenus.
> [4] S. Asia (Mirzapur, Punjab, India).

Subtribe **Papionina** Burnett, 1828:306.

> [= Papionidae Burnett, 1828:306; Papionina Szalay & Delson, 1979:335.] [Including Cercocebina Jolly, 1966:430.]
> E. Plioc.-R.; Af. ?E. Pleist.; Eu. R.; As.[1]
>
> [1] S.W. Asia (S.W. Arabia).

Papio Muller, 1773:119.[1] Baboons, chacma baboon, hamadryas baboon, mandrill, drill.
    [= *Pavianus* Frisch, 1775:19[2]; *Papio* Erxleben, 1777:15; *Cynocephalus* Cuvier &
    Geoffroy, 1795[3]; *Mandrillus* Ritgen, 1824:33; *Mormon* Wagner, 1839:164-168[4];
    Maimon Trouessart, 1904:21.[5]] [Including *Chaeropithecus* Gervais, 1839:90;
    Hamadryas Lesson, 1840:107-111[6]; *Drill* Reichenbach, 1862:162; *Comopithecus* J. A.
    Allen, 1925:290, 312.[7]]
    E. Plioc.-R.; Af. ?E. Pleist.; Eu. R.; As.[8]

[1] Described as a subgenus of *Cercopithecus*.
[2] Unavailable.
[3] Not Boddaert, 1768.
[4] Not *Mormon* Illiger, 1811, a bird.
[5] Replacement name for *Mormon* "Lesson," (actually Wagner, 1839). Not *Maimon* Wagner, 1839, a
    synonym of *Macaca*.
[6] Not *Hamadryas* Hübner, 1806, a lepidopteran.
[7] Replacement name for *Hamadryas* Lesson, 1840.
[8] S.W. Asia (Arabian Peninsula).

†*Parapapio* Jones, 1937:709-728.
    E.-L. Plioc.; Af.

†*Dinopithecus* Broom, 1936:750-768.
    L. Plioc. and/or E. Pleist.; Af.

Cercocebus É. Geoffroy Saint-Hilaire, 1812:97. Mangabeys.
    [Including *Semnocebus* Gray, 1870:27-28[1]; *Lophocebus* Palmer, 1903:873[2];
    Leptocebus Trouessart, 1904:15[2]; *Cercolophocebus* Matschie, 1914:341.[3]]
    L. Plioc.-E. Pleist., R.; Af.[4]

[1] Proposed as subgenus of *Cercocebus*; raised to generic rank by Lydekker 1990:595-596. Preoccupied by
    Semnocebus Lesson, 1840, a synonym of the indrid genus *Avahi*.
[2] Replacement name for *Semnocebus* Gray, 1870.
[3] Proposed as subgenus of *Cercocebus*.
[4] Tropical Africa.

†*Gorgopithecus* Broom & Robinson, 1949:379.
    E. Pleist.; Af.

Tribe **Cercopithecini** Gray, 1821:297.
    [= Cercopithecidae Gray, 1821:297; Cercopithecina Gray, 1825:338[1]; Cercopithecini
    Blanford, 1888:10; Cercopithecina Strasser & Delson, 1987:96.[2]] [Including
    Allenopithecina Strasser & Delson, 1987:93.]
    L. Plioc., M. Pleist.-R.; Af.[3]

[1] As tribe.
[2] As subtribe.
[3] Sub-Saharan Africa.

Cercopithecus Brunnich, 1772:34, 40-41. Guenons, white-nosed monkeys, cercopitheques, avem,
    osok, grivet, talapoin, vervet, green monkey, weissnases, blue monkey.
    [= or including *Lasiopyga* Illiger, 1811:68; *Rhinostigma* Elliot, 1913:273.] [Including
    Miopithecus I. Geoffroy Saint-Hilaire, 1842:720[1]; *Diademia* Reichenbach, 1862:107-
    109[2]; *Chlorocebus* Gray, 1870:24; *Diadema* Trouessart, 1878:122; *Diana* Trouessart,
    1878:124[3]; *Pogonocebus* Trouessart, 1904:14[4]; *Neopithecus* Elliot, 1913:lx[5];
    Allochrocebus Elliot, 1913:296[6]; *Insignicebus* Elliot, 1913:296[6]; *Melanocebus* Elliot,
    1913:296[7]; *Neocebus* Elliot, 1913:296.[6]]
    L. Plioc., M. Pleist.-R.; Af.[8]

[1] Sometimes as a genus.
[2] Supposedly preoccupied by *Diadema* Schumacher, 1817, a genus of crustaceans.
[3] Preoccupied by *Diana* Risso, 1826, a genus of fishes.
[4] Replacement for *Diana* Trouessart, 1878.
[5] = *Neocebus* Elliot, 1913, and in any case preoccupied by †*Neopithecus* Abel, 1902, a synonym of
    †*Dryopithecus*.
[6] As subgenus of *Lasiopyga* Illiger, 1811.
[7] As subgenus of *Lasiopyga* Illiger, 1811. Replacement for *Diademia* Reichenbach, 1862.
[8] Sub-Saharan Africa.

Allenopithecus Lang, 1923:1. Allen's monkey.
    R.; Af.[1]

[1] Congo Basin.

Erythrocebus Trouessart, 1897:19.[1] Patas monkey, red monkey.
R.; Af.[2]

[1] Described as subgenus of *Cercopithecus*; as distinct genus in Thorington & Groves, 1970.
[2] Tropical Africa.

Family **Hominidae** Gray, 1825:338. Humans, apes.

[= Hominina Bonaparte, 1850: unpaged chart; Hominoidea Gregory & Hellman, 1923:14; LIPOCERCA Haeckel, 1866:clx[1]; Humana Haeckel, 1866:clx[2]; Anthropidae Huxley, 1869:99; ANTHROPOMORPHA Huxley, 1872:402; Helacytidae Van Valen & Maiorana, 1991:73.] [Including Lipotyla Haeckel, 1866:clx[3]; Pongidae Elliot, 1913:181[4]; Pongoidae Hay, 1930:930[5]; ANTHROPIFORMES Hay, 1930:930[6]; Gorillidae Frechkop, 1943; †Proconsulidae L. S. B. Leakey, 1963; Panidae Schwartz, 1986:32.]

L. Olig.-R.; Af. E.-L. Mioc., Plioc., E. Pleist.-R.; As. M.-L. Mioc., M. Pleist.-R.; Eu. ?E. Pleist., M. Pleist.-R.; E. Indies. M. Pleist.-R.; Mediterranean. L. Pleist.-R.; N.A. L. Pleist.-R.; Aus. L. Pleist.-R.; New Guinea. R.; Madagascar. R.; Indian O. R.; Atlantic. R.; Arctic O. R.; Cent. A. R.; W. Indies. R.; S.A. R.; Antarctica. R.; New Zealand. R.; Pacific.

[1] Proposed as a sectio of subordo CATARRHINAE.
[2] Not based upon a generic name.
[3] Not based upon a generic name. It would otherwise be a valid senior synonym of Pongidae.
[4] The errata page, vol. 1, states that this work was published in 1913, not 1912 as printed. Pongidae was proposed on page 181 of the third volume.
[5] (Sic), as superfamily.
[6] As suborder.

Subfamily **Homininae** Gray, 1825:338.

[= Hominidae Gray, 1825:338; Hominina C. L. Bonaparte, 1841:250[1]; Homininae Delson & Andrews, in Luckett & Szalay, eds., 1975:441.] [Including †Australopithecinae Gregory & Hellman, 1939:370; Gorillidae Frechkop, 1943; †Ramapithecinae Simonetta, 1957:53-112; †Proconsulidae L. S. B. Leakey, 1963; †Sivapithecinae Pilbeam, in Pilbeam, Meyer, Badgley, Rose, Pickford, Behrensmeyer & Ibrahim Shah, 1977:689-695; Paninae Delson, 1977:450.]

L. Olig.-R.; Af. E.-L. Mioc., Plioc., E. Pleist.-R.; As. M.-L. Mioc., M. Pleist.-R.; Eu. ?E. Pleist., M. Pleist.-R.; E. Indies. M. Pleist.-R.; Mediterranean. L. Pleist.-R.; N.A. L. Pleist.-R.; Aus. L. Pleist.-R.; New Guinea. R.; Madagascar. R.; Indian O. R.; Atlantic. R.; Arctic O. R.; Cent. A. R.; W. Indies. R.; S.A. R.; Antarctica. R.; New Zealand. R.; Pacific.

[1] Used as subfamily.

†*Graecopithecus* G. H. R. von Koenigswald, 1972:390.
[= or including †*Ouranopithecus* de Bonis & Melentis, 1977:1395.]
L. Mioc.; Eu.

†*Langsonia* Schwartz, Long, Cuong. Kha & Tattersall, 1995:16.[1]
M. Pleist.; As.

[1] Validity dubious.

Tribe **Pongini** Elliot, 1913: vol. 1, errata.

[= Simiadae Fleming, 1822:xxxii, 172; Simiina Gray, 1825:338[1]; Simidae C. L. Bonaparte, 1837:7; Simiidae C. L. Bonaparte, 1850: unpaged chart; Pongidae Elliot, 1913: vol. 1, errata; Ponginae Allen, 1925:477; Pongina Delson & Andrews, in Luckett & Szalay, eds., 1975:441; Pongini Goodman, Tagle, Fitch, Bailey, Czelusniak, Koop, Benson & Slightom, 1990:265.] [Including †Dryopithecinae Gregory & Hellman, 1939:370; †Ramapithecinae Simonetta, 1957:53-112; †Sugrivapithecini Simonetta, 1957:53-112; †Proconsulidae L. S. B. Leakey, 1963; †Sivapithecinae Pilbeam, in Pilbeam, Meyer, Badgley, Rose, Pickford, Behrensmeyer & Ibrahim Shah, 1977; †Proconsuloidea Harrison, 1988:104.]

L. Olig.-L. Mioc.; Af. E.-L. Mioc., E.-M. Pleist.; As. M.-L. Mioc.; Eu. M. Pleist.-R.; E. Indies.

[1] Proposed as tribe.

†*Dryopithecus* Lartet, 1856:219-223.

[= or including †*Palaeopithecus* Lydekker, 1879:33[1]; †*Paidopithex* Pohlig, 1895:149, 151; †*Pliohylobates* Dubois, 1895:155; †*Anthropodus* Schlosser, 1901:261-271[2]; †*Griphopithecus* Abel, 1902:1177; †*Neopithecus* Abel, 1902:1172; †*Sivapithecus* Pilgrim, 1910:63; †*Palaeosimia* Pilgrim, 1915:29; †*Hylopithecus* Pilgrim, 1927:11; †*Sugrivapithecus* Lewis, 1934:168; †*Austriacopithecus* Ehrenberg, 1937; †*Hispanopithecus* de Villalta Comella & Crusafont Pairó, 1944:91-139; †*Udabnopithecus* Burchak-Abramovich & Gabashvili, 1945:457; †*Indopithecus* G. H. R. von Koenigswald, 1950; †*Rhenopithecus* G. H. R. von Koenigswald, 1956:330; †*Ankarapithecus* Ozansoy, 1957.] [Including †*Rangwapithecus* Andrews, 1974:188[3]; †*Chinjipithecus* G. H. R. von Koenigswald, 1981:511[4]; †*Ataxopithecus* Kretzoi, 1984:91; †*Sinopithecus* Wu, 1986.]

E.-L. Mioc.; Af. M.-L. Mioc.; As. M.-L. Mioc.; Eu.

[1] Objective synonym of †*Palaeosimia*. Not †*Palaeopithecus* Voigt, 1835, a *nomen dubium* (primate or other trackway?).
[2] Objective synonym of †*Neopithecus*. Not †*Anthropodus* de Lapouge, 1894, a synonym of †*Mesopithecus*.
[3] As subgenus.
[4] Objective synonym of †*Sivapithecus*?

†*Kamoyapithecus* M. G. Leakey, Ungar & Walker, 1995:520.

L. Olig.; Af.

†*Proconsul* Hopwood, 1933:96-98.

[= †*Xenopithecus* Hopwood, 1933:96-98.]

E.-M. Mioc.; Af.

†*Limnopithecus* Hopwood, 1933:96-98.

E.-M. Mioc.; Af. E. Mioc.; As.

†*Kalepithecus* Harrison, 1988:85.

E. Mioc.; Af.

†*Platodontopithecus* Gu & Lin, 1983:305.

M. Mioc.; As.

Pongo Lacépède, 1799:4. Orangutan, mia.

[= *Simia* Linnaeus, 1758:25-29[1]; *Macrobates* Billberg, 1828: before 1.[2]] [Including †*Sivasimia* Chopra, 1983:544.]

M. Mioc., E.-M. Pleist.; As. M. Pleist.-R.; E. Indies.[3,4,5]

[1] Suppressed because of ambiguity by the International Commission on Zoological Nomenclature, Opinion 114, 1929.
[2] Replacement name.
[3] Java in the middle Pleistocene.
[4] Sumatra in the late Pleistocene.
[5] Sumatra and Borneo in the Recent.

†*Ramapithecus* Lewis, 1934:162.

[= or including †*Bramapithecus* Lewis, 1934:173; †*Kenyapithecus* L. S. B. Leakey, 1962:690; †*Rudapithecus* Kretzoi, 1967; †*Bodvapithecus* Kretzoi, 1974:578-581.] [Including †*Mabokopithecus* G. H. R. von Koenigswald, 1969:50.]

M.-L. Mioc.; Af. M.-L. Mioc.; As. L. Mioc.; Eu.

†*Micropithecus* Fleagle & Simons, 1978:427.

M. Mioc.; Af.[1]

[1] E. Africa.

†*Lufengpithecus* R. Wu, 1987:269.

L. Mioc.; As.

Tribe †**Gigantopithecini** Gremyatskii, 1960.

[= †Gigantopithecinae Gremyatskii, 1960; †Gigantopithecinae Woo, 1962; †Gigantopithecina Delson & Andrews, in Luckett & Szalay, 1975:441; †Gigantopithecini Delson, 1977:450.[1]]

L. Mioc., Plioc., E.-M. Pleist.; As.

[1] Suggested alternative.

†*Gigantopithecus* G. H. R. von Koenigswald, 1935:874.

> [= †*Gigantanthropus* Weidenreich, 1946[1]; †*Giganthropus* Weinert, 1950.[1]]
> L. Mioc., Plioc., E.-M. Pleist.; As.

[1] Unjustified emendation.

Tribe **Hominini** Gray, 1825:338.

> [= Hominidae Gray, 1825:338; Hominina Gray, 1825:338[1]; Hominini
> Delson & P. Andrews, in Luckett & Szalay, eds., 1975:441.] [Including
> †Australopithecinae Gregory & Hellman, 1939:370; Gorillidae Frechkop, 1943.]
> ?L. Mioc., E. Plioc.-R.; Af. ?E. Pleist., M. Pleist.-R.; E. Indies.
> E. Pleist.-R.; As. M. Pleist.-R.; Mediterranean. M. Pleist.-R.; Eu.
> L. Pleist.-R.; N.A. L. Pleist.-R.; Aus. L. Pleist.-R.; New Guinea.
> R.; Madagascar. R.; Indian O. R.; Atlantic. R.; Arctic O. R.; Cent. A.
> R.; W. Indies. R.; S.A. R.; Antarctica. R.; New Zealand. R.; Pacific.

[1] Proposed as tribe.

Subtribe **Hominina** Gray, 1825:338.

> [= Hominidae Gray, 1825:338; Hominina Gray, 1825:338[1]; Hominina
> C. L. Bonaparte, 1838:111[2]; Hominina Delson & P. Andrews, in Luckett
> & Szalay, eds., 1975:441.[3]] [= or including Anthropini Huxley, 1864:153;
> Anthropidae Huxley, 1869:99.] [Including Paninae Delson, 1977:450;
> Panino Delson, 1977:450.[4]]
> ?L. Mioc., E. Plioc.-R.; Af. ?E. Pleist., M. Pleist.-R.; E. Indies.
> E. Pleist.-R.; As. M. Pleist.-R.; Mediterranean. M. Pleist.-R.; Eu.
> L. Pleist.-R.; N.A. L. Pleist.-R.; Aus. L. Pleist.-R.; New Guinea.
> R.; Madagascar. R.; Indian O. R.; Atlantic. R.; Arctic O. R.; Cent. A.
> R.; W. Indies. R.; S.A. R.; Antarctica. R.; New Zealand. R.; Pacific.

[1] Proposed as tribe.
[2] Used as subfamily.
[3] Used as subtribe.
[4] Used as an infratribe, with an otherwise rare or unused suffix. Some workers might regard Delson's 1977 proposals as conditional (see International Code of Zoological Nomenclature, Article 15).

Pan Oken, 1816:xi.[1] Chimpanzee, jocko, barris (of Pyrard de Laval), smitten (of Bosman), quimpese (of De la Brosse), quojas moras (of Tulpius).

> [= *Troglodytes* É. Geoffroy Saint-Hilaire, 1812:87[2]; *Mimetes* Leach, 1820:104[3];
> *Theranthropus* Brookes, 1828:28; *Chimpansee* Voigt, 1831:76; *Anthropopithecus*
> de Blainville, 1839:360; *Hylanthropus* Gloger, 1841:xxvii, 34; *Pseudanthropos*
> Reichenbach, 1860:191-194; *Pongo* Haeckel, 1866:cl[4]; *Engeco* Haeckel, 1866:cl, clx;
> *Anthropithecus* Haeckel, 1895:600[5]; *Fsihego* de Pauw, 1905[6]; *Pseudanthropus* Elliot,
> 1913:xxxvii.[7]] [Including *Bonobo* Tratz & Heck, 1954:97-101.]
> R.; Af.[8]

[1] A non-binominal name, but validated by the International Commission on Zoological Nomenclature, Opinion 1368, 1985.
[2] Not *Troglodytes* Viellot, 1806, a genus of birds.
[3] Not *Mimetes* Hübner, 1816, a lepidopteran.
[4] Not *Pongo* Lacepede, 1799, the orangutan.
[5] Contraction of *Anthropopithecus* Blainville, 1838.
[6] See Conisbee (1953:32) for comment.
[7] Emendation of *Pseudanthropos* Reichenbach, 1860.
[8] Tropical Africa.

†*Australopithecus* Dart, 1925.

> [Including †*Plesianthropus* Broom, 1938; †*Paranthropus* Broom, 1938;
> †*Praeanthropus* Senyürek, 1955[1]; †*Zinjanthropus* L. S. B. Leakey, 1959;
> †*Paraustralopithecus* Arambourg & Coppens, 1967:590;
> †*Ardipithecus* White, Suwa & Asfaw, 1995:88.]
> ?L. Mioc., E. Plioc.-E. Pleist.; Af. Pleist.; As.

[1] "†*Praeanthropus* Hennig, 1948" was based on the same material, but Hennig did not name a species.

Homo Linnaeus, 1758:20.[1] People, men, women, Heidelberg man, Java man, Peking man, Neanderthal man, Cro-Magnon man, Solo man.

[= *Epanthropos* Cope, 1879:194[2]; *Metanthropos* Cope, 1879:194[3]; †*Hemianthropus* Freudenberg, 1929:240; *Helacyton* Van Valen & Maiorana, 1991:73.]

[= or including *Euranthropus* Sergi, 1909; *Euranthropus* Arambourg, 1955[4]; *Heoanthropus* Sergi, 1909; *Notanthropus* Sergi, 1909; *Anthropus* Boyd-Dawkins, 1926; *Praehomo* Eickstedt, 1932; †*Maueranthropus* Montandon, 1943; *Nipponanthropus* Hasebe, 1948; *Praanthropus* Hennig, 1948[5]; *Europanthropus* Wust, 1950; *Homopithecus* Deraniyagala, 1960.]

[Including †*Pithecanthropus* Dubois, 1894:1-26, 31; †*Protanthropus* Haeckel, 1895:616-17, 644; †*Proanthropus* Wilser, 1900:551-576; †*Pseudohomo* Ameghino, 1909:195[6]; †*Palaeoanthropus* Bonarelli, 1910:26-31; †*Eoanthropus* Woodward, 1913[7]; †*Sinanthropus* D. Black & Zdansky, in Black, 1927; †*Cyphanthropus* Pycraft, 1928:412; †*Javanthropus* Oppenoorth, 1932:49-63; †*Metanthropus* Sollas, 1933:408; †*Africanthropus* Dreyer, 1935[8]; †*Africanthropus* Weinert, 1938:125-129[4]; †*Meganthropus* G. H. R. von Koenigswald, in Weidenreich, 1945:16; †*Telanthropus* Broom & Robinson, 1949:322-323; †*Atlanthropus* Arambourg, 1954:895; †*Hemianthropus* G. H. R. von Koenigswald, 1957:158[9]; †*Hemanthropus* G. H. R. von Koenigswald, 1957:416[10]; †*Tchadanthropus* Coppens, 1965:2869.]

L. Plioc.-R.; Af. ?E. Pleist., M. Pleist.-R.; E. Indies. E. Pleist.-R.; As. M. Pleist.-R.; Mediterranean. M. Pleist.-R.; Eu. L. Pleist.-R.; N.A. L. Pleist.-R.; Aus. L. Pleist.-R.; New Guinea. R.; Madagascar. R.; Indian O. R.; Atlantic. R.; Arctic O. R.; Cent. A. R.; W. Indies. R.; S.A. R.; Antarctica. R.; New Zealand. R.; Pacific.

[1] This genus of PRIMATES not only has been assaulted by taxonomic splitters, but also has been accorded various high monotypic ranks since the time of Aristotle's DIPODA. Storr, 1780 called it suborder PALMARES; Illiger, 1811:65 used order ERECTA; Gray, 1821:297 called it a class, BIMANES (probably an emendation of BIMANUS Blumenbach, 1779-1797); and Owen, 1859:26 dubbed it a subclass, ARCHENCEPHALA, and an order, BIMANA. Owen's ARCHENCEPHALA and J. S. Huxley's (1958, 1959) kingdom-level "psychozoan grade" are similar concepts.
[2] Provisional name.
[3] Provisional name. "*Metanthropus*" Sollas, 1933:408 was for Neanderthals.
[4] Preoccupied.
[5] As *Präanthropus*.
[6] Objective synonym of †*Palaeoanthropus*.
[7] As restricted by G. Miller, 1915:19. Objective synonym of †*Palaeoanthropus*.
[8] Named as subgenus of *Homo*.
[9] Preoccupied by †*Hemianthropus* Freudenberg, 1929:240.
[10] Replacement name for †*Hemianthropus* von Koenigswald.

Subtribe **Gorillina** Frechkop, 1943.

[= Gorillidae Frechkop, 1943; Gorillina Goodman, Tagle, Fitch, Bailey, Czelusniak, Koop, Benson & Slightom, 1990:265.]

R.; Af.

Gorilla I. Geoffroy Saint-Hilaire, 1853:933. Gorilla.

[= *Pseudogorilla* Elliot, 1913:224.]

R.; Af.[1]

[1] Tropical Africa.

Subfamily **Hylobatinae** Gray, 1870:4, 9.

[= Tylogluta Haeckel, 1866:clx[1]; Hylobatina Gray, 1870:4, 9; Hylobatinae Gill, 1872:2, 52; Hylobatidae Blyth, 1875:1; Hylobatoidea Thenius, 1969.]

M.-L. Mioc., M. Pleist.-R.; As. R.; E. Indies.

[1] Proposed as a family but not based upon a generic name.

Hylobates Illiger, 1811:67. Gibbons, hoolock, wau-wau, siamang.

[= *Satyrus* Oken, 1816:xi, 1225-7[1]; *Cheiron* Burnett, 1828:307; *Methylobates* Ameghino, 1884:365; †*Bunopithecus* Matthew & Granger, 1923:588.]

[Including *Symphalangus* Gloger, 1842:xxvii, 34[2]; *Brachytanytes* Schultz, 1932:369[2]; *Nomascus* Miller, 1933:158.[2]]

M.-L. Mioc., M. Pleist.-R.; As. R.; E. Indies.

[1] Replacement name for *Hylobates* Illiger, 1811. Supposedly preoccupied by *Satyra* Meigen, 1803, a dipteran.
[2] As subgenus.

Superfamily **Callitrichoidea** Gray, 1821:298, **new rank.**[1]
>[= Callitricidae Gray, 1821:298; Callitrichidae Thomas, 1903:457; PLATYRRHINI É. Geoffroy Saint-Hilaire, 1812:104[2]; PLATYRRINAE Haeckel, 1866:clx[3]; HESPEROPITHECI Haeckel, 1866:cliv; HESPEROPITHECA Haeckel, 1895:600, 609.] [Including Atelina Gray, 1825:338; Ateloidea Rosenberger, Setoguchi & Shigehara, 1990:211; Cebina Bonaparte, 1831:6, 14; Ceboidea Simpson, 1931:271.]

[1] An undescribed taxon has been reported from the early Miocene of Cuba.
[2] Proposed as group of family Singes. Used as suborder by Simpson, 1945:181 and as infraorder by McKenna, 1975:41.
[3] Used as suborder of order SIMIAE (PITHECI) Haeckel.

†*Carlocebus* Fleagle, 1990:67.
>E. Mioc.; S.A.

Subfamily †**Branisellinae** Hershkovitz, 1977.
>[= †Branisellidae Hershkovitz, 1977; †Branisellinae Szalay & Delson, 1979:286.]
>L. Olig.; S.A.

†*Branisella* Hoffstetter, 1969:434.
>[= †*Szalatavus* Rosenberger, Hartwig & Wolff, 1991:225.]
>L. Olig.; S.A.

Family **Callitrichidae** Gray, 1821:298.   Marmosets.
>[= Callitricidae Gray, 1821:298; Callithricina Gray, 1825:338[1]; Callitrichidae Thomas, 1903:457; Harpaladae Gray, 1821:298[2]; Hapalidae Wagner, 1840:238; Hapalida Haeckel, 1866:clx; ARCTOPITHECI Haeckel, 1866:clx[3]; Arctopithecini Huxley, 1872:392.] [Including Mididae Gill, 1872:2, 54; Sariguidae Gray, 1825:338.[4]]
>M. Mioc., R.; S.A. L. Pleist. and/or R.; W. Indies.[5] R.; Cent. A.[6]

[1] Used as tribe.
[2] Family Harpaladae was also called an order, GAMPSTONYCHAE Gray, 1821:298.
[3] Proposed as suborder.
[4] *Lapsus calami* for Saguinidae, a family of tamarins; not based on *Sarigua*, a synonym of *Didelphis*.
[5] An indet. subfossil callitrichid has been reported from Hispaniola.
[6] Panama and Costa Rica.

Subfamily **Callitrichinae** Gray, 1821:298.
>[= Callitricidae Gray, 1821:298; Callithricina Gray, 1825:338[1]; Callithricinae Thomas, 1903:457; Callitrichini Szalay & Delson, 1979:281; Callitrichina Rosenberger, Setoguchi & Shigehara, 1990:211; Harpaladae Gray, 1821:298; Hapalina Bonaparte, 1837:7[2]; Hapalidae Wagner, 1840:238; Hapalineae Lesson, 1840; Hapalida Haeckel, 1866:clx[3]; Arctopitheci É. Geoffroy Saint-Hilaire, 1812; Arctopithecina Gravenhorst, 1843; Arctopithecae Dahlbom, 1856; Arctopithecini Huxley, 1872; Jacchina Gray, 1849; Ouistidae Burnett, 1828; Ouistitidae Burnett, 1828.] [Including Saguinina Gray, 1825:338[4]; Saguina Schneider, Schneider, Sampaio, Harada, Stanhope, Czelusniak & Goodman, 1993:235; Mididae Gill, 1872:2, 54; Leontocebinae Miranda-Ribeiro, 1940; Leontocebina Rosenberger, Setoguchi & Shigehara, 1990:211; Leontocebinae Hill, 1959; Leontopithecina Schneider, Schneider, Sampaio, Harada, Stanhope, Czelusniak & Goodman, 1993:235.]
>M. Mioc., R.; S.A. R.; Cent. A.[5]

[1] As tribe.
[2] As subfamily.
[3] Used as a family.
[4] Proposed as tribe.
[5] Panama and Costa Rica.

†*Micodon* Setoguchi & Rosenberger, 1985:617.
>M. Mioc.; S.A.

Leontopithecus Lesson, 1840:184, 200-202. Lion-headed marmoset, golden marmoset, leoncito, maned tamarin.
>[= *Leontideus* Cabrera, 1956:52.]
>R.; S.A.[1]

[1] S.E. Brazil.

Saguinus Hoffmansegg, 1807:102. Tamarins.

[= *Leontocebus* Wagner, 1839:ix; *Tamarin* Gray, 1870:68.[1]] [= or including *Cercopithecus* Gronow, 1763:5[2]; *Midas* Geoffroy Saint-Hilaire, 1812:120, 121[3]; *Marikina* Lesson, 1840; *Marikina* Reichenbach, 1862:xviii[4]; *Tamarinus* Trouessart, 1904:29.[1]] [Including *Oedipus* Lesson, 1840:197-200[5]; *Oedipomidas* Reichenbach, 1862:5-6[6]; *Hapanella* Gray, 1870:65-66[7]; *Mystax* Gray, 1870:66[1]; *Seniocebus* Gray, 1870:68.]

R.; Cent. A.[8] R.; S.A.

[1] Proposed as subgenus of *Midas*.
[2] Not *Cercopithecus* Brunnich, 1772, eliminated by suspension of the Rules.
[3] Preoccupied by *Midas* Latreille, 1796, a genus of Diptera.
[4] Junior homonym of *Marikina* Lesson, 1840.
[5] Preoccupied by *Oedipus* Tschudi, 1838, a genus of Amphibia.
[6] New name for *Oedipus* Lesson.
[7] Proposed as subgenus of *Oedipus*.
[8] Panama and Costa Rica.

Callithrix Erxleben, 1777:xxxi, 44, 45. Marmosets.

[= *Callitrix* Boddaert, 1784:42; *Cercopithecus* Blumenbach, 1779:68[1]; *Sagoinus* Kerr, 1792:80; *Sagouin* Lacépède, 1799:4; *Saguin* Fischer de Waldheim, 1803:113; *Sagoin* Desmarest, 1804:8; *Saguinus* Illiger, 1811:71[2]; *Hapale* Illiger, 1811:71; *Harpale* Gray, 1821:298; *Hapales* F. Cuvier, 1829:401; *Jacchus* É. Geoffroy Saint-Hilaire, 1812:359; *Iacchus* Spix, 1823:32; *Jachus* Schlegel, 1876:254; *Arctopithecus* G. Cuvier, 1817:115; *Ouistitis* Burnett, 1828:307; *Cuistitis* Cabrera, 1958:185[3]; *Anthopithecus* F. Cuvier, 1829:401; *Mico* Lesson, 1840:184, 192; *Liocephalus* Wagner, 1840:ix, 244; *Micoella* Gray, 1870:130.]

R.; S.A.[4]

[1] Junior homonym of *Cercopithecus* Brunnich, 1772.
[2] Junior homonym of *Saguinus* Hoffmannsegg, 1807.
[3] Misprint for *Ouistitis*.
[4] Brazil, Boliva.

Cebuella Gray, 1866:734.[1]

R.; S.A.[2]

[1] Proposed as subgenus of *Hapale*; raised to generic rank by Gray, 1870:64.
[2] Amazon Basin.

Subfamily **Callimiconinae** Thomas, 1913:131.

[= Calimiconidae Dollman, 1931:908; Callimiconini Szalay & Delson, 1979:282; Callimiconina Schneider, Schneider, Sampaio, Harada, Stanhope, Czelusniak & Goodman, 1993:235.]

R.; S.A.[1]

[1] Amazon Basin.

Callimico Miranda-Ribeiro, 1911:21. Goeldi's marmoset, callimico.

R.; S.A.[1]

[1] Amazon Basin.

Family **Atelidae** Gray, 1825:338. New World monkeys.

[= Atelina Gray, 1825:338; Ateloidea Rosenberger, Setoguchi & Shigehara, 1990:211.] [Including Cebina Bonaparte, 1831:6, 14; Cebidae Swainson, 1835:76, 81, 350; Ceboidea Simpson, 1931:271; Aphyocerca Haeckel, 1866:clx[1]; Labidocerca Haeckel, 1866:clx[2]; Alouattidae Miller, 1912:380.]

E.-L. Mioc., ?E. and/or ?M. Pleist., L. Pleist.-R.; S.A. Pleist., R.; Cent. A. Pleist. and/or R.; W. Indies.[3] R.; N.A.[4]

[1] Not based upon a generic name. Used as a family, equivalent to Ceboidea.
[2] Not based on a generic name.
[3] Cuba, Hispaniola, and Jamaica; now extinct.
[4] Tropical North America.

Subfamily †**Xenotrichinae** Hershkovitz, 1970:3.[1]

[= †Xenothricidae Hershkovitz, 1970:3; †Xenothricinae Szalay, 1976:421; †Xenotrichidae Morgan, 1993:428.]

Pleist. and/or R.; W. Indies.[2]

[1] For spelling, see the International Code of Zoological Nomenclature, Article 35(d)(i) and Appendix D, 24.
[2] Jamaica.

†*Xenothrix* Williams & Koopman, 1952:12.
>Pleist. and/or R.; W. Indies.[1]
>
>[1] Jamaica.

Subfamily †**Homunculinae** Ameghino, 1894:265.
>[= †Homunculidae Ameghino, 1894:265; †Homunculinae Bordas, 1942; †Homunculini Rosenberger, Setoguchi & Shigehara, 1990; †Homunculina Rosenberger, Setoguchi & Shigehara, 1990:211.]
>E.-M. Mioc.; S.A.

†*Homunculus* Ameghino, 1891:217.
>[= †*Ecphantodon* Mercerat, 1891:73; †*Anthropops* Ameghino, 1891:387.] [Including †*Dolichocebus* Kraglievich, 1951:57.]
>E.-M. Mioc.; S.A.[1]
>
>[1] Argentina.

Subfamily **Cebinae** Bonaparte, 1831:6, 14.
>[= Cebina Bonaparte, 1831:6, 14; Cebinae Mivart, 1865:547.] [Including Nyctipithecinae Mivart, 1865:547; Chrysotrichinae Cabrera, 1900; Aotinae Poche, 1908; Aotidae Miller, 1912:379; Aotini Tattersall, Delson & Van Couvering, 1988:xxxiii; Saimiridae Miller, 1912:379; Saimiriini Rosenberger, 1981; Callicebinae Pocock, 1925; Callicebina Szalay & Delson, 1979:293; Callicebidae Groves, 1989.]
>E.-M. Mioc., ?L. Mioc., Pleist., R.; S.A.[1] Pleist., R.; Cent. A. Pleist. and/or R.; W. Indies.[2]
>
>[1] Cf. Cebinae in the late Miocene (Huayquerian) Rio Acre local fauna, Brazil.
>[2] Hispaniola and possibly Jamaica; now extinct.

†*Chilecebus* J. Flynn, Wyss, Charrier & Swisher, 1995:603.
>E. Mioc.; S.A.

†*Antillothrix* MacPhee, Horovitz, Arredondo & Jiménez Vasquez, 1995:3.
>L. Pleist. and/or R.; W. Indies.[1]
>
>[1] Hispaniola.

Tribe **Cebini** Bonaparte, 1831:6, 14.
>[= Cebina Bonaparte, 1831:6, 14; Cebini Rosenberger, Setoguchi & Shigehara, 1990:211.] [Including Nyctipithecinae Mivart, 1865.[1]]
>M. Mioc., Pleist., R.; S.A. Pleist., R.; Cent. A. Pleist. and/or R.; W. Indies.[2]
>
>[1] Not available.
>[2] Hispaniola and possibly Jamaica; now extinct.

†*Laventiana* Rosenberger, Setoguchi & Hartwig, 1991:2137.
>M. Mioc.; S.A.

Saimiri Voigt, in Cuvier, 1831:95.[1] Squirrel monkeys, saimiris, caymiri, cai.
>[= *Chrysothrix* Kaup, 1835:50-52; *Samiris* Blyth, 1840:61; *Saimiris* Geoffroy Saint-Hilaire, 1843:1151.[2]] [Including †*Neosaimiri* Stirton, 1951:326.]
>M. Mioc., R.; S.A. R.; Cent. A.[3]
>
>[1] Described as subgenus of *Simia*.
>[2] And many later authors.
>[3] Tropical Central America.

Cebus Erxleben, 1777:44. Capuchins, sapajous, caiarara, cay.
>[= *Calyptrocebus* Reichenbach, 1862:55.[1]] [Including *Sapajus* Kerr, 1792:74-79; *Sapajou* Lacépède, 1799:4; *Sapaju* Ritgen, 1824:33; *Eucebus* Reichenbach, 1862:56[2]; *Otocebus* Reichenbach, 1862:55, 56[2]; *Pseudocebus* Reichenbach, 1862:55.[2]]
>Pleist., R.; Cent. A.[3] Pleist. and/or R.; W. Indies.[4] Pleist., R.; S.A.[5]
>
>[1] Subgenus as defined, but the type species of *Cebus*, *C. capucinus*, was included.
>[2] Subgenus as defined.
>[3] Honduras southward.
>[4] Hispaniola and possibly Jamaica; now extinct.
>[5] Northern Argentina northward.

Aotus Humboldt, 1811:358. Douroucoulis.
>[= *Aotus* Humboldt, 1811:358; *Nyctipithecus* Spix, 1823:24-26; *Nocthora* F. Cuvier, 1824.]

M. Mioc., R.; S.A.[1] R.; Cent. A.[2]

[1] Tropical South America.
[2] Panama.

Tribe †**Tremacebini** Hershkovitz, 1974:1, **new rank**.
  [= †Tremacebinae Hershkovitz, 1974:1.]
  E. Mioc.; S.A.

†*Tremacebus* Hershkovitz, 1974:3.
  E. Mioc.; S.A.

Tribe †**Lagonimiconini** Kay, 1994:536.
  M. Mioc.; S.A.

†*Lagonimico* Kay, 1994:536.
  M. Mioc.; S.A.

Tribe **Callicebini** Pocock, 1925, **new rank**.
  [= Callicebina Szalay & Delson, 1979:293; Callicebidae Groves, 1989.]
  Pleist., R.; S.A.

Callicebus Thomas, 1903:457. Titis, titi monkeys, widow monkeys.
  [= *Callithrix* É. Geoffroy Saint-Hilaire, 1812:357.[1]]
  Pleist., R.; S.A.

[1] Not *Callithrix* Erxleben, 1777, a genus of marmosets.

Subfamily **Pitheciinae** Mivart, 1865:547.
  [= Pithecina Gray, 1870:37, 58; Pithecidae Ludwig, 1883; Pitheciini Szalay & Delson, 1979:293; Pitheciina Szalay & Delson, 1979:293.] [Including Chiropotina Schneider, Schneider, Sampaio, Harada, Stanhope, Czelusniak & Goodman, 1993:235; †Soriacebina Rosenberger, Setoguchi & Shigehara, 1990.]
  E.-M. Mioc., R.; S.A.

†*Soriacebus* Fleagle, Powers, Conroy & Watters, 1987:66.
  E. Mioc.; S.A.

†*Cebupithecia* Stirton & D. E. Savage, 1951:350.
  M. Mioc.; S.A.

†*Mohanamico* Luchterhand, Kay & Madden, 1986:1754.
  M. Mioc.; S.A.

Pithecia Desmarest, 1804:8. Sakis.
  R.; S.A.[1]

[1] Guianas and Amazon Basin.

Chiropotes Lesson, 1840:178. Red-backed sakis.
  R.; S.A.[1]

[1] Guianas and Amazon Basin.

Cacajao Lesson, 1840:181.[1] Uakaris.
  [= *Brachyurus* Spix, 1823:11-13[2]; *Ouakaria* Gray, 1849:cxc, 9-10; *Uakeria* Lydekker, in Flower & Lydekker, 1891:712; *Cothurus* Palmer, 1899:493[3]; *Neocothurus* Palmer, 1903:873.[4]]
  R.; S.A.[5]

[1] Described as subgenus of *Pithecia*; raised to generic rank by Reichenbach, 1862:75.
[2] Preoccupied by *Brachyurus* Fischer de Waldheim, 1813, a rodent.
[3] Not *Cothurus* Champion, 1891, a coleopteran.
[4] Replacement name for *Cothurus* Palmer, 1899.
[5] Amazon Basin.

Subfamily **Atelinae** Gray, 1825:338.
  [= Atelina Gray, 1825:338; Atelinae Miller, 1924:170; Atelidae Tattersall, Delson & Van Couvering, 1988:xxv, xxvii.] [Including Mycetina Gray, 1825:338; Alouatinae Trouessart, 1879:32; Alouattinae Elliot, 1904:725; Alouattidae Miller, 1912:380; Alouattini Szalay & Delson, 1979:291; †Stirtoninae Hershkovitz, 1970:6.]

M.-L. Mioc., ?E. and/or ?M. Pleist., L. Pleist.-R.; S.A. Pleist., R.; Cent. A.
Pleist. and/or R.; W. Indies.[1] R.; N.A.[2]

[1] Cuba (now extinct).
[2] Southern North America (Mexico).

Tribe **Atelini** Gray, 1825:338.

[= Atelina Gray, 1825:338[1]; Atelini Szalay & Delson, 1979:290;
Atelina Rosenberger, Setoguchi & Shigehara, 1990:211.]
[Including Lagotrichina Gray, 1870[1]; Lagothrichinae Cabrera, 1900.]
Pleist. and/or R.; W. Indies.[2] L. Pleist.-R.; S.A. R.; N.A.[3] R.; Cent. A.

[1] Proposed as a tribe.
[2] Cuba (now extinct).
[3] Southern North America (Mexico).

Ateles É. Geoffroy Saint-Hilaire, 1806:262-269. Spider monkeys, coaitas.

[= *Ateleus* Fischer de Waldheim, 1813:529-532[1]; *Paniscus* Rafinesque, 1815:53;
Mamatelesus Herrera, 1899:19[2]; †*Montaneia* F. Ameghino, 1911;
Ameranthropoides Montandon, 1929:817.[3]]
Pleist. and/or R.; W. Indies.[4] R.; N.A.[5] R.; Cent. A. R.; S.A.[6]

[1] Invalid emendation.
[2] See the International Code of Zoological Nomenclature, Article 1(b)(8), and Palmer (1904:25-26).
[3] Described as a platyrrhine.
[4] Cuba (now extinct).
[5] Southern North America (Mexico north to Tamaulipas).
[6] Amazon Basin.

†*Caipora* Cartelle & Hartwig, 1996:6405.

L. Pleist.; S.A.[1]

[1] S.E. Brazil.

Brachyteles Spix, 1823:36. Woolly spider monkeys, mirikis.

[= *Brachyteleus* Elliot, 1913.[1]]
[= or including *Eriodes* I. Geoffroy Saint-Hilaire, 1829:143-145.]
R.; S.A.[2]

[1] Invalid emendation.
[2] S.E. Brazil.

Lagothrix É. Geoffroy Saint-Hilaire, 1812:106-107. Woolly monkeys, belly monkeys,
maricamicos.

[= or including *Gastrimargus* Spix, 1823:39-42.]
[Including *Oreonax* Thomas, 1927:156.]
R.; S.A.[1]

[1] Tropical South America (chiefly Amazon Basin).

Tribe **Alouattini** Trouessart, 1897:32.

[= Mycetina Gray, 1825:338[1]; Alouatinae Trouessart, 1897:32;
Alouattinae Elliot, 1904:725; Alouattini Szalay & Delson, 1979:291.]
[Including †Stirtoninae Hershkovitz, 1970:6.]
M.-L. Mioc., ?E. and/or ?M. Pleist., L. Pleist.-R.; S.A. Pleist., R.; Cent. A.
Pleist. and/or R.; W. Indies.[2]

[1] Proposed as tribe.
[2] Cuba (now extinct).

†*Protopithecus* Lund, 1838:14.

L. Pleist.; S.A.

†*Paralouatta* Rivero & Arredondo, 1991:1.

Pleist. and/or R.; W. Indies.[1]

[1] Cuba.

†*Stirtonia* Hershkovitz, 1970:6.

[= †*Kondous* Setoguchi, 1985:96-101.]
M.-L. Mioc.; S.A.

Alouatta Lacépède, 1799:4. Howler monkeys, aluatas, carayas.
[= *Mycetes* Illiger, 1811:70.]
Pleist., R.; Cent. A.[1] Pleist., R.; S.A.[2]
[1] As far north as southern Mexico.
[2] Tropical South America.

Order **SCANDENTIA** Wagner, 1855:xix, 524.[1]
[= TUPAYAE Peters, 1864:20; TUPAII Broers, 1963:362[2];
SCANDENTIFORMES Kinman, 1994:37.]
[Including PRAESIMII Gingerich, 1984:60[3]; PROPRIMATES Gingerich, 1989:23.[3]]
[1] Used as a family of INSECTIVORA by Wagner, 1855: xix, 524; not Scandentia Wagner, 1855: xiv, 268, a family of MARSUPIALIA. Although the Code of International Nomenclature does not apply, Scandentia Fischer de Waldheim, 1817:372, proposed for *Myrmecophaga,* is actually a senior homonym. Newman 1843:34 also employed SCANDENTIA as a synonym of PRIMATES. Butler 1972:263 advocated the use of SCANDENTIA Wagner, 1855 as an order for the tupaiids; it was used as an infraorder of STREPSIRRHINI by Kalandadze & Rautian, 1992:79.
[2] Proposed as suborder.
[3] In part.

Family **Tupaiidae** Gray, 1825:339. Tree shrews.
[= Tupaina Gray, 1825:339; Tupaiadae Bell, 1839:994; Tupayae Peters, 1864;
Tupaiidae Mivart, 1868:145; Tupayidae Gill, 1872:19; Tupaioidea Dobson, 1882:4;
Tupajidae Schlosser, 1887:91, 114; Tupaioidea Van Valen, 1967:258;
Cladobatida Haeckel, 1866:clx.[1]]
M. Eoc., Olig., L. Mioc., Plioc., R.; As.[2] R.; E. Indies.
[1] *Nomen oblitum.*
[2] Tupaiids have been reported in the Oligocene of Asia.

Subfamily **Tupaiinae** Gray, 1825:339.
[= Tupaina Gray, 1825:339; Tupaiinae Lyon, 1913:4; Cladobatidina C. L. Bonaparte,
1838:113[1]; Cladobatina C. L. Bonaparte, 1845:5; Cladobatida Haeckel, 1866:clx.]
M. Eoc., L. Mioc., Plioc., R.; As. R.; E. Indies.
[1] As subfamily.

†*Eodendrogale* Tong, 1988:214.[1]
M. Eoc.; As.[2]
[1] (Tung).
[2] E. Asia.

†*Prodendrogale* Qiu, 1986:309.
L. Mioc.; As.

Tupaia Raffles, 1822:256. Tree shrews.
[= *Cladobates* F. Cuvier, 1825:251; *Tupaja* Haeckel, 1866:clx.] [= or including
Hylogale Temminck, 1827:xix[1]; *Hylogalea* Schlegel & Müller, 1843:159.]
[Including *Lyonogale* Conisbee, 1953:46; *Tana* Lyon, 1913:134[2];
†*Palaeotupaia* Chopra & Vasishat, 1979:214.]
?L. Mioc., Plioc., R.; As. R.; E. Indies.
[1] Or 1824.
[2] Objective synonym of *Lyonogale.* Not *Tana* Reed, 1888, a fly.

Anathana Lyon, 1913:120. Indian tree shrew.
R.; As.[1]
[1] S. Asia (India).

Dendrogale Gray, 1848:23. Small smooth-tailed tree shrews.
R.; E. Indies.[1] R.; As.[2]
[1] Borneo.
[2] S.E. Asia.

Urogale Mearns, 1905:435. Philippine tree shrew.
R.; E. Indies.[1]
[1] Philippines.

Subfamily **Ptilocercinae** Lyon, 1913:4.
> R.; E. Indies. R.; As.[1]

[1] S.E. Asia.

Ptilocercus Gray, 1848:24. Pen-tailed tree shrew.
> R.; E. Indies.[1] R.; As.[2]

[1] Sumatra, Borneo.
[2] S.E. Asia (Malay Peninsula).

Grandorder **UNGULATA** Linnaeus, 1766.[1] Ungulates.
> [= PYCNODERMA Haeckel, 1866:cxlv[2]; TERNATES Trouessart, 1879:2.[3]] [= or including DIPLARTHRA Cope, 1882:445.[4]] [Including CHELOPHORA Haeckel, 1866:clix[5]; PROTUNGULATA Weber, 1904:587; PANTOMESAXONIA Franz, 1924:823.[6]]

[1] This term goes back to John Ray's pre-Linnaean usage, but as used here the term reflects an expansion of Owen's 1868 usage, along the lines suggested by Flower & Lydekker, 1891. Used as a grandorder by McKenna, 1975:41 and as a superorder by Novacek, 1986.

[2] Now forgotten; used for (terrestrial) ungulates plus whales.

[3] In part. Cohort TERNATES (compare PRIMATES and Trouessart's SECONDATES) was more inclusive than UNGULATA. This obscure paraphyletic term for the ungulate-paenungulate tangle included three orders, PROBOSCIDEA, UNGULATA, and †AMBLYPODA. Thus it excluded whales and sirenians, but included proboscideans, hyraxes, perissodactyls, artiodactyls, uintatheres, and pantodonts.

[4] Coined for ARTIODACTYLA and PERISSODACTYLA alone.

[5] Order CHELOPHORA included four suborders, LAMNUNGIA (hyraxes), TOXODONTA, †GONYOGNATHA (deinotheres), and PROBOSCIDEA.

[6] Used for †CONDYLARTHRA, HYRACOIDEA, †DINOCERATA, PROBOSCIDEA, and PERISSODACTYLA, but not including ARTIODACTYLA, SIRENIA, South American ungulates, nor †DESMOSTYLIA. Revived from obscurity by Prothero, Manning & Fischer, 1988:212-213 to subsume TETHYTHERIA and PERISSODACTYLA (in which they include HYRACOIDEA), but not †CONDYLARTHRA, †EMBRITHOPODA, nor †DINOCERATA.

†*Protungulatum* Sloan & Van Valen, 1965:226.
> L. Cret.-E. Paleoc.; N.A.

Order **TUBULIDENTATA** Huxley, 1872:288.
> [= NEORYCTIDA Haeckel, 1895:516[1]; TUBULIDENTIFORMES Kinman, 1994:37.]

[1] Proposed as a "family" for *Orycteropus*.

Family **Orycteropodidae** Gray, 1821:305.
> [= Orycteropidae Gray, 1821:305; Orycteropina Gray, 1825:343[1]; Orycteropodina C. L. Bonaparte, 1837:8[2]; Orycteropodidae C. L. Bonaparte, 1850; Orycteropi Murray, 1866:227.]
> Olig., L. Mioc., Plioc.; Eu. E. Mioc.-R.; Af. M. Mioc.-L. Plioc.; As.

[1] As tribe.
[2] As subfamily.

†*Palaeorycteropus* Filhol, 1893:135-136.
> [= †*Palaeoryctoropus* Filhol, 1893:135-136[1]; *Palaeorycteropus* Lydekker, 1894:41.[2]]
> Olig.; Eu.

[1] Misprint.
[2] Correction.

†*Archaeorycteropus* Ameghino, 1905:223.[1]
> Olig.; Eu.

[1] Validity doubtful.

†*Myorycteropus* MacInnes, 1956:1.
> E. Mioc.; Af.

Orycteropus É. Geoffroy Saint-Hilaire, 1796:102.[1] Aardvark.
> M. Mioc.-R.; Af. M. Mioc.-L. Plioc.; As. L. Mioc., Plioc.; Eu.

[1] There is a possibility that the name dates from 1791 (Palmer, 1904:483).

†*Leptorycteropus* Patterson, 1975:186.
> L. Mioc.; Af.[1]

[1] E. Africa.

Order †**DINOCERATA** Marsh, 1873:117, 118.[1]

[= †DINOCEREA Marsh, 1872:344[2]; †UINTATHERIAMORPHA Schoch & Lucas, 1985:33[3]; †DINOCERATIFORMES Kinman, 1994:38.]

[1] "No animal is there that possesses both tusks and horns...." Aristotle, Historia Animalium Book II(1).
[2] Proposed as an order.
[3] In part. Proposed as Mirorder, it also included Order †XENUNGULATA and possibly †PYROTHERIA but did not deal with the †Eoastrapostylopidae.

Family †**Uintatheriidae** Flower, 1876:387.

[= †Tinoceridae Marsh, 1872:323; †Tinoceratidae Marsh, 1873:295; †Sphaleroceratinae Brandt, 1878:18; †Dinoceratidae Zittel, 1893:439.] [Including †Eobasileidae Cope, 1873:3 or 293; †Eobasiliidae Cope, 1873:563; †Bathyopsidae Osborn, 1898:182; †Prodinoceratidae Flerov, 1952:1029; †Gobiatheriidae Flerov, 1952:1032.]
L. Paleoc.-M. Eoc.; As. L. Paleoc.-M. Eoc.; N.A.

Subfamily †**Uintatheriinae** Flower, 1876:387.

[= †Uintatheriidae Flower, 1876:387[1]; †Uintatheriinae Wheeler, 1961.] [Including †Bathyopsidae Osborn, 1898:182; †Prodinoceratidae Flerov, 1952:1029; †Bathyopsinae Wheeler, 1961:19; †Prodinoceratinae Szalay & McKenna, 1971:312; †Bathyopsini Schoch & Lucas, 1985:33, 43.]
L. Paleoc.-M. Eoc.; As. L. Paleoc.-M. Eoc.; N.A.

[1] Revised and restricted by Flerov, 1952:1032.

†*Prodinoceras* Matthew, Granger & Simpson, 1929:10.

[= or including †*Bathyopsoides* Patterson, 1939:373; †*Prouintatherium* Dorr, 1958:507; †*Jiaoluotherium* Tong, 1978:92[1]; †*Houyanotherium* Tong, 1978:95[1]; †*Pyrodon* Zhai, 1978:104; †*Phenaceras* Tong, 1979:395[1]; †*Ganatherium* Tong, 1979:397.[1]] [Including †*Probathyopsis* Simpson, 1929:1; †*Eobathyopsis* Osborn, 1929:65[2]; †*Mongolotherium* Flerov, 1952:1030.]
L. Paleoc.-E. Eoc.; As. L. Paleoc.-E. Eoc.; N.A.

[1] (Tung).
[2] *Nomen nudum.*

†*Bathyopsis* Cope, 1881:75.
E.-M. Eoc.; N.A.

†*Uintatherium* Leidy, 1872:168-169.

[= †*Uintamastix* Leidy, 1872:169; †*Dinoceras* Marsh, 1872:344; †*Tinoceras* Marsh, 1872:504[1]; †*Paroceras* Marsh, 1885:200[1]; †*Platoceras* Marsh, 1885:213, 214[2]; †*Laoceras* Marsh, 1885:216[2]; †*Octotomus* Cope, 1885:44, 52-53; †*Ditetrodon* Cope, 1885:594; †*Elachoceras* Scott, 1886:304-307.]
M. Eoc.; As. M. Eoc.; N.A.

[1] Proposed as a subgenus of †*Dinoceras.*
[2] Proposed as a subgenus of †*Tinoceras.*

†*Eobasileus* Cope, 1872:485.[1]

[= †*Lefalaphodon* Cope, 1872:1[2]; †*Loxolophodon* Cope, 1872:1-2[3]; †*Uintacolotherium* Cook, 1926.]
M. Eoc.; N.A.

[1] August 20, 1872.
[2] August 19, 1872. Misprint for †*Loxolophodon.*
[3] August 22, 1872. In part. †*Loxolophodon* Cope, (February 16) 1872:420 may be a synonym of †*Coryphodon.*

†*Tetheopsis* Cope, 1885:594.
M. Eoc.; N.A.

Subfamily †**Gobiatheriinae** Flerov, 1952:1032.

[= †Gobiatheriidae Flerov, 1952:1032; †Gobiatheriinae Wheeler, 1961:56.]
M. Eoc.; As.

†*Gobiatherium* Osborn & Granger, 1932:10.
M. Eoc.; As.

Mirorder **EPARCTOCYONA** McKenna, 1975:41.

[Including CETUNGULATA Irwin & Wilson, 1993:264.]

Order †**PROCREODI** Matthew, 1915:5.

[= †ARCTOCYONIA Van Valen, 1969:123[1]; †ARCTOCYONIFORMES Kinman, 1994:37.]

[1] In part. Proposed as a suborder with a more inclusive content but later used in a more restricted sense but at the higher rank of order by McKenna, 1975:42 and in Stucky & McKenna, in Benton, ed., 1993:758, and by R. J. G. Savage, 1986:130, 132. †ARCTOCYONIA was restricted to the †Arctocyonidae by Szalay, 1977:371.

Family †**Oxyclaenidae** Scott, 1892:294.

[= †Oxyclaeninae Matthew, 1937:38.] [Including †Chriacidae Osborn & Earle, 1895:20; †Chriacinae Matthew, 1937:64; †Astigalidae Zhang & Tong, 1981:133[1]; †Petrolemuridae Szalay & Li, 1987:394, caption, fig. 3.[2]]

E.-L. Paleoc., L. Eoc.; As. E. Paleoc., ?E. Eoc.; Eu. E. Paleoc.-E. Eoc.; N.A.

[1] (Chang & Tung).
[2] Not 1986 as stated.

†*Oxyprimus* Van Valen, 1978:53.

E. Paleoc., ?L. Paleoc.; N.A.

†*Carcinodon* Scott, 1892:323.

E. Paleoc.; N.A.

†*Chriacus* Cope, 1883:80.

[= or including †*Tricentes* Cope, 1883:315; †*Epichriacus* Scott, 1892:296; †*Metachriacus* Simpson, 1935:235; †*Spanoxyodon* Simpson, 1935:236.]

E. Paleoc.-E. Eoc.; N.A. ?E. Eoc.; Eu.

†*Zhujegale* Zhang & Tong, 1981:137.[1]

E. Paleoc.; As.[2]

[1] (Chang & Tung).
[2] E. Asia.

†*Astigale* Zhang & Tong, 1981:133.[1]

E. Paleoc.; As.[2]

[1] (Chang & Tung).
[2] E. Asia.

†*Oxyclaenus* Cope, 1884:312-313, 324.

[Including †*Thangorodrim* Van Valen, 1978:55.]

E. Paleoc.; N.A.

†*Prolatidens* Sudre & D. E. Russell, 1982:180.

E. Paleoc.; Eu.

†*Oxytomodon* Gazin, 1941:27.

E. Paleoc.; N.A.

†*Prothryptacodon* Simpson, 1935:233.

[= †*Pantinomia* Van Valen, 1967:222.]

E. Paleoc.; N.A.

†*Princetonia* Gingerich, 1989:43.

L. Paleoc.-E. Eoc.; N.A.

†*Thryptacodon* Matthew, 1915:7.

L. Paleoc.-E. Eoc.; N.A.

†*Petrolemur* Tong, 1979:65.[1]

L. Paleoc.; As.[2]

[1] (Tung).
[2] E. Asia.

†*Khashanagale* Szalay & McKenna, 1971:305.

L. Paleoc.; As.[1]

[1] E. Asia.

†*Lantianius* Chow, 1964:1.

L. Eoc.; As.[1]

[1] E. Asia.

Family †**Arctocyonidae** Giebel, 1855:755.[1]
>   [= †Arctocyoninae Giebel, 1855:755[1]; †Arctocyonidae Murray, 1866:329;
>   †Arctocyonida Haeckel, 1866:clix[1]; †Arctocyonidae Gervais, 1870:150.]
>   E. Paleoc.-E. Eoc.; N.A. L. Paleoc.-E. Eoc.; Eu. ?E. Eoc.; Af.[2]
>   [1] Used as a family.
>   [2] A worn and broken premolar tooth from the Ypresian of Morocco has been referred to this taxon. The
>   reference is questionable.

Subfamily †**Loxolophinae** Van Valen, 1978:55.
>   E.-L. Paleoc.; N.A.

†*Platymastus* Van Valen, 1978:56.
>   E.-L. Paleoc.; N.A.

†*Desmatoclaenus* Gazin, 1941:34.
>   E.-L. Paleoc.; N.A.

†*Baioconodon* Gazin, 1941:292.
>   [Including †*Ragnarok* Van Valen, 1978:55.]
>   E. Paleoc.; N.A.

†*Loxolophus* Cope, 1885:386.
>   [= or including †*Protochriacus* Scott, 1892:296; †*Protogonodon* Scott, 1892:322;
>   †*Paradoxodon* Scott, 1892:322[1]; †*Paradoxodonta* Strand, 1943:94-114.]
>   E. Paleoc.; N.A.
>   [1] Objective synonym of †*Paradoxodonta*. Not *Paradoxodon* Wagner, 1855: 805, a synonym of *Suncus*; nor
>   †*Paradoxodon* Filhol, 1890, a synonym of †*Paradoxodonides*.

†*Mimotricentes* Simpson, 1937:203.
>   E. Paleoc., ?L. Paleoc.; N.A.

†*Deuterogonodon* Simpson, 1935:232.
>   E. Paleoc.; N.A.

Subfamily †**Arctocyoninae** Giebel, 1855:755.
>   [= †Arctocyoninae Giebel, 1855:755[1]; †Arctocyoninae Van Valen, 1978:55.]
>   E. Paleoc.-E. Eoc.; N.A. L. Paleoc.-E. Eoc.; Eu.
>   [1] Used as a family.

†*Neoclaenodon* Gidley, 1919:547.
>   E.-L. Paleoc.; N.A.

†*Claenodon* Scott, 1892:298.
>   E.-L. Paleoc.; N.A.

†*Colpoclaenus* Patterson & McGrew, 1962:2.
>   E.-L. Paleoc.; N.A.

†*Mentoclaenodon* Weigelt, 1960:31.
>   L. Paleoc.; Eu. L. Paleoc.; N.A.

†*Arctocyonides* Lemoine, 1891:275.
>   [= †*Procynictis* Lemoine, 1885:205[1]; †*Creodapis* Lemoine, 1893:353.]
>   L. Paleoc.; Eu.
>   [1] *Nomen oblitum.*

†*Landenodon* Quinet, 1968:18.[1]
>   L. Paleoc.-E. Eoc.; Eu.
>   [1] *Nomen nudum* in Quinet, 1966.

†*Arctocyon* de Blainville, 1841:73, 112, 121.
>   [= *Palaeocyon* de Blainville, 1841:73, 112, 121[1]; †*Hyodectes* Cope, 1880:79;
>   †*Heteroborus* Cope, 1880:79; †*Arctotherium* Lemoine, 1896:342.[2]]
>   L. Paleoc.; Eu.
>   [1] Alternative name for †*Arctocyon*.
>   [2] Not †*Arctotherium* Bravard, 1860 (or 1857), a synonym of the ursid †*Paractotherium*.

†*Anacodon* Cope, 1882:181.
>   ?L. Paleoc., E. Eoc.; N.A.

†*Lambertocyon* Gingerich, 1979:524.
   L. Paleoc.; N.A.

Order †**CONDYLARTHRA** Cope, 1881:1018.
   [= †PHENACODONTA McKenna, 1975:42; †PHENACODONTIA Kalandadze &
   Rautian, 1992:105.] [Including †LEMURAVIDA Haeckel, 1895:600, 603;
   †HYOPSODONTIFORMES Kinman, 1994:37; †PHENACODONTIFORMES
   Kinman, 1994:38; †DIDOLODONTIFORMES Kinman, 1994:38.]

  Family †**Hyopsodontidae** Trouessart, 1879:229.
    [= †Lemuravidae Marsh, 1875:239-240; †Hyopsodinae Trouessart, 1879:229;
    †Hyopsodidae Schlosser, 1887:43; †Hyopsodontidae Nicholson & Lydekker,
    1889:1465; †Hyopsodina Haeckel, 1895:600, 604[1]; †LEMURAVIDA Haeckel,
    1895:600, 603.[2]]
    E. Paleoc.-E. Eoc., ?E. Olig.; Eu. E. Paleoc.-M. Eoc.; N.A. E. Eoc.; Af.[3] E.-L.
    Eoc.; As.[4]
    [1] Used as a family.
    [2] Used as a suborder, but apparently based on †*Lemuravus*, a synonym of †*Hyopsodus.*
    [3] N. Africa.
    [4] E. Asia.

   †*Midiagnus* D. E. Russell, in Godinot, Crochet, Hartenberger, Lange-Badre, Russell & Sigé,
    1987:281.
    E. Eoc.; Eu.

  Subfamily †**Tricuspiodontinae** Simpson, 1929:16.
    [= †Tricuspiodontidae Simpson, 1929:16; †Tricuspiodontinae Van Valen, 1978:61.]
    E.-L. Paleoc.; N.A. L. Paleoc.; Eu.

   †*Litomylus* Simpson, 1935:243.
    E.-L. Paleoc.; N.A.

   †*Aletodon* Gingerich, 1977:238.
    L. Paleoc.; N.A.

   †*Paratricuspiodon* D. E. Russell, 1964:234.
    L. Paleoc.; Eu.

   †*Tricuspiodon* Lemoine, 1885:205.
    [= †*Conaspidotherium* Lemoine, 1891:275; †*Plesiphenacodus* Lemoine, 1896:343.]
    L. Paleoc.; Eu.

  Subfamily †**Hyopsodontinae** Trouessart, 1879:229.
    [= †Lemuravidae Marsh, 1875:239-240; †Hyopsodinae Trouessart, 1879:229;
    †Hyopsodontinae Matthew, 1937:194; †Hyopsodina Haeckel, 1895:600, 604.]
    E. Paleoc.-M. Eoc.; N.A. E.-L. Eoc.; As.[1] E. Eoc.; Eu.
    [1] E. Asia.

   †*Haplaletes* Simpson, 1935:243.
    E.-L. Paleoc.; N.A.

   †*Dorraletes* Gingerich, 1983:244.
    L. Paleoc.; N.A.

   †*Utemylus* Gingerich, 1983:227, 252.
    L. Paleoc.; N.A.

   †*Hyopsodus* Leidy, 1870:109-110.
    [= †*Stenacodon* Marsh, 1872:210; †*Lemuravus* Marsh, 1875:239.]
    L. Paleoc.-M. Eoc.; N.A. E.-M. Eoc.; As.[1] E. Eoc.; Eu.
    [1] E. Asia.

   †*Heptaconodon* Zdansky, 1930:58.
    L. Eoc.; As.[1]
    [1] E. Asia.

  Subfamily †**Louisininae** Sudre & D. E. Russell, 1982:175.
    E. Paleoc.-E. Eoc.; Eu. L. Paleoc.-E. Eoc.; N.A. E. Eoc.; Af.[1] ?M. Eoc.; As.
    [1] N. Africa.

†*Monshyus* Sudre & D. E. Russell, 1982:178.
  E. Paleoc.; Eu.

†*Khamsaconus* Sudre, Jaeger, Sigé & Vianey-Liaud, 1993:611.[1]
  E. Eoc.; Af.[2]
  [1] Subfamily (and even family) reference is questionable.
  [2] N. Africa (Morocco).

†*Paschatherium* D. E. Russell, 1964:237.
  L. Paleoc.-E. Eoc.; Eu.

†*Dipavali* Van Valen, 1978:61.
  L. Paleoc.; Eu.

†*Louisina* D. E. Russell, 1964:242.
  L. Paleoc.-E. Eoc.; Eu.

†*Haplomylus* Matthew, 1915:313.
  L. Paleoc.-E. Eoc.; N.A. ?M. Eoc.; As.

†*Chambius* Hartenberger, 1986:247.
  E. Eoc.; Af.[1]
  [1] N. Africa (Tunisia).

†*Microhyus* Teilhard de Chardin, 1927:24.
  E. Eoc.; Eu.

Subfamily †**Apheliscinae** Matthew, 1918:592.
  [= †Apheliscidae Matthew, 1918:592.]
  L. Paleoc.-E. Eoc.; N.A. ?E. Olig.; Eu.
†*Phenacodaptes* Jepsen, 1930:517.
  L. Paleoc.; N.A.

†*Apheliscus* Cope, 1875:13, 16-17.
  [Including †*Parapheliscus* Van Valen, 1967:247.]
  L. Paleoc.-E. Eoc.; N.A.

†*Epapheliscus* Van Valen, 1966:88.
  ?E. Olig.; Eu.

Family †**Mioclaenidae** Osborn & Earle, 1895:48.
  E.-L. Paleoc.; N.A. E. Paleoc.; S.A. L. Paleoc.; Eu.

†*Raulvaccia* Bonaparte, Van Valen & Kramartz, 1993:21.
  E. Paleoc.; S.A.

†*Escribania* Bonaparte, Van Valen & Kramartz, 1993:24.
  E. Paleoc.; S.A.

Subfamily †**Pleuraspidotheriinae** Zittel, 1892:222.
  [= †Pleuraspidotheridae Zittel, 1892:222; †Pleuraspidotheriinae Simpson, 1945:124.]
  [Including †Protoseleninae Rigby, 1980:131.]
  E. Paleoc., ?L. Paleoc.; N.A. L. Paleoc.; Eu.
†*Protoselene* Matthew, 1897:317.
  [Including †*Dracoclaenus* Gazin, 1939:281.]
  E. Paleoc., ?L. Paleoc.; N.A.

†*Pleuraspidotherium* Lemoine, 1878:104.
  L. Paleoc.; Eu.

†*Orthaspidotherium* Lemoine, 1885:205.
  L. Paleoc.; Eu.

Subfamily †**Mioclaeninae** Osborn & Earle, 1895:48.
  [= †Mioclaenidae Osborn & Earle, 1895:48; †Mioclaeninae Matthew, 1937:vi, 195.]
  [Including †Molinodinae Bonaparte, Van Valen & Kramartz, 1993:29.]
  E.-L. Paleoc.; N.A. E. Paleoc.; S.A.
†*Litaletes* Simpson, 1935:242.
  [Including †*Jepsenia* Gazin, 1939:285.]
  E. Paleoc.; N.A.

†*Ellipsodon* Scott, 1892:298.
>[Including †*Bomburia* Van Valen, 1978:59.]
>E. Paleoc.; N.A.

†*Choeroclaenus* Simpson, 1937:232.
>E. Paleoc.; N.A.

†*Bubogonia* Johnson, in Johnson & Fox, 1984:202.
>E. Paleoc.; N.A.

†*Molinodus* de Muizon & Marshall, 1987:771.
>E. Paleoc.; S.A.

†*Tiuclaenus* de Muizon & Marshall, 1987:947.
>E. Paleoc.; S.A.

†*Tiznatzinia* Simpson, 1936:7.
>E. Paleoc.; N.A.

†*Promioclaenus* Trouessart, 1904:43.
>E.-L. Paleoc.; N.A.

†*Pucanodus* de Muizon & Marshall, 1991:203.
>E. Paleoc.; S.A.

†*Mioclaenus* Cope, 1881:489.
>E. Paleoc.; N.A.

Family †**Phenacodontidae** Cope, 1881:1018.
>[= †Phenacodontida Haeckel, 1895:530, 534; †Phenacodontidea Kalandadze & Rautian, 1992:105[1]; †Eohyidae Marsh, 1894:260.] [Including †Meniscotheriidae Cope, 1882:334; †Almogaveridae Crusafont Pairó & Villalta, 1954:173.]
>E. Paleoc.-M. Eoc.; N.A. E.-M. Eoc.; Eu.
>[1] In part. Misspelled.

Subfamily †**Phenacodontinae** Cope, 1881:1018, **new rank**.
>[= †Phenacodontidae Cope, 1811:1018; †Eohyidae Marsh, 1894:260.] [Including †Almogaveridae Crusafont Pairó & Villalta, 1974:173.]
>E. Paleoc.-M. Eoc.; N.A. E.-M. Eoc.; Eu.

†*Tetraclaenodon* Scott, 1892:299-300.
>[= †*Protogonia* Cope, 1881:492[1]; †*Euprotogonia* Cope, 1893:378.]
>E. Paleoc.; N.A.
>[1] Suppressed by the International Commission on Zoological Nomenclature, Opinion 980, 1972. Not *Protogonius* Hübner, 1816, a genus of lepidopterans.

†*Phenacodus* Cope, 1873:3-4.
>[= †*Opisthotomus* Cope, 1875:13, 15-16; †*Theocodus* Cope, 1883:509[1]; †*Trispondylus* Cope, 1884:900; †*Eohyus* Marsh, 1894:259-260.[2]] [Including †*Almogaver* Crusafont Pairó & Villalta, 1954:173.]
>L. Paleoc.-M. Eoc.; N.A. E.-M. Eoc.; Eu.
>[1] Misspelling.
>[2] *Nomen nudum* in Cope, 1877:362, 1878:240.

†*Copecion* Gingerich, 1989:52.
>E. Eoc.; N.A.

Subfamily †**Meniscotheriinae** Cope, 1882:334.
>[= †Meniscotheriidae Cope, 1882:334; †MESODACTYLA Marsh, 1892:446[1]; †Meniscotheriinae Simpson, 1937:224.]
>L. Paleoc.-M. Eoc.; N.A.
>[1] Proposed as an order.

†*Ectocion* Cope, 1882:522.
>[Including †*Gidleyina* Simpson, 1935:240; †*Prosthecion* Patterson & West, 1973:2.]
>L. Paleoc.-M. Eoc.; N.A.

†*Meniscotherium* Cope, 1874:22-23.
>[= †*Hyracops* Marsh, 1892:446.]
>L. Paleoc.-E. Eoc.; N.A.

Family †**Periptychidae** Cope, 1882:832.
> [= †Catathleidae Ameghino, 1906:296; †Periptychoidea Kalandadze & Rautian, 1992:105.]
> ?E. Paleoc., L. Paleoc.; As. E.-L. Paleoc.; N.A. ?E. Paleoc.; S.A. ?E. Eoc.; Eu.

†*Pseudanisonchus* Zhang, Zheng & Ding, 1979:384.[1]
> L. Paleoc.; As.[2]
> [1] Alternatively spelled Chang, Cheng & Ting. Possibly referable to †Anisonchinae.
> [2] E. Asia.

Subfamily †**Anisonchinae** Osborn & Earle, 1895:58.
> E. Paleoc.; N.A. ?E. Paleoc.; S.A. ?E. Eoc.; Eu.

†*Mimatuta* Van Valen, 1978:62.
> [Including †*Earendil* Van Valen, 1978:63.]
> E. Paleoc.; N.A. ?E. Paleoc.; S.A.

†*Anisonchus* Cope, 1881:488-489.
> [= †*Zetodon* Cope, 1883:968.]
> E. Paleoc.; N.A.

†*Haploconus* Cope, 1882:417.
> E. Paleoc.; N.A.

†*Mithrandir* Van Valen, 1978:64.[1]
> [Including †*Gillisonchus* Rigby, 1981:97.]
> E. Paleoc.; N.A.
> [1] Described as subgenus of †*Anisonchus*.

†*Hemithlaeus* Cope, 1882:832.
> E. Paleoc.; N.A.

†*Lessnessina* Hooker, 1979:43, 50.[1]
> E. Eoc.; Eu.
> [1] Affinities questionable.

Subfamily †**Periptychinae** Cope, 1882:832.
> [= †Periptychidae Cope, 1882:832; †Periptychinae Osborn & Earle, 1895:53.]
> ?E. Paleoc.; As.[1] E.-L. Paleoc.; N.A.
> [1] S.E. Asia (S. China).

†*Ectoconus* Cope, 1884:795-796.
> [= †*Ectogonus* Trouessart, 1898:723.]
> ?E. Paleoc.; As.[1] E. Paleoc.; N.A.
> [1] S.E. Asia (S. China).

†*Maiorana* Van Valen, 1978:61.
> E. Paleoc.; N.A.

†*Periptychus* Cope, 1881:337.
> [Including †*Plagioptychus* Matthew, 1936:9[1]; †*Carsioptychus* Simpson, 1936:234[2]; †*Catathlaeus* Cope, 1881:487.]
> E.-L. Paleoc.; N.A.
> [1] Objective synonym of †*Carsioptychus*. Not *Plagioptychus* Matheron, 1843, a mollusk.
> [2] Sometimes as subgenus.

†*Tinuviel* Van Valen, 1978:61.
> E. Paleoc.; N.A.

Subfamily †**Conacodontinae** Archibald, Schoch & Rigby, 1983:3.
> E. Paleoc.; N.A.

†*Oxyacodon* Osborn & Earle, 1895:25.
> [= †*Escatepos* Reynolds, 1936:207.] [Including †*Fimbrethil* Van Valen, 1978:62.]
> E. Paleoc.; N.A.

†*Conacodon* Matthew, 1897:298.
> E. Paleoc.; N.A.

Family †**Peligrotheriidae** Bonaparte, Van Valen & Kramartz, 1993:16.
E. Paleoc.; S.A.

†*Peligrotherium* Bonaparte, Van Valen & Kramartz, 1993:16.
E. Paleoc.; S.A.

Family †**Didolodontidae** Scott, 1913:489.
[= †Didolodidae Scott, 1913:489; †Didolodontidae Simpson, 1934:6; †Didolodontoidea Cifelli, 1983:9.] [= or including †Bunolitopternidae Schlosser, 1923:585.]
L. Paleoc.-E. Eoc., L. Olig.; S.A.

†*Lamegoia* Paula Couto, 1952:363.
L. Paleoc.; S.A.

†*Paulacoutoia* Cifelli, 1983:3.
L. Paleoc.; S.A.

†*Didolodus* Ameghino, 1897:437.
[= †*Enneoconus* Ameghino, 1901:378; †*Lonchoconus* Ameghino, 1901:379; †*Paulogervaisia* Ameghino, 1901:389; †*Cephanodus* Ameghino, 1902:25.] [= or including †*Nephacodus* Ameghino, 1902:19; †*Argyrolambda* Ameghino, 1904:338.]
E. Eoc.; S.A.

†*Salladolodus* Soria & Hoffstetter, 1983:267.
L. Olig.; S.A.

Order †**ARCTOSTYLOPIDA** Cifelli, Schaff & McKenna, 1989:5.
[= †ARCTOSTYLOPIFORMES Kinman, 1994:38.[1]]
[1] Proposed with a query but the provisions of the International Code of Zoological Nomenclature do not apply above family-group rank.

Family †**Arctostylopidae** Schlosser, 1923:614.
[Including †Palaeostylopidae Hau, 1976:50, 52.[1]]
L. Paleoc.-E. Eoc.; As. L. Paleoc.; N.A.
[1] *Lapsus calami?*

†*Allostylops* Zheng, 1979:391.[1]
L. Paleoc.; As.[2]
[1] (Cheng).
[2] E. Asia.

Subfamily †**Arctostylopinae** Schlosser, 1923:614.
[= †Arctostylopidae Schlosser, 1923:614; †Arctostylopinae Zheng, 1979:391.[1]]
[Including †Palaeostylopidae Hau, 1976:50, 52[2]; †Synostylopinae Nessov, 1987:211[3]; †Kazachostylopinae Nessov, 1987:211.]
L. Paleoc.-E. Eoc.; As. L. Paleoc.; N.A.
[1] (Cheng).
[2] *Lapsus calami?*
[3] Misspelling.

Tribe †**Arctostylopini** Schlosser, 1923:614, **new rank**.
[= †Arctostylopidae Schlosser, 1923:614.]
L. Paleoc.-E. Eoc.; As. L. Paleoc.; N.A.

†*Gashatostylops* Cifelli, Schaff & McKenna, 1989:15.
L. Paleoc.; As.

†*Palaeostylops* Matthew & Granger, 1925:2.
L. Paleoc.; As.

†*Arctostylops* Matthew, 1915:429.
L. Paleoc.; N.A.

†*Anatolostylops* Zhai, 1978:109.
[= †*Anatostylops* Schaff, 1985:593.[1]]
E. Eoc.; As.
[1] Misprint.

Tribe †**Sinostylopini** Nessov, 1987:211, **new rank**.
[= †Synostylopinae Nessov, 1987:211.[1]]
L. Paleoc.; As.
[1] Misspelling.

†*Sinostylops* Tang & Yan, 1976:91.
L. Paleoc.; As.[1]
[1] E. Asia.

†*Bothriostylops* Zheng & Huang, 1986:121.
L. Paleoc.; As.[1]
[1] E. Asia.

†*Kazachostylops* Nessov, 1987:211.
L. Paleoc.; As.[1]
[1] Kazakhstan.

Subfamily †**Asiostylopinae** Zheng, 1979:387.[1]
L. Paleoc.; As.[2]
[1] (Cheng).
[2] E. Asia.

†*Asiostylops* Zheng, 1979:388.[1]
L. Paleoc.; As.[2]
[1] (Cheng).
[2] E. Asia.

Order **CETE** Linnaeus, 1758:75.[1]
[= CETINA Newman, 1843:148.]
[1] Proposed as an order. Used as mirorder by McKenna (1975:42).

Suborder †**ACREODI** Matthew, 1909:327, **new rank**.[1]
[= †MESONYCHIA Matthew, 1937:39[2]; †MESONYCHIFORMES Kinman, 1994:37.]
[1] Proposed as approximately an infraorder of creodonts. Raised to the rank of order by McKenna, 1975:42.
[2] Proposed, perhaps inadvertently, for †*Dissacus*, †*Pachyaena*, and †*Mesonyx*, but without specifying taxonomic rank. Used by Van Valen, 1969:123 as suborder.

†*Microclaenodon* Scott, 1892:302.
E. Paleoc.; N.A.

†*Olsenia* Matthew & Granger, 1925:3.[1]
L. Eoc.; As.[2]
[1] Possibly referable to †Mesonychidae.
[2] E. Asia.

Family †**Triisodontidae** Trouessart, 1904:161.
[= †Triisodontinae Trouessart, 1904:161; †Triisodontidae Scott, 1892:300.] [Including †Andrewsarchinae Szalay & Gould, 1966:154.]
E. Paleoc.; N.A. M. Eoc.; As.[1]
[1] E. Asia.

†*Goniacodon* Cope, 1888:320-321.[1]
E. Paleoc.; N.A.
[1] As subgenus of †*Mioclaenus*. Raised to generic rank by Scott, 1892:301-302.

†*Eoconodon* Matthew & Granger, 1921:6.
E. Paleoc.; N.A.

†*Triisodon* Cope, 1881:667.
[= †*Sarcothraustes* Cope, 1882:193.[1]]
E. Paleoc.; N.A.
[1] 1882-1883 (Dec. 1881).

†*Stelocyon* Gingerich, 1978:2.
E. Paleoc.; N.A.

†*Andrewsarchus* Osborn, 1924:1.
[= †*Paratriisodon* Chow, 1959:133.]
M. Eoc.; As.

Family †**Mesonychidae** Cope, 1875:444.
>>> [= †Mesonychidae Cope, 1875:444; †Mesonychida Haeckel, 1895:578, 583;
>>> †Mesonychinae Wortman, 1901:286; †Mesonychoidea Osborn, 1910:527.[1]]
>>> E. Paleoc.-L. Eoc., ?E. Olig.; As. E. Paleoc.-M. Eoc.; N.A. L. Paleoc.-M. Eoc.; Eu.

[1] Authorship questionable.

†*Yantanglestes* Ideker & Yan, 1980:138-140.
>>> [= †*Lestes* Yan & Tang, 1976:252[1]; †*Notodissacus* Yan & Tang, 1976: pl. 1.[2]]
>>> E.-L. Paleoc.; As.[3]

[1] Not *Lestes* Leach, 1815, an insect.
[2] Evidently, prior to proposing †*Lestes*, Yan & Tang had intended to call the animal †*Notodissacus*. On plate 1 of their paper, the type is figured with the legend, "†*Notodissacus conexus* gen. et sp. nov." We interpret the International Code of Zoological Nomenclature (especially Art. 13) to mean that †*Notodissacus* is not an available name.
[3] E. Asia.

†*Hukoutherium* Chow, Chang, Wang & Ting, 1973:32.
>>> E. Paleoc.; As.[1]

[1] S.E. Asia (S. China).

†*Dissacusium* Zhou, Zhang, Wang & Ding, 1977:25.[1]
>>> E. Paleoc.; As.[2]

[1] (Chow, Chang, Wang & Ting).
[2] S.E. Asia (S. China).

†*Ankalagon* Van Valen, 1980:266.
>>> [= †*Ancalagon* Van Valen, 1978:65.[1]]
>>> E. Paleoc.; N.A.

[1] Not †*Ancalagon* Conway Morris, 1977, a Cambrian priapulid.

†*Sinonyx* Zhou, Zhai, Gingerich & Chen, 1995:388.
>>> L. Paleoc.; As.

†*Dissacus* Cope, 1881:1018.
>>> [= or including †*Hyoenodictis* Lemoine, 1880:5[1]; †*Hyaenodictis* Lemoine, 1891:272.]
>>> [Including †*Plesidissacus* Lemoine, 1894:353-354, 363; †*Plagiocristodon* Chow & Qi, 1978:80.]
>>> E. Paleoc.-E. Eoc.; N.A. L. Paleoc.-E. Eoc.; As. L. Paleoc.-M. Eoc.; Eu.

[1] *Nomen nudum* in Lemoine, 1878:65.

†*Pachyaena* Cope, 1874:13.
>>> L. Paleoc., M. Eoc.; As.[1] ?L. Paleoc., E. Eoc.; N.A. E. Eoc.; Eu.

[1] E. Asia.

†*Jiangxia* Zhang, Zheng & Ding, 1979:382.[1]
>>> L. Paleoc.; As.[2]

[1] Alternatively, Chang, Cheng & Ting.
[2] E. Asia.

†*Mongolonyx* Szalay & Gould, 1966:134.
>>> M. Eoc.; As.[1]

[1] E. Asia.

†*Harpagolestes* Wortman, 1901:286.
>>> M.-L. Eoc., ?E. Olig.; As.[1] M. Eoc.; N.A.

[1] E. Asia.

†*Hessolestes* Peterson, 1931:338.
>>> M. Eoc.; N.A.

†*Synoplotherium* Cope, 1872:483.
>>> [Including †*Dromocyon* Marsh, 1876:403.]
>>> M. Eoc.; N.A.

†*Mesonyx* Cope, 1872:1.
>>> M. Eoc.; As. M. Eoc.; N.A.

†*Guilestes* Zheng & Chi, 1978:98.
>>> L. Eoc.; As.[1]

[1] S.E. Asia.

†*Mongolestes* Szalay & Gould, 1966:136.
　　　　L. Eoc.; As.

Family †**Hapalodectidae** Szalay & Gould, 1966:152.
　　　　[= †Hapalodectinae Szalay & Gould, 1966:152; †Hapalodectidae Ding & Li, 1987:162.[1]]
　　　　?L. Paleoc., E.-L. Eoc.; As.[2] E.-M. Eoc.; N.A.
　　　[1] (Ting & Li).
　　　[2] E. Asia.

†*Hapalodectes* Matthew, 1909:93.
　　　　?L. Paleoc., E.-L. Eoc.; As.[1] E. Eoc.; N.A.
　　　[1] E. Asia.

†*Hapalorestes* Gunnell & Gingerich, 1996:414.
　　　　M. Eoc.; N.A.

†*Metahapalodectes* Dashzeveg, 1976:26.
　　　　M. Eoc.; As.[1]
　　　[1] E. Asia.

†*Lohoodon* Chow, Li & Chang, 1973:176.
　　　　M. Eoc.; As.[1]
　　　[1] E. Asia.

†*Honanodon* Chow, 1965:286.
　　　　L. Eoc.; As.[1]
　　　[1] E. Asia.

Suborder **CETACEA** Brisson, 1762:3, 225, **new rank**.[1] Whales.
　　　　[= KETODE Aristotle, 330 B.C.[2]; MUTICA Linnaeus, 1766:24; CETACEAE Gray, 1821:309[3]; CARNIVORAE Gray, 1821:309[4]; NATANTIA Newman, 1843:35[5]; CETOMORPHA Haeckel, 1895:490, 562.] [Including †PHOCODONTIA Huxley, 1872:349.]
　　　[1] Proposed as an order.
　　　[2] Non-Linnean but included here for the sake of interest. Date approximate.
　　　[3] Named as a class and subdivided into two orders, HERBIVORAE (for sirenians) and CARNIVORAE (for true cetaceans).
　　　[4] Not to be confused with CARNIVORA.
　　　[5] In part; also including other aquatic animals. Newman's work put even that of the Quinarians to shame as a codification of illogic.

†*Chonecetus* L. S. Russell, 1968:929.
　　　　L. Olig.; N.A.[1]
　　　[1] Pacific North America.

†*Ferecetotherium* Mchedlidze, 1970:1-112.
　　　　L. Olig.; As.[1]
　　　[1] W. Asia.

Infraorder †**ARCHAEOCETI** Flower, 1883:182, **new rank**.[1]
　　　　[= †ZEUGLOCETA Haeckel, 1866:clix; †ZEUGLODONTA Haeckel, 1866:clix; †ZEUGLODONTIA Gill, 1871:122; †PHOCODONTIA Huxley, 1872:349[2]; †HYDROTHEREUTA Ameghino, 1889:44, 355, 895; †ARCHAEOCETIFORMES Kinman, 1994:37.]
　　　[1] Proposed as a suborder. Despite several prior synonyms, we maintain †ARCHAEOCETI in the interest of stability. The provisions of the International Code of Zoological Nomenclature do not govern taxa higher than those of the family-group.
　　　[2] In part. Huxley's †PHOCODONTIA was paraphyletic, used for "†*Zeuglodon*, †*Squalodon*, and other large extinct cetaceans of the Tertiary Epoch." Huxley divided all whales then known into three "groups": BALAENOIDEA, DELPHINOIDEA, and †PHOCODONTIA.

†*Pontobasileus* Leidy, 1873:337.[1]
　　　　?Eoc. and/or ?Olig. and/or ?Mioc.; N.A.[2]
　　　[1] Assignment to the ARCHAEOCETI is questionable.
　　　[2] Provenance not precise; supposedly from "some Eocene or Miocene formation of the Atlantic States" (Palmer, 1904:557).

†*Platyosphys* Kellogg, 1936:97.
E. Olig.; Eu.[1]
[1] E. Europe.

Family †**Basilosauridae** Cope, 1868:144.
[= †Zeuglodontidae Bonaparte, 1849:618[1]; †Zeuglodontida Haeckel, 1866:clix; †Hydrarchidae Bonaparte, 1850:1; †Hydrarchina Bonaparte, 1850; †Basilosauroidea Mitchell, 1989:2231.] [= or including †Stegorhinidae Brandt, 1873:344.] [Including †Dorudontidae Miller, 1923:13; †Prozeuglodontidae Moustafa, 1954:87.]
M.-L. Eoc.; Af.[2] M. Eoc.; Eu. M.-L. Eoc.; N.A. ?M. Eoc.; Antarctica.[3] L. Eoc.; New Zealand.
[1] Invalid.
[2] N. Africa.
[3] Possible basilosaurid vertebrae have been reported from the middle Eocene of Seymour Island.

Subfamily †**Basilosaurinae** Cope, 1868:144. Zeuglodonts.
[= †Basilosauridae Cope, 1868:144; †Zeuglodontinae Slijper, 1936:540; †Basilosaurinae Barnes & Mitchell, 1978:590.]
M.-L. Eoc.; Af.[1] M. Eoc.; Eu. L. Eoc.; N.A.
[1] N. Africa.

†*Basilosaurus* Harlan, 1834:397-403.
[= †Zeuglodon Owen, 1839:24-28; †Zygodon Owen, 1839:35-36; †Hydrargos Koch, 1845:1-16; †Hydrarchos Koch, 1845:1-24; †Hydrarchus Muller, 1849:3; †Alabamornis Abel, 1906:450-458.]
M.-L. Eoc.; Af.[1] M. Eoc.; Eu.[2] L. Eoc.; N.A.[3]
[1] N. Africa (Egypt).
[2] England.
[3] S.E. United States.

Subfamily †**Dorudontinae** Miller, 1923:13.
[= †Dorudontidae Miller, 1923:13; †Dorudontinae Slijper, 1936:540.]
M.-L. Eoc.; Af.[1] M. Eoc.; Eu. M.-L. Eoc.; N.A. L. Eoc.; New Zealand.
[1] N. Africa.

†*Dorudon* Gibbes, 1845:254-256.
[= †Doryodon Cope, 1868:144, 155; †Durodon Gill, 1872:93.[1]] [Including †Prozeuglodon Andrews, 1906:243; †Pachycetus Van Beneden, 1883:31-32.[2]]
M.-L. Eoc.; Af.[3] ?M. Eoc.; Eu.[4] M. Eoc.; N.A. ?L. Eoc.; New Zealand.
[1] Misspelling.
[2] Synonymy questionable.
[3] N. Africa (Egypt).
[4] Germany.

†*Ancalecetus* Gingerich & Uhen, 1996:363.
L. Eoc.; Af.[1]
[1] N. Africa (Egypt).

†*Saghacetus* Gingerich, 1992:73.
L. Eoc.; Af.[1]
[1] N. Africa (Egypt).

†*Zygorhiza* True, 1908:78.
M. Eoc.; Eu.[1] L. Eoc.; N.A.
[1] England.

†*Pontogeneus* Leidy, 1852:52.
?L. Eoc.; Af.[1] L. Eoc.; N.A.
[1] N. Africa (Egypt).

Family †**Protocetidae** Stromer, 1908:148.
[= †Protocetoidea Mitchell, 1989:2232.] [Including †Ambulocetidae Thewissen, Madar & Hussain, 1996:9.]
E.-M. Eoc.; As. M. Eoc.; Af. ?M. Eoc.; N.A.

†*Rodhocetus* Gingerich, Raza, Arif, Anwar & Zhou, 1994:844.
      M. Eoc.; As.[1]
    [1] S. Asia (Pakistan).

†*Eocetus* Fraas, 1904:374.
    [= †*Mesocetus* Fraas, 1904:21[1]; †*Eocetus* Abel, 1905:85.[2]]
      M. Eoc.; Af.[3]
    [1] Not †*Mesocetus* Moreno, 1892, a physeterid whale, nor †*Mesocetus* Van Beneden, 1880, a cetotheriid whale.
    [2] Abel settled on the same replacement name as Fraas had created.
    [3] N. Africa (Egypt).

†*Pappocetus* Andrews, 1920:309.
      M. Eoc.; Af.[1]
    [1] Nigeria.

†*Babiacetus* Trivedi & Satsangi, 1984:322-323.
      M. Eoc.; As.

†*Takracetus* Gingerich, Arif & Clyde, 1995:300.
      M. Eoc.; As.

†*Gaviacetus* Gingerich, Arif & Clyde, 1995:305.
      M. Eoc.; As.

Subfamily †**Pakicetinae** Gingerich & D. E. Russell, 1990:17.
    [Including †Ambulocetidae Thewissen, Madar & Hussain, 1996:9.]
      E. Eoc., ?M. Eoc.; As.

†*Gandakasia* Dehm & Oettingen-Spielberg, 1958:11.
      E. Eoc.; As.

†*Ambulocetus* Thewissen, Hussain & Arif, 1994:212.
      E. and/or M. Eoc.; As.[1]
    [1] S. Asia (Pakistan).

†*Pakicetus* Gingerich & D. E. Russell, 1981:238.
      E. Eoc.; As.

†*Ichthyolestes* Dehm & Oettingen-Spielberg, 1958:15.
      E. Eoc.; As.

Subfamily †**Protocetinae** Stromer, 1908:148.
    [= †Protocetidae Stromer, 1908:148; †Protocetinae Gingerich & Russell, 1990:18.]
    [Including †Indocetinae Gingerich, Raza, Arif, Anwar & Zhou, 1993:414.]
      M. Eoc.; Af.[1] M. Eoc.; As.[2] ?M. Eoc.; N.A.
    [1] N. Africa.
    [2] S. Asia.

†*Indocetus* Sahni & Mishra, 1975:1-48.
      M. Eoc.; As.[1]
    [1] S. Asia.

†*Protocetus* Fraas, 1904:5.
      M. Eoc.; Af.[1] ?M. Eoc.; N.A.
    [1] N. Africa (Egypt).

Family †**Remingtonocetidae** Kumar & Sahni, 1986:330.
    [= †Remingtonocetoidea Mitchell, 1989:2232.]
      M. Eoc.; As.[1]
    [1] S. Asia.

†*Dalanistes* Gingerich, Arif & Clyde, 1995:317.
      M. Eoc.; As.[1]
    [1] S. Asia (Pakistan).

†*Remingtonocetus* Kumar & Sahni, 1986:330.
      M. Eoc.; As.[1]
    [1] S. Asia.

†*Andrewsiphius* Sahni & Mishra, 1975:1-48.
  M. Eoc.; As.[1]
  [1] S. Asia (India).

Infraorder **AUTOCETA** Haeckel, 1866:clix.
  [= DELPHINOIDEA Huxley, 1872:336.[1]] [Including MYSTICETI Flower, 1864:388; ODONTOCETE Flower, 1864:388; ODONTOCETI Flower, 1867:110, 115.]
  [1] Used as a "group" of CETACEA, of approximately subordinal rank.

†*Iniopsis* Lydekker, 1893.
  L. Olig.; As.[1]
  [1] W. Asia.

†*Microzeuglodon* Stromer, 1903.
  L. Olig.; As.[1]
  [1] W. Asia.

†*Oligodelphis* Mchedlidze & Aslanova, 1968:181.
  L. Olig.; As.[1]
  [1] W. Asia.

†*Mirocetus* Mchedlidze, 1970.
  L. Olig.; As.[1]
  [1] W. Asia.

†*Atropatenocetus* Aslanova, 1977:61.
  L. Olig.; As.[1]
  [1] W. Asia.

†*Kharthlidelphis* Mchedlidze & Pilleri, 1988:9.
  L. Olig.; As.[1]
  [1] W. As. (Caucasus).

†*Archaeodelphis* Glover M. Allen, 1921:4.
  ?L. Olig.; N.A.

†*Xenorophus* Kellogg, 1923.
  L. Olig.; N.A.

†*Ceterhinops* Leidy, 1877.
  L. Olig. and/or Mioc.; N.A.

†*Miodelphis* L. E. Wilson, 1935:76.
  E. Mioc.; N.A.

†*Protodelphinus* Dal Piaz, 1922.
  E. Mioc.; Eu.

†*Scaptodon* Chapman, 1918.[1]
  ?Mioc.; Aus.[2]
  [1] Possibly referable to Physeteridae.
  [2] Ulverstone, Tasmania; age unknown.

†*Sinanodelphis* Makiyama, 1936.
  M. Mioc.; As.[1]
  [1] E. Asia (Japan).

†*Imerodelphis* Mchedlidze, 1959.
  M. Mioc.; As.[1]
  [1] W. Asia.

†*Stereodelphis* Gervais, 1848.[1]
  M. Mioc.; Eu.
  [1] 1848-1852.

†*Macrochirifer* Brandt, 1874.
  M. Mioc.; Eu.

†*Delphinopsis* J. Müller, 1853.
  M. Mioc.; Eu.

†*Heterodelphis* Brandt, 1873.
    M.-L. Mioc.; Eu.

†*Phocageneus* Leidy, 1869.
    M. Mioc.; Eu. M. Mioc.; N.A.

†*Hadrodelphis* Kellogg, 1966.
    M. Mioc.; Eu. M. Mioc.; N.A.

†*Nannolithax* Kellogg, 1931:386.
    M. Mioc.; Eu. M. Mioc.; N.A.

†*Platylithax* Kellogg, 1931:388.
    M. Mioc.; Eu. M. Mioc.; N.A.

†*Agabelus* Cope, 1875.
    M. Mioc.; N.A.[1]
    [1] Atlantic North America.

†*Priscodelphinus* Leidy, 1851.
    M. Mioc.; N.A.[1]
    [1] Atlantic North America.

†*Ixacanthus* Cope, 1868.
    M. Mioc.; N.A.[1]
    [1] Atlantic North America.

†*Tretosphys* Cope, 1868.
    M. Mioc.; N.A.[1]
    [1] Atlantic North America.

†*Belosphys* Cope, 1875.
    M. Mioc.; N.A.[1]
    [1] Atlantic North America.

†*Pelodelphis* Kellogg, 1955.
    M. Mioc.; N.A.[1]
    [1] Atlantic North America.

†*Araeodelphis* Kellogg, 1957.
    M. Mioc.; N.A.[1]
    [1] Atlantic North America.

†*Graphiodon* Leidy, 1870.[1]
    M. Mioc.; N.A.[2]
    [1] Possibly referable to Physeteridae.
    [2] Atlantic North America.

†*Oedolithax* Kellogg, 1931:378.
    M. Mioc.; N.A.[1]
    [1] Pacific North America (California).

†*Lamprolithax* Kellogg, 1931:381.
    ?M. and/or ?L. Mioc.; Eu. M. Mioc.; N.A.[1]
    [1] Pacific North America (California).

†*Loxolithax* Kellogg, 1931:390.
    M. Mioc.; N.A.[1]
    [1] Pacific North America (California).

†*Delphinavus* Lull, 1914.
    L. Mioc.; N.A.[1]
    [1] Pacific North America (California).

†*Hesperocetus* True, 1912.
    L. Mioc.; N.A.

†*Lonchodelphis* Allen, 1924.
    L. Mioc.; N.A.

†*Incacetus* Colbert, 1944.
    E. Plioc.; S.A.[1]
    [1] Peru.

Family †**Agorophiidae** Abel, 1913:720.
    L. Olig.; N.A.

†*Agorophius* Cope, 1895.
    L. Olig.; N.A.

Superfamily †**Squalodontoidea** Brandt, 1873:576.
    [= †Squalodontidae Brandt, 1873:576; †Squalodontoidea Simpson, 1945:100.]
    [Including †Eurhinodelphoidea de Muizon, 1988.]

†*Austrosqualodon* Climo & Baker, 1972:66.
    L. Olig.; New Zealand.

†*Tangaroasaurus* Benham, 1935:232.
    E. Mioc.; New Zealand.

Family †**Squalodontidae** Brandt, 1873:576. Shark-toothed whales.
    [= †Squalodontinae Rothausen, 1968:88.] [Including †Patriocetidae Abel, 1914:200;
    †Patriocetinae Rothausen, 1968:84, 87.]
    ?E. Olig., L. Olig.-E. Mioc.; Aus. L. Olig., Mioc.; As. L. Olig.-L. Mioc., Plioc.;
    Eu. L. Olig.-M. Mioc., ?L. Mioc.; N.A.[1] L. Olig.-E. Mioc.; New Zealand. E.-M.
    Mioc.; Mediterranean. Mioc.; S.A.
    [1] Including indet. squalodonts in the early Miocene Pyramid Hill l.f., California and the Astoria Fm., Oregon.

†*Parasqualodon* Hall, 1911:262.
    L. Olig.; Aus.

†*Metasqualodon* Hall, 1911:262.
    ?E. Olig., L. Olig.; Aus. L. Olig.; As.[1]
    [1] E. Asia (Japan).

†*Eosqualodon* Rothausen, 1968:88.
    L. Olig.; Eu.

†*Prosqualodon* Lydekker, 1894:125.
    ?L. Olig., E. Mioc.; New Zealand. Mioc.; S.A. E. Mioc.; Aus.[1]
    [1] Tasmania.

†*Patriocetus* Abel, 1913:28-34, 57-68.
    L. Olig.; Eu.

†*Agriocetus* Abel, 1913:28-34, 57-68.
    L. Olig.; Eu.

†*Saurocetus* Agassiz, 1848:4-5, 57.[1]
    L. Olig. and/or Mioc.; N.A.
    [1] Originally "*Sauro-cetus.*"

†*Colophonodon* Leidy, 1853:377.
    L. Olig. and/or Mioc.; N.A.

†*Sulakocetus* Mchedlidze, 1976:41.
    L. Olig.; Eu.[1]
    [1] S.E. Europe.

†*Squalodon* Grateloup, 1840:346.
    [Including †*Phocodon* Agassiz, 1841:236; †*Arionius* Meyer, 1841:315-331;
    †*Rhytisidon* Costa, 1852; †*Trirhizodon* Cope, 1890:603; †*Rhytisodon* Paolo,
    1897:49.]
    L. Olig.; As.[1] L. Olig.-L. Mioc., Plioc.; Eu. ?L. Olig. and/or ?E. Mioc., M. Mioc.;
    N.A. ?L. Olig.; New Zealand.[2] ?L. Olig.; Aus.[3] E.-M. Mioc.; Mediterranean.
    [1] E. Asia (Japan).
    [2] Nominal record probably not representing †*Squalodon* in the strict sense.
    [3] Nominal record probably not representing †*Squalodon* in the strict sense (see Pledge & Rothausen, 1977;
    Fordyce, 1980).

†*Kelloggia* Mchedlidze, 1976:57.
L. Olig.; As.[1]
[1] W. Asia.

†*Microcetus* Kellogg, 1923:5.
[Including †*Uncamentodon* Rothausen, 1970.[1]]
L. Olig.; As. L. Olig.; Eu. L. Olig.; N.A. L. Olig.; New Zealand.
[1] Possibly a separate genus but a *nomen nudum*.

†*Neosqualodon* Dal Piaz, 1904:1, 17.
[Including †*Microsqualodon* Abel, 1905:35.]
E.-M. Mioc.; Eu. M. Mioc.; Mediterranean.

†*Phoberodon* Cabrera, 1926:378.
Mioc.; S.A.

†*Sachalinocetus* Dubrovo, 1971:88.
Mioc.; As.[1]
[1] E. Asia (Sakhaline Island).

Family †**Rhabdosteidae** Gill, 1871:123-124, 126.
[Including †Eurhinodelphidae Abel, 1901:60; †Eurhinodelphinae Trouessart, 1905:761;
†Eurhinodelphinidae Rice, in Anderson & Jones, 1984:468.]
Mioc.; As.[1] E. Mioc., ?M. Mioc., L. Mioc.; Eu. E.-M. Mioc.; N.A. Mioc.; S.A.
E. Mioc.; New Zealand. ?M. Mioc.; Mediterranean. M. Mioc.; Aus.[2]
[1] E. Asia (Japan).
[2] Rhabdosteidae genus and sp. indet. from the Namba Fm.

†*Eurhinodelphis* Du Bus, 1867:568-569.
Mioc.; As.[1] E. Mioc., ?M. Mioc., L. Mioc.; Eu. ?M. Mioc.; Mediterranean. M.
Mioc.; N.A.
[1] E. Asia (Japan).

†*Macrodelphinus* L. E. Wilson, 1935:27.
E. Mioc.; N.A. ?M. Mioc.; Eu.

†*Ziphiodelphis* Dal Piaz, 1912:15, 18.[1]
E. Mioc., L. Mioc.; Eu.
[1] *Nomen nudum* in 1908.

†*Rhabdosteus* Cope, 1867:145.
Mioc.; Eu. M. Mioc.; N.A.

†*Argyrocetus* Lydekker, 1894:125.
[Including †*Doliodelphis* L. E. Wilson, 1935:97.]
E. Mioc.; N.A. Mioc.; S.A.

†*Phocaenopsis* Huxley, 1859:509-510.
E. Mioc.; New Zealand.

Parvorder **MYSTICETI** Cope, 1891:69, **new rank**.[1] Whalebone whales, baleen whales.
[= MYSTICETE Flower, 1864:388[1]; MYSTACOCETI Flower, 1869:110[2];
MYSTACOCETA Haeckel, 1895:566, 572; MYSTACOCETI Imamura, 1961:135;
BALAENOIDEA Flower, 1864:388[3]; ANODONTOCETE A. W. Scott, 1873:63;
MICROZOOPHAGA A. W. Scott, 1873:63; MYSTICETA Haeckel, 1895:566, 572;
MYSTICETIFORMES Kinman, 1994:38.] [Including †CRENATICETI Mitchell,
1989:2220[4]; CHAEOMYSTICETI Mitchell, 1989:2232.[4]]
[1] Proposed as suborder.
[2] As suborder.
[3] Alternative name for suborder MYSTICETE.
[4] Described as infraorder.

†*Mizuhoptera* Hatai, Hayasaka & Masuda, 1963:9.
Mioc., E. Plioc.; As.[1]
[1] E. Asia (Japan).

†*Siphonocetus* Cope, 1895:140-141.[1]
M. Mioc.; N.A.
[1] Assignment to MYSTICETI is questionable.

†*Ulias* Cope, 1895:141-143.[1]
        ?E. Plioc.; N.A.
    [1] Assignment to MYSTICETI is questionable.

†*Tretulias* Cope, 1895:143-145.[1]
        ?E. Plioc.; N.A.
    [1] Assignment to MYSTICETI is questionable.

†*Eucetites* Ameghino, 1901:80.[1]
        Mioc.; S.A.
    [1] Validity questionable. Probably best considered a *nomen nudum* or *nomen oblitum*.

Subfamily †**Kekenodontinae** Mitchell, 1989:2231.
        E. Olig.; Eu. L. Olig.; New Zealand.

†*Phococetus* Gervais, 1876:64-70.
        E. Olig.; Eu.[1]
    [1] France.

†*Kekenodon* Hector, 1881:434.
        L. Olig.; New Zealand.

Subfamily †**Llanocetidae** Mitchell, 1989:2220.
        M. Eoc.; Antarctica.

†*Llanocetus* Mitchell, 1989:2220.
        M. Eoc.; Antarctica.

Family †**Aetiocetidae** Emlong, 1966.
        L. Olig.; N.A.[1]
    [1] Pacific North America.

†*Aetiocetus* Emlong, 1966.
        L. Olig.; N.A.[1]
    [1] Pacific North America.

Family †**Mammalodontidae** Mitchell, 1989:2231.
        ?E. Olig., L. Olig.; Aus.

†*Mammalodon* Pritchard, 1939:155.
        ?E. Olig., L. Olig.; Aus.

Family †**Cetotheriidae** Brandt, 1872:116.
        [= †Cetotheriinae Brandt, 1872:116; †Cetotheriidae Miller, 1923:21, 41.]
        ?L. Olig. and/or ?E. Mioc., M. Mioc., ?L. Mioc.; As.[1] L. Olig.-L. Mioc., Plioc.; Eu.
        ?L. Olig., E. Mioc.-L. Plioc.; N.A.[2] ?L. Olig.; New Zealand. Mioc., E. Plioc.; S.A.
        ?E. Mioc.; Aus. M. Mioc.; Mediterranean.[3]
    [1] Including a cetothere from the Miocene of Sri Lanka, named †"*Mioceta*" Deraniyagala, 1967:50, an unavailable genus according to the 1961 International Code of Zoological Nomenclature, article 15.
    [2] Dwarf baleen whales, the last of the family Cetotheriidae, are said to have survived until 3 Ma. along the coast of California and Oregon.
    [3] Sardinia.

†*Cetotheriopsis* Brandt, 1871:566.
        [= †Stenodon Van Beneden, 1865:75-79[1]; †Aulocetus Van Beneden, 1875.]
        L. Olig.-L. Mioc., ?E. Plioc.; Eu. ?L. Olig.; N.A.[2] M. Mioc.; Mediterranean.[3]
    [1] Not *Stenodon* Rafinesque, 1818, a genus of mollusks; nor †*Steneodon* Croizet, 1833, a genus of felids (synonym of †*Megantereon*).
    [2] Atlantic North America (coastal South Carolina).
    [3] Sardinia.

†*Mauicetus* Benham, 1939.
        ?L. Olig.; As. ?L. Olig.; New Zealand.

†*Cophocetus* Packard & Kellogg, 1934.
        ?E. Mioc., M. Mioc.; N.A.

†*Cetotherium* Brandt, 1843.
        E.-M. Mioc., ?E. Plioc.; N.A. M. Mioc.; As.[1] M. Mioc.; Eu.
    [1] W. Asia.

†*Pelocetus* Kellogg, 1965.
    Mioc.; Eu. M. Mioc.; N.A.

†*Aglaocetus* Kellogg, 1934.
    Mioc.; S.A. ?E. Mioc.; Aus. M. Mioc.; N.A.

†*Diorocetus* Kellogg, 1968.
    M. Mioc.; N.A.

†*Pinocetus* Czyzewska & Ryziewicz, 1976:261.
    M. Mioc.; Eu.[1]
        [1] E. Europe.

†*Tiphyocetus* Kellogg, 1931:317.
    M. Mioc.; N.A.

†*Metopocetus* Cope, 1896:141-143.
    M. Mioc.; N.A. L. Mioc.; Eu.

†*Imerocetus* Mchedlidze, 1964.
    M. Mioc.; As.[1]
        [1] W. Asia.

†*Mesocetus* Van Beneden, 1880.
    M.-L. Mioc., Plioc.; Eu. ?M. and/or ?L. Mioc.; N.A.

†*Herpetocetus* Van Beneden, 1872:20.
    [= †*Erpetocetus* Van Beneden, 1880:25.]
    M. and/or L. Mioc.; Eu. ?L. Mioc. and/or ?Plioc.; N.A.

†*Isocetus* Van Beneden, 1880:24, 25.
    M. Mioc., ?L. Mioc.; Eu. M.-L. Mioc.; N.A.

†*Peripolocetus* Kellogg, 1931:338.
    M. Mioc.; N.A.

†*Halicetus* Kellogg, 1969.
    M. Mioc.; N.A.

†*Eucetotherium* Brandt, 1873.
    M. Mioc.; Eu.

†*Cetotheriomorphus* Brandt, 1873:161, 162.
    ?M. Mioc.; Eu.

†*Parietobalaena* Kellogg, 1924.
    M. Mioc.; N.A.

†*Thinocetus* Kellogg, 1969.
    M. Mioc.; N.A.

†*Otradnocetus* Mchedlidze, 1984.
    M. Mioc.; Eu.[1]
        [1] S.E. Europe (N. Caucasus).

†*Heterocetus* Van Beneden, 1880.
    L. Mioc.; Eu.

†*Nannocetus* Kellogg, 1929:450.
    L. Mioc.; N.A.

†*Mixocetus* Kellogg, 1934.
    L. Mioc.; N.A.

†*Amphicetus* Van Beneden, 1880.
    L. Mioc.; Eu.

†*Cephalotropis* Cope, 1896:880.
    L. Mioc.; Eu. E. Plioc.; N.A.

†*Piscocetus* G. Pilleri & O. Pilleri, in G. Pilleri, ed., 1989:14.
    E. Plioc.; S.A.[1]
        [1] Peru.

†*Rhegnopsis* Cope, 1896:145.[1]
    [= †*Protobalaena* Leidy, 1869:440-441.[2]]
    E. Plioc.; N.A.
[1] Replacement name for †*Protobalaena* Leidy, 1869.
[2] Not †*Protobalaena* Du Bus, 1867, a balaenid whale, nor "†*Protobalaena*" Haeckel, 1895, a hypothetical concept.

†*Piscobalaena* Pilleri & Siber, 1989:112.
    E. Plioc.; S.A.[1]
[1] Peru.

Family **Balaenopteridae** Gray, 1964:203.
    [= Balaenopteroidea Gray, 1868.] [Including †Palaeocetidae Gray, 1866; Megapteridae Gray, 1868:2; Agaphelidae Gray, 1870:391; Rachianectidae Weber, 1904:575; Eschrichtiidae Ellerman & Morrison-Scott, 1951:713; Eschrichtioidea Mitchell, 1989:2232.]
    Mioc., E.-L. Plioc., Pleist.; As. ?E. Mioc., M. Mioc.-L. Plioc., ?L. Pleist.; Eu. Mioc. and/or E. Plioc., Pleist.; S.A. M. Mioc.-R.; N.A. ?L. Mioc. and/or ?Plioc.; Aus. ?E. Plioc.; New Zealand. R.; Indian O. R.; Mediterranean. R.; Atlantic. R.; Arctic O. R.; Pacific.

Subfamily **Balaenopterinae** Gray, 1864:203.
    [= Balaenopteridae Gray, 1864:203; Balaenopterinae Brandt, 1872.] [Including Physalina Gray, 1864:211[1]; Megapterina Gray, 1864:205; Megapterinae Flower, 1864:391; Megapteridae Gray, 1868:2; †Palaeocetidae Gray, 1866.]
    Mioc., E.-L. Plioc., Pleist.; As. ?E. Mioc., M. Mioc.-L. Plioc., ?L. Pleist.; Eu. Mioc. and/or E. Plioc., Pleist.; S.A. M. Mioc.-L. Pleist.; N.A. ?L. Mioc. and/or ?Plioc.; Aus.[2] ?E. Plioc.; New Zealand.[3] R.; Indian O. R.; Mediterranean. R.; Atlantic. R.; Arctic O. R.; Pacific.
[1] Based on *Physalus* Gray, 1821, not Lacépède, 1804.
[2] Earbones possibly from balaenopterines have been collected from Flinders Island and from Beaumaris and Grange Burn, near Hamilton Victoria, Australia.
[3] A possible balaenopterine is known from the Opoitian near Taihape, New Zealand.

Balaenoptera Lacépède, 1804:114. Rorquals, finback whales, fin whale, blue whale, minke whale.
    [= *Catoptera* Rafinesque, 1815:61[1]; *Cetoptera* Rafinesque, 1815:219; *Physalus* Gray, 1821:310[2]; *Benedenia* Gray, 1864:211; *Swinhoia* Gray, 1866:382.] [Including *Boops* Gray, 1821:310; *Rorqual* G. Cuvier, 1829:298; *Rorqualus* F. Cuvier, 1836:303; *Mysticetus* Wagler, 1830:33; *Pterobalaena* Eschricht, 1849:108; *Sibbaldius* Flower, 1864:39; *Sibbaldus* Gray, 1864:222-223[3]; *Cuvierius* Gray, 1866:114; *Rudolphius* Gray, 1866:170[4]; *Fabricia* Gray, 1866:382[5]; *Flowerius* Lilljeborg, 1867:11; *Cetotheriophanes* Brandt, 1873:148-159.]
    Mioc., E.-L. Plioc., ?L. Pleist.; Eu. M. Mioc., E.-L. Plioc., Pleist.; As.[6] M. Mioc.-L. Pleist.; N.A. L. Mioc., ?E. Plioc., Pleist.; S.A. R.; Indian O. R.; Mediterranean. R.; Atlantic. R.; Arctic O. R.; Pacific.
[1] Replacement name.
[2] Not *Physalus* Lacépède, 1804.
[3] Sometimes as subgenus.
[4] Proposed as a subgenus of *Sibbaldus*.
[5] Described as subgenus of *Balaenoptera*. Not *Fabricia* de Blainville, 1828, a genus of "Vermes."
[6] Japan in the M. Mioc.

†*Plesiocetus* Van Beneden, 1859:139-141.
    [= †*Plesiocetopsis* Brandt, 1873:143-148.]
    ?E. Mioc., M.-L. Mioc., Plioc.; Eu. Mioc. and/or E. Plioc.; S.A. ?M. and/or L. Mioc. and/or ?Plioc.; N.A.

†*Idiocetus* Capellini, 1876:12, 13.
    Mioc.; As.[1] M.-L. Mioc., L. Plioc.; Eu.
[1] E. Asia (Japan).

Megaptera Gray, 1846:16. Humpback whale.
    [= *Kyphobalaena* Eschricht, 1849:108; *Cyphobalaena* Marschall, 1873:5; *Poescopia* Gray, 1864:207.[1]]

?L. Mioc., Pleist.; N.A. Plioc.; Eu. ?E. Plioc., Pleist.; S.A. Pleist.; As.[2] R.; Indian O. R.; Mediterranean. R.; Atlantic. R.; Arctic O. R.; Pacific.

[1] Described as subgenus of *Megaptera*; raised to generic rank by Gray, 1866.
[2] E. Asia (Japan).

†*Megapteropsis* Van Beneden, 1872.
Plioc.; Eu.

†*Burtinopsis* Van Beneden, 1872:19-20.
E. Plioc.; Eu.

†*Palaeocetus* Seeley, 1865:54-57.
Plioc.; Eu.

†*Notiocetus* Ameghino, 1891:167.
L. Pleist.; S.A.

Subfamily **Eschrichtiinae** Ellerman & Morrison-Scott, 1951:713, **new rank**.
[= Eschrichtiidae Ellerman & Morrison-Scott, 1951:713; Eschrichtioidea Mitchell, 1989:2232; Agaphelidae Gray, 1870:391; Rhachianectidae Weber, 1904:575.]
L. Pleist.-R.; N.A. R.; Atlantic.[1] R.; Pacific.[2]

[1] Formerly N. Atlantic.
[2] N. Pacific.

Eschrichtius Gray, 1864:350.[1] Gray whale.
[= *Cyphonotus* Rafinesque, 1815:61; *Agaphelus* Cope, 1868:159; *Rhachianectes* Cope, 1869:15.]
L. Pleist.-R.; N.A. R.; Atlantic.[2] R.; Pacific.[3]

[1] Described as subgenus of *Megaptera*; raised to generic rank by Gray, 1865.
[2] Formerly N. Atlantic.
[3] N. Pacific.

Family **Balaenidae** Gray, 1821:310.
[= Balanadae Gray, 1821:310; Balaenidae Gray, 1825:340; Balaenina Gray, 1825:340[1]; Balaenodea Giebel, 1855:76; Balaenida Haeckel, 1866:clix; Balaenoidea Gray, 1868.]
[Including Eubalaenida Haeckel, 1895:566; †Protobalaenida Haeckel, 1895; Neobalaenidae Miller, 1923:21.]
Mioc., Pleist.; S.A. M. Mioc.-E. Plioc., Pleist.; N.A.[2] M. and/or L. Mioc., ?E. Pleist.; New Zealand.[3,4] L. Mioc.-L. Plioc., Pleist.; Eu. L. Mioc., ?E. Plioc.; Aus.[5,6] ?Pleist.; As. R.; Indian O. R.; Atlantic. R.; Arctic O. R.; Pacific.

[1] As tribe.
[2] The left dentary of a Balaenidae, genus and sp. indet., is known from the Barstovian Sharktooth Hill l.f., California.
[3] Balaenidae genus and sp. indet. from the middle or late Miocene of Gore Bay.
[4] Possible Balaenidae genus and sp. indet. from the early Pleistocene (Nukumaruan).
[5] Undescribed belaenid periotics are known from the late Miocene (Cheltenhamian) of Beaumaris.
[6] Possible undescribed balaenid periotics are known from the early Pliocene Hamilton fauna.

†*Morenocetus* Cabrera, 1926:364.
Mioc.; S.A.

†*Balaenula* Van Beneden, 1872.
[Including †*Balaenotus* Van Beneden, 1872.]
L. Mioc., Plioc.; Eu. L. Mioc.-E. Plioc.; N.A.

†*Protobalaena* Du Bus, 1867.[1]
[= †*Probalaena* Van Beneden, 1872.[1]]
?L. Mioc.; Eu.

[1] Probably *nomen nudum*.

†*Mesoteras* Cope, 1870.
E. Plioc.; N.A.

Balaena Linnaeus, 1758:75-76. Right whales, bowhead.
[= *Leiobalaena* Eschricht, 1849:108.] [Including *Eubalaena* Gray, 1864:201[1]; *Hunterus* Gray, 1864:349; *Hunterius* Gray, 1866:78, 98-100; *Macleayius* Gray, 1864:589; *Halibalaena* Gray, 1873:140.]

E.-L. Plioc., Pleist.; Eu. E. Plioc., Pleist.; N.A. ?Pleist.; As. Pleist.; S.A. R.; Indian O. R.; Atlantic. R.; Arctic O. R.; Pacific.

[1] Sometimes as subgenus; sometimes as separate genus.

Caperea Gray, 1864. Pygmy right whale.
 [= *Neobalaena* Gray, 1870.]
 R.; Indian O. R.; Atlantic.[1] R.; Pacific.[2]
[1] S. Atlantic.
[2] S. Pacific.

Parvorder **ODONTOCETI** Flower, 1869:111, **new rank.**[1] Toothed whales.
 [= ODONTOCETE Flower, 1864:388[2]; DELPHINOIDEA Flower, 1864:388[3];
 DENTICETA Haeckel, 1895:566, 569; ODONTOCETIFORMES Kinman, 1994:38.]
 [Including PHYSETERIDA Haeckel, 1866:clix.]
[1] Proposed as suborder. Simpson (1945:100) gave the date as 1867, but the original publication is dated 1869.
[2] Proposed as suborder.
[3] As alternative name for the suborder.

†*Pachyacanthus* Brandt, 1871.
 M. Mioc.; Eu.

Superfamily **Physeteroidea** Gray, 1821:310.
 [= Physeteridae Gray, 1821:310; Physeteroidea Gill, 1872:15.]

Family **Physeteridae** Gray, 1821:310. Sperm whales.
 [= Physeterina Gray, 1825:340; Physeteroidea Gill, 1872:15; Catodontidae F. Cuvier, 1836:564.] [Including Physodontidae Lydekker, 1894:4.]
 E. Mioc.; As.[1] E. Mioc.-L. Plioc., Pleist.; Eu. Mioc., E. Plioc.; S.A. E. and/or M. Mioc., L. Mioc., ?E. Plioc.; Aus.[2,3] M. Mioc.-E. Plioc., ?L. Plioc., Pleist.; N.A. ?L. Plioc.; New Zealand.[4] R.; Indian O. R.; Mediterranean. R.; Atlantic. R.; Pacific.
[1] Japan.
[2] Undescribed physeterid material from the probable early or middle Miocene (Batesfordian or Bairnsdalian) of Fyansford.
[3] Undescribed physeterid material from the probable early Pliocene (Kalimnan) of Hamilton.
[4] Physeteridae genus and sp. indet. from the ?Mangapanian of Waipukuran.

†*Helvicetus* Pilleri, 1986:25.
 E. Mioc.; Eu.

Subfamily †**Hoplocetinae** Cabrera, 1926:408.
 E. Mioc.; As.[1] E. Mioc.-E. Plioc., Pleist.; Eu. Mioc.; S.A. M.-L. Mioc., Plioc., ?Pleist.; N.A. L. Mioc.; Aus.
[1] Japan.

†*Scaldicetus* Du Bus, 1867:567-568.
 [= †*Eucetus* Du Bus, 1867:571-572; †*Homoeocetus* Du Bus, 1867:572-573; †*Palaeodelphis* Du Bus, 1872:503-508; †*Dinoziphius* Van Beneden, 1880:344-345; †*Homocetus* Lydekker, 1887:14; †*Palacodelphis* Trouessart, 1898:1053[1]; †*Homoecetus* Van Beneden, 1977:855.] [Including †*Physodon* Gervais, 1872:101.[2]]
 E. Mioc.-E. Plioc.; Eu. M.-L. Mioc., Plioc., ?Pleist.; N.A. L. Mioc.; Aus.
[1] Misprint.
[2] Not †*Physodon* Haldeman, 1843.

†*Idiorophus* Kellogg, 1925:6.
 [= †*Apenophyseter* Cabrera, 1926:406.]
 Mioc.; S.A.

†*Diaphorocetus* Ameghino, 1894:181.
 [= †*Mesocetus* Moreno, 1892:395-397[1]; †*Hypocetus* Lydekker, 1894:7; †*Paracetus* Lydekker, 1894:8.]
 Mioc.; S.A.
[1] Not †*Mesocetus* Van Beneden, 1880, a cetotheriid.

†*Aulophyseter* Kellogg, 1927:4.
　　E. Mioc.; As.[1] M. Mioc.; N.A. L. Mioc.; S.A.
　　[1] Japan.

†*Hoplocetus* Gervais, 1848:161.[1]
　　M. Mioc.-E. Plioc.; Eu.
　　[1] 1848-1852.

†*Paleophoca* Van Beneden, 1859:255-258.
　　E. Plioc., Pleist.; Eu.

Subfamily **Physeterinae** Gray, 1821:310.
　　[= Physeteridae Gray, 1821:310; Physeterina Gray, 1825:340[1]; Physeterinae Flower,
　　1867:114.]
　　E. Mioc.-L. Plioc.; Eu. M. Mioc., ?L. Mioc. and/or ?Plioc., Pleist.; N.A. L. Mioc.;
　　Aus. R.; Indian O. R.; Mediterranean. R.; Atlantic. R.; Pacific.
　　[1] As tribe.

†*Placoziphius* Van Beneden, 1869:11-12.[1]
　　E. and/or M. Mioc.; Eu.
　　[1] *Nomen nudum* in 1864 and 1866.

Physeter Linnaeus, 1758:76. Sperm whale, cachalot.
　　[= *Catodon* Linnaeus, 1761:18; *Physalus* Lacépède, 1804:xl, 219; *Physeterus* Duméril,
　　1806:28; *Tursio* Fleming, 1822:211; *Megistosaurus* Harlan, 1828:186-187[1];
　　Physeteres F. Cuvier, 1829:518; *Meganeuron* Gray, 1865:440.[2]] [= or including
　　Physelus Rafinesque, 1815:60.]
　　E. Mioc.-L. Plioc.; Eu. ?Mioc. and/or ?Plioc., Pleist.; N.A. R.; Indian O. R.;
　　Mediterranean. R.; Atlantic. R.; Pacific.
　　[1] "Remains of the largest saurian fossil ever heard of" (Godman, in Harland, 1828).
　　[2] Described as subgenus of *Catodon*; raised to generic rank by Gray, 1866:387-389.

†*Orycterocetus* Leidy, 1853:378.
　　[= or including †*Gargantuodon* Ginsburg, 1969:995.[1]]
　　E.-M. Mioc.; Eu. M. Mioc., ?L. Mioc.; N.A.
　　[1] Thought originally to be the earliest known walrus.

†*Physeterula* Van Beneden, 1877:851-856.
　　Mioc., ?L. Plioc.; Eu.

†*Prophyseter* Abel, 1905:82.
　　M. and/or L. Mioc.; Eu.

†*Thalassocetus* Abel, 1905:70.
　　M. and/or L. Mioc.; Eu.

†*Idiophyseter* Kellogg, 1925:18.
　　M. Mioc.; N.A.

†*Physetodon* McCoy, 1879:vi, 19-20.
　　L. Mioc.; Aus.

†*Balaenodon* Owen, 1846:536-542.
　　?L. Mioc.; N.A. Plioc.; Eu.

†*Priscophyseter* Portis, 1886:315-321.
　　L. Plioc.; Eu.

Subfamily **Kogiinae** Gill, 1871:732.
　　[= Kogiidae Miller, 1923:33, 45.]
　　E.-M. Mioc., L. Plioc.; Eu. L. Mioc.-E. Plioc.; N.A.[1] L. Mioc.-E. Plioc.; S.A.[2]
　　R.; Indian O. R.; Mediterranean. R.; Atlantic. R.; Pacific.
　　[1] Periodics of a new kogiid taxon have been reported from the early Pliocene Yorktown Fm., North
　　Carolina.
　　[2] Peru. Periotics of a new kogiid taxon have been reported from the early Pliocene Sacaco level, Pisco Fm.

†*Miokogia* Pilleri, 1986:155, 157.
　　E.-M. Mioc.; Eu.

†*Kogiopsis* Kellogg, 1929:2.
  L. Mioc.; N.A.

†*Praekogia* Barnes, 1973:1, 4.
  L. Mioc.; N.A.

†*Scaphokogia* de Muizon, 1988.
  L. Mioc.; S.A.[1]
  [1] Peru.

Kogia Gray, 1846:22. Pygmy sperm whale.
  [= *Euphysetes* Wall, 1851:37; *Euphycetes* Gray, 1866:391; *Callignathus* Gill, 1871:737-738[1]; *Cogia* Wallace, 1876:208; *Callignathula* Strand, 1928:61.[2]]
  L. Plioc.; Eu. R.; Indian O. R.; Mediterranean. R.; Atlantic. R.; Pacific.
  [1] Not *Callignathus* Agassiz, 1846, a genus of beetles, nor *Calignathus* Costa, 1853, a genus of fishes.
  [2] Strand's paper was published in 1928, not 1926.

Superfamily **Hyperoodontoidea** Gray, 1846:24.
  [= Hyperoodontina Gray, 1846:24; Hyperoodontoidea Moore, 1968:276.] [Including Ziphioidea Gray, 1868:9.]

Family **Hyperoodontidae** Gray, 1846:24. Beaked whales.
  [= Hyperoodontina Gray, 1846:24; Hyperodontidae Gray, 1866:62; Hyperoodontidae Gray, 1866:327; Hyperoodonta Haeckel, 1866:clix; Hyperoodontini Moore, 1968:276; Anarnacinae Gill, 1871:124.[1]] [Including Ziphiina Gray, 1850:59, 61[2]; Ziphiidae Gray, 1865:528; Ziphiinae Fraser & Purves, 1960:fig. 16; Ziphini Moore, 1968:276; Squalodelphidae Dal Piaz, 1916:32; Berardiina Moore, 1968:276; Tasmacetina Moore, 1968:276; Indopacetina Moore, 1968:277; Squalodelphinae Ginsburg & Janvier, 1971:170; †Squaloziphiinae de Muizon, 1991:282.]
  L. Olig. and/or E. Mioc., Plioc.; Aus.[3] E. Mioc.; Af.[4] Mioc.; As.[5] E.-L. Mioc., ?E. Plioc., L. Plioc.; Eu. E. Mioc., ?M. Mioc., L. Mioc.-E. Plioc.; N.A.[6,7] Mioc., E. Plioc.; S.A. M. and/or L. Mioc.; New Zealand.[8] L. Mioc., R.; Mediterranean. R.; Indian O. R.; Atlantic. R.; Arctic O. R.; Pacific.
  [1] Synonymy questionable.
  [2] Fide Palmer (1904:776).
  [3] Nominal record from the late Oligocene or early Miocene.
  [4] Indet. ziphiid from the upper Turkana Grit, Kenya.
  [5] E. Asia (Japan).
  [6] ?Ziphiid genus indet. from the middle Miocene Sharktooth Hill l.f., California.
  [7] Ziphiid genus indet. from the late Miocene upper Monterey Fm., California.
  [8] Ziphiid mandible, genus indet.

†*Notocetus* Moreno, 1892:17.[1]
  [= †*Diochotichus* Ameghino, 1894:182; †*Argyrodelphis* Lydekker, 1894:12-13.]
  Mioc.; S.A.
  [1] †*Notocetus* Moreno, 1892, is not a homonym of †*Notiocetus* Ameghino, 1891.

†*Medocinia* de Muizon, 1988:82.
  E. Mioc.; Eu.

†*Belemnoziphius* Huxley, 1864:392-395.
  ?Mioc.; N.A. M. and/or L. Mioc. and/or Plioc.; Eu.

†*Pelycorhamphus* Cope, 1895:137-139.
  Mioc.; N.A.

†*Squalodelphis* Dal Piaz, 1916:5.
  E. Mioc., ?M. Mioc.; Eu.

†*Choneziphius* Duvernoy, 1851:43, 61-63.
  Mioc. and/or Plioc.; Eu. Mioc.; N.A.

Mesoplodon Gervais, 1850:16. Cow fish.
  [= *Aodon* Lesson, 1828:155[1]; *Anodon* Gray, 1850:71; *Nodus* Wagler, 1830:34[2]; *Micropterus* Wagner, 1846:281[3]; *Micropteron* Eschricht, 1849:97[4]; *Mesodiodon* Duvernoy, 1851:41.] [Including *Dioplodon* Gervais, 1850:512[5]; *Dolichodon* Gray, 1866:353[5]; *Neoziphius* Gray, 1871:101; *Callidon* Gray, 1871:368[6]; *Oulodon* von Haast, 1876:457; *Paikea* Oliver, 1922:574.]

Mioc.; As.[7] Mioc. and/or Plioc.; Eu. ?Mioc., E. Plioc.; N.A. L. Mioc.; Mediterranean. Plioc.; Aus. R.; Indian O. R.; Atlantic. R.; Pacific.

[1] Not *Aodon* Lacépède, 1798, a genus of fishes.
[2] *Nomen oblitum.*
[3] Not *Micropterus* Lacépède, 1802, a genus of fishes.
[4] Incorrect subsequent spelling of *Micropterus* Wagner, 1846.
[5] As subgenus.
[6] Objective synonym of *Dolichodon.*
[7] Japan.

†*Squaloziphius* de Muizon, 1991:282.
E. Mioc.; N.A.

†*Cetorhynchus* Gervais, 1861:122-124.
M.-L. Mioc.; Eu.

†*Ziphirostrum* Du Bus, 1868:622-625.[1]
[= †*Ziphirostris* Van Beneden, 1864:396[2]; †*Aporotus* Du Bus, 1868:626-627; †*Mioziphius* Abel, 1905:102.]
M. and/or L. Mioc.; Eu.

[1] Replacement name for "†*Ziphirostris.*"
[2] *Nomen nudum* in 1864 and again in 1868.

†*Messapicetus* Bianucci, Landini & Varola, 1992:261.
L. Mioc.; Eu.[1]

[1] S. Italy.

†*Palaeoziphius* Abel, 1905:90.
M. Mioc.; Eu.

†*Ziphioides* Probst, 1886:109-116.
M. Mioc.; Eu.

†*Anoplonassa* Cope, 1869:188-190.
L. Mioc. and/or E. Plioc.; N.A.

†*Proroziphius* Leidy, 1876:86-87.
[= †*Prozoziphius* Alston, 1879:15.[1]]
L. Mioc. and/or E. Plioc.; N.A.

[1] Misspelling.

†*Eboroziphius* Leidy, 1876:81.
L. Mioc. and/or E. Plioc.; N.A.

†*Ninoziphius* de Muizon, 1983:1.
E. Plioc.; N.A. E. Plioc.; S.A.[1]

[1] Peru.

Hyperoodon Lacépède, 1804:xliv, 319-24. Bottle-nosed whales.
[= *Anarnak* Lacépède, 1804:164[1]; *Anarnacus* Dumeril, 1806:28[1]; *Ancylodon* Illiger, 1811:142; *Uranodon* Illiger, 1811:143; *Bidens* Fischer de Waldheim, 1814:686; *Heterodon* de Blainville, 1817:151, 175-179; *Cetodiodon* Jacob, 1825:72; *Orca* Wagler, 1830:34; *Uperodon* Gray, 1843:xxiii; *Chaenodelphinus* Eschricht, 1843:651-655; *Chaenocetus* Eschricht, 1846:17; *Chenocetus* Gray, 1866:328-329; *Hyperhoodon* Gervais, 1850:6-13; *Hyperodon* Gray, 1863:200; *Hyperaodon* Cope, 1869:31; *Hyperoodus* Schulze, 1897:6; *Lagenocetus* Gray, 1863:200; *Lagocetus* Gray, 1866:82.]
[Including *Frasercetus* Moore, 1968:274.[2]]
?Plioc.; Eu. R.; Indian O. R.; Mediterranean. R.; Atlantic. R.; Arctic O. R.; Pacific.

[1] Synonymy questionable.
[2] As subgenus.

Ziphius G. Cuvier, 1823:352. Two-toothed whale, Cuvier's beaked whale, goose-beaked whale.
[= *Diodon* Lesson, 1828:124, 440[1]; *Hypodon* Haldeman, 1841:127[2]; *Xiphias* Murchison, 1843:560[3]; *Xiphius* Agassiz, 1846:389; *Aliama* Gray, 1864:242; *Petrorhynchus* Gray, 1865:524; *Ziphiorrhynchus* Burmeister, 1865[4]; *Ziphiorhynchus* Van Beneden, 1868:96.]

E. Plioc.; N.A. Plioc.; Aus. R.; Indian O. R.; Mediterranean. R.; Atlantic. R.; Pacific.

[1] Not *Diodon* Linnaeus, 1758, a genus of fishes, nor *Diodon* Storr, 1780, a synonym of *Monodon*.
[2] Replacement name for *Diodon* Lesson, 1828.
[3] Not *Xiphias* Linnaeus, 1758, a genus of fishes.
[4] Or 1866:94.

†*Berardiopsis* Portis, 1886:326-329.
    L. Plioc.; Eu.

Berardius Duvernoy, 1851:41. Fourtooth whales.
    [Including *Rostrifer* Zenkovicz, 1947:15.]
    R.; Indian O.[1] R.; Atlantic.[2] R.; Pacific.

[1] S. Indian Ocean.
[2] S. Atlantic.

Tasmacetus Oliver, 1937:372. Tasman beaked whale, Shepherd's beaked whale.
    R.; Atlantic.[1] R.; Pacific.[2]

[1] S. Atlantic.
[2] S. Pacific.

Indopacetus Moore, 1968:254. Indo-Pacific whale.
    R.; Indian O. R.; Pacific.[1]

[1] S. Pacific.

Superfamily **Platanistoidea** Gray, 1846:45. Fresh-water dolphins.
    [= Platanistina Gray, 1846:45; Platanistoidea Simpson, 1945:100; Susuoidea Hershkovitz, 1961:555.]

Family **Platanistidae** Gray, 1846:45.
    [= Platanistina Gray, 1846:45; Platanistinae Flower, 1867:114; Platanistidae Gray, 1868:199; Susuidae Hershkovitz, 1961:555.]
    E.-L. Mioc., L. Plioc.; N.A.[1,2] R.; As.[3]

[1] Platanistids, genus indet., have been reported from the late Miocene (Clarendonian) Santa Margarita Fm., California.
[2] Platanistids, genus indet., have been reported from the late Pliocene (Blancan) San Diego Fm., California.
[3] S. Asia.

†*Allodelphis* L. E. Wilson, 1935:13.
    E. Mioc.; N.A.

†*Zarhachis* Cope, 1868:186, 189.
    [= †*Zarachis* Van Beneden & Gervais, 1880:512.]
    M. Mioc.; N.A.

Platanista Wagler, 1830:35. Susu, Indian river dolphins, Ganges River dolphin.
    [= *Susu* Lesson, 1828:212, 440.[1]]
    R.; As.[2]

[1] *Nomen oblitum.*
[2] S. Asia.

Superfamily **Delphinoidea** Gray, 1821:310.
    [= Delphinidae Gray, 1821:310; Delphinoidea Flower, 1864:389; DELPHINOIDEA Huxley, 1872:336.[1]] [Including Monodontoidea Fraser & Purves, 1960:1.]

[1] Used as a "group" of CETACEA, at approximately subordinal rank.

†*Hesperoinia* Zei, 1956.
    L. Mioc.; Eu.

†*Palaeophocaena* Abel, 1905:376.
    M. Mioc.; Eu.[1]

[1] E. Europe.

†*Protophocaena* Abel, 1905.
    M. and/or L. Mioc.; Eu.

†*Prionodelphis* Frenguelli, 1922.[1]
    L. Mioc.; S.A.

[1] Based on a single tooth and perhaps therefore should be considered a *nomen dubium* (Simpson, 1945:27; de Muizon & Hendey, 1980:93).

Family **Delphinidae** Gray, 1821:310. Dolphins.
[= Delphinusidae Lesson, 1842:197; Delphinodea Giebel, 1855:86; Delphinida Haeckel, 1866:CLX.[1]] [Including Orcini Wagner, 1846:292; Orcadae Gray, 1871:85; Globiocephalidae Gray, 1850:313; Grampidae Gray, 1871:82.]
Olig., ?E. Mioc., M. Mioc., ?L. Mioc. and/or ?E. Plioc., L. Plioc., Pleist., R.; Eu.[2] Mioc., ?Plioc., E. Pleist., ?M. and/or ?L. Pleist., R.; As.[3] L. Mioc.-E. Plioc., ?L. Plioc., Pleist.; N.A. ?L. Mioc.; Aus.[4] E. Plioc.; Af. E. Plioc., L. Pleist.-R.; S.A.[5] E. Plioc.; Antarctica.[6] L. Plioc., ?E. Pleist.; New Zealand. R.; Indian O. R.; Mediterranean. R.; Atlantic. R.; Arctic O. R.; Pacific.
[1] As a family.
[2] Black Sea in the Recent.
[3] Mid-Holocene of China.
[4] "*Steno*" *cudmorei* Chapman, 1917, from Beaumaris.
[5] Amazon Basin in the Recent.
[6] Undescribed species of Delphinidae from Marine Plain, Vestfold Hills.

†*Anacharsis* Bogachev, 1956.[1]
M. Mioc.; Eu.[2]
[1] Assignment to the Delphinidae is tentative.
[2] E. Europe.

Subfamily **Orcininae** Wagner, 1846:292.
[= Orcini Wagner, 1846:292; Orcadina Gray, 1850:278; Orcadae Gray, 1871:85; Orcinae Slijper, 1936:556; Orcininae Kasuya, 1973:40.] [Including Pseudorcaina Gray, 1871:79.]
Olig., M. Mioc., L. Plioc., Pleist.; Eu. E. Plioc.; N.A. ?L. Plioc., ?E. Pleist.; New Zealand. E. Pleist., ?M. and/or ?L. Pleist.; As. L. Pleist.; S.A. R.; Indian O. R.; Mediterranean. R.; Atlantic. R.; Arctic O. R.; Pacific.

Orcinus Fitzinger, 1860:204. Killer whale.
[= *Orca* Gray, 1846:33[1]; *Ophysia* Gray, 1868:8[2]; *Gladiator* Gray, 1870:71.[2]]
Olig., M. Mioc., L. Plioc.; Eu. E. Pleist.; As.[3] ?E. Pleist.; New Zealand. L. Pleist.; S.A. R.; Indian O. R.; Mediterranean. R.; Atlantic. R.; Arctic O. R.; Pacific.
[1] Not *Orca* Wagler, 1830, a synonym of *Hyperoodon*.
[2] Described as subgenus of *Orca*.
[3] Japan.

Pseudorca Reinhardt, 1862:151. False killer whale.
[= *Neorca* Gray, 1871.]
E. Plioc.; N.A. ?L. Plioc.; New Zealand. Pleist.; As.[1] Pleist.; Eu. R.; Indian O. R.; Mediterranean. R.; Atlantic. R.; Pacific.
[1] E. Asia (Japan).

Subfamily **Delphininae** Gray, 1821:310.
[= Delphinidae Gray, 1821:310; Delphinina Gray, 1825:340[1]; Delphininae Flower, 1867:115; Delphinini Pavlinov & Rossolimo, 1987:92.] [Including Lagenorhynchi Wagner, 1846:317; Lagenorhynchina Gray, 1868:7; Stenonina Gray, 1868:5; Stenidae Fraser & Purves, 1960:59; Steninae Mead, 1975; Lissodelphinae Fraser & Purves, 1960:108.]
Mioc., ?Plioc., R.; As.[2] ?E. Mioc., M. Mioc., ?L. Mioc. and/or ?E. Plioc., L. Plioc., Pleist., R.; Eu.[3] L. Mioc., ?Plioc., Pleist.; N.A. E. Plioc.; Af. E. Plioc., L. Pleist.; S.A. L. Plioc.; New Zealand. R.; Indian O. R.; Mediterranean. R.; Atlantic. R.; Pacific.
[1] As tribe.
[2] Mid-Holocene of China.
[3] Black Sea in the Recent.

Delphinus Linnaeus, 1758:77. Common dolphin, cape dolphin.
[= *Rhinodelphis* Wagner, 1846:281, 316[1]; *Delphis* Gray, 1864:236[2]; *Eudelphinus* Van Beneden & Gervais, 1880:600; *Mamdelphinus* Herrera, 1899:27.[3]]
Mioc., ?Plioc.; As. ?Mioc., Plioc., R.; Eu.[4] ?E. Plioc.; N.A. L. Plioc.; New Zealand. R.; Indian O. R.; Mediterranean. R.; Atlantic. R.; Pacific.
[1] Proposed as subgenus.
[2] Proposed as subgenus of *Delphinus*. Not *Delphis* Wagler, 1830, a synonym of *Delphinapterus*.

[3] See the International Code of Zoological Nomenclature, Article 1(b)(8), and Palmer (1904:25-26).
[4] Black Sea in the Recent.

Tursiops Gervais, 1855:323. Bottle-nosed dolphins.
    [= *Tursio* Gray, 1843:xxiii, 105.[1]]
    M. Mioc., L. Plioc., ?Pleist., R.; Eu.[2] ?L. Mioc. and/or ?Plioc., Pleist.; N.A. E.
    Plioc., L. Pleist.; S.A. ?L. Plioc.; New Zealand. R.; Indian O. R.; As.[3] R.;
    Mediterranean. R.; Atlantic. R.; Pacific.
    [1] Not *Tursio* Fleming, 1822, a synonym of *Physeter*, nor *Tursio* Wagler, 1830, a synonym of *Lissodelphis*.
    [2] Black Sea in the Recent.
    [3] Mid-Holocene of China.

Lagenorhynchus Gray, 1846:84. Striped dolphins, white-sided dolphins, white-beaked dolphins.
    [= *Leucopleurus* Gray, 1866:216.] [Including *Sagmatias* Cope, 1866:294-295.]
    ?E. Plioc.; Af.[1] ?E. Plioc., L. Pleist.; N.A.[2] Pleist.; Eu. R.; Indian O. R.; Atlantic.
    R.; Pacific.
    [1] N. Africa.
    [2] Pacific North America.

Steno Gray, 1846:43. Rough-toothed dolphin.
    [= *Glyphidelphis* Gervais, 1859:301; *Stenopontistes* Miranda-Ribeiro, 1936:19.]
    L. Plioc.; Eu. R.; Indian O. R.; Mediterranean. R.; Atlantic. R.; Pacific.

Lissodelphis Gloger, 1841:169. Right whale dolphins.
    [= *Delphinapterus* Lesson & Garnot, 1826:179[1]; *Tursio* Wagler, 1830:34[2];
    Leucorhamphus Lilljeborg, 1861:5.[3]] [Including *Pachypleurus* Brandt, 1873:234-239;
    Archaeocetus Sinzow, 1898[4]; *Pristinocetus* Trouessart, 1898.[4]]
    ?L. Pleist.; N.A.[5] R.; Indian O. R.; Atlantic.[6] R.; Pacific.
    [1] Not *Delphinapterus* Lacépède, 1804.
    [2] Not *Tursio* Fleming, 1822, a synonym of *Physeter*.
    [3] Replacement name for *Delphinapterus* Lesson & Garnot, 1826.
    [4] Unnecessary replacement name for *Pachypleurus*.
    [5] Pacific North America.
    [6] S. Atlantic.

Lagenodelphis Fraser, 1956:496. Bornean dolphin, Fraser's dolphin.
    R.; Indian O. R.; Atlantic. R.; Pacific.[1]
    [1] S. Pacific.

Stenella Gray, 1866:213.[1] Spinner dolphin, spotted dolphin.
    [Including *Clymene* Gray, 1864:237[2]; *Clymenia* Gray, 1868:6[3]; *Euphrosyne* Gray,
    1866:214[4]; *Micropia* Gray, 1868:6[5]; *Prodelphinus* Gervais, in Van Beneden & Gervais,
    1880:604; *Fretidelphis* Iredale & Troughton, 1934:65.]
    R.; Indian O. R.; Mediterranean. R.; Atlantic. R.; Pacific.[6]
    [1] Described as subgenus of *Steno*.
    [2] Proposed as subgenus of *Delphinus*. Raised to generic rank by Gray, 1866:214. Not *Clymene* Oken, 1815,
    a genus of mollusks.
    [3] Not *Clymenia* Münster, 1839, a genus of mollusks.
    [4] Proposed as subgenus of *Clymene* Gray. Not *Euphrosyna* Von Siebold, 1843, a genus of "Vermes."
    [5] Proposed as subgenus of *Clymenia* Gray.
    [6] Including the Bering Sea.

Subfamily **Globicephalinae** Gray, 1850:313.
    [= Globicephalidae Gray, 1850:313; Globicephalina Gray, 1863:201; Globicephalinae
    Gill, 1872:15; Globicephalinae Van Bree, 1972:212.] [Including Grampidae Gray,
    1871:82; Grampini Pavlinov & Rossolimo, 1987:94.]
    Plioc.; Eu. E. Plioc., Pleist.; N.A. E. Plioc.; S.A.[1] R.; Indian O. R.;
    Mediterranean. R.; Atlantic. R.; Pacific.
    [1] Globicephalinae nov. gen., nov. sp. from Sud-Sacaco level, Pisco Fm., Peru.

Globicephala Lesson, 1828:441. Pilot whales, blackfish, caa'ing whale.
    [= *Globiocephalus* Gray, 1843:xxii; *Globicephalus* Van Beneden, 1880:554; *Globiceps*
    Flower, 1883:508.[1]] [= or including *Sphaerocephalus* Gray, 1864:244.[2]]
    Plioc.; Eu. E. Plioc., Pleist.; N.A. R.; Indian O. R.; Mediterranean. R.; Atlantic.
    R.; Pacific.
    [1] Flower was the first to select a type species, but his *Globiceps* was preoccupied by *Globiceps* Lepelletier &
    Serville, 1825, a genus of Hemiptera.
    [2] Described as subgenus of *Globiocephalus*.

Grampus Gray, 1828:2. Risso's dolphin, gray grampus.
[= *Grayius* A. W. Scott, 1873:104; *Grampidelphis* Iredale & Troughton, 1933:31.[1]]
R.; Indian O. R.; Mediterranean. R.; Atlantic. R.; Pacific.
[1] New name for *Grampus*.

Peponocephala Nishiwaki & Norris, 1966:95-99. Melon-headed whale.
[= *Electra* Gray, 1866:268-272.[1]]
R.; Indian O.[2] R.; Atlantic.[3] R.; Pacific.[4]
[1] Described as subgenus of *Lagenorhynchus*; raised to generic rank by Gray, 1868. Not *Electra* Lamouroux, 1816, a genus of "polyps", nor *Electra* Stephens, 1829, a genus of Lepidoptera.
[2] Tropical and warm temperate Indian Ocean.
[3] Tropical and warm temperate Atlantic Ocean.
[4] Tropical and warm temperate Pacific Ocean.

Feresa Gray, 1871. Pygmy killer whale.
R.; Indian O. R.; Atlantic. R.; Pacific.

Subfamily **Cephalorhynchinae** Fraser & Purves, 1960:108.
[Including Sotaliinae Kasuya, 1973:32.]
R.; Indian O. R.; Atlantic. R.; S.A.[1] R.; Pacific.
[1] Amazon Basin.

Cephalorhynchus Gray, 1846:36. Piebald dolphin.
[Including *Eutropia* Gray, 1862.]
R.; Indian O.[1] R.; Atlantic.[2] R.; Pacific.[3]
[1] S. Indian Ocean.
[2] S. Atlantic Ocean.
[3] S. Pacific Ocean.

Sotalia Gray, 1866. River dolphins, tucuxi.
[Including *Tucuxa* Gray, 1866.]
R.; Atlantic.[1] R.; S.A.[2]
[1] W. Atlantic (coastal waters of South America).
[2] Amazon Basin.

Sousa Gray, 1866. Freckled dolphin, plumbeous dolphin, white dolphins.
R.; Indian O.[1] R.; Atlantic.[2] R.; Pacific.[3]
[1] Coastal waters and rivers of Africa, S. Asia and E. Indies.
[2] E. Atlantic (coastal waters of Africa).
[3] W. Pacific (coastal waters and rivers from China to Australia).

Family **Pontoporiidae** Gray, 1870:393.
[= Pontoporiadae Gray, 1870:393; Pontoporidae Gill, 1871; Pontoporiidae Kasuya, 1973:61.]
M. Mioc.-E. Plioc., L. Pleist.; S.A. L. Mioc.-L. Plioc.; N.A.[1] R.; Atlantic.[2]
[1] Pontoporid periotics have been reported from the early Pliocene Yorktown Fm., North Carolina.
[2] Coastal waters of S.E. South America.

†*Brachydelphis* de Muizon, 1988.
M.-L. Mioc.; S.A.[1]
[1] Peru.

†*Piscorhynchus* G. Pilleri & H. J. Siber, in Pilleri, ed., 1989:199.[1]
L. Mioc.; S.A.[2]
[1] Assignment to Pontoporiidae is questionable.
[2] Peru.

Subfamily **Pontoporiinae** Gray, 1870:393.
[= Pontoporiadae Gray, 1870:393; Stenodelphininae Miller, 1923:40; Pontoporiinae Rice, 1967:315.]
L. Mioc.-E. Plioc., L. Pleist.; S.A. R.; Atlantic.[1]
[1] Coastal waters of S.E. South America.

†*Pontistes* Burmeister, 1885:138-144.
[= †*Palaeopontoporia* Doering, 1882:437, 455[1]; †*Pontivaga* Ameghino, 1891:165-166.]

L. Mioc.; S.A.[2]

[1] Palmer (1904:557) stated that †*Pontistes* is antedated by †*Palaeopontoporia*, based upon the same type species. Palmer's comment seems to have been long-forgotten. †*Palaeopontoporia*, if truly a senior synonym, probably should be officially suppressed.
[2] Argentina.

†*Pliopontos* de Muizon, 1983:625.
E. Plioc.; S.A.[1]

[1] Peru.

Pontoporia Gray, 1846:45.[1] La Plata River dolphin, franciscana.
[= *Stenodelphis* Gervais, in d'Orbigny, 1847.]
?L. Mioc., L. Pleist.; S.A.[2] R.; Atlantic.[3]

[1] *Pontoporia* Gray, 1846, is not a homonym of *Pontoporeia* Kroyer, 1842, a genus of Crustacea, nor is it antedated by *Pontoporia* Agassiz, 1846, an emendation of *Pontoporeia*.
[2] Argentina.
[3] La Plata River estuary and coastal waters of S.E. South America.

Subfamily †**Parapontoporiinae** Barnes, 1984:1, 6.
L. Mioc., L. Plioc.; N.A.[1]

[1] Pacific North America.

†*Parapontoporia* Barnes, 1984:1, 6.
L. Mioc., L. Plioc.; N.A.[1]

[1] Pacific North America.

Family **Lipotidae** Zhou, Qian & Li, 1979:72.
[= Lipotinae Barnes, 1985.]
Mioc., R.; As.[1]

[1] China.

†*Prolipotes* Zhou, Zhou & Zhao, 1984:173.
Mioc.; As.[1]

[1] China.

Lipotes Miller, 1918:2. Yangtze River dolphin, Chinese dolphin, white flag dolphin, baiji.
R.; As.[1]

[1] China.

Family **Iniidae** Gray, 1846:25, 45.
[= Iniina Gray, 1846:25, 45; Iniidae Gray, 1866:226; Iniinae Flower, 1867:114.]
?E. and/or ?M. Mioc., L. Mioc., ?Pleist., R.; S.A. M. Mioc.; Eu.[1] M. and/or L. Mioc.; N.A. ?E. Plioc.; Af.[2]

[1] An iniid periotic has reportly been found in the middle Miocene of Visiano.
[2] An iniine, genus indet., is known from the Sahabi Fm., Libya.

†*Plicodontinia* Miranda-Ribeiro, 1938.
?Pleist.; S.A.

†*Proinia* True, 1910.
Mioc.; S.A.

†*Goniodelphis* Allen, 1941.
M. and/or L. Mioc.; N.A.

†*Ischyrorhynchus* Ameghino, 1891:163.
[Including †*Anisodelphis* Rovereto, 1915.]
L. Mioc.; S.A.

†*Saurodelphis* Burmeister, 1891:161-162.[1]
[= †*Saurocetes* Burmeister, 1871:51-55[2]; †*Saurocetus* Coues, 1890:5355[2]; †*Pontoplanodes* Ameghino, 1891:255.[1]]
L. Mioc.; S.A.

[1] Replacement name for †*Saurocetes* Burmeister, 1871.
[2] Not †*Saurocetus* Agassiz, 1848, a squalodontid.

Inia d'Orbigny, 1834:31-36. Amazon River dolphin, boutu.
R.; S.A.[1]

[1] Amazon and Orinoco river systems.

Family †**Kentriodontidae** Slijper, 1936:556.
> [= †Kentriodontinae Slijper, 1936:556; †Kentriodontidae Slijper, 1958: figure 36.]
> ?L. Olig.; New Zealand.[1] Mioc.; As.[2] ?E. Mioc., M. Mioc., ?L. Mioc.; Eu.[3] M.-L. Mioc.; N.A. M. Mioc.; S.A.
>
> [1] A probable kentriodontid, genus and sp. indet., has been reported from the Waitakian of Port Waikato.
> [2] E. Asia (Japan).
> [3] Possible kentriodontid auditory bones have been reported from the early Miocene of Rosignano and Vignale, Italy.

†*Atocetus* de Muizon, 1988.
> M. Mioc.; S.A.[1]
>
> [1] Peru.

†*Belonodelphis* de Muizon, 1988.
> M. Mioc.; S.A.[1]
>
> [1] Peru.

Subfamily †**Kentriodontinae** Slijper, 1936:556.
> Mioc.; As.[1] M. Mioc., ?L. Mioc.; Eu. M. Mioc., ?L. Mioc.; N.A.
>
> [1] E. Asia (Japan).

†*Kentriodon* Kellogg, 1927.
> [Including †*Grypolithax* Kellogg, 1931:393.]
> ?Mioc.; As.[1] ?M. Mioc.; Eu. M. Mioc.; N.A.
>
> [1] E. Asia (Japan).

†*Sarmatodelphis* Kirpichnikov, 1954.
> M. Mioc.; Eu.[1]
>
> [1] E. Europe.

†*Leptodelphis* Kirpichnikov, 1954.
> M. Mioc.; Eu.[1]
>
> [1] E. Europe.

†*Pithanodelphis* Abel, 1905.
> M. and/or L. Mioc.; Eu. ?L. Mioc.; N.A.

†*Delphinodon* Leidy, 1869.
> M. Mioc.; N.A.

†*Microphocaena* Kudrin & Tatarinov, 1965.
> M. Mioc.; Eu.[1]
>
> [1] E. Europe.

Subfamily †**Lophocetinae** Barnes, 1978:11.
> M. and/or L. Mioc.; N.A.

†*Lophocetus* Cope, 1867.
> M. and/or L. Mioc.; N.A.

Subfamily †**Kampholophinae** Barnes, 1978:4.
> M. Mioc.; Eu. M.-L. Mioc.; N.A.

†*Kampholophus* Rensberger, 1969:2.
> M. Mioc., ?L. Mioc.; N.A.

†*Liolithax* Kellogg, 1931:375.
> M. Mioc.; Eu. M.-L. Mioc.; N.A.

Family **Monodontidae** Gray, 1821:310.
> [= Tachynicidae Brookes, 1828; Narvallidae Burnett, 1830:360; Narwalina Reichenbach, 1845:5; Monodonta Giebel, 1855:112.[1]]
> E. Mioc., ?E. and/or ?M. Pleist., L. Pleist.; Eu.[2] M. Mioc., R.; As. L. Mioc.-L. Plioc., ?E. and/or ?M. Pleist., L. Pleist.; N.A. E. Plioc.; S.A. R.; Indian O. R.; Atlantic.[3] R.; Arctic O. R.; Pacific.
>
> [1] Proposed as a family.
> [2] Monodontidae indet. in the early Miocene of Europe
> [3] N. Atlantic.

Subfamily **Delphinapterinae** Gill, 1871:124.
> [= Beluginae Flower, 1867:115; Belugidae Gray, 1868:9; Delphinapteridae Weber, 1904:577, 579.]
> M. Mioc.; As.[1] L. Mioc.-L. Plioc., L. Pleist.; N.A.[2] E. Plioc.; S.A.[3] L. Pleist.; Eu. R.; Atlantic.[4] R.; Arctic O. R.; Pacific.[5]
>
> [1] W. Asia.
> [2] Delphinapterinae, genus indet., from the Blancan of North America.
> [3] Delphinapterinae indet. from Pisco Fm., Peru.
> [4] N. Atlantic.
> [5] N. Pacific.

†*Denebola* Barnes, 1984:1, 13.
> L. Mioc.; N.A.

Delphinapterus Lacépède, 1804:xli. Beluga, white whale.
> [= *Beluga* Rafinesque, 1815:60[1]; *Delphis* Wagler, 1830; *Argocetus* Gloger, 1842:xxxiv, 169; *Leucas* Brandt, 1873:234.[2]]
> M. Mioc.; As.[3] E. Plioc., L. Pleist.; N.A. L. Pleist.; Eu. R.; Atlantic.[4] R.; Arctic O. R.; Pacific.[5]
>
> [1] New name for *Delphinapterus*. Not *Beluga* Gmelin, 1774.
> [2] Described as subgenus of *Delphinapterus*.
> [3] W. Asia.
> [4] N. Atlantic.
> [5] N. Pacific.

Subfamily **Monodontinae** Gray, 1821:310.
> [= Monodontidae Gray, 1821:310; Monodontina C. L. Bonaparte, 1837:7[1]; Monodonta Haeckel, 1866:clix[2]; Monodontinae Miller, 1923:34, 40, 51.]
> Pleist.; Eu. Pleist.; N.A. R.; Atlantic.[3] R.; Arctic O. R.; Pacific.[4]
>
> [1] As subfamily.
> [2] Used as family.
> [3] N. Atlantic.
> [4] Bering Sea.

Monodon Linnaeus, 1758:75. Narwhal.
> [= *Ceratodon* Brisson, 1762:218; *Diodon* Storr, 1780:42[1]; *Narwalus* Lacépède, 1804:xxxvii, 142-163; *Oryx* Oken, 1816:672[2]; *Tachynices* Brookes, 1828:40; *Narvallus* Burnett, 1830:361; *Monodus* Schulze, 1897:5.]
> Pleist.; Eu. Pleist.; N.A. R.; Atlantic.[3] R.; Arctic O. R.; Pacific.[4]
>
> [1] Not *Diodon* Linnaeus, 1758, a fish.
> [2] New name for *Monodon* Linnaeus. Not *Oryx* de Blaineville, 1816, a genus of Bovidae.
> [3] N. Atlantic.
> [4] Bering Sea.

Subfamily **Orcaellinae** Nishiwaki, 1972:11.
> [= Orcaelidae Nishiwaki, in Ridgway, 1972:11; Orcaellinae Kasuya, 1973:61.]
> R.; Indian O. R.; As.[1] R.; Pacific.
>
> [1] Irrawaddy River, Burma.

Orcaella Gray, 1866. Irrawaddy dolphin.
> R.; Indian O.[1] R.; As.[2] R.; Pacific.[3]
>
> [1] Coastal waters of India, S.E. Asia, E. Indies and N. Australia.
> [2] Irrawaddy River, Burma.
> [3] Coastal waters of S.E. Asia, E. Indies, N. Australia and New Guinea.

Family †**Odobenocetopsidae** de Muizon, 1993:746.
> Plioc.; S.A.

†*Odobenocetops* de Muizon, 1993:746.
> Plioc.; S.A.[1]
>
> [1] Peru.

Family †**Dalpiazinidae** de Muizon, 1988:66.
> E. Mioc.; Eu.

†*Dalpiazina* de Muizon, 1988:66.
> E. Mioc.; Eu.

Family †**Acrodelphinidae** Abel, 1905:41.

[= †Acrodelphidae Abel, 1905:41; †Acrodelphinidae Rice, in Anderson & Jones, 1984:466.] [Including †Champsodelphidae Scott, 1873:67[1]; †Eoplatanistidae de Muizon, 1988.]

E. Mioc.; Af.[2] Mioc.; As.[3] E.-M. Mioc.; Mediterranean. E.-L. Mioc.; Eu. M.-L. Mioc.; N.A.

[1] A. W. Scott's family †Champsodelphidae has many years priority, but is a *nomen oblitum*.
[2] N. Africa.
[3] W. Asia.

†*Pomatodelphis* G. M. Allen, 1921:148.

?E. Mioc., M. Mioc.; Eu. L. Mioc.; N.A.

†*Champsodelphis* Gervais, 1848.[1]

E.-M. Mioc.; Eu.

[1] 1848-1852.

†*Schizodelphis* Gervais, 1861:125-126.

[= †*Cyrtodelphis* Abel, 1900:850.]

E. Mioc.; Af.[1] Mioc.; As.[2] E.-M. Mioc., ?L. Mioc.; Eu. M.-L. Mioc.; N.A.

[1] N. Africa.
[2] W. Asia.

†*Eoplatanista* Dal Piaz, 1916:5.

E.-M. Mioc.; Mediterranean. E.-M. Mioc.; Eu.

†*Acrodelphis* Abel, 1900.

M.-L. Mioc.; Eu. M. Mioc.; N.A.

Family **Phocoenidae** Gray, 1825:340. Porpoises.

[= Phocaenina Gray, 1825:340; Phocaenidae Bravard, 1885:144.]

L. Mioc. and/or E. Plioc., ?Pleist.; As.[1] L. Mioc., ?Plioc.; N.A.[2] L. Mioc.-E. Plioc.; S.A.[3] R.; Indian O. R.; Eu.[4] R.; Atlantic. R.; Arctic O. R.; Pacific.

[1] Phocoenids have been listed from the Pleistocene of Asia.
[2] Pacific North America.
[3] Pacific South America.
[4] Black Sea.

†*Lomacetus* de Muizon, 1986.

L. Mioc.; S.A.[1]

[1] Peru.

†*Australithax* de Muizon, 1988.

L. Mioc.; S.A.[1]

[1] Pacific South America (Peru).

Subfamily **Phocoenoidinae** Barnes, 1984:1, 17.

L. Mioc. and/or E. Plioc.; As.[1] L. Mioc.; N.A.[2] ?L. Mioc., E. Plioc.; S.A.[3] ?R.; Indian O. R.; Atlantic.[4] R.; Pacific.

[1] Phocoenoidinae gen. and sp. indet. in the late Miocene or early Pliocene of Japan.
[2] Pacific North America.
[3] Pacific South America (Peru).
[4] S. Atlantic.

†*Piscolithax* de Muizon, 1983.

L. Mioc.; N.A.[1] ?L. Mioc., E. Plioc.; S.A.[2]

[1] Pacific North America.
[2] Pacific South America (Peru).

†*Salumiphocaena* Barnes, 1985:149.

L. Mioc.; N.A.[1]

[1] Pacific N.A. (California).

Phocoenoides Andrews, 1911:31. Dall porpoise.

[= *Phocaenoides* Simpson, 1945:104.[1]]

R.; Pacific.[2]

[1] Probably a *lapsus calami* but possibly an attempted emendation.
[2] N. Pacific.

Australophocaena Barnes, 1985:153. Spectacled porpoise.
　　　　?R.; Indian O. R.; Atlantic.[1] R.; Pacific.[2]
[1] S. Atlantic.
[2] S. Pacific.

Subfamily **Phocoeninae** Gray, 1825:340.
　　　　[= Phocaenina Gray, 1825:340[1]; Phocoeninae Barnes, 1984:17.]
　　　　L. Mioc.; N.A.[2] R.; Indian O. R.; Eu.[3] R.; Atlantic. R.; Arctic O. R.; Pacific.
[1] Proposed as tribe.
[2] Phoecoeninae sp. indet. from the latest Miocene Purisima Fm., California.
[3] Black Sea.

Neophocaena Palmer, 1899:23.[1] Black finless porpoise.
　　　　[= *Neomeris* Gray, 1846:30.[2]]
　　　　R.; Indian O. R.; Pacific.[3]
[1] Proposed as new name for *Neomeris* Gray.
[2] Not *Neomaris* Lamouroux, 1816, a genus of "polyps," nor *Neomeris* Costa, 1844.
[3] W. Pacific.

Phocoena Cuvier, 1817:279. Common porpoises, harbor porpoise, vaquita.
　　　　[= *Phocaena* G. Cuvier, 1817:163; *Phocena* G. Cuvier, 1821:310.] [Including
　　　　Acanthodelphis Gray, 1866:304-305.[1]]
　　　　R.; Eu.[2] R.; Atlantic. R.; Arctic O. R.; Pacific.
[1] Described as subgenus of *Phocaena*; raised to generic rank by Gray, 1868:8.
[2] Black Sea.

Family †**Albireonidae** Barnes, 1984:1, 29.
　　　　L. Mioc.; N.A.[1]
[1] Pacific North America.

†*Albireo* Barnes, 1984:1, 31.
　　　　L. Mioc.; N.A.[1]
[1] Baja California.

Family †**Hemisyntrachelidae** Slijper, 1936:550.
　　　　E. Plioc.; Eu.

†*Hemisyntrachelus* Brandt, 1873:239-242.
　　　　E. Plioc.; Eu.

Order **ARTIODACTYLA** Owen, 1848:131.[1] Artiodactyls, even-toed ungulates.
　　　　[= AMBULANTIA Newman, 1843:35[2]; ARTIODACTYLI Gill, 1872:290; PORCINA
　　　　Trouessart, 1879:2; PARAXONIA Marsh, 1884:9, 177; CLINODACTYLA Marsh,
　　　　1884:177[3]; PARIDIGITATA Depéret, 1909:321; ARTIODACTYLIFORMES
　　　　Kinman, 1994:37.] [Including BISULCA Giebel, 1855:248; NEOSELENODONTIA
　　　　Webb & Taylor, 1980:151.[4]]
[1] The term ARTIODACTYLA dates from Owen, but the concept was first expressed (informally) by de
　Blainville in 1816.
[2] In large part.
[3] In part. CLINODACTYLA was for perissodactyls (MESAXONIA) and artiodactyls (PARAXONIA), but
　not other ungulates as then understood.
[4] To include suborders TYLOPODA and RUMINANTIA.

Suborder **SUIFORMES** Jaeckel, 1911:233.[1]
　　　　[= TESSERACHENAE Gray, 1821:306[2]; OMNIVORA Owen, 1859:52[3]; PORCINA
　　　　Trouessart, 1879:2.[4]] [Including Obesa Illiger, 1811:97[5]; SUINA Gray, 1868:20[6];
　　　　†DENTATA Gill, 1872:290[7]; BUNODONTA Kovalevskii, 1893:152[8];
　　　　BUNODONTIA Schlosser, 1923; †OREODONTA Osborn, 1910:549[6];
　　　　†HYPOCONIFERA Stehlin, 1910:1135; †PALAEODONTA Matthew, 1929:406[6];
　　　　HYODONTA Matthew, 1929:406; †BUNOSELENODONTIA Schlosser, 1923;
　　　　ANCODONTA Matthew, 1929:406[6]; †AGRIOCHOERIFORMES Hay, 1930:776;
　　　　†CAENOTHERIA Golpe-Posse, 1972:57.[9]]
[1] "Essentially = non-RUMINANTIA of various authors" (Simpson 1945:143).

[2] Proposed as order to include Hippopotamidae, Suidae (including the tayassuids), and †Anoplotheriidae. The term has long been ignored and is diagnosed by plesiomorphy in any case.

[3] Proposed as a family. *Nomen oblitum*; in brief prior use but later forgotten. For reasons of stability it should probably stay that way. Listed here for historical purposes.

[4] Proposed as suborder. In brief prior use but later forgotten. For reasons of stability it should probably stay that way. Listed here for historical purposes.

[5] Proposed as a family based on *Hippopotamus* but not based on a generic name. Used by Gill, 1872:290 not only for the Hippopotamidae but also for the †Merycopotamidae.

[6] Often used as an infraorder of SUIFORMES.

[7] Used for anoplotheres and oreodonts.

[8] Sometimes spelled "Kowalewsky."

[9] Proposed as suborder.

†*Aksyiria* Gabunia, 1973:741.
   M. and/or L. Eoc.; As.

Family †**Raoellidae** Sahni, Bhatia, Hartenberger, Jaeger, Kumar, Sudre & Vianey-Liaud, 1981:692.
   E.-M. Eoc.; As.

†*Indohyus* Rao, 1971:126.
   ?E. Eoc., M. Eoc.; As.

†*Khirtharia* Pilgrim, 1940:141.
   [Including †*Bunodentus* Rao, 1972:3.]
   E.-M. Eoc.; As.[1]
   [1] S. Asia.

†*Kunmunella* Sahni & Khare, 1971:50.
   E. and/or M. Eoc.; As.[1]
   [1] S. Asia.

†*Raoella* Sahni & Khare, 1971:47.
   E. and/or M. Eoc.; As.[1]
   [1] S. Asia.

†*Metkatius* Kumar & Sahni, 1985:157.
   M. Eoc.; As.

†*Haqueina* Dehm & Oettingen-Spielberg, 1958:26.[1]
   M. Eoc.; As.[2]
   [1] Assignment to †Raoellidae is questionable.
   [2] S. Asia.

Family †**Choeropotamidae** Owen, 1845:559.
   [= †Chaeropotamina Bonaparte, 1845:4.]
   M.-L. Eoc.; Eu. L. Eoc.; As.[1]
   [1] Choeropotamid indet. from the Naduo Fm., Kwangsi.

†*Choeropotamus* G. Cuvier, 1821:9.[1]
   M.-L. Eoc.; Eu.
   [1] Fide Desmarest, 1822:544. Simpson (1945:144) considered the original spelling †*Chaeropotamus* to be a *lapsus* and, therefore, subject to valid correction, noting that "Cuvier himself later spelled it †*Choeropotamus*, as have most authors since then." This problem crops up occasionally in the Nineteenth Century literature (e.g., "*Platychaerops*"), but in many publications "ae" and "oe" can be shown to have been used interchangeably.

Superfamily **Suoidea** Gray, 1821:306.
   [= Suidae Gray, 1821:306; Suoidea Cope, 1887:381; Setigera Illiger, 1811:99[1]; Suillida Haeckel, 1866:clviii[2]; Setifera Gill, 1872:82.[3]] [Including Hippopotamoidea Gray, 1821:306.]
   [1] Proposed as a family for *Sus* [s.l., suids and tayassuids] but not based on a generic name.
   [2] Proposed as a synonym of family Setigera.
   [3] Proposed as a superfamily for Phacochoeridae, Suidae, and Dicotylidae, but not based on a generic name.

†*Xenohyus* Ginsburg, 1980:863.
   E. Mioc.; Eu.

Family **Suidae** Gray, 1821:306. Pigs.
    ?L. Eoc., E. Olig.-R.; Eu. Olig., E. Plioc., ?L. Plioc., E. Pleist.; Mediterranean.[1]
    ?L. Olig., E. Mioc.-R.; As. E. Mioc.-R.; Af. L. Plioc.-R.; E. Indies. R.;
    Madagascar.[2]
   [1] Suid n. sp. in the early Pleistocene of Sardinia.
   [2] Also Recent practically worldwide in domestication.

†*Hemichoerus* Filhol, 1882:106-111.[1]
    L. Eoc.; Eu.
   [1] Assignment to Suidae is questionable.

†*Paradoxodonides* Strand, 1943:114.[1]
    [= †*Paradoxodon* Filhol, 1890:133, 134.[2]]
    L. Eoc. and/or Olig.; Eu.
   [1] Assignment to Suidae is questionable.
   [2] Not *Paradoxodon* Wagner, 1855, a synonym of *Suncus*.

†*Cainochoerus* Pickford, 1988:231.
    E. Plioc.; Af.

Subfamily †**Palaeochoerinae** Matthew, 1924:176.
    Olig.; Mediterranean.[1] E. Olig.-M. Mioc.; Eu. E. Mioc.; Af.
   [1] Balearic Islands.

†*Palaeochoerus* Pomel, 1847:381.[1]
    [= †*Palaeocherus* Pomel, 1847:381; †*Palaeochoerus* Pomel, 1847:392; †*Amphichoerus*
    Gore, from Bravard (MS), 1874:6.] [Including †*Propalaeochoerus* Stehlin, 1899:9.]
    Olig.; Mediterranean.[2] E. Olig.-E. Mioc.; Eu. E. Mioc.; Af.
   [1] Simpson (1945:145) took "the first spelling '*Paleocherus*' to be a *lapsus* legally corrigible, since it is
    manifestly contrary to Pomel's own and contemporary etymological usage and was at once corrected by
    Pomel."
   [2] Balearic Islands.

†*Yunnanochoerus* van der Made & Han, 1994:34.
    L. Mioc.; As.

†*Aureliachoerus* Ginsburg, 1974:70, 73, 76.
    E.-M. Mioc.; Eu.

Subfamily †**Tetraconodontinae** Lydekker, 1876:60.
    [= †Tetraconodontidae Lydekker, 1876:60; †Tetraconodontinae Simpson, 1945:145.]
    ?L. Olig., E. Mioc.-L. Plioc.; As. M. Mioc.-E. Pleist.; Af. M.-L. Mioc.; Eu.

†*Conohyus* Pilgrim, 1925:207.
    [Including †*Adaetontherium* Lewis, 1934:174.[1]]
    ?L. Olig., E.-L. Mioc.; As. M. Mioc.; Af. M. Mioc.; Eu.
   [1] Synonymy questionable.

†*Tetraconodon* Falconer, 1868:149-156.
    [Including †*Choerotherium* Cautley & Falconer, 1835:707.[1]]
    ?M. Mioc., L. Mioc., ?Plioc.; As.
   [1] Probably a *nomen nudum*; synonymy questionable.

†*Parachleuastochoerus* Golpe-Posse, 1972:125.
    M.-L. Mioc.; Eu.

†*Nyanzachoerus* L. S. B. Leakey, 1958:4.
    L. Mioc.-L. Plioc.; Af. L. Mioc.; As.[1]
   [1] S.W. Asia (Abu Dhabi).

†*Lophochoerus* Pilgrim, 1926:28.
    L. Mioc.; As.

†*Sivachoerus* Pilgrim, 1926:19.
    L. Mioc.-L. Plioc.; As.

†*Notochoerus* Broom, 1925:308.
    [Including †*Gerontochoerus* L. S. B. Leakey, 1943:47.]
    E. Plioc.-E. Pleist.; Af.

Subfamily †**Hyotheriinae** Cope, 1888:1087.
>> [= †Hyotherida Haeckel, 1895:522.]
>> L. Olig.-L. Mioc.; Eu. E.-L. Mioc., Plioc.; As.

†*Dubiotherium* Hellmund, 1992:23.
>> L. Olig.; Eu.

†*Hyotherium* von Meyer, 1834:30-31.
>> [= †*Hypotherium* Ginsburg, de Broin, Crouzel, Duranthon, Esquillié, Juillard & Lassaube, 1991:31.[1]]
>> E.-L. Mioc.; As. E.-L. Mioc.; Eu.
>> [1] *Lapsus calami.*

†*Chleuastochoerus* Pearson, 1928:11.
>> L. Mioc., Plioc.; As.

Subfamily †**Kubanochoerinae** Gabunia, 1958:1189.
>> E.-M. Mioc.; Af. E.-M. Mioc.; As. M. Mioc.; Eu.[1]
>> [1] E. Europe.

†*Nguruwe* Pickford, 1986:15.
>> E. Mioc.; Af.

†*Kenyasus* Pickford, 1986:24.
>> E. Mioc.; Af.

†*Libycochoerus* Arambourg, 1961:108.
>> E.-M. Mioc.; Af. E.-M. Mioc.; As.

†*Megalochoerus* Pickford, 1993:251.
>> E. Mioc.; As. M. Mioc.; Af.

†*Kubanochoerus* Gabunia, 1955:1203.
>> M. Mioc.; As. M. Mioc.; Eu.[1]
>> [1] E. Europe (N. Caucasus).

Subfamily †**Listriodontinae** Lydekker, 1884:100.
>> [= †Listriodontidae Lydekker, 1884:100; †Listriodontinae Simpson, 1945:145.]
>> E.-M. Mioc.; Af. E.-M. Mioc., ?L. Mioc.; As. E.-M. Mioc., ?L. Mioc.; Eu.

†*Bunolistriodon* Arambourg, 1963:904.[1]
>> [Including †*Eurolistriodon* Pickford & Moya Sola, 1995:344.]
>> E. Mioc., ?M. Mioc.; Af. E.-M. Mioc.; As.[2] E.-M. Mioc.; Eu.
>> [1] †*Bunolistriodon* Arambourg 1933:137, is a *nomen nudum.*
>> [2] W. Asia.

†*Listriodon* Meyer, 1846:466.
>> M. Mioc.; Af. M. Mioc., ?L. Mioc.; As. M. Mioc., ?L. Mioc.; Eu.

†*Lopholistriodon* Pickford & Wilkinson, 1975:128-137.
>> M. Mioc.; Af. ?M. Mioc.; As.[1]
>> [1] S.W. Asia (Saudi Arabia).

Subfamily †**Namachoerinae** Pickford, 1995:319.
>> E.-M. Mioc.; Af.

†*Namachoerus* Pickford, 1995:319.
>> E.-M. Mioc.; Af.

Subfamily **Suinae** Gray, 1821:306.
>> [= Suidae Gray, 1821:306; Suina Gray, 1825:343[1]; Suinae Zittel, 1893:343.]
>> [Including Phacochoerinae Van Hoepen & Van Hoepen, 1932:61.]
>> ?E. and/or ?M. Mioc., L. Mioc.-R.; Af. M. Mioc.-R.; As. M. Mioc.-R.; Eu. E. Plioc., L. Plioc. and/or E. Pleist.; Mediterranean.[2] L. Plioc.-R.; E. Indies. R.; Madagascar.[3]
>> [1] Used as tribe.
>> [2] Sardinia.
>> [3] Also Recent practically worldwide in domestication.

Tribe **Suini** Gray, 1821:306.

[= Suidae Gray, 1821:306; Suina Gray, 1825:343[1]; Suini L. S. B. Leakey, 1965:28; Suini Thenius, 1970:327.] [Including †Dicoryphochoerini Schmidt-Kittler, 1971:160.] ?Mioc., Plioc., E. Pleist.-R.; Af. ?M. Mioc., L. Mioc.-R.; As. M. Mioc.-R.; Eu. E. Plioc., L. Plioc. and/or E. Pleist.; Mediterranean.[2] E. Pleist.-R.; E. Indies.[3]

[1] As tribe.
[2] Sardinia.
[3] Also Recent practically worldwide in domestication.

Sus Linnaeus, 1758:49. Pigs, wild boars.

[= *Scrofa* Gray, 1868:38.[1]] [Including *Capriscus* Gloger, 1841:130[2]; *Porcula* Hodgson, 1847:423-428[3]; *Centuriosus* Gray, 1862:17; *Gyrosus* Gray, 1862:278[4]; *Ptychochoerus* Fitzinger, 1864[5]; *Eusus* Gray, 1868:32[3]; *Euhys* Gray, 1869:339[6]; *Porculia* Jerdon, 1874:243-245[7]; *Caprisculus* Strand, 1928:61.[8]]
?Mioc., Plioc., E. Pleist.-R.; Af.[9] ?M. Mioc., L. Mioc.-R.; As. L. Mioc.-R.; Eu. E. Plioc., L. Plioc. and/or E. Pleist.; Mediterranean.[10] E. Pleist.-R.; E. Indies.[11]

[1] Antedated by *Scrofa* Gistel, 1848, a fish.
[2] Not *Capriscus* Rafinesque, 1810, a fish.
[3] As subgenus.
[4] Objective synonym of *Centuriosus*.
[5] Replacement name for *Centuriosus*.
[6] As subgenus. Objective synonym of *Eusus*.
[7] Replacement name for *Porcula* Hodgson, 1847.
[8] Replacement name for *Capriscus* Gloger, 1841.
[9] N. Africa in the Recent.
[10] Sardinia.
[11] Also Recent practically worldwide in domestication. Feral populations on New Guinea, on Pacific islands (Solomons, Fiji, the Galapagos, and Hawaii), in North America, the West Indies, Central America, and South America, on the Mediterranean islands of Corsica and Sardinia, on the Indian Ocean island of Mauritius, and in Australia.

†*Korynochoerus* Schmidt-Kittler, 1971:155.

M. Mioc.-E. Plioc.; Eu. L. Mioc.; As.

†*Hippopotamodon* Lydekker, 1877:81.

[Including †*Dicoryphochoerus* Pilgrim, 1925:207.]
L. Mioc., Plioc., E. Pleist.; As.

†*Eumaiochoerus* Hürzeler, 1982:421.

L. Mioc.; Eu.

†*Microstonyx* Pilgrim, 1926:8.

[Including †*Limnostonyx* Ginsburg, 1988:58.[1]]
M.-L. Mioc.; Eu. L. Mioc.; As.

[1] Subgenus as defined.

Tribe **Potamochoerini** Gray, 1873:434.

[= Potamochoerina Gray, 1873:434; Potamochoerini Thenius, 1970:327.]
M. Mioc.-E. Pleist., M. and/or L. Pleist.; As. L. Mioc.-R.; Af. L. Mioc.; Eu. L. Plioc., Pleist.; E. Indies.[1] R.; Madagascar.

[1] Sulawesi.

†*Propotamochoerus* Pilgrim, 1925:207.

M.-L. Mioc., Plioc.; As. ?L. Mioc.; Af.

Potamochoerus Gray, 1854:129-132. African river hog, bush pig.

[= *Choiropotamus* Gray, 1843:185[1]; †*Potamochaerus* Murray, 1866:164.] [Including †*Postpotamochoerus* Thenius, 1950:32.[2]]
M. Mioc.-E. Pleist., M. and/or L. Pleist.; As. L. Mioc.; Eu. E.-L. Plioc., M. Pleist.-R.; Af. R.; Madagascar.

[1] Not †*Chaeropotamus* G. Cuvier, 1821 (= †*Choeropotamus* G. Cuvier, 1822), a dichobunoid. In this case Simpson (1945:145) chose not to follow the rule that one different letter always makes a different name [International Code of Zoological Nomenclature, Article 56(b)]. We agree with Simpson in following this rule when it helps to preserve a widely used name and to prevent confusion, but not in following it here, where the name would be *Choiropotamus*, for that would be to discard the universally used name *Potamochoerus* and promote confusion. [See also ICZN, Appendix D(3).]
[2] Proposed as subgenus.

†*Kolpochoerus* van Hoepen & van Hoepen, 1932:59.
        [= †*Mesochoerus* Shaw & Cooke, 1941:293.[1]] [Including †*Omochoerus* Arambourg, 1943:474-475; †*Promesochoerus* L. S. B. Leakey, 1965:28; †*Ectopotamochoerus* L. S. B. Leakey, 1965:30.]
        L. Mioc.-M. Pleist.; Af.
[1] Not "†*Mesochaerus* Jourdan" Depéret, 1887:236 (= "†*Mesochoerus* Jourdan" Trouessart, 1898).

Hylochoerus Thomas, 1904:577. Forest hog.
        ?E. Pleist., L. Pleist.-R.; Af.[1]
[1] Tropical Africa.

†*Celebochoerus* Hooijer, 1948:1025.
        L. Plioc., Pleist.; E. Indies.[1]
[1] Sulawesi.

Tribe †**Hippohyini** Thenius, 1970:327.
        L. Mioc.-E. Pleist.; As.

†*Sivahyus* Pilgrim, 1926:52.
        [Including †*Hyosus* Pilgrim, 1926:56.]
        L. Mioc., Plioc.; As.

†*Sinohyus* G. H. R. Von Koenigswald, 1963:196.
        ?L. Mioc.; As.

†*Hippohyus* Falconer & Cautley, in Owen, 1840:562-563.[1]
        L. Mioc.-E. Pleist.; As.
[1] 1840-1845.

Tribe **Phacochoerini** Gray, 1868:47.
        [= Phacochoeridae Gray, 1868:47; Phacochoerinae van Hoepen & van Hoepen, 1932:61; Phacochoerini L. S. B. Leakey, 1965:33.]
        L. Plioc.-R.; Af. M.-L. Pleist.; As.[1]
[1] W. Asia (Israel).

†*Potamochoeroides* Dale, 1948:116.
        L. Plioc., ?E. Pleist.; Af.

Phacochoerus F. Cuvier, 1817:236, 237.[1] Wart hog.
        [= *Macrocephalus* Frisch, 1775:3[2]; *Eureodon* G. Fischer de Waldheim, 1817:373, 417; †*Phocochorus* Voigt, 1819:422[3]; *Phascochaeres* Rüppell, 1826:61[3]; *Phascochaerus* Griffith, 1827:289[3]; *Phacochaeres* F. Cuvier, 1829:506[3]; *Phacocherus* Smuts, 1832:60, 61[3]; *Phascochoerus* Agassiz, 1842:25[3]; *Phacellochoerus* Hemprich & Ehrenberg, 1832[3]; *Phacellochaerus* Hemprich & Ehrenberg, 1832; *Dinochoerus* Gloger, 1841:131.] [= or including *Aper* Pallas, 1766:16-29.] [Including †*Tapinochoerus* van Hoepen & van Hoepen, 1932:58; *Potamochoerops* Ewer, 1956:527.[4]]
        L. Plioc.-R.; Af. M.-L. Pleist.; As.[5]
[1] F. Cuvier spelled it "*Phaco-choerus,*" later corrected by Fleming, 1822:200.
[2] Invalid. Not published in a consistently binominal work. See the International Code of Zoological Nomenclature, Article 11(c).
[3] Replacement name.
[4] Proposed as subgenus.
[5] W. Asia (Israel).

†*Metridiochoerus* Hopwood, 1926:267.
        [Including †*Pronotochoerus* L. S. B. Leakey, 1943:53.]
        L. Plioc.-M. Pleist.; Af. M. Pleist.; As.[1]
[1] W. Asia (Israel).

†*Stylochoerus* van Hoepen & van Hoepen, 1932:53.
        [Including †*Synaptochoerus* van Hoepen & van Hoepen, 1932:56; †*Afrochoerus* L. S. B. Leakey, 1942:187; †*Orthostonyx* L. S. B. Leakey, 1958:48.]
        E.-M. Pleist.; Af.

Tribe **Babyrousini** Gray, 1868:21.

>[= Babirussina Gray, 1868:21[1]; Babirussinae Hilzheimer & Heck, 1925; Babyrousinae Thenius, 1970:326.]
>
>Pleist., R.; E. Indies.
>
>[1] Rank not specified but Gray's names with the ending"ina" were usually tribes.

Babyrousa Perry, 1811. Babirusa.

>[= *Babirussa* Frisch, 1775[1]; *Babiroussus* Gray, 1821:306; *Babiroussa* F. Cuvier, 1825:257; *Porcus* Wagler, 1830:17[2]; *Babyrussa* Burnett, 1830:352; *Babirusa* Lesson, 1842:162; *Elaphochoerus* Gistel, 1848:x.[3]]
>
>Pleist., R.; E. Indies.
>
>[1] Invalid. Not published in a consistently binominal work. See the International Code of Zoological Nomenclature, Article 11(c).
>[2] Preoccupied by *Porcus* Geoffroy Saint-Hilaire, 1829, a genus of fishes.
>[3] Replacement name for *Porcus* Wagler, 1830.

Family **Tayassuidae** Palmer, 1897:174. Peccaries.

>[= Dicotylina Turner, 1849:157; Dicotylidae Gray, 1868:21.[1]] [Including †Cynorcidae Cope, 1867:144.[1]]
>
>L. Eoc.-E. Olig., E.-L. Mioc.; As. L. Eoc.-R.; N.A. Olig.; Mediterranean. E. Olig.-L. Mioc.; Eu. M. Mioc.; Af. L. Mioc., L. Pleist.-R.; Cent. A. L. Plioc.-R.; S.A.[2]
>
>[1] The priority of names here is a complex issue that should be submitted to the International Commission on Zoological Nomenclature.
>[2] Also Cuba by introduction.

†*Odoichoerus* Tong & Zhao, 1986:129.[1]

>E. Olig.; As.[2]
>
>[1] (Tung & Zhao).
>[2] E. Asia.

Subfamily **Tayassuinae** Palmer, 1897:174.

>[= Dicotylinae Leidy, 1853:323; Tayassuidae Palmer, 1897:174; Tayassuinae Hay, 1902:658.] [Including †Cynorcidae Cope, 1867:144.]
>
>L. Eoc.; As.[1] L. Eoc.-R.; N.A. L. Mioc., L. Pleist.-R.; Cent. A. L. Plioc.-R.; S.A.[2]
>
>[1] S.E. Asia (Thailand).
>[2] Also Cuba by introduction.

†*Egatochoerus* Ducrocq, 1994:766.

>L. Eoc.; As.[1]
>
>[1] S.E. Asia (Thailand).

†*Perchoerus* Leidy, 1869:194-197, 389.

>[Including †*Bothrolabis* Cope, 1888:63, 66-79.]
>
>L. Eoc.-E. Mioc.; N.A.

†*Thinohyus* Marsh, 1875:248.

>E. Olig.-E. Mioc.; N.A.

†*Hesperhys* Douglass, 1903:174.

>[Including †*Desmathyus* Matthew, 1907:217; †*Pediohyus* Loomis, 1910:381; †*Floridachoerus* White, 1941:96.]
>
>E. Mioc.-E. Plioc.; N.A.

†*Chaenohyus* Cope, 1879:4.

>E. Mioc.; N.A.

†*Cynorca* Cope, 1867:144, 151.

>[= or including †*Thinotherium* Cope, 1870:292-293.[1]]
>
>E.-M. Mioc.; N.A.
>
>[1] Probably best regarded as a *nomen dubium*, but the name antedates †*Thinotherium* Marsh, 1872.

†*Dyseohyus* Stock, 1937:398.

>E.-M. Mioc., ?L. Mioc.; N.A.

†*Prosthennops* Gidley, in Matthew & Gidley, 1904:265.

>[= or including †*Macrogenis* Matthew, 1924:179.[1]]

M. Mioc.-E. Plioc.; N.A. L. Mioc.; Cent. A.

[1] Proposed as subgenus of †*Prosthennops*.

†*Platygonus* Le Conte, 1848:103.[1]

[= †*Hyops* Le Conte, 1848:104; †*Protochoerus* Le Conte, 1848:105-106[2]; †*Euchoerus* Leidy, 1853:340; †*Coyametla* Duges, 1887:16.] [Including †*Antaodon* Ameghino, 1886:149; †*Parachoerus* Rusconi, 1930:150[3]; †*Selenogonus* Stirton, 1947:322.]

L. Mioc.-L. Pleist.; N.A. L. Plioc.-L. Pleist.; S.A.

[1] "The first spelling was actually '*Platigonus*,' but there is sufficient indication that this was a misprint. The name is differently misprinted '*Platydonus*' in another paper by Le Conte, published almost simultaneously. '*Platygonus*' is now universally used and may be retained" (Simpson, 1945:146). See also the International Code of Zoological Nomenclature, Articles 32 and 33.
[2] *Nomen dubium.*
[3] Proposed as subgenus.

†*Mylohyus* Cope, 1889:134. Longnosed peccary.

E. Plioc.-L. Pleist.; N.A.

Tayassu G. Fischer de Waldheim, 1814:284. White-lipped peccary, collared peccary, javelina.

[= *Notophorus* Fischer de Waldheim, 1817:418; *Tayassus* Trouessart, 1905:658.] [Including *Dicotyles* G. Cuvier, 1817:237-238; *Adenonotus* Brookes, 1828:11[1]; *Pecari* Reichenbach, 1835:1[2]; *Mamdicotylesus* Herrera, 1899:17.[3]]

?L. Plioc., Pleist., R.; S.A. ?L. Pleist., R.; N.A. L. Pleist.-R.; Cent. A.[4]

[1] Replacement name for *Dicotyles* G. Cuvier, 1817.
[2] Described as subgenus of *Sus*. Objective synonym of *Dicotyles*.
[3] See the International Code of Zoological Nomenclature, Article 1(b)(8) and Palmer (1904:25-26).
[4] Also Cuba by introduction.

†*Argyrohyus* Kraglievich, 1959:230.

L. Plioc.; S.A.

Catagonus Ameghino, 1904:73.[1] Roman-nosed peccary.

[Including †*Interchoerus* Rusconi, 1930:168.[2]]

M. Pleist.-R.; S.A.

[1] Originally discovered as fossils.
[2] Proposed as subgenus.

†*Brasiliochoerus* Rusconi, 1930:160.[1]

M.-L. Pleist.; S.A.

[1] Proposed as a subgenus of †*Platygonus*.

Subfamily †**Doliochoerinae** Simpson, 1945:146.

Olig.; Mediterranean. E. Olig.-L. Mioc.; Eu. E.-L. Mioc.; As. M. Mioc.; Af.

Tribe †**Doliochoerini** Simpson, 1945:146.

[= †Doliochoerini Pickford, 1986:74.] [Including †Doliochoerinae Simpson, 1945:146.]

Olig.; Mediterranean. E. Olig.-M. Mioc.; Eu. E.-L. Mioc.; As. ?M. Mioc.; Af.

†*Doliochoerus* Filhol, 1882:1259-1260.

Olig.; Mediterranean. E.-L. Olig.; Eu.

†*Taucanamo* Simpson, 1945:146.

[= †*Choerotherium* Lartet, 1851:32-33.[1]] [Including †*Albanohyus* Ginsburg, 1974:73-75.]

E.-M. Mioc.; Eu. ?M. Mioc.; Af. M.-L. Mioc.; As.

[1] Not †*Choerotherium* Cautley & Falconer, 1835, a possible synonym of †*Tetraconodon*.

†*Pecarichoerus* Colbert, 1933:2.

E.-M. Mioc.; As.

†*Barberahyus* Golpe-Posse, 1977:33.

M. Mioc.; Eu.

Tribe †**Schizochoerini** Golpe-Posse, 1972:18.

[= †Schizochoerinae Thenius, 1979:6.]

M. Mioc.; Af. M.-L. Mioc.; As. L. Mioc.; Eu.

†*Schizochoerus* Crusafont Pairó & Lavocat, 1954:89.

M. Mioc.; Af. M.-L. Mioc.; As. L. Mioc.; Eu.

Family †**Sanitheriidae** Simpson, 1945:145.
> [= †Sanitheriinae Simpson, 1945:145; †Sanitheriidae Pickford, 1984:133, 134.]
> [Including †Xenochoerinae Thenius, 1979:1, 5.]
> E.-M. Mioc.; Af. E.-M. Mioc.; As. M. Mioc.; Eu.

†*Diamantohyus* Stromer, 1922:332.
> E. Mioc., ?M. Mioc.; Af. E. Mioc.; As.

†*Sanitherium* von Meyer, 1866:15-17.
> [Including †*Xenochoerus* Zdarsky, 1909:264.]
> M. Mioc.; Af. M. Mioc.; As. M. Mioc.; Eu.

Family **Hippopotamidae** Gray, 1821:306. Hippopotamuses.
> [= Obesa Illiger, 1811:97[1]; Hippopotamida Haeckel, 1895:530, 555; Hippopotamoidea
> Kalandadze & Rautian, 1992:122.]
> M. Mioc.-R.; Af. L. Mioc.-L. Pleist., ?R.; As.[2] L. Mioc.-E. Plioc., E.-L. Pleist.;
> Eu. L. Plioc.-L. Pleist.; E. Indies.[3] Pleist.; Mediterranean. R.; Madagascar.[4]
>
> [1] Proposed as a family based on *Hippopotamus* but not based on a generic name.
> [2] Possibly Recent but now extinct.
> [3] Java.
> [4] Extinct.

†*Trilobophorus* Geze, 1985.
> L. Plioc.; Af.

Subfamily †**Kenyapotaminae** Pickford, 1983:195.
> M.-L. Mioc.; Af.

†*Kenyapotamus* Pickford, 1983:195.
> M.-L. Mioc.; Af.

Subfamily **Hippopotaminae** Gray, 1821:306.
> [= Hippopotamidae Gray, 1821:306; Hippopotamina Gray, 1825:343[1];
> Hippopotaminae Gill, 1872:10.] [Including Choeropsinae Gill, 1872:10.]
> L. Mioc.-R.; Af. L. Mioc.-L. Pleist., ?R.; As.[2] L. Mioc.-E. Plioc., E.-L. Pleist.;
> Eu. L. Plioc.-L. Pleist.; E. Indies.[3] Pleist.; Mediterranean. R.; Madagascar.[4]
>
> [1] As tribe.
> [2] Possibly Recent but now extinct.
> [3] Java.
> [4] Extinct.

Hexaprotodon Falconer & Cautley, 1836:51.[1] Pygmy hippopotamus.
> [Including *Diprotodon* Duvernoy, 1849:277[2]; *Choerodes* Leidy, 1852:52[3]; *Choeropsis*
> Leidy, 1853:213[4]; *Ditomeodon* Gratiolet, 1869:250.]
> L. Mioc.-R.; Af. L. Mioc.-L. Pleist.; As. L. Mioc.-E. Plioc.; Eu. L. Plioc.-L.
> Pleist.; E. Indies.[5]
>
> [1] Described as a subgenus of *Hippopotamus*.
> [2] Described as a subgenus of *Hippopotamus*. Not †*Diprotodon* Owen, 1838, a genus of marsupials.
> [3] Not *Chaerodes* White, 1846, a genus of Coleoptera. Suppressed under the plenary powers by the
> International Commission on Zoological Nomenclature, Opinion 1393, 1986.
> [4] New name for †*Choerodes* Leidy, 1852. Conserved and placed on the official list of generic names by the
> International Commission on Zoological Nomenclature, Opinion 1393, 1986.
> [5] Java.

Hippopotamus Linnaeus, 1758:74. Hippopotamus, dwarf hippopotamuses.
> [= *Hippotamus* Rafinesque, 1815:56.[1]] [Including *Tetraprotodon* Falconer & Cautley,
> 1836:51; †*Hippoleakius* Deraniyagala, 1947:226-229; †*Prechoeropsis* Deraniyagala,
> 1948:22; †*Phanourios* Boekshoten & Sondaar, 1972:326.]
> L. Plioc.-R.; Af. E.-L. Pleist., ?R.; As.[2] Pleist.; Mediterranean. E.-L. Pleist.; Eu.
> R.; Madagascar.[3]
>
> [1] New name for *Hippopotamus*.
> [2] W. Asia; possibly Recent but now extinct in Israel.
> [3] Now extinct.

Superfamily †**Dichobunoidea** Turner, 1849:158.
> [= †Dichobunina Turner, 1849:158; †Dichobunoidea Weber, 1904:644, 688.]
> [Including †Trigonolestoidea Cope, 1898:130.]

†*Acoessus* Cope, 1881:380, 397.[1]
  M. Eoc.; Eu.
  [1] Validity and placement doubtful.

Family †**Dichobunidae** Turner, 1849:158.
  [= †Dichobunina Turner, 1849:158; †Dichobunidae Gill, 1872:10.] [Including
  †Homacodontidae Marsh, 1894:263; †Leptochoeridae Marsh, 1894:273;
  †Trigonolestoidea Cope, 1898:130; †Trigonolestidae Schlosser, 1899:349;
  †Diacodexidae Viret, 1961:891; †Diacodexeidae Krishtalka & Stucky, 1985:420;
  †Antiacodontidae Krishtalka & Stucky, 1985:413; †Bunomerycidae Gentry & Hooker,
  in Benton, ed., 1988:266.]
  E.-M. Eoc.; As. E. Eoc.-L. Olig.; Eu. E. Eoc.-L. Olig.; N.A.

†*Paraphenacodus* Gabunia, 1971:233.[1]
  ?E. Eoc., M. Eoc.; As.
  [1] Assignment to †Dichobunidae is questionable.

†*Dulcidon* Van Valen, 1965:393.[1]
  M. Eoc.; As.[2]
  [1] Assignment to †Dichobunidae is questionable.
  [2] S. Asia.

†*Chorlakkia* Gingerich, D. E. Russell, Sigogneau-Russell & Hartenberger, 1979:118.[1]
  E. and/or M. Eoc.; As.[2]
  [1] Assignment to †Dichobunidae is questionable.
  [2] S. Asia.

†*Pakibune* Thewissen, Gingerich & D. E. Russell, 1987:253.
  ?E. Eoc., M. Eoc.; As.

Subfamily †**Dichobuninae** Turner, 1849:158.
  [= †Dichobunina Turner, 1849:158; †Dichobuninae Zittel, 1893:374.] [Including
  †Hyperdichobuninae Gentry & Hooker, in Benton, ed., 1988:266.]
  E. Eoc.-L. Olig.; Eu. M. Eoc.; As.[1]
  [1] E. Asia.

Tribe †**Hyperdichobunini** Gentry & Hooker, in Benton, ed., 1988:266, **new rank**.
  [= †Hyperdichobuninae Gentry & Hooker, in Benton, ed., 1988:266.]
  E.-L. Eoc.; Eu.

†*Mouillacitherium* Filhol, 1882:139.
  E.-L. Eoc.; Eu.

†*Hyperdichobune* Stehlin, 1910:1096.
  M.-L. Eoc.; Eu.

Tribe †**Dichobunini** Turner, 1849:158, **new rank**.
  [= †Dichobunina Turner, 1849:158.]
  E. Eoc.-L. Olig.; Eu. M. Eoc.; As.[1]
  [1] E. Asia.

†*Aumelasia* Sudre, 1980:203.
  E.-M. Eoc.; Eu.

†*Meniscodon* Rütimeyer, 1888:50-52.
  M. Eoc.; Eu.

†*Messelobunodon* Franzen, in Franzen & Krumbiegel, 1980:1553.
  M. Eoc.; Eu.

†*Dichobune* Cuvier, 1822:64, 70-71.
  M. Eoc.; As.[1] M. Eoc.-E. Olig.; Eu.
  [1] E. Asia.

†*Buxobune* Sudre, 1978:21.
  M. Eoc.; Eu.

†*Neufferia* Franzen, 1994:196.
  M. Eoc.; Eu.

†*Metriotherium* Filhol, 1882:99-103.
> [= †*Mesotherium* Filhol, 1880:1571-1580.[1]]
> L. Eoc.-L. Olig.; Eu.
> [1] Not †*Mesotherium* Serres, 1867: 140, a notoungulate.

†*Synaphodus* Pomel, 1848:325.
> E. Olig.; Eu.

Subfamily †**Diacodexeinae** Gazin, 1955:10, 22.
> [= †Trigonolestoidea Cope, 1898:130[1]; †Trigonolestidae Schlosser, 1899:349; †Diacodexinae Gazin, 1955:10, 22; †Diacodexidae Viret, 1961:891; †Diacodectidae Romer, 1966:388; †Diacodexeidae Krishtalka & Stucky, 1985:420.] [Including †Bunophorinae Stucky & Krishtalka, 1990:150.]
> E.-M. Eoc.; As. E.-M. Eoc.; Eu. E.-M. Eoc.; N.A.
> [1] †*Trigonolestes*, on which this was based, was long placed in the †Pantolestidae.

†*Diacodexis* Cope, 1882:1029.
> [Including †*Trigonolestes* Cope, 1894:868; †*Simpsonodus* Krishtalka & Stucky, 1986:186.]
> E.-M. Eoc.; As. E. Eoc.; Eu. E.-M. Eoc.; N.A.

†*Bunophorus* Sinclair, 1914:273.
> [Including †*Wasatchia* Sinclair, 1914:268.]
> E. Eoc.; Eu. E.-M. Eoc.; N.A.

†*Protodichobune* Lemoine, 1891:287-288.
> E. Eoc.; Eu.

†*Lutzia* Franzen, 1994:192.
> M. Eoc.; Eu.

†*Tapochoerus* McKenna, 1959:126.
> M. Eoc.; N.A.

†*Neodiacodexis* Atkins, in West & Atkins, 1970:16.
> M. Eoc.; N.A.

Subfamily †**Homacodontinae** Marsh, 1894:263.
> [= †Homacodontinae Peterson, 1919:66.] [Including †Antiacodontinae Gazin, 1958:1; †Antiacodontidae Romer, 1966:274; †Bunomerycidae Gentry & Hooker, in Benton, ed., 1988:266.]
> E.-M. Eoc.; N.A. M. Eoc.; Eu.

Tribe †**Bunomerycini** Gentry & Hooker, in Benton, ed., 1988:266, **new rank**.
> [= †Bunomerycidae Gentry & Hooker, in Benton, ed., 1988:266.]
> M. Eoc.; N.A.

†*Bunomeryx* Wortman, 1898:97.
> M. Eoc.; N.A.

†*Hylomeryx* Peterson, 1919:67.
> [Including †*Sphenomeryx* Peterson, 1919:71.]
> M. Eoc.; N.A.

†*Mesomeryx* Peterson, 1919:73.
> M. Eoc.; N.A.

†*Mytonomeryx* Gazin, 1955:35.
> M. Eoc.; N.A.

†*Pentacemylus* Peterson, 1931:72.
> M. Eoc.; N.A.

Tribe †**Homacodontini** Marsh, 1894:263, **new rank**.
> [= †Homacodontinae Marsh, 1894:263.]
> E.-M. Eoc.; N.A. M. Eoc.; Eu.

†*Hexacodus* Gazin, 1952:73.
> E. Eoc., ?M. Eoc.; N.A.

†*Antiacodon* Marsh, 1872:210-212.
[Including †*Sarcolemur* Cope, 1875:256.]
E.-M. Eoc.; N.A.

†*Eygalayodon* Sudre & Marandat, 1993:158.
M. Eoc.; Eu.

†*Homacodon* Marsh, 1872:126.
[Including †*Nanomeryx* Marsh, 1894:263-264.]
M. Eoc.; N.A.

†*Auxontodon* Gazin, 1958:2.
M. Eoc.; N.A.

†*Microsus* Leidy, 1870:113.
M. Eoc.; N.A.

†*Texodon* West, 1982:13.
M. Eoc.; N.A.

Subfamily †**Leptochoerinae** Marsh, 1894:273.
[= †Leptochoeridae Marsh, 1894:273; †Leptochoerinae Van Valen, 1971:525.]
M. Eoc.-L. Olig.; N.A.

†*Stibarus* Cope, 1873:3.
[= †*Menotherium* Cope, 1874:22-23.]
?M. Eoc., L. Eoc.-E. Olig.; N.A.

†*Ibarus* Storer, 1984:76.
M.-L. Eoc.; N.A.

†*Laredochoerus* Westgate, 1994:296.
M. Eoc.; N.A.

†*Leptochoerus* Leidy, 1856:88.
[= †*Laopithecus* Marsh, 1875:240.] [Including †*Nanochoerus* Macdonald, 1955:452.]
L. Eoc.-L. Olig.; N.A.

Family †**Cebochoeridae** Lydekker, 1883:146.
[= †Cebochoerinae Simpson, 1945:144.] [Including †Acotherulidae Lydekker, 1883:146.]
E. Eoc.-E. Olig.; Eu.

†*Cebochoerus* Gervais, 1848:4.[1]
[Including †*Choeromorus* Gervais, 1848:7[1]; †*Gervachoerus* Sudre, 1978:50.[2]]
E.-L. Eoc.; Eu.
[1] 1848-1852.
[2] As subgenus.

†*Acotherulum* Gervais, 1850:604.
[Including †*Leptacotherulum* Filhol, 1877:53-54[1]; †*Metadichobune* Filhol, 1877:53.]
M. Eoc.-E. Olig.; Eu.
[1] As subgenus.

†*Moiachoerus* Golpe-Posse, 1972:18.
L. Eoc.; Eu.

Family †**Mixtotheriidae** Lydekker, 1883:146.
[= †Mixtotheriodontidae Lydekker, 1883:146[1]; †Mixtotheriidae Pearson, 1927:431; †Mixtotheriinae Simpson, 1945:144.]
E.-L. Eoc.; Eu. ?L. Eoc.; Af.[2]
[1] "Based on †*Mixtotherium* despite the erroneous form" (Simpson, 1945:144).
[2] N. Africa.

†*Mixtotherium* Filhol, 1880:1580.
[Including †*Adrotherium* Filhol, 1883:94-96.]
E.-L. Eoc.; Eu. ?L. Eoc.; Af.[1]
[1] N. Africa.

Family †**Helohyidae** Marsh, 1877:364.
> [= †Helohyinae Gazin, 1955:37.] [Including †Achaenodontinae Zittel, 1893:334; †Achaenodontida Haeckel, 1895:522; †Achaenodontidae Matthew, 1899:34.]
> M. Eoc.; As. M. Eoc.; N.A.

†*Apriculus* Gazin, 1956:23.
> M. Eoc.; N.A.

†*Ithygrammodon* Osborn, Scott & Speir, 1878:56.[1]
> M. Eoc.; N.A.
> [1] Assignment to this family is questionable.

†*Helohyus* Marsh, 1872:207-208.
> [Including †*Thinotherium* Marsh, 1872:208.[1]]
> M. Eoc.; N.A.
> [1] Not †*Thinotherium* Cope, 1870: 292-293, a synonym of †*Cynorca* (Tayassuidae).

†*Lophiohyus* Sinclair, 1914:276.
> M. Eoc.; N.A.

†*Parahyus* Marsh, 1876:402.
> M. Eoc.; N.A.

†*Achaenodon* Cope, 1873:2.[1]
> [Including †*Protelotherium* Osborn, 1895:105.]
> ?M. Eoc.; As. M. Eoc.; N.A.
> [1] Cope originally spelled the name †*"Archaenodon"*, but all subsequent authors including Cope himself have spelled it †*"Achaenodon."* Palmer (1904:74) gives the rationale.

†*Gobiohyus* Matthew & Granger, 1925:7.
> M. Eoc.; As.

†*Pakkokuhyus* Holroyd & Ciochon, 1995:177.
> M. Eoc.; As.[1]
> [1] S.E. Asia (Burma).

Superfamily †**Anthracotherioidea** Leidy, 1869:202.
> [= †Anthracotheridae Leidy, 1869:202; †Anthracotherioidea Gill, 1872:11.] [Including †Merycopotamoidea Gill, 1872:10.]

Family †**Haplobunodontidae** Sudre, 1978:2, 90.[1]
> [= †Haplobunodontinae Kalandadze & Rautian, 1992:122.]
> ?E. Eoc., M.-L. Eoc.; Eu. M. Eoc.; As.[2] ?M. Eoc.; N.A. ?L. Eoc.; Af.[3] ?L. Eoc.; E. Indies.[4]
> [1] Referred to Pilgrim, 1941, by Sudre, but we have not found any mention of this family in Pilgrim's publication.
> [2] W. Asia (Turkey).
> [3] N. Africa (Egypt).
> [4] A lost specimen possibly of a haplobunodontid has been reported from Borneo.

†*Masillabune* Tobien, 1980:14.
> M. Eoc.; Eu.

†*Hallebune* Erfurt & Sudre, 1995:87.
> M. Eoc.; Eu.

†*Lophiobunodon* Depéret, 1908.
> M. Eoc.; Eu.

†*Haplobunodon* Depéret, 1908:3-4.
> M.-L. Eoc.; Eu.

†*Parabunodon* Ducrocq & Sen, 1991:13.
> M. Eoc.; As.[1]
> [1] W. Asia (Turkey).

†*Rhagatherium* Pictet & Humbert, 1855: plate III, fig. 1.[1]
> [= †*Rhogatherium* Gervais, 1867:255[2]; †*Ragatherium* Filhol, 1877:53.[3]]
> ?E. Eoc., M.-L. Eoc.; Eu. ?M. Eoc.; N.A. ?L. Eoc.; Af.[4]
> [1] 1855-1857.
> [2] 1867-1869. Misprint.

[3] Misprint.
[4] N. Africa (Egypt).

†*Anthracobunodon* Heller, 1934:247.
M. Eoc.; Eu.

†*Amphirhagatherium* Depéret, 1908:158-162.
L. Eoc.; Eu.

Family †**Anthracotheriidae** Leidy, 1869:202.
[= †Anthracotherida Haeckel, 1866:clviii[1]; †Anthracotheridae Leidy, 1869:202; †Anthracotheriidae Gill, 1872:11; †Anthracotheriinae Gill, 1872:11.] [Including †Merycopotamidae Gill, 1872:10; †Merycopotaminae Gaziry, in Boaz et al., 1987:297; †Hyopotaminae Gill, 1872:11[2]; †Hyopotamidae Kovalevskii, 1873:147[2]; †Ancodontidae Marsh, 1894:178[2]; †Bothriodontinae Scott, 1940:443.]
M. Eoc.-L. Mioc., Plioc., ?E. Pleist.; As. M. Eoc.-E. Mioc., ?M. Mioc., L. Mioc.; Eu. M. Eoc.-E. Mioc.; N.A. L. Eoc., E.-L. Mioc., ?E. Plioc.; Af. L. Eoc.; E. Indies.[3] L. Olig.; Mediterranean.[4]
[1] Haeckel's family-group name has priority, but should be suppressed. See the International Code of Zoological Nomenclature, Article 23(b).
[2] Objective synonym of †Bothriodontinae.
[3] Timor, possibly salted.
[4] Mallorca, Balearic Islands.

†*Probrachyodus* Xu & Chiu, 1962.
M. Eoc., ?L. Eoc.; As. ?E. Olig.; Eu.

†*Anthracotherium* G. Cuvier, 1822:336-337.
?M. Eoc., L. Olig.-L. Mioc.; As. E. Olig.-E. Mioc.; Eu. L. Olig.; Mediterranean.[1]
[1] Mallorca, Balearic Islands.

†*Brachyodus* Depéret, 1895:397-408.
M. Eoc.-E. Olig., E.-M. Mioc.; As. L. Olig.; Mediterranean.[1] E. Mioc.; Af. E. Mioc., ?M. Mioc.; Eu.
[1] Mallorca, Balearic Islands.

†*Atopotherium* Ducrocq, Chaimanee, Suteethorn & Jaeger, 1996:391.
L. Eoc.; As.[1]
[1] S.E. Asia (Thailand).

†*Anthracokeryx* Pilgrim & Cotter, 1916:61.
M. Eoc.-E. Olig.; As.

†*Bakalovia* Nikolov & Heissig, 1985:69.
M.-L. Eoc.; As. L. Eoc.; Eu.[1]
[1] E. Europe (Bulgaria).

†*Prominatherium* Teller, 1884:133.
M. Eoc., E. Olig.; Eu.

†*Ulausuodon* Hu, 1963:310-317.
M. Eoc.; As.

†*Anthracosenex* Zdansky, 1930:14.
M. Eoc.; As.

†*Siamotherium* Suteethorn, Buffetaut, Helmcke-Ingavat, Jaeger & Jongkanjanasoontorn, 1988:566.
M. Eoc.; As.[1]
[1] S.E. Asia (S. Thailand).

†*Anthracothema* Pilgrim, 1928.
M.-L. Eoc.; As. L. Eoc.; E. Indies.[1]
[1] Timor, possibly salted.

†*Diplopus* Kowalevsky, 1873:149.
L. Eoc.; Eu.

†*Heptacodon* Marsh, 1894:409.
[Including †*Octacodon* Marsh, 1894:92.]
M. Eoc.-E. Olig.; N.A.

†*Huananothema* Tang, 1978:13.
      L. Eoc.; As.[1]
    [1] E. Asia (China).

†*Elomeryx* Marsh, 1894:176-177.
      L. Eoc.-E. Mioc.; Eu. E.-L. Olig.; N.A. L. Olig.; As.[1]
    [1] W. Asia.

†*Thaumastognathus* Filhol, 1890:34-38.[1]
      L. Eoc.; Eu.
    [1] "Spelled *'Taumastognathus'* and *'Taumastognatus'* in the original publication, but the fact that one must be a misprint suggests that both are and that the now universal emended spelling is permissible" (Simpson 1945:147).

†*Heothema* Tang, 1978:14.
      L. Eoc.-E. Olig.; As.[1]
    [1] E. Asia (China).

†*Anthracohyus* Pilgrim & Cotter, 1916:48.
      M. Eoc.; As.[1] L. Eoc.; Eu.
    [1] S.E. Asia (Burma).

†*Bothriogenys* Schmidt, 1913:6.[1]
      L. Eoc., E. Mioc.; Af.[2]
    [1] Proposed as subgenus of †*Brachyodus.*
    [2] N. Africa.

†*Bothriodon* Aymard, 1846:239, 246-247.
      [= †*Bothryodon* Gaudry, 1866:355; †*Ancodon* Pomel, 1847:207[1]; †*Ancodus* Pomel, 1848:324-325.[2]] [Including †*Hyopotamus* Owen, 1848:103-126.[3]]
      L. Eoc.-E. Mioc.; As. L. Eoc.-E. Olig.; Eu. L. Eoc.; N.A.
    [1] Proposed as subgenus of †*Palaeotherium.*
    [2] Proposed as genus.
    [3] Not †*Hyopotamus* Kaup, 1844, another artiodactyl.

†*Aepinacodon* Troxell, 1921.
      L. Eoc.-E. Olig.; N.A.

†*Pachychoerops* Kretzoi, 1941:349.
      [= †*Isodactylus* Dal Piaz, 1927.[1]]
      Olig.; Eu.
    [1] Proposed as subgenus of †*Anthracotherium.* Not *Isodactylus* Gray, 1845, a reptile.

†*Bunobrachyodus* Depéret, 1908:162.
      E.-L. Olig.; Eu.

†*Anthracochoerus* Dal Piaz, 1930.
      E. Olig.; Eu.

†*Microbunodon* Depéret, 1908:2.
      [Including †*Microselenodon* Depéret, 1908:2.]
      L. Olig.; Eu. ?E. Mioc.; As.[1]
    [1] S. Asia.

†*Kukusepasutanka* Macdonald, 1956:641.
      L. Olig.; N.A.

†*Sivameryx* Lydekker, 1883:169.[1]
      [= †*Hyoboops* Trouessart, 1904:651.] [Including †*Merycops* Pilgrim, 1910:68.]
      L. Olig.-M. Mioc.; As. E. Mioc.; Af.
    [1] †*Sivameryx* was suggested by Lydekker, 1878:80, but the name was not proposed adequately until 1883.

†*Hemimeryx* Lydekker, 1883:167-169.[1]
      [= †*Gelasmodon* Forster-Cooper, 1913:515.]
      L. Olig.-M. Mioc., ?L. Mioc.; As. L. Mioc.; Eu.
    [1] *Nomen nudum* in Lydekker, 1878:79-80.

†*Arretotherium* Douglass, 1901:269-278.
      L. Olig.-E. Mioc.; N.A.

†*Telmatodon* Pilgrim, 1907:51.
    L. Olig.-M. Mioc.; As.

†*Parabrachyodus* Forster Cooper, 1915:404.
    L. Olig.-E. Mioc.; As.

†*Bugtitherium* Pilgrim, 1908:148, 153.
    E. Mioc.; As.[1]
    [1] S. Asia (Pakistan).

†*Gonotelma* Pilgrim, 1908:148.
    E. Mioc.; As.

†*Masritherium* Fourtau, 1918:195-196.
    E. Mioc.; Af. E. Mioc.; As.[1]
    [1] S.W. Asia (Saudi Arabia).

†*Choeromeryx* Pomel, 1848:687.
    [= †*Chaeromeryx* Lydekker, 1885:37.]
    E.-L. Mioc.; As.[1]
    [1] S. Asia.

†*Merycopotamus* Falconer & Cautley, in Owen, 1845:566.[1]
    [Including †*Libycosaurus* Bonarelli, 1947:26.]
    M.-L. Mioc., ?E. Plioc.; Af. M.-L. Mioc., Plioc., ?E. Pleist.; As.[2]
    [1] Owen, who wrote the manuscript, stated that Cautley and Falconer had found and "determined" †*Merycopotamus*, but no species was mentioned. Lydekker, 1885:209, made †*Hippopotamus dissimilis* Falconer & Cautley, 1836:51, the type species, citing Falconer & Cautley, in Owen, 1845 as the authors of †*Merycopotamus*.
    [2] S. Asia.

†*Afromeryx* Pickford, 1991:1499.
    E.-M. Mioc.; Af. E. and/or M. Mioc.; As.[1]
    [1] S.W. Asia.

Superfamily †**Anoplotherioidea** Gray, 1821:306.
    [= †Anoplotheriadae Gray, 1821:306; †Anoplotheroidea Romer, 1966:389; †Anoplotherioidea Sudre, 1977:213, 216.]

Family †**Dacrytheriidae** Depéret, 1917:114.
    [Including †Tapirulidae Cope, 1879:228[1]; †Dacrytheriinae Viret, 1961:936.]
    E. Eoc.-E. Olig.; Eu.
    [1] †Tapirulidae is the valid name for this family according to the Principle of Priority [International Code of Zoological Nomenclature, Article 23 (a) and (d)]; however, the continued use of †Dacrytheriidae promotes stability, the purpose of the Principle of Priority [Article 23(b)]. This issue should be referred to the Commission for a ruling.

†*Cuisitherium* Sudre, D. E. Russell, Louis & D. E. Savage, 1983:281, 349.
    E. Eoc.; Eu.

†*Dacrytherium* Filhol, 1876:288.
    M.-L. Eoc.; Eu.

†*Tapirulus* Gervais, 1850:604.
    M. Eoc.-E. Olig.; Eu.

†*Catodontherium* Depéret, 1908.[1]
    [= †*Catodus* Depéret, 1905.]
    M.-L. Eoc.; Eu.
    [1] "Proposed on the grounds that †*Catodus* is preoccupied by *Catodon* Linnaeus, 1761. This is not preoccupation, but †*Catodus* was a *nomen nudum* in its earlier publication (1905) so that †*Catodontherium* may be retained" (Simpson 1945:147).

†*Leptotheridium* Stehlin, 1910:912.
    M.-L. Eoc.; Eu.

Family †**Anoplotheriidae** Gray, 1821:306.
    [= †Anoplotheriadae Gray, 1821:306; †Anoplotheriidae Bonaparte, 1850; †Anoplotherida Haeckel, 1866:clviii.]
    M. and/or L. Eoc.; Mediterranean. M. Eoc.-L. Olig.; Eu.

Subfamily †**Robiaciinae** Sudre, 1977:213, 216.
  M.-L. Eoc.; Eu.

†*Robiacina* Sudre, 1969:132.
  M.-L. Eoc.; Eu.

Subfamily †**Anoplotheriinae** Gray, 1821:306.
  [= †Anoplotheriadae Gray, 1821:306; †Anoplotherina C. L. Bonaparte, 1837:8[1];
  †Anoplotheriinae Viret, 1961:938.]
  M. and/or L. Eoc.; Mediterranean. M. Eoc.-L. Olig.; Eu.
  [1] As subfamily.

†*Robiatherium* Sudre, 1988:141.
  M. Eoc.; Eu.

†*Anoplotherium* G. Cuvier, 1804:370-382.
  [= †*Oplotherium* Laizer & Parieu, 1838:276-277; †*Pleregnathus* Laizer & Parieu,
  1838:341; †*Hoplotherium* Meyer, 1841:461.]
  M. and/or L. Eoc.; Mediterranean. L. Eoc.-E. Olig.; Eu.

†*Diplobune* Rütimeyer, 1862:74.[1]
  [= or including †*Thylacomorphus* Gervais, 1876:52.[2]] [Including †*Hyrocodon* Filhol,
  1873:88; †*Hyracodon* Filhol, 1876:288-289[3]; †*Hyracodontherium* Filhol, 1877:153-
  156.[4]]
  M. and/or L. Eoc.; Mediterranean. L. Eoc.-E. Olig.; Eu.
  [1] Proposed as subgenus of †*Dichobune*.
  [2] Synonymy of this *nomen oblitum* is questionable.
  [3] Emendation of †*Hyrocodon* Filhol, 1873. Not †*Hyracodon* Leidy, 1856, a perissodactyl, nor *Hyracodon*
  Tomes, 1863, a marsupial.
  [4] Replacement name for †*Hyracodon* Filhol, 1876.

†*Ephelcomenus* Hürzeler, 1938:318.
  E.-L. Olig.; Eu.

†*Deilotherium* Filhol, 1882:112-113.[1]
  Olig.; Eu.
  [1] Assignment to the anoplotheres is questionable, but we follow Palmer (1904:219). If an anoplothere, then it
  is probably an anoplotheriine. †*Deilotherium* is not a prior homonym of †*Delotherium* Ameghino,
  1889:655, 920.

Family †**Cainotheriidae** Cope, 1881:378.
  [= †Caenotheriidae Cope, 1881:378; †Cainotheridae Rütimeyer, 1891:87;
  †Caenotherida Haeckel, 1895:552; †Caenotherioidea Weber, 1928:623[1];
  †Cainotherioidea Camp & VanderHoof, 1940:339; †Cainotheriidae Camp &
  VanderHoof, 1940:339.]
  L. Eoc.-M. Mioc.; Eu. Olig.; Mediterranean.[2]
  [1] As suborder.
  [2] An unnamed taxon has been reported from the Oligocene of the Balearic Islands.

Subfamily †**Oxacroninae** Hürzeler, 1936:94.
  [= †Oxacronide Hürzeler, 1936:94; †Oxacroninae Viret, 1961:957.]
  L. Eoc.-E. Olig.; Eu.

†*Paroxacron* Hürzeler, 1936:105.
  L. Eoc.-E. Olig.; Eu.

†*Oxacron* Filhol, 1884:64-65.
  L. Eoc.-E. Olig.; Eu.

Subfamily †**Cainotheriinae** Cope, 1881:378.
  [= †Caenotheriidae Cope, 1881:378; †Caenotheriinae Osborn, 1910:548;
  †Cainotheriinae Viret, 1961:958.]
  E. Olig.-M. Mioc.; Eu.

†*Procaenotherium* Hürzeler, 1936:95.
  E. Olig.; Eu.

†*Plesiomeryx* Gervais, 1873:369.
  [= †*Plesiomaeryx* Gervais, 1873:369[1]; †*Plesiomeryx* Gervais, 1876:45.[2]]

E.-L. Olig.; Eu.

[1] Misprint.
[2] Correction.

†*Caenomeryx* Hürzeler, 1936:100.
    E.-L. Olig.; Eu.

†*Cainotherium* Bravard, 1828:90, 113.
    [= †*Caenotherium* Agassiz, 1846:57; †*Crinotherium* Filhol, 1882:42.[1]]
    E. Olig.-M. Mioc.; Eu.

[1] Misprint.

Superfamily †**Oreodontoidea** Leidy, 1869:7, 379.
    [= †Oreodontidae Leidy, 1869:7, 379; †Oreodontoidea Gill, 1872:10.] [= or including
    †Merycoidodontoidea Thorpe, 1937:23.]

†*Hadrohyus* Leidy, 1872:248.[1]
    L. Olig. and/or E. Mioc.; N.A.

[1] Possibly best regarded as a *nomen oblitum*.

†*Eucrotaphus* Leidy, 1850:90.[1]
    Olig.; N.A.

[1] Possibly should be regarded as a *nomen dubium*.

Family †**Agriochoeridae** Leidy, 1869:131.[1]
    [= †Agriochoerinae Gill, 1872:10; †Agriochoeroidae Hay, 1930:776.] [Including
    †Protoreodontinae Scott, 1890:320, 361; †Protoreodontidae Scott, 1890:503.]
    M. Eoc.-E. Mioc.; N.A.

[1] "Spelled Agriochaeridae in 1869, Agriochoeridae by Leidy in 1871, but since Leidy used the generic
    spelling *Agriochoerus* the former must be a misprint" (Simpson 1945:148).

†*Diplobunops* Peterson, 1919:76.
    M. Eoc.; N.A.

†*Agriochoerus* Leidy, 1850:121-122.[1]
    [Including †*Coloreodon* Cope, 1879:6; †*Merycopater* Cope, 1879:197; †*Artionyx*
    Osborn & Wortman, 1893:5; †*Agriomeryx* Marsh, 1894:270-271.]
    M. Eoc.-E. Mioc.; N.A.

[1] 1850-1851.

†*Protoreodon* Scott & Osborn, 1887:257-258.
    [= †*Hyomeryx* Marsh, 1894:268.] [Including †*Eomeryx* Marsh, 1894:364, 365[1];
    †*Agriotherium* Scott, 1898:79-81[2]; †*Chorotherium* Berg, 1899:79[3]; †*Protagriochoerus*
    Scott, 1899:100-111; †*Mesagriochoerus* Peterson, 1934:377.]
    M.-L. Eoc.; N.A.

[1] "Thorpe credits †*Eomeryx* to Marsh in 1875, and on this showing he should have used this name instead of
    †*Protoreodon*. The fact, however, appears to be that Marsh did not publish †*Eomeryx* until 1877 and that it
    remained essentially a *nomen nudum* until 1894, so that it is antedated by †*Protoreodon*" (Simpson
    1945:148).
[2] Not †*Agriotherium* Wagner, 1837:335, a carnivoran.
[3] Replacement for †*Agriotherium* Scott, 1898.

Family †**Oreodontidae** Leidy, 1869:7, 379.[1]
    [= †Cotylopidae Lydekker, 1889:1326; †Oreodontida Haeckel, 1895:530, 558;
    †Merycoidodontinae Hay, 1902:665; †Merycoidodontidae Thorpe, 1923:239.]
    M. Eoc.-L. Mioc.; N.A.   ?Mioc.; Cent. A.

[1] For use of †Oreodontidae rather than †Merycoidodontidae see the International Code of Zoological
    Nomenclature, Articles 23(d), 36, 40, 41, 65, 70. Article 40 does not apply if †*Oreodon* and
    †*Merycoidodon* cannot be shown to be synonymous. The matter should be decided by the International
    Commission on Zoological Nomenclature. These animals have always been called oreodonts.

Subfamily †**Oreonetinae** Schultz & Falkenbach, 1956:453.
    L. Eoc.; N.A.

†*Bathygenys* Douglass, 1901:20.
    L. Eoc.; N.A.

†*Megabathygenys* Schultz & Falkenbach, 1968:387.
    L. Eoc.; N.A.

†*Oreonetes* Loomis, 1924:370.
    L. Eoc.; N.A.

†*Parabathygenys* Schultz & Falkenbach, 1968:388.
    L. Eoc.; N.A.

Subfamily †**Leptaucheniinae** Schultz & Falkenbach, 1940:215.
    M. Eoc.-L. Olig.; N.A.

Tribe †**Leptaucheniini** Schultz & Falkenbach, 1968:258.
    M. Eoc.-L. Olig.; N.A.

†*Limnenetes* Douglass, 1901:259.
    M.-L. Eoc.; N.A.

†*Pseudocyclopidius* Schultz & Falkenbach, 1968:323.
    E.-L. Olig.; N.A.

†*Hadroleptauchenia* Schultz & Falkenbach, 1968:303.
    E.-L. Olig.; N.A.

†*Leptauchenia* Leidy, 1856:88.
    E.-L. Olig.; N.A.

†*Pithecistes* Cope, 1878:219.
    E.-L. Olig.; N.A.

†*Cyclopidius* Cope, 1878:221.
    [= †*Brachymeryx* Cope, 1878:220.] [Including †*Chelonocephalus* Thorpe, 1921:415.[1]]
    L. Olig.; N.A.
    [1] Proposed as subgenus.

Tribe †**Sespiini** Schultz & Falkenbach, 1968:238.
    L. Olig.; N.A.

†*Megasespia* Schultz & Falkenbach, 1968:255.
    L. Olig.; N.A.

†*Sespia* Stock, 1930:38.
    L. Olig.; N.A.

Subfamily †**Oreodontinae** Leidy, 1869:7, 379.
    [= †Oreodontidae Leidy, 1869:7, 379; †Oreodontinae Gill, 1872:81; †Cotylopinae Flower & Lydekker, 1891:763; †Merycoidodontinae Hay, 1902:665.]
    L. Eoc.-E. Mioc.; N.A.

†*Aclistomycter* J. A. Wilson, 1971:40.
    L. Eoc.; N.A.

†*Merycoidodon* Leidy, 1848:47.
    [= †*Oreodon* Leidy, 1851:237[1]; †*Cotylops* Leidy, 1851:237.[1]] [Including †*Anomerycoidodon* Schultz & Falkenbach, 1968:73[2]; †*Blickohyus* Schultz & Falkenbach, 1968:79.[2]]
    L. Eoc.-E. Olig.; N.A.
    [1] Synonymy questionable.
    [2] Proposed as subgenus.

†*Otionohyus* Schultz & Falkenbach, 1968:106.
    [Including †*Otarohyus* Schultz & Falkenbach, 1968:116.[1]]
    L. Eoc.-E. Olig.; N.A.
    [1] Proposed as subgenus.

†*Paramerycoidodon* Schultz & Falkenbach, 1968:86.
    [Including †*Barbourochoerus* Schultz & Falkenbach, 1968:90[1]; †*Gregoryochoerus* Schultz & Falkenbach, 1968:98.[1]]
    E.-L. Olig.; N.A.
    [1] Proposed as subgenus.

†*Genetochoerus* Schultz & Falkenbach, 1968:135.
    [Including †*Osbornohyus* Schultz & Falkenbach, 1968:142.[1]]

E. Olig.; N.A.
[1] Proposed as subgenus.

†*Pseudogenetochoerus* Schultz & Falkenbach, 1968:156.
E. Mioc.; N.A.

†*Epigenetochoerus* Schultz & Falkenbach, 1968:163.
E. Mioc.; N.A.

Subfamily †**Miniochoerinae** Schultz & Falkenbach, 1956:391.
L. Eoc.-E. Olig.; N.A.

†*Stenopsochoerus* Schultz & Falkenbach, 1956:435.
[Including †*Pseudostenopsochoerus* Schultz & Falkenbach, 1956:443.[1]]
L. Eoc.-E. Olig.; N.A.
[1] Proposed as subgenus.

†*Platyochoerus* Schultz & Falkenbach, 1956:425.
E. Olig.; N.A.

†*Parastenopsochoerus* Schultz & Falkenbach, 1956:450.
E. Olig.; N.A.

†*Miniochoerus* Schultz & Falkenbach, 1956:391.
[Including †*Paraminiochoerus* Schultz & Falkenbach, 1956:402.[1]]
E. Olig.; N.A.
[1] Proposed as subgenus.

Subfamily †**Desmatochoerinae** Schultz & Falkenbach, 1954:163.
L. Eoc.-E. Mioc.; N.A.

†*Prodesmatochoerus* Schultz & Falkenbach, 1954:225.
L. Eoc.-E. Olig.; N.A.

†*Subdesmatochoerus* Schultz & Falkenbach, 1954:217.
E. Olig.; N.A.

†*Desmatochoerus* Thorpe, 1921:241.[1]
[Including †*Paradesmatochoerus* Schultz & Falkenbach, 1954:193.[2]]
L. Olig.-E. Mioc.; N.A.
[1] Proposed as subgenus of †*Promerycochoerus*.
[2] Proposed as subgenus of †*Desmatochoerus*.

†*Megoreodon* Schultz & Falkenbach, 1954:163.
[Including †*Superdesmatochoerus* Schultz & Falkenbach, 1954:213.]
L. Olig.-E. Mioc.; N.A.

†*Pseudodesmatochoerus* Schultz & Falkenbach, 1954:203.
L. Olig.-E. Mioc.; N.A.

Subfamily †**Promerycochoerinae** Schultz & Falkenbach, 1949:84.
E. Olig.-E. Mioc.; N.A.

†*Promesoreodon* Schultz & Falkenbach, 1949:152.
E. Olig.; N.A.

†*Promerycochoerus* Douglass, 1901:82.
[= †*Pseudopromerycochoerus* Schultz & Falkenbach, 1949:121.[1]] [Including
†*Paracotylops* Matthew, in Merriam, 1901:296; †*Hypselochoerus* Loomis, 1933:728;
†*Parapromerycochoerus* Schultz & Falkenbach, 1949:114.[1]]
L. Olig.-E. Mioc.; N.A.
[1] Proposed as subgenus.

†*Merycoides* Douglass, 1907:101.
L. Olig.-E. Mioc.; N.A.

†*Mesoreodon* Scott, 1893:659.
L. Olig.; N.A.

Subfamily †**Merychyinae** Simpson, 1945:149.
L. Olig.-M. Mioc.; N.A.

†*Oreodontoides* Thorpe, 1921:107.
    L. Olig.-E. Mioc.; N.A.

†*Paroreodon* Thorpe, 1921:109.
    L. Olig. and/or E. Mioc.; N.A.

†*Paramerychyus* Schultz & Falkenbach, 1947:247.
    E. Mioc.; N.A.

†*Merychyus* Leidy, 1858:25.
    [Including †*Metoreodon* Matthew & Cook, 1909:391.[1]]
    E.-M. Mioc.; N.A.
    [1] As subgenus.

Subfamily †**Eporeodontinae** Schultz & Falkenbach, 1940:215.
    L. Olig. and/or E. Mioc.; N.A.

†*Dayohyus* Schultz & Falkenbach, 1968:215.
    L. Olig. and/or E. Mioc.; N.A.

†*Eporeodon* Marsh, 1875:249.
    [Including †*Paraeporeodon* Schultz & Falkenbach, 1968:204.[1]]
    L. Olig. and/or E. Mioc.; N.A.
    [1] Proposed as subgenus.

Subfamily †**Phenacocoelinae** Schultz & Falkenbach, 1950:100.
    E. Mioc.; N.A.

†*Phenacocoelus* Peterson, 1906:24, 29.
    E. Mioc.; N.A.

†*Submerycochoerus* Schultz & Falkenbach, 1950:124.
    E. Mioc.; N.A.

†*Pseudomesoreodon* Schultz & Falkenbach, 1950:128.
    E. Mioc.; N.A.

†*Hypsiops* Schultz & Falkenbach, 1950:113.
    E. Mioc.; N.A.

Subfamily †**Ticholeptinae** Schultz & Falkenbach, 1940:215.[1]
    E.-L. Mioc.; N.A.
    [1] Named, 1940. First diagnosed, 1941.

†*Mediochoerus* Schultz & Falkenbach, 1941:92.
    E.-M. Mioc.; N.A.

†*Ticholeptus* Cope, 1878:129.
    [Including †*Poatrephes* Douglass, 1903:176.]
    E.-M. Mioc.; N.A.

†*Ustatochoerus* Schultz & Falkenbach, 1941:10.
    M.-L. Mioc.; N.A.

Subfamily †**Merycochoerinae** Schultz & Falkenbach, 1940:216.
    E.-M. Mioc.; N.A.   ?Mioc.; Cent. A.

†*Merycochoerus* Leidy, 1858:24.
    E. Mioc.; N.A.

†*Brachycrus* Matthew, 1901:397.[1]
    [= †*Pronomotherium* Douglass, 1907:94.]
    E.-M. Mioc.; N.A.   ?Mioc.; Cent. A.
    [1] Proposed as subgenus of †*Merycochoerus*.

Superfamily †**Entelodontoidea** Lydekker, 1883:146.
    [= †Entelodontidae Lydekker, 1883:146; †Entelodontoidea Colbert, 1938:105.]

Family †**Entelodontidae** Lydekker, 1883:146.
    [= †Entelodontinae Osborn, 1909:61; †Elotheriidae Alston, 1878:18; †Elotheriinae
    Cope, 1888:1089.]
    M. Eoc.-E. Mioc.; As.   M. Eoc.-E. Mioc.; N.A.   E.-L. Olig.; Eu.

†*Eoentelodon* Chow, 1958:30.
M.-L. Eoc.; As.[1]

[1] E. Asia.

†*Brachyhyops* Colbert, 1937:87.
[Including †*Dyscritochoerus* Gazin, 1956:26.]
M.-L. Eoc.; N.A.

†*Archaeotherium* Leidy, 1850:92-93.
[Including †*Arctodon* Leidy, 1851:278; †*Pelonax* Cope, 1874:504-505[1]; †*Megachoerus* Troxell, 1920:433[1]; †*Choerodon* Troxell, 1920:442[1]; †*Scaptohyus* Sinclair, 1921:480.[1]]
L. Eoc.-E. Mioc.; N.A.

[1] As subgenus.

†*Paraentelodon* Gabunia, 1964:109.
E. Olig.; Eu.[1] L. Olig.-E. Mioc.; As.

[1] E. Europe.

†*Entelodon* Aymard, 1846:240-242.
[= †*Elotherium* Pomel, 1847:307-308.[1]] [Including †*Oltinotherium* Delfortrie, 1874:261-263; †*Elodon* Kretzoi, 1941:272, 347; †*Brachyodon* Trofimov, 1952:146[2]; †*Ergilobia* Trofimov, 1958:119[3]; †*Gobielodon* Kretzoi, 1968:164.[3]]
L. Eoc.-E. Olig.; As. E.-L. Olig.; Eu. ?E. Olig.; N.A.

[1] "The name †*Elotherium* probably has priority and is often used (necessitating changes of subfamily, family, and superfamily names as well), but recently †*Entelodon* is more common in the literature. As Peterson has shown, it is not certain that †*Elotherium* is prior, and the type specimen was inadequate, poorly described, unfigured, and is lost. No one but Pomel ever saw it, and it is fair to say that his genus was not recognizably established. †*Entelodon* was firmly established, and the name can legitimately continue in use" (Simpson, 1945: 144).
[2] Not †*Brachyodon* Lartet, 1868: 1121, an ungulate; nor *Brachyodon* Jukes-Browne, 1905: 222, a mollusk.
[3] New name for †*Brachyodon* Trofimov, 1952.

†*Daeodon* Cope, 1878:15.
[= †*Daledon* Zittel, 1892:304; †*Dalodon* Zittel, 1893:308.] [Including †*Boochoerus* Cope, 1879:59-67[1]; †*Ammodon* Marsh, 1893:409-410; †*Dinochoerus* Peterson, 1905:212[2]; †*Dinohyus* Peterson, 1905:719.[3]]
?E. Olig., L. Olig.-E. Mioc.; N.A.

[1] Synonymy questionable.
[2] Not *Dinochoerus* Gloger, 1841: 141, a synonym of *Phacochoerus*.
[3] Replacement name for †*Dinochoerus* Peterson, 1905.

†*Neoentelodon* Aubekerova, 1969:47-52.
L. Olig.; As.

Suborder **TYLOPODA** Illiger, 1811:102.
[= HYDROPHORAE Gray, 1821:307.[1]] [Including CAMELIDA Kalandadze & Rautian, 1992:123.[2]]

[1] Named as an order.
[2] Used as infraorder of TYLOPODA. Not the same concept as Haeckel's family Camelida.

Family †**Xiphodontidae** Flower, 1884:xxviii, 335.
[= †Xiphodonta Haeckel, 1866:clviii[1]; †Xiphodontoidea Viret, 1961:933.] [Including †Dichodontidae Cope, 1874:26.]
M.-L. Eoc., ?E. Olig.; Eu.

[1] Described as a family. Haeckel's name has priority over that proposed by Flower, but the matter should be taken up by the International Commission on Zoological Nomenclature according to the International Code of Zoological Nomenclature, Article 23(b).

†*Dichodon* Owen, 1848:36-42.
[Including †*Tetraselenodon* Schlosser, 1886:134.]
M.-L. Eoc.; Eu.

†*Haplomeryx* Schlosser, 1886:96.
M.-L. Eoc.; Eu.

†*Xiphodon* G. Cuvier, 1822:60-62.[1]
M.-L. Eoc., ?E. Olig.; Eu.

[1] Described as subgenus of †*Anoplotherium*.

†*Paraxiphodon* Sudre, 1978:129.
L. Eoc.; Eu.

Superfamily **Cameloidea** Gray, 1821:307.
[= Camelidae Gray, 1821:307; Cameloidea Cope, 1891:87, 89.]

Family **Camelidae** Gray, 1821:307. Camels.
[= Camelina Gray, 1825:342[1]; Camelida Haeckel, 1866:clviii.[2]] [Including
†Protolabididae Cope, 1884:16; †Eschatiidae Cope, 1887:379, 394.]
M. Eoc.-L. Pleist.; N.A. L. Mioc., ?E. Plioc., L. Plioc.-R.; Eu.[3] L. Mioc., L.
Pleist.; Cent. A. E. Plioc.-R.; As. L. Plioc.-R.; Af.[4] L. Plioc.-R.; S.A.[5]
[1] Used as tribe of Bovidae.
[2] Used as family.
[3] Extant only in domestication.
[4] Extant only in domestication.
[5] Also feral in Australia by introduction.

†*Miotylopus* Schlaikjer, 1935:174.
L. Olig.-E. Mioc.; N.A.

Subfamily †**Poebrodontinae** J. A. Wilson, 1974:26.
[= †Poebrodoninae Wilson, 1974:26.]
M.-L. Eoc.; N.A.

†*Poebrodon* Gazin, 1955:78.
M. Eoc.; N.A.

†*Hidrosotherium* J. A. Wilson, 1974:26.
L. Eoc.; N.A.

Subfamily †**Poebrotheriinae** Cope, 1874:26.
[= †Poebrotheriidae Cope, 1874:26; †Poebrotherinae Zittel, 1893:361.]
L. Eoc.-E. Olig., E. Mioc.; N.A.

†*Poebrotherium* Leidy, 1847:322-326.
[Including †*Protomeryx* Leidy, 1856:164.[1]]
L. Eoc.-E. Olig.; N.A.
[1] Possibly a *nomen dubium.*

†*Paralabis* Lull, 1921:392.[1]
E. Olig.; N.A.
[1] Proposed as a subgenus of †*Pseudolabis,* †*Paralabis* is either valid at the generic level or should be merged
with †*Poebrotherium.*

†*Paratylopus* Matthew, 1904:211.[1]
E. Olig.; N.A.
[1] Described as subgenus of †*Miolabis.*

†*Gentilicamelus* Loomis, 1936:59.
[= †*Gomphotherium* Cope, 1886:618-620.[1]]
E. Mioc.; N.A.
[1] Not †*Gomphotherium* Burmeister, 1837, a proboscidean, nor †*Gomphotherium* Filhol, 1884: 62-63, a
synonym of the lipotyphlan †*Tetracus.*

Subfamily †**Pseudolabidinae** Simpson, 1945:150.
E. Olig.; N.A.

†*Pseudolabis* Matthew, 1904:211.
E. Olig.; N.A.

Subfamily **Camelinae** Gray, 1821:307.
[= Camelidae Gray, 1821:307; Camelinae Zittel, 1893:364.] [Including †Protolabinae
Zittel, 1893:363.]
L. Olig.-L. Pleist.; N.A. L. Mioc., ?E. Plioc., L. Plioc.-R.; Eu.[1] L. Mioc., L.
Pleist.; Cent. A. E. Plioc.-R.; As. L. Plioc.-R.; Af.[2] L. Plioc.-R.; S.A.[3]
[1] Extant only in domestication.
[2] Extant only in domestication.
[3] Also feral in Australia by introduction.

†*Dyseotylopus* Stock, 1935:121.
L. Olig.; N.A.

†*Nothotylopus* Patton, 1969:160.
M. Mioc.; N.A.

Tribe †**Protolabidini** Cope, 1884:16.
[= †Protolabididae Cope, 1884:16; †Protolabinae Zittel, 1893:363; †Protolabidini Webb, 1965:44.]
E.-L. Mioc.; N.A. L. Mioc.; Cent. A.

†*Tanymykter* Honey & Taylor, 1978:375.
E. Mioc.; N.A.

†*Michenia* Frick & Taylor, 1971:5.
E.-L. Mioc.; N.A.

†*Protolabis* Cope, 1876:144.
E.-L. Mioc.; N.A. L. Mioc.; Cent. A.

Tribe **Lamini** Webb, 1965:44.
[= Auchenina Bonaparte, 1845:4; Aucheniina Bonaparte, 1850: unpaged chart; Lamina Harrison, 1979:8.] [Including †Eschatiidae Cope, 1887:379, 394; Camelopini Webb, 1965:44.]
M. Mioc.-L. Pleist.; N.A. L. Plioc.-R.; S.A. L. Pleist.; Cent. A.

†*Hemiauchenia* H. Gervais & Ameghino, 1880:120-123.
[Including †*Tanupolama* Stock, 1928:29; †*Prochenia* Frick, 1929:107.[1]]
M. Mioc.-L. Pleist.; N.A. E.-L. Pleist.; S.A.
[1] Simpson (1945:150) considered †*Prochenia* a *nomen nudum*, perhaps a subjective synonym of †*Tanupolama*.

†*Pliauchenia* Cope, 1875:258.
L. Mioc.; N.A.

†*Alforjas* Harrison, 1979:12.
L. Mioc.-E. Plioc.; N.A.

†*Blancocamelus* Dalquest, 1975:37.
[= †*Leptotylopus* Meade, 1945:538.[1]]
L. Plioc.; N.A.
[1] Not †*Leptotylopus* Matthew, 1924 (unpublished MS), which would have been a synonym of †*Hemiauchenia*.

†*Camelops* Leidy, 1854:172-173.
[Including †*Palauchenia* del Castillo, 1869:481; †*Holomeniscus* Cope, 1884:16; †*Eschatius* Cope, 1884:18[1]; †*Eschatinus* Sclater, 1886:43.[2]]
L. Plioc.-L. Pleist.; N.A. L. Pleist.; Cent. A.
[1] A *nomen dubium* based on deciduous teeth, a possible synonym of †*Camelops*.
[2] Possibly a *lapsus calami* rather than an attempted correction.

†*Palaeolama* P. Gervais, 1867:281.
[Including †*Protauchenia* Branco, 1883:110-126[1]; †*Astylolama* Churcher, 1965:199.[1]]
L. Plioc.-L. Pleist.; S.A. E.-L. Pleist.; N.A.
[1] As subgenus.

Lama G. Cuvier, 1800: Table I. Llama, alpaca, guanaco.
[= *Lama* Frisch, 1775:4[1]; *Lacma* Tiedemann, 1808:420-421; *Llacma* Illiger, 1815:48; *Llama* Gray, 1872:101; *Auchenia* Illiger, 1811:103[2]; *Auchenias* Wagner, 1843:349; *Vicunia* Rafinesque, 1815:55[3]; *Dromedarius* Wagler, 1830:31; *Neoauchenia* Ameghino, 1891:242.] [Including *Pacos* Gray, 1872:101[4]; †*Mesolama* Ameghino, 1884:199; †*Stilauchenia* Ameghino, 1889:591-593.]
L. Plioc.-R.; S.A.
[1] Invalid. Not published in a consistently binominal work. See the International Code of Zoological Nomenclature, Article 11(c).
[2] Not *Auchenia* Thunberg, 1789, a coleopteran.
[3] New name for *Lama*.
[4] Proposed as a subgenus of *Llama*.

Vicugna Lesson, 1842:167.[1] Vicunia, vicuña.
>   M. Pleist.-R.; S.A.
>   [1] Used as a subgenus of *Llama* by Gray, 1872 101; used as a genus by Miller, 1924:1.

†*Eulamaops* Ameghino, 1889:594.
>   [= †*Eulamops* Lydekker, 1890:44.]
>   L. Pleist.; S.A.

Tribe **Camelini** Gray, 1821:307.
>   [= Camelidae Gray, 1821:307; Camelina Gray, 1825:342[1]; Camelini Webb, 1965:44;
>   Camelina Harrison, 1979:8.]
>   M. Mioc.-E. Pleist.; N.A. L. Mioc., ?E. Plioc., L. Plioc.-R.; Eu.[2] L. Mioc.; Cent.
>   A. E. Plioc.-R.; As. L. Plioc.-R.; Af.[3]
>   [1] As tribe.
>   [2] Extant only in domestication.
>   [3] Extant only in domestication.
>   Also feral in Australia by introduction.

†*Procamelus* Leidy, 1858:23-24.
>   [Including †*Homocamelus* Leidy, 1869:158-159, 382.]
>   M.-L. Mioc.; N.A. L. Mioc.; Cent. A.

†*Megatylopus* Matthew & Cook, 1909:396.[1]
>   L. Mioc.-E. Plioc.; N.A.
>   [1] As subgenus of †*Pliauchenia*.

Camelus Linnaeus, 1758:65-66. Camels, dromedary, Bactrian camel.
>   [= *Dromedarius* Gloger, 1841:134.[1]] [Including †*Paracamelus* Schlosser, 1903:95;
>   †*Neoparacamelus* Khaveson, 1950:919-920[2]; †*Cameliscus* Kretzoi, 1954:235.[3]]
>   L. Mioc., L. Plioc.-R.; Eu.[4] E. Plioc.-R.; As. L. Plioc.-R.; Af.[5]
>   [1] Not *Dromedarius* Wagler, 1830, a synonym of *Lama*.
>   [2] Proposed as subgenus of †*Paracamelus*.
>   [3] Proposed as subgenus of *Camelus*.
>   [4] Extant only in domestication.
>   [5] Extant only in domestication.
>   Also feral in Australia by introduction.

†*Titanotylopus* Barbour & Schultz, 1934:291-294.
>   [Including †*Gigantocamelus* Barbour & Schultz, 1939:20.]
>   ?Plioc.; As. ?E. Plioc.; Eu. E. Plioc.-E. Pleist.; N.A.

†*Megacamelus* Frick, 1929:107.
>   E. Plioc.; N.A.

Subfamily †**Aepycamelinae** Webb, 1965:37.[1]
>   [Including †Alticamelinae Simpson, 1945:150; †Nothokemadidae White, 1947:515.[2]]
>   L. Olig.-L. Mioc.; N.A.
>
>   [1] Proposed as a replacement for †Alticamelinae, whose type genus, †*Alticamelus*, was found to be based on
>   a *nomen dubium* (see footnote under †*Aepycamelus*).
>   [2] Under the rules of the International Code of Zoological Nomenclature, Article 23(d), †Nothokemadinae,
>   predating †Aepycamelinae, should be the name of this group. However, because the systematic position of
>   †Nothokemadidae is poorly understood, †Aepycamelinae is used here as it has been in recent literature in
>   order to preserve the original taxonomic concept of the subfamily.

†*Oxydactylus* Peterson, 1904:434.
>   L. Olig.-E. Mioc., ?M. Mioc.; N.A.

†*Australocamelus* Patton, 1969:145.
>   E. Mioc.; N.A.

†*Aepycamelus* Macdonald, 1956:198.
>   [Including †*Alticamelus* Matthew, 1901:426, 429[1]; †*Altomeryx* Frick, 1929:107.[2]]
>   E.-L. Mioc.; N.A.
>   [1] The type species of †*Alticamelus* Matthew, 1901, †*A. altus* (Marsh, 1894), is a *nomen dubium*. For this
>   reason Macdonald proposed †*Aepycamelus* as a new name for this group, designating, as the type species,
>   †*A. giraffinus* (Matthew, 1909), a species clearly central to Matthew's concept of †*Alticamelus*.
>   [2] A *nomen nudum* possibly separable from †*Aepycamelus* but probably belonging in the same subfamily
>   (Simpson 1945:150).

†*Priscocamelus* M. S. Stevens, in M. S. Stevens, J. B. Stevens & M. R. Dawson, 1969:30.
>   E. Mioc.; N.A.

†*Delahomeryx* M. S. Stevens, 1969:40.
E. Mioc.; N.A.

†*Hesperocamelus* Macdonald, 1949:186.
?E. Mioc., M.-L. Mioc.; N.A.

†*Nothokemas* White, 1947:508.
E. Mioc.; N.A.

Subfamily †**Miolabinae** Hay, 1902:676.
E.-L. Mioc.; N.A.

†*Miolabis* Hay, 1899:593.[1]
E.-L. Mioc.; N.A.

[1] Actually first published by Matthew, March 31, 1899:24, 74, but credited to Hay, whose paper was in press. Hay's paper proposing the genus was published April 21, 1899:593.

†*Paramiolabis* Kelly, 1992:7.
M. Mioc.; N.A.

Subfamily †**Stenomylinae** Matthew, Gregory & Mosenthal, in Osborn, 1910:550.[1]
E.-M. Mioc.; N.A.

[1] Peterson, 1908:300, first suggested that †*Stenomylus* might represent a then unnamed camelid subfamily. Simpson (1945:150) attributed the subfamily to Frick, 1937:656. W. D. Matthew was the principal author of most of the content of the classification that appeared in Osborn, 1910:511-563.

†*Stenomylus* Peterson, 1906:35, 41.
[Including †*Pegomylus* Frick & Taylor, 1968:19.[1]]
E. Mioc.; N.A.

[1] Proposed as subgenus.

†*Blickomylus* Frick & Taylor, 1968:22.
E. Mioc.; N.A.

†*Rakomylus* Frick, 1937:657.
M. Mioc.; N.A.

Subfamily †**Floridatragulinae** Maglio, 1966:3.
E.-M. Mioc.; N.A.

†*Floridatragulus* White, 1940:34.
[= †*Hypermekops* White, 1942:11.]
E.-M. Mioc.; N.A.

†*Aguascalientia* M. S. Stevens, 1977:52.
E. Mioc.; N.A.

†*Cuyamacamelus* Kelly, 1992:13.[1]
M. Mioc.; N.A.

[1] Assignment to the †Floridatragulinae somewhat doubtful.

Family †**Oromerycidae** Gazin, 1955:68.
[= †Oromerycinae J. A. Wilson, 1974:18.[1]]
M.-L. Eoc.; N.A.

[1] As subfamily of Camelidae.

†*Camelodon* Granger, 1910:248.
M. Eoc.; N.A.

†*Protylopus* Wortman, 1898:104-110.
M. Eoc.; N.A.

†*Oromeryx* Marsh, 1894:269.
M.-L. Eoc.; N.A.

†*Merycobunodon* Golz, 1976:41.
M. Eoc.; N.A.

†*Eotylopus* Matthew, 1910:36.
M.-L. Eoc.; N.A.

†*Malaquiferus* Gazin, 1955:76.
M:-L. Eoc.; N.A.

†*Montanatylopus* Prothero, 1986:459.
　　L. Eoc.; N.A.

Superfamily †**Protoceratoidea** Marsh, 1891:82, **new rank**.
　　[= †Protoceratidae Marsh, 1891:82.]

Family †**Protoceratidae** Marsh, 1891:82.
　　M. Eoc.-E. Plioc.; N.A. E. Mioc.; Cent. A.

Subfamily †**Leptotragulinae** Zittel, 1893:361.
　　[Including †Leptotragulini Simpson, 1945:151.]
　　M.-L. Eoc., ?E. Olig.; N.A.

†*Toromeryx* J. A. Wilson, 1974:7.
　　M. Eoc.; N.A.

†*Leptotragulus* Scott & Osborn, 1887:258.
　　[= or including †*Parameryx* Marsh, 1894:269.[1]]
　　M. Eoc., ?L. Eoc.; N.A.
　　[1] "A *nomen nudum* for a *genus coelebs* in 1877, based on scraps of several genera, and really a *nomen vanum* that should be discarded entirely" (Simpson 1945:151).

†*Trigenicus* Douglass, 1903:162.
　　L. Eoc., ?E. Olig.; N.A.

†*Poabromylus* Peterson, 1931:75.
　　M.-L. Eoc.; N.A.

†*Leptoreodon* Wortman, 1898:95-97.[1]
　　[= †*Merycodesmus* Scott, 1898:75-77[2]; †*Camelomeryx* Scott, 1898:77-78.[2]]
　　[Including †*Hesperomeryx* Stock, 1936:178.[3]]
　　M. Eoc.; N.A.
　　[1] April 9, 1898.
　　[2] April 15, 1898.
　　[3] As subgenus of †*Leptoreodon.*

†*Heteromeryx* Matthew, 1905:23.
　　L. Eoc.; N.A.

Subfamily †**Protoceratinae** Marsh, 1891:82.
　　[= †Protoceratidae Marsh, 1891:82; †Protoceratinae Zittel, 1893:405.]
　　L. Eoc.-L. Mioc.; N.A. E. Mioc.; Cent. A.

†*Pseudoprotoceras* Cook, 1934:149.
　　L. Eoc.; N.A.

†*Protoceras* Marsh, 1891:81-82.
　　[Including †*Calops* Marsh, 1894:94.]
　　E. Olig.-E. Mioc.; N.A.

†*Paratoceras* Frick, 1937:608.
　　E. Mioc.; Cent. A.[1] M.-L. Mioc.; N.A.
　　[1] Panama.

Subfamily †**Synthetoceratinae** Frick, 1937:595-6, 602.
　　[Including †Syndyoceratinae Frick, 1937:595-6, 607.]
　　E. Mioc.-E. Plioc.; N.A.

Tribe †**Synthetoceratini** Frick, 1937:595-6, 602.
　　[= †Synthetoceratinae Frick, 1937:595-6, 602; †Synthetoceratini Webb, 1981:357, 362.]
　　E.-L. Mioc.; N.A.

†*Prosynthetoceras* Frick, 1937:605.[1]
　　[Including †*Lambdoceras* Stirton, 1967:4.[2]]
　　E.-M. Mioc.; N.A.
　　[1] Described as subgenus of †*Synthetoceras*; elevated to rank of genus by Patton, 1967:35.
　　[2] Described as genus; made subgenus of †*Prosynthetoceras* by Patton & Taylor, 1973:410.

†*Synthetoceras* Stirton, 1932:147.
L. Mioc.; N.A.

Tribe †**Syndyoceratini** Frick, 1937:595-6, 607.
[= †Syndyoceratinae Frick, 1937:595-6, 607.] [Including †Kyptoceratini Webb, 1981:357-8, 362.]
E. Mioc., E. Plioc.; N.A.

†*Syndyoceras* Barbour, 1905:797.
E. Mioc.; N.A.

†*Kyptoceras* Webb, 1981:358.
E. Plioc.; N.A.

Suborder **RUMINANTIA** Scopoli, 1777:493-496.[1]
[= RUMINANTES Gray, 1821:307.[2]] [Including PECORA Linnaeus, 1758:65[3]; ELAPHIA Haeckel, 1866:clviii[4]; TRAGULINA Flower, 1883:184[5]; MOSCHINA Webb & Taylor, 1980:152[6]; EUPECORA Webb & Taylor, 1980:152.[7]]

[1] Often used in early literature at ordinal rank, coordinate with ARTIODACTYLA.
[2] Named as an order.
[3] One of Linnaeus' (1758) eight mammalian orders, differing from his earlier (1735) concept only by the addition of *Moschus*. However, Linnaeus regarded *Camelus* as a pecoran. Later authors have restricted PECORA and it has often been used as an infraorder of RUMINANTIA. The International Code of Zoological Nomenclature does not pertain to names above the family-group rank.
[4] In part. Also included †*Poebrotherium*. Used as a subsectio of sectio RUMINANTIA.
[5] Often used as an infraorder of RUMINANTIA.
[6] Proposed as division of PECORA. Defined to include †Gelocidae and Moschidae, it was therefore paraphyletic.
[7] Proposed as division of PECORA.

†*Palaeohypsodontus* Trofimov, 1957:138.
?E. Mioc.; As.

†*Gobiocerus* Sokolov, 1952:155.
E. Mioc.; As.

†*Georgiomeryx* Paraskevaidis, 1940:405, 408.
M. Mioc.; As.[1]
[1] W. Asia (Chios).

Family †**Amphimerycidae** Stehlin, 1910:1143.
[= †Amphimerycoidea Colbert, 1941:21.]
E. Eoc.-E. Olig.; Eu.

†*Pseudamphimeryx* Stehlin, 1910.
E. Eoc.-E. Olig.; Eu.

†*Amphimeryx* Pomel, 1849:72.[1]
[= †*Amphimerix* Pomel, 1849:72[2]; †*Amphimeryx* Pictet, 1853:341; †*Amphimoerix* Gervais, 1859:162-163; †*Hyaegulus* Pomel, 1851:218.]
L. Eoc.-E. Olig.; Eu.
[1] Always now so spelled (Simpson 1945:151).
[2] As originally spelled. Possibly a correctable *lapsus* (Simpson 1945:151).

Family †**Hypertragulidae** Cope, 1879:66.
[= †Hypertraguloidea Scott, 1940:507.] [Including †Andegamerycidae Ginsburg & Morales, 1989.]
M. Eoc.-E. Mioc.; N.A. E.-M. Mioc.; Eu.

†*Simimeryx* Stock, 1934:625.[1]
M. Eoc.; N.A.
[1] Assignment to the †Hypertragulidae is questionable.

Subfamily †**Hypertragulinae** Cope, 1879:66.
[= †Hypertragulidae Cope, 1879:66; †Hypertragulinae Matthew, 1908:561; †Hypertragulini Frick, 1937:618.] [Including †Andegamerycidae Ginsburg & Morales, 1989.]
M. Eoc.-E. Mioc.; N.A. E.-M. Mioc.; Eu.

†*Hypertragulus* Cope, 1874:26-27.
      [Including †*Allomeryx* Sinclair, 1905:129.]
      M. Eoc.-E. Olig., ?L. Olig., E. Mioc.; N.A.

†*Parvitragulus* Emry, 1978:1005.
      L. Eoc.; N.A.

†*Nanotragulus* Lull, 1922:111, 116.
      E. Olig.-E. Mioc.; N.A.

†*Andegameryx* Ginsburg, 1971:997.
      E.-M. Mioc.; Eu.

Subfamily †**Hypisodontinae** Cope, 1887:389.
      [= †Hypisodontini Frick, 1937:618.]
      L. Eoc.-E. Olig.; N.A.

†*Hypisodus* Cope, 1873:7.
      L. Eoc.-E. Olig.; N.A.

Family **Tragulidae** Milne-Edwards, 1864:157.
      [= Traguloidea Gill, 1872:9, 73, 88.[1]] [Including Hyemoschidae Gray, 1872:5, 99;
      †Dorcatheriinae Matthew, Gregory & Mosenthal, in Osborn, 1910:551.]
      E.-M. Mioc., Pleist., R.; Af. E.-L. Mioc., Plioc., E. Pleist., ?M. and/or ?L. Pleist.,
      R.; As. E.-L. Mioc.; Eu. M. and/or L. Mioc.; Mediterranean.[2] R.; E. Indies.
      [1] The Traguloidea of most authors is paraphyletic.
      [2] Crete.

†*Dorcatherium* Kaup, 1833:419.
      E.-M. Mioc.; Af. E.-L. Mioc., Plioc.; As. E.-L. Mioc.; Eu. M. and/or L. Mioc.;
      Mediterranean.[1]
      [1] Crete.

†*Dorcabune* Pilgrim, 1910:68.
      M.-L. Mioc., E. Pleist.; As.

†*Siamotragulus* Thomas, Ginsburg, Hintong & Suteethorn, 1990:991.
      M. Mioc.; As.[1]
      [1] S.E. Asia (N. Thailand).

†*Yunnanotherium* Han, 1986:68.
      L. Mioc.; As.

Tragulus Pallas, 1779:27-28. Chevrotain, mouse deer.
      [= *Tragulus* Brisson, 1762:65-68.[1]] [Including *Moschiola* Hodgson, 1843:292.]
      ?Plioc., Pleist., R.; As. R.; E. Indies.[2]
      [1] Brisson's 1762 names were inconsistently binominal and are therefore all invalid. However, see the
      International Code of Zoological Nomenclature, Article 11(c)(i) for confusion.
      [2] Sumatra, Java, Borneo, Balabac Island (S.W. Philippines).

Hyemoschus Gray, 1845:350. Water chevrotain.
      [= *Hyeomoschus* Turner, 1849:158; *Hyomoschus* Blyth, 1864:483; *Hyaemoschus*
      Zittel, 1893:387.]
      Pleist., R.; Af.

Family †**Leptomerycidae** 1893:389.
      [= †Leptomerycinae Zittel, 1893:389; †Leptomerycidae Scott, 1899:15.]
      M. Eoc.-E. Olig.; As. M. Eoc.-M. Mioc.; N.A.

Subfamily †**Archaeomerycinae** Simpson, 1945:151.
      M. Eoc.-E. Olig.; As.

†*Archaeomeryx* Matthew & Granger, 1925:9.
      M. Eoc.; As.

†*Indomeryx* Pilgrim, 1928:33.
      M.-L. Eoc.; As.

†*Xinjiangmeryx* Zheng, 1978:120.
L. Eoc.; As.[1]
[1] N.W. China.

†*Miomeryx* Matthew & Granger, 1925:10.
L. Eoc.-E. Olig.; As.

Subfamily †**Leptomerycinae** Zittel, 1893:389.
[= †Leptomerycini Frick, 1937:618.]
M. Eoc.-M. Mioc.; N.A. ?E. Olig.; As.

†*Leptomeryx* Leidy, 1853:394.
[Including †*Trimerodus* Cope, 1873:8; †*Hendryomeryx* Black, 1978:254.]
M. Eoc.-L. Olig., ?E. Mioc.; N.A. ?E. Olig.; As.

†*Pronodens* Koerner, 1940:842.
L. Olig.; N.A.

†*Pseudoparablastomeryx* Frick, 1937:244.[1]
E.-M. Mioc.; N.A.
[1] Proposed as subgenus of †*Parablastomeryx*. The record length of this unimaginative genus-group name is of critical interest mainly to fabricators of classificatory software.

Family †**Bachitheriidae** Janis, 1987:214.[1]
E.-L. Olig.; Eu.
[1] Regarded as informal by Janis, 1978:210. Whether this is the same as a conditional proposal (International Code of Zoological Nomenclature, Article 15) is questionable. †Bachitheriidae was therefore not available on page 10, but on page 215 of the same paper Janis stated that figure 10 on page 214 was her final conclusion. In that figure there is no uncertainty expressed. We judge that Janis is the author of †Bachitheriidae and that it does not need rechristening.

†*Bachitherium* Filhol, 1882:138-139.
[= †*Pachitherium* Filhol, 1882:42.]
E.-L. Olig.; Eu.

Family †**Lophiomerycidae** Janis, 1987:213.
L. Eoc.-L. Olig.; As. E.-L. Olig.; Eu.

†*Cryptomeryx* Schlosser, 1886:74, 93-94.
E. Olig.; Eu.

†*Iberomeryx* Gabunia, 1964:179.
L. Olig.; As.[1]
[1] W. Asia.

†*Lophiomeryx* Pomel, 1854:97-98.
L. Eoc.-L. Olig.; As. E.-L. Olig.; Eu.

Family †**Gelocidae** Schlosser, 1886:41.
[Including †Pseudoceratinae Frick, 1937:649.]
L. Eoc.-L. Olig.; As. L. Eoc.-L. Olig.; Eu. E. Mioc.; Af. L. Mioc.; N.A. L. Mioc.; Cent. A.

†*Gelocus* Aymard, 1855.[1]
[= †*Gelaucus* Bonney (?), 1880:296.]
E. Olig., ?L. Olig.; Eu. E. Mioc.; Af.
[1] Or 1856:233. See Palmer (1904:292).

†*Notomeryx* Qiu, 1978:9.
L. Eoc.; As.[1]
[1] E. Asia.

†*Gobiomeryx* Trofimov, 1957:138.
?L. Eoc., E. Olig.; As.

†*Prodremotherium* Filhol, 1877:228-236.
?E. Olig., L. Olig.; As. E.-L. Olig.; Eu.

†*Phaneromeryx* Schlosser, 1886:62, 95.
L. Eoc.; Eu.

†*Pseudogelocus* Schlosser, 1893:387.[1]
    [= †*Protomeryx* Schlosser, 1886:95-96.[2]]
    E. Olig.; Eu.
    [1] New name for †*Protomeryx* Schlosser, 1886.
    [2] Not †*Protomeryx* Leidy, 1856, a synonym of †*Poebrotherium* (Camelidae).

†*Pseudomeryx* Trofimov, 1957:138.
    E. Olig.; As.

†*Paragelocus* Schlosser, 1902.
    E. Olig.; Eu.

†*Pseudoceras* Frick, 1937:649.
    L. Mioc.; N.A. L. Mioc.; Cent. A.

Superfamily **Cervoidea** Goldfuss, 1820:xx, 374.
    [= Cervina Goldfuss, 1820:xx, 374; Cervoidea Hay, 1930:815; Cervoidea Simpson,
    1931:284; Eucervoidea Janis & Scott, 1987:78.[1]]
    [1] In part.

†*Eumeryx* Matthew & Granger, 1924:3.[1]
    E. Olig.; As.
    [1] *Nomen nudum* in Matthew & Granger, 1923:4.

†*Rutitherium* Filhol, 1876:289.
    L. Olig.; Eu.

†*Walangania* Whitworth, 1958.
    [= †*Kenyameryx* Ginsburg & Heintz, 1966:980.]
    E. Mioc.; Af. ?E. Mioc.; As.[1]
    [1] S.W. Asia (Israel).

Family **Moschidae** Gray, 1821:307.
    [= Moschina Gray, 1825:342; Moschifera Haeckel, 1866:clviii.] [Including
    †Dremotheriidae Ginsburg & Heintz, 1966; †Hispanomerycidae Moyà Solà, 1986:267.]
    E. Olig.-L. Mioc.; Eu. E.-L. Mioc., ?E. Plioc., Pleist., R.; As. E. Mioc.-L. Plioc.;
    N.A.

†*Hispanomeryx* Morales, Moyà-Solà & Soria, 1981:467.
    ?M. Mioc.; As. M. and/or L. Mioc.; Eu.

Subfamily †**Dremotheriinae** Haeckel, 1895:552.
    [= †Dremotherida Haeckel, 1895:552; †Dremotheriinae Crusafont Pairó, 1952:186.]
    E. Olig.-E. Mioc.; Eu. E. Mioc.; As.

†*Dremotherium* É. Geoffroy Saint-Hilaire, 1833:622.
    E. Olig.-E. Mioc.; Eu. E. Mioc.; As.

Subfamily †**Blastomerycinae** Frick, 1937:215.
    [= †Blastomerycini Frick, 1937:215; †Blastomerycinae Taylor & Webb, 1976:1.]
    [Including †Longirostromerycinae Frick, 1937:217; †Parablastomerycinae Frick,
    1937:217; †Pseudoblastomerycinae Frick, 1937:253.]
    E. Mioc.-L. Plioc.; N.A.

†*Blastomeryx* Cope, 1877:350, 360.
    [Including †*Parablastomeryx* Frick, 1937:217[1]; †*Pseudoblastomeryx* Frick, 1937:251[1];
    †*Problastomeryx* Frick, 1937:251.[1]]
    E.-L. Mioc., Plioc.; N.A.
    [1] Proposed as subgenus.

†*Machaeromeryx* Matthew, 1926:1.
    E. Mioc.; N.A.

†*Longirostromeryx* Frick, 1937:217.
    M. Mioc.-L. Plioc.; N.A.

Subfamily **Moschinae** Gray, 1821:307.
    [= Moschidae Gray, 1821:307; Moschina Gray, 1825:342[1]; Moschinae Sclater,
    1870:115.]

E.-L. Mioc.; Eu. M.-L. Mioc., ?E. Plioc., Pleist., R.; As.

[1] As tribe.

†*Micromeryx* Lartet, 1851:36.
> [Including †*Orygotherium* von Meyer, 1838:413.]
> E.-L. Mioc.; Eu. M. Mioc.; As.

Moschus Linnaeus, 1758:66. Musk deer.
> [= *Moschifer* Frisch, 1775; *Odontodorcus* Gistel, 1848:82.]
> L. Mioc., ?E. Plioc., Pleist., R.; As.

Family **Antilocapridae** Gray, 1866:326.
> E. Mioc.-R.; N.A.

Subfamily †**Cosorycinae** Cope, 1887:396.
> [Including †Merycodontidae Matthew, 1904:102, 103; †Merycodontinae Matthew, 1909:114, 115; †Ramocerotinae Frick, 1937:35, 271.]
> E.-L. Mioc.; N.A.

†*Merycodus* Leidy, 1854:90.
> [Including †*Subparacosoryx* Frick, 1937:271, 292; †*Meryceros* Frick, 1937:293; †*Submeryceros* Frick, 1937:293.]
> E.-L. Mioc.; N.A.

†*Merriamoceros* Frick, 1937.[1]
> E.-M. Mioc.; N.A.

[1] Described as subgenus of †*Ramoceros*; raised to generic rank by Gregory, 1942:407.

†*Paracosoryx* Frick, 1937, **new rank**.[1]
> E.-L. Mioc.; N.A.

[1] Described as subgenus of †*Cosoryx*; raised to generic rank on advice of B. Taylor and E. Manning (June 1974, personal communication).

†*Ramoceros* Frick, 1937.
> [Including †*Paramoceros* Frick, 1937:291.]
> M. Mioc.; N.A.

†*Cosoryx* Leidy, 1869.
> [Including †*Subcosoryx* Frick, 1937.]
> M.-L. Mioc.; N.A.

Subfamily **Antilocaprinae** Gray, 1866:326.
> [= Antilocapridae Gray, 1866:326; Antilocaprinae Brooke, 1876:223.] [Including †Ilingoceratinae Frick, 1937:35, 469; †Pliocerotinae Frick, 1937:36, 469; †Stockocerotinae Frick, 1937:36, 469, 521.]
> L. Mioc.-R.; N.A.

†*Subantilocapra* Webb, 1973.
> L. Mioc.; N.A.

†*Osbornoceros* Frick, 1937.
> L. Mioc.; N.A.

†*Ilingoceros* Merriam, 1909:320.
> L. Mioc.; N.A.

†*Sphenophalos* Merriam, 1909:325.
> L. Mioc.; N.A.

†*Texoceros* Frick, 1937:501.
> L. Mioc.; N.A.

†*Hayoceros* Frick, 1937.
> ?L. Mioc., M. Pleist.; N.A.

†*Hexobelomeryx* Furlong, 1941.
> L. Mioc.; N.A.

†*Ottoceros* Miller & Downs, 1974:13.
> L. Mioc.; N.A.

†*Plioceros* Frick, 1937.
  L. Mioc.; N.A.

†*Hexameryx* White, 1941.
  L. Mioc.; N.A.

†*Proantilocapra* Barbour & Schultz, 1934.
  L. Mioc.; N.A.

†*Ceratomeryx* Gazin, 1935:390.
  E. Plioc.; N.A.

†*Capromeryx* Matthew, 1902.
  [Including †*Breameryx* Furlong, 1946.]
  Plioc., E.-L. Pleist.; N.A.

†*Tetrameryx* Lull, 1921.
  L. Plioc.-L. Pleist.; N.A.

†*Stockoceros* Frick, 1937.
  M.-L. Pleist.; N.A.

Antilocapra Ord, 1818. Pronghorn, American antelope.
  [= *Dicranocerus* H. Smith, 1827.] [Including †*Neomeryx* Parks, 1925.]
  L. Pleist.-R.; N.A.

Family †**Palaeomerycidae** Lydekker, 1883:174.
  [Including †Triceromerycidae Crusafont Pairó, 1952:49; †Dromomerycidae Crusafont
  Pairó, 1952:187.]
  ?E. Olig., L. Olig.-M. Mioc.; Eu. L. Olig.; Mediterranean. E. and/or M. Mioc.; Af.
  E.-M. Mioc., ?L. Mioc.; As. E.-L. Mioc., ?E. Plioc.; N.A.

†*Prolibytherium* Arambourg, 1961.
  E. and/or M. Mioc.; Af.[1]
  [1] N. Africa.

†*Amphitragulus* Pomel, 1846:369-371.
  [Including †*Hydropotopsis* Jehenne, 1985; †*Pomelomeryx* Ginsburg & Morales, 1989.]
  ?E. Olig., L. Olig.-M. Mioc.; Eu. L. Olig.; Mediterranean. E. Mioc.; As.

Subfamily †**Palaeomerycinae** Lydekker, 1883:174.
  [= †Palaeomerycidae Lydekker, 1883:174; †Palaeomerycinae Matthew, 1904:102;
  †Palaeomerycini Simpson, 1945:152.] [Including †Triceromerycidae Crusafont Pairó,
  1952:49; †Triceromerycinae Kalandadze & Rautian, 1992:125.]
  E.-M. Mioc.; Eu. M. Mioc., ?L. Mioc.; As.

†*Palaeomeryx* von Meyer, 1834:31, 92-102.
  [Including †*Bedenomeryx* Jehenne, 1988:1991-1996; †*Sinomeryx* Duranthon, Moya
  Sola, Astibia & Köhler, 1995:344.]
  E.-M. Mioc.; Eu. M. Mioc.; As.

†*Ampelomeryx* Duranthon, Moya Sola, Astibia & Köhler, 1995:344.
  E. Mioc.; Eu.

†*Oriomeryx* Ginsburg, 1985:1076.
  E. Mioc.; Eu.

†*Triceromeryx* de Villalta Comella, Crusafont Pairó & Lavocat, 1946:1-4.[1]
  [= †*Hispanocervus* Viret, 1946:48-49.[2]]
  E. Mioc., ?M. Mioc.; Eu.[3] ?M. Mioc., ?L. Mioc.; As.[4,5]
  [1] A fuller description by the same authors, also stating the genus to be new, was published in 1949.
  [2] February 4, 1946.
  [3] Spain.
  [4] Turkey in the M. Mioc.?
  [5] China in the L. Mioc.?

Subfamily †**Dromomerycinae** Frick, 1937:46, 99.
  [= †Dromomerycinae Frick, 1937:46, 99; †Dromomerycini Frick, 1937:51[1];
  †Dromomerycidae Crusafont Pairó, 1952:187.] [Including †Barbouromerycinae Frick,

1937:32, 127; †Yumaceratinae Frick, 1937:33, 142; †Drepanomerycinae Frick, 1937:46, 137; †Aletomerycini Frick, 1937:51.]
E. Mioc.; As.   E.-L. Mioc., ?E. Plioc.; N.A.
[1] Proposed as division of Cervidae with rank higher than subfamily.

†*Asiagenes* Vislobokova, 1983:34.
E. Mioc.; As.

Tribe †**Aletomerycini** Frick, 1937:33, 147.
[= †Aletomerycini Frick, 1937:32, 47[1]; †Aletomerycini Simpson, 1945:153.[2]]
E. Mioc.; N.A.
[1] Proposed as "division" of Cervidae, higher than the rank of subfamily.
[2] As tribe of †Dromomerycinae.

†*Aletomeryx* Lull, 1920:85.
[Including †*Dyseomeryx* Matthew, 1924:196.[1]]
E. Mioc.; N.A.
[1] As subgenus.

†*Sinclairomeryx* Frick, 1937:153-155.
E. Mioc.; N.A.

Tribe †**Dromomerycini** Frick, 1937:46, 99, **new rank**.
[= †Dromomerycinae Frick, 1937:46, 99; †Dromomerycini Frick, 1937:51.[1]]
[Including †Drepanomerycinae Frick, 1937:46, 137.]
E.-M. Mioc.; N.A.
[1] Proposed as division of Cervidae with rank higher than subfamily.

†*Dromomeryx* Douglass, 1909:461.
[Including †*Subdromomeryx* Frick, 1937:49, 123.[1]]
E.-M. Mioc.; N.A.
[1] Proposed as subgenus.

†*Rakomeryx* Frick, 1937:48, 99.
M. Mioc.; N.A.

†*Drepanomeryx* Sinclair, 1915.
[Including †*Matthomeryx* Frick, 1937:50, 138.[1]]
M. Mioc.; N.A.
[1] Proposed as subgenus.

Tribe †**Cranioceratini** Frick, 1937:32, 75, **new rank**.
[= †Cranioceratinae Frick, 1937:32, 75.] [Including †Barbouromerycinae Frick, 1937:32, 127; †Yumaceratinae Frick, 1937:33, 142.]
E.-L. Mioc., ?E. Plioc.; N.A.

†*Barbouromeryx* Frick, 1937:49, 134.
[Including †*Probarbouromeryx* Frick, 1937:49, 135[1]; †*Protobarbouromeryx* Frick, 1937:49, 136.[1]]
E. Mioc.; N.A.
[1] Proposed as subgenus.

†*Bouromeryx* Frick, 1937:49, 127.[1]
E.-M. Mioc.; N.A.
[1] Described as a subgenus of †*Barbouromeryx*.

†*Cranioceras* Matthew, 1918:223.
[Including †*Procranioceras* Frick, 1937:32, 75.[1]]
M.-L. Mioc.; N.A.
[1] Proposed as subgenus.

†*Yumaceras* Frick, 1937:50, 142.
L. Mioc.; N.A.

†*Pediomeryx* Stirton, 1936:644.
[Including †*Procoileus* Frick, 1937:191.]
L. Mioc., ?E. Plioc.; N.A.

Family †**Hoplitomerycidae** Leinders, 1983:4.
E.-L. Mioc.; Eu.

†*Amphimoschus* Bourgeois, 1873:235-236.[1]
E.-M. Mioc.; Eu.
[1] Not *Amphimoschus* (Falconer MS.) Gray, 1852:247, 248, a *nomen nudum.*

†*Hoplitomeryx* Leinders, 1983:4.
L. Mioc.; Eu.

Family **Cervidae** Goldfuss, 1820:374. Deer.
[= Cervina Goldfuss, 1820:374; Cervidae Gray, 1821:307.] [Including Elaphidae
Brookes, 1828:61; Rangiferinidae Brookes, 1828:61[1]; Capreolidae Brookes, 1828:62[2];
Mazamadae Brookes, 1828:62[3]; Rusadae Brookes, 1828:62[4]; Axidae Brookes,
1828:62[4]; Alcedae Brookes, 1828:61[5]; Alcadae Gray, 1872:66; Alceidae Kashin,
1974:215; †Procervulidae Bubenik, 1966:23.]
E. Mioc.-R.; As. E. Mioc.-R.; Eu. E. Plioc.-R.; N.A. ?L. Plioc., E. Pleist.-R.;
S.A. Pleist., R.; E. Indies. Pleist., R.; Cent. A. M. Pleist.-R.; Mediterranean. L.
Pleist.-R.; Af.[6]
[1] Fide Gray, 1852:188.
[2] Fide Gray, 1852:221.
[3] Fide Gray, 1852:228.
[4] Fide Gray, 1852:202.
[5] Fide Gray, 1852:186.
[6] N. Africa.

†*Cervavitulus* Kretzoi, 1951:394, 412.
L. Mioc.; Eu.

†*Muva* Deraniyagala, 1958:137.
Pleist.; As.[1]
[1] S. Asia (Ceylon).

†*Lucentia* Azanza & Montoya, 1995:1165.
L. Mioc.; Eu.

Subfamily **Hydropotinae** Trouessart, 1898:865.
[= Hydropotini Simpson, 1945:155.]
R.; As.

Hydropotes Swinhoe, 1870:90.[1] Chinese water deer.
[= *Hydrelaphus* Lydekker, 1898:219-222.[2]]
R.; As.[3]
[1] Not preoccupied by *Hydropota* Rondani, 1861, a fly.
[2] Replacement name for *Hydropotes* Swinhoe, 1870, thought to be preoccupied.
[3] E. Asia.

Subfamily **Muntiacinae** Pocock, 1923:207.[1]
[= Cervulinae Sclater, 1870:115.]
E. Mioc.-R.; As. E. Mioc.-E. Plioc.; Eu. R.; E. Indies.
[1] Although predated by Cervulinae, Muntiacinae has been the generally accepted subfamily name and was
validated by the Commission on Zoological Nomenclature, Opinion 544, 1959.

Tribe †**Lagomerycini** Pilgrim, 1941:176.
[= †Lagomerycidae Pilgrim, 1941:176[1]; †Lagomerycinae Crusafont Pairó, 1952:187;
†Lagomerycini Kalandadze & Rautian, 1992:126.] [Including †Procervulidae Bubenik,
1966:23.]
E.-L. Mioc.; As. E.-M. Mioc.; Eu.
[1] "The family was proposed by Teilhard de Chardin in 1939, but he neglected to apply any name to it"
(Simpson 1945:155).

†*Procervulus* Gaudry, 1877:87.
[Including †*Heterocemas* Young, 1937:228.]
E.-L. Mioc.; As. E.-M. Mioc.; Eu.

†*Lagomeryx* Roger, 1904.
E.-M. Mioc.; Eu. M.-L. Mioc.; As.

Tribe †**Dicrocerini** Simpson, 1945:153.
> [= †Dicrocerina Kalandadze & Rautian, 1992:126.]
> E.-L. Mioc.; Eu. M.-L. Mioc., ?Plioc.; As.

†*Acteocemas* Ginsburg, 1985:68.
> E. Mioc.; Eu.

†*Stephanocemas* Colbert, 1936: erratum.
> [= †*Stephanoceras* Colbert, 1936:1.[1]] [Including †*Stehlinocemas* Ginsburg, 1985:68[2];
> †*Stehlinoceros* Azanza & Menendez, 1990:78.[3]]
> E.-M. Mioc.; Eu. M. Mioc., ?L. Mioc.; As.

[1] Not *Stephanoceras* Waagen, 1869, a genus of cephalopods.
[2] Subgenus as proposed.
[3] Objective junior synonym of †*Stehlinocemas* but originally given generic rank.

†*Dicrocerus* Lartet, 1837:418.
> [= †*Dicroceros* Agassiz, 1846:4; †*Dicrocemas* Kretzoi, 1976:428.]
> E.-L. Mioc.; Eu. M.-L. Mioc.; As.

†*Palaeoplatycerus* Pacheco, 1915.
> M.-L. Mioc.; Eu.

†*Euprox* Stehlin, 1928:248.
> M. Mioc.; As. M.-L. Mioc.; Eu.

†*Paradicrocerus* Gabunia, 1959:114.
> M. Mioc.; Eu.[1]

[1] E. Europe.

†*Heteroprox* Stehlin, 1928:255.
> M. Mioc.; Eu.

†*Oschinotherium* Belyaeva, 1974:80.
> M. and/or L. Mioc.; As.

†*Amphiprox* Haupt, 1935:50-55.
> L. Mioc.; Eu.

†*Platycemas* Teilhard de Chardin & Trassaert, 1937:25.
> ?Plioc.; As.

Tribe **Muntiacini** Pocock, 1923:207.
> [= Muntiacinae Pocock, 1923:207; Muntiacini Weber, 1928:572; Cervulini Kalandadze
> & Rautian, 1992:126; Cervulina Kalandadze & Rautian, 1992:127.]
> L. Mioc.-R.; As. L. Mioc.-E. Plioc.; Eu. R.; E. Indies.

†*Paracervulus* Teilhard de Chardin & Trassaert, 1937:15.
> [= or including †*Diglochis* Gervais, 1859:149-150.[1]] [Including †*Praecapreolus* Portis,
> 1920:133.[2]]
> ?L. Mioc., L. Plioc.-E. Pleist.; As. L. Mioc.-E. Plioc.; Eu.

[1] As subgenus. Preoccupied by *Diglochis* Forster, 1856, a genus of Hymenoptera.
[2] Described as subgenus of *Cervus*.

Muntiacus Rafinesque, 1815:56.[1] Muntjak, barking deer.
> [= *Cervulus* de Blainville, 1816:74; *Prox* Ogilby, 1837:135.] [Including *Procops*
> Pocock, 1923:207; *Megamuntiacus* Tuoc, Dung, Dawson, Arctander & MacKinnon,
> 1994:4-13.[2]]
> L. Mioc., L. Plioc.-R.; As. L. Mioc.-E. Plioc.; Eu.[3] R.; E. Indies.

[1] Conserved by the International Commission on Zoological Nomenclature, Opinion 460, 1957.
[2] Publication of taxonomic names such as this one in an obscure journal unlikely to be in many libraries is
acceptable under the Code but is not helpful to the scientific community at large.
[3] E. Europe.

†*Eostyloceros* Zdansky, 1925:3.
> L. Mioc.-E. Plioc., E. Pleist.; As. L. Mioc.-E. Plioc.; Eu.

†*Metacervulus* Teilhard de Chardin & Trassaert, 1937:13.
> L. Mioc.-E. Plioc., E. Pleist., L. Pleist.; As.

Elaphodus Milne-Edwards, 1871:93. Tufted deer, Tibetan muntjak.
> M. Pleist., R.; As.

Subfamily **Cervinae** Goldfuss, 1820:xx, 374.
>   [= Cervina Goldfuss, 1820:xx, 374; Cervinae Baird, 1857:630.] [Including
>   †Pliocervinae Khomenko, 1913:107, 108.]
>   ?M. Mioc., L. Mioc.-R.; As. L. Mioc.-R.; Eu. ?E. and/or ?M. Pleist., L. Pleist.-R.;
>   Af.[1] Pleist., R.; E. Indies. E. Pleist.-R.; N.A. M. Pleist.-R.; Mediterranean.
>   [1] N. Africa.

†*Tamanalces* Vereshchagin, 1957:55.
>   Plioc.; Eu.[1]
>   [1] E. Europe (Taman Peninsula).

†*Pseudalces* Flerov, 1962:373.
>   ?Plioc.; As.[1] Plioc.; Eu.[2]
>   [1] W. Asia (Georgia).
>   [2] E. Europe (Stavropol).

†*Torontoceros* Churcher & Peterson, 1982:190. Toronto subway deer.
>   L. Pleist.; N.A.

Tribe †**Pliocervini** Symeonidis, 1974:311.
>   [= †Pliocervinae Symeonidis, 1974:308-316.] [= or including †Pliocervinae
>   Khomenko, 1913:108[1]; †Pliocervini Azzaroli, 1948:47.[1]]
>   ?M. Mioc., L. Mioc.-E. Pleist.; As. L. Mioc.; Eu.
>   [1] Not based on the name of a genus contained within this subfamily.

†*Cervavitus* Khomenko, 1913:108.
>   [Including †*Cervocerus* Khomenko, 1913:108; †*Damacerus* Khomenko, 1913:109;
>   †*Procervus* Alexejev, 1914:1.[1]]
>   ?M. Mioc., L. Mioc.-E. Pleist.; As. L. Mioc.; Eu.[2]
>   [1] Not *Procervus* de Blainville, 1840:392, an emendation of *Procerus* Serres, 1838:143, 204, 230,
>   preoccupied, a synonym of *Rangifer*; not *Procervus* Hodgson, 1847, a synonym of *Cervus*; not †*Procervus*
>   Alexejev, 1913, a *nomen nudum*.
>   [2] E. Europe.

†*Pliocervus* Hilzheimer, 1922:743.[1]
>   [= †*Ctenocerus* Kretzoi, 1941:350[2]; †*Ctenocervus* Kretzoi, 1968:164.[3]]
>   L. Mioc., Plioc.; As. L. Mioc.; Eu.
>   [1] Said to be preoccupied by "†*Pliocervus* Alexejew," but we have not been able to verify this.
>   [2] Replacement name for †*Pliocervus* Hilzheimer, 1922, but itself preoccupied by *Ctenocerus* Dahlbom,1845,
>   a hymenopteran.
>   [3] Replacement name for †*Ctenocerus* Kretzoi, 1941.

Tribe †**Megacerini** Viret, 1961:1018.
>   [= †Megalocerini Kalandadze & Rautian, 1992:128.]
>   L. Mioc., L. Plioc.-L. Pleist.; Eu. E. Plioc.-L. Pleist.; As. ?M. Pleist., L. Pleist.;
>   Mediterranean. L. Pleist.; Af.[1]
>   [1] N. Africa.

†*Praesinomegaceros* Vislobokova, 1983:58.
>   E. Plioc.; As.

†*Arvernoceros* Heintz, 1970:241.[1] Arde deer.
>   L. Plioc.; Eu.
>   [1] Listed but not described by Heintz, 1968:2184.

†*Neomegaloceros* Korotkevitsch, 1971:59-62.
>   L. Mioc.; Eu.[1]
>   [1] E. Europe.

†*Orchonoceros* Vislobokova, 1979:31.
>   L. Plioc.; As.[1]
>   [1] E. Asia.

†*Sinomegaceros* Dietrich, 1933.
>   [Including †*Sinomegaceroides* Shikama, 1949:107[1]; †*Mongolomegaceros* Shikama &
>   Okafuji, 1958:83; †*Megaceraxis* Matsumoto, 1963:346.[2]]
>   L. Plioc.-L. Pleist.; As.[3]
>   [1] Proposed as subgenus of *Cervus*.

[2] Proposed as subgenus of †*Megaceros*.
[3] E. Asia.

†*Psekupsoceros* Radulesco & Samson, 1967:332.[1]
     E. Pleist.; Eu.[2]

[1] (Rădulescu).
[2] E. Europe.

†*Praemegaceros* Portis, 1920:136.[1]
     [= †*Praerangifer* Portis, 1920:137[1]; †*Orthogonoceros* Kahlke, 1956:62.]
     E.-M. Pleist.; Eu.

[1] Proposed as subgenus of *Cervus*.

†*Candiacervus* Kuss, 1975:28.
     ?M. Pleist., L. Pleist.; Mediterranean.[1]

[1] Crete, Kasos and Karpathos.

†*Megaloceros* Brookes, 1828:61.
     [= †*Megaceros* Owen, 1844:237.[1]] [Including †*Praedama* Portis, 1920:136[2];
     †*Dolichodoryceros* Kahlke, 1956:62.]
     M.-L. Pleist.; As. M.-L. Pleist.; Eu.

[1] Proposed as subgenus of *Cervus*.
[2] In part. Proposed as subgenus of *Cervus*.

†*Allocaenelaphus* Radulesco & Samson, 1967:324.[1]
     M. Pleist.; Eu.[2]

[1] (Rădulescu).
[2] E. Europe.

†*Megaceroides* Joleaud, 1914:737.[1]
     E. Pleist.; Eu. L. Pleist.; Af.[2]

[1] Proposed as subgenus of *Cervus*; often as subgenus of †*Megaloceros*.
[2] N. Africa.

†*Nesoleipoceros* Radulesco & Samson, 1967:334.[1]
     [Including †*Notomegaceros* Gliozzi & Malatesta, 1982:311, 315.[2]]
     L. Pleist.; Mediterranean.[3]

[1] (Rădulescu).
[2] Proposed as subgenus of †*Praemegaceros*.
[3] Corsica and Sicily.

Tribe **Cervini** Goldfuss, 1820:xx, 374.
     [= Cervina Goldfuss, 1820:xx, 374; Cervina Gray, 1825:342[1]; Cervini Weber,
     1928:572.]
     ?L. Mioc., E. Plioc.-R.; As. E. Plioc.-R.; Eu. ?Pleist., R.; Af.[2] Pleist., R.; E.
     Indies. E. Pleist.-R.; N.A. M. Pleist.-R.; Mediterranean.

[1] As tribe.
[2] N. Africa.

Cervus Linnaeus, 1758:66. Red deer, stag, wapiti, American elk, maral, sambar, barasingha,
     thamin, sika.
     [= *Strongyloceros* Owen, 1846:470; *Elaphus* C. H. Smith, in Griffith, Smith &
     Pidgeon, 1827:307; *Harana* Hodgson, 1838:154; *Pseudocervus* Hodgson, 1841:914;
     Eucervus Acloque, 1899:71.[1]] [Including *Procervus* Hodgson, 1847:689-690[2]; *Rusa* C.
     H. Smith, in Griffith, Smith & Pidgeon, 1827:309-312[3]; *Hippelaphus* Boneparte,
     1836:4[4]; *Rucervus* Hodgson, 1838:154[5]; *Panolia* Gray, 1843:xxvii, 180; *Sika* Sclater,
     1870:115[6]; *Pseudaxis* Gray, 1872:70; *Elaphoceros* Fitzinger, 1874:347, 352;
     Melanaxis Heude, 1888:8, 19; *Sambur* Heude, 1888:8; *Sikaillus* Heude, 1898:98; *Sica*
     Trouessart, 1898:878; †*Praeelaphus* Portis, 1920:133-134[7]; †*Epirusa* Zdansky,
     1925:53; *Przewalskium* Flerov, 1930:115[3]; †*Deperetia* Shikama, 1936:251-254[8];
     †*Metacervocerus* Dietrich, 1938:263, 265[5]; †*Nipponicervus* Kretzoi, 1941[3];
     †*Thaocervus* Pocock, 1943:554[6]; †*Cervodama* Pidoplichko & Flerov, 1952:1239.[5]]
     E. Plioc.-R.; As. E. Plioc.-R.; Eu. E. Pleist.-R.; N.A. M. Pleist.-R.;
     Mediterranean.[9,10] R.; Af.[11] R.; E. Indies.[12]

[1] Not *Eucervus* Gray, 1866, a synonym of *Odocoileus*.
[2] Not *Procervus* de Blainville, 1840:392, an emendation of *Procerus* Serres, 1838:143, 204, 230,
     preoccupied, a synonym of *Rangifer*.

³ As subgenus.
⁴ Not *Hippelaphus* Reichenbach, 1835, a genus of bovids.
⁵ Subgenus as described.
⁶ Sometimes as subgenus.
⁷ Proposed as subgenus of *Cervus*.
⁸ Sometimes as subgenus, but actually a preoccupied objective synonym of subgenus †*Nipponicervus*. Not *Deperetia* Teppner, 1921, a pectinid, nor †*Deperetia* Schaub, 1923, a synonym of †*Megalovis*.
⁹ Malta and Sicily in the middle-late Pleistocene.
¹⁰ Corsica and Sardinia in the Recent.
¹¹ N. Africa.
¹² Moluccas, Timor.

Axis C. H. Smith in Griffith, Smith & Pidgeon, 1827:312-313.¹ Spotted deer, chital, axis deer, hog deer.
> [Including *Hyelaphus* Sundevall, 1846:180-181²; †*Cervavus* Schlosser, 1903:116.]
> ?L. Mioc., L. Plioc.-M. Pleist., R.; As. Pleist., R.; E. Indies.³

¹ Described as subgenus of *Cervus*; raised to generic rank by Gray, 1843:xxvii, 178.
² As subgenus.
³ Bawean Island, Indonesia; Calamian Islands, Philippines.

†*Eucladoceros* Falconer, 1868:472-480.
> [= †*Eucladocerus* Waterhouse, 1902:132.] [Including †*Euctenoceros* Trouessart, 1898:880¹; †*Praedama* Portis, 1920:136²; †*Kosmelaphus* Kretzoi, 1954:253.]
> L. Plioc.-M. Pleist.; As. L. Plioc.-M. Pleist.; Eu.

¹ Proposed as a subgenus of *Cervus*.
² Portis did not designate a type species for †*Praedama*.

†*Croizetoceros* Heintz, 1970:93.
> [= or including †*Anoglochis* Croizet & Jobert, 1826¹; †*Polycladus* Pomel, 1853:107.²]
> L. Plioc.-E. Pleist.; Eu.

¹ Validity dubious. See Palmer (1904:105).
² Proposed as a section of †*Cervus (Anoglochis)*. Not †*Polycladus* Blanchard, 1847, a genus of "Vermes."

Elaphurus Milne-Edwards, 1866:1090-1091. Mi-lu, Pere David's deer.
> [Including †*Metaplatyceros* Shikama, 1941:1158; †*Elaphuroides* Otsuka, 1972:201.¹]
> E. Pleist.-R.; As.²

¹ Proposed as subgenus.
² E. Asia, now surviving only in captivity.

Dama Frisch, 1775:table.¹ Fallow deer.
> [= *Platyceros* Zimmermann, 1780; *Dactyloceros* Wagner, 1855:352²; *Palmatus* Lydekker, 1898:125.] [= or including *Machlis* Zittel, 1893:402.]
> ?Pleist.; Af.³ E. Pleist., L. Pleist.-R.; As.⁴,⁵ E. Pleist.-R.; Eu.⁶ M. and/or L. Pleist.; Mediterranean.⁷

¹ Validated under the plenary powers of the International Commission on Zoological Nomenclature, Opinion 581, 1960.
² Proposed as subgenus of *Cervus*.
³ N. Africa.
⁴ E. Asia (China) in the early Pleistocene.
⁵ S.W. Asia in the late Pleistocene and Recent.
⁶ Extant in Europe by reintroduction.
⁷ Sicily.

†*Sangamona* Hay, 1920:91.
> L. Pleist.; N.A.

Subfamily **Odocoileinae** Pocock, 1923:206.¹
> [= or including Neocervinae Kalandadze & Rautian, 1992:127.²] [Including Rangiferidae Brookes, 1828:61³; Rangerinae Gray, 1852:188; Rangiferinae Pocock, 1923:206; Alcedae Brookes, 1828:61⁴; Alcinae Jerdon, 1874:253.]
> L. Mioc.-R.; As. L. Mioc.-R.; Eu. E. Plioc.-R.; N.A. ?L. Plioc., E. Pleist.-R.; S.A. Pleist., R.; Cent. A.

¹ "But, as frequently in the classification and not usually noted, the present subfamily is far more extensive than Gray's, or than Brookes' 'family'" (Simpson 1945:154).
² We are not aware of a genus *"Neocervus,"* upon which a family-group name such as this would have to be based.
³ Fide Gray, 1852:188.
⁴ Fide Gray, 1852:186.

Tribe **Capreolini** Brookes, 1828:62.

[= Capreolidae Brookes, 1828:62[1]; Capreolini Simpson, 1945:155.]
L. Mioc.-R.; As. L. Mioc.-E. Plioc., ?E. Pleist., M. Pleist.-R.; Eu.
[1] Fide Gray, 1852:221.

†*Procapreolus* Schlosser, 1924:637.

L. Mioc.-E. Pleist.; As. L. Mioc.-E. Plioc.; Eu.

Capreolus Gray, 1821:307. Roe deer, roebuck.

[= *Capreolus* Frisch, 1775: table[1]; *Caprea* Ogilby, 1837:135.]
L. Plioc., ?E. Pleist., M. Pleist.-R.; As. ?E. Pleist., M. Pleist.-R.; Eu.
[1] Not published in a consistently binominal work.

Tribe **Alceini** Brookes, 1828:61.

[= Alcedae Brookes, 1828:61[1]; Alcadae Gray, 1872:66[2]; Alcinae Jerdon, 1874:253[3];
Alcini Simpson, 1945:155; Alceidae Kashin, 1974:215; Alceini Pavlinov &
Rossolimo, 1987:113.]
?L. Plioc., E. Pleist.-R.; As. L. Plioc.-R.; Eu. M. Pleist.-R.; N.A.
[1] Fide Gray, 1852:186.
[2] Not Alcadae Anon., 1820, a family of birds.
[3] Not Alcidae, emendation of Alcadae Anon., 1820, a family of birds.

†*Cervalces* Scott, 1885:422. Stag-moose, elk-moose.

[Including.†*Praealces* Portis, 1920:137[1]; †*Libralces* Azzaroli, 1952:134.[2]]
?L. Plioc., E.-L. Pleist.; As. L. Plioc.-L. Pleist.; Eu. M.-L. Pleist.; N.A.
[1] Proposed as subgenus of *Cervus*.
[2] Proposed as genus; made subgenus of †*Cervalces* by Azzaroli, 1981:147-148.

Alces Gray, 1821:307. Elk, European elk, moose.

[= *Alce* Frisch, 1775:3[1]; *Paralces* J. Allen, 1902:160.[2]]
?M. Pleist., L. Pleist.-R.; Eu. L. Pleist.-R.; As. L. Pleist.-R.; N.A.
[1] Not published in a consistently binominal work. See the International Code of Zoological Nomenclature,
Article 11(c).
[2] New name for *Alces* Gray.

Tribe **Odocoileini** Pocock, 1923:206.

[= Odocoileinae Pocock, 1923:206; Odocoileini Simpson, 1945:154.] [= or including
Neocervinae Kalandadze & Rautian, 1992:127[1]; Neocervini Kalandadze & Rautian,
1992:127.[1]] [Including Mazamadae Brookes, 1828:62[2]; Rangiferinidae Brookes,
1828:61[3]; Rangiferini Simpson, 1945:155.]
L. Mioc., E. Pleist., M. and/or L. Pleist., R.; As. E. Plioc.-R.; N.A. ?L. Plioc., E.
Pleist.-R.; S.A. Pleist., R.; Cent. A. M. Pleist.-R.; Eu.
[1] We are not aware of a genus "*Neocervus*," upon which a family-group name such as this would have to be
based.
[2] Fide Palmer (1904:752).
[3] Fide Palmer (1904:767).

†*Pavlodaria* Vislobokova, 1980:98.

L. Mioc.; As.

†*Bretzia* Fry & Gustafson, 1974:378.

E. Plioc., L. Pleist. and/or R.; N.A.

†*Morenelaphus* Carette, 1922:393-472.

[Including †*Pampaeocervus* Carette, 1922:393-472; †*Paraceros* Castellanos, 1924[1];
†*Habromeryx* Cabrera, 1929:53-64[2]; †*Rohnia* Castellanos, 1957:9.[3]]
?L. Plioc., Pleist.; S.A.
[1] Not †*Paraceros* Ameghino, 1889, a synonym of †*Antifer*.
[2] Objective synonym of †*Paraceros* Castellanos.
[3] Probably a *nomen nudum*.

Odocoileus Rafinesque, 1832:109-110. American deer, white-tailed deer, black-tailed deer, mule
deer, waiking.

[= *Cariacus* Lesson, 1842:173; *Reduncina* Wagner, 1844:363-384[1]; *Gymnotis*
Fitzinger, 1879:343-350[2]; *Mamcariacus* Herrera, 1899:26[3]; *Odocoelus* G. M. Allen,
1901:449; *Odontocoelus* Sclater, 1902:290; *Palaeoodocoileus* Spillmann, 1931:30-32;
Protomazama Spillmann, 1931:42.] [= or including *Dorcelaphus* Gloger, 1841:140.]

[Including *Macrotis* Wagner, 1855:368-372[4]; *Eucervus* Gray, 1866:338-339; *Otelaphus* Fitzinger, 1874:347-348, 356.[5]]

E. Plioc.-R.; N.A.   Pleist., R.; Cent. A.   Pleist., R.; S.A.

[1] Proposed as subgenus of *Cervus*.
[2] *Nomen nudum* in 1878.
[3] See the International Code of Zoological Nomenclature, Article 1(b)(8), and Palmer (1904:25-26).
[4] Not *Macrotis* Dejean, 1833, a genus of Coleoptera, nor *Macrotis* Reid, 1837, a living peramelid marsupial.
[5] New name for *Macrotis* Wagner, 1855.

Ozotoceros Ameghino, 1891:243.[1] Pampas deer, guazuy, suasutinga, veado dos pampas, veado campeiro.

[= *Blastoceros* Fitzinger, 1860:176[2]; *Ozotoceras* Palmer, 1904:492[3]; *Ozelaphus* Knottnerus-Meyer, 1907:19.[4]]

?L. Plioc., Pleist., R.; S.A.

[1] New name for "*Blastoceros* Gray, 1872".
[2] Not *Blastocerus* Wagner, 1844, 1855, nor Gray, 1850, another odocoileine genus; not *Blastocera* Gerstaecker, 1856, a genus of Diptera.
[3] *Lapsus* for *Ozotoceros*.
[4] New name for *Ozotoceros* Ameghino. Described as a subgenus of *Dama*.

Blastocerus Wagner, 1844:366.[1] Marsh deer, swamp deer, suasupucu, guazu-puco, veado galheiro, veado dos pantanos.

[= *Blastoceri* Sundevall, 1846:182-183[2]; *Bezoarticus* Marelli, 1932:57[3]; *Edocerus* Avila-Pires, 1957:5.] [Including †*Epieuryceros* Ameghino, 1889:613-614.]

E. Pleist., M. and/or L. Pleist., R.; S.A.

[1] Proposed as a category of *Cervus (Elaphus)*; considered a genus by Gray, 1850:68; considered a subgenus of *Cervus* by Wagner, 1955.
[2] As a division of *Cervus*.
[3] Proposed as subgenus.

†*Antifer* Ameghino, 1889:610.

[Including †*Paraceros* Ameghino, 1889:605-607.]

?L. Plioc., E. Pleist., M. and/or L. Pleist.; S.A.

†*Charitoceros* Hoffstetter, 1963:201.

E. Pleist., L. Pleist.; S.A.[1,2]

[1] Andean Bolivia in the early Pleistocene.
[2] Andean Peru in the late Pleistocene.

†*Agalmaceros* Hoffstetter, 1952:361-362.

L. Pleist.; S.A.[1]

[1] Andean Ecuador and Peru.

Mazama Rafinesque, 1817:363. Brocket, gouazou.

[= *Homelaphus* Gray, 1872:90.[1]] [Including *Subulo* H. Smith, 1827:318-319[2]; *Passalites* Gloger, 1841:xxxiii, 140; *Coassus* Gray, 1843:xxvii, 174; *Doryceros* Fitzinger, 1874:360.]

Pleist., R.; S.A. R.; N.A.[3] R.; Cent. A.

[1] Provisional name.
[2] Proposed as subgenus of *Cervus*.
[3] Southern North America (Mexico north to Tamaulipas).

Pudu Gray, 1852:242. Pudu.

[= *Pudua* Brooke, 1878:926-927.] [Including *Pudella* Thomas, 1913:588.]

Pleist., R.; S.A.

†*Navahoceros* Kurtén, 1975:507. Mountain deer.

L. Pleist.; N.A.

Rangifer C. H. Smith, in Griffith, Smith & Pidgeon, 1827. Caribou, reindeer.

[= *Rangifer* Frisch, 1775:3[1]; *Tarandus* Billberg, 1828:22; *Achlis* Reichenbach, 1845:12.] [= or including *Procerus* Serres, 1838:143, 204, 230[2]; *Procervus* de Blainville, 1840:392.[3]]

E. Pleist., M. and/or L. Pleist., R.; As. E. Pleist.-R.; N.A. M. Pleist.-R.; Eu.

[1] Not published in a consistently binominal work. See International Code of Zoological Nomenclature, Article 11(c).
[2] Not preoccupied by *Proceros* Rafinesque, a fish; not *Procerus* Megerle in Dejean, 1821, a coleopteran.
[3] New name for *Procerus* Serres.

Hippocamelus Leuckart, 1816:24. Andean deer, huemul.

[= *Cervequus* Lesson, 1842:173[1]; *Huamela* Gray, 1872:445.] [Including *Furcifer* Wagner, 1844:384[2]; *Anomolocera* Gray, 1869:385-386[3]; *Xenelaphus* Gray, 1869:496-498[4]; *Creagroceros* Fitzinger, 1874:348, 358.[5]]
?E. Pleist., M. Pleist., ?L. Pleist., R.; S.A.

[1] Proposed as subgenus of *Cervus*.
[2] Proposed as category of *Cervus (Elaphus)*; raised to generic rank by Gray, 1852; made subgenus of *Cervus* by Wagner, 1855. Not *Furcifer* Fitzinger, 1843, a genus of Reptilia.
[3] Not *Anomalocera* Templeton, 1837, a genus of Crustacea.
[4] New name for *Anomolocera* Gray, 1869.
[5] New name for *Furcifer* Wagner.

Superfamily **Giraffoidea** Gray, 1821:307.

[= Giraffidae Gray, 1821:307; Giraffoidea Simpson, 1931:284; Camelopardina Gray, 1825:342.]

†*Teruelia* Moya-Sola, 1987:249.
E. Mioc.; Eu.

†*Lorancameryx* Morales, Pickford & D. Soria, 1993:209.
E. Mioc.; Eu.

†*Propalaeoryx* Stromer, 1926:117.
E. Mioc.; Af.

Family †**Climacoceratidae** Hamilton, 1978:168.

[= †Climacoceridae Hamilton, 1978:168; †Climacoceratidae Gentry, 1994:134.]
M. Mioc.; Af.

†*Nyanzameryx* Thomas, 1984:64-89.
M. Mioc.; Af.

†*Climacoceras* MacInnes, 1936:521-530.
M. Mioc.; Af.

Family **Giraffidae** Gray, 1821:307.

[= Giraffae Haeckel, 1866:clviii; Camelopardina Gray, 1825:342; Camelopardalidae Bonaparte, 1831:24.] [= or including Devexa Haeckel, 1866:clviii.[1]] [Including †Sivatheriina Bonaparte, 1850: chart; †Sivatheriidae Gill, 1872:9; †Sivatheriidae Hamilton, 1973:103; †Helladotheridae Dawkins, 1868:4; †Progiraffinae Pilgrim, 1911:26; †Canthumerycidae Hamilton, 1978:178.]
E. Mioc.-R.; Af. E. Mioc.-L. Plioc., Pleist.; As. M. Mioc.-E. Pleist., ?M. and/or ?L. Pleist.; Eu.

[1] As a synonym of Giraffae Haeckel but apparently not based on a generic name.

†*Canthumeryx* Hamilton, 1973:81.

[= †*Zarafa* Hamilton, 1973:85.]
E.-M. Mioc.; Af. E. and/or M. Mioc.; As.[1]

[1] S.W. Asia (Saudi Arabia).

†*Injanatherium* Heintz, Brunet & Sen, 1981:423.
L. Mioc.; As.[1]

[1] W. Asia (Iraq).

†*Propalaeomeryx* Lydekker, 1883:173-174.[1]
[Including †*Progiraffa* Pilgrim, 1908:148, 155.]
E. Mioc., M. and/or L. Mioc.; As.

[1] Proposed as a provisional name.

Subfamily **Giraffinae** Gray, 1821:307.

[= Giraffidae Gray, 1821:307; Giraffinae Zittel, 1893:407; Camelopardalina C. L. Bonaparte, 1837:8.[1]] [Including Palaeotraginae Pilgrim, 1911:29; Okapinae Bohlin, 1926:133.]
E. Mioc.-R.; Af. E. Mioc.-L. Plioc., Pleist.; As. M. Mioc.-E. Pleist.; Eu.

[1] As subfamily.

Tribe **Palaeotragini** Pilgrim, 1911:29, **new rank**.

[= Palaeotraginae Pilgrim, 1911:29.]

E. Mioc.-E. Plioc., E. Pleist.-R.; Af. E. Mioc.-L. Plioc.; As. M.-L. Mioc., L. Plioc.-E. Pleist.; Eu.

Subtribe †**Palaeotragina** Pilgrim, 1911:29, **new rank**.
[= †Palaeotraginae Pilgrim, 1911:29.]
E. Mioc.-E. Plioc.; Af. E. Mioc.-L. Plioc.; As. M.-L. Mioc., L. Plioc.-E. Pleist.; Eu.

†*Praepalaeotragus* Godina, Vislobokova & Abdrachmanova, 1993:92.
E. Mioc.; As.

†*Palaeotragus* Gaudry, 1861:239-240.
[= †*Palaeotragoceros* Lydekker, 1891:349.[1]] [Including †*Achtiaria* Borissiak, 1914:4; †*Orlovia* Godina, 1975:73[2]; †*Yuorlovia* Godina, 1979:47.[3]]
E.-L. Mioc.; Af. ?E. Mioc., M. Mioc.-L. Plioc.; As. M.-L. Mioc.; Eu.
[1] *Lapsus calami?*
[2] Subgenus as defined.
[3] As subgenus. Replacement name for †*Orlovia?*

†*Samotherium* Forsyth Major, 1889:1181.
[Including †*Alcicephalus* Rodler & Weithofer, 1890:154, 155; †*Chersonotherium* Alexejev, 1915:138; †*Schansitherium* Killgus, 1923:251.]
M. Mioc.-E. Plioc.; Af. M. Mioc.-E. Plioc.; As. L. Mioc.; Eu.

†*Giraffokeryx* Pilgrim, 1910:69.
M. Mioc.-E. Plioc.; As. M. Mioc., ?L. Mioc.; Eu. L. Mioc.; Af.

†*Sogdianotherium* Sharapov, 1974:86-91.
L. Plioc.; As.

†*Macedonitherium* Sickenberg, 1967.
L. Plioc.-E. Pleist.; Eu.

Subtribe **Okapiina** Bohlin, 1926:133, **new rank**.
[= Okapinae Bohlin, 1926:133.]
L. Mioc.; Eu. E. Pleist.-R.; Af.

†*Csakvarotherium* Kretzoi, 1930.
L. Mioc.; Eu.[1]
[1] E. Europe.

Okapia Lankester, 1901. Okapi.
[= *Ocuapia* Gatti, 1936:295.]
E. Pleist.-R.; Af.

Tribe **Giraffini** Gray, 1821:307. Giraffes.
[= Giraffidae Gray, 1821:307; Giraffina Gray, 1852:180[1]; Camelopardina Gray, 1825:342.[1]]
L. Mioc.-L. Plioc., Pleist.; As. L. Mioc.-E. Plioc., E. Pleist.; Eu. E. Plioc.-R.; Af.
[1] As tribe.

Giraffa Brunnich, 1771. Giraffe.
[= *Giraffa* Brisson, 1756[1]; *Ovifera* Frisch, 1775: table; *Camelopardalis* Schreber, 1784; *Orasius* Oken, 1816.]
L. Mioc.-L. Plioc., Pleist.; As. L. Mioc.; Eu.[2] E. Plioc.-R.; Af.
[1] See the International Code of Zoological Nomenclature, Article 11(c)(i) for confusion.
[2] E. Europe.

†*Bohlinia* Matthew, 1929:546.
[= †*Orasius* Wagner, 1861.[1]]
L. Mioc.-E. Plioc.; Eu.
[1] Not †*Orasius* Oken, 1816. "A name given to the giraffe in the 13th century by Vincentus Bellovacensis (who died about 1264), and by Albertus Magnus (1193-1280)" (Palmer 1904:478).

†*Honanotherium* Bohlin, 1927.
L. Mioc. and/or Plioc.; As.

†*Mitilanotherium* Samson & Radulesco, 1966.
>   E. Pleist.; Eu.[1]
>   [1] E. Europe.

Subfamily †**Sivatheriinae** Bonaparte, 1850: chart.
>   [= †Sivatheriina Bonaparte, 1850:chart; †Sivatheriidae Gill, 1872:9; †Sivatheriinae
>   Zittel, 1893:409; †Sivatheriidae Hamilton, 1973:103.] [Including Helladotheridae
>   Dawkins, 1868:4.]
>   M. Mioc.-L. Plioc., Pleist.; As.   M.-L. Mioc., ?E. Plioc., Pleist.; Eu.   ?L. Mioc., E.
>   Plioc.-L. Pleist.; Af.

†*Helladotherium* Gaudry, 1860:804.
>   [Including †*Panotherium* Wagner, 1861:79-80.; †*Maraghatherium* de Mecquenem,
>   1908:27-79.]
>   ?L. Mioc.; Af.[1]   L. Mioc.; As.[2]   L. Mioc.; Eu.[3]
>   [1] N. Africa.
>   [2] W. Asia.
>   [3] E. Europe.

†*Karsimatherium* Meladze, 1962:51.
>   L. Mioc.; As.[1]
>   [1] W. Asia.

†*Vishnutherium* Lydekker, 1876:91, 103.
>   L. Mioc.; As.

†*Decennatherium* Crusafont Pairó, 1952.
>   M. Mioc.; As.[1]   M.-L. Mioc.; Eu.
>   [1] S.W. Asia (Iran).

†*Birgerbohlinia* Crusafont Pairó, 1952:100.[1]
>   L. Mioc., ?E. Plioc.; Eu.
>   [1] *Nomen nudum* in Crusafont Pairó & de Villalta Comella, 1951.

†*Bramatherium* Falconer, 1845:363-365.
>   L. Mioc.; As.

†*Hydaspitherium* Lydekker, 1876:154.
>   [= †*Hydaspidotherium* Lydekker, 1876:154[1]; †*Hydaspitherium* Lydekker, 1878:159[2];
>   †*Hydraspotherium* Beddard, 1902:306.]
>   L. Mioc.; As.
>   [1] Incorrect spelling.
>   [2] Emendation.

†*Sivatherium* Cautley & Falconer, 1835:706.
>   [= †*Thaumatherium* Gloger, 1841:138.[1]] [Including †*Libytherium* Pomel, 1892:100-
>   102; †*Indratherium* Pilgrim, 1910:69; †*Griquatherium* Haughton, 1922:11-16[2];
>   †*Orangiatherium* van Hoepen, 1932:63.]
>   E. Plioc.-L. Pleist.; Af.   E.-L. Plioc., Pleist.; As.   Pleist.; Eu.[3]
>   [1] Replacement name.
>   [2] Name questionable.
>   [3] E. Europe.

Superfamily **Bovoidea** Gray, 1821:308.
>   [= Bovidae Gray, 1821:308; Bovina Gray, 1825:342; Booidea Gill, 1872:8, 9; Bovoidea
>   Simpson, 1931:264, 284.]

Family **Bovidae** Gray, 1821:308.
>   [= Bovina Gray, 1825:342; Bovesideae Lesson, 1842:184; Tauroda Haeckel,
>   1866:clviii.[1]] [= or including Cavicornia Illiger, 1811:106[0]; Cavicornidae Reichenow,
>   1886:132.] [Including Antilopidae Gray, 1821:307; Capridae Gray, 1821:307;
>   Rupicapradae Brookes, 1828:63; Damalidae Brookes, 1828:64; Hircidae Brookes,
>   1828:72; Ovidae Brookes, 1828:72; Ovesideae Lesson, 1842:182; Oegosceridae
>   Cobbold, 1859; Probatoda Haeckel, 1866:clviii[2]; Cephalophoridae Gray, 1871:588;
>   Cephalophidae Gray, 1872:3, 21; Nesotragidae Gray, 1872:3, 30; Ovibovidae Gray,
>   1872:3, 31; Pantholopidae Gray, 1872:3, 33; Heleotragidae Gray, 1872;
>   Strepsicerotidae Gray, 1872:3, 46; Connochetidae Gray, 1872:4, 42; Aepycerotidae

Gray, 1872:4, 42; Saigadae Gray, 1872:7, 32; Orygidae Rochebrune, 1883:125, 155; Cervicapridae Rochebrune, 1883:126, 156; Alcelaphidae Rochebrune, 1883:132, 156; Tragelaphidae Rochebrune, 1883; Aegodontia Schlosser, 1923:589; Boodontia Schlosser, 1923:592; †Urmiatheriidae Dietrich, 1937:323, 324; †Hypsodontinae Köhler, 1987:133-246.]

Olig., E. Mioc.-R.; As. E. Mioc.-R.; Af. E. Mioc.-R.; Eu. L. Mioc., L. Plioc.-R.; Mediterranean. L. Mioc., ?E. Pleist., M. Pleist.-R.; N.A. Pleist., R.; E. Indies. L. Pleist.; Cent. A.[3]

[1] Used as a family.
[2] Used as a family but not based on a generic name.
[3] Also Recent worldwide in domestication.

†*Hanhaicerus* Huang, 1985:155.
  Olig.; As.

†*Namibiomeryx* Morales, Soria & Pickford, 1995:1211.
  E. Mioc.; Af.

†*Hypsodontus* Sokolov, 1949:1103.[1]
  M. Mioc.; As.[2] M. Mioc.; Eu.[3]
  [1] Questionably referred to the Bovidae.
  [2] W. Asia.
  [3] E. Europe.

†*Kubanotragus* Gabunia, 1973:112.
  M. Mioc.; As. M. Mioc.; Eu.[1]
  [1] E. Europe.

†*Pseudoeotragus* van der Made, 1989:218.
  E.-M. Mioc.; Eu.

†*Procobus* Khomenko, 1913.
  L. Mioc.; Eu.[1]
  [1] E. Europe.

†*Prodamaliscus* Schlosser, 1904:29.
  L. Mioc.; As.[1]
  [1] W. Asia (Samos).

†*Samodorcas* Bouvrain & de Bonis, 1985:257, 287.
  L. Mioc.; As.[1]
  [1] W. Asia (Samos).

†*Shaanxispira* Liu, Li & Zhai, 1978:181.
  [= †*Shensispira* Chiu, Li & Chiu, 1979:267.[1]]
  L. Mioc.; As.[2]
  [1] Apparently a new spelling of †*Shaanxispira*.
  [2] E. Asia (China).

†*Torticornis* Dmitrieva, 1977:48.
  L. Mioc. and/or E. Plioc.; As.

†*Kabulicornis* Heintz & Thomas, 1981.
  E. Plioc.; As.

†*Damalacra* Hendey, 1981:52.
  E. Plioc.; Af.

†*Pontoceros* Vereshchagin, Alekseeva, David & Baigusheva, 1969:167.
  L. Plioc.-M. Pleist.; Eu.[1]
  [1] E. Europe.

†*Vishnumeryx* Pilgrim, 1939:29.
  Plioc. and/or Pleist.; As.

†*Fenhoryx* Pei, 1958.
  L. Plioc. and/or E. Pleist.; As.

†*Parabubalis* Gromova, 1931.
  Pleist.; As.

Tribe **Peleini** Gray, 1872.
>   [= Peleadae Gray, 1872; Peleini Sokolov, 1953:46, 267.]
>   E. Pleist.-R.; Af.[1]
>   [1] S. Africa.

Pelea Gray, 1851:126.[1] Rhebok.
>   E. Pleist.-R.; Af.
>   [1] As subgenus of *Eleotragus*. Raised to generic rank by Gray, 1852:90.

Subfamily **Antilopinae** Gray, 1821:307. Antelopes.
>   [= Antilopidae Gray, 1821:307; Antilopina C. L. Bonaparte, 1837:8[1]; Antilopinae Baird, 1857:664; Antilopida Haeckel, 1866:clviii.[2]] [Including Gazellae Haeckel, 1866:clviii[2]; Gazellinae Coues, 1889:2474; Nesotragidae Gray, 1872:3, 30; Neotraginae Sclater & Thomas, 1894:2; Madoquinae Pocock, 1910.]
>   E. Mioc.-R.; Af. E. Mioc.-R.; As. M. Mioc.-R.; Eu. M.-L. Pleist.; N.A.[3]
>   [1] As subfamily.
>   [2] Used as a family.
>   [3] Alaska and Canadian Arctic.

Tribe **Antilopini** Gray, 1821:307.
>   [= Antilopidae Gray, 1821:307; Antilopina Pilgrim, 1939:23; Antilopini Simpson, 1945:161.] [Including Saigadae Gray, 1872:7, 32; Saigiidae Gill, 1872:8, 72; Saigina Pilgrim, 1939:23; Saigini Simpson, 1945:161; Gazellina Pilgrim, 1939:23; Gazellini Sokolov, 1953:46, 231; Pantholopidae Gray, 1872:3, 33; Pantholopina Pilgrim, 1939:23; Pantholopini Sokolov, 1953:187; Antidorcatini Haltenorth, 1963:106; Procaprini Haltenorth, 1963:107; Ammodorcini Haltenorth, 1963:113; Lithocraniini Haltenorth, 1963:114.]
>   E. Mioc.-R.; Af. E. Mioc.-R.; As. M. Mioc.-R.; Eu. M.-L. Pleist.; N.A.[1]
>   [1] Alaska and Canadian Arctic.

Gazella de Blainville, 1816:75.[1] Gazelles, chinkara, korin, mhorr, goa, zeren.
>   [= *Dorcas* Gray, 1821:307; *Leptoceros* Wagner, 1844:422, 423.[2]] [Including *Procapra* Hodgson, 1846:334[3]; *Tragops* Hodgson, 1847:695, 696[4]; *Tragopsis* Fitzinger, 1869:157[5]; *Eudorcas* Fitzinger, 1869:159; *Korin* Gray, 1872:39; *Nanger* Lataste, 1885:173[3]; †*Protetraceros* Schlosser, 1903:136; *Matschiea* Knottnerus-Meyer, 1907:57; *Prodorcas* Pocock, 1918:130; *Trachelocele* Ellerman & Morrison-Scott, 1951:389[3]; *Rhinodorcas* von Boetticher, 1953:89[3]; †*Vetagazella* Dmitrieva, 1970:141-151[3]; †*Miogazella* Korotkevich, 1976:149.[3]]
>   E. Mioc.-R.; Af. E. Mioc.-R.; As. M. Mioc.-M. Pleist.; Eu.
>   [1] "*Gazella* was previously used by Pallas, 1769, and by Lichtenstein, 1814, in conflicting senses. By suspension of the Rules, the International Commission on Zoological Nomenclature has authorized its use for the present genus as of De Blainville, 1816" (Simpson 1945:161).
>   [2] Preoccupied by *Leptoceros* Leach, 1817, a neuropteran.
>   [3] As subgenus.
>   [4] Preoccupied by *Tragops* Wagler, 1830, a reptile.
>   [5] Replacement name for *Tragops* Hodgson, 1847.

†*Prostrepsiceros* Forsyth Major, 1891:608-609.
>   [Including †*Helicoceras* Weithofer, 1888:288[1]; †*Helicophora* Weithofer, 1889:79[2]; †*Helicotragus* Palmer, 1903:873; †*Hemistrepsiceros* Pilgrim & Hopwood, 1928:94.]
>   M.-L. Mioc.; As.[3] L. Mioc.; Eu.[4]
>   [1] Objective synonym of †*Helicotragus* Palmer, 1903. Not †*Helicoceras* d'Orbigny, 1842, a genus of mollusks.
>   [2] Objective synonym of †*Helicotragus* Palmer, 1903. Not †*Helicophora* Gray, 1842, a genus of mollusks.
>   [3] W. Asia.
>   [4] E. Europe.

†*Ouzocerus* Bouvrain & de Bonis, 1986:661, 664.
>   L. Mioc.; As.[1] L. Mioc.; Eu.[2]
>   [1] W. Asia (Iran).
>   [2] E. Europe (Greece).

†*Dorcadoryx* Teilhard de Chardin & Trassaert, 1938:32.
>   L. Mioc.; As.

Antilope Pallas, 1766:1. Blackbuck, Indian antelope.
    [= *Cervicapra* Sparrman, 1780:275-281.]
    L. Mioc.-R.; As. L. Plioc.; Af.

†*Protragelaphus* Dames, 1883:95-97.
    L. Mioc.; As.[1] L. Mioc.; Eu.[2]
    [1] W. Asia.
    [2] E. Europe.

†*Qurliqnoria* Bohlin, 1937:34.
    L. Mioc.; As.

†*Hispanodorcas* Thomas, Morales & Heintz, 1982:211.
    L. Mioc.; Eu.

†*Nisidorcas* Bouvrain, 1979:507, 510.
    L. Mioc.; As. L. Mioc.; Eu.[1]
    [1] E. Europe (Macedonia).

†*Sinapocerus* Tekkaya, 1974:174, 178.
    L. Mioc.; As.[1]
    [1] W. Asia.

†*Antilospira* Teilhard de Chardin & Young, 1931:1-67.
    E.-L. Plioc.; As.

†*Sinoreas* Teilhard de Chardin & Trassaert, 1938:73.
    E. Plioc., ?E. Pleist.; As.

†*Parastrepsiceros* Vekua, in Gabunia & Vekua, 1968:49-55.
    E. Plioc.; As.[1]
    [1] W. Asia.

Antidorcas Sundevall, 1847:271. Springbuck.
    [Including †*Phenacotragus* Schwarz, 1937:53.]
    L. Plioc.-R.; Af.

†*Spirocerus* Boule & Teilhard de Chardin, 1928:66.
    L. Plioc.-L. Pleist.; As.

†*Gazellospira* Pilgrim & Schaub, 1939:6.
    L. Plioc.-M. Pleist.; Eu. E. Pleist.; As.[1]
    [1] W. Asia.

†*Tragospira* Kretzoi, 1954.
    E.-M. Pleist.; Eu.

Pantholops Hodgson, 1834:81. Chiru, Tibetan antelope.
    Pleist., R.; As.

Litocranius Kohl, 1886. Gerenuk, Waller's gazelle, giraffe gazelle.
    [= *Lithocranius* Thomas, 1891.]
    M. Pleist., R.; Af.

Saiga Gray, 1843:xxvi, 160. Saiga.
    [= *Colus* Wagner, 1844:419.[1]]
    M. Pleist.-R.; Eu. M.-L. Pleist.; N.A.[2] L. Pleist.-R.; As.
    [1] Described as subgenus of *Antilope*. Not *Colus* Humphrey, 1767, a genus of mollusks.
    [2] Alaska and Canadian Arctic.

Ammodorcas Thomas, 1891. Dibatag.
    R.; Af.

Tribe **Neotragini** Sclater & Thomas, 1894:2.
    [= Neotraginae Sclater & Thomas, 1894:2; Neotragina Pilgrim, 1939:23; Neotragini
    Simpson, 1945:160.] [Including Nesotragidae Gray, 1872:3, 30; Madoquinae Pocock,
    1910; Madoquina Pilgrim, 1939:23; Madoquini Haltenorth, 1963:73; Oreotragina
    Pilgrim, 1939:23; Oreotragini Haltenorth, 1963:75; Raphicerini Haltenorth, 1963:77;
    Dorcatragini Haltenorth, 1963:76.]
    M. Mioc.-R.; Af.[1] ?M. Mioc.; As.[2] L. Mioc.; Eu.
    [1] Neotragini indet. in the late Miocene of N. Africa.
    [2] S.W. Asia.

†*Homoiodorcas* Thomas, 1981:404.[1]
  M. Mioc.; Af. ?M. Mioc.; As.[2]
  [1] Assignment to Neotragini is questionable.
  [2] S.W. Asia (Saudi Arabia).

†*Tyrrhenotragus* Hürzeler & Engesser, 1976:334.
  L. Mioc.; Eu.

Raphicerus Hamilton Smith, 1827:342-343. Steinbok, grysboks.
  [= *Pediotragus* Fitzinger, 1860:396; *Rhaphicerus* Lonnberg, 1908:40.[1]] [Including
  Nototragus Thomas & Schwann, 1906:10; *Grysbock* Knottnerus-Meyer, 1907:55.[2]]
  E. Plioc.-R.; Af.
  [1] Variant spelling.
  [2] Objective synonym of *Nototragus*.

Madoqua Ogilby, 1837:137. Dik-diks.
  [Including †*Praemadoqua* Dietrich, 1950:34; *Rhynchotragus* Neumann, 1905:88.[1]]
  E.-L. Plioc., M. Pleist., R.; Af.
  [1] As subgenus.

†*Palaeotragiscus* Broom, 1934:471-480.
  ?L. Plioc.; Af.

Oreotragus A. Smith, 1834:212. Klipspringer.
  [= *Oritragus* Gloger, 1841:154.]
  L. Plioc.-R.; Af.

Ourebia Laurillard, 1841:622-623. Oribi.
  ?E. Pleist., M. Pleist.-R.; Af.

Neotragus H. Smith, 1827:269. Pygmy antelopes, royal antelope, suni.
  [Including *Nesotragus* Dueben, 1847:221[1]; *Hylarnus* Thomas, 1906:149.]
  Pleist., R.; Af.
  [1] As subgenus.

Dorcatragus Noack, 1894. Beira.
  R.; Af.

Subfamily **Caprinae** Gray, 1821:307. Goats, sheep.
  [= Capridae Gray, 1821:307; Caprina Haeckel, 1866:clviii[1]; Caprinae Gill, 1872:9.]
  [= or including Aegomorpha Haeckel, 1866:clviii[2]; Ovicaprina Noack, 1887:202[3];
  Ovicaprinae Schlosser, 1911:591.[3]] [Including Ovidae Brookes, 1828:72[4]; Ovinae
  Baird, 1857:xxxi, 664; Ovina Haeckel, 1866:clviii[1]; Hircidae Brookes, 1828:72[5];
  Probatoda Haeckel, 1866:clviii[6]; †Pseudotraginae Schlosser, 1904:87; Rupicaprinae
  Trouessart, 1905:734.]
  E. Mioc.-R.; As. M. Mioc., ?L. Mioc., E. Plioc.-E. Pleist., ?M. Pleist., L. Pleist.-
  R.; Af. M.-L. Mioc., ?E. Plioc., L. Plioc.-R.; Eu. L. Mioc., ?E. Plioc., M. Pleist.-
  R.; N.A. L. Plioc.-R.; Mediterranean. R.; E. Indies.[7]
  [1] Used as a family .
  [2] Used as a family but not based on a generic name.
  [3] Not based on a generic name.
  [4] Fide Gray, 1852:160.
  [5] Fide Gray, 1852:143.
  [6] Used as a family, synonym of Ovina Haeckel, but not based on a generic name.
  [7] Sumatra.
  Also introduced and feral populations in New Zealand and the Kerguelen Islands of the S. Indian Ocean.
  Recent worldwide in domestication.

†*Pachygazella* Teilhard de Chardin & Young, 1931.
  L. Mioc.; As.

†*Samotragus* Sickenberg, 1936.
  L. Mioc.; As.[1] L. Mioc.; Eu.[2]
  [1] W. Asia (Samos).
  [2] E. Europe (Greece).

†*Myotragus* Bate, 1909. Cave goat.
  L. Plioc.-R.; Mediterranean.[1]
  [1] Balearic Islands.

†*Capraoryx* Alekseeva, 1977.
     E. Pleist.; Eu.[1]
     [1] E. Europe.

Tribe **Caprini** Gray, 1821:307.
     [= Capridae Gray, 1821:307; Caprini Simpson, 1945:162; Caprina Kalandadze &
     Rautian, 1992:131.] [= or including Ovicaprina Noack, 1887:202.[1]] [Including
     Tragina Haeckel, 1895; †Pseudotraginae Schlosser, 1904:87; †Pseudotragini Solounias
     & Moelleken, 1991:797, 805; †Oiocerina Pilgrim, 1939:23; †Oiocerini Sokolov,
     1953:161; Rupicapradae Brookes, 1828:63[2]; Rupicaprini Simpson, 1945:161;
     Rupicaprina Pilgrim, 1939:23; Ovina Haeckel, 1866:clviii[3]; Probatoda Haeckel,
     1866:clviii.[4]]
     E.-L. Mioc., ?E. Plioc., L. Plioc.-R.; As. M. Mioc., ?L. Mioc., L. Plioc.-E. Pleist.,
     ?M. Pleist., L. Pleist.-R.; Af.[5] M.-L. Mioc., ?E. Plioc., L. Plioc.-R.; Eu. L. Mioc.,
     M. Pleist.-R.; N.A. R.; Mediterranean.[6]
     [1] Not based on a generic name.
     [2] Fide Gray, 1852:115.
     [3] Used as a family.
     [4] Used as a family and synonym of Ovina Haeckel. Not based on a generic name.
     [5] Caprini? genus and sp. indet. in the middle Pleistocene of Ternifine, Algeria.
     [6] Sardinia and Corsica, possibly by introduction.
     Also introduced and feral populations in New Zealand and the Kerguelen Islands of the S. Indian Ocean.
     Recent worldwide in domestication.

†*Sinopalaeoceros* Chen, 1988:166.
     E. Mioc.; As.[1]
     [1] China.

†*Oioceros* Gaillard, 1902.
     E.-L. Mioc.; As. M. Mioc.; Af. M.-L. Mioc.; Eu.[1]
     [1] E. Europe.

†*Pachytragus* Schlosser, 1904:56.
     M. Mioc., ?L. Mioc.; Af. M.-L. Mioc.; As.[1]
     [1] W. Asia (including Samos).

†*Turcocerus* Köhler, 1987:133-246.
     [= †*Sinomioceros* Chen, 1988:166.]
     M. Mioc.; As.[1]
     [1] N. China and Mongolia.

†*Benicerus* Heintz, 1973:245.
     M. Mioc.; Af.[1]
     [1] N. Africa.

†*Caprotragoides* Thenius, 1979:9, 11.
     M. Mioc.; As.

†*Tethytragus* Azanza & Morales, 1994:256.
     M. Mioc.; As.[1] M. Mioc.; Eu.
     [1] W. Asia (Turkey).

†*Gentrytragus* Azanza & Morales, 1994:270.
     M. Mioc.; Af. M. Mioc.; As.[1]
     [1] S.W. Asia (Arabian Peninsula).

†*Tossunnoria* Bohlin, 1937:37.
     M.-L. Mioc.; As. ?L. Plioc.; Af.[1]
     [1] E. Africa.

†*Protoryx* Forsyth Major, 1891:608, 609.
     ?M. and/or L. Mioc. and/or ?Plioc.; As. L. Mioc.; Eu.[1]
     [1] E. Europe.

†*Norbertia* Köhler, Moya-Sola & Morales, 1995:172.
     L. Mioc. and/or E. Plioc.; Eu.

†*Sinocapra* Chen, 1991:230.
     L. Plioc.; As.[1]
     [1] China.

†*Mesembriacerus* Bouvrain, 1975:1357.
  L. Mioc.; Eu.[1]
  [1] E. Europe.

†*Pseudotragus* Schlosser, 1904:51.
  [Including †*Leptotragus* Bohlin, 1936:1-22.]
  L. Mioc.; As.[1]
  [1] W. Asia (Samos, Turkey).

Capra Linnaeus, 1758:68. Goats, ibexes, tur, markhor.
  [= *Mamcapraus* Herrera, 1899:8.[1]] [Including *Hircus* Brisson, 1762:38-48; *Ibex* Pallas,
  1776:52; *Tragus* Schrank, 1798:78; *Aegoceros* Pallas, 1811:224; *Orthaegoceros*
  Trouessart, 1905:738[2]; *Turus* Hilzheimer, 1916:273; *Euibex* Camerano, 1916:338;
  Eucapra Camerano, 1916:338; *Turocapra* de Beaux, 1949:17.]
  ?L. Mioc., Pleist., R.; As. ?L. Mioc., M. Pleist.-R.; Eu. L. Plioc., R.; Af.[3]
  [1] See the International Code of Zoological Nomenclature, Article 1(b)(8), and Palmer (1904:25-26).
  [2] Sometimes as subgenus.
  [3] N. Africa.
  Also Recent worldwide in domestication.

†*Sporadotragus* Kretzoi, 1968:164.
  [= †*Microtragus* Andree, 1926:150.[1]]
  L. Mioc.; As.[2]
  [1] Not †*Microtragus* White, 1846, a coleopteran.
  [2] W. Asia (Samos).

†*Sinotragus* Bohlin, 1935:133.
  L. Mioc.; As.

†*Neotragocerus* Matthew & Cook, 1909:413.
  L. Mioc.; N.A.

†*Parapseudotragus* Sokolov, 1961:112-118.
  L. Mioc.; As. L. Mioc.; Eu.[1]
  [1] E. Europe.

†*Olonbulukia* Bohlin, 1937:30.
  L. Mioc.; As.

†*Prosinotragus* Bohlin, 1935:130.
  L. Mioc.; As.

†*Paraprotoryx* Bohlin, 1935:126.
  L. Mioc.; As.

†*Procamptoceras* Schaub, 1923:282.
  L. Plioc.-E. Pleist.; Eu.

Ovis Linnaeus, 1758:70-71. Sheep, mouflons, urial, argali, bighorn, thinhorns, snow sheep,
  Dall's sheep, range maggot, hoofed locust.
  [= *Aries* Brisson, 1762:48-51; *Aires* Storr, 1780: Table C; *Aries* Link, 1795:96-97;
  Ammon de Blainville, 1816:76.] [Including *Musimon* Pallas, 1776:8; *Caprovis*
  Hodgson, 1847:702; *Argali* Gray, 1850:37; *Pachyceros* Gromova, 1936:84.[1]]
  L. Plioc.-R.; As. E. Pleist.-R.; Eu.[2] M. Pleist.-R.; N.A. L. Pleist.; Af. R.;
  Mediterranean.[3]
  [1] As subgenus.
  [2] Recent of Europe by introduction (probably from Sardinia and Corsica).
  [3] Sardinia and Corsica (possibly primitive domestic populations, now feral).
  Also introduced and feral populations in New Zealand and the Kerguelen Islands of the S. Indian Ocean.
  Recent worldwide in domestication.

Hemitragus Hodgson, 1841:218. Tahrs.
  L. Plioc.-L. Pleist.; Eu. R.; As.

†*Numidocapra* Arambourg, 1949:290.
  E. Pleist.; Af.[1]
  [1] N. Africa.

†*Sivacapra* Pilgrim, 1939:49.
  E. Pleist.; As.

Ammotragus Blyth, 1840:13. Barbary sheep, aoudad, arui.
>> Pleist., R.; Af.

Rupicapra de Blainville, 1816:75. Chamois.
>> [= *Rupicapra* Frisch, 1775[1]; *Capella* Keyserling & Blasius, 1840:iv, 9, 28; *Cemas* Gloger, 1841:153.[2]]
>> M. Pleist.-R.; Eu. R.; As.[3]

>> [1] Invalid. Not published in a consistently binominal work. See the International Code of Zoological Nomenclature, Article 11(c).
>> [2] New name for *Rupicapra* de Blainville. Not *Cemas* Oken, 1816, a gnu.
>> [3] W. Asia.

Oreamnos Rafinesque, 1817:44. Rocky Mountain goat.
>> [= *Aplocerus* Hamilton Smith, 1827:354-355; *Haplocerus* Wagner, 1844:462; *Haploceros* Lydekker, 1891:351.]
>> L. Pleist.-R.; N.A.

Pseudois Hodgson, 1846:343. Blue sheep, bharal.
>> [= *Pseudovis* Gill, 1872:79.]
>> R.; As.

Tribe **Ovibovini** Gill, 1872:9, 77. Musk oxen.
>> [= Ovibovinae Gill, 1872:9, 77; Ovibovini Simpson, 1945:162.] [Including †Urmiatheriinae Sickenberg, 1932:2; †Euceratheriinae Frick, 1937:xvi, 548; †Euceratherini Hibbard, 1955:80; †Megalovinae Dietrich, 1950:21; Budorcadina Pilgrim, 1939:23.]
>> M. Mioc., E. Plioc.-E. Pleist.; Af.[1] M. Mioc.-E. Plioc., E. Pleist.-R.; As. L. Mioc., ?E. Plioc., L. Plioc.-L. Pleist.; Eu. ?E. Pleist., M. Pleist.-R.; N.A.

>> [1] Ovibovini sp. indet. in the early Pliocene 'E' Quarry, Langebaanweg.

†*Damalavus* Arambourg, 1959.
>> M. Mioc.; Af.[1]

>> [1] N. Africa.

†*Urmiatherium* Rodler, 1888:114-115.
>> [Including †*Pseudobos* Schlosser, 1903:154; †*Chilinotherium* Wiman, 1922.]
>> M.-L. Mioc.; As.

†*Tsaidamotherium* Bohlin, 1935.
>> L. Mioc.; As.

†*Parurmiatherium* Sickenberg, 1932:1.
>> [Including †*Paraboselaphus* Schlosser, 1903; †*Plesiaddax* Schlosser, 1903.]
>> L. Mioc.; As.

†*Palaeoreas* Gaudry, 1861:298, 299.
>> L. Mioc.; As.[1] L. Mioc.; Eu.

>> [1] W. Asia.

†*Palaeoryx* Gaudry, 1861:240-241.
>> [Including †*Sinoryx* Teilhard de Chardin & Trassaert, 1938:38.]
>> L. Mioc.; As. L. Mioc., ?Plioc.; Eu.

†*Criotherium* Forsyth Major, 1891:608-610.
>> L. Mioc.; As.[1]

>> [1] Including W. Asia (Samos).

†*Lyrocerus* Teilhard de Chardin & Trassaert, 1938:77.
>> E. Plioc.; As.

†*Makapania* Wells & Cooke, 1957:26.
>> L. Plioc.-E. Pleist.; Af.

†*Megalovis* Schaub, 1923.
>> [Including †*Deperetia* Schaub, 1923:287[1]; †*Pliotragus* Kretzoi, 1941:349[2]; †*Hesperoceras* de Villalta Comella & Crusafont Pairó, 1953:6[3]; †*Hesperidoceras* Crusafont Pairó, 1965.[4]]
>> L. Plioc.-E. Pleist.; Eu. E. Pleist.; As.

>> [1] Not †*Deperetia* Teppner, 1921, a pectinid.

[2] Replacement name for †*Deperetia* Schaub, 1923.
[3] Not †*Hesperoceras* Miller & Youngquist, 1947.
[4] Replacement name for †*Hesperoceras* Villalta & Crusafont Pairó, 1953.

Budorcas Hodgson, 1850:65. Takin.
Pleist., R.; As.

†*Praeovibos* Staudinger, 1908.
[Including †*Parovibos* Kretzoi, 1942:362.]
E.-M. Pleist.; As.   E.-M. Pleist.; Eu.   ?E. Pleist., M. Pleist.; N.A.

†*Bootherium* Leidy, 1852:71.
M.-L. Pleist.; N.A.

Ovibos de Blainville, 1816:76. Musk ox.
[= *Bosovis* Kowarzik, 1911:107.]
M.-L. Pleist.; Eu.   M. Pleist.-R.; N.A.   L. Pleist.; As.

†*Symbos* Osgood, 1905:223.
[= †*Scaphoceros* Osgood, 1905:174[1]; †*Liops* Gidley, 1906:165[2]; †*Gidleya* Cossman, 1907:64; †*Lissops* Gidley, 1908:684.[3]]
M.-L. Pleist.; N.A.

[1] Not *Scaphocera* Saalmueller, 1884: 181, a genus of Lepidoptera.
[2] Not *Liops* Fieber, 1870: 254, a genus of Hemiptera.
[3] Possibly *lapsus* or replacement for †*Liops*.

†*Soergelia* Schaub, 1951.
M. Pleist.; As.[1]   M. Pleist.; Eu.   M. Pleist.; N.A.
[1] N.E. Asia.

†*Boopsis* Teilhard de Chardin, 1936.
M. Pleist.; As.

†*Euceratherium* Furlong & Sinclair, 1904.
[Including †*Preptoceras* Furlong, 1905; †*Aftonius* Hay, 1913.]
M.-L. Pleist.; N.A.

Tribe **Naemorhedini** Pilgrim, 1939:23.[1]
[= Nemorhaedina Pilgrim, 1939:23; Nemorhaedini Sokolov, 1953:162.]
L. Plioc.-E. Pleist.; Eu.   Pleist., R.; As.   E. Pleist.; Mediterranean.[2]   R.; E. Indies.[3]
[1] Emendation.
[2] Sardinia.
[3] Sumatra.

†*Gallogoral* Guérin, 1965:5.
L. Plioc.-E. Pleist.; Eu.

†*Nesogoral* Gliozzi & Malatesta, 1980:295-347.
E. Pleist.; Mediterranean.[1]
[1] Sardinia.

Naemorhedus C. H. Smith, in Griffith, Smith & Pidgeon, 1827:352. Gorals.
[= *Nemorhaedus* Hodgson, 1841:913; *Naemorhaedus* Jardine, 1836:97[1]; *Nemorhedus* Agassiz, 1842:22[1]; *Nemorrhedus* Gray, 1843:166[1]; *Kemas* Ogilby, 1837:138; *Hemitragus* Van der Hoeven, 1855:943.[2]] [Including *Urotragus* Gray, 1871:372.]
Pleist., R.; As.
[1] Emendation.
[2] Replacement name but preoccupied by *Hemitragus* Hodgson, 1841.

Capricornis Ogilby, 1837. Serows.
[Including *Capricornulus* Heude, 1898:13[1]; *Lithotragus* Heude, 1898:13; *Nemotragus* Heude, 1898:13; *Austritragus* Heude, 1898:14.]
Pleist., R.; As.   R.; E. Indies.[2]
[1] As subgenus.
[2] Sumatra.

Subfamily **Bovinae** Gray, 1821:308.
[= Bovidae Gray, 1821:308; Bovina Gray, 1825:342; Bovinae Gill, 1872:8.] [Including Bisontinae Rutimeyer, 1865:320; †Parabovinae Dietrich, 1942; Bubalinae Dietrich, 1950:47; Tragelaphidae Rochebrune, 1883; Tragelaphinae Jerdon, 1874.]

E. Mioc.-R.; Af. ?E. Mioc., M. Mioc.-R.; As. E. Mioc.-R.; Eu. ?E. Pleist., M.
Pleist., ?L. Pleist., R.; E. Indies. ?E. Pleist., M. Pleist.-R.; N.A. L. Pleist.;
Mediterranean.[1] L. Pleist.; Cent. A.[2]

[1] Sicily.
[2] Also Recent practically worldwide in domestication.

†*Gona* Deraniyagala, 1958:146.
   Pleist.; As.[1]
   [1] S. Asia (Sri Lanka).

Pseudoryx Dung, Giao, Chinh, Tuoc, Arctander & MacKinnon, 1993:443. Sao la, Vu Quang ox.
   R.; As.[1]
   [1] S.E. Asia (Vietnam, ?Laos).

Tribe **Boselaphini** Knottnerus-Meyer, 1907:38, 98, 116.
   [= Boselaphinae Knottnerus-Meyer, 1907:38, 98, 116; Boselaphina Pilgrim, 1939:23;
   Boselaphini Simpson, 1945:158.] [Including Tetracerosidae Brookes, 1828[1];
   †Helicoportacina Pilgrim, 1939:23; †Strepsiportacina Pilgrim, 1939:23; Tetracerina
   Pilgrim, 1939:23; †Tragocerina Pilgrim, 1939:23; †Eotragini Viret, 1961:1044.]
   E. Mioc.-L. Plioc.; Af. ?E. Mioc., M.-L. Mioc., Plioc., Pleist., R.; As. E.-L.
   Mioc., Plioc.; Eu. Pleist.; E. Indies.[2]
   [1] Long forgotten.
   [2] Java.

†*Eotragus* Pilgrim, 1939:137.
   [= †*Eocerus* Schlosser, 1911:502.[1]]
   E. Mioc., ?M. Mioc.; Af. ?E. Mioc., M. Mioc.; As. E.-M. Mioc.; Eu.
   [1] Not †*Eocerus* Sharp, 1907.

†*Protragocerus* Depéret, 1887:381.
   [Including †*Helicoportax* Pilgrim, 1937:746; †*Strepsiportax* Pilgrim, 1937:756;
   †*Austroportax* Kretzoi, 1941; †*Paratragocerus* Sokolov, 1949.]
   M. Mioc.; As. M. Mioc.; Eu.

†*Miotragocerus* Stromer, 1928:36.
   [= †*Tragocerus* Gaudry, 1861:297-298[1]; †*Graecoryx* Pilgrim & Hopwood, 1928:54[2];
   †*Tragoceridus* Kretzoi, 1968:165.[2]] [Including †*Sivaceros* Pilgrim, 1937:792;
   †*Dystychoceras* Kretzoi, 1941:262, 336; †*Gazelloportax* Kretzoi, 1941:266, 341;
   †*Pontoportax* Kretzoi, 1941:266, 341; †*Indotragus* Kretzoi, 1941:267, 342;
   †*Pikermicerus* Kretzoi, 1941:267, 342; †*Mirabilocerus* Gadzhiev, 1961.]
   M.-L. Mioc.; As. M.-L. Mioc.; Eu. L. Mioc.; Af.
   [1] Not †*Tragocerus* Dejean 1821, a coleopteran. Not preoccupied by *Tragocera* Billberg, 1820, a
   lepidopteran.
   [2] Replacement name for †*Tragocerus* Gaudry, 1861.

†*Sivoreas* Pilgrim, 1939:130.
   ?M. Mioc.; Af. M. Mioc.; As.

†*Samokeros* Solounias, 1981:127.
   ?M. Mioc.; Af.[1] L. Mioc.; As.[2] ?L. Mioc.; Eu.[3]
   [1] N. Africa.
   [2] W. Asia (Samos and Iran).
   [3] E. Europe (Greece).

†*Elachistoceras* Thomas, 1977:375.
   M.-L. Mioc.; As.

†*Kipsigicerus* Thomas, 1984:46.
   M. Mioc.; Af.

†*Strogulognathus* Filhol, 1890:265.
   [= †*Platuprosopos* Filhol, 1888:30-32; †*Platyprosopos* Lydekker, 1890:52[1];
   †*Strongylognathus* Lydekker, 1892:46.[2]]
   M. Mioc.; Eu.
   [1] Not *Platyprosopus* Mannerheim, 1830, a supposed senior homonym based on a living coleopteran.
   [2] Preoccupied by *Strongylognathus* Mayer, 1853, an insect.

†*Selenoportax* Pilgrim, 1937:737.
L. Mioc.; As.

†*Phronetragus* Gabunia, 1955.
L. Mioc.; As.[1]
[1] W. Asia.

†*Tragoportax* Pilgrim, 1937.
L. Mioc.; As. L. Mioc.; Eu.

†*Pachyportax* Pilgrim, 1937:766.
L. Mioc., Plioc.; As.

†*Tragoreas* Schlosser, 1904:34.
L. Mioc.; As.[1]
[1] W. Asia (including Samos).

†*Pliodorcas* Kretzoi, 1975.
L. Mioc.; Eu.

†*Perimia* Pilgrim, 1939:165.
L. Mioc.; As.

†*Ruticeros* Pilgrim, 1939:240.
L. Mioc.; As.

†*Mesembriportax* Gentry, 1974.
E.-L. Plioc.; Af.[1]
[1] S. Africa.

†*Sivaportax* Pilgrim, 1939:163.
Plioc.; As.

†*Plioportax* Korotkevich, 1975.
Plioc.; Eu.[1]
[1] E. Europe.

†*Duboisia* Stremme, 1911:115.
Pleist.; E. Indies.[1] Pleist.; As.
[1] Java.

Boselaphus de Blainville, 1816:75. Nilgai, blue bull.
[= *Buselaphus* Reichenbach, 1845:142; *Bosephalus* Horsfield, 1851:169.] [Including
Portax C. H. Smith, 1827:366-367.]
Pleist., R.; As.

Tetracerus Leach, 1825:524. Four-horned antelope, chousingha, bekra, bhokra, doda.
[= *Tetraceros* Voigt, 1831:314-315.]
Pleist., R.; As.

†*Proboselaphus* Matsumoto, 1915.
Pleist.; As.

Tribe **Bovini** Gray, 1821:308.
[= Bovidae Gray, 1821:308; Bovina Gray, 1825:342[1]; Bovini Simpson, 1945:158;
Taurina Rütimeyer, 1865; Tauroda Haeckel, 1866:clviii.[2]] [Including Bibovina
Rütimeyer, 1865; Bubalina Rütimeyer, 1865:320, 329; Bubalinae Dietrich, 1950:47;
Bisontina Rütimeyer, 1865:320, 335; Bisonina Dubrovo & Burchak-Abramovich,
1986:20; †Leptobovina Pilgrim, 1939:23; Syncerina Pilgrim, 1939:23; †Parabovinae
Dietrich, 1942; †Ioritragini Vekua, 1968:50; †Adjiderebovina Dubrovo & Burchak-
Abramovich, 1986:20.]
?L. Mioc., E. Plioc.-R.; Af. L. Mioc.-R.; As. L. Mioc.-R.; Eu. ?E. Pleist., M.
Pleist., ?L. Pleist., R.; E. Indies. ?E. Pleist., M. Pleist.-R.; N.A. L. Pleist.;
Mediterranean.[3] L. Pleist.; Cent. A.[4]
[1] As tribe.
[2] Used as a family.
[3] Sicily.
[4] Also Recent practically worldwide in domestication.

†*Urmiabos* Burchak-Abramovich, 1950.
> L. Mioc.; As.[1]
>> [1] W. Asia.

†*Proleptobos* Pilgrim, 1939:309.[1]
> L. Mioc. and/or Plioc.; As.
>> [1] "A *nomen nudum* since 1913, but defined in 1939" (Simpson 1945:158).

†*Proamphibos* Pilgrim, 1939:270.
> L. Mioc.-E. Pleist.; As.

†*Leptobos* Rütimeyer, 1877: pls. I, IV, VI, VII.[1]
> [Including †*Smertiobos* Duvernois, 1992:160.[2]]
> ?L. Mioc., ?Pleist.; Af.[3] ?L. Mioc., L. Plioc.-M. Pleist.; Eu.   ?Plioc., E.-M. Pleist.; As.
>> [1] Or 1878:157.
>> [2] As subgenus.
>> [3] N. Africa.

†*Parabos* Arambourg & Piveteau, 1929.
> L. Mioc.-L. Plioc., ?E. Pleist.; Eu.

†*Eosyncerus* Vekua, 1972:277.
> E. Plioc.; As.[1]
>> [1] W. Asia.

†*Probison* Sahni & Khan, 1968.
> Plioc.; As.

†*Alephis* Gromolard, 1980:770.
> E. Plioc.; Eu.

†*Protobison* Burchak-Abramovich, Gadzhiev & Vekua, 1980:486.
> E. Plioc.; As.[1]
>> [1] W. Asia (Azerbaijan).

†*Ugandax* H. B. S. Cooke & Coryndon, 1970:205.
> E.-L. Plioc.; Af.[1]
>> [1] E. Africa.

†*Brabovus* Gentry, 1987:382.
> E. Plioc.; Af.

†*Simatherium* Dietrich, 1941.
> E.-L. Plioc.; Af.

†*Pelorovis* Reck, 1925.
> [= †*Bularchus* Hopwood, 1936.]
> ?E. Plioc., L. Plioc.-L. Pleist.; Af.

Syncerus Hodgson, 1847:709. African buffalo, cape buffalo.
> [Including †*Homoioceras* Bate, 1949:397.]
> L. Plioc.-R.; Af.

†*Ioribos* Vekua, 1968:255.
> E. Plioc.; As.[1]
>> [1] W. Asia.

†*Yakopsis* Kretzoi, 1954:254.
> L. Plioc.; Eu.

†*Hemibos* Falconer, in Rutimeyer, 1865:330.
> [Including †*Amphibos* Falconer, in Rutimeyer, 1865:331; †*Peribos* Lydekker, 1878:141.]
> ?L. Plioc.; Af.[1] ?L. Plioc., E. Pleist.; As.
>> [1] E. Africa.

Bos Linnaeus, 1758:71. Cattle, oxen, aurochs, yak, gaur, gayal, kouprey, banteng.
> [= *Taurus* Rafinesque, 1814:30[1]; *Platatherium* Gervais & Ameghino, 1880:130-133.]
> [Including *Bisonus* Hodgson, 1835:525; *Bissonius* Gray, 1843:153; *Urus* Frisch, 1775: unpaged plate[2]; *Urus* C. H. Smith, in Griffith, Smith & Pidgeon, 1827:417; *Bibos*

Hodgson, 1837:499[3]; *Poephagus* Gray, 1843:xxvi[3]; *Gaveus* Hodgson, 1847:705-706; *Gauribos* Heude, 1901:3[4]; *Uribos* Heude, 1901:5[4]; *Bubalibos* Heude, 1901:6[4]; *Novibos* Coolidge, 1940:425; *Colombibos* Hernandez Camacho & Porta, 1960:46.]
?L. Plioc. and/or ?E. Pleist., M. Pleist.-R.; Af.[5] E. Pleist.-R.; As. Pleist.; N.A.[6] M. Pleist.-R.; Eu.[7] L. Pleist.; Mediterranean.[8] R.; E. Indies.[9]

[1] Substitute name for *Bos*.
[2] Invalid because not published in a consistently binominal work.
[3] As subgenus.
[4] Objective synonym of the subgenus *Bibos*.
[5] N. Africa, now extinct.
[6] Alaska.
[7] Now extinct in the wild.
[8] Sicily.
[9] Java, Borneo.
Also Recent practically worldwide in domestication.

Bubalus C. H. Smith, in Griffith, Smith & Pidgeon, 1827:371. Water buffalo, Asiatic buffalo.
[= *Bubalus* Frisch, 1775: unpaged plate[1]; *Buffelus* Rütimeyer, 1865:332-334.]
L. Plioc.-R.; As. M. Pleist.; Eu. R.; E. Indies.[2]

[1] Not published in a consistently binominal work. See the International Code of Zoological Nomenclature, Article 11(c).
[2] Also introduced and feral in Australia and the Pacific (Bismarcks).
Recent of N. Africa, S. Europe, and eastern South America in domestication.

†*Adjiderebos* Dubrovo & Burchak-Abramovich, 1984:717-720.
L. Plioc.; As.

Bison C. H. Smith, in Griffith, Smith & Pidgeon, 1827:373.[1] Bison, American buffalo, wisent.
[= *Bonasus* Wagner, 1844:515.[2]] [= or including *Harlanus* Owen, 1846:94.] [Including †*Simobison* Hay & Cook, 1930:33; †*Stelabison* Figgins, 1933:18; †*Superbison* Frick, 1937:567; †*Platycerobison* Skinner & Kaisen, 1947:197; †*Gigantobison* Skinner & Kaisen, 1947:203; †*Parabison* Skinner & Kaisen, 1947:219; †*Leptobison* Matsumoto & Mori, 1956:243; †*Eobison* Flerov, 1972:82.[3]]
E.-L. Pleist.; As. E. Pleist.-R.; Eu. M. Pleist.-R.; N.A. L. Pleist.; Mediterranean.[4] L. Pleist.; Cent. A.

[1] Proposed as subgenus of *Bos*; raised to generic rank by Turner, 1850:177.
[2] Proposed as subgenus of *Bos*.
[3] Sometimes as subgenus.
[4] Sicily.

†*Platycerabos* Barbour & Schultz, 1942:223.
[= †*Parabos* Barbour & Schultz, 1941:63.[1]]
Pleist.; N.A.

[1] Not †*Parabos* Arambourg & Piveteau, 1929, another member of the Bovini.

†*Platybos* Pilgrim, 1939:327.
E. Pleist.; As.

Anoa C. H. Smith, in Griffith, Smith & Pidgeon, 1827:355.[1] Dwarf buffalo, tamarau.
[= †*Probubalus* Rütimeyer, 1865:331.]
Pleist., R.; E. Indies.[2]

[1] Ex Leach MS. Proposed as subgenus of *Antilope*; raised to generic rank by Gray, 1830.
[2] Philippines, Sulawesi.

†*Bucapra* Rütimeyer, 1877: table 2.[1]
E. Pleist.; As.

[1] Or 1878:105.

†*Epileptobos* Hooijer, 1956:239.
M. Pleist.; E. Indies.[1]

[1] Java.

Tribe **Tragelaphini** Jerdon, 1874:271.[1]
[= Strepsicereae Gray, 1846:230; Strepsicerotidae Gray, 1872:vi, 4, 46; Strepsicerotini Simpson, 1945:151; Tragelaphinae Jerdon, 1874:271[1]; Tragelaphidae Rochebrune, 1883; Tragelaphini Ellerman, Morrison-Scott & Hayman, 1953:174.] [Including Taurotragini L. S. B. Leakey, 1965:37.]
L. Mioc.-R.; Af.

[1] Attributed to "Blyth."

Tragelaphus de Blainville, 1816:75. Bushbuck, sitatunga, kudus, nyalas.
>[Including *Strepsiceros* Frisch, 1775: table[1]; *Strepsiceros* C. H. Smith, 1827:365-366; *Limnotragus* Sclater & Thomas, 1900:90; *Strepsicerastes* Knottnerus-Meyer, 1903:113; *Strepsicerella* Zukowsky, 1910:206[2]; *Ammelaphus* Heller, 1912:15[2]; *Nyala* Heller, 1912:16.]
>
>L. Mioc.-R.; Af.

[1] Not published in a consistently binominal work. See the International Code of Zoological Nomenclature, Article 11(c).

[2] Objective synonym of *Strepsicerastes*.

Taurotragus Wagner, 1855:438-439. Elands.
>[= *Oreas* Desmarest, 1822:471[1]; *Orias* Lydekker, 1894:267-273.]
>
>L. Plioc.-R.; Af.

[1] Described as subgenus of *Antilope*. Not *Oreas* Hübner, 1806, a genus of lepidopterans.

Boocercus Thomas, 1902:309-310. Bongo.
>[= *Euryceros* Gray, 1850:27.[1]]
>
>R.; Af.

[1] Not *Eurycerus* Illiger, 1807, a genus of Coleoptera.

Tribe †**Udabnocerini** Burchak-Abramovich & Gabashvili, 1968:144.
>L. Mioc.; As.[1]

[1] W. Asia.

†*Udabnocerus* Burchak-Abramovich & Gabashvili, 1968:145.
>[= †*Udabnoceras* Burchak-Abramovich & Gabashvili, 1964:27-36.[1]]
>
>L. Mioc.; As.[2]

[1] *Nomen nudum.*

[2] W. Asia (Georgia).

Subfamily **Hippotraginae** Retzius & Loven, 1845:445.[1]
>[= Hippotragina Retzius & Loven, 1845:445[1]; Hippotraginae Brooke, 1876:223; Oegosceridae Cobbold, 1859.] [Including Heleotragidae Gray, 1872; Orygidae Rochebrune, 1883:125, 155[2]; Oryginae Lydekker & Blaine, 1914:117.]
>
>?M. Mioc., L. Mioc.-R.; Af. L. Mioc., Plioc., E. Pleist., R.; As.

[1] Attributed to "Sundevall."

[2] Attributed to "Gray."

Tribe **Reduncini** Lydekker & Blaine, 1914:197.
>[= Cervicapridae Rochebrune, 1883:126, 156; Reduncinae Lydekker & Blaine, 1914:197; Reduncini Simpson, 1945:159.] [Including Heleotragidae Gray, 1872; Adenotinae Jerdon, 1874:271[1]; †Menelikinae Arambourg, 1947:391; †Menelikini Viret, 1961:1060.]
>
>?M. Mioc., L. Mioc.-R.; Af.[2,3] L. Mioc., Plioc., E. Pleist.; As.

[1] Attributed to "Blyth."

[2] ?Reduncini genus and sp. indet., from the middle Miocene Ngorora Fm., Kenya.

[3] Reduncines reported from the late Miocene Lukeino Fm., Kenya.

†*Dorcadoxa* Pilgrim, 1939:44.
>L. Mioc.; As.

Redunca C. H. Smith, 1827:337-340. Reedbucks.
>[= *Cervicapra* de Blainville, 1816:75.[1]] [Including *Nagor* Laurillard, in d'Orbigny, 1841:621-622; *Eleotragus* Gray, 1843:165; *Heleotragus* Kirk, 1865:657-658[2]; *Oreodorcas* Heller, 1912:13.]
>
>?L. Mioc., L. Plioc.-R.; Af. ?L. Mioc.; As.[3]

[1] Not *Cervicapra* Sparrman, 1780.

[2] Objective synonym of *Eleotragus*.

[3] N.W. Iran.

†*Cambayella* Pilgrim, 1939:121.
>L. Mioc.; As.

†*Kobikeryx* Pilgrim, 1939:125.
L. Mioc.; As.

Kobus A. Smith, 1840. Waterbucks, lechwes, kob, puku.
[= *Cobus* Buckley, 1876.] [Including *Adenota* Gray, 1847[1]; *Hydrotragus* Fitzinger, 1866[1]; *Onotragus* Gray, 1872.[2]]
E. Plioc.-R.; Af.
[1] As subgenus.
[2] As subgenus. Objective synonym of *Hydrotragus*.

†*Vishnucobus* Pilgrim, 1939:102.
[= †*Indoredunca* Pilgrim, 1939:113.]
?Plioc., E. Pleist.; As.

†*Hydaspicobus* Pilgrim, 1939:115.
Plioc.; As.

†*Menelikia* Arambourg, 1941.
L. Plioc.-E. Pleist.; Af.

†*Sivadenota* Pilgrim, 1939:105.
E. Pleist.; As.

†*Gangicobus* Pilgrim, 1939:109.
E. Pleist.; As.

†*Sivacobus* Pilgrim, 1939:99.
E. Pleist.; As.

†*Thaleroceros* Reck, 1935.
M. Pleist.; Af.

Tribe **Hippotragini** Retzius & Lovén, 1845:445.[1]
[= Hippotragina Retzius & Loven, 1845:445[1]; Hippotragini Simpson, 1945:159; Oegosceridae Cobbold, 1859.] [Including Orygidae Rochebrune, 1883:125, 155[2]; Orygini Haltenorth, 1963:87; Addacina Pilgrim, 1939:23.]
L. Mioc.-R.; Af.[3] ?L. Plioc., E. Pleist., R.; As.
[1] Attributed to "Sundevall."
[2] Attributed to "Gray."
[3] Hippotragine horncore reported from the late Miocene of Sahabi, Libya.

†*Praedamalis* Dietrich, 1950:30.
[= †*Aeotragus* Dietrich, 1950:38.]
E.-L. Plioc.; Af.

†*Wellsiana* Vrba, 1987:53.[1]
L. Plioc.; Af.
[1] Tentatively referred to Hippotragini.

Hippotragus Sundevall, 1846:196. Blaauwbok, roan antelope, sable antelope.
[= *Egocerus* Desmarest, 1822:475-476[1]; *Aigocerus* C. H. Smith, 1827:324-325; *Oegocerus* Lesson, 1842:179-180.] [Including *Ozanna* Reichenbach, 1845:126-131; †*Sivatragus* Pilgrim, 1939:80; †*Hippotragoides* H. B. S. Cooke, 1947.]
L. Plioc.-R.; Af. E. Pleist.; As.
[1] Not *Aegoceros* Pallas, 1811. See the International Commission on Zoological Nomenclature, Opinion 90, 1922.

Oryx de Blainville, 1816:75. Gemsbok, oryxes.
[= *Onyx* Gray, 1821:307.] [Including *Aegoryx* Pocock, 1918:221; †*Sivoryx* Pilgrim, 1939:73.]
L. Plioc.-R.; Af. ?L. Plioc., E. Pleist., R.; As.

Addax Laurillard, in d'Orbigny, 1841:619-621. Addax.
[= *Addax* Rafinesque, 1815:56.[1]]
R.; Af.
[1] Not published in a consistently binominal work. See the International Code of Zoological Nomenclature, Article 11(c).

Subfamily **Alcelaphinae** Rochebrune, 1883:132, 156.
> [= Damalidae Brookes, 1828:64; Alcelaphidae Rochebrune, 1883:132, 156;
> Alcelaphinae Pilgrim, 1939:i, 63; Bubalidinae Schlater & Thomas, 1894:2, 3;
> Bubalinae Trouessart, 1898:904.] [Including Connochetidae Gray, 1872:4, 42.]
> ?L. Mioc., E. Plioc.-R.; Af.[1] L. Mioc.; Mediterranean.[2] L. Mioc.; Eu.[3] L. Plioc.-E.
> Pleist., L. Pleist.-R.; As.[4]
>
> [1] Possible alcelaphine molar from the late Miocene of Wadi Natrun, N. Africa.
> [2] Sardinia.
> [3] Italy (Tuscany).
> [4] Now extinct.

†*Maremmia* Hürzeler, 1983:243.
> L. Mioc.; Mediterranean.[1] L. Mioc.; Eu.[2]
>
> [1] Sardinia.
> [2] Italy (Tuscany).

Tribe **Alcelaphini** Rochebrune, 1883:132, 156.
> [= Alcelaphidae Rochebrune, 1883:132, 156; Alcelaphini Simpson, 1945:160.]
> [Including Connochaetini Sokolov, 1953:46, 200.]
> E. Plioc.-R.; Af. L. Plioc.-E. Pleist., L. Pleist.-R.; As.[1]
>
> [1] Now extinct.

†*Parmularius* Hopwood, 1934.
> E. Plioc.-L. Pleist.; Af.

†*Oreonagor* Pomel, 1895:45.[1]
> L. Plioc.; Af.
>
> [1] Described as subgenus of *Antilope*.

Alcelaphus de Blainville, 1816:143-144. Hartebeest, bubal, tora.
> [= *Bubalis* Frisch, 1775:2[1]; *Bubalis* Goldfuss, 1820; *Damalis* C. H. Smith, 1827:343[2];
> *Acronotus* C. H. Smith, 1827:346-354.[3]]
> L. Plioc., M. Pleist.-R.; Af. L. Pleist.-R.; As.[4]
>
> [1] Not published in a consistently binominal work. See the International Code of Zoological Nomenclature,
>   Article 11(c).
> [2] Not *Damalis* Fabricius, 1805.
> [3] Proposed as subgenus of *Damalis*.
> [4] S.W. Asia. Now extinct.

†*Damalops* Pilgrim, 1939:67.
> ?L. Plioc.; Af. L. Plioc.-E. Pleist.; As.

†*Rhynotragus* Reck, 1925.
> [= †*Pelorocerus* van Hoepen, 1932:65; †*Lunatoceras* Hoffman, 1953.] [= or including
> †*Megalotragus* van Hoepen, 1932:63.] [Including †*Xenocephalus* L. S. B. Leakey,
> 1965.]
> L. Plioc.-L. Pleist.; Af.

Connochaetes Lichtenstein, 1814:152.[1] Gnus, wildebeests.
> [= *Cemas* Oken, 1816:727-744; *Catablepas* Gray, 1821:307; *Catoblepas* H. Smith,
> 1827:366-372.] [Including *Gorgon* Gray, 1850[2]; †*Pultiphagonides* Hopwood, 1934.]
> L. Plioc.-R.; Af.
>
> [1] As subgenus of *Antilope*.
> [2] As subgenus.

Damaliscus Sclater & Thomas, 1894:51-91. Blesbok, bontebok, korrigum, topi, tiang, tsessebe,
> Hunter's antelope, herola.
> [= *Damalis* Gray, 1846:233.[1]] [Including *Beatragus* Heller, 1912:8.]
> L. Plioc.-R.; Af.
>
> [1] Not *Damalis* H. Smith, 1827.

†*Parestigorgon* Dietrich, 1950:30.
> L. Plioc.; Af.

Sigmoceros Heller, 1912:4. Lichtenstein's hartebeest.
> Pleist., R.; Af.

†*Rabaticeras* Ennouchi, 1953:128.
> E.-M. Pleist.; Af.

†*Rusingoryx* Pickford & Thomas, 1984:443.[1]
    L. Pleist.; Af.
    [1] Referral to Alcelaphini is questionable.

Tribe **Aepycerotini** Gray, 1872:4, 42.
    [= Aepycerotidae Gray, 1872:4, 42; Aepycerina Pilgrim, 1939:23; Aepycerini Haltenorth, 1963:96.]
    E. Plioc.-R.; Af.

Aepyceros Sundevall, 1847:271. Impala.
    E. Plioc.-R.; Af.

Subfamily **Cephalophinae** Gray, 1872:v, 3, 21. Duikers.
    [= Cephalophidae Gray, 1872:v, 3, 21; Cephalophoridae Gray, 1871:588; Cephalophinae Brooke, 1876:224; Cephalophini Simpson, 1945:159.] [Including Sylvicaprina Sundevall, 1846.[1]]
    ?L. Mioc., L. Plioc., M. Pleist.-R.; Af.[2]
    [1] Long unused.
    [2] Possible cephalophine molar in the late Miocene Lukeino Fm.

Cephalophus H. Smith, 1827:258.[1] Yellow-backed duiker, forest duiker, blue duiker, bloubokkie.
    [= *Cephalophora* Gray, 1842:266.[2]] [Including *Philantomba* Blyth, 1840:140[1]; *Guevei* Gray, 1852:86[3]; *Cephalophella* Knottnerus-Meyer, 1907:45; *Cephalophidium* Knottnerus-Meyer, 1907:45; *Cephalophops* Knottnerus-Meyer, 1907:46; *Cephalophula* Knottnerus-Meyer, 1907:46; *Cephalopia* Knottnerus-Meyer, 1907:44; *Cephalophia* Knottnerus-Meyer, 1907:99, 120.[4]]
    L. Plioc., M. Pleist.-R.; Af.
    [1] As subgenus.
    [2] Emendation.
    [3] As subgenus. Objective synonym of *Philantomba*.
    [4] Correction of *Cephalopia*.

Sylvicapra Ogilby, 1837. Grey duiker, duikerbok.
    M. Pleist.-R.; Af.

Mirorder †**MERIDIUNGULATA** McKenna, 1975:42. South American ungulates.
    [Including †NOTOTHERIA Kalandadze & Rautian, 1992:103.]

†*Andinodus* de Muizon & Marshall, 1987:949.
    E. Paleoc.; S.A.

†*Anagonia* Ameghino, 1904:185.
    E. Eoc.; S.A.

†*Lambdaconus* Ameghino, 1897:441.[1]
    ?E. Eoc.; S.A.
    [1] *Nomen dubium.*

†*Prostylophorus* Roth, 1902:252.
    M. Eoc.; S.A.

†*Trilobodon* Roth, 1902:253.
    M. Eoc.; S.A.

†*Heterolophodon* Roth, 1903:147.
    M. Eoc.; S.A.

†*Acamana* Simpson, Minoprio & Patterson, 1962:276.
    M. Eoc.; S.A.

Family †**Perutheriidae** Van Valen, 1978:65.
    [= †Perutheriinae Van Valen, 1978:65; †Perutheriidae Marshall, de Muizon & Sigé, 1983:14.]
    L. Cret.; S.A.

†*Perutherium* Thaler, 1967:707-710.[1]
    L. Cret.; S.A.
    [1] Possibly a didolodontid or a primitive notoungulate.

Family †**Amilnedwardsiidae** Soria, 1984:19.
[= †Amilnedwardsidae Soria, 1984:19.]
E. Eoc.; S.A.

†*Amilnedwardsia* Ameghino, 1901:386.
[= or including †*Ernestohaeckelia* Ameghino, 1901:382.]
E. Eoc.; S.A.

†*Rutimeyeria* Ameghino, 1901:385.
E. Eoc.; S.A.

Order †**LITOPTERNA** Ameghino, 1889:492.[1]
[= †LOPHIOLIPTERNA Cifelli, 1983:15; †LITOPTERNIFORMES Kinman, 1994:38.] [Including †NOTOPTERNA Soria, 1984:23.[2]]
[1] Proposed as an order; used as grandorder by McKenna, in Stucky & McKenna, in Benton, ed., 1993.
[2] Order †NOTOPTERNA was created for †Amilnedwardsiidae, †Indaleciinae (as "†Indalecidae"), and a then undescribed Paleocene mammal from Tucumán, Argentina.

†*Miguelsoria* Cifelli, 1983:14.
L. Paleoc.; S.A.

†*Asmithwoodwardia* Ameghino, 1901:379.
[= or including †*Ernestokokenia* Ameghino, 1901:380; †*Notoprotogonia* Ameghino, 1904:41; †*Archaeohyracotherium* Ameghino, 1906:307, 467.]
L. Paleoc.-M. Eoc.; S.A.

†*Eoauchenia* Ameghino, 1887:16-17.
L. Mioc.-E. Plioc.; S.A.

†*Megamys* d'Orbigny & Laurillard, 1837: pl. 8.[1]
[= †*Megalomys* Trouessart, 1903:387.[2]]
?Pleist.; S.A.
[1] Based on a tibia and patella from Ensenada de Ross, south of Río Negro, Argentina. Age actually unknown but possibly Pleistocene. At best a *nomen dubium*, but of interest because, if a litoptern, it was published a year before †*Macrauchenia*.
[2] Emendation of †*Megamys*. Not †*Megalomys* Trouessart, 1881.

Family †**Protolipternidae** Cifelli, 1983:10.[1]
L. Paleoc.; S.A.
[1] Validity doubtful.

†*Protolipterna* Cifelli, 1983:11.
L. Paleoc.; S.A.

Superfamily †**Macrauchenioidea** Gervais, 1855:36.
[= †Macrauchénidés Gervais, 1855:36; †Macrauchenioidea Cifelli, 1983:27.]

Family †**Macraucheniidae** Gervais, 1855:36.[1]
[= †Macrauchénidés Gervais, 1855:36 (or 335); †Macrauchenidae Huxley, 1871:343, 366; †Macraucheniidae Gill, 1872:12.] [Including †Mesorhinidae Ameghino, 1891; †Sparnotheriodontidae Soria, 1980:194.[2]]
L. Paleoc.-M. Eoc., L. Olig.-L. Pleist.; S.A. M. Eoc.; Antarctica.
[1] See the International Code of Zoological Nomenclature, Article 11(f)(iii).
[2] Described as notoungulates.

Subfamily †**Sparnotheriodontinae** Soria, 1980:194, **new rank**.
[= †Sparnotheriodontidae Soria, 1980:194.]
L. Paleoc.-M. Eoc.; S.A. M. Eoc.; Antarctica.

†*Victorlemoinea* Ameghino, 1901:383.
[= or including †*Guilielmofloweria* Ameghino, 1901:397.]
L. Paleoc.-E. Eoc.; S.A. M. Eoc.; Antarctica.

†*Phoradiadius* Simpson, Minoprio & Patterson, 1962:249.
M. Eoc.; S.A.

†*Sparnotheriodon* Soria, 1980:194.
E. Eoc.; S.A.

Subfamily †**Polymorphinae** Simpson, 1945:124.
M. Eoc.; S.A.

†*Polymorphis* Roth, 1899:385.
[= †*Polyacrodon* Roth, 1899:382[1]; †*Oroacrodon* Ameghino, 1904:335.[2]] [= or
including †*Megacrodon* Roth, 1899:384; †*Decaconus* Ameghino, 1901:378;
†*Periacrodon* Ameghino, 1904:334.]
M. Eoc.; S.A.
[1] Supposedly preoccupied by *Polyacrodus* Jaeckel, 1889, a fish.
[2] Replacement name for †*Polyacrodon* Roth, 1898.

Subfamily †**Cramaucheniinae** Ameghino, 1902:90.
[Including †Theosodontinae Ameghino, 1902:90.]
L. Olig.-L. Mioc.; S.A.

†*Pternoconius* Cifelli & Soria, 1983:148.
L. Olig.-E. Mioc.; S.A.

†*Caliphrium* Ameghino, 1895:633-634.
[= †*Coniopternium* Ameghino, 1895:632; †*Notodiaphorus* Loomis, 1914:31.]
L. Olig.; S.A.

†*Cramauchenia* Ameghino, 1902:90-93.
E. Mioc.; S.A.

†*Theosodon* Ameghino, 1887:19.
[= †*Pseudocoelosoma* Ameghino, 1891:294.]
E.-L. Mioc.; S.A.

†*Phoenixauchenia* Ameghino, 1904:57.
M. Mioc.; S.A.

Subfamily †**Macraucheniinae** Gervais, 1855:36.
[= †Macraucheniidae Gervais, 1855:36; †Macraucheniidae Gill, 1872:12;
†Macraucheniinae Ameghino, 1894:277; †Macraucheniinae Bordas, 1939:416.]
[Including †Mesorhinidae Ameghino, 1891.]
L. Mioc.-L. Pleist.; S.A.

†*Paranauchenia* Ameghino, 1904:55.
L. Mioc.; S.A.

†*Oxyodontherium* Ameghino, 1883:284-288.
[Including †*Mesorhinus* Ameghino, 1885:94-97.]
L. Mioc.; S.A.

†*Scalabrinitherium* Ameghino, 1883:108-112.
[= †*Scalabrinia* Lydekker, 1894:122.]
L. Mioc.; S.A.

†*Cullinia* Cabrera & Kraglievich, 1931:113.
L. Mioc.; S.A.

†*Macrauchenidia* Cabrera, 1939:33-34.
L. Mioc.; S.A.

†*Promacrauchenia* Ameghino, 1904:58.
[Including †*Pseudomacrauchenia* Kraglievich, 1930:160.[1]]
L. Mioc.-L. Plioc.; S.A.
[1] Subgenus as described.

†*Windhausenia* Kraglievich, 1930:159.
L. Plioc. and/or E. Pleist.; S.A.

†*Macrauchenia* Owen, 1838:35.
[= †*Macraucheniopsis* Paula Couto, 1945:239.]
E.-L. Pleist.; S.A.

†*Xenorhinotherium* Cartelle & Lessa, 1988:4.
L. Pleist.; S.A.

Family †**Notonychopidae** Soria, 1989:261.
>> L. Paleoc.; S.A.
> †*Notonychops* Soria, 1989:261.
>> L. Paleoc.; S.A.

Family †**Adianthidae** Ameghino, 1891:134.
>> [= †Adiantidae Ameghino, 1894:283.]
>> E.-M. Eoc., L. Eoc. and/or E. Olig., L. Olig.-M. Mioc.; S.A.
> †*Proectocion* Ameghino, 1904:83.
>> [= or including †*Oxybunotherium* Pascual, 1965:59.]
>> E. Eoc.; S.A.

Subfamily †**Indaleciinae** Bond & Vucetich, 1983:109.
>> [= †Indalecidae Soria, 1984:19.]
>> E.-M. Eoc., L. Eoc. and/or E. Olig., L. Olig.; S.A.[1,2]

[1] A new genus and species of †Indaleciinae has been reported from the late Eocene and/or early Oligocene Tinguiririca Fauna, Chile.
[2] An †"indalecid" has been reported from the late Oligocene (Deseadan) at Cabeza Blanca, Chubut, Argentina.

> †*Indalecia* Bond & Vucetich, 1983:109.
>> E. Eoc.; S.A.
> †*Adiantoides* Simpson & Minoprio, 1949:6.
>> E.-M. Eoc.; S.A.

Subfamily †**Adianthinae** Ameghino, 1891:134.
>> [= †Adianthidae Ameghino, 1891:134; †Adiantinae Bordas, 1939:417; †Adianthinae Patterson, 1940:13.]
>> L. Olig.-M. Mioc.; S.A.

Tribe †**Proadiantini**, **new**.
>> L. Olig.; S.A.
> †*Proadiantus* Ameghino, 1897:455.
>> [= †*Proadianthus* Loomis, 1914:51.]
>> L. Olig.; S.A.
> †*Thadanius* Cifelli & Soria, 1983:15.
>> L. Olig.; S.A.
> †*Tricoelodus* Ameghino, 1897:454.
>> L. Olig.; S.A.

Tribe †**Adianthini** Ameghino, 1891:134, **new rank**.
>> [= †Adianthidae Ameghino, 1891:134.]
>> E.-M. Mioc.; S.A.
> †*Proheptaconus* Bordas, 1936:110.
>> E. Mioc.; S.A.
> †*Adianthus* Ameghino, 1891:143-135.
>> E.-M. Mioc.; S.A.

Superfamily †**Proterotherioidea** Cifelli, 1983:27.

Family †**Proterotheriidae** Ameghino, 1887:19.
>> [= †Prototheridae Ameghino, 1887:19; †Proterotheriidae Cope, 1889:876.]
>> L. Paleoc.-M. Eoc., ?E. Olig., L. Olig.-L. Plioc.; S.A.

Subfamily †**Anisolambdinae** Cifelli, 1983:15.
>> L. Paleoc.-M. Eoc., ?E. Olig., L. Olig.; S.A.
> †*Paranisolambda* Cifelli, 1983:15.
>> L. Paleoc.; S.A.
> †*Ricardolydekkeria* Ameghino, 1901:397.
>> [= or including †*Lopholambda* Ameghino, 1904:36; †*Heterolambda* Ameghino, 1904:338.]
>> L. Paleoc.-E. Eoc.; S.A.

†*Anisolambda* Ameghino, 1901:383.
>> [= †*Josepholeidya* Ameghino, 1901:384; †*Eulambda* Ameghino, 1904:340; †*Anissolambda* Ameghino, 1906:467.]
>> L. Paleoc.-E. Eoc.; S.A.

†*Wainka* Simpson, 1935:9.
>> L. Paleoc.; S.A.

†*Xesmodon* Berg, 1899:79.
>> [= †*Glyphodon* Roth, 1899:383.[1]]
>> M. Eoc.; S.A.
>> [1] Not †*Glyphodon* Gunther, 1858: 210, a reptile.

†*Protheosodon* Ameghino, 1897:453.
>> [= or including †*Heteroglyphis* Roth, 1899:387.]
>> M. Eoc., ?E. Olig., L. Olig.; S.A.

Subfamily †**Proterotheriinae** Ameghino, 1885.
>> [= †Proterotheridae Ameghino, 1885; †Proterotheriinae Ameghino, 1891.]
>> L. Olig.-L. Plioc.; S.A.

†*Deuterotherium* Ameghino, 1895:633.
>> L. Olig.; S.A.

†*Megadolodus* McKenna, 1956:737.
>> M. Mioc.; S.A.[1]
>> [1] Colombia.

†*Eoproterotherium* Ameghino, 1904:59.
>> [= †*Eoprotherotherium* Ameghino, 1904:59.[1]]
>> L. Olig.; S.A.
>> [1] *Lapsus calami.*

†*Prolicaphrium* Ameghino, 1902:86-88.[1]
>> E. Mioc.; S.A.
>> [1] *Nomen nudum* in 1901.

†*Prothoatherium* Ameghino, 1902:88-89.
>> [= †*Licaphrops* Ameghino, 1904:64; †*Paramacrauchenia* Bordas, 1939:413-434.]
>> [Including †*Neodolodus* Hoffstetter & Soria, 1986:1620.]
>> E.-M. Mioc.; S.A.

†*Licaphrium* Ameghino, 1887:20.
>> E. Mioc., L. Mioc., L. Plioc.; S.A.

†*Diadiaphorus* Ameghino, 1887:20.
>> [= or including †*Coelosoma* Ameghino, 1891:137.[1]]
>> E.-M. Mioc., ?L. Mioc.; S.A.
>> [1] Or perhaps best regarded as a *nomen dubium.*

†*Thoatherium* Ameghino, 1887:19-20.
>> E. Mioc., L. Mioc.; S.A.

†*Brachytherium* Ameghino, 1883:289-291.
>> [Including †*Epitherium* Ameghino, 1889:569-572; †*Lophogonodon* Ameghino, 1904:62[1]; †*Lophogododon* Ameghino, 1904:62.[2]]
>> L. Mioc.-L. Plioc.; S.A.
>> [1] Objective synonym of †*Epitherium.*
>> [2] *Lapsus calami.*

†*Epecuenia* Cabrera, 1939:13-15.
>> L. Mioc.; S.A.

†*Diplasiotherium* Rovereto, 1914:129.
>> E. Plioc.; S.A.

†*Proterotherium* Ameghino, 1883:291-293.
>> L. Mioc.; S.A.

Order †**NOTOUNGULATA** Roth, 1903:11, 12.
[= †TOXODONTIA Owen, 1853:309[1]; †NOTOUNGULIFORMES Kinman, 1994:38.]
[1] Sensu Lydekker, 1894:2 and Scott, 1904:590, not †TOXODONTIA Ameghino, 1906, or of many other authors.

†*Acoelodus* Ameghino, 1897:454.[1]
E. Eoc.; S.A.
[1] Family †Acoelodidae Ameghino, 1901:364, was created for this nearly indeterminate genus.

†*Pleurystylops* Ameghino, 1901:394.
E. Eoc.; S.A.

†*Lophiodonticulus* Ameghino, 1902:17.
E. Eoc.; S.A.

†*Tonostylops* Ameghino, 1902:32.
E. Eoc.; S.A.

†*Isotypotherium* Ameghino, 1904:421.
E. Eoc.; S.A.

†*Epitypotherium* Ameghino, 1904:422.
E. Eoc.; S.A.

†*Carolodarwinia* Ameghino, 1901:406.
M. Eoc.; S.A.

†*Ortholophodon* Roth, 1902:253.
M. Eoc.; S.A.

†*Procolpodon* Ameghino, 1904:438.
M. Eoc.; S.A.

†*Senodon* Ameghino, 1895:628.
L. Olig.; S.A.

†*Loxocoelus* Ameghino, 1895:653.
L. Olig.; S.A.

†*Pyralophodon* Ameghino, 1904:85.
L. Olig.; S.A.

†*Orthogeniops* Ameghino, 1902:33.[1]
[= †*Orthogenium* Roth, 1901:255.[2]]
Eoc. and/or Olig.; S.A.
[1] Probably a *nomen dubium* and best forgotten.
[2] Not *Orthogenium* Chaudoir, 1835, a coleopteran.

†*Pyramidon* Roth, 1901:255.[1]
L. Eoc. and/or E. Olig.; S.A.
[1] Probably a *nomen dubium* and best forgotten.

†*Archaeotypotherium* Roth, 1903:152-153.[1]
L. Eoc. and/or E. Olig.; S.A.
[1] Probably a *nomen dubium* and best forgotten.

Suborder †**NOTIOPROGONIA** Simpson, 1934:7, 10.[1]
[1] The original contents were †Arctostylopidae, †Henricosborniidae, and †Notostylopidae.

†*Satshatemnus* Soria, 1990:146.
L. Paleoc.; S.A.

†*Seudenius* Simpson, 1935:14.[1]
L. Paleoc.; S.A.
[1] Possibly referable to †Notostylopidae.

Family †**Henricosborniidae** Ameghino, 1901:357.
[= †Pantostylopidae Ameghino, 1901:423; †Selenoconidae Ameghino, 1902:20.]
L. Paleoc.-E. Eoc.; S.A.

†*Othnielmarshia* Ameghino, 1901:358.
> [= or including †*Postpithecus* Ameghino, 1901:358.] [Including †*Camargomendesia* Paula Couto, 1978:222.]
> L. Paleoc.-E. Eoc.; S.A.

†*Peripantostylops* Ameghino, 1904:37.
> ?L. Paleoc., E. Eoc.; S.A.

†*Henricosbornia* Ameghino, 1901:357.
> [= or including †*Selenoconus* Ameghino, 1901:381; †*Pantostylops* Ameghino, 1901:423; †*Microstylops* Ameghino, 1901:426; †*Prohyracotherium* Ameghino, 1902:15; †*Monolophodon* Roth, 1903:143; †*Hemistylops* Ameghino, 1904:38; †*Polystylops* Ameghino, 1904:40.]
> L. Paleoc.-E. Eoc.; S.A.

†*Simpsonotus* Pascual, Vucetich & Fernandez, 1978:374.
> L. Paleoc.; S.A.

Family †**Notostylopidae** Ameghino, 1897:488.
> E.-M. Eoc., L. Eoc. and/or E. Olig.; S.A.[1]
> [1] †Notostylopidae, new genus and species, in the late Eocene and/or early Oligocene Tinguiririca Fauna, Chile.

†*Homalostylops* Ameghino, 1901:422.
> [= †*Acrostylops* Ameghino, 1901:421.]
> E. Eoc.; S.A.

†*Notostylops* Ameghino, 1897:488.
> [= †*Anastylops* Ameghino, 1897:490; †*Catastylops* Ameghino, 1901:421; †*Pliostylops* Ameghino, 1901:421; †*Entelostylops* Ameghino, 1901:425; †*Isostylops* Ameghino, 1902:33.] [= or including †*Eostylops* Ameghino, 1901:424.]
> E. Eoc.; S.A.

†*Edvardotrouessartia* Ameghino, 1901:401.
> E. Eoc.; S.A.

†*Boreastylops* Vucetich, 1980:364.
> E. Eoc.; S.A.

†*Otronia* Roth, 1901:255.
> M. Eoc.; S.A.

Suborder †**TOXODONTIA** Owen, 1853:319.
> [= †TOXODONTA Haeckel, 1866:clix[1]; †MULTIDIGITATA Burmeister, 1866:xvii; †POLYDACTYLA Burmeister, 1885:168; †PENTADACTYLA Ameghino, 1885:72[2]; †TOXODONTOIDEA Noack, 1894:541[3]; †LIOPTERNA Haeckel, 1895:490, 536.[4]] [Including †ENTELONYCHIA Ameghino, 1894:312; †HOMALODONTOTHERIA Loomis, 1914:134.]
> [1] Proposed as suborder of order CHELOPHORA.; also used as a family by Haeckel.
> [2] In part.
> [3] Given ordinal rank.
> [4] Not to be confused with †LITOPTERNA.

†*Brandmayria* Cabrera, 1935:13.[1]
> L. Paleoc.; S.A.
> [1] Possibly referable to †Isotemnidae.

†*Colhuelia* Roth, 1902:254.[1]
> M. Eoc.; S.A.
> [1] Possibly referable to †Isotemnidae.

†*Lafkenia* Roth, 1902:254.[1]
> M. Eoc.; S.A.
> [1] Possibly referable to †Isotemnidae.

†*Colhuapia* Roth, 1902:255.[1]
> M. Eoc.; S.A.
> [1] Possibly referable to †Isotemnidae.

†*Allalmeia* Rusconi, 1946.[1]
    M. Eoc.; S.A.
    [1] Unnumbered page. Possibly referable to †Oldfieldthomasiidae.

†*Brachystephanus* Simpson, Minoprio & Patterson, 1962:254.[1]
    M. Eoc.; S.A.
    [1] Possibly referable to †Oldfieldthomasiidae.

†*Xenostephanus* Simpson, Minoprio & Patterson, 1962:262.[1]
    M. Eoc.; S.A.
    [1] Possibly referable to the †Oldfieldthomasiidae.

†*Phanotherus* Ameghino, 1889:900.[1]
    L. Mioc.; S.A.
    [1] *Nomen dubium*, possibly synonymous with some advanced toxodont.

†*Hermoseodon* Mercerat, 1917:6.[1]
    L. Plioc. and/or Pleist.; S.A.
    [1] *Nomen dubium*, possibly a synonym of some advanced toxodont.

Family †**Isotemnidae** Ameghino, 1897:479.
    L. Paleoc.-M. Eoc., L. Eoc. and/or E. Olig., L. Olig.; S.A.[1]
    [1] Specimens referred to †Isotemnidae in the late Eocene and/or early Oligocene Tinguiririca Fauna, Chile.

†*Isotemnus* Ameghino, 1897:480.
    [= †*Eochalicotherium* Ameghino, 1901:417; †*Dimerostephanos* Ameghino, 1902:30; †*Amphitemnus* Ameghino, 1904:234.] [Including †*Prostylops* Ameghino, 1897:486; †*Lelfunia* Roth, 1902:255.]
    L. Paleoc.-E. Eoc.; S.A.

†*Hedralophus* Ameghino, 1901:406-407.[1]
    E. Eoc.; S.A.
    [1] *Nomen dubium.*

†*Thomashuxleya* Ameghino, 1901:409-410.
    E. Eoc.; S.A.

†*Pampatemnus* Vucetich & Bond, 1982:8.
    E. Eoc.; S.A.

†*Coelostylodon* Simpson, 1970:8.
    E. Eoc.; S.A.

†*Pleurostylodon* Ameghino, 1897:485.
    [= †*Parastylops* Ameghino, 1897:491; †*Anchistrum* Ameghino, 1901:369; †*Tychostylops* Ameghino, 1901:369; †*Dialophus* Ameghino, 1901:415.] [Including †*Coelostylops* Ameghino, 1901:422; †*Porotemnus* Ameghino, 1902:28; †*Paratemnus* Ameghino, 1904:242.]
    E. Eoc.; S.A.

†*Anisotemnus* Ameghino, 1902:25.
    [= †*Toxotemnus* Ameghino, 1904:235.]
    E. Eoc.; S.A.

†*Periphragnis* Roth, 1899:387.
    [= †*Proasmodeus* Ameghino, 1902:23; †*Tehuelia* Roth, 1902:253; †*Lemudeus* Roth, 1903:144.] [Including †*Calodontotherium* Roth, 1903:148; †*Eurystephanodon* Roth, 1903:158.]
    M. Eoc.; S.A.

†*Rhyphodon* Roth, 1899:388.
    [= †*Setebos* Roth, 1902:253; †*Pehuenia* Roth, 1902:254.]
    M. Eoc.; S.A.

†*Distylophorus* Ameghino, 1902:19.
    [= †*Stylophorus* Roth, 1902:252.[1]]
    M. Eoc.; S.A.
    [1] Not *Stylophorus* Hesse, 1870, a genus of crustaceans.

†*Lophocoelus* Ameghino, 1904:245.[1]
L. Olig.; S.A.
[1] Probably best regarded as a *nomen dubium* or *nomen oblitum*.

†*Pleurocoelodon* Ameghino, 1895:645.
L. Olig.; S.A.

Family †**Leontiniidae** Ameghino, 1895:646.
M. Eoc., L. Olig., M. Mioc.; S.A.

†*Martinmiguelia* Bond & Lopez, 1995:303.
M. Eoc.; S.A.

†*Ancylocoelus* Ameghino, 1895:650.
[= †*Rodiotherium* Ameghino, 1895:653.]
L. Olig.; S.A.

†*Leontinia* Ameghino, 1895:647.
[Including †*Scaphops* Ameghino, 1895:629; †*Stenogenium* Ameghino, 1895:654.]
L. Olig.; S.A.

†*Scarrittia* Simpson, 1934:2.
L. Olig.; S.A.

†*Taubatherium* Soria & Alvarenga, 1989:159.
L. Olig.; S.A.

†*Huilatherium* Villarroel & Guerrero Diaz, 1985:35-40.
M. Mioc.; S.A.

Family †**Notohippidae** Ameghino, 1894:283.
[= or including †Protequidae Ameghino, 1891.[1]] [Including †Colpodontidae Ameghino, 1906:469; †Rhynchippidae Loomis, 1914:86.]
E.-M. Eoc., L. Eoc. and/or E. Olig., L. Olig.-E. Mioc., ?M. and/or ?L. Mioc.; S.A.
[1] Not based upon a generic name.

†*Edvardocopeia* Ameghino, 1901:395.
E. Eoc.; S.A.

†*Acoelohyrax* Ameghino, 1902:10.
E. Eoc.; S.A.

†*Trimerostephanos* Ameghino, 1895:646.
M. Eoc., L. Olig.; S.A.

Subfamily †**Rhynchippinae** Loomis, 1914:88.
[= †Rhynchippidae Loomis, 1914:88; †Rhynchippinae Simpson, 1945:127.]
E.-M. Eoc., L. Eoc. and/or E. Olig., L. Olig.; S.A.

†*Pampahippus* Bond & Lopez, 1993:60.
E. Eoc.; S.A.

†*Plexotemnus* Ameghino, 1904:236.
E. Eoc.; S.A.

†*Puelia* Roth, 1902:252.
M. Eoc.; S.A.

†*Eomorphippus* Ameghino, 1901:373.
[= †*Pseudostylops* Ameghino, 1901:395; †*Pleurystomus* Ameghino, 1902:14; †*Eurystomus* Roth, 1902:256[1]; †*Lonkus* Roth, 1902:256.]
M. Eoc., L. Eoc. and/or E. Olig.; S.A.
[1] Objective synonym of †*Pleurystomus*. Not *Eurystomus* Vieillot, 1816, a genus of birds.

†*Morphippus* Ameghino, 1897:459.
L. Olig.; S.A.

†*Eurygenium* Ameghino, 1895:655.[1]
[= †*Eurygeniops* Ameghino, 1897:464.]
L. Olig.; S.A.
[1] Not preoccuped by *Eurygenius* La Ferte, 1849, a coleopteran.

†*Rhynchippus* Ameghino, 1897:462.
　　L. Olig.; S.A.

Subfamily †**Notohippinae** Ameghino, 1894:283.
　　[= †Notohippidae Ameghino, 1894:283; †Notohippinae Simpson, 1945:127.]
　　[Including †Colpodontidae Ameghino, 1906:469.]
　　?E. Olig., L. Olig.-E. Mioc., ?M. and/or ?L. Mioc.; S.A.

†*Nesohippus* Ameghino, 1904:34.[1]
　　L. Olig.; S.A.
　　[1] Pagination of reprint.

†*Stilhippus* Ameghino, 1904:35.
　　E. Mioc.; S.A.

†*Perhippidion* Ameghino, 1904:36.
　　[= †*Perhippidium* Simpson, 1945:128.[1]]
　　E. Mioc.; S.A.
　　[1] Misspelling? Ameghino clearly intended "†*Perhippidion*."

†*Notohippus* Ameghino, 1891:135-136.
　　[= †*Entocasmus* Ameghino, 1891:139.]
　　E. Mioc.; S.A.

†*Argyrohippus* Ameghino, 1902:81-85.[1]
　　E. Mioc.; S.A.
　　[1] Nomen nudum in 1901.

†*Purperia* Paula Couto, 1982:69.[1]
　　[= †*Megahippus* Paula Couto, 1981:461-477.[2]]
　　Olig. and/or Mioc.; S.A.
　　[1] New name for †*Megahippus* Paula Couto, 1981.
　　[2] Not †*Megahippus* McGrew, 1938, a genus of horses.

†*Colpodon* Burmeister, 1885:161.
　　E. Mioc.; S.A.

Family †**Toxodontidae** Owen, 1845:121.
　　[= †Toxodonta Haeckel, 1866:clix.[1]] [= or including †Atrypteridae Ameghino, 1889:375, 482.] [Including †Nesodontidae Murray, 1866:xiii, 168, 388; †Protoxodontidae Ameghino, 1889:375; †Xotodontidae Ameghino, 1889:375, 402; †Haplodontidae Ameghino, 1906:481; †Haplodontheriidae Ameghino, 1907:89.]
　　L. Olig.-L. Pleist.; S.A.   ?E. Pleist., L. Pleist.; Cent. A.
　　[1] Used as a family.

Subfamily †**Nesodontinae** Murray, 1866:xiii, 168.
　　[= †Nesodontidae Murray, 1866:xiii, 168; †Atryptheridae Ameghino, 1889:375; †Protoxodontidae Ameghino, 1889:375; †Nesodontinae Simpson, 1945:128.]
　　L. Olig.-M. Mioc., E. Plioc.; S.A.

†*Proadinotherium* Ameghino, 1895:625.
　　[Including †*Pronesodon* Ameghino, 1895:626-628; †*Coresodon* Ameghino, 1895:630; †*Interhippus* Ameghino, 1902:13.[1]]
　　L. Olig.-E. Mioc.; S.A.
　　[1] Possibly an objective synonym of †*Coresodon*.

†*Posnanskytherium* Liendo Lazarte, 1943:7.
　　E. Plioc.; S.A.[1]
　　[1] Bolivia. Originally thought to be much older.

†*Palyeidodon* Roth, 1898:189-190.
　　M. Mioc.; S.A.

†*Adinotherium* Ameghino, 1887:17.
　　[Including †*Phobereotherium* Ameghino, 1887:18; †*Noaditherium* Ameghino, 1907:84.]
　　E.-M. Mioc.; S.A.

†*Nesodon* Owen, 1846:67.
> [= or including †*Lithops* Ameghino, 1887:15[1]; †*Adelphotherium* Ameghino, 1887:16;
> †*Gronotherium* Ameghino, 1887:17; †*Atryptherium* Ameghino, 1887:18;
> †*Rhadinotherium* Ameghino, 1887:18; †*Protoxodon* Ameghino, 1887:62;
> †*Acrotherium* Ameghino, 1891:133; †*Palaeolithops* Ameghino, 1891:240[2];
> †*Xotoprodon* Ameghino, 1891:241; †*Nesotherium* Mercerat, 1891:386.]
> E.-M. Mioc.; S.A.

[1] Thought to be preoccupied by *Lithopsis* Scudder, 1878, an insect.
[2] Replacement name for †*Lithops*.

Subfamily †**Toxodontinae** Owen, 1845:121.
> [= †Toxodontidae Owen, 1845:121; †Toxodontinae Trouessart, 1898:688.] [Including
> †Xotodontidae Ameghino, 1889:375, 402; †Xotodontinae Trouessart, 1898:687;
> †Dinotoxodontinae Madden, 1991:83.]
> E. Mioc.-L. Pleist.; S.A.

†*Minitoxodon* Paula Couto, 1982:16.
> ?Mioc.; S.A.

†*Stereotoxodon* Ameghino, 1904:31.
> M. Mioc.; S.A.

†*Stenotephanos* Ameghino, 1886:107.
> [Including †*Xotodon* Ameghino, 1887:53; †*Zotodon* Lydekker, 1894:30.]
> ?E. Mioc., M.-L. Mioc., Plioc.; S.A.

†*Hyperoxotodon* Mercerat, 1895:305-306.
> E. Mioc., ?M. Mioc.; S.A.

†*Nesodonopsis* Roth, 1898:181-188.
> M. Mioc.; S.A.

†*Pericotoxodon* Madden, 1991:83.
> M. Mioc.; S.A.

†*Andinotoxodon* Madden, 1991:233.
> M. Mioc.; S.A.

†*Eutomodus* Ameghino, 1889:403.
> [= †*Tomodus* Ameghino, 1886:111-112.[1]]
> L. Mioc.; S.A.

[1] Not *Tomodus* Trautschold, 1879, a fish.

†*Neotoxodon* Paula Couto, 1982:16.
> L. Mioc. and/or E. Plioc.; S.A.

†*Gyrinodon* Hopwood, 1928:536.
> L. Mioc. and/or E. Plioc.; S.A.[1]

[1] Venezuela.

†*Mesenodon* Paula Couto, 1982:13.
> L. Mioc. and/or E. Plioc.; S.A.

†*Pisanodon* Zetti, 1972.
> L. Mioc.; S.A.

†*Neoadinotherium* Bordas, 1941:25.
> L. Mioc.; S.A.

†*Hemixotodon* Cabrera & Kraglievich, 1931:110.
> L. Mioc.; S.A.

†*Dinotoxodon* Mercerat, 1895:208, etc..
> [= †*Palaeotoxodon* Ameghino, 1904:401.] [Including †*Alitoxodon* Rovereto,
> 1914:124.]
> L. Mioc.-E. Plioc.; S.A.

†*Plesiotoxodon* Paula Couto, 1982:14.
> ?L. Mioc.; S.A.

†*Chapalmalodon* Mercerat, 1917.[1]
> L. Plioc.; S.A.
> [1] "A very dubious genus. Based originally on a femur and a skull, Kraglievich selected the femur as type, and this is doubtful as to origin and affinities" (Simpson 1945: 128).

†*Toxodon* Owen, 1837:541-542.
> [= or including †*Dilobodon* Ameghino, 1886:109-111.[1]]
> L. Plioc.-L. Pleist.; S.A.
> [1] Sometimes said to be a *nomen nudum*, but this is apparently not correct.

†*Nonotherium* Castellanos, 1942:50.
> E. Pleist.; S.A.

†*Ceratoxodon* Ameghino, 1907:64.[1]
> Pleist.; S.A.
> [1] Validity dubious.

Subfamily †**Haplodontheriinae** Ameghino, 1907:89.
> [= †Haplodontidae Ameghino, 1906:481; †Haplodontheriidae Ameghino, 1907:89; †Haplodontheriinae Kraglievich, 1934:96.] [Including †Paratrigodontinae Kraglievich, 1934:96.]
> E. Mioc.-E. Plioc., ?L. Plioc., Pleist.; S.A.   ?E. Pleist., L. Pleist.; Cent. A.

†*Paratrigodon* Cabrera & Kraglievich, 1931:110.
> E. Mioc., L. Mioc.; S.A.

†*Prototrigodon* Kraglievich, 1930:133.
> M. Mioc.; S.A.

†*Pachynodon* Burmeister, 1891:433-440.
> L. Mioc.; S.A.

†*Haplodontherium* Ameghino, 1885:79-81.
> [= †*Haplodontotherium* Sclater, 1886:5.]
> L. Mioc.; S.A.

†*Abothrodon* Paula Couto, 1944:1.
> [= †*Neotrigodon* Spillmann, 1949:25.[1]]
> L. Mioc. and/or E. Plioc.; S.A.
> [1] Synonymy questionable.

†*Toxodontherium* Ameghino, 1883:105-107.
> L. Mioc.-E. Plioc.; S.A.

†*Mesotoxodon* Paula Couto, 1982:21.
> L. Mioc. and/or E. Plioc.; S.A.

†*Ocnerotherium* Pascual, 1954:113.
> L. Mioc.; S.A.

†*Trigodon* Ameghino, 1887:8-9.[1]
> [= †*Trigonodon* Ameghino, 1891:240.[2]]
> E. Plioc.; S.A.
> [1] *Nomen nudum* in 1882.
> [2] Preoccupied by *Trigonodon* Sismonda, 1849, a fish, and *Trigonodon* Conrad, 1852, a mollusk.

†*Trigodonops* Kraglievich, 1930:227-228.
> ?Plioc., Pleist.; S.A.

†*Mixotoxodon* van Frank, 1957:6.
> ?E. Pleist., L. Pleist.; Cent. A.[1]   Pleist.; S.A.[2]
> [1] Central America north to Nicaragua and Guatemala.
> [2] Northern South America.

Family †**Homalodotheriidae** Gregory, 1910:466.
> [= †Homalodontotheridae Ameghino, 1889:523, 551; †ENTELONYCHIA Ameghino, 1894:312[1]; †HOMALODONTOTHERIA Loomis, 1914:134.] [Including †Chasicotheriinae Cabrera & Kraglievich, 1931:108.]
> L. Eoc. and/or E. Olig., L. Olig.-L. Mioc.; S.A.
> [1] Content expanded by various subsequent authors.

†*Trigonolophodon* Roth, 1903:16.[1]
  L. Eoc. and/or E. Olig.; S.A.
  [1] Possibly best considered a *nomen dubium*.

†*Asmodeus* Ameghino, 1895:643.
  L. Olig.; S.A.

†*Homalodotherium* Flower, 1873:383.
  [= †*Homalodontotherium* Flower, 1874:173-182; †*Homalodon* Burmeister, 1891:389.]
  [= or including †*Diorotherium* Ameghino, 1891:296.]
  E.-M. Mioc.; S.A.

†*Chasicotherium* Cabrera & Kraglievich, 1931:108.
  [= †*Puntanotherium* Bordas, 1941:23.]
  L. Mioc.; S.A.

Suborder †**TYPOTHERIA** Zittel, 1892:62, 212, 490.
  [= †Glires hebetidentati Alston, 1876:74; †TYPODONTIA Haeckel, 1895:505.]

†*Nesciotherium* Roth, 1898:41.[1]
  ?M. Mioc.; S.A.
  [1] Probably best considered a *nomen dubium* or forgotten.

Family †**Archaeopithecidae** Ameghino, 1897:422.
  E. Eoc.; S.A.

†*Archaeopithecus* Ameghino, 1897:422.
  E. Eoc.; S.A.

†*Acropithecus* Ameghino, 1904:4.
  E. Eoc.; S.A.

Family †**Oldfieldthomasiidae** Simpson, 1945:126.
  L. Paleoc.-M. Eoc.; S.A.

†*Colbertia* Paula Couto, 1952:4.
  L. Paleoc.-E. Eoc.; S.A.

†*Itaboraitherium* Paula Couto, 1970:79.
  L. Paleoc.; S.A.

†*Kibenikhoria* Simpson, 1935:16.
  L. Paleoc.; S.A.

†*Oldfieldthomasia* Ameghino, 1901:366.
  E. Eoc.; S.A.

†*Maxschlosseria* Ameghino, 1901:413.
  [Including †*Paracoelodus* Ameghino, 1904:46.]
  E. Eoc.; S.A.

†*Paginula* Ameghino, 1901:415.
  E. Eoc.; S.A.

†*Ultrapithecus* Ameghino, 1901:359.
  E. Eoc.; S.A.

†*Tsamnichoria* Simpson, 1936:84.
  M. Eoc.; S.A.

Family †**Interatheriidae** Ameghino, 1887:63.
  [= †Interatheridae Ameghino, 1887:63; †Interatheriidae Sinclair, 1909:2.] [Including
  †Tembotheridae Ameghino, 1887:65[1]; †Protypotheridae Ameghino, 1891:393[2];
  †Notopithecidae Ameghino, 1897:415.]
  ?L. Paleoc., E.-M. Eoc., L. Eoc. and/or E. Olig., L. Olig.-L. Mioc.; S.A.
  [1] Sic. Should be spelled †Tembotheriidae.
  [2] Sic. Should be spelled †Protypotheriidae.

Subfamily †**Notopithecinae** Ameghino, 1897:415.
  [= †Notopithecidae Ameghino, 1897:415; †Notopithecinae Simpson, 1945:128.]

?L. Paleoc., E.-M. Eoc., L. Eoc. and/or E. Olig.; S.A.[1]

[1] †"Notopithecinae", new genus and species, in the late Eocene and/or early Oligocene Tinguiririca Fauna, Chile.

†*Transpithecus* Ameghino, 1901:356.
   ?L. Paleoc., E. Eoc.; S.A.

†*Antepithecus* Ameghino, 1901:356.
   [= †*Infrapithecus* Ameghino, 1901:357; †*Pseudadiantus* Ameghino, 1901:372; †*Patriarchippus* Ameghino, 1904:135.]
   E. Eoc.; S.A.

†*Notopithecus* Ameghino, 1897:419.
   [= †*Adpithecus* Ameghino, 1901:355; †*Epipithecus* Ameghino, 1904:193; †*Gonopithecus* Ameghino, 1904:196.]
   E.-M. Eoc.; S.A.

†*Guilielmoscottia* Ameghino, 1901:360.
   M. Eoc.; S.A.

Subfamily †**Interatheriinae** Ameghino, 1887:63.
   [= †Interatheridae Ameghino, 1887:63; †Interatheriinae Simpson, 1945:129.]
   [Including †Tembotheridae Ameghino, 1887:65[1]; †Protypotheridae Ameghino, 1891:393.[2]]
   L. Eoc. and/or E. Olig., L. Olig.-L. Mioc.; S.A.[3]

[1] Sic. Should be spelled †Tembotheriidae.
[2] Sic. Should be spelled †Protypotheriidae.
[3] †Interatheriinae, new genera and species, in late Eocene and/or early Oligocene Tinguiririca Fauna, Chile.

†*Archaeophylus* Ameghino, 1897:6, 17.
   [= or including †*Progaleopithecus* Ameghino, 1904:28.[1]]
   L. Olig.; S.A.

[1] Synonymy uncertain.

†*Cochilius* Ameghino, 1902:75-77.[1]
   L. Olig.-E. Mioc.; S.A.

[1] *Nomen nudum* in 1901.

†*Medistylus* Stirton, 1952:531.
   [= †*Phanophilus* Ameghino, 1904:12.[1]]
   L. Olig.; S.A.

[1] Not *Phanophilus* Sharp, 1886: 380, a coleopteran.

†*Plagiarthrus* Ameghino, 1896:92.
   [= †*Clorinda* Ameghino, 1895:624[1]; †*Argyrohyrax* Ameghino, 1897:435-436.]
   L. Olig.; S.A.

[1] Not †*Clorinda* Barrande, a brachiopod.

†*Paracochilius* Bordas, 1939:413-434.
   E. Mioc.; S.A.

†*Protypotherium* Ameghino, 1882.[1]
   [= †*Patriarchus* Ameghino, 1889:480-481.]
   E.-L. Mioc.; S.A.

[1] Fide Ameghino, 1889:474-480.

†*Epipatriarchus* Ameghino, 1904:13.
   M. Mioc.; S.A.

†*Miocochilius* Stirton, 1953:268.
   M. Mioc.; S.A.

†*Caenophilus* Ameghino, 1904:16.
   M. Mioc.; S.A.

†*Interatherium* Ameghino, 1887:63.[1]
   [Including †*Tembotherium* Ameghino, 1887:65[2]; †*Icochilus* Ameghino, 1889:469.]
   E. Mioc.; S.A.

[1] *Nomen nudum* in Moreno, 1882:117.
[2] Attributed by Ameghino to Moreno, 1882:117.

Subfamily †**Munyiziinae** Kraglievich, 1931:261.
[= †Muñizinae Kraglievich, 1931:261; †Muniziinae Simpson, 1945:130; †Munyiziinae Cifelli, 1993:209n.]
L. Mioc.; S.A.

†*Munyizia* Kraglievich, 1931:261.
[= †*Muñizia* Kraglievich, 1931; †*Munizia* Cifelli, 1985; †*Munyizia* Cifelli, 1993:209n.[1]]
L. Mioc.; S.A.

[1] Emendation.

Family †**Campanorcidae** Bond, Vucetich & Pascual, 1984:20.
E. Eoc.; S.A.

†*Campanorco* Bond, Vucetich & Pascual, 1984:20.
E. Eoc.; S.A.

Family †**Mesotheriidae** Alston, 1876:75, 98.
[= †HEBETIDENTATA Gill, 1883:72[1]; †Typotheriidae Lydekker, 1886:170.]
[Including †Trachytheridae Ameghino, 1894:276; †Eutrachytheriidae Ameghino, 1897:427.]
M. Eoc., L. Olig., ?E. Mioc., M. Mioc.-M. Pleist.; S.A.

[1] Proposed as a suborder of RODENTIA, based on suborder "Glires hebetidentati" Alston, 1876:74, coordinate with suborders "Glires Simplicidentati" Lilljeborg, 1866, and "Glires Duplicidentati" Lilljeborg, 1866.

Subfamily †**Trachytheriinae** Ameghino, 1894:276.
[= †Trachytheridae Ameghino, 1894:276; †Eutrachytheriidae Ameghino, 1897:427; †Trachytheriinae Simpson, 1945:129.]
M. Eoc., L. Olig.; S.A.[1]

[1] An unquestioned trachytheriine, perhaps not †*Trachytherus*, occurs in the middle Eocene Divisadero Largo fauna.

†*Trachytherus* Ameghino, 1889:918.[1]
[= †*Eutrachytherus* Ameghino, 1897:427-429.] [= or including †*Ameghinotherium* Podestá, 1898.]
?M. Eoc., L. Olig.; S.A.

[1] Not preoccupied by †*Trachytherium* Gervais, 1849, a fossil sirenian.

†*Proedium* Ameghino, 1895:623.[1]
[= †*Proedrium* Ameghino, 1897:17n; †*Isoproedrium* Ameghino, 1903:18.]
L. Olig.; S.A.

[1] Not preoccupied by *Proedrus* Foerster, 1888.

Subfamily †**Fiandraiinae** Roselli, 1976:155.
Mioc.; S.A.[1]

[1] Uruguay.

†*Fiandraia* Roselli, 1976:155.
Mioc.; S.A.[1]

[1] Uruguay.

Subfamily †**Mesotheriinae** Alston, 1876:75, 98.
[= †Mesotheriidae Alston, 1876:75, 98; †Mesotheriinae Simpson, 1945:129.]
M. Mioc.-M. Pleist.; S.A.

†*Eutypotherium* Roth, 1901:256.[1]
[= †*Tachytypotherium* Roth, 1903:26; †*Trachytypotherium* Ameghino, 1904:165[2]; †*Typothericulus* Kraglievich, 1930:148.]
M. Mioc.; S.A.

[1] "Rejected by most students, including Roth himself, on grounds of preoccupation by †*Eutypotherium* Haeckel, 1895, but, as Patterson has pointed out, the latter name was given to a purely hypothetical genus, hence has no standing in nomenclature and cannot preoccupy a name based on a real animal" (Simpson 1945:129).
[2] Misspelling.

†*Microtypotherium* Villarroel, 1974:551.
  M. and/or L. Mioc.; S.A.[1]
  [1] Bolivia.

†*Plesiotypotherium* Villarroel, 1974:255, 279.
  L. Mioc.; S.A.[1]
  [1] Bolivia.

†*Typotheriopsis* Cabrera & Kraglievich, 1931:111.
  [= †*Acrotypotherium* Rusconi, 1936:1.[1]]
  L. Mioc.; S.A.
  [1] Synonymy questionable.

†*Pseudotypotherium* F. Ameghino, 1904:163.
  [Including †*Typotheriodon* C. Ameghino, 1919:148.]
  L. Mioc.-L. Plioc., E. and/or M. Pleist.; S.A.

†*Hypsitherium* Anaya & MacFadden, 1995:129.
  E. Plioc.; S.A.[1]
  [1] Bolivia.

†*Mesotherium* Serres, 1867:140.[1]
  [= †*Typotherium* Gervais, 1867[2]; †*Uptiodon* Serres, 1867:593-599; †*Entelomorphus* Ameghino, 1889:421; †*Typotheridion* Cabrera, 1939:371; †*Bravardia* Cattoi, 1941:2.]
  [= or including †*Xenotherium* Ameghino, 1904:24.[3]]
  E.-M. Pleist.; S.A.
  [1] "†*Mesotherium*" Serres, 1857, was a *nomen nudum*.
  [2] Published later than †*Mesotherium* Serres, 1867. †*Typotherium* Bravard, 1857, was a *nomen nudum*; †*Typotherium* Bravard, 1858, was a *genus coelebs* and, moreover, the publication was doubtful; Bravard, 1860, was printed but not published. †*Typotherium* Gervais, 1859, ex Bravard, was a *genus coelebs*.
  [3] Synonymy doubtful, as is its validity. Assumed to be a valid senior homonym of †*Xenotherium* Douglass, 1906, when that animal was renamed †*Epoicotherium* by Simpson, 1927.

Suborder †**HEGETOTHERIA** Simpson, 1945:130.

†*Eohegetotherium* Ameghino, 1901:370.[1]
  M. Eoc.; S.A.
  [1] Assignment to †HEGETOTHERIA is questionable.

†*Eopachyrucos* Ameghino, 1901:370.[1]
  M. Eoc.; S.A.
  [1] Assignment to †HEGETOTHERIA is questionable.

†*Pseudopachyrucos* Ameghino, 1901:371.[1]
  M. Eoc.; S.A.
  [1] Assignment to †HEGETOTHERIA is questionable.

†*Getohetherium* Ameghino, 1904:17.
  E. Mioc.; S.A.

†*Tegehotherium* Ameghino, 1904:18.
  Mioc.; S.A.

Family †**Archaeohyracidae** Ameghino, 1897:431.
  ?L. Paleoc., E.-M. Eoc., L. Eoc. and/or E. Olig., L. Olig.; S.A.

†*Eohyrax* Ameghino, 1901:363.
  ?L. Paleoc., E.-M. Eoc.; S.A.

†*Pseudhyrax* Ameghino, 1901:362.
  [= †*Degonia* Roth, 1902:251; †*Pseudopithecus* Roth, 1902:251; †*Rankelia* Roth, 1902:252.]
  M. Eoc., L. Eoc. and/or E. Olig.; S.A.

†*Bryanpattersonia* Simpson, 1967:112.
  M. Eoc., L. Eoc. and/or E. Olig.; S.A.

†*Archaeohyrax* Ameghino, 1897:3-9.
  [= or including †*Notohyrax* Ameghino, 1901:362.]
  L. Olig.; S.A.

Family †**Hegetotheriidae** Ameghino, 1894:275.
> [= †Hegetotheridae Ameghino, 1894:275; †Hegetotheriidae Sinclair, 1909:2;
> †Hegetotheroidea Romer, 1966:387.[1]]
> M. Eoc., L. Olig.-L. Plioc., ?E. Pleist.; S.A.

[1] Proposed for hegetotheriids and archaeohyracids, but as a family-group name. The term is equivalent to †HEGETOTHERIA.

Subfamily †**Pachyrukhinae** Lydekker, 1894:3.
> [= †Pachyrucidae Lydekker, 1894:3; †Pachyrukhinae Kraglievich, 1934:96.]
> L. Olig.-L. Plioc., ?E. Pleist.; S.A.

†*Propachyrucos* Ameghino, 1897:425.
> L. Olig.; S.A.

†*Prosotherium* Ameghino, 1897:426-427.
> L. Olig.; S.A.

†*Pachyrukhos* Ameghino, 1885:160-162.
> [= †*Pachyrucos* Ameghino, 1889:422-436, 918.]
> E.-M. Mioc., ?E. Pleist.; S.A.[1]

[1] A very improbable range for a genus, although of course not impossible.

†*Paedotherium* Burmeister, 1888:179.
> L. Mioc.-L. Plioc., ?E. Pleist.; S.A.

†*Tremacyllus* Ameghino, 1891:241.
> ?L. Mioc., E.-L. Plioc.; S.A.

†*Raulringueletia* Zetti, 1972:53.
> L. Mioc.; S.A.

Subfamily †**Hegetotheriinae** Ameghino, 1894:275.
> [= †Hegetotheridae Ameghino, 1894:275; †Hegetotherinae Ameghino, 1894:277;
> †Hegetotheriinae Simpson, 1945:130.]
> M. Eoc., L. Olig.-L. Mioc.; S.A.

†*Ethegotherium* Simpson, Minoprio & Patterson, 1962:273.
> M. Eoc.; S.A.

†*Prohegetotherium* Ameghino, 1897:424-425.
> L. Olig.; S.A.

†*Hegetotherium* Ameghino, 1887:14.
> [= †*Selatherium* Ameghino, 1894:19-20.]
> E.-M. Mioc.; S.A.

†*Pseudohegetotherium* Cabrera & Kraglievich, 1931:111.
> L. Mioc.; S.A.

†*Hemihegetotherium* Rovereto, 1914:38.
> L. Mioc.; S.A.

Order †**ASTRAPOTHERIA** Lydekker, 1894:42.[1]
> [= †ASTRAPOTHEROIDEA Ameghino, 1894:268, 303; †ASTRAPOTHERIOIDEA
> Simpson, 1934:7; †AMEGHINIDA Van Valen, 1988:15[2];
> †ASTRAPOTHERIFORMES Kinman, 1994:38.[3]] [Including
> †TRIGONOSTYLOPOIDEA Simpson, 1934:4, 19[4]; †TRIGONOSTYLOPIDA
> Cifelli, 1985:253.[5]]

[1] Ameghino's †ASTRAPOTHEROIDEA has actual priority (Palmer 1904:931) but the International Code of Zoological Nomenclature does not currently apply at levels above the family-group of names. Moreover, Lydekker's †ASTRAPOTHERIA is entrenched.

[2] In part. Van Valen's †AMEGHINIDA was for both astrapotheres and notoungulates (exclusive of arctostylopids), but excluded litopterns and other meridiungulates. If evidence is provided to back up Van Valen's claim, the term may prove useful. Van Valen's usage of †ASTRAPOTHERIA included the (hypothetical) common ancestor of notoungulates and astrapotheres; therefore we synonymize †AMEGHINIDA with †ASTRAPOTHERIA, but we do not subsume †NOTOUNGULATA within †ASTRAPOTHERIA.

[3] Proposed with a query but the provisions of the International Code of Zoological Nomenclature do not apply above family-group rank.

[4] Proposed as suborder of order †ASTRAPOTHERIA; used as order by Simpson, 1967:209.

[5] *Lapsus?*

†*Notorhinus* Roth, 1903:136.[1]
> [= †*Tonorhinus* Ameghino, 1904:82.]
> M. Eoc.; S.A.

[1] Both †*Notorhinus* and its anagram †*Tonorhinus* are *nomina oblita*. Ameghino believed †*Notorhinus* to be preoccupied by *Notorhina*, a genus of beetles described in 1848.

†*Blastoconus* Roth, 1903:137.[1]
> M. Eoc.; S.A.

[1] Assignment to †ASTRAPOTHERIA is questionable.

Family †**Eoastrapostylopidae** Soria & Powell, 1981:160.[1]
> L. Paleoc.; S.A.

[1] Assignment to †ASTRAPOTHERIA is probable. Date of publication possibly 1982.

†*Eoastrapostylops* Soria & Powell, 1982:160.
> L. Paleoc.; S.A.

Family †**Trigonostylopidae** Ameghino, 1901:390.
> L. Paleoc.-M. Eoc.; S.A. M. Eoc.; Antarctica.

†*Tetragonostylops* Paula Couto, 1963:339.
> L. Paleoc., ?E. Eoc.; S.A.

†*Shecenia* Simpson, 1935:19.
> L. Paleoc., ?E. Eoc.; S.A.

†*Trigonostylops* Ameghino, 1897:492.
> [Including †*Chiodon* Berg, 1899:79; †*Staurodon* Roth, 1899:386.[1]]
> E.-M. Eoc.; S.A. ?M. Eoc.; Antarctica.

[1] Objective synonym of †*Chiodon*. Not †*Staurodon* Lowe, 1854, a genus of mollusks.

Family †**Astrapotheriidae** Ameghino, 1887:19.
> [= †Astrapotheridae Ameghino, 1887:19; †Astrapotheriinae Simpson, 1945:130.]
> [Including †Albertogaudryidae Ameghino, 1901:398; †Uruguaytheriinae Kraglievich, 1928:12; †Albertogaudryinae Simpson, 1945:130.]
> E.-M. Eoc., L. Eoc. and/or E. Olig., L. Olig.-L. Mioc.; S.A.

†*Scaglia* Simpson, 1957:11.
> E. Eoc.; S.A.

†*Albertogaudrya* Ameghino, 1901:399.
> [= †*Scabellia* Ameghino, 1901:400.]
> E. Eoc.; S.A.

†*Astraponotus* Ameghino, 1901:401.
> [= or including †*Notamynus* Roth, 1903:3; †*Megalophodon* Roth, 1903:6; †*Grypolophodon* Roth, 1903:9.[1]]
> M. Eoc.; S.A.

[1] Synonymy questionable.

†*Isolophodon* Roth, 1903:142.
> ?M. Eoc., L. Eoc. and/or E. Olig.; S.A.

†*Xenastrapotherium* Kraglievich, 1928:11.
> M.-L. Mioc.; S.A.[1]

[1] Venezuela, Colombia, Bolivia.

†*Parastrapotherium* Ameghino, 1895:635.
> [= or including †*Liarthrus* Ameghino, 1895:641; †*Traspoatherium* Ameghino, 1895:641; †*Henricofilholia* Ameghino, 1901:404; †*Helicolophodon* Roth, 1903:11.[1]]
> L. Olig.; S.A.

[1] Synonymy questionable.

†*Uruguaytherium* Kraglievich, 1928:40.
> [= or including †*Synastrapotherium* Paula Couto, 1976:241.]
> Olig. and/or Mioc.; S.A.

†*Astrapothericulus* Ameghino, 1901:73.
> L. Olig.-E. Mioc.; S.A.

†*Astrapotherium* Burmeister, 1879:517-520.
E.-M. Mioc.; S.A.

†*Monoeidodon* Roth, 1898:191.[1]
M. Mioc.; S.A.

[1] Validity doubtful and, in any case, a *nomen oblitum*.

Order †**XENUNGULATA** Paula Couto, 1952:370.
[= †XENUNGULIFORMES Kinman, 1994:38.]

Family †**Carodniidae** Paula Couto, 1952:370.
[Including †Etayoidae Villarroel, 1987:242.]
L. Paleoc.; S.A.

†*Etayoa* Villarroel, 1987:242.
?L. Paleoc.; S.A.[1]

[1] Colombia.

†*Carodnia* Simpson, 1935:20.
[= †*Ctalecarodnia* Simpson, 1935:22.]
L. Paleoc.; S.A.[1]

[1] Brazil and Patagonia.

Order †**PYROTHERIA** Ameghino, 1895:608.[1]
[= †PYROTHERIFORMES Kinman, 1994:38.]

[1] Used by Ameghino as a suborder (of ungulates), related to PROBOSCIDEA.

Family †**Pyrotheriidae** Ameghino, 1889:894.
[Including †Carolozittelidae Ameghino, 1901:387, 388; †Colombitheriidae Hoffstetter, 1970:155.]
E.-M. Eoc., L. Eoc. and/or E. Olig., L. Olig.; S.A.

†*Proticia* Patterson, 1977:403.
?E. Eoc.; S.A.[1]

[1] Venezuela.

†*Griphodon* Anthony, 1924:1.
Eoc. and/or Olig.; S.A.

†*Carolozittelia* Ameghino, 1901:388.
[= †*Archaeolophus* Ameghino, 1897:447.[1]]
E. Eoc.; S.A.

[1] Synonymy questionable. This name appears to be a *nomen oblitum*. It was based on a small pyrotherian incisor from the Casamayoran, not the Deseadan as described. Ultimately, suppression may be in order.

†*Colombitherium* Hoffstetter, 1970:149, 155.
M. and/or L. Eoc. and/or E. Olig.; S.A.[1]

[1] Colombia.

†*Propyrotherium* Ameghino, 1901:387.
[= or including †*Promoeritherium* Ameghino, 1906:333.]
M. Eoc.; S.A.

†*Pyrotherium* Ameghino, 1888:10.
[= †*Ricardowenia* Ameghino, 1901:390; †*Parapyrotherium* Ameghino, 1902:28.]
L. Olig.; S.A.

Mirorder **ALTUNGULATA** Prothero & Schoch, 1989:510, **new rank**.[1]
[Including TRICHENAE Gray, 1821:306.[2]]

[1] Proposed as a grandorder.
[2] Long forgotten. Created to include rhinos, hyraxes, tapirs, and †*Palaeotherium*, obviously meant to contrast with order MONOCHENAE (for *Equus*) and order TESSERACHENAE (for hippos, suids, tayassuids, and †*Anoplotherium*).

†*Radinskya* McKenna, Chow, Ting & Luo, in Prothero & Schoch, 1989:25.[1]
L. Paleoc.; As.

[1] Assignment to this position is questionable.

Order **PERISSODACTYLA** Owen, 1848:131.[1] Odd-toed ungulates, perissodactyls.

[= JUMENTA Trouessart, 1879:2; MESAXONIA Marsh, 1884:9, 177; CLINODACTYLA Marsh, 1884:177[2]; STEREOPTERNA F. Ameghino, 1889:492; †IMPARIDIGITATA Depéret, 1909:322; PERISSODACTYLIFORMES Kinman, 1994:38; TRIDACTYLA Kalandadze & Rautian, 1992:110.[3]] [Including MONOCHENAE Gray, 1821:306.[4]]

[1] Not November 3, 1847, when the paper was given orally. We continue to use the name PERISSODACTYLA in preference to MESAXONIA or its prior synonym JUMENTA, but exclude hyraxes and other taxa put in the PERISSODACTYLA by Owen in 1848 but not in 1852.
[2] In part.
[3] Named as a superorder.
[4] Named as an order and including only *Equus*.

†*Hallensia* Franzen & Haubold, 1986:37.

E.-M. Eoc.; Eu.

†*Schizotheroides* Hough, 1955:34.[1]

[= †*Schizotheriodes* Hough, 1955:34.[2]]

L. Eoc.; N.A.

[1] Emended. See the International Code of Zoological Nomenclature, Article 32(c)(ii). While it is indeed true that Radinsky accepted the spelling "†*Schizotheriodes*" in 1964, he did not cite any alternative names and cannot be a "first revisor," because he did no revision (see the International Code of Zoological Nomenclature, Article 24). The spelling is simply a result of inadequate proofreading by Hough. Common sense, as well as the International Code of Zoological Nomenclature, supports emendation to "†*Schizotheroides*."
[2] Misspelling. And as "†*Schizotherioides*" in the abstract (page 22).

Suborder **HIPPOMORPHA** Wood, 1937:106.

[= SOLIDUNGULA Blumenbach, 1779:109; SOLIPEDA Meckel, 1809: table 1[1]; MONOCHENAE Gray, 1821:306.[2]] [Including †HYRACOTHERIA Kalandadze & Rautian, 1992:110.[3]]

[1] 1809-1810.
[2] When Gray based this order on Equidae, *Equus* was the only known equid. According to rigid advocates of crown groups, it still is.
[3] Named as an infraorder; paraphyletic.

Family **Equidae** Gray, 1821:307. Horses.

[= Aschidae Aristotle, 330 B.C.[1]; Equuina Haeckel, 1866:clviii; Equinae Steinmann & Döderlein, 1890:767; Equoidea Hay, 1902:608; Equini Quinn, 1955:43; Equina Hulbert & MacFadden, 1991:6; Hippoidea Osborn, 1898:79[2]; Hippidae Schultz, 1900:197.] [Including †Hippotheriina Bonaparte, 1850; †Hippotheriinae Cope, 1881:398[3]; †Hippotherida Haeckel, 1895:530, 547[4]; †Hippotheriini Prothero & Schoch, 1989:532; †Anchitheridae Leidy, 1869:402; †Anchitheriinae Osborn, 1910:555; †Pliolophoidea Gill, 1872:88; †Pliolophidae Gill, 1872:88; †Hyracotheriinae Cope, 1881:381; †Protohippinae Gidley, 1907:868; †Protohippini Quinn, 1955:13; †Protohippina Hulbert, 1988; †Hippariinae Dietrich, 1942; †Hipparionini Quinn, 1955:63; †Hipparioninae Sondaar, 1969; †Hippariini Samson, 1975:192; †Calippini Quinn, 1955:27; †Pliohippina Prado & Alberdi, 1996:676.]

E. Eoc., E. Mioc.-R.; Eu. E. Eoc.-L. Pleist.; N.A. E. Mioc., L. Mioc., L. Pleist.; Cent. A. M. Mioc.-R.; As. L. Mioc.-R.; Af. L. Mioc.; Mediterranean.[5] L. Plioc.-R.; S.A.[6]

[1] Date approximate. Pre-Linnean but included here for the sake of interest.
[2] In part. Proposed as a superfamily for Equidae and †Palaeotheriidae but not based upon a generic name.
[3] Objective synonym of †Protohippinae.
[4] Used as a family.
[5] Crete.
[6] Extinct in South America by about 8000 BP.
  Also introduced and now feral in North America, in the West Indies (Hispaniola), on the Chagos Islands in the Indian Ocean, in Australia, in New Zealand, in the Pacific (Hawaii and the Galapagos Islands), and probably on other oceanic islands.
  Recent practically worldwide in domestication.

†*Hyracotherium* Owen, 1840:163.

[= †*Syotherium* von Meyer, 1848:603.] [= or including †*Eopithecus* Owen, 1858:544; †*Lophiodochoerus* Lemoine, 1880:589.] [Including †*Pliolophus* Owen, 1858:55;

†*Eohippus* Marsh, 1876:401; †*Systemodon* Cope, 1881:1018; †*Protorohippus* Wortman, 1896:104.]

E. Eoc.; Eu. E. Eoc.; N.A.

†*Orohippus* Marsh, 1872:207.

[= †*Helohippus* Marsh, 1892:353.] [Including †*Helotherium* Cope, 1872:466; †*Orotherium* Marsh, 1872:217[1]; †*Oligotomus* Cope, 1873:606; †*Aminippus* Granger, 1908:254.[2]]

E.-M. Eoc.; N.A.

[1] Not †*Orotherium* Aymard, 1850, an indeterminate artiodactyl.
[2] Proposed as subgenus of †*Orohippus* as replacement name for †*Orotherium* Marsh.

†*Epihippus* Marsh, 1878:236.

[= †*Ephippus* Winge, 1906.[1]] [Including †*Duchesnehippus* Peterson, 1931:67, fig. 4.[2]]

M. Eoc.; N.A.

[1] Attempted emendation.
[2] Proposed as subgenus of †*Epihippus* in the figure caption, at least insofar as the format used is indicative of the author's intention, but hypothetically proposed to be raised to generic rank on the next page, "when more satisfactorily determined."

†*Haplohippus* McGrew, 1953:167.

L. Eoc.; N.A.

†*Mesohippus* Marsh, 1875:248.

[Including †*Pediohippus* Schlaikjer, 1935:141.]

M. Eoc.-E. Olig.; N.A.

†*Miohippus* Marsh, 1874:249.

L. Eoc.-E. Mioc.; N.A.

†*Archaeohippus* Gidley, 1906:385.

E.-L. Mioc.; N.A. E. Mioc.; Cent. A.

†*Anchitherium* von Meyer, 1844:298.

[Including †*Paranchitherium* Borissiak, 1937:789.]

E.-L. Mioc.; Eu. E.-M. Mioc.; N.A. E. Mioc.; Cent. A. M. Mioc.; As.

†*Kalobatippus* Osborn, in Cope & Matthew, 1915: pl. cviii.

E. Mioc.; N.A.

†*Hypohippus* Leidy, 1858:26.[1]

[= †*Hyphippus* Winge, 1906.[2]] [Including †*Drymohippus* Merriam, 1913:420.[3]]

E.-L. Mioc.; N.A.

[1] Proposed as subgenus of †*Anchitherium*.
[2] Attempted emendation.
[3] As subgenus.

†*Sinohippus* Zhai, 1962:49.

L. Mioc.; As.

†*Megahippus* McGrew, 1938:315.

M. Mioc.; N.A.

†*Parahippus* Leidy, 1858:26.

[= †*Parippus* Winge, 1906.[1]] [= or including †*Anchippus* Leidy, 1868:231.] [Including †*Desmatippus* Scott, 1893:661; †*Desmathippus* Winge, 1906[1]; †*Altippus* Douglass, 1908:271.]

?L. Olig., E.-M. Mioc.; N.A.

[1] Attempted emendation.

†*Merychippus* Leidy, 1857:311.

[Including †*Stylonus* Cope, 1879:76.]

E.-M. Mioc.; N.A.

†*Acritohippus* Kelly, 1995:9.

M. Mioc.; N.A.

†*Pseudhipparion* F. Ameghino, 1904:262.

[= †*Pseudhipparion* F. Ameghino, 1904:536, index[1]; †*Pseudohipparion* Winge, 1906.[2]] [Including †*Griphippus* Quinn, 1955:42.]

M.-L. Mioc.; N.A.

[1] Misprint.
[2] Attempted emendation.

†*Nannippus* Matthew, 1926:165.[1]

[= †*Nannipus* Hay, 1930:694[2]; †*Nannithehippus* Hay, 1930:691[2]; †*Nannihippus* Hay, 1930:1019.[2]]

M. Mioc.-L. Plioc.; N.A.

[1] Described as subgenus of †*Hipparion*.
[2] Misspelling.

†*Neohipparion* Gidley, 1903:467.

[Including †*Hesperohipparion* Dalquest, 1981:506.]

M. Mioc.-E. Plioc.; N.A.

†*Hipparion* de Christol, 1832:180.

[= †*Hippotherium* Kaup, 1833:327.[1]] [= or including †*Eurygnathohippus* van Hoepen, 1930:23; †*Notohipparion* Haughton, 1931:421; †*Libyhipparion* Joleaud, 1933:7; †*Hemihipparion* Wehrli, 1941:374.] [Including †*Hippodactylus* Cope, 1888:449; †*Parahipparion* Kretzoi, 1954:251[2]; †*Perihipparion* Kretzoi, 1965:134[3]; †*Plesiohipparion* Qiu, Huang & Guo, 1987:52[4]; †*Cremohipparion* Qiu, Huang & Guo, 1987:63[4]; †*Baryhipparion* Qiu, Huang & Guo, 1987:68[4]; †*Gremohipparion* Qiu, Huang & Guo, 1987:238.[5]]

M. Mioc.-L. Plioc.; Eu. M.-L. Mioc.; N.A. L. Mioc.-M. Pleist.; Af. L. Mioc.-E. Pleist.; As. L. Mioc.; Mediterranean.[6] L. Mioc.; Cent. A.

[1] Described as subgenus of *Equus*.
[2] Not †*Parahipparion* Ameghino, 1904.
[3] Replacement name for †*Parahipparion* Kretzoi.
[4] As subgenus.
[5] Misspelling.
[6] Crete.

†*Proboscidipparion* Sefve, 1927:55.

L. Plioc.-E. Pleist.; As.

†*Stylohipparion* van Hoepen, 1932:31.

Plioc. and/or Pleist.; Af.

†*Cormohipparion* Skinner & MacFadden, 1977:917.

[= or including †*Sivalhippus* Lydekker, 1877:31.[1]] [Including †*Notiocradohipparion* Hulbert, 1988:246.[2]]

M. Mioc.-L. Plioc.; N.A. L. Mioc.; As. L. Mioc.; Eu. L. Mioc.; Cent. A.

[1] Technically a *nomen oblitum*, for a species of what now is called †*Cormohipparion*. It might be useful to suppress †*Sivalhippus*.
[2] Proposed as subgenus.

†*Protohippus* Leidy, 1858:26.[1]

[= †*Prohippus* Heude, 1894:167[2]; †*Prothippus* Winge, 1906.[3]] [= or including †*Eoequus* Quinn, 1955:54.]

M.-L. Mioc.; N.A.

[1] Described as subgenus of *Equus*; raised to generic rank by Leidy, 1869.
[2] Invalid emendation or misprint.
[3] Attempted emendation.

†*Parapliohippus* Kelly, 1995:3.

M. Mioc.; N.A.

†*Heteropliohippus* Kelly, 1995:5.

L. Mioc.; N.A.

†*Pliohippus* Marsh, 1874:252.

M.-L. Mioc.; N.A. L. Mioc.; Cent. A.

†*Calippus* Matthew & Stirton, 1930:354.

[Including †*Grammohippus* Hulbert, 1988:254.[1]]

M.-L. Mioc.; N.A. L. Mioc.; Cent. A.

[1] Described as subgenus of †*Calippus*.

†*Astrohippus* Stirton, 1940:190.[1]

L. Mioc.-E. Plioc.; N.A.

[1] Proposed as a subgenus of †*Pliohippus*.

†*Hippidion* Owen, 1869:268.

[= †*Hippidium* Burmeister, 1875:5[1]; †*Rhinippus* Burmeister, 1875:15n.] [= or including †*Hipphaplous* F. Ameghino, 1885:98[2]; †*Hipphaplus* F. Ameghino, 1889:502[3]; †*Stereohippus* F. Ameghino, 1904:275.] [Including †*Plagiohippus* F. Ameghino, 1908:423; †*Parahipparion* F. Ameghino, 1904:273; †*Hyperhippidium* Sefve, 1910:5.[4]]

L. Mioc., L. Plioc. and/or E. Pleist.; N.A. L. Plioc.-R.; S.A.[5]

[1] And almost all later writers.
[2] *Nomen nudum* in Ameghino, 1882.
[3] Correction of †*Hipphaplous*.
[4] Considered a subgenus of †*Parahipparion* by Sefve, 1912.
[5] Extinct by about 8000 BP.

†*Onohippidium* Moreno, 1891:65.

[= †*Onohippus* Burmeister, 1891:470[1]; †*Onohippidion* Boule & Thevin, 1920:79n.[2]]

E. Plioc.; N.A. L. Plioc. and/or Pleist.; S.A.

[1] *Lapsus calami*.
[2] Replacement name for †*Onohippidium*, which did not require replacement.

†*Dinohippus* Quinn, 1955:43.

L. Mioc.-E. Plioc.; N.A.

Equus Linnaeus, 1758:73. Horses, tarpan, asses, donkeys, kiang, onager, quagga, zebras, takh.

[= *Caballus* Rafinesque, 1815:55.[1]] [Including *Asinus* Frisch, 1775:244-248[2]; *Asinus* Gray, 1824:244[3]; *Onager* Boddaert, 1785; *Hemionus* F. Cuvier, 1823[3]; *Hippotigris* H. Smith, 1841:321[3]; †*Tomolabis* Cope, 1892:125; *Zebra* Allen, 1909:163[4]; *Dolichohippus* Heller, 1912:1[3]; *Megacephalon* Hilzheimer, 1912:95[5]; *Grevya* Hilzheimer, 1912:476[6]; †*Neohippus* Abel, 1913:754a; *Ludolphozecora* Griffini, 1913:332[6]; †*Euhippus* von Reichenau, 1915:150[3]; †*Microhippus* von Reichenau, 1915:152[3]; *Microhippus* Matschie, 1924:68[7]; †*Plesippus* Matthew, 1924:2; †*Sterrohippus* van Hoepen, 1930:6; †*Kraterohippus* van Hoepen, 1930:7; †*Kolpohippus* van Hoepen, 1930:8; †*Quagga* Shortridge, 1934:397; †*Allohippus* Kretzoi, 1938:93[3]; †*Macrohippus* Kretzoi, 1938:149; †*Hypsohipparion* Dietrich, 1941:219; *Megacephalonella* Strand, 1943:216[6]; †*Amerhippus* Hoffstetter, 1950:433[3]; *Pseudoquagga* Hoffstetter, 1950:690[3]; †*Hesperohippus* Hibbard, 1955:66[8]; †*Allozebra* Trumler, 1961:115; *Asinohippus* Trumler, 1961:118[9]; †*Praehemionus* Trumler, 1961:119[9]; †*Hydruntinus* Radulesco & Samson, 1962:174[10]; †*Quaggoides* Willoughby, 1974:35[3]; †*Parastilidequus* Mooser & Dalquest, 1975:807.[3]]

E. Plioc.-L. Pleist.; N.A. L. Plioc.-R.; Af. L. Plioc.-R.; As. L. Plioc.-R.; Eu. M.-L. Pleist.; S.A. L. Pleist.; Cent. A.[11]

[1] New name for *Equus*.
[2] Not published in a consistently binominal work. See the International Code of Zoological Nomenclature, Article 11(c).
[3] Sometimes as subgenus.
[4] Not *Zebra* Shuttleworth, 1856, a mollusk.
[5] Not *Megacephalon* Temminck, 1844, a bird.
[6] Replacement name for *Megacephalon* Hilzheimer. Antedated by *Dolichohippus*.
[7] Sometimes as subgenus. Not *Microhippus* von Reichenau, 1915.
[8] Proposed as a subgenus of *Equus*.
[9] Described as subgenus of *Hemionus*.
[10] (Rădulescu).
[11] Also introduced and now feral in North America, W. Indies (Hispaniola), South America (Colombia), Pacific (Galapagos Islands, Hawaii), Australia, New Zealand, the Chagos Islands of the Indian Ocean, and probably on other oceanic islands.
Recent practically worldwide in domestication.

Family †**Palaeotheriidae** Bonaparte, 1850: unpaged chart.

[= †Palaeotheriina Bonaparte, 1850: unpaged chart; †Palaeotheridae Girard, 1852:328; †Palaeotherida Haeckel, 1866:clviii; †Palaeotheriidae Gill, 1872:12; †Palaeotheriinae Remy, 1976:122; †Palaeotheriini Remy, 1976:122.] [Including †Paloplotheriinae Osborn, 1892:93; †Pachynolophynae Pavlow, 1877[1]; †Pachynolophidae Pavlow, 1888:136, 145.]

E. Eoc., M. and/or L. Eoc., ?E. Olig.; As. E. Eoc.-E. Olig.; Eu. M. and/or L. Eoc.; Mediterranean.

[1] (Paulow).

†*Propachynolophus* Lemoine, 1891:285, 286.
    [= or including †*Hyracotherhyus* Lemoine, 1891:266.]
    E. Eoc.; As. E.-M. Eoc.; Eu.

†*Palaeotherium* G. Cuvier, 1804:275-303.
    [= or including †*Monacrum* Aymard, 1853:309, 311.[1]] [Including †*Franzenitherium* Remy, 1992:212.[2]]
    M. Eoc.-E. Olig.; Eu.
    [1] Named as subgenus of †*Palaeotherium*; raised to generic rank by Aymard, 1854:674.
    [2] As subgenus.

†*Propalaeotherium* Gervais, 1849:383.
    E. and/or M. Eoc.; As. E.-L. Eoc.; Eu.

†*Plagiolophus* Pomel, 1847:586.
    [Including †*Paloplotherium* Owen, 1848:20-36.]
    M. and/or L. Eoc.; Mediterranean. M. Eoc.-E. Olig.; Eu.

†*Pachynolophus* Pomel, 1847:327.[1]
    M.-L. Eoc.; Eu.
    [1] Described as subgenus of †*Lophodon*; raised to generic rank by Gervais, 1849.

†*Leptolophus* Remy, 1965:4362.
    M. Eoc.; Eu.

†*Gobihippus* Dashzeveg, 1979:15.
    M. and/or L. Eoc.; As.[1]
    [1] E. Asia.

†*Paraplagiolophus* Depéret, 1917:60.
    M. Eoc.; Eu.

†*Cantabrotherium* Casanovas-Cladellas & Santafé-Llopis, 1987:245.
    L. Eoc.; Eu.

†*Franzenium* Casanovas-Cladellas & Santafé-Llopis, 1989:45.
    M.-L. Eoc.; Eu.

†*Qianohippus* Miao, 1982:526.
    ?E. Olig.; As.

†*Pseudopalaeotherium* Franzen, 1972:317.
    E. Olig.; Eu.

Suborder **CERATOMORPHA** Wood, 1937:106.[1]
    [= TRICHENAE Gray, 1821:306; TRIDACTYLA Latreille, 1825:61.]
    [1] We continue to use Wood's taxon for reasons of stability.

Infraorder †**SELENIDA, new.**[1]
    [Including †TITANOTHERIOMORPHA Hooker, 1989:98; †BRONTOTHERIA Kalandadze & Rautian, 1992:111.[2]]
    [1] Definition: for the most recent common ancestor of †Brontotherioidea and Chalicotherioidea, and all its descendants.
    [2] Named as an infraorder.

Superfamily †**Brontotherioidea** Marsh, 1873:486.
    [= †Brontotheriidae Marsh, 1873:486; †Brontotherioidea Hay, 1902:629; †BRONTOTHERIA Kalandadze & Rautian, 1992:111.[1]] [Including †Titanotherioidea Osborn, 1898:79.]
    [1] Named as an infraorder.

Family †**Brontotheriidae** Marsh, 1873:486.
    [= †Brontotheridae Marsh, 1873:486.] [Including †Titanotheridae Flower, 1876:327; †Titanotheriidae Flower, 1876:109; †Lambdotheriidae Cope, 1889:152; †Palaeosyopidae Osborn, 1910:556.]
    ?E. Eoc., M. Eoc.-E. Olig.; As. E.-L. Eoc.; N.A. ?M. Eoc., L. Eoc., ?E. Olig.; Eu.

†*Pakotitanops* West, 1980:521.
M. Eoc.; As.[1]
[1] S. Asia (Pakistan).

†*Nanotitan* Qi & Beard, 1996:578.
M. Eoc.; As.

Subfamily †**Lambdotheriinae** Cope, 1889:152.
[= †Lambdotheriidae Cope, 1889:152; †Lambdotheriinae Hay, 1902:629.]
E. Eoc.; N.A.

†*Lambdotherium* Cope, 1880:746-747.
E. Eoc.; N.A.

†*Xenicohippus* Bown & Kihm, 1981:258.
E. Eoc.; N.A.

Subfamily †**Palaeosyopinae** Steinmann & Döderlein, 1890:777.
[= †Eotitanopinae Osborn, 1914:403.]
?E. Eoc., M. Eoc.; As.[1] E.-M. Eoc.; N.A.
[1] The possible occurrence of a member of this subfamily in the early or middle Eocene of the Zaisan Basin, Kazakhstan has been noted.

†*Palaeosyops* Leidy, 1870:113.
[Including †*Limnohyops* Marsh, 1890:525; †*Eotitanops* Osborn, 1907:242[1]; †*Eometarhinus* Osborn, 1919:568.]
E.-M. Eoc.; N.A. ?M. Eoc.; As.[2]
[1] Possibly a *nomen nudum*. Adequately described by Osborn, 1908:600-601.
[2] Pakistan and China.

†*Mulkrajanops* Kumar & Sahni, 1985:153.
M. Eoc.; As.[1]
[1] S. Asia.

Subfamily †**Dolichorhininae** Riggs, 1912:25, 40.[1]
[= †Dolichorhinae Riggs, 1912:25, 40; †Dolichorhininae Osborn, 1929:245.]
[Including †Rhadinorhininae Osborn, 1929:429.]
M. Eoc.; N.A.
[1] See the International Code of Zoological Nomenclature, Article 40.

†*Metarhinus* Osborn, 1908:609.
[Including †*Rhadinorhinus* Riggs, 1912:36; †*Heterotitanops* Peterson, 1914:53.]
M. Eoc.; N.A.

†*Sphenocoelus* Osborn, 1895:98.[1]
[= or including †*Dolichorhinus* Hatcher, 1895:1090[2]; †*Tanyorhinus* Cook, 1926:13.]
M. Eoc.; N.A.
[1] Between January and June, 1895.
[2] December 1895.

†*Mesatirhinus* Osborn, 1908:608.
M. Eoc.; N.A.

Subfamily †**Brontotheriinae** Marsh, 1873:486.
[= †Brontotheridae Marsh, 1873:486; †Brontotheriinae Steinmann & Döderlein, 1890:777.] [Including †Megaceropinae Osborn, 1914:405.]
M.-L. Eoc.; N.A.

†*Duchesneodus* Lucas & Schoch, 1982:1022.
M.-L. Eoc.; N.A.

†*Brontotherium* Marsh, 1873:486-487.
L. Eoc.; N.A.

†*Megacerops* Leidy, 1870:1-2.[1]
L. Eoc.; N.A.
[1] "In the literature previous to Osborn, 1929, the genus here called . . . †*Megacerops* was usually called †*Symborodon*, and the genus here called †*Brontops* was usually called †*Megacerops*" (Simpson 1945:139).

Subfamily †**Embolotheriinae** Osborn, 1929:9.
M. Eoc.-E. Olig.; As.

†*Titanodectes* Granger & Gregory, 1943:369.
M.-L. Eoc.; As.

†*Embolotherium* Osborn, 1929:9.
L. Eoc.-E. Olig.; As.

†*Protembolotherium* Yanovskaya, 1954:11.
L. Eoc.; As.

Subfamily †**Brontopinae** Osborn, 1914:405.
[Including †Manteoceratinae Osborn, 1914:403; †Epimanteoceratinae Granger & Gregory, 1943:357.]
M.-L. Eoc., ?E. Olig.; As. M. and/or L. Eoc., ?E. Olig.; Eu. M.-L. Eoc.; N.A.

†*Brachydiastematherium* Bockh & Maty, 1876:125-150.
M. and/or L. Eoc.; Eu.

†*Pachytitan* Granger & Gregory, 1943:366.
M. Eoc.; As.

†*Dianotitan* Chow, Chang & Ting, 1974:262.
M.-L. Eoc.; As.[1]
[1] E. Asia.

†*Gnathotitan* Granger & Gregory, 1943:363.
M. Eoc.; As.

†*Microtitan* Granger & Gregory, 1943:361.
M. Eoc.; As.

†*Epimanteoceras* Granger & Gregory, 1943:357.
M.-L. Eoc.; As.

†*Protitan* Granger & Gregory, 1943:358.
M. Eoc.; As.

†*Rhinotitan* Granger & Gregory, 1943:364.
M. Eoc.; As.

†*Metatitan* Granger & Gregory, 1943:367.
M.-L. Eoc.; As.

†*Dolichorhinoides* Granger & Gregory, 1943:363.
M. Eoc.; As.

†*Protitanotherium* Hatcher, 1895:1084.
M. Eoc.; N.A.

†*Parabrontops* Granger & Gregory, 1943:366.
L. Eoc.; As.

†*Oreinotherium* L. S. Russell, 1934:62.
L. Eoc.; N.A.

†*Brontops* Marsh, 1887:326-328.[1]
[Including †*Diploclonus* Marsh, 1890:523-524; †*Teleodus* Marsh, 1890:524.]
L. Eoc.; N.A. ?E. Olig.; Eu.
[1] "In the literature previous to Osborn, 1929, the genus here called . . . †*Brontops* was usually called †*Megacerops*" (Simpson 1945:139).

†*Protitanops* Stock, 1936:656.
L. Eoc.; N.A.

†*Pygmaetitan* Miao, 1982:528.
?E. Olig.; As.

Subfamily †**Telmatheriinae** Osborn, 1914:403.
[Including †Metatelmatheriinae Granger & Gregory, 1943:355.]
M.-L. Eoc.; As. M. Eoc., ?L. Eoc.; N.A. ?L. Eoc.; Eu.

†*Acrotitan* Ye, 1983:112.
    M. Eoc.; As.[1]
    [1] E. Asia.

†*Desmatotitan* Granger & Gregory, 1943:356.
    M. Eoc.; As.

†*Arctotitan* Wang, 1978:118.
    M.-L. Eoc.; As.[1]
    [1] E. Asia (China).

†*Hyotitan* Granger & Gregory, 1943:357.
    M. Eoc.; As.

†*Sthenodectes* Gregory, 1912:545.
    M. Eoc.; N.A.

†*Telmatherium* Marsh, 1872:123-124.
    [= †*Telmatotherium* Marsh, 1880:10.[1]] [= or including †*Manteoceras* Hatcher, 1895:1090.] [Including †*Leurocephalus* Osborn, Scott & Speir, 1878:42; †*Metatelmatherium* Granger & Gregory, 1938:435.]
    M. Eoc.; As. M. Eoc., ?L. Eoc.; N.A.
    [1] Emendation, in a privately printed list of genera. If not available, the emendation dates from Scudder, 1882:328. Preferred by Earle, 1891 and other authors prior to Hay (1902).

†*Sivatitanops* Pilgrim, 1925:3.
    M. Eoc.; As. ?L. Eoc.; Eu.

Subfamily †**Menodontinae** Cope, 1881:378.
    [= †Menodontidae Cope, 1881:378; †Menodontinae Osborn, 1914:405.] [Including †Diplacodontinae Osborn, 1914:403.]
    M.-L. Eoc.; N.A. L. Eoc.; Eu.

†*Diplacodon* Marsh, 1875:246.
    M. Eoc.; N.A.

†*Eotitanotherium* Peterson, 1914:220.[1]
    [= †*Diploceras* Peterson, 1914:30.[2]]
    M. Eoc.; N.A.
    [1] Replacement name for †*Diploceras* Peterson.
    [2] Preoccupied by *Diploceras* Conrad, 1844, a mollusk.

†*Notiotitanops* Gazin & Sullivan, 1942:4.
    M. Eoc.; N.A.

†*Menodus* Pomel, 1849:73-75.[1]
    [= †*Titanotherium* Leidy, 1852:551-552; †*Symborodon* Cope, 1873:2-3.] [Including †*Anisacodon* Marsh, 1875:246[2]; †*Diconodon* Marsh, 1876:339[3]; †*Allops* Marsh, 1887:331.]
    L. Eoc.; Eu. L. Eoc.; N.A.
    [1] "In the literature previous to Osborn, 1929, the genus here called †*Menodus* was usually called †*Titanotherium*, the genus here called †*Megacerops* was usually called †*Symborodon*" (Simpson 1945:139).
    [2] Preoccupied by †*Anisacodon* Marsh, 1872, a synonym of †*Pantolestes*.
    [3] Replacement name for †*Anisacodon* Marsh, 1875.

†*Ateleodon* Schlaikjer, 1935:83.
    L. Eoc.; N.A.

Family †**Anchilophidae, new**.
    M.-L. Eoc.; Mediterranean. M.-L. Eoc.; Eu.

†*Anchilophus* Gervais, 1852.[1]
    M.-L. Eoc.; Eu.
    [1] 1848-1852.

†*Paranchilophus* Casanovas-Cladellas & Santafé-Llopis, 1989:38.
    L. Eoc.; Eu.

†*Lophiotherium* Gervais, 1850:381.[1]
>   M.-L. Eoc.; Mediterranean. M.-L. Eoc.; Eu.
>   [1] Not †*Lophiotherium* Fischer de Waldheim, 1829 (suppressed by the International Commission on Zoological Nomenclature); nor †*Lophiotherium* Gervais, 1849 (a *nomen nudum*).

Superfamily †**Chalicotherioidea** Gill, 1872:8.
>   [= †Chalicotheriidae Gill, 1872:8; †CHALICOTHERIA Marsh, 1892:448[1]; †Chalicotheroidea Osborn, 1907:184; †Chalicotherioidae Hay, 1930:660; †ANCYLOPODA Cope, 1889:153.[2]]
>   [1] Given ordinal rank by Marsh, who stated that Cope's ANCYLOPODA was preoccupied.
>   [2] Proposed as an order; used as an infraorder by Kalandadze & Rautian, 1992:112.

†*Pernatherium* Gervais, 1876:425-432.
>   M. Eoc.; Eu.

Family †**Eomoropidae** Matthew, 1929:519.
>   [= †Eomoropinae Matthew, 1929:519; †Eomoropidae Viret, in Piveteau, 1958:417.]
>   E.-L. Eoc.; As. E.-M. Eoc.; Eu. E.-M. Eoc.; N.A.

†*Paleomoropus* Radinsky, 1964:2.[1]
>   E. Eoc.; N.A.
>   [1] Chalicothere affinities challenged by Fischer in 1977.

†*Lophiaspis* Depéret, 1910.[1]
>   E.-M. Eoc.; Eu.
>   [1] Chalicothere affinities challenged by Fischer in 1977.

†*Danjiangia* Wang, 1995:138.
>   E. Eoc.; As.

†*Eomoropus* Osborn, 1913:262.
>   M.-L. Eoc.; As. M. Eoc.; N.A.

†*Lunania* Chow, 1957:205.
>   M. Eoc.; As.[1]
>   [1] E. Asia.

†*Litolophus* Radinsky, 1964:17.
>   M. Eoc.; As.

†*Grangeria* Zdansky, 1930:67.
>   M. Eoc.; As. M. Eoc.; N.A.

Family †**Chalicotheriidae** Gill, 1872:8.
>   [= †Macrotheriidae Alston, 1878:23; †Chalicotherida Haeckel, 1895:530, 546.]
>   [Including †Ancylotheridae Dawkins, 1868:3[1]; †Moropodidae Marsh, 1877:249.]
>   L. Eoc.-E. Plioc., ?L. Plioc., Pleist.; As. ?E. Olig., L. Olig.-E. Plioc.; Eu. E. Mioc., L. Mioc.-E. Pleist.; Af. E.-L. Mioc.; N.A.
>   [1] Or, according to Dawkins as quoted by Palmer (1904:727), Gaudry, 1867. Either reference would have priority over Gill's paper. †Ancylotheridae (†Ancylotheriidae) should be suppressed (see the International Code of Zoological Nomenclature, Articles 23 and 79).

Subfamily †**Schizotheriinae** Holland & Peterson, 1914:203.
>   [= †Schizotherini Colbert, 1934:354.]
>   L. Eoc.-L. Mioc.; As. ?E. Olig., L. Olig.-L. Mioc.; Eu. E.-L. Mioc.; N.A. L. Mioc.-E. Pleist.; Af.

†*Schizotherium* Gervais, 1876:58-59.
>   [= †*Limognitherium* Filhol, 1880:1580.] [= or including †*Kyzylkakhippus* Gabunia & Belyaeva, 1964:129.]
>   L. Eoc.-L. Olig.; As. Olig.; Eu.

†*Phyllotillon* Pilgrim, 1910:67.
>   [= †*Metaschizotherium* G. H. R. von Koenigswald, 1932:1-24.]
>   ?L. Olig., E. Mioc.; As. L. Olig.-L. Mioc.; Eu.

†*Borissiakia* Butler, 1965:225.
>   L. Olig.; As.

†*Moropus* Marsh, 1877:249.
> E. Mioc.; Eu. E.-L. Mioc.; N.A.

†*Tylocephalonyx* Coombs, 1979:7.
> E.-M. Mioc.; N.A.

†*Ancylotherium* Gaudry, 1863:129-142.
> [= †*Gansuodon* Wu & Chen, 1976:194.] [= or including †*Huanghotherium* Tung, Huang & Qiu, 1975:38.]
> M.-L. Mioc.; As. M.-L. Mioc.; Eu. L. Mioc.-E. Pleist.; Af.

†*Chemositia* Pickford, 1979:83-91.
> L. Mioc. and/or E. Plioc.; Af.

Subfamily †**Chalicotheriinae** Gill, 1872:8.
> [= †Chalicotheriidae Gill, 1872:8; †Chalicotheriini Winge, 1906:153; †Chalicotheriinae Matthew, 1929:518; †Chalicotherini Colbert, 1934:354; †Macrotheriinae Holland & Peterson, 1913:202, 209.]
> E. Mioc.; Af. E. Mioc.-E. Plioc., ?L. Plioc., Pleist.; As. E. Mioc.-E. Plioc.; Eu.

†*Chalicotherium* Kaup, 1833:4-8, 30-31.
> [= †*Macrotherium* Pictet, 1844:232; †*Anisodon* Pomel, 1848:686.] [Including †*Butleria* de Bonis, Bouvrain, Koufos & Tassy, 1995:169.]
> E. Mioc.; Af. E. Mioc.-E. Plioc.; As. E. Mioc.-E. Plioc.; Eu.

†*Nestoritherium* Kaup, 1859:3.
> [= †*Circotherium* Holland & Peterson, 1914:211.]
> ?Plioc., Pleist.; As.

Infraorder **TAPIROMORPHA** Haeckel, 1866:clviii, **new rank**.[1]
> [1] Proposed as a "sectio" of the suborder PERISSODACTYLA.

Superfamily **Rhinocerotoidea** Gray, 1825:343.
> [= Nasicornia Illiger, 1811:97[1]; Rhinocerina Gray, 1825:343; Rhinocerotoidea Gill, 1872:12.]
> [1] Used as a family but not based upon a generic name.

†*Meschotherium* Gabunia, 1964:67.
> L. Olig.; As.[1]
> [1] W. Asia.

Family †**Hyracodontidae** Cope, 1879:288.
> [Including †Forstercooperiidae Kretzoi, 1940:93; †Hyrachyidae H. E. Wood, 2nd, 1927:8.]
> E.-M. Eoc., ?E. Olig., L. Olig., Mioc.; Eu. E. Eoc.-L. Olig.; N.A. M. Eoc.-E. Mioc.; As.

†*Praeaceratherium* Abel, 1910:15.[1]
> Olig.; Eu.
> [1] Possibly referable to †Paraceratheriinae.

Subfamily †**Hyrachyinae** Osborn, 1892:93.
> [= †Hyrachyidae H. E. Wood, 2nd, 1927:8.]
> E.-M. Eoc.; Eu. E.-M. Eoc.; N.A. M.-L. Eoc.; As.

†*Hyrachyus* Leidy, 1871:229.
> [Including †*Colonoceras* Marsh, 1873:407-408; †*Metahyrachyus* Troxell, 1922:38[1]; †*Ephyrachyus* H. E. Wood, 2nd, 1934:232.]
> E.-M. Eoc.; Eu. E.-M. Eoc.; N.A. M.-L. Eoc.; As.
> [1] *Nomen nudum.*

†*Fouchia* Emry, 1989:794.
> M. Eoc.; N.A.

†*Dilophodon* Scott, 1883:46.
> [Including †*Heteraletes* Peterson, 1931:68.]
> M. Eoc.; N.A.

Subfamily †**Paraceratheriinae** Osborn, 1923:13.[1]

> [= †Indricotheriinae Borissiak, 1923:123; †Baluchitheriinae Osborn, 1923:13.]
> [Including †Forstercooperiidae Kretzoi, 1940:93; †Forstercooperiinae H. E. Wood, 2nd, 1963:2.]
> M. Eoc.-E. Mioc.; As. M. Eoc.; N.A. ?E. Olig., L. Olig., Mioc.; Eu.[2]

[1] This group of large perissodactyls is the subject of an extensive and complex literature in several languages and is in need of yet another revision. Some of the taxa are dubious and others apparently composite.
[2] E. Europe.

†*Forstercooperia* H. E. Wood, 2nd, 1939.

> [= †*Cooperia* H. E. Wood, 2nd, 1938:1.[1]] [Including †*Pappaceras* H. E. Wood, 2nd, 1963:2.]
> M.-L. Eoc.; As. M. Eoc.; N.A.

[1] Not †*Cooperia* Ransom, 1907, a nematode.

†*Ilianodon* Chow & Xu, 1961:296.

> L. Eoc.; As.[1]

[1] E. Asia.

†*Imequincisoria* Wang, 1976:104.

> L. Eoc.; As.[1]

[1] E. Asia.

†*Juxia* Chow & Chiu, 1964:264.

> L. Eoc.; As.[1]

[1] E. Asia.

†*Paraceratherium* Forster Cooper, 1911:711.

> [= †*Thaumastotherium* Forster Cooper, 1913:376[1]; †*Baluchitherium* Forster Cooper, 1913:504.[2]]
> E. Mioc.; As.

[1] Not †*Thaumastotherium* Kirkaldy, 1908.
[2] Replacement name for †*Thaumastotherium* Forster Cooper.

†*Indricotherium* Borissiak, 1916:131.[1]

> [= or including †*Aralotherium* Borissiak, 1939:271; †*Pristinotherium* Birjukov, 1953:1-8.]
> L. Eoc.-L. Olig., ?E. Mioc.; As. ?E. Olig., L. Olig., ?E. Mioc.; Eu.[2]

[1] Often given as "Borissiak, 1915" but cited by Borissiak himself as having been published in April 1916. Borissiak named no type species until †*Indricotherium asiaticum* Borissiak, 1923, but in 1922 Pavlova described †*Indricotherium transouralicum*, validating the genus [see the International Code of Zoological Nomenclature, Article 11(c)(i)]. †*I. asiaticum* is apparently a junior synonym of †*I. transouralicum*. †*Indricotherium* is the largest known land mammal, but is not so large as has often been claimed. In 1993 Fortelius & Kappelman estimated an absolute maximum weight of 15-20 tons and a minimum of about 6.7 tons for species of †*Indricotherium*. They believed the size of indricotheres to be not much larger than that of the largest proboscideans, reaching about 15 feet at the shoulder. The neck was long, not short as sometimes depicted by those who use modern rhinoceroses as a model. It had to drink, after all, although a proboscis may have been present.
[2] E. Europe.

†*Urtinotherium* Chow & Chiu, 1963:230.

> L. Eoc.; As. ?E. Olig.; Eu.[1]

[1] E. Europe (Transylvania).

†*Benaratherium* Gabunia, 1955:177-182.

> L. Olig.; Eu.[1]

[1] E. Europe (Caucasus and Romania).

†*Dzungariotherium* Chiu, 1973:182.

> E.-L. Olig.; As.

†*Caucasotherium* Vereshchagin, 1960:107-112.

> Mioc.; Eu.[1]

[1] E. Europe.

Subfamily †**Hyracodontinae** Cope, 1879:288.

> [= †Hyracodontidae Cope, 1879:288; †Hyracodontinae Steinmann & Döderlein, 1890:768, 772; †Hyracodontini Kalandadze & Rautian, 1992:115.] [Including †Rhodopaginae Reshetov, 1975:35; †Rhodopagidae Hooker, 1989:94.]
> M. Eoc.-L. Olig.; As. M. Eoc.-L. Olig.; N.A.

†*Rhodopagus* Radinsky, 1965:207.
M.-L. Eoc.; As.

†*Pataecops* Radinsky, 1966:222.
[= †*Pataecus* Radinsky, 1965:212.[1]]
M.-L. Eoc.; As.
[1] Not †*Pataecus* Richardson, 1844, a perciform fish.

†*Veragromovia* Gabunia, 1961.
M. and/or L. Eoc.; As.[1]
[1] Kazakhstan.

†*Yimengia* Wang, 1988:20.
M. Eoc.; As.

†*Hyracodon* Leidy, 1856:91-92.
M. Eoc.-L. Olig.; N.A.

†*Triplopides* Radinsky, 1967:30.
L. Eoc.; N.A.

†*Ardynia* Matthew & Granger, 1923:2.
[Including †*Ergilia* Gromova, 1952:99-119; †*Parahyracodon* Belyaeva, 1952:120-143, plate 1.]
M. Eoc.-L. Olig.; As.

†*Epitriplopus* H. E. Wood, 2nd, 1927:19.
M. Eoc.; N.A.

Subfamily †**Triplopodinae** Cope, 1881:340.
[= †Triplopodidae Cope, 1881:340; †Triplopodinae Osborn, 1892:93; †Triplopini Kalandadze & Rautian, 1992:115.]
M.-L. Eoc., ?L. Olig.; As. M. Eoc.; N.A.

†*Triplopus* Cope, 1880:382-383.
[Including †*Prothyracodon* Scott & Osborn, 1887:260; †*Eotrigonias* H. E. Wood, 2nd, 1927:28.]
M. Eoc., ?L. Olig.; As. M. Eoc.; N.A.

†*Ulania* Qi, 1990:218.
L. Eoc.; As.

Subfamily †**Eggysodontinae** Breuning, 1923:119.
[= †Allaceropinae H. E. Wood, 2nd, 1932:170; †Eggysodontinae Viret, 1958:441.]
?E. Olig., L. Olig.; Eu. L. Olig.; As.

†*Eggysodon* Roman, 1910:1558-1560.[1]
[= †*Engyodon* Stehlin, 1930:644-648[2]; †*Allacerops* H. E. Wood, 2nd, 1932:170.]
?E. Olig., L. Olig.; Eu.
[1] Possibly referable to †Paraceratheriinae.
[2] Attempted emendation.

†*Tenisia* Reshetov, Spassov & Baishashov, 1993:716.
L. Olig.; As.

Family **Rhinocerotidae** Gray, 1821:306.
[= Nasicornia Illiger, 1811:97[1]; Rhynocerotidae Gray, 1821:306[2]; Rhinocerina Gray, 1825:343; Rhinocerosidae Burnett, 1830:352; Rhinocerotidae Owen, 1845:587; Rhinocerida Haeckel, 1866:clviii.[3]] [Including †Elasmotheriidae Gill, 1872:12; Amynodontidae Scott & Osborn, 1883:4, 12, etc.; †Caenopidae Cope, 1887:926.]
M. Eoc.-R.; As. M. Eoc.-L. Pleist.; Eu. M. Eoc.-E. Plioc.; N.A. E. Mioc.-R.; Af.
E. Mioc., L. Mioc.; Cent. A. ?M. and/or ?L. Mioc.; Mediterranean. R.; E. Indies.[4]
[1] Used as a family but not based upon a generic name.
[2] Gray spelled the Linnaeus's type genus correctly in 1821.
[3] Used as a family by Haeckel (1866:clviii).
[4] Java, Sumatra, Borneo.

†*Teletaceras* Hanson, 1989:380.
M.-L. Eoc.; N.A. L. Eoc.; As.

†*Itanzatherium* Belyaeva, 1966:92-143.
    L. Plioc. and/or E. Pleist.; As.

Subfamily **Rhinocerotinae** Gray, 1821:306. Rhinoceroses.
    [= Rhynocerotidae Gray, 1821:306; Rhinocerontina C. L. Bonaparte, 1837:8[1];
    Rhinocerotinae Dollo, 1885:295.] [Including †Elasmotheriinae Dollo, 1885:295;
    †Aceratheriinae Dollo, 1885:295; Ceratorhinae Osborn, 1898:121; †Atelodinae
    Breuning, 1923:120; Dicerorhinae Ringström, 1924:5; Dicerorhininae Simpson,
    1945:142; †Caenopinae Breuning, 1924:16; †Menoceratinae Prothero, Manning &
    Hanson, 1986:348-349.]
    M. Eoc.-R.; As. M. Eoc.-L. Pleist.; Eu. L. Eoc.-E. Plioc.; N.A. E. Mioc.-R.; Af.
    E. Mioc., L. Mioc.; Cent. A. ?M. and/or ?L. Mioc.; Mediterranean. R.; E. Indies.[2]
    [1] As subfamily.
    [2] Java, Sumatra, Borneo.

†*Huananodon* You, 1977:46.
    L. Eoc.; As.[1]
    [1] S.E. Asia.

†*Guixia* You, 1977:48.
    L. Eoc.; As.[1]
    [1] S.E. Asia.

Tribe †**Diceratheriini** Dollo, 1885:295, **new rank**.
    [= †Diceratheriinae Dollo, 1885:295.] [Including †Trigoniadini Heissig, in Prothero &
    Schoch, eds., 1989:405, 406; †Caenopini Kalandadze & Rautian, 1992:116;
    †Menoceratinae Prothero, Manning & Hanson, 1986:348, 349.]
    M. Eoc.-E. Olig., E.-L. Mioc., ?E. Plioc.; As. M. Eoc.-L. Mioc., ?E. Plioc.; Eu. L.
    Eoc.-M. Mioc.; N.A. E. Mioc.; Cent. A.

Subtribe †**Caenopina** Cope, 1887:992, **new rank**.
    [= †Caenopidae Cope, 1887:992; †Caenopinae Breuning, 1924:16; †Caenopini
    Kalandadze & Rautian, 1992:116.] [Including †Ronzotheriinae Kretzoi, 1940:87-98;
    †Trigoniadini Heissig, in Prothero & Schoch, eds., 1989:405, 406.[1]]
    M. Eoc.-E. Olig.; As. M. Eoc.-L. Olig.; Eu. L. Eoc.-E. Olig., E. Mioc.; N.A.
    [1] Polyphyletic or paraphyletic as described.

†*Prohyracodon* Koch, 1897:481-490.
    [= or including †*Meninatherium* Abel, 1910:26.]
    M. Eoc.-E. Olig.; As.[1] M.-L. Eoc.; Eu.[2]
    [1] E. Asia.
    [2] E. Europe.

†*Trigonias* Lucas, 1900:221.
    L. Eoc.; N.A.

†*Amphicaenopus* H. E. Wood, 2nd, 1927:72.
    L. Eoc.-E. Olig.; N.A.

†*Subhyracodon* Brandt, 1878:30-32.
    [= or including †*Anchisodon* Cope, 1879:233; †*Anchirodon* Forbes, 1881:19.]
    [Including †*Caenopus* Cope, 1880:611; †*Leptaceratherium* Osborn, 1898:132.]
    L. Eoc.-E. Olig.; N.A.

†*Epiaceratherium* Abel, 1910:20.
    E.-L. Olig.; Eu.

†*Ronzotherium* Aymard, 1854:676.
    [= †*Badactherium* Croizet, 1842:49.[1]] [Including †*Paracaenopus* Breuning, 1923:120;
    †*Symphysorrhachis* Belyaeva, 1954:192, 195; †*Tongriceros* Misonne, 1957:9.]
    E. Olig.; As. E.-L. Olig.; Eu.
    [1] *Nomen nudum.*

†*Moschoedestes* Stevens, 1969:23.
    E. Mioc.; N.A.

Subtribe †**Diceratheriina** Dollo, 1885:295, **new rank**.

[= †Diceratheriinae Dollo, 1885:295.] [Including †Menoceratinae Prothero, Manning & Hanson, 1986:348-349.]

E. Olig.-L. Mioc., ?E. Plioc.; Eu. L. Olig.-M. Mioc.; N.A. E.-L. Mioc., ?E. Plioc.; As. E. Mioc.; Cent. A.

†*Protaceratherium* Abel, 1910:10.

[Including †*Proaceratherium* Ginsburg & Hugueney, 1980:274.[1]]

E. Olig.-E. Mioc.; Eu. E. Mioc.; As.

[1] Misspelling.

†*Diceratherium* Marsh, 1875:242.

[Including †*Metacoenopus* Cook, 1909:245; †*Brachydiceratherium* Lavocat, 1951:113.[1]]

L. Olig.; Eu. L. Olig.-M. Mioc.; N.A. E. Mioc.; Cent. A.

[1] Described as a subgenus.

†*Pleuroceros* Roger, 1898:25, 26.[1]

L. Olig.-L. Mioc., ?E. Plioc.; Eu. E.-L. Mioc., ?E. Plioc.; As.

[1] "As H. E. Wood has noted, this name is not preoccupied by †*Pleuroceras* Hyatt, 1868. The later Tertiary species referred to this genus probably do not belong to it" (Simpson 1945:142).

†*Menoceras* Troxell, 1921:206-207.

E.-M. Mioc.; N.A.

Tribe **Rhinocerotini** Gray, 1821:306.

[= Rhynocerotidae Gray, 1821:306; Rhinocerina Gray, 1825:343[1]; Rhinocerotini Heissig, 1972:16.] [Including Dicerotini Loose, 1975:7; †Aceratheriini Heissig, 1972:59; †Teleoceratini Heissig, 1972:71; †Elasmotherini Heissig, 1972:50; †Elasmotheriini Kalandadze & Rautian, 1992:116; Dicerorhinini Kalandadze & Rautian, 1992:116; †Iranotheriini Kalandadze & Rautian, 1992:116.]

L. Olig.-R.; As. L. Olig.-L. Pleist.; Eu. E. Mioc.-R.; Af. E. Mioc.-E. Plioc.; N.A. ?M. and/or ?L. Mioc.; Mediterranean.[2,3] L. Mioc.; Cent. A. R.; E. Indies.[4]

[1] As tribe.
[2] Crete in the ?middle Miocene.
[3] Sicily in the late Miocene?
[4] Java, Sumatra, Borneo.

Subtribe †**Aceratheriina** Dollo, 1885:295, **new rank**.

[= †Aceratherinae Dollo, 1885:295; †Aceratherini Heissig, 1972:59; †Aceratheriini Kalandadze & Rautian, 1992:115; †Turkanatheriinae Deraniyagala, 1951:24.]

L. Olig.-L. Plioc., ?Pleist.; As. L. Olig.-E. Plioc.; Eu. E.-L. Mioc.; Af. E. Mioc.-E. Plioc.; N.A. ?M. Mioc.; Mediterranean.[1]

[1] Crete.

†*Aceratherium* Kaup, 1832:898-904.[1]

[= †*Turkanatherium* Deraniyagala, 1951:24.] [Including †*Mesaceratherium* Heissig, 1969:90[2]; †*Alicornops* Ginsburg & Guérin, 1979:114.[2]]

L. Olig.-L. Mioc., ?E. Plioc., ?Pleist.; As. L. Olig.-L. Mioc.; Eu. E.-L. Mioc.; Af. ?M. Mioc.; Mediterranean.[3]

[1] Proposed as subgenus of *Rhinoceros*.
[2] Proposed as subgenus of †*Aceratherium*.
[3] Crete.

†*Subchilotherium* Heissig, 1972.

M.-L. Mioc.; As.

†*Chilotherium* Ringström, 1924:26.

E. Mioc.-L. Plioc.; As. L. Mioc.-E. Plioc.; Eu.

†*Chilotheridium* Hooijer, 1971:342.

E.-M. Mioc.; Af.

†*Acerorhinus* Kretzoi, 1942:311.

Mioc.; Eu. L. Mioc.; As.

†*Floridaceras* H. E. Wood, 2nd, 1964:364.

E. Mioc.; N.A.

†*Plesiaceratherium* Young, 1937:214.
> [= †*Dromoceratherium* Crusafont Pairó, de Villalta Comella & Truyols, 1955:152-156.[1]]
> E.-L. Mioc.; Eu. M.-L. Mioc.; As.
> [1] *Nomen nudum* in de Villalta Comella, 1954:107-114.

†*Diaceratherium* Dietrich, 1931:203.
> L. Olig.-E. Mioc.; Eu.

†*Aphelops* Cope, 1873:1-2.
> M. Mioc.-E. Plioc.; N.A.

†*Hoploaceratherium* Ginsburg & Heissig, 1989:418.
> M. Mioc.; As. M.-L. Mioc.; Eu.

†*Peraceras* Cope, 1880:540.
> M. and/or L. Mioc.; N.A.

Subtribe **Rhinocerotina** Gray, 1821:306.
> [= Rhynocerotidae Gray, 1821:306; Rhinocerina Gray, 1825:343[1]; Rhinocerotina Prothero & Schoch, 1989:536.] [Including †Elasmotherina Bonaparte, 1845:4; †Elasmotheriina Prothero & Schoch, 1989:536; †Elasmotheriini Kalandadze & Rautian, 1992:116; Ceratorhinae Osborn, 1898:121[2]; †Atelodinae Breuning, 1923:120; Dicerorhinae Ringström, 1924:5; Dicerorhinina Prothero & Schoch, 1989:536; Dicerinae Ringström, 1924:97; Dicerotina Prothero & Schoch, 1989:536; †Iranotheriinae Kretzoi, 1942:315; †Iranotheriini Kalandadze & Rautian, 1992:116; †Iranotheriina Kalandadze & Rautian, 1992:117; †Coelodontina Kalandadze & Rautian, 1992:116.]
> L. Olig.-R.; As. L. Olig.-L. Pleist.; Eu. E. Mioc.-R.; Af. M. Mioc.-E. Plioc.; N.A. ?L. Mioc.; Mediterranean.[3] L. Mioc.; Cent. A. R.; E. Indies.[4]
> [1] As tribe.
> [2] Objective synonym of Dicerorhininae Ringstrom.
> [3] Sicily.
> [4] Java, Sumatra, Borneo.

Infratribe †**Teleocerati** Hay, 1902:646, **new rank.**
> [= †Teleoceratinae Hay, 1902:646; †Teleoceratini Heissig, 1972:71.]
> L. Olig.-L. Mioc.; As. E. Mioc.-L. Plioc.; Af. E.-L. Mioc.; Eu. M. Mioc.-E. Plioc.; N.A. L. Mioc.; Cent. A.

†*Aprotodon* Forster Cooper, 1915:408.
> [= or including †*Indotherium* Kretzoi, 1942:315.]
> ?L. Olig., L. Mioc.; As.

†*Brachypotherium* Roger, 1904:13.
> L. Olig.-M. Mioc.; As. E. Mioc.-L. Plioc.; Af. E.-L. Mioc.; Eu. ?M. Mioc.; N.A.[1]
> [1] The North American reference is apparently based on a long-legged rhino that may not be true †*Brachypotherium.*

†*Teleoceras* Hatcher, 1894:149-150.
> [Including †*Mesoceras* Cook, 1930:49.[1]]
> M. Mioc.-E. Plioc.; N.A. L. Mioc.; Cent. A.
> [1] Described as subgenus.

†*Prosantorhinus* Heissig, 1974:37.[1]
> [= †*Brachypodella* Heissig, 1972:65.[2]]
> E.-M. Mioc.; Eu.
> [1] New name for †*Brachypodella* Heissig.
> [2] Not *Brachypodella* Beck, 1837: 89, a genus of gastropods.

Infratribe †**Elasmotherii** Bonaparte, 1845:4, **new rank.**
> [= †Elasmotherina Bonaparte, 1845:4; †Elasmotheriina Prothero & Schoch, 1989:536; †Elasmotheriini Kalandadze & Rautian, 1992:116.] [Including †Iranotheriinae Kretzoi, 1942:237, 315; †Iranotheriini Kalandadze & Rautian, 1992:116; †Iranotheriina Kalandadze & Rautian, 1992:117; †Begertheriinae Belyaeva, 1971:80.]
> E. Mioc.-E. Plioc., E.-M. Pleist.; As. M. Mioc., E. Pleist.; Eu. L. Mioc.; Af.

†*Shennongtherium* Huang & Yan, 1983:223.
>   Mioc.; As.[1]
>   [1] China.

†*Hispanotherium* Crusafont Pairó & Villalta, 1947:867.
>   [Including †*Begertherium* Beliajeva, 1971:80.]
>   M.-L. Mioc.; As.[1] M. Mioc.; Eu.
>   [1] W. Asia (Turkey).

†*Caementodon* Heissig, 1972:50.
>   M.-L. Mioc.; As.

†*Beliajevina* Heissig, 1974:27.
>   L. Mioc.; As.[1]
>   [1] W. Asia.

†*Tesselodon* Yan, 1979:189-199.
>   Mioc.; As.[1]
>   [1] E. Asia (Hubei).

†*Ninxiatherium* Chen, 1977:143.
>   L. Mioc.; As.[1]
>   [1] E. Asia.

†*Iranotherium* Ringström, 1924:147.
>   [Including †*Gobitherium* Kretzoi, 1943:270.]
>   E.-M. Mioc., ?E. Plioc.; As.

†*Kenyatherium* Aguirre & Guérin, 1974:232.
>   L. Mioc.; Af.

†*Sinotherium* Ringström, 1922.
>   [Including †*Parelasmotherium* Killgus, 1923.]
>   L. Mioc.-E. Plioc.; As.

†*Elasmotherium* Fischer de Waldheim, 1808:23-28.
>   [= †*Stereoceros* Duvernoy, 1853:109.]
>   E.-M. Pleist.; As. E. Pleist.; Eu.

Infratribe **Rhinoceroti** Gray, 1821:306, **new rank**.
>   [= Rhynocerotidae Gray, 1821:306; Rhinocerotina Prothero & Schoch, 1989:536.]
>   [Including Ceratorhinae Osborn, 1898:121; †Atelodinae Breuning, 1923:120;
>   Dicerorhinae Ringström, 1924:5; Dicerorhinina Groves, 1983:312;
>   Dicerinae Ringström, 1924:97; Dicerina Kalandadze & Rautian, 1992:116;
>   Dicerotina Prothero & Schoch, 1989:536; †Lartetotheriina Groves, 1983:312;
>   †Coelodontina Kalandadze & Rautian, 1992:116.]
>   L. Olig.-R.; As. L. Olig.-L. Pleist.; Eu. E. Mioc.-R.; Af.
>   ?L. Mioc.; Mediterranean.[1] R.; E. Indies.[2]
>   [1] Sicily.
>   [2] Java, Sumatra, Borneo.

Dicerorhinus Gloger, 1841:125. Steppe rhinoceros, Asiatic two-horned rhinoceros,
>   Sumatran rhinoceros.
>   [= *Didermocerus* Brookes, 1828:75[1]; *Ceratorhinus* Gray, 1867:1021.]
>   [Including †*Dihoplus* Brandt, 1878:48-51; †*Procerorhinus* Kretzoi, 1942:314;
>   †*Stephanorhinus* Kretzoi, 1942:348; †*Lartetotherium* Ginsburg, 1974:597.]
>   L. Olig.-R.; As.[2] L. Olig.-L. Pleist.; Eu. E.-L. Mioc., L. Plioc., L. Pleist.; Af.
>   R.; E. Indies.[3]
>   [1] Rejected by the International Commission on Zoological Nomenclature, Opinion 1080, 1977.
>   [2] Extant in S.E. Asia.
>   [3] Sumatra, Borneo.

†*Coelodonta* Bronn, 1831:51-61. Woolly rhinoceros.
>   [= †*Gryphus* Schubert, 1823:718[1]; †*Tichorhinus* Brandt, 1849:393;
>   †*Coelorhinus* Frech, 1904: table 10[2].] [= or including †*Hysterotherium*
>   Giebel, 1847:54, 456; †*Atelodus* Pomel, 1853:114.[3]]

L. Plioc.-R.; As. M.-L. Pleist.; Eu.

[1] Preoccupied by *Gryphus* Humphries, 1797, a genus of mollusks. Also used by Brisson, 1760, for a genus of birds.
[2] A *lapsus.*
[3] Proposed as subgenus of *Rhinoceros.*

†*Gaindatherium* Colbert, 1934:1.
E. and/or M. Mioc.; Eu. M.-L. Mioc.; As.

Rhinoceros Linnaeus, 1758:56. One-horned rhinoceros, Indian rhinoceros, Javanese rhinoceros.
[= *Eurhinoceros* Gray, 1867:1009-1015; *Naricornis* Frisch, 1775: table.[1]] [= or including †*Shansirhinus* Kretzoi, 1942:311; †*Monocerorhinus* Wust, 1922:654.] [Including †*Sinorhinus* Kretzoi, 1942:317.]
?L. Mioc.; Af. E. Plioc.-R.; As. R.; E. Indies.[2]

[1] Substitute name. Invalid because not published in a consistently binominal work.
[2] Sumatra (now extinct) and Java.

†*Punjabitherium* Kahn, 1971:105-109.
L. Plioc.; As.[1]

[1] S. Asia (India).

†*Paradiceros* Hooijer, 1968:78.
M.-L. Mioc.; Af.

Ceratotherium Gray, 1868:1027. African white rhinoceros.
L. Mioc.-R.; Af. L. Mioc.; As.[1] ?L. Mioc.; Mediterranean.[2] L. Mioc.; Eu.

[1] W. Asia.
[2] Sicily.

Diceros Gray, 1821:306. African black rhinoceros.
[= *Opsiceros* Gloger, 1841:xxxii, 125; *Rhinaster* Gray, in Gerrard, 1862:282[1]; *Keitloa* Gray, 1867:1025; †*Colobognathus* Brandt, 1878:51.]
?L. Mioc., E. Plioc.-R.; Af.

[1] Not *Rhinaster* Wagler, 1830, a condylurine mole.

Subfamily †**Amynodontinae** Scott & Osborn, 1883:4, 12.
[= †Amynodontidae Scott & Osborn, 1883:4, 12; †Amynodontinae Kretzoi, 1942:144.] [Including †Cadurcotheriinae Kretzoi, 1942:144; †Rostriamynodontinae Wall & Manning, 1986:911; †Cadurcodontinae Prothero & Schoch, 1989:534.]
M. Eoc.-E. Olig., ?L. Olig.; As. M. Eoc.-E. Olig.; N.A. L. Eoc., ?E. Olig., L. Olig., ?E. Mioc.; Eu.

†*Rostriamynodon* Wall & Manning, 1986:911.
M. Eoc.; As.

†*Gigantamynodon* Gromova, 1954:161.
L. Eoc.-E. Olig., ?L. Olig.; As.

†*Teilhardia* Matthew & Granger, 1926:4.[1]
M. Eoc.; As.

[1] Assignment to †Amynodontinae is questionable.

†*Caenolophus* Matthew & Granger, 1925:6.[1]
L. Eoc.; As.

[1] Assignment to †Amynodontinae is questionable.

†*Hypsamynodon* Gromova, 1954:165.
E. Olig.; As.

Tribe †**Amynodontini** Scott & Osborn, 1883:4, 12.
[= †Amynodontidae Scott & Osborn, 1883:4, 12; †Amynodontini Wall, 1982:564.]
M. Eoc.-E. Olig.; As. M.-L. Eoc.; N.A.

†*Amynodontopsis* Stock, 1933:762.
M. Eoc.; N.A. L. Eoc.; As.[1]

[1] E. Asia (Inner Mongolia).

†*Lushiamynodon* Chow & Xu, 1965:190.
M.-L. Eoc.; As.

†*Amynodon* Marsh, 1877:251-252.
    [= †*Orthocynodon* Scott & Osborn, 1882:223-225.]
    M. Eoc.-E. Olig.; As. M. Eoc.; N.A.

†*Euryodon* Xu, Yan, Zhou, Han & Zhang, 1979:428.
    M. Eoc.; As.[1]
    [1] E. Asia.

†*Armania* Gabuniya & Dashzeveg, 1988:244.
    L. Eoc.; As.

†*Sharamynodon* Kretzoi, 1942:145.
    L. Eoc.; As.

†*Mesamynodon* Peterson, 1931:71.
    M. Eoc.; N.A.

†*Toxotherium* H. E. Wood, 2nd, 1961:1.
    M.-L. Eoc.; N.A.

†*Penetrigonias* Tanner & Martin, 1976:212.
    L. Eoc.; N.A.

Tribe †**Cadurcotheriini** Kretzoi, 1942:144, **new rank**.
    [= †Cadurcotheriinae Kretzoi, 1942:144.] [Including †Cadurcodontini Kalandadze & Rautian, 1992:115.]
    M.-L. Eoc., Olig.; As. M. Eoc.-E. Olig.; N.A. L. Eoc., ?E. Olig., L. Olig., ?E. Mioc.; Eu.

Subtribe †**Cadurcotheriina** Kretzoi, 1942:144, **new rank**.
    [= †Cadurcotheriinae Kretzoi, 1942:144.] [Including †Cadurcodontini Wall, 1982:564; †Cadurcodontinae Prothero & Schoch, 1989:534.]
    L. Eoc., Olig.; As. L. Eoc., ?E. Olig., L. Olig., ?E. Mioc.; Eu. L. Eoc. and/or E. Olig.; N.A.

†*Cadurcodon* Kretzoi, 1942:146.
    [Including †*Paracadurcodon* Xu, 1966:145.]
    L. Eoc., Olig.; As. L. Eoc., L. Olig.; Eu.[1]
    [1] E. Europe.

†*Procadurcodon* Gromova, 1960:129.
    L. Eoc. and/or E. Olig.; As.

†*Sianodon* Xu, 1965:83.
    L. Eoc. and/or E. Olig.; As.

†*Cadurcopsis* Kretzoi, 1942:140.
    L. Eoc. and/or E. Olig.; N.A.

†*Cadurcotherium* Gervais, 1873:106.
    Olig.; As. Olig., ?E. Mioc.; Eu.

†*Cadurcamynodon* Kretzoi, 1942:146.
    L. Olig.; Eu.

Subtribe †**Metamynodontina** Kretzoi, 1942:145, **new rank**.
    [= †Metamynodontinae Kretzoi, 1942:145; †Metamynodontini Wall, 1982:564.] [Including †Paramynodontinae Kretzoi, 1942:145.]
    M.-L. Eoc., Olig.; As. M. Eoc.-E. Olig.; N.A.

†*Metamynodon* Scott & Osborn, 1887:165-169.
    M. Eoc.-E. Olig.; N.A. Olig.; As.

†*Megalamynodon* H. E. Wood, 2nd, in Scott, 1945:250.
    M. Eoc.; N.A.

†*Paramynodon* Matthew, 1929:512.
    M. Eoc.; As.[1]
    [1] S.E. Asia (Burma).

†*Zaisanamynodon* Belyaeva, 1971:39-61.
　　　　L. Eoc., ?Olig.; As.[1,2]
　　　　[1] Kazakhstan in the late Eocene.
　　　　[2] Mongolia in the Oligocene.

Superfamily **Tapiroidea** Gray, 1825:343.
　　　　[= Tapirina Gray, 1825:343; Tapirina Van der Hoeven, 1858; Tapiroidea Gill, 1872:12.] [Including †Lophiodontoidea Gill, 1872:83.]

†*Sastrilophus* Sahni & Khare, 1971:45.
　　　　M. and/or L. Eoc.; As.[1]
　　　　[1] S. Asia (India).

†*Indolophus* Pilgrim, 1925:22.
　　　　M. Eoc.; As.[1]
　　　　[1] S.E. Asia (Burma).

Family †**Helaletidae** Osborn, 1892:127.
　　　　[= †Helaletinae Wortman & Earle, 1893:173.] [Including †Colodontinae Wortman & Earle, 1893:173.]
　　　　E. Eoc.-E. Olig., ?L. Olig.; As. E. Eoc.; Eu. E. Eoc.-E. Olig.; N.A.

†*Cymbalophus* Hooker, 1984:230.
　　　　?E. Eoc.; As. E. Eoc.; Eu.

†*Heptodon* Cope, 1882:1029.
　　　　E.-M. Eoc.; As.[1] E. Eoc.; N.A.
　　　　[1] E. Asia.

†*Selenaletes* Radinsky, 1966:740.
　　　　E. Eoc.; N.A.

†*Helaletes* Marsh, 1872:218.
　　　　[Including †*Desmatotherium* Scott, 1883:46; †*Chasmotheroides* H. E. Wood, 2nd, 1934:187.]
　　　　M.-L. Eoc.; As. M. Eoc.; N.A.

†*Colodon* Marsh, 1890:524.
　　　　[= †*Mesotapirus* Osborn, 1889:470, 524.] [Including †*Paracolodon* Matthew & Granger, 1925:4.]
　　　　?M. Eoc., L. Eoc.-E. Olig., ?L. Olig.; As. M. Eoc.-E. Olig.; N.A.

†*Plesiocolopirus* Schoch, 1989:306-7.
　　　　L. Eoc. and/or E. Olig.; N.A.

Family †**Isectolophidae** Peterson, 1919:115.
　　　　[= †Isectolophinae Peterson, 1919:116; †Isectolophomorpha Prothero & Schoch, 1989:532.[1]] [Including †Homogalaxinae Peterson, 1919:116.]
　　　　E.-M. Eoc.; As. E.-M. Eoc.; N.A.
　　　　[1] Proposed as subinfraorder of infraorder MOROPOMORPHA (PERISSODACTYLA, MESAXONIA).

†*Homogalax* Hay, 1899:593.[1]
　　　　E.-M. Eoc.; As.[2] E. Eoc.; N.A.
　　　　[1] "In the older literature this genus is called †*Systemodon* Cope, 1881, but the type of †*Systemodon* belongs to †*Hyracotherium*" (Simpson 1945:140).
　　　　[2] E. Asia.

†*Orientolophus* Ding, 1993:202.[1]
　　　　E. Eoc.; As.
　　　　[1] (Ting).

†*Cardiolophus* Gingerich, 1991:194.
　　　　E. Eoc.; N.A.

†*Isectolophus* Scott & Osborn, 1887:260.
　　　　[Including †*Parisectolophus* Peterson, 1919:121; †*Schizolophodon* Peterson, 1919:122.]
　　　　M. Eoc.; N.A.

Family †**Lophiodontidae** Gill, 1872:12.
    [= †Lophiodonta Haeckel, 1866:clviii.[1]]
    E.-L. Eoc.; Eu. M. Eoc.; As.[2]
    [1] Haeckel's term should be suppressed. See the International Code of Zoological Nomenclature, Article 23
    (b).
    [2] S. Asia.

Subfamily †**Lophiodontinae** Gill, 1872:12.
    [= †Lophiodonta Haeckel, 1866:clviii[1]; †Lophiodontidae Gill, 1872:12;
    †Lophiodontinae Viret, in Piveteau, 1958:464.]
    E.-L. Eoc.; Eu.
    [1] Haeckel's term should be suppressed.

†*Lophiodon* Cuvier, 1822:161.
    [= †*Hypsolophiodon* Kretzoi, 1940:88-98.]
    E.-M. Eoc.; Eu.

†*Atalonodon* Dal Piaz, 1929:5.
    M. Eoc.; Eu.

Subfamily †**Chasmotheriinae** Viret, 1958:466.
    E.-L. Eoc.; Eu. M. Eoc.; As.[1]
    [1] S. Asia (India).

†*Chasmotherium* Rütimeyer, 1862:63-67.
    E.-L. Eoc.; Eu. M. Eoc.; As.[1]
    [1] S. Asia (India).

Subfamily †**Paralophiodontinae** Dedieu, 1977:2221.
    M. Eoc.; Eu.

†*Paralophiodon* Dedieu, 1977:2221.[1]
    [= †*Rhinocerolophiodon* Fischer, 1977:910.[2]]
    M. Eoc.; Eu.
    [1] June 1977.
    [2] July 1977.

Family †**Deperetellidae** Radinsky, 1965:214.
    M. Eoc.-E. Olig.; As. E. Olig.; Eu.

†*Teleolophus* Matthew & Granger, 1925:3.
    [Including †*Pachylophus* Tong & Lei, 1984:275.[1]]
    M. Eoc.-E. Olig.; As.
    [1] (Tung & Lei).

†*Deperetella* Matthew & Granger, 1925:4.
    [Including †*Cristidentinus* Zdansky, 1930:32.]
    L. Eoc.; As.

†*Diplolophodon* Zdansky, 1930:35.
    M.-L. Eoc.; As.

†*Haagella* Heissig, 1978:266.
    E. Olig.; Eu.

Family †**Lophialetidae** Matthew & Granger, 1925:7.
    [= †Lophialetinae Matthew & Granger, 1925:7; †Lophialetidae Radinsky, 1965:188;
    †Lophialetini Kalandadze & Rautian, 1992:113.] [Including †Breviodontinae Reshetov,
    1975:19-53.]
    M.-L. Eoc.; As.

†*Kalakotia* Ranga Rao, 1972:7.
    [= †*Aulaxolophus* Ranga Rao, 1972:13.]
    M. Eoc.; As.[1]
    [1] S. Asia (India).

†*Lophialetes* Matthew & Granger, 1925:5.
    M.-L. Eoc.; As.

†*Parabreviodon* Reshetov, 1975:19-53.
    M. Eoc.; As.

†*Schlosseria* Matthew & Granger, 1926:3.
    M.-L. Eoc.; As.

†*Breviodon* Radinsky, 1965:203.
    M.-L. Eoc.; As.

†*Simplaletes* Qi, 1980:215.
    M.-L. Eoc.; As.[1]
    [1] E. Asia.

†*Eoletes* Biryukov, 1974:57.
    M. Eoc.; As.

Family **Tapiridae** Gray, 1821:306. Tapirs.
    [= Nasuta Illiger, 1811:98[1]; Taperidae Gray, 1821:306; Tapirina Gray, 1825:343[2]; Tapiridae Burnett, 1830:352; Tapirida Haeckel, 1866:clviii.[3]] [Including Elasmognathinae Gray, 1867:885; †Tapiravini Schoch, 1984:3; †Miotapirini Schoch, 1984:8.]
    E. Olig.-L. Plioc., Pleist.; Eu. E. Olig.-L. Pleist.; N.A. M. Mioc.-R.; As. ?E. Plioc., L. Plioc.-R.; S.A. Pleist., R.; Cent. A. R.; E. Indies.[4]

[1] Used as a family but not based on a generic name.
[2] As tribe.
[3] As family.
[4] Sumatra.

†*Eotapirus* Cerdeno & Ginsburg, 1988:84.
    Olig., E. Mioc.; Eu.

†*Protapirus* Filhol, 1877:131-135.
    [Including †*Tanyops* Marsh, 1894:348.]
    E. Olig.-M. Mioc.; Eu. E. Olig.-E. Mioc.; N.A.

†*Palaeotapirus* Filhol, 1888:55-58.
    [Including †*Paratapirus* Depéret & Douxami, 1902:34.]
    E. Mioc.; Eu.

†*Miotapirus* Schlaikjer, 1937:231.
    E. Mioc.; N.A.

†*Plesiotapirus* Qiu, Yan & Sun, 1991:119.
    M. Mioc.; As.

Tapirus Brunnich, 1772:32, 44-45. Tapirs, Brazilian tapir, woolly tapir, mountain tapir, pinchaque, danta, huagra, Asiatic tapir.
    [= *Tapirus* Brisson, 1762:81-82[1]; *Tapirussa* Frisch, 1775:4[2]; *Syspotamus* Billberg, 1828:before 1[3]; *Rhinochoerus* Wagler, 1830:17[4]; *Tapyra* Liais, 1872:397.] [= or including *Tapir* Gmelin, 1778; †*Tapiralum* Rusconi, 1934:9.[5]] [Including *Tapir* Blumenbach, 1779:129; *Elasmognathus* Gill, 1865:183[6]; *Cinchacus* Gray, 1873:34[7]; *Pinchacus* Hershkovitz, 1954:469[8]; *Tapirella* Palmer, 1903:873[9]; *Acrocodia* Goldman, 1913:65.[10]]
    L. Mioc.-R.; As. L. Mioc.-L. Plioc., Pleist.; Eu. L. Mioc.-L. Pleist.; N.A. ?E. Plioc., L. Plioc.-R.; S.A. Pleist., R.; Cent. A. R.; E. Indies.[11]

[1] Brisson's paper was not consistently binominal. However, see the International Code of Zoological Nomenclature, Article 11(c)(i) for confusion.
[2] Unavailable.
[3] Replacement name for *Tapir* Gmelin, 1778.
[4] Replacement name for *Tapirus* Brisson, 1762, invalid.
[5] Described as subgenus of *Tapirus*.
[6] Not *Elasmognathus* Fieber, 1844, an hemipteran.
[7] Proposed as subgenus.
[8] Emendation of *Cinchacus* Gray. As subgenus.
[9] Replacement name for *Elasmognathus* Gill, 1865. Sometimes as subgenus.
[10] Sometimes as subgenus.
[11] Sumatra.

†*Tapiravus* Marsh, 1877:252.
    L. Mioc.; N.A.

†*Tapiriscus* Kretzoi, 1951:393, 411.
L. Mioc.; Eu.[1]
[1] E. Europe.

†*Megatapirus* Matthew & Granger, 1923:588.[1]
M. Pleist.-R.; As.
[1] Described as subgenus of *Tapirus*.

Order **URANOTHERIA, new**.[1]

[= PAENUNGULATA Simpson, 1945:131.[2]]
[1] Definition: for the most recent common ancestor of HYRACOIDEA, †EMBRITHOPODA, and TETHYTHERIA, and all its descendants.
[2] In part. Proposed as a superorder to include †PANTODONTA, †DINOCERATA, and †PYROTHERIA, as well as HYRACOIDEA, †EMBRITHOPODA, SIRENIA, †DESMOSTYLIA (as †DESMOSTYLIFORMES), and PROBOSCIDEA.

Suborder **HYRACOIDEA** Huxley, 1869:101, **new rank**.[1] Hyraxes, dassies.

[= Lamnunguia Illiger, 1811:95[2]; LAMNUNGIA Van der Hoeven, 1858[3]; LAMINUNGULA Gray, 1869:278[4]; HYRACEA Haeckel, 1895:490, 534; PROCAVIATA Imamura, 1961:133; HYRACIFORMES Kinman, 1994:38.]
[Including PROCAVIAMORPHA Whitworth, 1954:54, 55[5]; †PSEUDHIPPOMORPHA Whitworth, 1954:54, 55.[5]]

[1] Proposed as order.
[2] In part. *Nomen oblitum*, although used by Fischer de Waldheim, 1817:373. Proposed as a family but not based on a generic name, it contained two taxa, *Lipura* (a synonym of *Marmota*) and *Hyrax* (a synonym of *Procavia*). We note the priority of LAMNUNGUIA, but we attempt to maintain stability by the continued use of HYRACOIDEA.
[3] As suborder. *Nomen oblitum*, although used by Haeckel, 1866:cxlvii as a family.
[4] Misspelling.
[5] As suborder.

†*Seggeurius* Mahboubi, Ameur, Crochet & Jaeger, 1986:24.
E. Eoc.; Af.[1]
[1] N. Africa.

Family †**Pliohyracidae** Osborn, 1899:172.

E. Eoc.-E. Olig., E.-L. Mioc.; Af. E. Mioc.-E. Pleist.; As. M. Mioc.; Mediterranean.[1] M. Mioc.-E. Plioc.; Eu.
[1] Crete.

Subfamily †**Saghatheriinae** Andrews, 1906:86.

[= †Saghatheriidae Andrews, 1906:86; †Saghatheriinae Whitworth, 1954:40.]
[Including †Titanohyracidae Matsumoto, 1926:259.]
E. Eoc.-E. Olig., E.-M. Mioc.; Af. M. Mioc.; As.[1]
[1] S.W. Asia (Arabian Peninsula).

†*Pachyhyrax* Schlosser, 1910:502.
L. Eoc.-E. Olig., E.-M. Mioc.; Af. M. Mioc.; As.[1]
[1] S.W. Asia (Saudi Arabia).

†*Bunohyrax* Schlosser, 1910:502.
?M. Eoc., L. Eoc.; Af.

†*Titanohyrax* Matsumoto, 1921:844.
E.-L. Eoc.; Af.[1]
[1] N. Africa.

†*Microhyrax* Sudre, 1979:95.
M. Eoc.; Af.[1]
[1] N. Africa.

†*Megalohyrax* Andrews, 1903:341.
[Including †*Mixohyrax* Schlosser, 1910:502.]
M.-L. Eoc., ?M. Mioc.; Af.

†*Saghatherium* Andrews & Beadnell, 1902:5.
L. Eoc.; Af.[1]
[1] N. Africa.

†*Selenohyrax* Rasmussen & Simons, 1988:76.
  L. Eoc.; Af.[1]
   [1] N. Africa (Egypt).

†*Thyrohyrax* Meyer, 1973:3.
  L. Eoc.; Af.[1]
   [1] N. Africa.

Subfamily †**Geniohyinae** Andrews, 1906:193.
  [= †Geniohyidae Matsumoto, 1926:259.]
  L. Eoc.; Af. E. Mioc.; As.

†*Geniohyus* Andrews, 1904:160.
  L. Eoc.; Af. E. Mioc.; As.

Subfamily †**Pliohyracinae** Osborn, 1899:172.
  [= †Pliohyracidae Osborn, 1899:172; †Pliohyracinae Whitworth, 1954:54, 55.]
  E.-L. Mioc.; Af. M. Mioc.-E. Pleist.; As. M. Mioc.; Mediterranean.[1] M. Mioc.-E.
  Plioc.; Eu.
   [1] Crete.

†*Meroehyrax* Whitworth, 1954:40.
  E. Mioc.; Af.[1]
   [1] E. Africa.

†*Parapliohyrax* Lavocat, 1961:87.
  M.-L. Mioc.; Af.

†*Pliohyrax* Osborn, 1899:172-173.
  [= †*Leptodon* Gaudry, 1860:927-929.[1]] [Including †*Neoschizotherium* Viret,
  1947:354[2]; †*Neoschizotherium* Viret & Mazenot, 1948:17-59.]
  M.-L. Mioc.; As. M. Mioc.; Mediterranean.[3] M. Mioc.-E. Plioc.; Eu.
   [1] Not †*Leptodon* Sundevall, 1835, a bird.
   [2] *Nomen nudum.* Objective synonym of †*Neoschizotherium* Viret & Mazenot.
   [3] Crete.

†*Sogdohyrax* Dubrovo, 1978:99.
  L. Mioc.; As.

†*Postschizotherium* G. H. R. von Koenigswald, 1932:21-22.
  ?L. Mioc., E. Plioc.-E. Pleist.; As.

†*Kvabebihyrax* Gabunia & Vekua, 1966:643.
  E. Plioc.; As.[1]
   [1] W. Asia.

Family **Procaviidae** Thomas, 1892:51.
  [= Hyracidae Gray, 1821:306[1]; Hyracida Haeckel, 1866:clix[2]; Procaviinae Whitworth,
  1954:54; Procavioidea Kalandadze & Rautian, 1992:106.]
  E. Mioc.-R.; Af. L. Pleist.-R.; As.[3]
   [1] Invalid.
   [2] As a family.
   [3] S.W. Asia.

†*Prohyrax* Stromer, 1926.
  E.-M. Mioc.; Af.

Procavia Storr, 1780:40, table B. Rock hyraxes, cape hyrax, shafan.
  [= *Procauia* Storr, 1780: Table B; *Hyrax* Hermann, 1783:115; *Euhyrax* Gray, 1868:46.]
  E. Plioc.-R.; Af. L. Pleist.-R.; As.[1]
   [1] S.W. Asia.

Heterohyrax Gray, 1868:50. Yellow spotted hyraxes, gray hyraxes.
  L. Mioc., L. Plioc., M. and/or L. Pleist., R.; Af.

†*Gigantohyrax* Kitching, 1965:91.
  L. Plioc.; Af.

Dendrohyrax Gray, 1868:48. Tree hyraxes.
  M. and/or L. Pleist., R.; Af.

Suborder †**EMBRITHOPODA** Andrews, 1906:224, **new rank**.[1]
> [= †BARYPODA Andrews, 1904:481[2]; †ARSINOITHERIA Schlosser, 1911:155; †EMBRITHOPODIFORMES Kinman, 1994:38.]
> [1] Proposed as an order.
> [2] Proposed as an order. Not order †BARYPODA Haeckel, 1866: clvii, an agglomeration of †*Stereognathus*, †*Pliolophus*, †*Nototherium*, and †*Diprotodon*.

Family †**Phenacolophidae** South China Red Bed Team, 1977:235.
> [= †Phenacolophidae Van Valen, 1978[1]; †Phenacolophidae Zhang, 1978:267.[2]]
> ?E. Paleoc., L. Paleoc.; As.
> [1] December 1978.
> [2] (Chang).

†*Phenacolophus* Matthew & Granger, 1925:8.
> [= †*Procoryphodon* Flerov, 1957:73; †*Tienshanilophus* Tong, 1978:84[1]; †*Yuelophus* Zhang, 1978:274[2]; †*Ganilophus* Zhang, 1978:269[3]; †*Ganolophus* Zhang, 1979:373.[2]]
> ?E. Paleoc., L. Paleoc.; As.
> [1] (Tung).
> [2] (Chang).
> [3] (Chang). *Nomen nudum*.

†*Minchenella* Zhang, 1980:257.[1]
> [= †*Conolophus* Zhang, 1978:268.[2]]
> L. Paleoc.; As.
> [1] (Chang).
> [2] (Chang). Not *Conolophus* Fitzinger, 1843, an iguana.

Family †**Arsinoitheriidae** Andrews, 1904:160.
> M. Eoc., ?L. Eoc.; As.[1] L. Eoc.; Af. L. Eoc. and/or E. Olig.; Eu.[2]
> [1] W. Asia.
> [2] E. Europe.

Subfamily †**Palaeoamasiinae** Sen & Heintz, 1979:76.[1]
> [= †Palaeoamasinae Sen & Heintz, 1979:76.]
> M. Eoc.; As.[2] L. Eoc. and/or E. Olig.; Eu.[3]
> [1] Corrected spelling.
> [2] W. Asia.
> [3] E. Europe.

†*Palaeoamasia* Ozansoy, 1966:41.
> M. Eoc.; As.[1]
> [1] W. Asia (Turkey).

†*Crivadiatherium* Radulesco, Iliesco & Iliesco, 1976:694.[1]
> L. Eoc. and/or E. Olig.; Eu.[2]
> [1] (Rădulescu).
> [2] E. Europe.

Subfamily †**Arsinoitheriinae** Andrews, 1904:160, **new rank**.
> [= †Arsinoitheriidae Andrews, 1904:160.]
> L. Eoc.; Af. ?L. Eoc.; As.[1]
> [1] S.W. Asia (Arabian Peninsula).

†*Arsinoitherium* Beadnell, 1902:494-495.
> L. Eoc.; Af. ?L. Eoc.; As.[1]
> [1] S.W. Asia (Oman).

Suborder **TETHYTHERIA** McKenna, 1975:42, **new rank**.[1]
> [= BRUTA Linnaeus, 1758:33[2]; Gravigrades de Blainville, 1844.[3]] [Including HETEROGNATHA Gray, 1869:358.[4]]
> [1] Proposed as mirorder.
> [2] In part. Linnaeus' BRUTA was based upon *Elephas*, *Trichechus*, *Bradypus*, and *Myrmecophaga*.
> [3] 1839-1864, probably 1844.
> [4] Long forgotten.

Infraorder **SIRENIA** Illiger, 1811:140, **new rank**.[1]

> [= HERBIVORAE Gray, 1821:309; SIRENOIDEA van Beneden, 1855[2];
> PHYCOCETA Haeckel, 1866:clix[3]; HALOBIOIDEA Ameghino, 1889:652[4];
> TRICHECHIFORMES Hay, 1923:109[5]; SIRENIFORMES Kinman, 1994:38.]

[1] Proposed as a family of order NATANTIA; used as an order by Gill, 1872:13.
[2] Date uncertain. See Agassiz (1859:360).
[3] As a suborder of CETACEA.
[4] Proposed at an unspecified superordinal level, but containing only the SIRENIA and a hypothetical order †PROSIRENIA.
[5] Proposed as suborder.

†*Florentinoameghinia* Simpson, 1932:18.[1]

> E. Eoc.; S.A.

[1] Assignment to SIRENIA is questionable.

†*Lophiodolodus* Stirton, 1947:332.[1]

> E. Olig.; S.A.

[1] Compared with didolodont condylarths by Stirton, who also noted a similarity of the type specimen to a peccary tooth. Assignment to SIRENIA is questionable.

†*Sirenavus* Kretzoi, 1941:147.[1]

> M. Eoc.; Eu.[2]

[1] Possibly referable to †*Halitherium*; not close to †*Prorastomus*.
[2] E. Europe.

†*Anisosiren* Kordos, 1977:313.[1]

> M. Eoc.; Eu.[2]

[1] Possibly referable to Dugongidae.
[2] E. Europe.

†*Prohalicore* Flot, 1887:134-138.[1]

> Plioc.; Eu.

[1] Possibly referable to Dugongidae.

Family †**Prorastomidae** Cope, 1889:876.

> [= †Prorastomatidae Flower & Lydekker, 1891:224[1]; †Prorastominae Reinhart, 1959:5.]
> E.-M. Eoc.; W. Indies.[2] ?M. Eoc., ?L. Eoc.; N.A.[3,4]

[1] Unjustified emendation.
[2] Jamaica.
[3] †Prorastomidae genus et sp. indet., probably from the middle Eocene Inglis Fm., Florida.
[4] †Prorastomidae genus et sp. indet., probably from the late Eocene Crystal River Fm., Florida.

†*Prorastomus* Owen, 1855:543.

> [= †*Prorastoma* Lydekker, 1892:83.[1]]
> E.-M. Eoc.; W. Indies.[2]

[1] Unjustified emendation.
[2] Jamaica.

Family **Dugongidae** Gray, 1821:309. Dugongs.

> [= Halicoridae Gray, 1825:341.] [Including †Rytinadae Gray, 1843:xxiii, 107;
> †Rhytinidae Gill, 1872:14, 91, 92.; †Halitherida Carus, 1868:168; †Halitheriidae Gill, 1872:13; †Hydrodamalidae Palmer, 1895:450; †Eotheroidinae Kretzoi, 1941:154; †Prototheriidae Kretzoi, 1941:155.]
> ?E. Eoc., M. Eoc., L. Olig.-E. Mioc., ?M. Mioc., L. Mioc., ?E. Plioc., L. Plioc., ?M. Pleist.; As.[1] E. Eoc.-E. Plioc., ?L. Plioc.; Eu. M. Eoc.-E. Olig., E. Mioc., ?M. Mioc., L. Mioc. and/or E. Plioc.; Af.[2] M. Eoc., ?E. Olig., L. Olig.-E. Pleist., L. Pleist.; N.A. Olig.; Madagascar. L. Olig.-M. Mioc.; W. Indies.[3] Mioc.; Mediterranean.[4] E.-M. Mioc., E. Plioc.; S.A. L. Mioc.; E. Indies.[5] ?L. Mioc., E. Plioc.; Cent. A. E. Plioc.; New Zealand. L. Pleist.-R.; Pacific. R.; Indian O.

[1] A single dugongid vertebra has been reported from the early Eocene of India.
[2] N. Africa.
[3] Greater Antilles.
[4] Mallorca and Sardinia.
[5] Java.

†*Eotheroides* Palmer, 1899:494.

[= †*Eotherium* Owen, 1875:100-105[1]; †*Masrisiren* Kretzoi, 1941:152.] [Including †*Eosiren* Andrews, 1902:293-295; †*Archaeosiren* Abel, 1913.]
E.-M. Eoc., ?L. Eoc.; Eu. M. Eoc.-E. Olig.; Af.[2]

[1] Not †*Eotherium* Leidy, 1853, a titanothere.
[2] N. Africa (Egypt).

†*Sirenotherium* Paula Couto, 1967.

[= or including †*Tracypleurotherium* Dilg, 1909:90.[1]]
E. Mioc.; S.A.[2]

[1] *Nomen nudum.* A possible synonym.
[2] Brazil.

†*Miodugong* Deraniyagala, 1969:97.[1]

Mioc.; As.[2]

[1] Validity dubious.
[2] S. Asia (Ceylon).

Subfamily †**Protosireninae** Sickenberg, 1934:193.

[= †Protosirenidae Sickenberg, 1934:193; †Protosireninae Reinhart, 1959:62.]
M. Eoc.; Af.[1] M. Eoc.; As.[2] M. Eoc.; Eu. M. Eoc.; N.A.[3]

[1] N. Africa.
[2] S. Asia.
[3] Atlantic North America.

†*Protosiren* Abel, 1907:29.[1]

M. Eoc.; Af.[2] M. Eoc.; As.[3] M. Eoc.; Eu.[4] M. Eoc.; N.A.[5]

[1] *Nomen nudum* in Abel, 1904:214 and Abel, 1906:51.
[2] N. Africa (Egypt).
[3] S. Asia (Kutch).
[4] France.
[5] Atlantic North America.

Subfamily †**Halitheriinae** Abel, 1913:358.

[= †Halitherida Carus, 1868:168.] [Including †Prototheriidae Kretzoi, 1941:155.]
M. Eoc.-M. Mioc.; Eu. Olig.; Madagascar. ?E. Olig., L. Olig.; N.A. L. Olig.-E. Mioc., L. Mioc.; As.[1]

[1] †Halitheriinae genus et sp. indet. from the late Miocene Aoso Fm., northern Japan.

†*Prototherium* de Zigno, 1887:731.

[Including †*Mesosiren* Abel, 1906; †*Paraliosiren* Abel, 1906.]
M.-L. Eoc.; Eu.[1]

[1] Italy and Spain.

†*Paralitherium* Kordos, 1977:350.

L. Eoc.; Eu.[1]

[1] E. Europe.

†*Halitherium* Kaup, 1838:319.[1]

[= †*Pugmeodon* Kaup, 1838:319.] [= or including †*Trachytherium* Gervais, 1849:644-645.] [Including †*Manatherium* Hartlaub, 1886:369-378.]
Olig.; Madagascar. E. Olig.-E. Mioc.; Eu. ?E. Olig., L. Olig.; N.A. L. Olig.-E. Mioc.; As.[2]

[1] "'†*Halytherium*' in first citation, but this was probably a typographical error and was changed to †*Halitherium* by the same author in the same volume" (Simpson 1945:135).
[2] S. Asia.

†*Crenatosiren* Domning, 1991:398.

L. Olig.; N.A.

†*Thalattosiren* Sickenberg, 1928.

M. Mioc.; Eu.

Subfamily †**Hydrodamalinae** Palmer, 1895:450.

[= †Rytinadae Gray, 1843:xxiii, 107; †Rhytinea Brandt, 1849:141; †Rhytinida Haeckel, 1866:clix; †Rhytinidae Gill, 1872:14, 91, 92; †Hydrodamalidae Palmer, 1895:450; †Hydrodamalinae Simpson, 1932:424.] [Including †Metaxytheriinae Kretzoi, 1941:155; †Halianassinae Reinhart, 1959:8, 62, 63.]

L. Olig.-M. Mioc.; W. Indies.[1] E. Mioc., L. Mioc. and/or E. Plioc.; Af.[2] E. Mioc., L. Mioc., ?E. Plioc., L. Plioc., ?M. Pleist.; As. Mioc.; Mediterranean.[3] E. Mioc.-E. Plioc., ?L. Plioc.; Eu. E. Mioc.-E. Pleist., L. Pleist.; N.A.[4] E.-M. Mioc., E. Plioc.; S.A. E. Plioc.; New Zealand. L. Pleist.-R.; Pacific.[5]

[1] Greater Antilles.
[2] N. Africa.
[3] Mallorca and Sardinia.
[4] Hydrodamaline rib fragments have been reported from the early Pleistocene Lomita Marl, California.
[5] N. Pacific (Bering Sea).

†*Caribosiren* Reinhart, 1959:8.
L. Olig.; W. Indies.[1]

[1] Puerto Rico.

†*Metaxytherium* Christol, 1840:322-323.
[= or including †*Crassitherium* Van Beneden, 1871:164-171; †*Haplosiren* Kretzoi, 1951:438.] [Including †*Cheirotherium* Bruno, 1839:143-160[1]; †*Felsinotherium* Capellini, 1865:281-283; †*Halianassa* Studer, 1887:1-20[2]; †*Halysiren* Kretzoi, 1941:153.]
E. Mioc., L. Mioc. and/or E. Plioc.; Af.[3] E. Mioc.; As.[4] Mioc.; Mediterranean.[5] E. Mioc.-E. Plioc., ?L. Plioc.; Eu. ?E. Mioc., M.-L. Mioc.; N.A.[6] E.-M. Mioc.; W. Indies.[7,8] E.-M. Mioc., E. Plioc.; S.A. E. Plioc.; New Zealand.

[1] Not †*Cheirotherium* Kaup, 1835, a reptile.
[2] The type species of "†*Halianassa*" Meyer, 1838, is a *nomen nudum*. Studer's is the first valid usage.
[3] N. Africa.
[4] S. Asia.
[5] Mallorca and Sardinia.
[6] Atlantic North America.
[7] Puerto Rico in the E. Mioc.
[8] Cuba in the M. Mioc.

†*Dusisiren* Domning, 1978:13.
E.-L. Mioc.; N.A.[1] L. Mioc.; As.[2]

[1] Pacific North America.
[2] Japan.

†*Hesperosiren* Simpson, 1932:426.
E. and/or M. Mioc.; N.A.[1]

[1] Atlantic North America.

†*Hydrodamalis* Retzius, 1794:292. Steller's sea cow.
[= †*Manati* Zimmermann, 1780:426[1]; †*Sirene* Link, 1794:67; †*Rytina* Illiger, 1811:141; †*Rhytina* Berthold, in Latreille, 1827:62[2]; †*Rhytina* Gloger, 1841[3]; †*Nepus* Fischer de Waldheim, 1814:640; †*Stellera* Bowdich, 1821:86[4]; †*Stellerus* Desmarest, 1882:510; †*Haligyna* Billberg, 1828:33.]
L. Mioc.-L. Plioc., L. Pleist.; N.A.[5] ?E. Plioc., L. Plioc., ?M. Pleist.; As.[6] L. Pleist.-R.; Pacific.[7]

[1] Not *Manatus* Brunnich, 1771. Suppressed by the International Commission on Zoological Nomenclature, Opinion 1320, 1985.
[2] Unjustified emendation of †*Rytina* Illiger.
[3] Invalid emendation.
[4] Bowdich incorrectly cited Cuvier as the author.
[5] Pacific North America.
[6] Japan.
[7] N. Pacific (Bering Sea).

Subfamily †**Miosireninae** Abel, 1919:835.
?L. Olig., Mioc.; Eu.

†*Anomotherium* Siegfried, 1965.[1]
L. Olig.; Eu.

[1] Assignment to †Miosireninae is questionable.

†*Miosiren* Dollo, 1890:415-421.
Mioc.; Eu.

Subfamily †**Rytiodontinae** Abel, 1914:217.[1]
>> [= †Rhytiodinae Abel, 1914:217; †Rytiodontinae Kretzoi, 1914:155;
>> †Thelriopiinae Pilleri, 1987:65-66.]
>> Mioc.; Af.[2] E. Mioc.; Eu. E.-M. Mioc., E. Plioc.; N.A. E. Mioc.; S.A.[3]
>> ?L. Mioc., E. Plioc.; Cent. A.
>> [1] See the International Code of Zoological Nomenclature, Article 11(f)(ii).
>> [2] N. Africa.
>> [3] Brazil.

†*Dioplotherium* Cope, 1883:52.
>> E.-M. Mioc.; N.A. E. Mioc.; S.A.[1]
>> [1] Brazil.

†*Rytiodus* Lartet, 1866:682.
>> [= †*Rhytiodus* Delfortrie, 1872:282[1]; †*Thelriope* Pilleri, 1987:65.[2]]
>> Mioc.; Af.[3] E. Mioc.; Eu. ?E. Mioc.; S.A.[4]
>> [1] Invalid emendation of †*Rytiodus* Lartet, 1866. Preoccupied by *Rhytiodus* Kner, 1858:78, a genus of fishes.
>> [2] Replacement name for †*Rhytiodus* "Lartet."
>> [3] N. Africa (Libya).
>> [4] Brazil.

†*Xenosiren* Domning, 1989:429.
>> L. Mioc. and/or E. Plioc.; Cent. A.

†*Corystosiren* Domning, 1990:361.
>> E. Plioc.; N.A. E. Plioc.; Cent. A.

Subfamily **Dugonginae** Simpson, 1932:424.
>> [= Halicoridae Gray, 1825:341.]
>> L. Mioc.; E. Indies.[1] R.; Indian O. R.; Pacific.[2]
>> [1] Java.
>> [2] W. Pacific.

†*Indosiren* G. H. R. von Koenigswald, 1952:611.
>> L. Mioc.; E. Indies.[1]
>> [1] Java.

Dugong Lacépède, 1799:17. Dugong, sea cow.
>> [= *Platystomus* Fischer de Waldheim, 1803:353[1]; *Dugungus* Tiedemann, 1808:554;
>> *Halicore* Illiger, 1811:140; *Dugongidus* Gray, 1821:309.]
>> R.; Indian O.[2] R.; Pacific.[3]
>> [1] Not *Platystoma* Meigen, 1803, a genus of Diptera.
>> [2] Coastal waters of E. Africa, Madagascar, the Red Sea, S. and S.E. Asia, the East Indies, and Australia.
>> [3] W. Pacific (including coastal waters of S. Japan, S.E. Asia, the East Indies, New Guinea, and Australia).

Family **Trichechidae** Gill, 1872:14, 91. Manatees.
>> [= Manatidae Gray, 1821:309; Manatina C. L. Bonaparte, 1838:111[1];
>> Manatida Haeckel, 1866:clix[2].]
>> M.-L. Mioc., ?Plioc., Pleist., R.; S.A. E. Plioc., Pleist., R.; N.A. R.; Af.
>> R.; Atlantic. R.; Cent. A. R.; W. Indies.
>> [1] Used as a subfamily.
>> [2] Used as a family.

†*Potamosiren* Reinhart, 1951:203.
>> M. Mioc.; S.A.

†*Ribodon* Ameghino, 1883:112.
>> L. Mioc.; S.A. E. Plioc.; N.A.

Trichechus Linnaeus, 1758:34.[1] Manatees.
>> [= *Manatus* Brunnich, 1771:34, 38[2]; *Oxystomus* Fischer de Waldheim, 1803:353;
>> *Trichecus* Oken, 1816:685-690[2]; *Halipaedisca* Gistel, 1848:83.]
>> ?Plioc., Pleist., R.; S.A.[3] Pleist., R.; N.A.[4] R.; Af.[5] R.; Atlantic.[6] R.; Cent. A.[7]
>> R.; W. Indies.[7]
>> [1] Not Linnaeus, 1766, a synonym of *Odobenus*.
>> [2] A rejected name; however, see the International Code of Zoological Nomenclature, Article 11(c)(i) for confusion.

[3] Coastal rivers of N.E. South America and the Amazon Basin in the Recent.
[4] Coastal rivers of S.E. North America in the Recent.
[5] Rivers of West Africa.
[6] Caribbean coastal waters of S.E. North America, Central America, West Indies, and N.E. South America. Coastal waters of West Africa.
[7] Coastal areas and river systems.

Infraorder **BEHEMOTA**, new.[1]

[1] Authors: McKenna, Bell & Shoshani. Definition: for the most recent common ancestor of †DESMOSTYLIA and PROBOSCIDEA, and all its descendants.

Parvorder †**DESMOSTYLIA** Reinhart, 1953:187, **new rank**.[1]

[= †DESMOSTYLIFORMES Hay, 1923:109[2]; †DESMOSTYLIFORMES Kinman, 1994:38[1]; †DESMOSTYLOIDEA Abel, 1933:875.[3]]

[1] Proposed as an order.
[2] Proposed as suborder.
[3] As order. It would serve no purpose to resurrect Abel's prior but forgotten name.

Family †**Desmostylidae** Osborn, 1905:109.

[= †Desmostylinae Kalandadze & Rautian, 1992:110.] [Including †Cornwalliusidae Shikama, 1957:16-21; †Cornwalliidae Shikama, 1966:153[1]; †Cornwalliinae Kalandadze & Rautian, 1992:110; †Paleoparadoxidae Reinhart, 1959:94; †Paleoparadoxiidae Reinhart, 1959:94[2]; †Behemotopsidae Inuzuka, 1987:16.]
L. Olig., ?E. Mioc., M. and/or L. Mioc.; As.[3] L. Olig.-L. Mioc.; N.A.[4]

[1] Emendation of †Cornwalliusidae.
[2] Emendation of †Paleoparadoxidae.
[3] E. Asia.
[4] Pacific North America.

†*Cornwallius* Hay, 1923:107.
L. Olig.; N.A.[1]
[1] Pacific North America.

†*Behemotops* Domning, Ray & McKenna, 1986:6.
L. Olig.; As.[1] L. Olig.; N.A.[2]
[1] E. Asia (Hokkaido, Japan).
[2] Pacific North America (Washington and Oregon).

†*Desmostylus* Marsh, 1888:95.
[Including †*Desmostylella* Nagao, 1937:82-85.]
Mioc.; As.[1] ?E. Mioc., M.-L. Mioc.; N.A.[2]
[1] E. Asia (Japan, Sakhalin Island).
[2] Pacific North America. Supposed occurrences in Florida are so far incorrect.

†*Paleoparadoxia* Reinhart, 1959:94.
E.-L. Mioc.; N.A.[1] M. and/or L. Mioc.; As.[2]
[1] Pacific North America (California).
[2] E. Asia (Japan).

†*Kronokotherium* Pronina, 1957:312.
M. and/or L. Mioc.; As.[1]
[1] E. Asia (Kamchatka).

†*Vanderhoofius* Reinhart, 1959:90.
M. Mioc.; N.A.[1]
[1] Pacific North America (California).

Parvorder **PROBOSCIDEA** Illiger, 1811:96, **new rank**.[1]

[= PROBOSCIDIAE Gray, 1821:305[2]; PROBOSCIFORMES Kinman, 1994:38.[3]]
[Including ELEPHANTIFORMES Tassy, 1988:46[0]; ELEPHANTOTHERIA Kalandadze & Rautian, 1992:107.[4]]

[1] Proposed as a "family" of order MULTUNGULA; used as an order by Owen, 1859:52; used as a suborder of order UNGULATA by Lydekker, 1886:1.

The sequence of superfamilies in the PROBOSCIDEA is not hierarchically arranged here, but does generally proceed from the more plesiomorphous to the more apomorphous taxa. We have chosen not to insert additional ranks between parvorder and superfamily, although this can easily be done in the future if deemed useful.

[2] Used as an order.
[3] Proposed as an order.
[4] Proposed as a suborder.

Family †**Anthracobunidae** Wells & Gingerich, 1983:117, 118.
E.-M. Eoc.; As.[1]
[1] S. Asia.

†*Pilgrimella* Dehm & Oettingen-Spielberg, 1958:33.
E.-M. Eoc.; As.[1]
[1] S. Asia.

†*Anthracobune* Pilgrim, 1940:129.
E.-M. Eoc.; As.[1]
[1] S. Asia.

†*Ishatherium* Sahni & Kumar, 1980:133, 134.[1]
E. Eoc.; As.[2]
[1] Questionably distinct from †*Anthracobune*.
[2] S. Asia.

†*Lammidhania* Gingerich, 1977:199.
E.-M. Eoc.; As.[1]
[1] S. Asia.

†*Jozaria* Wells & Gingerich, 1983:125.
M. Eoc.; As.[1]
[1] S. Asia.

Family †**Moeritheriidae** C. W. Andrews, 1906:99.
[= †Moeritheriini Winge, 1906:169, 172[1]; †Moeritherioidea Osborn, 1921:2[2];
†Moeritheriinae "Winge-Osborn", in Osborn, 1923:1; †MOERITHERIA Deraniyagala,
1955:15.[3]]
?M. Eoc., L. Eoc.-E. Olig.; Af.
[1] December 1906.
[2] Proposed as either suborder or superfamily.
[3] Proposed as an order.

†*Moeritherium* C. W. Andrews, 1901:4.
?M. Eoc., L. Eoc.-E. Olig.; Af.

Family †**Numidotheriidae** Shoshani & Tassy, in Shoshani, ed., 1992:23.
L. Paleoc.-E. Eoc.; Af.[1]
[1] N. Africa.

†*Phosphatherium* Gheerbrant, Sudre & Cappetta, 1996:69.
L. Paleoc.; Af.[1]
[1] N. Africa (Morocco).

†*Numidotherium* Mahboubi, Ameur, Crochet & Jaeger, 1986:25.
E. Eoc.; Af.[1]
[1] N. Africa.

Family †**Barytheriidae** C. W. Andrews, 1906:172.
[= †BARYTHERIA C. W. Andrews, 1904:482[1]; †BARYTHERIOIDEA Simpson,
1945:134[2]; †Barytheriinae Sarwar, 1977:9.]
L. Eoc.-E. Olig.; Af.[3]
[1] Not †BARYTHERIA Cope, 1898, used for toxodont notoungulates.
[2] Proposed as suborder.
[3] N. Africa.

†*Barytherium* C. W. Andrews, 1901:577.[1]
[= †*Bradytherium* C. W. Andrews, 1901:407.[2]]
L. Eoc.-E. Olig.; Af.[3]
[1] October 1901.
[2] September 1901. *Nomen nudum* in C. W. Andrews (August 16) 1901:4. Preoccupied by †*Bradytherium*
Grandidier, (March) 1901, a synonym of †*Palaeopropithecus*.
[3] N. Africa (Egypt).

Family †**Deinotheriidae** C. L. Bonaparte, 1841:253.
[= †Curtognati Kaup, 1833:516; †Curtognathidae Osborn, 1936:81, 735; †Dinotherina
C. L. Bonaparte, 1841:253[1]; †Dinotheridae C. L. Bonaparte, 1845:4; †Dinotheriidae C.

L. Bonaparte, 1850; †Dinotherida Haeckel, 1866:clix[2]; †Dinotheridae Murray, 1866:xiii; †DEINOTHERIA A. W. Scott, 1873:44[3]; †Deinotherioidae A. W. Scott, 1873:44; †Dinotheriinae Osborn, 1910:558; †Deinotherioidea Osborn, 1921:2[4]; †DEINOTHERIOIDEA Osborn, 1936:81; †GONYOGNATHA Haeckel, 1866:cxlvii.[5]]
E. Mioc.-E. Pleist.; Af. E.-L. Mioc.; As. E.-L. Mioc.; Eu.

[1] Proposed as subfamily.
[2] Used as a family.
[3] *Nomen oblitum.*
[4] Proposed as suborder or superfamily.
[5] Proposed as suborder. *Nomen oblitum.*

†*Prodeinotherium* Éhik, 1930:14.
[= †*Prodinotherium* Éhik, 1930:14; †*Prodeinotherium* Harris, 1973.[1]]
E.-M. Mioc., L. Mioc. and/or E. Plioc.; Af. E.-M. Mioc.; As. E.-M. Mioc.; Eu.

[1] A justified emendation of †*Prodinotherium* Ehik, 1930 [see the International Code of Zoological Nomenclature, Article 33(b)(ii)].

†*Deinotherium* Kaup, 1829:401.
[= †*Dinotherium* Kaup, 1831.[1]]
E.-L. Mioc.; Eu. M.-L. Mioc.; As. L. Mioc.-E. Pleist.; Af.

[1] Or at least by 1832.

Family †**Palaeomastodontidae** Andrews, 1906:130.
[= †Palaeomastodontinae Osborn, 1936:691; †Palaeomastodontoidea Madden, 1983:59.]
L. Eoc.-E. Olig.; Af.[1]

[1] N. Africa.

†*Palaeomastodon* Andrews, 1901:319.
L. Eoc.-E. Olig.; Af.[1]

[1] N. Africa.

Family †**Phiomiidae** Kalandadze & Rautian, 1992:109, **new rank**.[1]
[= †Phiomyini Kalandadze & Rautian, 1992:109.]
L. Eoc.-E. Olig.; Af.

[1] Spelling corrected.

†*Phiomia* Andrews & Beadnell, 1902:1-9.
L. Eoc.-E. Olig.; Af.

Family †**Hemimastodontidae, new**.[1]
E. Mioc.; As.

[1] McKenna, Bell, Shoshani & Tassy.

†*Hemimastodon* Pilgrim, 1912:17.
E. Mioc.; As.

Superfamily †**Mammutoidea** Hay, 1922:101.[1]
[= †Mastodontoidea Osborn, 1921:2[2]; †Mammutinae Hay, 1922:101; †Mammutoidea Simpson, 1945:247.]

[1] †Mammutoidea plus Elephantoidea actually represents a clade, but we refrain from naming it here.
[2] Proposed as either suborder or superfamily.

Family †**Mammutidae** Hay, 1922:101.
[= †Mastodonadae Gray, 1821:306; †Mastodontidae Girard, 1852:328; †Mammutinae Hay, 1922:101; †Mammutidae Cabrera, 1929:74.]
E. Mioc., ?M. Mioc., L. Mioc., Plioc. and/or E. Pleist.; Af. E.-L. Mioc., Plioc.; As. E. Mioc.-E. Pleist.; Eu. M. Mioc.-R.; N.A. M. and/or L. Mioc., Plioc., Pleist.; Cent. A.

Subfamily †**Eozygodontinae, new**.[1]
E. Mioc.; Af.

[1] McKenna, Bell & Shoshani.

†*Eozygodon* Tassy & Pickford, 1983:58.
E. Mioc.; Af.[1]

[1] E. Africa.

Subfamily †**Mammutinae** Hay, 1922:101.
>
> [= †Mastodontina Bonaparte, 1845:4; †Mastodontinae Osborn, 1910:558.] [Including †Zygolophodontinae Osborn, 1923:1.]
>
> E. and/or M. Mioc., L. Mioc., Plioc. and/or E. Pleist.; Af. E.-L. Mioc., Plioc.; As. E. Mioc.-E. Pleist.; Eu. M. Mioc.-R.; N.A. M. and/or L. Mioc., Pleist.; Cent. A.

†*Zygolophodon* Vacek, 1877:45.
>
> [= †*Turicius* Osborn, 1926:3.] [= or including †*Mastolophodon* Chakravarti, 1957:84.] [Including †*Miomastodon* Osborn, 1922:4.]
>
> E. and/or M. Mioc., L. Mioc.; Af. E.-L. Mioc., Plioc.; As. E.-L. Mioc.; Eu. M.-L. Mioc.; N.A. M. and/or L. Mioc.; Cent. A.

†*Mammut* Blumenbach, 1799:697-698. Mastodons.
>
> [= †*Harpagmotherium* Fischer de Waldheim, 1808:19[1]; †*Mastodon* Rafinesque, 1814:182[2]; †*Mastotherium* Fischer de Waldheim, 1814:337-341; †*Mastodontum* de Blainville, 1817:276; †*Tetracaulodon* Godman, 1830:478-485; †*Missourium* Koch, in Oken, 1840:905-906; †*Leviathan* Koch, 1841:13.] [Including †*Pliomastodon* Osborn, 1926:1.]
>
> L. Mioc. and/or Plioc. and/or E. Pleist.; Af. L. Mioc., Plioc.; As. L. Mioc.-E. Pleist.; Eu. E. Plioc.-R.; N.A. Pleist.; Cent. A.

[1] Publication not seen. Possibly spelled "†*Harpagonotherium.*"
[2] Not G. Cuvier, 1817:232-233 as frequently cited.

Superfamily **Elephantoidea** Gray, 1821:305.
>
> [= Elephantidae Gray, 1821:305; Elephantoidea Osborn, 1921:2.[1]] [= or including †Bunomastodontoidea Moustafa, 1974:423.[2]] [Including †Stegodontoidea Osborn, 1935:407-408.[3]]

[1] Proposed as either a suborder or superfamily. Contrary to Osborn's intent, we also add the gomphotheres.
[2] Not based on a generic name.
[3] Proposed as suborder; used as superfamily by Osborn, 1936:22, 25.

Subfamily †**Choerolophodontinae** Gaziry, 1976:105.
>
> [= †Choerolophinae Sarwar, 1977.]
>
> E.-L. Mioc.; Af. E. Mioc.-E. Plioc.; As. L. Mioc.; Eu.

†*Choerolophodon* Schlesinger, 1917:181.[1] Pig-toothed mastodons.
>
> [Including †*Synconolophus* Osborn, 1929:9.]
>
> E.-L. Mioc.; Af. E. Mioc.-E. Plioc.; As. L. Mioc.; Eu.

[1] Described as subgenus of †*Mastodon.*

Family †**Gomphotheriidae** Hay, 1922:101. Gomphotheres.
>
> [= †Gomphotheriinae Hay, 1922:101; †Gomphotheriidae Cabrera, 1929:74; †Trilophodontidae Simpson, 1931:281.] [= or including †Bunomastodontinae Osborn, 1918:134[1]; †Bunomastodontidae Osborn, 1921:2.[2]] [Including †Gnathalodontinae Barbour & Sternberg, 1935:396[3]; †Dibunodontidae Hopwood, 1935:55; †Serridentidae Osborn, 1936:381, 729; †Humboldtidae Osborn, 1936:575.[4]]
>
> E.-M. Mioc., E. Plioc.; Af. E. Mioc.-M. Pleist.; As. E.-L. Mioc., ?E. Plioc.; Eu. Mioc., ?Plioc., E.-L. Pleist.; S.A. M. Mioc.-L. Pleist.; N.A. L. Mioc., ?E. and/or ?M. Pleist., L. Pleist.; Cent. A.

[1] In part. Not based on a generic name.
[2] Not based on a generic name.
[3] Based on the genus †*Gnathabelodon.*
[4] Based not on a generic name but on a specific name, †*Mastodon humboldtii* Cuvier, the type species of †*Cuvieronius* Osborn.

†*Gnathabelodon* Barbour & Sternberg, 1935:396.[1]
>
> M.-L. Mioc.; N.A.

[1] Possibly referable to †Amebelodontinae.

Subfamily †**Gomphotheriinae** Hay, 1922:101.
>
> [= †Longirostrinae Osborn, 1918:136.[1]] [Including †Serridentinae Osborn, 1921:232.]
>
> E.-M. Mioc., E. Plioc.; Af. E.-L. Mioc.; As. E.-L. Mioc., ?E. Plioc.; Eu. M. Mioc.-E. Plioc.; N.A.

[1] Not based on a generic name.

Tribe †**Gomphotheriini** Hay, 1922:101. Longirostrines.
[= †Longirostrinae Osborn, 1918:136[1]; †Gomphotheriinae Hay, 1922:101;
†Gomphotherini Madden, 1983:60; †Gomphotherina Madden, 1983:60;
†Gomphotheriina Kalandadze & Rautian, 1992:108.]
E.-M. Mioc.; Af. E.-L. Mioc.; As. E.-L. Mioc., ?E. Plioc.; Eu. M. Mioc.-E. Plioc.;
N.A.

[1] Not based on a generic name.

†*Gomphotherium* Burmeister, 1837:795.
[= †*Gamphotherium* Gloger, 1841:119[1]; †*Trilophodon* Falconer, 1857:316[2];
†*Bunolophodon* Vacek, 1877:45; †*Tetrabelodon* Cope, 1884:4.] [Including
†*Megabelodon* Barbour, 1914:217[3]; †*Genomastodon* Barbour, 1917:512; †*Serridentinus*
Osborn, 1923:2; †*Ocalientinus* Frick, 1933:579; †*Trobelodon* Frick, 1933:580;
†*Tatabelodon* Frick, 1933:581; †*Hemilophodon* Kretzoi, 1942:139; †*Kunatia* Sarwar &
Akhtar, 1992:219.]
E.-M. Mioc.; Af. E.-L. Mioc.; As. E.-L. Mioc., ?E. Plioc.; Eu. M. Mioc.-E. Plioc.;
N.A.

[1] Misprint of †*Gomphotherium*.
[2] As "section" of †*Mastodon* in Falconer & Cautley, 1846:54; as subgenus of †*Mastodon* in Falconer, 1857.
[3] Described as subgenus of †*Tetrabelodon*.

Tribe †**Amebelodontini** Barbour, 1927:131, **new rank**.
[= †Amebelodontidae Barbour, 1927:131; †Amebelodontinae Barbour, 1929:139.]
[Including †Platybelodontinae Borissiak, 1928:119.]
E.-M. Mioc., E. Plioc.; Af. E.-M. Mioc.; Eu. M.-L. Mioc.; As. L. Mioc., ?E.
Plioc.; N.A.

Subtribe †**Protanancina, new**.[1]
E.-M. Mioc.; Af. E.-M. Mioc.; Eu. M. Mioc., ?L. Mioc.; As.

[1] McKenna, Bell & Shoshani.

†*Archaeobelodon* Tassy, 1984:462.
E. Mioc., ?M. Mioc.; Af. E.-M. Mioc.; Eu.

†*Protanancus* Arambourg, 1945:493.
M. Mioc.; Af. M. Mioc., ?L. Mioc.; As.

Subtribe †**Amebelodontina** Barbour, 1927:131, **new rank**. Shovel-tuskers, shovel-tusked
gomphotheres.
[= †Amebelodontidae Barbour, 1927:131.]
E. Mioc., E. Plioc.; Af. M.-L. Mioc.; As. M. Mioc.; Eu.[1] L. Mioc., ?E. Plioc.;
N.A.

[1] E. Europe.

†*Serbelodon* Frick, 1933:594.
M. Mioc.; As.[1] L. Mioc.; N.A.

[1] E. Asia (China).

†*Platybelodon* Borissiak, 1928:120.
[Including †*Torynobelodon* Barbour, 1929:147; †*Selenolophodon* Chang & Zhai,
1978:136.]
E. Mioc.; Af. M.-L. Mioc.; As. M. Mioc.; Eu.[1] L. Mioc., ?E. Plioc.; N.A.

[1] E. Europe.

†*Amebelodon* Barbour, 1927:131.
[Including †*Konobelodon* Lambert, 1990:1033.[1]]
M. Mioc.; As.[2] L. Mioc., ?E. Plioc.; N.A. E. Plioc.; Af.[3]

[1] Described as subgenus.
[2] E. Asia (China).
[3] N. Africa.

Subfamily †**Rhynchotheriinae** Hay, 1922:101.
[= †Rhynchorostrinae Osborn, 1918:136.[1]] [Including †Humboldtinae Osborn,
1934:180.[2]]

Mioc., ?Plioc., E.-L. Pleist.; S.A.[3]   M. Mioc.-L. Pleist.; N.A.   L. Mioc.-M. Pleist.; As.   L. Mioc., ?E. and/or ?M. Pleist., L. Pleist.; Cent. A.

[1] Not based on a generic name.
[2] Based not on a generic name but on a specific name, †*Mastodon humboldtii* Cuvier, the type species of †*Cuvieronius* Osborn.
[3] A new taxon has been reported from the Miocene of eastern Peru.

†*Sinomastodon* Tobien, Chen & Li, 1986:160.
L. Mioc.-M. Pleist.; As.

†*Eubelodon* Barbour, 1914:186.
M. Mioc.; N.A.

Tribe †**Rhynchotheriini** Hay, 1922:101, **new rank**. Rhynchorostrines, beak-jawed gomphotheres.
[= Rhynchorostrinae Osborn, 1918:136[1]; †Rhynchotheriinae Hay, 1922:101.]
M.-L. Mioc., Plioc.; N.A. L. Mioc.; Cent. A.

[1] Not based on a generic name.

†*Rhynchotherium* Falconer, 1868:74-75.
[= †*Dibelodon* Cope, 1884:2.]   [Including †*Blickotherium* Frick, 1933:529; †*Aybelodon* Frick, 1933:532.]
M.-L. Mioc., Plioc.; N.A. L. Mioc.; Cent. A.

Tribe †**Cuvieroniini** Cabrera, 1929:76.
[= †Notorostrinae Osborn, 1921:330[1]; †Cuvieroniinae Cabrera, 1929:76; †Humboldtinae Osborn, 1934:180.[2]] [Including †Brevirostrinae Osborn, 1918:136[1]; †Notiomastodontinae Osborn, 1936:590, 730; †Notiomastodontina Madden, 1983:60; †Notiomastodonta Kalandadze & Rautian, 1992:108.[3]]
?L. Mioc., E. Plioc.-L. Pleist.; N.A. ?Plioc., E.-L. Pleist.; S.A. ?E. and/or ?M. Pleist., L. Pleist.; Cent. A.

[1] Not based on a generic name.
[2] Based not on a generic name but on a specific name, †*Mastodon humboldtii* Cuvier, the type species of †*Cuvieronius* Osborn.
[3] We are unsure of what the authors meant by this redundant echo of Osborn's †Notiomastodontinae, and we are only slightly more certain of the intended rank.

†*Stegomastodon* Pohlig, 1912:193.
[= †*Rhabdobunus* Hay, 1914:373.]
E. Plioc.-M. Pleist.; N.A.

†*Haplomastodon* Hoffstetter, 1950:24.[1]
[Including †*Aleamastodon* Hoffstetter, 1952:208.[2]]
?Plioc. and/or ?E. Pleist., M.-L. Pleist.; S.A. Pleist.; Cent. A.[3]

[1] Proposed as subgenus of †*Stegomastodon*; raised to generic rank by Hoffstetter, 1952.
[2] Proposed as subgenus of †*Haplomastodon*.
[3] Costa Rica.

†*Notiomastodon* Cabrera, 1929:90.
E. Pleist.; S.A.

†*Cuvieronius* Osborn, 1923:1.
[= †*Cordillerion* Osborn, 1926:15; †*Teleobunomastodon* Revilliod, 1931:21.]
?L. Mioc., E. Plioc.-L. Pleist.; N.A. ?L. Plioc., Pleist.; S.A. ?E. Pleist., L. Pleist.; Cent. A.

Family **Elephantidae** Gray, 1821:305.
[= Elephantida Haeckel, 1866:clix.[1]] [Including †Anancinae Hay, 1922:101; †Anancini Madden, 1983:60; †Anancina Madden, 1983:60; †Tetralophodontinae Van der Maarel, 1932:126; †Tetralophodontidae Gaziry, 1987:201.]
E. Mioc.-R.; As. M. Mioc.-R.; Af. M. Mioc.-L. Pleist.; Eu. M.-L. Mioc., ?Plioc., E.-L. Pleist.; N.A. L. Plioc.-R.; E. Indies. Pleist.; Mediterranean. L. Pleist.; Cent. A.

[1] As a family.

†*Tetralophodon* Falconer, 1857:316.[1]
[Including †*Lydekkeria* Osborn, 1924:2[2]; †*Geisotodon* Bergounioux & Crouzel, 1955:1488.]

M.-L. Mioc., ?E. Plioc.; Af. M.-L. Mioc., E. Pleist.; As. M.-L. Mioc.; Eu.

[1] Described as subgenus of †*Mastodon*.
[2] Described as subgenus of †*Tetralophodon*.

†*Morrillia* Osborn, 1924:1.[1]

?Plioc., E. Pleist.; N.A.

[1] Described as a subgenus of †*Tetralophodon*; used in Osborn, 1936 as both a subgenus (pp. 349-352, 377-379) and a genus (pp. 690, 739).

†*Anancus* Aymard, 1855:507.

[= †*Dibunodon* Schlesinger, 1917:124.[1]] [Including †*Pentalophodon* Falconer, 1857:314.[1]]

M.-L. Mioc.; N.A. L. Mioc.-E. Pleist.; Af. L. Mioc.-E. Pleist.; As. L. Mioc.-E. Pleist.; Eu.

[1] Described as subgenus of †*Mastodon*.

†*Paratetralophodon* Tassy, 1983:273.

L. Mioc.; As.[1]

[1] S. Asia.

Subfamily †**Stegodontinae** Osborn, 1918:135.

[= †Stegodontidae Hopwood, 1935:71; †Cryptomastodontidae Von Koenigswald, 1933:112.] [Including †Stegolophodontinae Osborn, 1936:700.]

E. Mioc.-R.; As. L. Mioc.-L. Plioc.; Af. ?Plioc.; Eu. L. Plioc., Pleist.; E. Indies.

†*Stegolophodon* Schlesinger, 1917:228.[1]

[= †*Prostegodon* Matsumoto, in Osborn, 1923:2.] [Including †*Eostegodon* Yabe, 1950:65; †*Tetrazygodon* Tobien, 1978:244[2]; †*Antelephas* Sarwar, 1979:133; †*Rulengchia* Zhou & Zhang, 1983:57.[3]]

E. Mioc.-L. Plioc., ?E. Pleist.; As.[4] ?Plioc.; Eu.

[1] Described as subgenus of †*Mastodon*.
[2] Sometimes as subgenus.
[3] (Chow & Chang).
[4] Including Japan.

†*Stegodon* Falconer, 1857:318.[1]

[= †*Emmenodon* Cope, 1889:194.] [Including †*Parastegodon* Matsumoto, 1929:13[2]; †*Cryptomastodon* G. H. R. von Koenigswald, 1933:112; †*Platystegodon* Deraniyagala, 1954:25[3]; †*Sulcicephalus* Deraniyagala, 1955:216.[4]]

L. Mioc.-L. Plioc.; Af. L. Mioc.-R.; As.[5] L. Plioc., Pleist.; E. Indies.[6]

[1] Described as subgenus of *Elephas*.
[2] *Nomen nudum* in Matsumoto, 1924:256-257, 1926:1.
[3] Described as subgenus of †*Stegodon*.
[4] Described as subgenus of †*Stegodon*. *Nomen nudum* in Deraniyagala, 1954:25-26.
[5] Including Japan, Taiwan, and Holocene of Shuanlong Cave, Zhejiang Province, PRC.
[6] Java, Sulawesi, Flores, Timor, Philippines.

Subfamily **Elephantinae** Gray, 1821:305.

[= Elephantidae Gray, 1821:305; Elephantina Bonaparte, 1838:112[1]; Elephantinae Gill, 1872:13.] [Including Loxodontinae Osborn, 1918:135; †Mammontinae Osborn, 1921:234; †Dicyclotheriinae Deraniyagala, 1955:34; †Stegotetrabelodontinae Aguirre, 1969:1370; †Stegotetrabelodonta Kalandadze & Rautian, 1992:108[2]; †Palaeoloxodontinae Zhang & Zong, 1983:302, 304.[3]]

L. Mioc.-R.; Af. L. Plioc.-R.; E. Indies. L. Plioc.-R.; As. L. Plioc.-L. Pleist.; Eu. Pleist.; Mediterranean. E.-L. Pleist.; N.A. L. Pleist.; Cent. A.

[1] As subfamily.
[2] We are unsure what rank was intended by the authors of this redundant term.
[3] (Chang & Zong).

†*Stegotetrabelodon* Petrocchi, 1941:110.

L. Mioc.-E. Plioc.; Af.

†*Stegodibelodon* Coppens, 1972:2964.

L. Mioc.-E. Plioc.; Af.

Tribe **Elephantini** Gray, 1821:305.[1] Elephants.

[= Elephantidae Gray, 1821:305; Elephantina Gray, 1825:343.[2]]

L. Mioc.-R.; Af. L. Plioc.-R.; E. Indies. L. Plioc.-R.; As. L. Plioc.-L. Pleist.; Eu.
Pleist.; Mediterranean. E.-L. Pleist.; N.A. L. Pleist.; Cent. A.

[1] New suffix.
[2] As tribe.

†*Primelephas* Maglio, 1970:10.
L. Mioc.-L. Plioc.; Af.

Subtribe **Loxodontina** Osborn, 1918:135, **new rank**.[1]
[= Loxodontinae Osborn, 1918:135; Loxodontini Kalandadze & Rautian, 1992:109.]
L. Mioc.-R.; Af.
[1] Shoshani.

Loxodonta Anonymous, 1827:140. African elephant, African forest elephant, African savannah
elephant.
L. Mioc.-R.; Af.

Subtribe **Elephantina** Gray, 1821:305, **new rank**.[1]
[= Elephantidae Gray, 1821:305; Elephantina Gray, 1825:343[2]; Elephantina Bonaparte,
1838:112.[3]]
E. Plioc.-L. Pleist.; Af. L. Plioc.-R.; E. Indies. L. Plioc.-R.; As. L. Plioc.-L.
Pleist.; Eu. Pleist.; Mediterranean. E.-L. Pleist.; N.A. L. Pleist.; Cent. A.
[1] Shoshani.
[2] As tribe.
[3] As a subfamily.

†*Mammuthus* Brookes, 1828:73-74.[1] Mammoths, woolly mammoth, steppe elephant, dwarf
mammoths.
[= †*Dicyclotherium* É. Geoffroy Saint-Hilaire, 1837:119, 120; †*Cheirolites* von Meyer,
1848:286.] [= or including †*Mammonteum* de Blainville, 1864:237.[2]] [Including
†*Archidiskodon* Pohlig, 1888:138[3]; †*Parelephas* Osborn, 1924:4; †*Metarchidiskodon*
Osborn, 1934:12.]
E. Plioc.-M. Pleist.; Af. L. Plioc.-R.; As.[4] L. Plioc.-L. Pleist.; Eu. E.-L. Pleist.;
N.A. L. Pleist.; Cent. A.
[1] Not Burnett, 1830.
[2] †*"Mammonteus* Camper, 1788" was used by Osborn, 1924. This was reconstructed from "Mammonteum"
used in the vernacular by Camper and not Linnaean (Simpson 1945:134, 249).
[3] Described as subgenus of *Elephas*.
[4] Wrangel Island, Siberian Arctic in the Holocene.

Elephas Linnaeus, 1758:33. Asian elephant, dwarf island elephants,[1] forest elephant.
[= *Elephantus* É. Geoffroy Saint-Hilaire & Cuvier, 1795:189.] [Including †*Euelephas*
Falconer, 1857:318[2]; †*Palaeoloxodon* Matsumoto, 1924:257[3]; †*Sivalikia* Osborn,
1924:2; †*Pilgrimia* Osborn, 1924:2; †*Leith-Adamsia* Matsumoto, 1928:214;
†*Hesperoloxodon* Osborn, 1931:21; †*Hypselephas* Osborn, 1936:12[4]; †*Platelephas*
Osborn, 1936:12[5]; †*Stegoloxodon* Kretzoi, 1950:405; †*Omoloxodon* Deraniyagala,
1955:25; †*Protelephas* Garutt, 1957:189; †*Phanagoroloxodon* Garutt, 1957:333;
†*Wanoloxodon* Liu & Zhen, 1981:220.]
E. Plioc.-L. Pleist.; Af. L. Plioc.-R.; E. Indies.[6,7] L. Plioc.-R.; As.[8,9] ?L. Plioc.,
E.-L. Pleist.; Eu. Pleist.; Mediterranean.[10]
[1] The cyclops legend may be based upon interpretation of the external narial opening of a dwarf elephant
skull.
[2] Described as subgenus of *Elephas*.
[3] Sometimes as subgenus of *Elephas*.
[4] *Nomen nudum* in Osborn, 1934:285; full description in Osborn, 1942:1340.
[5] *Nomen nudum* in Osborn, 1934:285; full description in Osborn, 1942:1358.
[6] Including Java and Sulawesi in the Pliocene and Pleistocene.
[7] Extant in Sumatra and Borneo.
[8] Including Japan and Taiwan in the late Pleistocene.
[9] Recent formerly from E. Asia to W. Asia, including Rhodes and Tilos; now extant only in S. and S. E. Asia.
[10] Cyprus, Crete, Malta, Sicily, Sardinia.

APPENDIX A

Article 36 of the International Code of Zoological Nomenclature

Article 36 of the 1961 International Code of Zoological Nomenclature (International Commission on Zoological Nomenclature 1961:39) reads as follows:

Article 36. Categories co-ordinate.— All categories in the family-group are of coordinate status in nomenclature, that is, they are subject to the same rules and recommendations, and a name established for a taxon in any category in the group, and based on a given type-genus, is thereupon available with its original date and author for a taxon based on the same type-genus in each of the other categories, with appropriate change of suffix.

Example.—The proposal of HESPERIIDAE Latreille, 1809 (as HESPERIDES), based on *Hesperia* Fabricius, 1793, thereupon makes available, from the year 1809, the superfamily name HESPERIOIDEA and the subfamily name HESPERIINAE, even though the former was first used by Comstock, J. H. & A. B., 1904, and the latter by Watson, 1893.

In the presently operative 3rd edition of the ICZN, Article 36 (International Commission on Zoological Nomenclature 1985:77) reads:

Article 36. Principle of Coordination.—
(a) **Statement of the Principle of Coordination.—**A name established for a taxon at any rank in the family group is deemed to be simultaneously established with the same author and date for taxa based upon the same name-bearing type (type genus) at other ranks in the family group, with appropriate mandatory change of suffix [Art. 34a].

Example.—The family name HESPERIIDAE, based on *Hesperia* Fabricius, was established in 1809 by Latreille (as Hesperides). For authorship and priority Latreille, 1809, is considered also to have simultaneously established the coordinate superfamily name HESPERIOIDEA and the coordinate subfamily name HESPERIINAE, even though the former was first used by Comstock and Comstock,1904, and the latter by Watson, 1893.

(b) **Type genus.—**When a nominal taxon is raised or lowered in rank in the family group, its type genus remains the same [Art. 61b (ii)].

Comment

The aim of Article 36 is mainly to discourage irresponsible taxonomists from changing the ranks of taxa to previously unused levels in order to be the author of the taxon in its new status. When this rule came into effect, however, it was a time consuming task to modify all mammalian family-group taxa so as to list the author who first proposed any of them as author of all. This was a major factor in Simpson's choice not to publish an update of his 1945 classification (personal communication to McKenna from Simpson; see also Bradley 1962:178).

APPENDIX B

Mammaliaform and Mammalian Characters

Some recent papers, including those cited here, have wrestled mightily with the problem of what taxa to include in or exclude from the definition of MAMMALIA. Rowe (1988:247) suggested, "Mammalia may be defined as comprising the most recent common ancestor of living Monotremata (Ornithorhynchidae and Tachyglossidae) and Theria (Marsupialia and Placentalia), and all of its descendants." This is a crown group definition that in our opinion would also include multituberculates, but Wible (1991) suggested modifications that would remove multituberculates from the crown group. Rowe (1987, 1988, 1996) is followed here regarding mammaliaform and mammalian content. Rowe (1988, 1996) used MAMMALIAFORMES for the mammalian crown group plus †Morganucodontidae, a view differing from those of Kemp (1988b), Hopson (1991), Miao (1991), Lucas (1992), Lucas and Luo (1993), Luo (1994), and Jenkins et al. (1997), authors who included the morganucodonts in MAMMALIA. Other definitions and diagnoses have been proposed that result in a slightly different mammalian content.

Gauthier, Kluge, and Rowe (1988a:208-209) provided a list of attributes (not necessarily diagnostic) occurring in crown group MAMMALIA, i.e, monotremes and therians in the traditional sense (Hennig 1969; Jefferies 1979; de Queiroz 1995). Their taxonomic definition of MAMMALIA excluded morganucodonts as well as other extinct therapsid clades.

Proposed diagnostic characters for MAMMALIA and the more inclusive clade MAMMALIAFORMES have been given by a number of authors. Those provided by Gauthier, Kluge, and Rowe (1988a:205) are listed here with their numbering system placed in parentheses:

MAMMALIAFORMES
- Loss of tabular (25). [Gauthier, Kluge, and Rowe (1988a:171) scored this character as true for †Morganucodontidae (contra Rowe 1988) and either true or false for what they defined as MAMMALIA. This was, no doubt, a reference to the supposed presence of separate tabulars in multituberculates (Kielan-Jaworowska 1970), but in agreement with Kielan-Jaworowska et al. (1986:600), Rowe (1988: 251), and Wible (1991:5) we doubt that multituberculates had separate tabular bones.]
- Maxillary ventral margin bowed ventrally (-30). [This is a reversal of uncertain significance. See also Wible (1991:13) and, for †Haldanodon, see Lillegraven and Krusat (1991:52-55).]
- Cochlear promontorium (72).
- Dentary/squamosal articulation (91).

- Vertebral anapophyses (125) [uncertain significance].
- Lumbar vertebrae faces inclined (140).

MAMMALIA
- Appearance of independent centers of ossification in epiphyses of all longbones (G36a (2)). [This is a homoplasy with lepidosaurs and is a weak character in any case because of missing data.]
- Internasal process of premaxilla lost (A32). [This is a homoplasy with TESTUDINES and crocodiles. †Haldanodon (and presumably other docodonts) retained an internasal process (Rowe 1988; Miao 1988; Lillegraven and Krusat 1991: fig. 4.]
- Craniomandibular joint anterior to occiput (70 (2)). [But also true of †Haldanodon (Lillegraven and Krusat 1991:51).]
- Atlas arch-intercentrum in contact (-135). [This is a reversal from condition seen in dicynodonts, therocephalians, †Procynosuchus, †Thrinaxodon, †Diademodon, †Exaeretodon, †Tritylodontidae, and †Morganucodontidae, but a similarity to TESTUDINES, †Captorhinidae, †ARAEOSCELIDIA, and various reptiles.]
- Interclavicle T-shaped anteriorly (154). [Inasmuch as this was scored as "?" for †Morganucodontidae, this character might well apply to MAMMALIAFORMES.]

We question the usefulness of characters G36a (2), 70 (2), and 154. Therefore we cull from Gauthier, Kluge, and Rowe (1988a) the following hard tissue anatomical characters as approximately diagnostic for a clade (including †Morganucodontidae) that we continue to term MAMMALIAFORMES:

MAMMALIAFORMES
- Loss or fusion of tabular (25). [Not known in †Morganucodon, even though suggested by Kermack and Kermack (1984: fig. 4.4). Not a separate bone in †Haldanodon (Lillegraven and Krusat 1991:80).]
- Occipital condyles expanded dorsally to enclose ventral two-thirds of foramen magnum, and traverse wide arc of abduction [and adduction]. [We list this character, even though it is apparently not true for morganucodonts or multituberculates (Kielan-Jaworowska 1971: fig. 2; Miao 1988: fig. 28; Wible 1991:5). It does, however, characterize †Haldanodon (Lillegraven and Krusat 1991) so we reassign the character, with doubt, to MAMMALIAFORMES.]

•Cochlear promontorium (72).

•Tegmen tympani present. [See Wible (1991:10) for discussion of this character.]

•Dentary/squamosal articulation (91). [Additional bones of the craniomandibular joint still plesiomorphously present initially.]

•Lumbar vertebrae faces inclined (140).

•Ossified cribriform plate separating nasal cavity from brain cavity. [As noted by Wible (1991: 5), the presence or absence of a cribriform plate is still unknown in trithelodonts, †Sinoconodon, †Dinnetherium, †Kuehneotherium, and multituberculates. However, Lillegraven and Krusat (1991:120) identified a cribriform plate in †Haldanodon. In monotremes, adult echidnas possess it, but in the platypus it occurs only briefly in ontogeny (Zeller 1988). Previously, the presence of a cribriform plate has been thought diagnostic of MAMMALIA, but it is evidently pre-mammalian in origin and has been secondarily reduced or remained unossified in a few mammals (e.g., †Anagale?).]

Reworking the lists of mammalian crown group characters given by Gauthier, Kluge, and Rowe (1988a), as well as by Rowe (1988, 1996), and Wible (1991), we renumber the characters and segregate them into separate lists dealing with hard parts (H, accessible in fossils) and soft tissue anatomy (S, crown group only, inaccessible or only indirectly accessible and uncertain in nearly all fossils). Finally, in agreement with Rowe (1996), we select one of them (H1) as the most practical morphological rubicon between MAMMALIA and other MAMMALIAFORMES. Such a selection is for the purpose of minimally ambiguous approximate diagnosis, but we realize that future discoveries might show that some character other than H1 would allow a closer match of the diagnosis with the definition. Moreover, as Gauthier and his co-authors noted, some of these characters occur as well in other amniote groups. Indeed, there is enough homoplasy to have misled Gardiner (1982, 1993) and Løvtrup (1985) in a way that has proven quite instructive concerning outgroup, ingroup, and morphocline polarity analysis (Donoghue et al. 1989).

MAMMALIA

H1. Accessory jaw bones [except coronoid bone and splenial bone if it were simply lost] shifted away from the craniomandibular joint to become associated with the cranium alone. [This process probably proceeded in stages, although the stages may have been short-lived. In addition to the stapes (derived from the columella auris), the middle ear bones consist of at least five separate ossifications formerly associated with the jaw apparatus: incus (quadrate), malleus (articular), ectotympanic (angular), os goniale (prearticular), and ossiculum accessorium mallei (surangular), that are suspended from the skull.]

H2. Absence of notches in squamosal for quadrate and quadratojugal. [In addition to taxa analyzed in Wible's (1991: Appendix 1) matrix, these notches are lost in an early Cretaceous triconodontid (Crompton and Sun 1985: fig. 7A; Wible 1991:3) and in †Vincelestes (Bonaparte and Rougier 1987: fig. 4).]

H3. Quadratojugal not a separate ossification. [This is uncertain for †Haldanodon and †Morganucodon. See Kermack et al. (1981), Lillegraven and Krusat (1991:99), and Wible (1991:13). Possibly this character should be shifted to MAMMALIAFORMES.]

H4. Coronoid bone small and/or fused to medial surface of dentary. [Independently fused and/or absent in some multituberculates, possibly monotremes (Rowe 1988), and therian mammals (Wible 1991:15).]

H5. Stapes very small relative to skull size.[†Haldanodon has a relatively very large, perforate stapes (Lillegraven and Krusat 1991:99). In †Morganucodon the stapes was also plesiomorphously relatively larger than in the more apomorphous mammals (Kermack et al. 1981: fig. 4; Allin and Hopson 1992: fig. 28.8; Wible and Hopson 1993:53).]

H6. Meckelian sulcus enclosed by dentary to form Meckelian canal.

H7. Lateral pterygoid flange (pterygoid transverse process) vestigial (hamulus pterygoidei) and widely separated from mandible.

H8. Proatlas absent postembryonically.

H9. Atlas intercentrum and neural arches fused to form single, ring-shaped osseous structure. [See discussion above.]

H10. Atlantal rib absent or fused to form part of transverse process.

H11. Axial prezygapophysis absent.

H12. Postaxial cervical ribs fused to their centra, forming transverse foramina.

H13. Epiphyses on long bones and girdles. [Weak character; see above.]

H14. Presence of parafibular flabellum.

H15. Flexor sesamoids in manus and pes.

H16. Styloid process of radius, tibia, and fibula.

H17. Saddle-shaped articulation between entocuneiform and first metatarsal.

S1. Single aortic trunk.

S2. Pulmonary artery with three semilunar valves.

S3. Endothermy.

S4. Incubation of eggs (monotremes).

S5. Erector muscles and dermal papillae.

S6. Three meninges.

S7. Folded cerebellum, pons variolii, inferior olive and pontine nuclei.

S8. Adventitious cartilage.

S9. Vascularized islets of pancreas.

S10. Renal macula densa.

S11. Loop of Henle.

S12. Completely divided heart with thick, compact myocardium.

S13. Loss of the sinus venosus and development of a sinu-venosi septum (ambiguous character).

S14. Absence of tendon to lower eyelid (ambiguous character).

S15. Intrinsic eyelid muscles derived from facial platysma.

S16. Subclavian arteries positioned anteriorly, near the carotids (ambiguous character).

S17. Auriculo-ventricular node and Purkinje fibers.

S18. Three neurofilament polypeptides.

S19. Most of stomach lumen lies posterior to pyloris.

S20. Processus recessus encloses anteriormost remnant of fissura metotica.

S21. M. panniculus carnosus forms continuous sheath of muscle wrapping trunk and neck.

S22. Muscular diaphragm encloses pleural cavities, and consequent development of diaphragmatic breathing.

S23. Superficial musculature expanded onto face and differentiated into muscle groups associated with eye, ear, and snout.

S24. Elaborate development of greater omental bursa.

S25. Epiglottis.

S26. Well-developed hippocampus.

S27. Dorsal or hippocampal commissure and anterior commissure interconnect pallial structures of two cerebral hemispheres.

S28. Motor nucleus of facial nerve expanded and divided into nucleus facialis dorsalis and nucleus facialis ventralis.

S29. Strong representation of facial nerve field in motor cortex.

S30. Restriction of sensory field of facial nerve and great expansion of cutaneous field of trigeminal nerve over face.

S31. Chorda tympani passes below stapes.

S32. Divided optic lobes.

S33. Well-developed specific motor nuclei that receive afferents from cerebellum or basal ganglia, project to restricted regions of telencephalon, and are situated rostrally in ventral half of thalamus.

S34. Central region of telencephalic pallium is isocortex.

S35. Hindbrain overlies fenestrae vestibuli.

S36. Thrombocytes take form of blood platelets.

S37. Erythrocytes lack nuclei at maturity.

S38. Adult liver and spleen play only minor role in erythrocyte formation.

S39. Hair (sometimes preserved, e.g., in multituberculates; see Meng and Wyss 1997).

S40. Sebaceous glands.

S41. Sweat glands.

S42. Mammary glands.

S43. Parotid, submaxillary and sublingual glands

S44. Tympanic membrane with middle layer or membrana propria.

S45. Thymus differentiates from ventral part of gill pouch of second postspiracular gill cleft.

S46. Cervical thymus gland forms from invagination of ectoderm of neck.

S47. Lungs expanded ventrally, surrounding heart and almost meeting on ventral midline, leaving only median strand of tissue, the ventral mediastinum, connecting pericardial sac with ventral body wall.

S48. Complex lung structure with division of lungs into lobes, bronchioles and alveoli.

As is clear from the work of Gauthier, Kluge, and Rowe (1988a), the closest extant outgroup to the synapsid crown group is REPTILIA, possibly TESTUDINES among extant sauropsid reptiles [Hopson 1991; but see Gaffney (1980:596, 605), who favored DIAPSIDA]. The branch point between these major amniote divisions would have been at least as long ago as the late Carboniferous in the Paleozoic Era. For this reason, the 48 soft tissue characters are interesting as diagnosing extant synapsids, but we doubt that all of these soft tissue anatomical characters diagnose MAMMALIA alone. Many of them could have diagnosed various nonmammalian synapsids as well, long before the origin of the mammalian clade. The chance that all of the characters listed arose at once in the first mammal are infinitesimally small.

Strangely, if more than one "independent" character is employed to diagnose a node-based taxon, the diagnosis can only be decreasingly accurate as characters are added. This is because, in our experience, such characters generally arise sequentially rather than synchronously. They occur in some unknown sequence in the stem between the diagnosed group and the common ancestor it shares with the closest known outgroup, before cladogenesis occurs in the group that is diagnosed. Otherwise, they would not occur in the diagnosed group's branches unless by homoplasy. Diagnoses employing more than one character are supported by ignorance of morphocline sequence within the stem of the diagnosed taxon rather than by mere knowledge of the presence of "diagnostic" characters among the members of the taxon for which a diagnosis is attempted. The problem is similar to determining the sequence of arrival of the various shuffled letters in a week's accumulation of pigeonholed mail. Nevertheless, one can specify various sequential cladogenetic events on which taxa can be based. For known nested taxa these can be diagnosed approximately.

Seemingly endless prose has been devoted to the various pros and cons of supplying a nonterminal taxon with a diagnosis. The aim of diagnosing in addition to defining high level taxa is to have a convenient method of "identification", some way to link higher taxa to morphological reality. For example, perissodactyls and artiodactyls each have distinctive astragali in the hind limbs that can serve as handy guides to "identification" of members of those taxa, with a very high degree of reliability. But when did the acquisition of these astragalar characters occur? The crown group definition of, say, PERISSODACTYLA refers to the most immediate common ancestor of the subsumed extant perissodactyl subclades and all its descendants, which would have been represented by a lineage of mammals immediately before it split into isolated and then genetically fixed sublineages leading to both fossil and extant perissodactyls. The distinctive astragalus hypothesized to be a synapomorphy that characterizes all perissodactyls would have had to have originated when the parent lineage was as yet unbranched; otherwise, the character would not appear in all resulting subclades unless the synapomorphy hypothesis is false. The parent lineage

at the time of acquisition of the synapomorphy would not have been a member of the later crown group it spawned because it was not the most immediate common ancestor of that group. Thus the diagnosis of a crown group could, and probably would, apply to some organisms that were not included by the crown group's definition. The problem would be more complex if more than one independent character were relied upon for diagnosis, inasmuch as the characters almost certainly would not have been acquired simultaneously, let alone at the moment of lineage splitting. We conclude that diagnoses have great practical utility, but can only be approximate.

We anticipate that the sequence of origin of mammalian characters will become better known with further research. In an attempt to be as unequivocal as possible, given present knowledge, we have arbitrarily settled on a single diagnostic character that is sufficiently complex to make the occurrence of homoplasy unlikely, or at least easily recognized should homoplasy have occurred. We take this as the closest practical approach that we can presently make to a true diagnosis of the taxon MAMMALIA. The following hard part character complex (H1 in the list given on page 508) would appear adequate to diagnose crown group mammals (monotremes, therians, their common ancestor, and all descendants of that ancestor, including multituberculates): accessory jaw bones (except coronoid bone and possibly the splenial bone if it were simply lost) shifted away from the craniomandibular joint to become associated with the cranium alone. The remaining 16 hard part characters that we list are less accessible and therefore less practical in the formation of an approximate diagnosis.

APPENDIX C

Some Close Mammalian Outgroups (Nonmammalian Mammaliaforms)

Unnamed rank **MAMMALIAFORMES** Rowe, 1988:250.[1]

    [1] Using the rank of superclass here would highlight a major problem of using named ranks in cladistic hierarchies. If the remaining nonmammalian synapsids are to be placed in the same hierarchy as MAMMALIA, then a whole series of ranks between that of MAMMALIA and that of SYNAPSIDA would theoretically be needed to express the cladistic structure.

    †*Tricuspes* E. von Huene, 1933.[1]
        L. Trias.; Eu.
        [1] Status questionable.

    †*Adelobasileus* Lucas & A. Hunt, 1990:42.[1]
        L. Trias.; N.A.
        [1] Status questionable.

    Family †**Theroteinidae** Sigogneau-Russell, Frank & Hemmerle, 1986:107.
        [= †THEROTEINIDA Hahn, Sigogneau-Russell & Wouters, 1989:211.[1]]
        L. Trias.; Eu.
        [1] Utilized at ordinal rank.

    †*Theroteinus* Sigogneau-Russell, Frank & Hemmerle, 1986:107.
        L. Trias.; Eu.

    Family †**Kollikodontidae** Flannery, Archer, Rich & Jones, 1995:418.
        E. Cret.; Aus.

    †*Kollikodon* Flannery, Archer, Rich & Jones, 1995:418.
        E. Cret.; Aus.

    Family †**Sinoconodontidae** Mills, 1971:45, 62.
        [= †SINOCONODONTIFORMES Kinman, 1994:37.]
        E. Juras.; As.[1]
        [1] E. Asia.

    †*Sinoconodon* Patterson & Olson, 1961:133.
        E. Juras.; As.[1]
        [1] E. Asia.

    †*Lufengoconodon* Yang, 1982:23.[1]
        E. Juras.; As.[2]
        [1] (C. C. Young, posthumous.)
        [2] E. Asia.

Order †**MORGANUCODONTA** Kermack, Mussett & Rigney, 1973:87, 110.[1]
        [= †MORGANUCODONTIFORMES Kinman, 1994:37.]
        [1] Proposed as a suborder. Elevated to rank of order by McKenna, in Stucky & McKenna, in Benton, ed., 1993:740.

    Family †**Morganucodontidae** Kühne, 1958:222.
        [= †Eozostrodontidae Hopson, 1969:211[1]; †Eozostrodontoidea Kalandadze & Rautian, 1992:48.]
        ?L. Trias., E. Juras.; As. L. Trias.-M. Juras.; Eu. E. Juras.; Af. E. Juras.; N.A.
        [1] Hopson's formal term is in violation of Article 40 of the International Code of Zoolgical Nomenclature. The term "eozostrodont" has often been used informally, however.

    †*Morganucodon* Kühne, 1949:345-350.
        L. Trias.-E. Juras.; Eu. E. Juras.; As.[1] E. Juras.; N.A.
        [1] E. Asia.

    †*Eozostrodon* Parrington, 1941:141.
        ?L. Trias.; As. L. Trias.; Eu.

    †*Brachyzostrodon* Sigogneau-Russell, 1983:238.
        L. Trias.; Eu.

    †*Helvetiodon* Clemens, 1980:81.
        [= †*Helveticodon* Tatarinov, 1985:139.[1]]
        L. Trias.; Eu.
        [1] Misspelling or attempted correction.

†*Erythrotherium* Crompton, 1964:442.
  E. Juras.; Af.

†*Wareolestes* Freeman, 1979:158.
  M. Juras.; Eu.

Order †**DOCODONTA** Kretzoi, 1946:111.
  [= †DOCODONTIFORMES Kinman, 1994:37.] [Including †TEGOTHERIDIA Tatarinov, 1994:129.[1]]
  [1] Described as an order of superorder †SYMMETRODONTA.

†*Dinnetherium* Jenkins, Crompton & Downs, 1983:1233.
  E. Juras.; N.A.

Family †**Megazostrodontidae** Gow, 1986:22.
  E. Juras.; Af.

†*Megazostrodon* Crompton & Jenkins, 1968:428.
  E. Juras.; Af.

Family †**Docodontidae** Simpson, 1929:84.
  [= †Diplocynodontidae Marsh, 1887:338; †Dicrocynodontidae Osborn, 1888:263; †Docodontoidea Butler, 1939:353; †Peraiocynodontidae Kretzoi, 1946:111.] [Including †Tegotheriidae Tatarinov, 1994:129.]
  M.-L. Juras.; Eu. L. Juras.; As. L. Juras.; N.A.

†*Borealestes* Waldman & R. J. G. Savage, 1972:122.
  M. Juras.; Eu.

†*Simpsonodon* Kermack, Lee, Lees & Mussett, 1987:1, 5.
  M. Juras.; Eu.

†*Tegotherium* Tatarinov, 1994:129.
  L. Juras.; As.

†*Haldanodon* Kühne & Krusat, 1972:300.[1]
  L. Juras.; Eu.
  [1] *Nomen nudum* in Kühne, 1968.

†*Docodon* Marsh, 1881:512-513.
  [Including †*Dicrocynodon* Osborn, 1888:263[1]; †*Diplocynodon* Marsh, 1880:235[2]; †*Ennacodon* Marsh, 1890:15; †*Enneodon* Marsh, 1887:339[3]; †*Peraiocynodon* Simpson, 1928:125.]
  L. Juras.; Eu. L. Juras.; N.A.
  [1] From Marsh (MS).
  [2] Objective synonym of †*Dicrocynodon*. Not †*Diplocynodon* Pomel, 1846: 372, a reptile.
  [3] Objective synonym of †*Ennacodon*. Not †*Enneodon* Prangner, 1845, a reptile.

Order †**HARAMIYOIDEA** Hahn, 1973:3, **new rank**.[1]
  [= †HARAMIYIDA Hahn, Sigogneau-Russell & Wouters, 1989:211.]
  [1] Proposed as suborder of †MULTITUBERCULATA; used as infraclass (as "HARAMYOIDEA") by McKenna, in Stucky & McKenna, in Benton, ed., 1993:742. Taxonomic position somewhat uncertain.

Family †**Haramiyidae** Simpson, 1947:497.[1]
  [= †Microlestidae Murray, 1866:xvi, 364; †Thomasiidae Poche, 1908; †Microcleptidae Simpson, 1928:52.]
  L. Trias.-E. Juras., ?M. Juras.; Eu. ?L. Trias., E. Juras.; N.A.
  [1] See ICZN, Art. 40.

†*Mojo* Hahn, Lepage & Wouters, 1987:40.[1]
  L. Trias.; Eu.
  [1] Assignment to the †Haramiyidae is questionable.

†*Hypsiprymnopsis* Dawkins, 1864:406.
  L. Trias.; Eu.

†*Haramiyavia* Jenkins, Gatesy, Shubin & Amaral, 1997:715.
  L. Trias. and/or E. Juras.; N.A.[1]
  [1] E. Greenland.

†*Thomasia* Poche, 1908:269.
  [= †*Microlestes* Plieninger, 1847:165[1]; †*Plieningeria* Krausse, 1919:50.] [= or including †*Microcleptes* Simpson, 1928:55[2]; †*Haramiya* Simpson, 1947:497.]
  L. Trias.-E. Juras., ?M. Juras.; Eu. E. Juras.; N.A.
  [1] Not *Microlestes* Schmidt-Goebel, 1846: 41, a coleopteran.
  [2] Not *Microcleptes* Newman, 1840, a coleopteran.

APPENDIX D
Taxonomic Suffixes Denoting Rank

We believe that taxonomic prefixes and suffixes are becoming less useful as computers rather than paper provide the structured storage systems we call classifications. Nonetheless, such verbal clues are still broadly useful although their implications are not always well understood. For those who wish to use them, various schemes of prefixes and suffixes already exist; they do not need re-invention. We provide information about them here.

Convenient suffixes have come to be employed for the names of taxonomic levels of the family-group, so that one can instantly recognize their relative rank [see Article 29(a) of the ICZN]. Attempts to extend that system higher up the ladder have met with resistance and those endings are not particularly standardized, even now (Brown 1957; Hendler and Pawson 1988; Steyskal 1988). Readers can gain entry into this legalistic world by reading Stenzel (1950), Simpson (1952, 1961), Brothers (1983), Bour and Dubois (1984), Telford and Mooi (1986), Starobogatov (1991), and Bock (1994). Some common sense about changing the names of higher taxa has been offered by Ghiselin (1977), Minelli (1991), and Christofferson (1995).

If they are to be used at all above the family-group, then at the very least suffixes should be standardized. That would be helpful in remembering the relative position of ranked levels. Inasmuch as Bour and Dubois' (1984) paper, mentioned above, is not readily available, we repeat in translation here some of their suggested levels and appropriate suffixes within the family-group, latinized to meet the requirements of Article 11(f) of the ICZN:

	OID-	ID-	IN-	IT-	IL-	IS-
ES	OIDES hyperfamily					
EA	OIDEA superfamily					
AE	OIDAE epifamily	IDAE family	INAE subfamily			
EI			INEI infrafamily			
I			INI tribe			
A			INA subtribe	ITA infratribe		
OI				ITOI clan	ILOI subclan	ISOI infraclan
OA						ISOA chaste

Unfortunately, the suffix proposed for hyperfamily by Bour and Dubois is sometimes used for genera (e.g., †*Deltatheroides*), as is the suffix for subtribe (e.g., *Blarina*).

For proposed uniform suffixes above the family-group level, one should consult the tables presented by Starobogatov (1991). Starobogatov (following Stys and Kerzhner 1975) distinguished between "typified" and "descriptive" names, proposing that the latter should be gradually phased out in favor of an extension upward of the rules governing the family-group, coupled with unique suffixes recommended for various levels. He proposed "order-groups," class-groups," and "phylum-groups," but the ICZN has yet to grapple with the problem (Savage 1990). Ideally, one might wish for uniform rules for all taxa (Scudder 1872:348), but inertia, familiarity, and practicality must also be considered. For some of the levels used by us here, the following suffixes were recommended by Starobogatov:

Class	iodes
Subclass	iones
Infraclass	ioni
Cohort	omorphi
Superorder	iformii
Order	iformes
Suborder	oidei
Infraorder	oinei

In our view, mammalogists and paleomammalogists are not ready for such regimentation (Pearse 1936; Levine 1958). We list these suffixes here in order to point out that, if they are needed, the work has already been done and there is no need to invent still more suffixes in competition with those already proposed. However, in the future, suffixes may not be needed at all (de Queiroz and Gauthier 1994). As Minelli (1991: 187) noted, a change in rank that requires a change in suffix generates instability. Unique tags without regimented suffixes above the ICZN's self-limited family-group level have the advantage of not loading down systematists with additional onerous housekeeping chores, a fact that we believe outweighs the convenience of learning relative rank of higher categories from the suffixes.

REFERENCES

Abe, J. M. and N. Papavero. 1992. *Teoria Intuitiva dos Conjuntos*. São Paulo: McGraw-Hill, Makron Books.

Adams, E. N., III. Consensus techniques and the comparison of taxonomic trees. *Systematic Zoology* 21(4): 390-397.

Adanson, M. 1757. *Histoire Naturelle du Sénégal. Coquillages. Avec la relation abrégée d'un voyage fait en ce pays, pendant les années 1749, 50, 51, 52 & 53*. Paris: Claude-Jean-Baptiste Bauche.

_____. 1763. *Familles des plantes: Partie 2*. Paris: Vincent.

_____. 1764. *Familles des Plantes: Partie 1*. Paris: Vincent.

[Both parts reprinted in 1966 in the same volume, with an introduction by F. A. Stafleu. Historiae Naturalis Classica, vol. 46. Codicote, Herts: Wheldon and Wesley; New York: Stechert-Hafner.]

Agassiz, A. 1872-1874. Illustrated Catalogue of the Museum of Comparative Zoology at Harvard College. No. 7. Revision of the Echini, Parts 1-4. *Memoirs of the Museum of Comparative Zoology* 3:i-xii, 1-762.

Agassiz, J. L. R. [Louis]. 1848. *Nomenclatoris Zoologici index universalis, continens nomina systematica classium, ordinum, familiarum et generum animalium omnium, tam viventium quam fossilium, secundum ordinem alphabeticum unicum disposita, adjectis Homonymiis Plantarum*. Solothurn: Jent et Gassmann.

_____. 1857. *Contributions to the Natural History of the United States of America*, vol. 1. Boston: Little, Brown. [Reprinted in 1978. New York: Arno Press.]

_____. 1859. *An Essay on Classification*. London: Longman, Brown, Green, Longmans, & Roberts, and Trübner.

Allin, E. F. 1975. Evolution of the mammalian middle ear. *Journal of Morphology* 147(4): 403-437.

Allin, E. F. and J. A. Hopson. 1992. Evolution of the auditory system in Synapsida ("mammal-like reptiles" and primitive mammals) as seen in the fossil record. In D. B. Webster, R. R. Fay, and A. N. Popper, eds., *The Evolutionary Biology of Hearing*, pp. 587-614. New York: Springer Verlag.

Amadon, D. 1966. Another suggestion for stabilizing nomenclature. *Systematic Zoology* 15(1): 54-58.

American Ornithologists' Union. 1886. *The Code of Nomenclature and Check-List of North American Birds, adopted by the American Ornithologists' Union*. New York: American Ornithologists' Union.

_____. 1892. *The Code of Nomenclature adopted by the American Ornithologists' Union*. New York: American Ornithologists' Union. [Identical to the 1886 Code but published separately from the check-list and supplied with an index.]

Anderson, E. 1940. The concept of the genus. 2. A survey of modern opinion. *Bull. Torrey Botanical Club* 67(5): 363-369.

Anderson, L. 1976. Charles Bonnet's taxonomy and chain of being. *Journal of the History of Ideas* 37(1): 45-58.

_____. 1982. *Charles Bonnet and the order of the known*. Studies in the History of Modern Science 11. Dordrecht: D. Riedel.

Anderson, S. 1974. Some suggested concepts for improving taxonomic dialogue. *Systematic Zoology* 23(1): 58-70.

_____. 1975. On the number of categories in biological classifications. *American Museum Novitates* 2584:1-9.

Archibald, J. D. 1994. Metataxon concepts and assessing possible ancestry using phylogenetic systematics. *Systematic Biology* 43(1): 27-40.

Arthur, J. C., J. H. Barnhart, N. L. Britton, F. E. Clements, O. F. Cook, F. V. Coville, F. S. Earle, A. W. Evans, T. E. Hazen, A. Hollick, M. A. Howe, F. H. Knowlton, G. T. Moore, H. H. Rushby, C. L. Shear, L. M. Underwood, D. White, and W. W. Wight. 1907. American Code of Botanical Nomenclature. *Bulletin of the Torrey Botanical Club* 34:167-178.

Ashlock, P. D. 1971. Monophyly and associated terms. *Systematic Zoology* 20(1): 63-69.

_____. 1973. Monophyly again. *Systematic Zoology* 21(4): 430-438. [Dated 1972 but mailed 16 January 1973.]

_____. 1980. An evolutionary systematist's view of classification. *Systematic Zoology* 28(4): 441-450. [Dated 1979 but mailed 23 January 1980.]

Augier, A. 1801. *Essai d'une nouvelle classification des végétaux conforme à l'ordre que la Nature paroit avoir suivi dans le règne végétal: D'ou resulte une Méthode qui conduit à la conoissance des plantes & de leur rapports naturels*. Lyon: Bruyset Ainé. [Not seen, but see Candolle 1813:74 and Stevens 1983: fig. 1.]

Ax, P. 1985. Stem species and the stem lineage concept. *Cladistics* 1(3): 279-287.

_____. 1987. *The Phylogenetic System: The Systematization of Organisms on the Basis of their Phylogenesis*. Translated from the original German [Ax 1984] by R. P. S. Jefferies. Chichester and New York: Wiley .

_____. 1988. *Systematik in der Biologie*. Stuttgart: Gustav Fischer.

Bacskai, J. A., L. J. Bryant, J. T. Gregory, G. V. Shkurkin, and M. C. Winans, eds. 1983. *Bibliography of Fossil Vertebrates 1973-1977*. 2 vols.

Falls Church, Va.: American Geological Institute.

Ball, I. R. 1983. On groups, existence and the ordering of nature. *Systematic Zoology* 32(4): 446-451.

Balme, D. M. 1961. Aristotle's use of differentiae in zoology. In S. Mansion, ed., *Aristote et les problèmes de méthode, Symposium Aristotelicum 2nd, 1960*, pp. 195-212. Louvain. [Revised 1975. In J. Barnes, M. Schofield and R. Sorabji, eds., *Articles on Aristotle.* Vol. 1: *Science*, pp. 183-193. London: Duckworth.].

———. 1962. Genos and Eidos in Aristotle's biology. *Classical Quarterly*, n. ser., 12: 81-98.

Banks, N. and A. N. Caudell. 1912. *The Entomological Code.* Washington, D.C.: Judd and Detweiller.

Barghusen, H. R. and J. A. Hopson. 1970. Dentary-squamosal joint and the origin of mammals. *Science* 168:573-575.

Bartlett, H. H. 1940. The concept of the genus. Part 1. History of the generic concept in botany. *Bulletin of the Torrey Botanical Club* 67(5): 349-362.

Bather, F. A. 1927. Biological classification: Past and future. Anniversary Address of the President. *Quarterly Journal of the Geological Society of London* 83(2): lxii-civ.

Beatty, J. 1982. Classes and cladists. *Systematic Zoology* 31(1): 25-34.

Benton, M. J. 1996. Testing the time axis of phylogenies. In P. H. Harvey, A. J. L. Brown, J. Maynard Smith, and S. Nee, eds., *New Uses for New Phylogenies*, pp. 217-233. Oxford: Oxford University Press.

Berggren, W. A., D. V. Kent, C. C. Swisher, III, and M.-P. Aubry. 1995. A revised Cenozoic geochronology and chronostratigraphy. In W. A. Berggren, D. V. Kent, M.-P. Aubry, and J. Hardenbol, eds., *Geochronology, Time Scales and Global Stratigraphic Correlation*, pp. 129-212. Tulsa: SEPM (Society for Sedimentary Geology), Special Publication 54.

Berggren, W. A., D. V. Kent, and J. A. van Couvering. 1985. The Neogene: Part 2. Neogene geochronology and chronostratigraphy. In N. J. Snelling, ed., *The Chronology of the Geological Record*, pp. 211-260. Geological Society Memoir No. 10. London: Blackwell.

Berggren, W. A. and D. R. Prothero. 1992. Eocene-Oligocene climatic and biotic evolution: An overview. In D. R. Prothero and W. A. Berggren, eds., *Eocene-Oligocene Climatic and Biotic Evolution*, pp. 1-28. Princeton: Princeton University Press.

Berlin, B. 1973. Folk systematics in relation to biological classification and nomenclature. *Annual Review of Ecology and Systematics* 4:259-271.

———. 1992. *Ethnobiological Classification: Principles of Categorization of Plants and Animals in Traditional Societies.* Princeton: Princeton University Press.

Berlin, B., D. E. Breedlove, and P. H. Raven. 1973. General principles of classification and nomenclature in folk biology. *American Anthropologist* 75(1): 214-242.

Bicheno, J. E. 1827. On systems and methods in natural history. *Transactions of the Linnaean Society of London* 15:479-496. [Reprinted in 1828, *Philosophical Magazine*, ser. 2, 3:213-219, 265-271.]

Bigelow, R. S. 1961. Higher categories and phylogeny. *Systematic Zoology* 10(2): 86-91.

Blackburn, D. G. 1991. Evolutionary origins of the mammary gland. *Mammal Review* 21(2): 81-96.

Blackwelder, R. E. 1967. *Taxonomy: A Text and Reference Book.* New York: Wiley.

Blanchard, R. 1890. *Règles de la Nomenclature des êtres Organisés adoptées par le Congres Internationale de Zoologie, 1889.* Paris. [Not seen].

———. 1905. *Règles internationales de la nomenclature zoologique adoptées par les Congrès Internationaux de Zoologie.* Paris: F. R. Rudeval. [English language section reprinted in 1926, with cited opinions, as: International rules of zoological nomenclature. *Proceedings of the Biological Society of Washington* 39:75-104.]

Bleeker, P. 1859. Enumeratio specierum piscium hucusque in Archipelago Indico observatorum. Verhandelingen der Natuurkundige Vereeniging in Nederlandsch Indie 6(3): i-xxxvi, 1-276.

Bock, W. J. 1974. Philosophical foundations of classical evolutionary classification. *Systematic Zoology* 22(4): 375-392. [Dated December 1973 but mailed 29 January 1974.]

———. 1977. Foundations and methods of evolutionary classification. In M. K. Hecht, P. C. Goody, and B. M. Hecht, eds., *Major Patterns in Vertebrate Evolution*, pp. 851-895. New York: Plenum.

———. 1994. History and nomenclature of avian family-group names. *Bulletin of the American Museum of Natural History* 222:1-281.

Bogan, A. E. and E. E. Spamer. 1995. Comment on *Towards a harmonized bionomenclature for life on Earth* (Hawksworth *et al.*, 1994). *Bulletin of Zoological Nomenclature* 52(2): 126-136.

Böger, H. 1989. The stem-group problem. In N. Schmidt-Kittler and R. Willmann, eds., *Phylogeny and the Classification of Fossil and Recent Organisms: Proceedings of a Symposium Organized by the Deutsche Forschungsgemeinschaft*, pp. 45-52. Abhandlungen des Naturwissenschaftlichen Vereins in Hamburg (NF) 28. Hamburg: Verlag Paul Parey.

Bonaparte, J. F. and G. Rougier. 1987. Mamíferos del Cretácico inferior de Patagonia. *IV Congreso Latinoamericano de Paleontólogia, Bolivia* 1:343-359.

Bonde, N. 1977. Cladistic classification as applied to vertebrates. In M. K. Hecht, P. C. Goody, and B. M. Hecht, eds., *Major Patterns in Vertebrate*

Evolution, pp. 741-804. NATO Advanced Studies Institute Symposium 14. New York: Plenum.

Bonnet, C. 1745. *Traité d'Insectologie, ou Observations sur les Pucerons*. Première Partie. Paris: Durand. [Not seen. See O'Hara 1992:158; Daudin 1926:237.]

_____. 1764. *Contemplation de la Nature*. 2 vols. Amsterdam: Marc-Michel Rey.

Boole, G. 1854. *An Investigation of the Laws of Thought, on Which Are Founded the Mathematical Theories of Logic and Probabilities*. London: Macmillan. [Reprinted in 1958. New York: Dover.]

Boucot, A. J. 1979. Cladistics: Is it really different from classical taxonomy? In J. Cracraft and N. Eldredge, eds., *Phylogenetic Analysis and Paleontology*, pp. 199-210. New York: Columbia University Press.

_____. 1995. The reality of higher taxa and the question of intermediate forms. *Revista Española de Paleontología* 10(1): 2-8.

Boudreaux, H. B. 1987. *Arthropod Phylogeny with Special Reference to Insects*. Malabar, Fla.: Krieger. [Reprint of 1st edition, 1979. New York: Wiley.]

Bouquet, M. 1996. Family trees and their affinities: The visual imperative of the genealogical diagram. *Journal of the Royal Anthropological Institute* 2(1): 43-66.

Bour, R. and A. Dubois. 1984. Nomenclature ordinale et familiale des tortues (Reptilia). *Stvdia Geologica Salmanticensia*, vol. especial 1 (Stvdia Palaeocheloniologica 1): 77-86.

_____. 1986. Nomenclature ordinale et familiale des tortues (Reptilia). Note Complémentaire. *Bulletin mensuel de la Société Linnéenne de Lyon* 55(3): 87- 90.

Bradley, J. C. 1962. Difficulty arising from compulsory application of the law of priority to family-group names, with proposed amendments. *Systematic Zoology* 11(4): 178-179.

Brady, R. H. 1982. Theoretical issues and "pattern cladistics." *Systematic Zoology* 31(3): 286-291.

Bremekamp, C. E. B. 1953a. A re-examination of Cesalpino's classification. *Acta Botanica Neerlandica* 1(4): 580-593.

_____. 1953b. Linné's views on the hierarchy of the taxonomic groups. *Acta Botanica Neerlandica* 2(2): 242-253.

Brickell, C. D., E. G. Voss, A. F. Kelly, F. Schneider, and R. H. Richens, eds. 1980. International Code of Nomenclature of Cultivated Plants, 1980. *Regnum Vegetabile* 104:1-31.

Briquet, J. 1905. Texte Synoptique des Documents Destinés a Servir de Base aux Débats du Congrès International de Nomenclature Botanique de Vienne, 1905, présenté au nom de la Commission internationale de Nomenclature botanique. Berlin: R.

Friedländer.

Broadfield, A. 1946. *The Philosophy of Classification*. London: Grafton.

Brothers, D. J. 1983. Nomenclature at the ordinal and higher levels. *Systematic Zoology* 32(1): 34-42.

Brown, W. L., Jr. 1958. The ending for subtribal names in zoology. *Systematic Zoology* 6(4): 193-194.

Brundin, L. 1966. Transantarctic relationships and their significance, as evidenced by chironomid midges. *Kungliga Svenska Vetenskapsakademiens Handlingar*, ser. 4, 11(1): 1-472.

_____. 1972. Evolution, causal biology, and classification. *Zoologica Scripta* 1(3-4): 107-120.

Bryant, H. N. 1994. Comments on the phylogenetic definition of taxon names and conventions regarding the naming of crown clades. *Systematic Biology* 43(1): 124-130.

Buchanan, R. E. and R. S. Breed, eds. 1948. International Bacteriological Code of Nomenclature. *Journal of Bacteriology* 55:287-306.

Buchanan, R. E., et al., eds. 1958. *International Code of Nomenclature of Bacteria and Viruses: Bacteriological Code*. Revised ed. Ames: Iowa State University Press. [Approved by Plenary Session of VII International Congress of Microbiology, Rome, Italy, September, 1953; edited by the editorial board of the International Committee on Bacteriological Nomenclature.]

Buck, R. C. and D. L. Hull. 1966. The logical structure of the Linnaean hierarchy. *Systematic Zoology* 15(2): 97-111.

Burkhardt, F., ed. 1996. *Charles Darwin's Letters: A Selection 1825-1859*. Cambridge: Cambridge University Press.

Burma, B. H. 1954. Reality, existence, and classification. *Madroño* 12:193-209.

Burtt, B. L. 1966. Adanson and modern taxonomy. *Notes from the Royal Botanic Garden, Edinburgh* 26(4): 427-431.

Cain, A. J. 1956. The genus in evolutionary taxonomy. *Systematic Zoology* 5(3): 97-109.

_____. 1958. Logic and memory in Linnaeus's system of taxonomy. *Proceedings of the Linnean Society of London* 169(1-2): 144-163.

_____. 1959a. Deductive and inductive methods in post-Linnaean taxonomy. *Proceedings of the Linnean Society of London* 170(2): 185-217.

_____. 1959b. Taxonomic concepts. *Ibis* 101:302-318.

_____. 1959c. The post-Linnaean development of taxonomy. *Proceedings of the Linnean Society of London* 170(3): 234-244.

_____. 1993. Linnaeus's *Ordines naturales*. *Archives of Natural History* 20(3): 405-415.

Cain, A. J. and G. A. Harrison. 1960. Phyletic weighting. *Proceedings of the Zoological Society of London* 135(1): 1-31.

Camp, C. L. and H. J. Allison, eds. 1961. *Bibliography of Fossil Vertebrates 1949-1953.* Geological Society of America, Memoir 84.

Camp, C. L., H. J. Allison, and R. H. Nichols, eds. 1964. *Bibliography of Fossil Vertebrates 1954-1958.* Geological Society of America, Memoir 92.

Camp, C. L., H. J. Allison, R. H. Nichols, and H. McGinnis, eds. 1968. *Bibliography of Fossil Vertebrates 1959-1963.* Geological Society of America, Memoir 117.

Camp, C. L., B. Brajnikov, E. Fulton, and J. A. Bacskai, eds. 1972. *Bibliography of Fossil Vertebrates 1964-1968.* Geological Society of America, Memoir 134.

Camp, C. L., D. N. Taylor, and S. P. Welles, eds. 1942. *Bibliography of Fossil Vertebrates 1934-1938.* Geological Society of America, Special Paper 42.

Camp, C. L. and V. L. Vanderhoof, eds. 1940. *Bibliography of Fossil Vertebrates 1928-1933.* Geological Society of America, Special Paper 27.

Camp, C. L., S. P. Welles, and M. Green, eds. 1949. *Bibliography of Fossil Vertebrates 1939-1943.* Geological Society of America, Memoir 37.

_____. 1953. *Bibliography of Fossil Vertebrates 1944-1948.* Geological Society of America, Memoir 57.

Candolle, A[lphonse]-L.-P.-P. de. 1867. *Lois de la nomenclature botanique adoptées par le Congrès International de Botanique... en... 1867.* Geneva and Basel: H. Géorg.

Candolle, A[ugustin]-P. de. 1813. *Théorie Élémentaire de la Botanique, ou exposition des principes de la classification naturelle et de l'art de décrire et d'étudier les végétaux.* Paris: Déterville. [Not seen. See 3rd edition, 1844. Paris: Roret.]

_____. 1818-1821. *Regni Vegetabilis Systema Naturale, sive ordines, genera et species plantarum, secundum methodi naturalis normas digestarum et descriptarum.* 2 vols. Paris.

Cannatella, D. C. 1991. [Review of] N. Schmidt-Kittler and R. Willmann, eds., 1989, Phylogeny and classification of fossil and Recent organisms. *Systematic Zoology* 40(3): 376-378.

Carroll, R. L. 1988. *Vertebrate Paleontology and Evolution.* New York: W. H. Freeman.

Cesalpino, A. [Andreae Caesalpini] 1583. *De Plantis Libri XVI.* Florence: Georgium Marescottum.

Charig, A. J. 1982. Systematics in biology: A fundamental comparison of some major schools of thought. In K. A. Joysey and A. E. Friday, eds., *Problems of Phylogenetic Reconstruction,* pp. 363-440. The Systematics Association Special Volume No. 21. London: Academic Press.

Choate, H. A. 1912. The origin and development of the binomial system of nomenclature. *The Plant World* 15:257-263.

Christoffersen, M. L. 1995. Cladistic taxonomy, phylogenetic systematics, and evolutionary ranking. *Systematic Biology* 44(3): 440-454.

Clark, R. B. 1956. Species and systematics. *Systematic Zoology* 5(1): 1-10.

Clerck, C. A. 1757. Svenska Spindlar. *Uti sina hufvud-slägter indelte samt under några och sextio särskildte arter beskrefne: Och med illuminerade figurer uplyste . . .* Stockholm: Literis Laurentii Salvii.

Colless, D. H. 1977. A cornucopia of categories. *Systematic Zoology* 26(3): 349-352.

_____. 1985. On "character" and related terms. *Systematic Zoology* 34(2): 229-233.

Condorcet, M.-J.-A.-N. de C., Marquis de. 1777. Sur les familles naturelles des plantes, et en particulier sur celle des Renoncules. *Histoire de l'Académie Royal des Sciences (Paris)* 1773:34-36.

Conisbee, L. R. 1953. *A List of the Names Proposed for Genera and Subgenera of Recent Mammals from the Publication of T. S. Palmer's 'Index Generum Mammalium' 1904 to the end of 1951.* London: British Museum (Natural History).

_____. 1960. Newly proposed genera, 1952-56. *Journal of Mammalogy* 41(1): 112-113.

_____. 1964. Newly proposed genera, 1957-1961. *Journal of Mammalogy* 45(3): 474-475.

Conklin, H. C. 1962. Lexicographical treatment of folk taxonomies. *International Journal of American Linguistics* 28:119-141.

_____. 1980. *Folk classification. A topically arranged bibliography of contemporary and background references through 1971.* Revised reprinting and author index. New Haven: Yale University Department of Anthropology.

Corbet, G. B. and J. E. Hill. 1980. *A World List of Mammalian Species.* London: British Museum (Natural History). [2nd edition published in 1986; 3rd edition, 1991.]

Cracraft, J. 1974. Phylogenetic models and classification. *Systematic Zoology* 23(1): 71-90.

_____. 1981. Pattern and process in paleobiology: The role of cladistic analysis in systematic paleontology. *Paleobiology* 7:456-468.

Craske, A. J. and R. P. S. Jefferies. 1989. A new mitrate from the Upper Ordovician of Norway, and a new approach to subdividing a plesion. *Palaeontology* 32(1): 69-99.

Craw, R. 1992. Margins of cladistics: Identity, difference and place in the emergence of phylogenetic systematics, 1864-1975. In P. Griffiths, ed., *Trees of Life: Essays in Philosophy of Biology,* pp. 65-107. Dordrecht: Kluwer.

Croizat, L. 1945. History and nomenclature of the higher units of classification. *Bulletin of the Torrey Botanical Club* 72(1): 52-75.

Crompton, A. W. and A. Sun. 1985. Cranial structure and relationships of the Liassic mammal *Sinoconodon. Zoological Journal of the Linnean Society*

85:99-119.

Croneis, C. 1939. A military classification for fossil fragments. *Science* 89:314-315.

Crowson, R. A. 1970. *Classification and Biology.* London: Heinemann. [2nd printing, 1972. Chicago: Aldine.]

Crusafont Pairó, M. 1962. Constitución de una nueva clase (Ambulatilia) para los llamados "reptiles mamiferoides." *Notas y Comunicaciones del Instituto Geologico y Minero de España* 66:259-266

Cuvier, Baron G. L. C. F. D. 1798. *Tableau Élémentaire de l'Histoire Naturelle des Animaux.* Paris: Baudouin.

———. 1817. *Le Règne animal distribué d'après son Organization. Tome I, contenant l'introduction, les mammifères et les oiseaux.* Paris: Deterville. [Possibly published in part in late 1816; see Whitehead 1967:300 and Bock 1994:233, 247.] [Also published in English as *The Animal Kingdom Arranged in Conformity with its Organization.* 15 vols. (1817-1835). London: William S. Orr.]

Dall, W. H. 1878. Report of the Committee on Zoological Nomenclature to Section B, of the American Association for the Advancement of Science, at the Nashville Meeting, August 31, 1877. *Proceedings of the American Association for the Advancement of Science* 26:7-56. [This publication is sometimes erroneously cited as Dall 1877.]

Dana, J. D., S. S. Haldeman, D. H. Storer, A. A. Gould, A. Binney, C. U. Shepard, C. Dewey, J. D. Whelpley, and E. C. Herrick. 1846. Report on scientific nomenclature, made to the Association of American Geologists and Naturalists, New Haven, May 1845. *Association of American Geologists and Naturalists, Proceedings for 1845*: 3-7.

Dantzig, T. 1954. *Number, the Language of Science.* Paperback reprint of 4th ed. Garden City, N.Y.: Doubleday Anchor.

Darwin, C. 1859. *On the Origin of Species by Means of Natural Selection, or the Preservation of Favoured Races in the Struggle for Life.* London: John Murray.

Daudin, H. 1926. *De Linné a Jussieu. Méthodes de la Classification et Idée de Série en Botanique et en Zoologie (1740-1790).* Études d'Histoire des Sciences Naturelles, vol. 1. Paris: Librairie Félix Alcan.

Davis, J. I. and K. Nixon. 1992. Populations, genetic variation, and the delimitation of phylogenetic species. *Systematic Biology* 41(4): 421-435.

Dawkins, R. 1986. *The Blind Watchmaker.* New York and London: W. W. Norton.

Delson, E. 1977. Catarrhine phylogeny and classification: Principles, methods and comments. *Journal of Human Evolution* 6:433-459.

de Queiroz, K. 1988. Systematics and the Darwinian revolution. *Philosophy of Science* 55(2): 238-259.

———. 1992. Phylogenetic definitions and taxonomic philosophy. *Biology and Philosophy* 7(3): 295-313.

———. 1994. Replacement of an essentialistic perspective on taxonomic definitions as exemplified by the definition of "Mammalia." *Systematic Biology* 43(4): 497-510.

de Queiroz, K. and M. J. Donoghue. 1988. Phylogenetic systematics and the species problem. *Cladistics* 4(4): 317-338.

de Queiroz, K. and J. A. Gauthier. 1990. Phylogeny as a central principle in taxonomy: Phylogenetic definitions of taxon names. *Systematic Zoology* 39(4): 307-322.

———. 1992. Phylogenetic taxonomy. *Annual Review of Ecology and Systematics* 23:449-480.

———. 1994. Toward a phylogenetic system of biological nomenclature. *Trends in Ecology and Evolution* 9(1): 27-31.

Donoghue, M. J. 1985. A critique of the biological species concept and recommendations for a phylogenetic alternative. *Bryologist* 88:172-181.

Donoghue, M. J., J. A. Doyle, J. Gauthier, A. G. Kluge, and T. Rowe. 1989. The importance of fossils in phylogeny reconstruction. *Annual Review of Ecology and Systematics* 20:431-460.

Douvillé, H. 1882. Règles proposées par le Comité de la Nomenclature paléontologique. *Congrès Géologique International, Compte Rendu de la 2me Session, Bologne, 1881*: 594-595.

Doyle, J. A. and M. J. Donoghue. 1987. The importance of fossils in elucidating seed plant phylogeny and macroevolution. *Review of Palaeontology and Palynology* 50:63-95.

Duarte Rodrigues, P. 1986. On the term Character. *Systematic Zoology* 35(1): 140-141.

DuBois, A. 1988. The genus in zoology: A contribution to the theory of evolutionary systematics. *Mémoires du Muséum National d'Histoire Naturelle*, sér. A. Zoologie, 140:6-122.

Duffin, K. E. 1976. *The search for a natural classification: Perspectives on the quinarian episode in ornithology.* Unpublished B.A. thesis, Harvard University.

DuPraw, E. J. 1964. Non-Linnean taxonomy. *Nature* 202:849-852.

Durkheim, É. and M. Mauss. 1903. De quelques formes primitives de classification. *Année sociologique, 6e année*: 1-72.

———. 1963. *Primitive Classification.* Chicago: University of Chicago Press. [Translated from the French by Rodney Needham.]

Dybowski, B. 1926. Synoptisches Verzeichnis mit kurzer Besprechung der Gattungen und Arten dieser Abteilung der Baikalflohkrebse. *Bulletin International de l'Académie Polonaise des Sciences et des Lettres*, sér. B, 1926:1-77.

Edwards, M. A. and A. R. Hopwood, eds. 1966. *Nomenclator Zoologicus.* Vol. 6: *1946-1955.*

London: Zoological Society of London.

Edwards, M. A. and H. G. Vevers, eds. 1975. *Nomenclator Zoologicus.* Vol. 7: *1956-1965.* London: Zoological Society of London.

Eernisse, D. J. and A. G. Kluge. 1993. Taxonomic congruence versus total evidence, and amniote phylogeny inferred from fossils, molecules, and morphology. *Molecular Biology and Evolution* 10(6): 1170-1195.

Ehlers, U. 1985. *Das phylogenetische System der Plathelminthes.* Stuttgart: Gustav Fischer.

Eldredge, N. and J. Cracraft. 1980. *Phylogenetic Patterns and the Evolutionary Process: Method and Theory in Comparative Biology.* New York: Columbia University Press.

Eldredge, N. and S. N. Salthe. 1984. Hierarchy and evolution. In R. Dawkins and M. Ridley, eds., *Oxford Surveys in Evolutionary Biology,* vol. 1, pp. 184-208. Oxford: Oxford University Press.

Ellen, R. 1993. *The Cultural Relations of Classification: An Analysis of Nuavlu Animal Categories from Central Seram.* Cambridge: Cambridge University Press.

Ellerman, J. R. and T. C. S. Morrison-Scott. 1951. *Checklist of Palaearctic and Indian Mammals, 1758 to 1946.* London: British Museum (Natural History). [2nd edition published in 1966.]

Engelmann, G. F. and E. O. Wiley. 1977. The place of ancestor-descendant relationships in phylogeny reconstruction. *Systematic Zoology* 26(1): 1-11.

Ereshefsky, M. 1991. Species, higher taxa, and the units of evolution. *Philosophy of Science* 58:84-101.

Ereshefsky, M., ed. 1992. *The Units of Evolution: Essays on the Nature of Species.* Cambridge, Mass.: MIT Press.

Estes, R., K. de Queiroz, and J. Gauthier. 1988. Phylogenetic relationships within the Squamata. In R. Estes and G. Pregill, eds., *Phylogenetic Relationships of the Lizard Families: Essays Commemorating Charles L. Camp,* pp. 119-281. Stanford: Stanford University Press.

Fahlbusch, V. 1976. Report on the International Symposium on mammalian stratigraphy of the European Tertiary. *Newsletters on Stratigraphy* 5(2/3): 160-167.

Farris, J. S. 1967a. The meaning of relationship and taxonomic procedure. *Systematic Zoology* 16(1): 44-51.

_____. 1967b. Definitions of taxa. *Systematic Zoology* 16(2): 174-175.

_____. 1968. Categorical ranks and evolutionary taxa in numerical taxonomy. *Systematic Zoology* 17(2): 151-159.

_____. 1975. Formal definitions of paraphyly and polyphyly. *Systematic Zoology* 23(4): 548-554. [Dated 1974 but mailed 19 February 1975.]

_____. 1976. Phylogenetic classification of fossils with recent species. *Systematic Zoology* 25(3): 271-282.

_____. 1980. The information content of the phylogenetic system. *Systematic Zoology* 28(4): 483-519. [Dated 1979, but mailed 23 January 1980.]

_____. 1985. The pattern of cladistics. *Cladistics* 1(2): 190-201.

_____. 1991. Hennig defined paraphyly. *Cladistics* 7(3): 297-304.

Flynn, J. J. and C. C. Swisher, III. 1995. Cenozoic South American Land Mammal Ages: Correlation to global geochronologies. In W. A. Berggren, D. V. Kent, M.-P. Aubry, and J. Hardenbol, eds., *Geochronology, Time Scales and Global Stratigraphic Correlation,* pp. 317-333. Tulsa: SEPM (Society for Sedimentary Geology), Special Publication 54.

Forey, P. L. 1992. Fossils and cladistic analysis. In P. L. Forey, C. J. Humphries, I. J. Kitching, R. W. Scotland, D. J. Siebert, and D. M. Williams, *Cladistics: A Practical Course in Systematics,* pp. 124-136. The Systematics Association Publication 10. Oxford: Clarendon Press.

Forey, P. L., C. J. Humphries, I. J. Kitching, R. W. Scotland, D. J. Siebert, and D. M. Williams. 1992. *Cladistics: A Practical Course in Systematics.* The Systematics Association Publication 10. Oxford: Clarendon Press. [Reprinted in 1993, 1995.]

Francki, R. I. B., C. M. Fauquet, D. L. Knudson, and F. Brown. 1990. Classification and nomenclature of viruses. *Archives of Virology,* Suppl. 2:1-445.

Fries, E. M. 1821-1832. *Systema Mycologicum sistens fungorum ordines, genera et species, hucusque cognitas, quas ad normam methodi naturalis determinavit, disposuit atque descripsit Elias Fries.* 4 vols. Lundae: Greifswald (Ernesti Mauritii for vol. 3).

Fries, T. M. 1903. *Linné. Lefnadstekning I-II.* Stokholm: Fahlcrantz. [Swedish.]

Gadow, H. 1898. *A Classification of Vertebrata, Recent and Extinct.* London: Adam and Charles Black.

Gaffney, E. S. 1979. An introduction to the logic of phylogenetic reconstruction. In J. Cracraft and N. Eldredge, eds., *Phylogenetic Analysis and Paleontology,* pp. 79-111. New York: Columbia University Press.

_____. 1980. Phylogenetic relationships of the major groups of amniotes. In A. L. Panchen, ed., *The Terrestrial Environment and the Origin of Land Vertebrates,* pp. 593-610. London: Academic Press.

_____. 1984. Historical analysis of theories of chelonian relationship. *Systematic Zoology* 33(3): 283-301.

Gaffney, E. S. and P. A. Meylan. 1988. A phylogeny of turtles. In M. J. Benton, ed., *The Phylogeny and Classification of the Tetrapods.* Vol. 1: *Amphibians, Reptiles, Birds,* pp. 157-219. The Systematics Association Special Volume 35A. Oxford: Clarendon Press.

Gardiner, B. G. 1982. Tetrapod classification. *Zoological Journal of the Linnean Society* 74(3): 207-232.

———. 1993. Haematothermia: Warm-blooded amniotes. *Cladistics* 9(4): 369-395.

Gauthier, J. 1984. *A cladistic analysis of the higher systematic categories of the Diapsida.* Ph.D. dissertation, Department of Paleontology, University of California, Berkeley. [University Microfilms, Ann Arbor, Michigan, #85-12825.]

———. 1986. Saurischian monophyly and the origin of birds. In K. Padian, ed., *The Origin of Birds and the Evolution of Flight,* pp. 1-55. Memoirs of the California Academy of Sciences 8.

Gauthier, J., D. Cannatella, K. de Queiroz, A. G. Kluge, and T. Rowe. 1989. Tetrapod phylogeny. In B. Fernholm, K. Bremer, and H. Jörnvall, eds., *The Hierarchy of Life,* pp. 337-353. Amsterdam: Elsevier .

Gauthier, J., R. Estes, and K. de Queiroz. 1988. A phylogenetic analysis of Lepidosauromorpha. In R. Estes and G. Pregill, eds., *Phylogenetic Relationships of the Lizard Families: Essays Commemorating Charles L. Camp,* pp. 15-98. Stanford: Stanford University Press.

Gauthier, J., A. G. Kluge, and T. Rowe. 1988a. Amniote phylogeny and the importance of fossils. *Cladistics* 4(2): 105-209.

———. 1988b. The early evolution of the Amniota. In Benton, M. J., ed., *The Phylogeny and Classification of the Tetrapods.* Vol. 1: *Amphibians, Reptiles, Birds,* pp. 103-155. The Systematics Association Special Volume 35A. Oxford: Clarendon Press.

Gauthier, J. and K. Padian. 1985. Phylogenetic, functional, and aerodynamic analyses of the origin of birds and their flight. In M. K. Hecht, J. H. Ostrom, G. Viohl, and P. Wellnhofer, eds., *The Beginnings of Birds,* pp. 185-197. Proceedings of the International *Archaeopteryx* Conference, Eichstätt, 1984.

Gazin, C. L. 1956. The upper Paleocene Mammalia from the Almy Formation in western Wyoming. *Smithsonian Miscellaneous Collections* 131(7): 1-18.

Geoffroy Saint-Hilaire, É. 1812. Tableau des Quadrumanes, ou des Animaux composant le premier Ordre de la Class des Mammifères. *Annales du Muséum d'Histoire Naturelle* 19: 85-122, 156-170.

Geoffroy Saint-Hilaire, É., and G. Cuvier. 1795. Mémoire sur une nouvelle division des mammifères, et sur les principes qui doivent servir de base dans cette sorte de travail. *Magasin Encyclopédique* 2:164-190. [Not seen.]

Geoffroy Saint-Hilaire, I. 1845. Classification parallélique des Mammifères. *Comptes Rendus Hebdomadaires des Seances, Academie des Sciences, Paris* 20:757-761. [Not seen.]

Ghiselin, M. T. 1966a. An application of the theory of definitions to systematic principles. *Systematic Zoology* 15(2): 127-130.

———. 1966b. On psychologism in the logic of taxonomic controversies. *Systematic Zoology* 15(3): 207-215.

———. 1975. A radical solution to the species problem. *Systematic Zoology* 23(4): 536-544. [Dated 1974 but mailed 19 February 1975. Reprinted in Ereshefsky 1992.]

———. 1977. On changing the names of higher taxa. *Systematic Zoology* 26(3): 346-349.

———. 1981. Categories, life, and thinking. *The Behavioral and Brain Sciences* 4:269-313. [With commentaries by M. Bunge, A. L. Caplan, P. A. Corning, W. L. Fink, H. Heise, D. L. Hull, T. D. Johnston, R. K. Jones and A. D. Pick, F. C. Keil, J. B. Kruskal, F. J. Odling-Smee and H. C. Plotkin, A. Packard, E. S. Reed, A. Rosenberg, M. Ruse, S. N. Salthe, S. P. Schwartx, and E. O. Wiley, and a response by M. T. Ghiselin.]

———. 1984a. "Definition," "character," and other equivocal terms. *Systematic Zoology* 33(1): 104-110.

———. 1984b. Narrow approaches to phylogeny: A review of nine books on cladism. In R. Dawkins and M. Ridley eds., *Oxford Surveys in Evolutionary Biology,* vol. 1, pp. 209-222. Oxford: Oxford University Press.

———. 1985. Can Aristotle be reconciled with Darwin? *Systematic Zoology* 34(4): 457-460.

Gill, T. 1871. On the characteristics of the primary groups of the class of mammals. *American Naturalist* 5:526-533.

———. 1872a. Arrangement of the families of mammals with analytical tables. *Smithsonian Miscellaneous Collections* 11(230): i-vi, 1-98.

———. 1872b. On the characteristics of the primary groups of the class of mammals. *Proceedings of the American Association for the Advancement of Science,* Twentieth Meeting, held at Indianapolis, Indiana, August, 1871:284-306.

———. 1896. Some questions of nomenclature. *Proceedings of the American Association for the Advancement of Science* 45:1-31.

———. 1898. Some questions of nomenclature. *Smithsonian Report* for 1896:457-483. [Expanded from paper published in 1896.]

———. 1902. The story of a word--mammal. *Popular Science Monthly* 61:434-438.

———. 1908. Systematic Zoology: Its progress and purpose. *Smithsonian Report* for 1907:449-472.

Gilmour, J. S. L. 1937. A taxonomic problem. *Nature* 139:1040-1042.

_____. 1940. Taxonomy and philosophy. In J. Huxley, ed., *The New Systematics*, pp. 461-474. London: Oxford University Press.

_____. 1951. The development of taxonomic theory since 1851. *Nature* 168:400-402.

_____. 1976. Two early papers on classification (with a Forward by S. M. Walters). *Classification Society Bulletin* 3(4): 2-15. [Reprinted in 1989 in *Plant Systematics and Evolution* 167:97-107.]

Gingerich, P. D. 1977. [Review of] C. Patterson and D. E. Rosen, Review of ichthyodectiform and other Mesozoic fishes and the theory and practice of classifying fossils. *Systematic Zoology* 26(3): 358-359.

_____. 1979. The stratophenetic approach to phylogeny reconstruction in vertebrate paleontology. In J. Cracraft and N. Eldredge, eds., *Phylogenetic Analysis and Paleontology*, pp. 41-77. New York: Columbia University Press.

_____. 1980. Paleontology, phylogeny, and classification: An example from the mammalian fossil record. *Systematic Zoology* 28(4): 451-464. [Dated 1979 but mailed 23 January 1980.]

Good, J. M. 1826. *The Book of Nature*. 3 vols. London: Longden, Rees, Orme, Brown, and Green.

Gould, G. C. 1995. Hedgehog phylogeny (Mammalia, Erinaceidae)—the reciprocal illumination of the quick and the dead. *American Museum Novitates* 3131:1-45.

Gould, S. J. 1991. Fall in the house of Ussher. *Natural History* 100(11): 12-21.

Gould, S. J. and N. Eldredge. 1977. Punctuated equilibria: The tempo and mode of evolution reconsidered. *Paleobiology* 3(2): 115-151.

Gould, S. W. 1954. Permanent numbers to supplement the binomial system of nomenclature. *American Scientist* 42(2): 269-274.

Gow, C. E. 1981. *Pachygenelus, Diarthrognathus* and the double jaw articulation. *Palaeontologia Africana* 24:15.

Gradstein, F. M., F. P. Agterberg, J. G. Ogg, J. Hardenbol, P. Van Veen, J. Thierry, and Z. Huang. 1995. A Triassic, Jurassic and Cretaceous time scale. In W. A. Berggren, D. V. Kent, M.-P. Aubry, and J. Hardenbol, eds., *Geochronology, Time Scales and Global Stratigraphic Correlation*, pp. 95-126. Tulsa: SEPM (Society for Sedimentary Geology), Special Publication 54.

Gray, J. R. 1841-1842. *A list of the genera of birds with their synonyma, and an indication of the typical species of each genus.* 2nd ed. London: Richard and John Taylor.

Gregg, J. R. 1950. Taxonomy, language and reality. *American Naturalist* 84:419-435.

_____. 1954. *The Language of Taxonomy*. New York: Columbia University Press.

_____. 1967. Finite Linnaean structures. *Bulletin of Mathematical Biophysics* 29(2): 191-206.

_____. 1968. Buck and Hull: A critical rejoinder. *Systematic Zoology* 17(3): 342-344.

Gregory, J. T., J. A. Bacskai, B. Brajnikov, and K. Munthe, eds. 1973. *Bibliography of Fossil Vertebrates 1969-1972.* Geological Society of America, Memoir 141.

Gregory, J. T., J. A. Bacskai, and G. V. Shkurkin, eds. 1981. *Bibliography of Fossil Vertebrates 1978.* Falls Church, Va.: American Geological Institute.

Gregory, J. T., J. A. Bacskai, G. V. Shkurkin, and L. J. Bryant, eds. 1981. *Bibliography of Fossil Vertebrates 1979.* Falls Church, Va.: American Geological Institute.

_____. 1983. *Bibliography of Fossil Vertebrates 1980.* Falls Church, Va.: American Geological Institute.

Gregory, J. T., J. A. Bacskai, G. V. Shkurkin, and B. H. Rauscher, eds. 1993. *Bibliography of Fossil Vertebrates 1990.* Lincoln, Neb.: Society of Vertebrate Paleontology.

_____. 1994. *Bibliography of Fossil Vertebrates 1991.* Chicago: Society of Vertebrate Paleontology.

Gregory, J. T., J. A. Bacskai, G. V. Shkurkin, and M. C. Winans, eds. 1984. *Bibliography of Fossil Vertebrates 1981.* Los Angeles: Society of Vertebrate Paleontology.

_____. 1985. *Bibliography of Fossil Vertebrates 1982.* Los Angeles: Society of Vertebrate Paleontology.

_____. 1986. *Bibliography of Fossil Vertebrates 1983.* Los Angeles: Society of Vertebrate Paleontology.

Gregory, J. T., J. A. Bacskai, G. V. Shkurkin, M. C. Winans, and L. J. Bryant, eds. 1987. *Bibliography of Fossil Vertebrates 1984.* Los Angeles: The Society of Vertebrate Paleontology.

Gregory, J. T., J. A. Bacskai, G. V. Shkurkin, M. C. Winans, and B. H. Rauscher, eds. 1988. *Bibliography of Fossil Vertebrates 1985.* Los Angeles: Society of Vertebrate Paleontology.

_____. 1989a. *Bibliography of Fossil Vertebrates 1986.* Los Angeles: Society of Vertebrate Paleontology.

_____. 1989b. *Bibliography of Fossil Vertebrates 1987.* Lincoln, Neb.: Society of Vertebrate Paleontology.

_____. 1991. *Bibliography of Fossil Vertebrates 1988.* Lincoln, Neb.: Society of Vertebrate Paleontology.

_____. 1992. *Bibliography of Fossil Vertebrates 1989.* Lincoln, Neb.: Society of Vertebrate Paleontology.

Gregory, W. K. 1910. The orders of mammals. *Bulletin of the American Museum of Natural History* 27:1-524.

Gregory, W. K. and J. K. Mosenthal. 1910. Outline classification of the Mammalia, Recent and extinct.

In H. F. Osborn, *The Age of Mammals in Europe, Asia and North America*, pp. 511-563. New York: Macmillan.

Greuter, W., F. R. Barrie, H. M. Burdet, W. G. Chaloner, V. Demoulin, D. L. Hawksworth, P. M. Jørgensen, D. H. Nicolson, P. C. Silva, P. Trehane, and J. McNeill. 1994. *International Code of Botanical Nomenclature (Tokyo Code) Adopted by the Fifteenth International Botanical Congress, Yokohama, August-September 1993*. Regnum Vegetabile 131. Königstein, Germany: Koeltz.

Griffiths, G. C. D. 1974. On the foundations of biological systematics. *Acta Biotheoretica* 23(3-4): 85-131.

_____. 1976. The future of Linnaean nomenclature. *Systematic Zoology* 25(2): 168-173.

Günther, A. C. L. G., ed. 1865-1870. *The Record of Zoological Literature*. London: Van Voorst. [Superseded by *Zoological Record*.]

Haeckel, E. 1866. Generelle Morphologie der Organismen. *Allgemeine Grundzüge der Organischen Formen-Wissenschaft, Mechanisch Begründet durch die von Charles Darwin Reformirte Descendenz-Theorie*. Band l: *Allgemeine Anatomie der Organismen*. Berlin: Georg Reimer.

_____. 1874. The gastraea-theory, the phylogenetic classification of the animal kingdom, and the homology of the germ-lamellae. *Quarterly Journal of Microscopical Science* (n.s.) 14:142-165, 223-247. [Translated by E. P. Wright.]

Hagberg, K. 1939. *Carl Linnaeus*. Stockholm. [English edition, 1952, translated by A. Blair. London: Jonathan Cape.]

Hargis, W. J., Jr. 1956. A suggestion for the standardization of the higher systematic categories. *Systematic Zoology* 5(1): 42-46.

Hawksworth, D. L. 1992. The need for a more effective biological nomenclature for the 21st century. *Botanical Journal of the Linnean Society* 109:543-567.

_____. 1995. Steps along the road to a harmonized bionomenclature. *Taxon* 44(3): 447-456.

Hay, O. P. 1902. Bibliography and catalogue of the fossil Vertebrata of North America. *Bulletin of the United States Geological Survey* 179:1-868.

_____. 1929. *Second Bibliography and Catalogue of the Fossil Vertebrata of North America*, vol. 1. Washington, D.C.: Carnegie Institution of Washington, Publication No. 390.

_____. 1930. *Second Bibliography and Catalogue of the Fossil Vertebrata of North America*, vol. 2. Washington, D.C.: Carnegie Institution of Washington, Publication No. 390.

Hedgpeth, J. 1961. *Taxonomy: Man's Oldest Profession*. 11th Annual University of the Pacific Lecture.

Heller, J. L. 1964. The early history of binomial nomenclature. *Huntia* 1:33-70.

Hemming, F. 1944. A discussion on the differences in observance between zoological and botanical nomenclature. 2. The case for the zoologists. *Proceedings of the Linnean Society of London* 156:134-137.

Hendler, G. and D. L. Pawson. 1988. Echinoderm nomenclature: Not in need of repair. *Systematic Zoology* 36(4): 395-396. [Dated 1987 but published 10 March 1988.]

Hennig, W. 1950. *Grundzüge einer Theorie der Phylogenetischen Systematik*. Berlin; Deutscher Zentralverlag.

_____. 1965. Phylogenetic systematics. *Annual Review of Entomology* 10:97-116.

_____. 1966. *Phylogenetic Systematics*. Urbana: University of Illinois Press. [Translated by D. D. Davis and R. Zangerl.]

_____. 1969. *Die Stammesgeschichte der Insekten*. Herausg. von der Senkenbergischen Naturforschenden Gesellschaft zu Frankfurt am Mein. Frankfurt am Main: W. Kramer.

_____. 1974. Kritische Bemerkungen zur Frage "Cladistic analysis or cladistic classification?" *Zeitschrift für Zoologische Systematik und Evolutionsforschung* 12(4): 279-294.

_____. 1975. "Cladistic analysis or cladistic classification?": A reply to Ernst Mayr. *Systematic Zoology* 24(2): 244-256. [Translation of Hennig 1974 by G. C. D. Griffiths, edited by G. Nelson, authorized by W. Hennig.]

Heppell, D. 1981. The evolution of the Code of Zoological Nomenclature. In A. Wheeler and J. H. Price, eds., *History in the Service of Systematics. Papers from the Conference to celebrate the Centenary of the British Museum (Natural History) 13-16 April, 1981*, pp. 135-141. Society for the Bibliography of Natural History Special Publication number 1.

_____. 1991. Names without number? In D. L. Hawksworth, ed., *Improving the Stability of Names: Needs and Options*, pp. 191-196. Proceedings of an International Symposium, Kew, 20-23 February 1991. Regnum Vegetabile 123. Königstein, Germany: Koeltz.

Heywood, V. H. 1988. The structure of systematics. In D. L. Hawksworth, ed., *Prospects in Systematics*, pp. 44-56. The Systematics Association Special Volume 36. Oxford: Clarendon Press.

Hoffman, A. 1985. Patterns of family extinction depend on definition and geological timescale. *Nature* 315:659-662.

Holman, E. W. 1985. Evolutionary and psychological effects in pre-evolutionary classifications. *Journal of Classification* 2:29-39.

_____. 1992. Statistical properties of large published classifications. *Journal of Classification* 9:187-210.

Holmes, E. B. 1980. Reconsideration of some systematic concepts and terms. *Evolutionary Theory* 5(1): 35-87.

———. 1981. Erratum: Reconsideration of some systematic concepts and terms. *Evolutionary Theory* 5(3): 188.

Honacki, J. H., K. E. Kinman, and J. W. Koeppl, eds. 1982. *Mammal Species of the World.* Lawrence, Kansas: Allen Press and the Association of Systematics Collections.

Hopson, J. A. 1967. Mammal-like reptiles and the origin of mammals. *Discovery* 2(2): 25-33.

———. 1991. Systematics of the nonmammalian Synapsida and implications for patterns of evolution in synapsids. In H.-P. Schultze and L. Trueb, eds., *Origins of the Higher Groups of Tetrapods: Controversy and Consensus,* pp. 635-693. Ithaca, N.Y.: Comstock Publishing Associates of Cornell University Press.

———. 1994. Synapsid evolution and the radiation of non-eutherian mammals. In D. R. Prothero and R. M. Schoch, eds., *Major Features of Vertebrate Evolution,* pp. 190-219. Short Courses in Paleontology 7. Lawrence, Kansas: The Paleontological Society.

Hopson, J. A. and A. W. Crompton. 1969. Origin of mammals. *Evolutionary Biology* 3:15-72.

Hopwood, A. T. 1950a. Animal classification from the Greeks to Linnaeus. Linnean Society of London, *Lectures on the Development of Taxonomy,* pp. 24-32. London: Linnaean Society.

———. 1950b. Animal classification from Linnaeus to Darwin. Linnean Society of London, *Lectures on the Development of Taxonomy,* pp. 46-59. London: Linnaean Society.

———. 1959. The development of pre-Linnaean taxonomy. *Proceedings of the Linnean Society of London* 170(3): 230-234.

Huelsenbeck, J. P. [1991] 1992. When are fossils better than extant taxa in phylogenetic analysis? *Systematic Zoology* 40(4): 458-469. [Dated 1991 but mailed 15 July 1992.]

Huene, F. R. von. 1948. Short review of the lower tetrapods. In A. L. du Toit, ed., *Robert Broom Commemorative Volume,* pp. 65-106. Special Publication of the Royal Society of South Africa. Cape Town: Royal Society of South Africa.

Hull, D. L. 1964. Consistency and monophyly. *Systematic Zoology* 13(1): 1-11.

———. 1965. The effect of essentialism on taxonomy—Two thousand years of stasis. *British Journal of the Philosophy of Science* 15:314-326; 16:1-18. [Reprinted in Ereshefsky 1992].

———. 1966. Phylogenetic numericlature. *Systematic Zoology* 15(1): 14-17.

———. 1968. The syntax of numericlature. *Systematic Zooology* 17(4): 472-474.

———. 1970. Contemporary systematic philosophies. *Annual Review of Ecology and Systematics* 1:19-54.

———. 1976. Are species really individuals? *Systematic Zoology* 25(2): 174-191.

———. 1978. A matter of individuality. *Philosophy of Science* 45:335-360.

———. 1980. The limits of cladism. *Systematic Zoology* 28(4): 416-440. [Dated 1979 but mailed 23 January 1980.]

———. 1984. Cladistic theory: Hypotheses that blur and grow. In T. Duncan and T. F. Stuessy, eds., *Cladistics: Perspectives on the Reconstruction of Evolutionary History,* pp. 5-23. New York: Columbia University Press. [Reprinted in Hull 1989:162-178.]

———. 1989. *The Metaphysics of Evolution.* Albany: State University of New York Press.

Huxley, J. S. 1958. Evolutionary processes and taxonomy with special reference to grades. *Uppsala Universitets Årsskrift* 6:21-39.

———. 1959. Clades and grades. In A. J. Cain, ed., *Function and Taxonomic Importance,* pp. 21-22. The Systematics Association Publication 3. London: British Museum (Natural History).

Huxley, T. H. 1869. *An Introduction to the Classification of Animals.* London: John Churchill.

———. 1880. On the application of the laws of evolution to the arrangement of the Vertebrata, and more particularly of the Mammalia. *Proceedings of the Zoological Society of London* 43:649-662.

Illiger, C. 1811. *Prodromus systematis mammalium et avium additis terminis zoographicis utriusque classis.* Berlin: C. Salfeld.

Inglis, W. G. 1991. Characters: The central mystery of taxonomy and systematics. *Biological Journal of the Linnean Society* 44:121-139.

International Code of Nomenclature of Bacteria. 1966. *International Journal of Systematic Bacteriology* 16(4): 459-490.

International Commission on Zoological Nomenclature. 1902. Rules of zoological nomenclature [translation of "Règles internationales de la nomenclature zoologique"]. In P. Matschie, ed., *Verhandlungen des V Internationalen Zoologen Congress zur Berlin, 12-16 August 1901,* pp. 963-972. Jena: G. Fischer.

———. 1961. *International Code of Zoological Nomenclature adopted by the XV International Congress of Zoology, London, July 1958.* London: International Trust for Zoological Nomenclature.

———. 1964. *International Code of Zoological Nomenclature adopted by the XV International Congress of Zoology, London, July 1958.* 2nd ed. London: International Trust for Zoological Nomenclature.

———. 1985. *International Code of Zoological Nomenclature.* 3rd ed. Adopted by the XX General Assembly of the International Union of Biological

Sciences, Helsinki, August 1979. London: International Trust for Zoological Nomenclature.

Jardine, N. 1969. A logical basis for biological classification. *Systematic Zoology* 18(1): 37-52.

Jefferies, R. P. S. 1979. The origin of chordates—a methodological essay. In M. R. House, ed., *The Origin of Major Invertebrate Groups*, pp. 443-477. The Systematics Association Special Volume 12. London and New York: Academic Press.

Jeffrey, C. 1973. *Biological Nomenclature*. London: Edward Arnold.

———. 1977. *Biological Nomenclature*. 2nd ed. New York: Crane, Russak.

———. 1989. *Biological Nomenclature*. 3rd ed. London: Edward Arnold.

Jenkins, F. A., Jr., S. M. Gatesy, N. H. Shubin, and W. W. Amaral. 1997. Haramiyids and Triassic mammalian evolution. *Nature* 385:715-718.

Jussieu, A. de. 1848. Taxonomie. In Ch. d'Orbigny, ed., *Dictionnaire Universel d'Histoire Naturelle*, vol. 12, pp. 368-434. Paris: Renard.

Jussieu, A. L. de. 1777. Examen de la famille des Renoncules. Mémoires de Mathématiques et de Physique présentés à l'Academie Royale des Sciences par divers savants et lus dans ses assemblies. *Mémoires de l'Académie des Sciences*, année 1773:214-240.

———. 1789. *Genera Plantarum, secundum ordines naturales disposita, juxta methodum in horto regio parisiensi exaratam anno MDCCLXXIV*. Paris: Hérissant and Barrois. [Reprinted in 1964. Weinheim: J. Cramer.]

Just, T. 1953. Generic synopses and modern taxonomy. *Chronica Botanica* 14:103-114.

Kalandadze, N. N. and S. A. Rautian. 1992. [The system of mammals and historical zoogeography.] In O. L. Rossolimo, ed., *Phylogenetics of Mammals*, pp. 44-152. Archives of the Zoological Museum, Moscow State University, 29. [Russian.]

Keen, A. M. and S. W. Muller, revisors. 1948. *Procedure in Taxonomy*. Stanford: Stanford University Press. [See Schenk & McMasters 1936, 1948.]

Kemp, T. S. 1982. *Mammal-like Reptiles and the Origin of Mammals*. London: Academic Press.

———. 1983. The relationships of mammals. *Zoological Journal of the Linnean Society* 77(4): 353-384.

———. 1988a. Interrelationships of the Synapsida. In M. J. Benton, ed., *The Phylogeny and Classification of the Tetrapods*. Vol. 2: *Mammals*, pp. 1-22. The Systematics Association Special Volume 35B. Oxford: Clarendon Press.

———. 1988b. A note on the Mesozoic mammals, and the origin of therians. In M. J. Benton, ed., *The Phylogeny and Classification of the Tetrapods*. Vol. 2: *Mammals*, pp. 23-29. The Systematics Association Special Volume 35B. Oxford: Clarendon Press.

Kermack, D. M. and K. A. Kermack. 1984. *The Evolution of Mammalian Characters*. London: Croom Helm.

Kermack, K. A. 1972. The origin of mammals and the evolution of the temporomandibular joint. *Proceedings of the Royal Society of Medicine* 65(4): 389-392.

Kermack, K. A. and F. Mussett. 1958. The jaw articulation of the Docodonta and the classification of Mesozoic mammals. *Proceedings of the Royal Society* B 149:204-215.

Kermack, K. A., F. Mussett, and H. W. Rigney. 1973. The lower jaw of *Morganucodon*. *Zoological Journal of the Linnean Society* 53(2): 87-175.

———. 1981. The skull of *Morganucodon*. *Zoological Journal of the Linnean Society* 71:1-158.

Kielan-Jaworowska, Z. 1970. Unknown structures in multituberculate skull. *Nature* 226:974-976.

———. 1971. Skull structure and affinities of the Multituberculata. *Palaeontologia Polonica* 25:5-41.

Kielan-Jaworowska, Z., R. Presley, and C. Poplin. 1986. The cranial vascular system in taeniolabidoid multituberculate mammals. *Philosophical Transactions of the Royal Society of London* B313:525-602.

Kirby, W. 1815. Strepsiptera, a new order of insects proposed; and the characters of the order, with those of its genera, laid down. *Transactions of the Linnean Society of London* 11:86-122.

Kluge, A. G. 1989. Metacladistics. *Cladistics* 5(3): 291-294.

———. 1990. On the special treatment of fossils and taxonomic burden: A response to Loconte. *Cladistics* 6(2): 191-193.

Kowalevsky, V. 1873. Sur l'*Anchitherium aurelianense* Cuv. et sur l'histoire paléontologique des chevaux. *Mémoires de l'Académie Impériale des Sciences de St.-Pétersbourg*, sér. 7, 20(5): 1-73.

Krishtalka, L. 1993. Anagenetic angst: Species boundaries in Eocene primates. In W. H. Kimbel and L. B. Martin, eds., *Species, Species Concepts, and Primate Evolution*, pp. 331-344. New York: Plenum.

Lam, H. J. 1936. Phylogenetic symbols, past and present. *Acta Biotheoretica* 2:153-194.

Lamarck, J. B. P. A. de Monet de. 1786. Classes. In J.-B.-P.-A. de M. de Lamarck and J.-L.-M. Poiret, *Encyclopédie méthodique. Botanique*, vol. 2(1), pp. 29-36. Paris: Panckoucke.

———. 1799. Prodrome d'une nouvelle classification des coquilles, comprenant une rédaction appropriée des caractères génériques, et l'établissement d'un grand nombre de genres nouveaux. *Memoires Société d'Histoire Naturelle de Paris* 1:63-91.

———. 1802. *Recherches sur l'Organisation des Corps Vivants*. Paris: Maillard.

———. 1809. *Philosophie Zoologique, ou Exposition des Considérations relative à l'histoire naturelle des*

Animaux. 2 vols. Paris: Dentu. [Reprinted and translated in 1914 as *Zoological Philosophy, an Exposition with Regard to the Natural History of Animals*, with an introduction by H. Elliot, London: Macmillan; in 1963, New York: Hafner; and in 1984 as *Zoological Philosophy: An Exposition with Regard to the Natural History of Animals*, Chicago: University of Chicago Press.]

_____. 1835-1845. *Histoire Naturelle des Animaux sans Vertèbres*. 2eme ed. 11 vols. Paris: J. B. Baillière. [1st edition, 1815-22. 7 vols. Paris: Verdière.]

Lamarck, J. B. P. A. de Monet de and C. F. Brisseau de Mirbel. 1803. *Histoire naturelle des végétaux, classés par familles*. 15 vols. Paris: Deterville.

Langius [Lang], C. N. 1722. *Methodus Nova Testacea marina in suas Classes, Genera, et Species distribuendi*. Lucerne. [Not seen.]

Lapage, S. P., P. H. A. Sneath, E. F. Lessel, V. B. D. Skerman, H. P. R. Seeliger, and W. A. Clark, eds. 1975. *International Code of Nomenclature of Bacteria, and Statutes of the International Committee on Systematic Bacteriology and Statutes of the Bacteriology Section of The International Association of Microbiological Societies: Bacteriological Code, 1976* [sic] *Revision*. Washington, D.C.: American Society for Microbiology. [A new version of the Bacterial Code appeared under the editorship of Sneath in 1992, but we have not seen it.]

Larson, J. L. 1971. *Reason and Experience: The Representation of Natural Order in the work of Carl von Linné*. Berkeley: University of California Press.

Latreille, P. A. 1796. *Précis des Caractères Génériques des Insectes, Disposés dans un Ordre Naturel*. Paris: A. Brive by F. Bourdeaux. [Reprinted in 1907.]

Lauterbach, K.-E. 1989. Trilobites and phylogenetic systematics: A reply to G. Hahn. In N. Schmidt-Kittler and R. Willmann, eds., *Phylogeny and the Classification of Fossil and Recent Organisms: Proceedings of a Symposium Organized by the Deutsche Forschungsgemeinschaft*, pp. 201-211. Abhandlungen des Naturwissenschaftlichen Vereins in Hamburg (NF) 28. Hamburg: Verlag Paul Parey.

Lee, M. S. Y. 1995. Species concepts and the recognition of ancestors. *Historical Biology* 10(4): 329-339.

Levine, N. D. 1958. Uniform endings for the names of higher taxa. *Systematic Zoology* 7(3): 134-135.

Lewis, J. E. 1963. A short history of taxonomy from Aristotle to Linnaeus. *Medical Arts and Sciences* 17:106-123.

Ley, W. 1968. *The Dawn of Zoology*. Englewood Cliffs, N.J.: Prentice Hall.

Li, C. and S. Ting. 1983. The Paleogene mammals of China. *Bulletin of Carnegie Museum of Natural History* 21:1-98.

Lillegraven, J. A. and G. Krusat. 1991. Craniomandibular anatomy of *Haldanodon exspectatus* (Docodonta; Mammalia) from the Late Jurassic of Portugal and its implications to the evolution of mammalian characters. *Contributions to Geology, University of Wyoming* 28(2): 39-138.

Lillegraven, J. A., M. C. McKenna, and L. Krishtalka. 1981. Evolutionary relationships of middle Eocene and younger species of *Centetodon* (Mammalia, Insectivora, Geolabididae) with a description of the dentition of *Ankylodon* (Adapisoricidae). *University of Wyoming Publications* 45:i-viii, 1-115.

Linnaeus (Linné), C. 1735. *Systema Naturae, sive Regna Tria Naturae Systematice Proposita per Classes, Ordines, Genera, & Species*. Lugduni Batavorum [Leiden]: Theodorum Haak. [Reprinted in 1964 in Dutch Classics on the History of Science, no. 83. Nieukoop: B. De Graff.]

_____. 1737a. *Critica Botanica in qua nomina Plantarum generica, specifica, & variantia examini subjiciuntur...Seu Fundamentorum Botanicorum pars IV*. Lugduni Batavorum [Leiden].

_____. 1737b. *Genera Plantarum eorumque characteres naturales secundum numerum, figuram, situm & proportionum omnium fructificationis partium*. Lugduni Batavorum [Leiden]: Conrad Wishoff.

_____. 1740. *Systema Naturae in quo Naturae Regna tria, secundum. Classes, Ordines, Genera, Species, systematice proponuntur*. Stockholm: Kiesewetter. [2nd edition of Linnaeus 1735. Reprinted as "fifth ed." in 1745, but retitled *Systema Naturae, sive Regna tria Naturae systematice proposita per Classes, Ordines, Genera & Species*. Lugduni Batavorum (Leiden).]

_____. 1745 . *Öländska och gothländska resa på riksens högloflige stånders befallning förrättad åhr 1741....* Stockholm and Upsala: Kiesewetter.

_____. 1747. *Wästgötha Resa förrättad år 1746*. Stockholm.

_____. 1748. *Systema naturae sistens regna tria naturae, in classes et ordines genera et species, redacta tabulisque aeneis illustrata*. Lipsiae [Leipzig]: Kiesewetter. [6th edition of Linnaeus 1735. Fide J. L. R. Agassiz (1857, 1859:303), Linnaeus' 3rd edition is a reprint of the first; the 4th (1744) and 5th are reprints of the 2nd. Editions 7, 8, & 9 are reprints of the sixth. The eleventh is a reprint of the 10th. The "13th," published after Linnaeus's death (1778) by Gmelin, is only a compilation.]

_____. 1749. *Pan Svecicus*. Upsaliae. [Not seen. See Gill 1896:5, 1898:461; Choate 1912:260. The work, also known by the title *Pan Suecus*, was repeatedly reprinted and translated.]

_____. 1751. *Philosophia Botanica, in qua explicantur fundamenta botanica cum definitionibus partium, exemplis terminorum, observationibus rariorum, adiectis figuris aeneis*. Stockholm: Kiesewetter.

_____. 1753a. *Incrementa botanices proxime praeter-lapsi semiseculi....* Holmiae [Stockholm]. [Not seen.]

_____. 1753b. *Species Plantarum. Exhibentes plantas rite cognitas. Ad genera relatas. Cum differentiis specificis, nominibus trivialibus, synonymis selectis, locis natalibus, secundum systema sexuale digestas....* 2 vols. Holmiae [Stockholm]: Laurentii Salvii. [Reprinted in 1957 by the Ray Society. London: Bernard Quaritch Ltd.]

_____. 1758. *Systema naturae per regna tria naturae, secundum classes, ordines, genera, species, cum characteribus, differentiis, synonymis, locis.* Vol. 1: *Regnum animale.* Editio decima, reformata. Stockholm: Laurentii Salvii. [Facsimile reprinted in 1956 by the British Museum (Natural History).]

_____. 1766. *Systema naturae per regna tria naturae, secundum classes, ordines, genera, species, cum characteribus, differentiis, synonymis, locis.* Vol. 1: *Regnum animale.* Editio decima, reformata. Stockholm: Laurentii Salvii. [12th edition of Linnaeus 1735.]

Little, F. J., Jr. 1964. The need for a uniform system of biological nomenclature. *Systematic Zoology* 13(4): 191-194.

Lloyd, G. E. R. 1961. The development of Aristotle's theory of the classification of animals. *Phronesis* 6:59-85.

Loconte, H. 1990. Cladistic classification of AMNIOTA: A response to Gauthier et al. *Cladistics* 6(2): 187-190.

Lorch, J. 1961. The natural system in biology. *Philosophy of Science* 28:282-295.

Løvtrup, S. 1973. Classification, convention and logic. *Zoologica Scripta* 2(2-3): 49-61.

_____. 1977. *The Phylogeny of Vertebrata.* London and New York: Wiley.

_____. 1979. The evolutionary species concept: Fact or fiction? *Systematic Zoology* 28(3): 386-392.

_____. 1985. On the classification of the taxon Tetrapoda. *Systematic Zoology* 34(4): 463-470.

_____. 1986. On the existence and definition of taxa. *Rivista di Biologia, Perugia* 79:265-268.

_____. 1987. On the species problem and some other taxonomic issues. *Environmental Biology of Fishes* 20(1): 3-9.

Lucas, S. G. 1990. The extinction criterion and the definition of the class Mammalia. *Journal of Vertebrate Paleontology* 10(Supplement to 3): 33A.

_____. 1992. Extinction and the definition of the class Mammalia. *Systematic Biology* 41(3): 370-371.

Lucas, S. G. and Z. Luo. 1993. *Adelobasileus* from the Upper Triassic of West Texas: The oldest mammal. *Journal of Vertebrate Paleontology* 13(3): 309-334.

Luo, Z. 1994. Sister-group relationships of mammals and transformations of diagnostic mammalian characters. In N. C. Fraser and H.-D. Sues, eds., *In the Shadow of the Dinosaurs: Early Mesozoic Tetrapods*, pp. 98-128. Cambridge: Cambridge University Press.

Macdonald, J. R. 1963. The Miocene faunas from the Wounded Knee area of Western South Dakota. *Bulletin of the American Museum of Natural History* 125(3): 139-238.

MacIntyre, G. T. 1967. Foramen pseudovale and quasi-mammals. *Evolution* 21(4): 834-841.

MacLeay, W. S. 1829. A letter to J. E. Bicheno, Esq., F. R. S., in examination of his paper "On systems and methods" in the Linnean Transactions. *Philosophical Magazine*, ser. 2, 6:199-212.

_____. 1830. On the dying struggle of the dichotomous system. *Philosophical Magazine*, ser. 2, 7:431-445; 8:53-57, 134-140, 200-207.

Maglio, V. J. and H. B. S. Cooke. 1978. *Evolution of African Mammals.* Cambridge, Mass.: Harvard University Press.

Magnol, P. 1689. *Prodromus Historae generalis plantarum, in quo Familiae Plantarum per Tabulas disponuntur.* Montpellier: Pech.

Marschall, A. 1873. *Nomenclator Zoologicus. Continens nomina systematica generum animalium tam viventium quam fossilium, secundum ordinem alphabeticum disposita.* Vindobonae [Vienna]: Caroli Ueberreuter (M. Salzer).

Marshall, L. G., A. Berta, R. Hoffstetter, R. Pascual, O. A. Reig, M. Bombin, and A. Mones. 1984. Mammals and stratigraphy: Geochronology of the continental mammal-bearing Quaternary of South America. *Palaeovertebrata*, Mémoire Extraordinaire: 1-76.

Marshall, L. G., R. Hoffstetter, and R. Pascual. 1983. Mammals and stratigraphy: Geochronology of the mammal-bearing Tertiary of South America. *Palaeovertebrata*, Mémoire Extraordinaire: 1-93.

Martin, R. D. 1981. Phylogenetic reconstruction versus classification: The case for clear demarcation. *Biologist*: 28(3): 127-132.

Mayr, E. 1949. The species concept: Semantics versus semantics. *Evolution* 3(4): 371-372.

_____. 1953. Concepts of classification and nomenclature in higher organisms and microorganisms. *Annals of the New York Academy of Sciences* 56(3): 391-397.

_____. 1965. Classification and phylogeny. *American Zoologist* 5(1): 165-174.

_____. 1968. Theory of biological classification. *Nature* 220:545-548.

_____. 1969. *Principles of Systematic Zoology.* New York: McGraw Hill.

_____. 1974. Cladistic analysis or cladistic classification? *Zeitschrift für Zoologische Systematik und Evolutionsforschung* 12(2): 94-128.

_____. 1981. Biological classification: Toward a synthesis of opposing methodologies. *Science* 214:510-516.

_____. 1982. *The Growth of Biological Thought: Diversity, Evolution, and Inheritance*. Cambridge, Mass.: Belknap Press of Harvard University Press.

_____. 1988. Recent historical developments. In D. L. Hawksworth, ed., *Prospects in Systematics*, pp. 31-43. The Systematics Association Special Volume 36. Oxford: Clarendon Press.

_____. 1995. Systems of ordering data. *Biology and Philosophy* 10(4): 419-434.

Mayr, E. and P. D. Ashlock. 1991. *Principles of Systematic Zoology*. 2nd ed. New York: McGraw-Hill.

Mayr, E., E. G. Linsley, and R. L. Usinger. 1953. *Methods and Principles of Systematic Zoology*. New York: McGraw-Hill.

McKenna, M. C. 1975. Toward a phylogenetic classification of the Mammalia. In W. P. Luckett and F. S. Szalay, eds., *Phylogeny of the Primates*, pp. 21-46. New York: Plenum.

_____. 1984. Holarctic landmass rearrangement, cosmic events, and Cenozoic terrestrial organisms. *Annals of the Missouri Botanical Garden* 70:459-489.

_____. 1987. Molecular and morphological analysis of high-level mammalian interrelationships. In C. Patterson, ed., *Molecules and Morphology in Evolution: Conflict or Compromise?*, pp. 55-93. Cambridge: Cambridge University Press.

McKenna, M. C., G. F. Engelmann, and S. F. Barghoorn. 1977. [Review of] P. D. Gingerich, Cranial Anatomy and Evolution of Early Tertiary Plesiadapidae (Mammalia, Primates). *Systematic Zoology* 26(2): 233-238.

Meier, R. and S. Richter. 1992. Suggestions for a more precise usage of proper names of taxa. Ambiguities related to the stem lineage concept. *Zeitschrift für systematische Zoologie und Evolutionsforschung* 30:81-88.

Melville, R. V. 1995. *Towards Stability in the Names of Animals: A History of the International Commission on Zoological Nomenclature 1895-1995*. London: International Trust for Zoological Nomenclature.

Meng, J. and A. R. Wyss. 1997. Multituberculate and other mammal hair recovered from Palaeogene excreta. *Nature* 385: 712-714

Miao, D. 1988. Skull morphology of *Lambdopsalis bulla* (Mammalia, Multituberculata) and its implications to mammalian evolution. *Contributions to Geology, University of Wyoming*, Special Paper 4:1-104.

_____. 1991. On the origins of mammals. In H.-P. Schultze and L. Trueb, eds., *Origins of the Higher Groups of Tetrapods: Controversy and Consensus*, pp. 579-597. Ithaca, N.Y.: Comstock Publishing Associates of Cornell University Press.

Miao, D. and J. A. Lillegraven. 1986. Discovery of three ear ossicles in a multituberculate mammal.

National Geographic Research 2(4): 500-507.

Michener, C. D. 1963, Some future developments in taxonomy. *Systematic Zoology* 12(4): 151-172.

Mickevich, M. F. and S. J. Weller. 1990. Evolutionary character analysis: Tracing character change on a cladogram. *Cladistics* 6(2): 137-170.

Minelli, A. 1991. Names for the system and names for the classification. In D. L. Hawksworth, ed., *Improving the Stability of Names: Needs and Options*, pp. 183-189. Proceedings of an International Symposium, Kew, 20-23 February 1991. Regnum Vegetabile 123. Königstein, Germany: Koeltz.

_____. 1993. *Biological Systematics: The State of the Art*. London: Chapman and Hall.

_____. 1995. The changing paradigms of biological systematics: New challenges to the principles and practice of biological nomenclature. *Bulletin of Zoological Nomenclature* 52(4): 303-309.

Minelli, A., G. Fusco, and S. Sartori. 1991. Self-similarity in biological classifications. *Biosystems* 26:89-97.

Mishler, B. D. and B. D. Brandon. 1987. Individuality, pluralism, and the phylogenetic species concept. *Biology and Philosophy* 2:397-414.

Moravcsik, J. M. E. 1967. Aristotle's theory of categories. In J. M. E. Moravcsik, ed., *Aristotle: A Collection of Critical Essays*, pp. 125-145. Garden City, N.Y.: Doubleday Anchor.

Morison, R. 1672. *Plantarum umbelliferarum distributio nova, per tabulas cognationis et affinitatis ex libro naturae observata et detecta*. Oxford: Sheldon Theatre.

Morrison, P. and P. Morrison. 1996. The physics of binary numbers. *Scientific American* 274(2): 130-131.

Muir, J. W. 1968. The definition of taxa. *Systematic Zoology* 17(3): 345.

Nannfeldt, J. A. 1958. Presidential address. In O. Hedberg, ed., *Systematics of Today: Proceedings of a Symposium held at the University of Uppsala in Commemoration of the 250th Anniversary of the Birth of Carolus Linnaeus*, pp. 7-12. Uppsala Universitets Årsskrift 1958(6). Uppsala: Lundequistska Bokhandeln.

Neave, S. A., ed. 1939. *Nomenclator Zoologicus: A List of the Names of Genera and Subgenera in Zoology from the Tenth Edition of Linnaeus 1758 to the end of 1935*, vols. 1-2. London: Zoological Society of London.

_____. 1940. *Nomenclator Zoologicus: A List of the Names of Genera and Subgenera in Zoology from the Tenth Edition of Linnaeus 1758 to the end of 1935*, vols. 3-4. London: Zoological Society of London.

_____. 1950. *Nomenclator Zoologicus*. Vol. 5: *1936-1945*. London: Zoological Society of London.

Needham, J. G. 1910. Practical nomenclature. *Science*, n. ser., 32:295-300.

Neff, N. A. 1986. A rational basis for a priori character weighting. *Systematic Zoology* 35(1): 110-123.

Nelson, G. J. 1969. Gill arches and the phylogeny of fishes, with notes on the classification of vertebrates. *Bulletin of the American Museum of Natural History* 141:475-552.

_____. 1971. Paraphyly and polyphyly: Redefinitions. *Systematic Zoology* 20(4): 471-472.

_____. 1972. Phylogenetic relationship and classification. *Systematic Zoology* 21(2): 227-231.

_____. 1973. "Monophyly again?"—a reply to P. D. Ashlock. *Systematic Zoology* 22(3): 310-312.

_____. 1974. Classification as an expression of phylogenetic relationships. *Systematic Zoology* 22(4): 344-359. [Dated 1973 but mailed 29 January 1974.]

_____. 1979. Cladistic analysis and synthesis: Principles and definitions, with a historical note on Adanson's *Familles des Plantes* (1763-1764). *Systematic Zoology* 28(1): 1-21.

_____. 1985. Class and individual: A reply to M. Ghiselin. *Cladistics* 1(4): 386-389.

Nelson, G. J. and N. Platnick. 1981. *Systematics and Biogeography: Cladistics and Vicariance.* New York: Columbia University Press.

Newman, E. 1833. Observations on the nomenclature of divisions in systematical arrangements of the subjects of natural history, more particularly in reference to "Some remarks on genera and subgenera, and on the principles on which they should be established; by the Rev. Leonard Jenyns, A.M. F.L.S.;" published in p. 385-390. *The Magazine of Natural History*, ser. 1, 6(36): 481-485.

Norell, M. A. 1989. Late Cenozoic lizards of the Anza Borrego Desert, California. *Contributions in Science, Natural History Museum of Los Angeles County* 414:1-31.

O'Hara, R. J. 1991. Representations of the natural system in the nineteenth century. *Biology and Philosophy* 6(2): 255-274.

_____. 1992. Telling the tree: Narrative representation and the study of evolutionary history. *Biology and Philosophy* 7:135-160.

_____. 1993. Systematic generalization, historical fate, and the species problem. *Systematic Biology* 42(3): 231-246.

Olson, E. C. 1959. The evolution of mammalian characters. *Evolution* 13(3): 344-353.

Oosterbroek, P. 1987. More appropriate definitions of paraphyly and polyphyly, with a comment on the Farris 1974 model. *Systematic Zoology* 36(2): 103-108.

Osborn, H. F. 1894. *From the Greeks to Darwin: An Outline of the Development of the Evolution Idea.* New York: Macmillan. [Reprinted several times.]

_____. 1910. *The Age of Mammals in Europe, Asia and North America.* New York: Macmillan.

Osborne, D. V. 1963. Some aspects of the theory of dichotomous keys. *New Phytologist* 62(2): 144-160.

Owen, R. 1841. Report on British fossil reptiles. *Reports of the British Association for the Advancement of Science* 11(2): 60-204.

_____. 1859. *On the Classification and Geographical Distribution of the Mammalia . . . , to which is added an appendix "On the Gorilla," and "On the extinction and transmutation of species."* London: John W. Parker.

Owens, J. 1960-1961. Aristotle on categories. *Review of Metaphysics* 14:73-90.

Palmer, T. S. 1904. *Index Generum Mammalium: A list of the Genera and Families of Mammals.* U.S. Department of Agriculture, North American Fauna 23. Washington, D.C.: Government Printing Office.

Panchen, A. L. 1982. The use of parsimony in testing phylogenetic hypotheses. *Zoological Journal of the Linnaean Society* 74(3): 305-328.

_____. 1992. *Classification, Evolution, and the Nature of Biology.* Cambridge: Cambridge University Press.

Pankhurst, R. J. 1991. *Practical Taxonomic Computing.* Cambridge: Cambridge University Press.

Papavero, N. and J. Llorente-Bousquets, eds. 1993. *Principia Taxonomica: Una introducción a los fundamentos lógicos, filosóficos y metodológicos de las escuelas de taxonomía biológica.* Vol. 1: *Conceptos básicos de la taxonomía: Una formalización.* Mexico, D.F.: Universidad Nacional Autónoma de México.

_____. 1994a. *Principia Taxonomica: Una introducción a los fundamentos lógicos, filosóficos y metodológicos de las escuelas de taxonomía biológica.* Vol. 2: *Las teorías clasificatorias de Éuritos de Taranto, Platón, Espeusipo y Arisóteles.* Mexico, D.F.: Universidad Nacional Autónoma de México.

_____. 1994b. *Principia Taxonomica: Una introducción a los fundamentos lógicos, filosóficos y metodológicos de las escuelas de taxonomía biológica.* Vol. 4: *El Sistema Natural y otros sistemas, reglas, mapas de afinidades y el advenimiento del tiempo en las clasificaciones: Buffon, Adanson, Maupertuis, Lamarck y Cuvier.* Mexico, D.F.: Universidad Nacional Autónoma de México.

_____. 1994c. *Principia Taxonomica: Una introducción a los fundamentos ógicos, filosóficos y metodológicos de las escuelas de taxonomía biológica.* Vol. 5: *Wallace y Darwin.* Mexico, D.F.: Universidad Nacional Autónoma de México.

Papavero, N. and J. Llorente-Bousquets. 1994d. La 'Isagoge' de Porfirio. In N. Papavero, J. Llorente-Bousquets, and A. Bueno-Hernández, eds., *Principia Taxonomica*, vol. 3, pp. 11-16. Mexico, D.F.: Universidad Nacional Autónoma de México.

Papavero, N., J. Llorente-Bousquets, and A. Bueno-

Hernández, eds. 1994. *Principia Taxonomica: Una introducción a los fundamentos lógicos, filosóficos y metodológicos de las escuelas de taxonomía biológica.* Vol. 3: *De Hsun Tzu a Kant.* Mexico, D.F.: Universidad Nacional Autónoma de México.

Parker-Rhodes, A. F. 1957. Review of Gregg's "The language of taxonomy." *Philosophical Review* 66:124-125.

Parkinson, P. G. 1990. *A Reformed Code of Botanical Nomenclature.* Wellington: The Plant Press. [Not seen.]

Patterson, C. 1978. *Evolution.* London: British Museum (Natural History).

_____. 1980. Cladistics. *Biologist* 27(5): 234-240.

_____. 1981a. Methods of paleobiogeography. In G. Nelson and D. E. Rosen, eds., *Vicariance Biogeography: A Critique,* pp. 446-489. New York: Columbia University Press.

_____. 1981b. Significance of fossils in determining evolutionary relationships. *Annual Review of Ecology and Systematics* 12:195-223.

_____. 1982a. Morphological characters and homology. In K. A. Joysey and A. E. Friday, eds., *Problems of Phylogenetic Reconstruction,* pp. 21-74. The Systematics Association Special Volume 21. London: Academic Press.

_____. 1982b. Classes and cladists or individuals and evolution. *Systematic Zoology* 31(3): 284-286.

_____. 1988. The impact of evolutionary theories on systematics. In D. L. Hawksworth, ed., *Prospects in Systematics,* pp. 59-91. The Systematics Association Special Volume 36. Oxford: Clarendon Press.

_____. 1989. Phylogenetic relations of major groups: Conclusions and prospects. In B. Fernholm, K. Bremer, and H. Jörnvall, eds., *The Hierarchy of Life,* pp. 471-488. Nobel Symposium 70. Excerpta Medica, International Congress Series 824. Amsterdam: Elsevier.

Patterson, C. and D. E. Rosen. 1977. Review of ichthyodectiform and other Mesozoic teleost fishes and the theory and practice of classifying fossils. *Bulletin of the American Museum of Natural History* 158(2): 81-172.

Patterson, C. and A. B. Smith. 1987. Is the periodicity of extinctions a taxonomic artefact? *Nature* 330:248-251.

Patterson, D. J. 1988. The evolution of protozoa. *Memorias do Instituto Oswaldo Cruz, Rio de Janeiro,* Supplemento 1, 83:580-600.

Payer, J.-B. 1844. *Des classifications et des méthodes en histoire naturelle.* Paris: Lacour et Maistrasse.

Pearse, A. S., ed. 1936. *Zoological Names: A list of Phyla, Classes, and Orders.* Durham, N.C.: Duke University Press.

Pellegrin, P. 1982. *La classification des animaux chez Aristote.* Paris: Les Belles Lettres. [Not seen.]

_____. 1986. *Aristotle's Classification of Animals: Biology and the Conceptual Unity of the Aristotelian Corpus.* Berkeley: University of California Press. [Translation of Pellegrin 1982 by A. Preus.]

Persoon, C. H. 1801. *Synopsis Methodica Fungorum.* Göttingen. [Reprinted in 1952. New York: Johnson Reprint Corporation.]

Piveteau, J., ed. 1958. *Traité de Paléontologie. Tome 6. L'Origine des Mammifères et les Aspects Fondamentaux de luer Évolution,* vol. 2. Paris: Masson et Cie.

_____. 1961. *Traité de Paléontologie. Tome 6. L'Origine des Mammifères et les Aspects Fondamentaux de leur Évolution,* vol. 1. Paris: Masson et Cie.

Planchon, J. É. 1869. *Pierre Richer de Belleval, Fondateur du Jardin des Plantes de Montpellier.* Montpellier: J. Martel Aine. [Not seen.]

Platnick, N. I. 1977a. Paraphyletic and polyphyletic groups. *Systematic Zoology* 26(2): 195-200.

_____. 1977b. Monotypy and the origin of higher taxa: A reply to E. O. Wiley. *Systematic Zoology* 26(3): 355-357.

_____. 1980. Philosophy and the transformation of cladistics. *Systematic Zoology* 28(4): 537-546. [Dated 1979 but mailed 23 January 1980.]

_____. 1985. Philosophy and the transformation of cladistics revisited. *Cladistics* 1(1): 87-94.

_____. 1986. On justifying cladistics. *Cladistics* 2(1): 83-85.

Poche, F. 1911. Die Klassen und höherer Gruppen des Tierreichs. *Archiv für Naturgeschichte* 77(1, Suppl. 1): 63-136.

_____. 1937. Ueber eine Neubearbeitung der Internationalen Nomenklaturregeln zwecks Erzielung einer eindeutigen, möglichst rationellen, einheitlichen und stabilen Benennung der Tiere. *12th International Congress of Zoology, Lisbon, 1933*:2405-2416.

_____. 1938. Neubearbeitung der Internationalen Regeln der Zoologischen Nomenklatur, zwecks Erzielung einer eindeutigen, möglichst rationellen, einheitlichen und stabilen Benennung der Tiere von der Nomenclaturkommission des Verbandes Deutschsprachlicher Entomologen-Vereine der Internationalen Nomenclaturkommission und dem Internationalen Zoologenkongress vorgeschlagen. *Konowia* 17(1): 45-121; 17(2): 138-243.

Qiu, Z. and Z. Qiu. 1995. Chronological sequence and subdivision of Chinese Neogene mammalian faunas. *Palaeogeography, Palaeoclimatology, Palaeoecology* 116(1-2): 41-70.

Queiroz, K. de, *see* de Queiroz, K.

Rasnitsyn, A. P. 1982. Proposal to regulate the names of taxa above the family group. Z.N.(S.) 2381. *Bulletin of Zoological Nomenclature* 39:200-207.

Raup, D. M. and S. M. Stanley. 1971. *Principles of*

Paleontology. San Francisco: W. H. Freeman.

Raven, P. H., B. Berlin, and D. E. Breedlove. 1971. The origins of taxonomy. *Science* 174:1210-1213.

Ray, J. 1660. *Catalogus plantarum circa Cantabrigiam nascentium.* Cambridge.

_____. 1682. *Methodus Plantarum Nova, Brevitatis & Perspicuitatis causa Synoptice in Tabulis Exhibita; Cum notis Generum tum summorum tum subalternorum Charactersticus, Observationibus nonnullis de seminibus Plantarum & Indice Copioso.* London: Henrici Faitborne & Joannis Kersey. [Reprinted in 1962. Historiae Naturalis Classica, vol. 26. Codicote, Herts: Wheldon & Wesley; New York: Hafner.] [See also the 1703 and 1733 revised editions. London: Smith & Walford.]

_____. 1693. *Synopsis methodica Animalium Quadrupedum et Serpentini generis.* London: Robert Southwell. [Reprinted in 1978. New York: Arno Press.]

Reed, C. A. 1960. Polyphyletic or monophyletic ancestry of mammals, or: What is a class? *Evolution* 14(3): 314-322.

Remane, A. 1989. Critical remarks to cladistic analysis and cladistic classification. In N. Schmidt-Kittler and R. Willmann, eds., *Phylogeny and the Classification of Fossil and Recent Organisms: Proceedings of a Symposium Organized by the Deutsche Forschungsgemeinschaft*, pp. 111-124. Abhandlungen des Naturwissenschaftlichen Vereins in Hamburg (NF) 28. Hamburg: Verlag Paul Parey.

Richter, S. and R. Meier. 1994. The development of phylogenetic concepts in Hennig's early theoretical publications (1947-1966). *Systematic Biology* 43(2): 212-221.

Ride, W. D. L. 1985. Introduction. In International Commission on Zoological Nomenclature, *International Code of Zoological Nomenclature*, pp. xiii-xix. 3rd ed. Adopted by the XX General Assembly of the International Union of Biological Sciences, Helsinki, August 1979. London: International Trust for Zoological Nomenclature.

_____. 1988. Towards a unified system of biological nomenclature. In D. L. Hawksworth, ed., *Prospects in Systematics*, pp. 332-353. The Systematics Association Special Volume 36. Oxford: Clarendon Press.

Ridley, M. 1986. *Evolution and Classification: The Reformation of Cladism.* London: Longman.

_____. 1989. The cladistic solution to the species problem. *Biology and Philosophy* 4:1-16.

Rieppel, O. C. 1988. *Fundamentals of Comparative Biology.* Basel and Boston: Birkhäuser Verlag.

Roger, O. 1887. Verzeichniss der bisher bekannten fossilen Säugethiere. *Bericht des naturwissenschaftliches Vereines für Schwaben und Neuburg* 29:1-162. [Additional volumes were published in 1894, 1896, and 1898, but were not seen.]

Rohdendorf, B. B. 1977. [The rationalization of names of higher taxa in zoology.] *Paleontologicheskiy Zhurnal* 1977(2): 14-22. [Russian.] [1977 English translation in *Paleontological Journal* 11(2): 149-155.]

Rollins, R. C. 1959. Linnaeus: Codes and nomenclature in biology. *Systematic Zoology* 8(1): 2-3.

Romer, A. S. 1966. *Vertebrate Paleontology.* 3rd ed. Chicago: University of Chicago Press.

_____. 1968. *Notes and Comments on Vertebrate Paleontology.* Chicago: University of Chicago Press.

Romer, A. S., N. E. Wright, T. Edinger, and R. Van Frank, eds. 1962. *Bibliography of Fossil Vertebrates Exclusive of North America, 1509-1927.* 2 vols. Geological Society of America, Memoir 87.

Ross, H. H. 1974. *Biological Systematics.* Reading, Mass.: Addison-Wesley.

Rougier, G. W., J. R. Wible, and J. A. Hopson. Basicranial anatomy of *Priacodon fruitaensis* (Triconodontidae, Mammalia) from the Late Jurassic of Colorado, and a reappraisal of mammaliaform interrelationships. *American Museum Novitates* 3183:1-38.

Rowe, T. 1987. Definition and diagnosis in the phylogenetic system. *Systematic Zoology* 36(2): 208-211.

_____. 1988. Definition, diagnosis and origin of Mammalia. *Journal of Vertebrate Paleontology* 8(3): 241-264.

_____. 1996. Coevolution of the mammalian middle ear and neocortex. *Science* 273:651-654.

Rowe, T. and J. Gauthier. 1992. Ancestry, paleontology, and definition of the name Mammalia. *Systematic Biology* 41(3): 372-378.

Rowley, G. 1956. Caconymy . . . or, a few short words against many long ones. *National Cactus and Succulent Journal* 11:3-4.

Rumpf (Rumphius), G. E. 1705. *D'Amboinische Rariteitkamer.* Amsterdam. [Reprinted in 1741.]

Russell, D. E., J. L. Hartenberger, C. Pomerol, S. Sen, N. Schmidt-Kittler, and M. Vianey-Liaud. 1982. Mammals and stratigraphy: The Paleogene of Europe. *Palaeovertebrata*, Mémoire Extraordinaire: 1-77.

Russell, D. E. and R. Zhai. 1987. The Paleogene of Asia: Mammals and stratigraphy. *Mémoires du Muséum National d'Histoire Naturelle, sér. C, Sciences de la Terre* 52:1-488.

Sabrosky, C. 1954. Nomenclature of families and superfamilies. *Journal of Paleontology* 28(4): 489-490.

Salthe, S. N. 1988. Notes toward a formal history of the levels concept. In G. Greenberg and E. Tobach, eds., *Evolution of Social Behavior and Integrative Levels*, pp. 53-64. Hillsdale, N.J.: Lawrence Erlbaum.

_____. 1989. Self-organization of/in hierarchically structured systems. *Systems Research* 6(3): 199-

208.

_____. 1991. Two forms of hierarchy theory in western discourses. *International Journal of General Systems* 18:251-264.

Savage, D. E. and D. E. Russell. 1983. *Mammalian Paleofaunas of the World.* Reading, Mass.: Addison-Wesley.

Savage, J. M. 1990. Meetings of the International Commission on Zoological Nomenclature. *Systematic Zoology* 39(4): 424-425.

Savory, T. H. 1962. *Naming the Living World: An Introduction to the Principles of Biological Nomenclature.* New York: Wiley.

Schander, C. and M. Thollesson. 1995. Phylogenetic taxonomy—some comments. *Zoologica Scripta* 24:263-268.

Schenk, E. T. and J. H. McMasters. 1936. *Procedure in Taxonomy, including a reprint of the International Rules of Zoölogical Nomenclature with summaries of opinions rendered to the present date.* [Revised edition, 1948, by A. M. Keen and S. W. Muller. Stanford: Stanford University Press. See also enlarged and in part rewritten 3rd edition, 1956.]

Schmidt-Kittler, N., ed. 1987. International Symposium on Mammalian Biostratigraphy and Paleoecology of the European Paleogene, Mainz, February 18-21, 1987. *Müncher Geowissenschaften Abhandlungen* A 10:1-312.

Schoch, R. M. 1986. *Phylogeny Reconstruction in Paleontology.* New York: Van Nostrand Reinhold.

Schulze, F. E., W. Kükenthal, and K. Heider. 1926-1954. *Nomenclator Animalium Generum et Subgenerum.* 5 vols. Berlin: Preussischen Akademie der Wissenschaften.

Scudder, S. H. 1872. Canons of systematic nomenclature for the higher groups. *American Journal of Science and Arts*, ser. 3, 3(17): 348-351.

_____. 1882. *Nomenclator zoologicus.* An alphabetical list of all generic names that have been employed by naturalists for recent and fossil animals from the earliest times to the close of the year 1879. In two parts: 1. Supplemental List; 2. Universal Index. *Bulletin of the United States National Museum* 19:i-xix , 1-376, 1-340.

Sepkoski, J. J. 1987. [Reply to Patterson and Smith 1987.] *Nature* 330:251-252.

Sherborn, C. D. 1902. *Index Animalium sive Index Nominum quae ab A. D. MDCCLVIII Generibus et Speciebus Animalium imposita sunt.* Sect. 1, 1758-1800, vol. 1. Cambridge: Cambridge University Press.

_____. 1922-1933. *Index Animalium sive Index Nonimum quae ab A.D. MDCCLVIII Generibus et Speciebus Animalium imposita sunt.* Sect. 2, 1800-1850, pts. 1-33. London: Trustees of the British Museum (Natural History).

Simonetta, A. M. 1995. Some remarks on the influ-

ence of historical bias in our approach to systematics and the so called "species problem." *Bollettino di Zoologia* 62(1): 37-44.

Simpson, G. G. 1931. A new classification of mammals. *Bulletin of the American Museum of Natural History* 59:259-293.

_____. 1937. Notes on the Clark Fork, upper Paleocene, fauna. *American Museum Novitates* 954:1-24.

_____. 1943. Criteria for genera, species, and subspecies in zoology and paleozoology. *Annals of the New York Academy of Sciences* 44(2): 145-178.

_____. 1945. The principles of classification and a classification of mammals. *Bulletin of the American Museum of Natural History* 85:i-xvi, 1-350.

_____. 1947. A Continental Tertiary time chart. *Journal of Paleontology* 21(5): 480-483.

_____. 1952. For and against uniform endings in zoological nomenclature. *Systematic Zoology* 1(1): 20-23.

_____. 1953. *The Major Features of Evolution.* New York: Columbia University Press.

_____. 1959a. Anatomy and morphology: Classification and evolution: 1859 and 1959. *Proceedings of the American Philosophical Society* 103(2): 286-306.

_____. 1959b. Mesozoic mammals and the polyphyletic origin of mammals. *Evolution* 13(3): 405-414.

_____. 1959c. The nature and origin of supraspecific taxa. *Cold Spring Harbor Symposia on Quantitative Biology* 24:255-271.

_____. 1960. Diagnosis of the classes Reptilia and Mammalia. *Evolution* 14(3): 388-392.

_____. 1961. *Principles of Animal Taxonomy.* New York: Columbia University Press.

_____. 1964. The meaning of taxonomic statements. In S. L. Washburn, ed., *Classification and Human Evolution*, pp. 1-31. Viking Fund Publications in Anthropology 37. Chicago: Aldine. [Dated 1963 but issued 1964.]

_____. 1971a. Status and problems of vertebrate phylogeny. In R. Alvarado, E. Gadea, and A. de Haro, eds., *Actas del I Simposio Internacional de Zoofilogenia, Salamanca, 13-17 de octubre de 1969*, pp. 353-368. Salamanca: Facultad de Ciencias, Universidad de Salamanca.

_____. 1971b. Concluding remarks: Mesozoic mammals revisited. In D. M. Kermack and K. M. Kermack, eds., *Early Mammals*, pp. 181-198. Zoological Journal of the Linnean Society 50, Supplement 1. London: Academic Press.

_____. 1976. The compleat palaeontologist? *Annual Review of Earth and Planetary Sciences* 4:1-13.

_____. 1983. *Fossils and the History of Life.* New York: Scientific American Books.

Slaughter, M. M. 1982. *Universal Languages and Sci-*

entific Taxonomy in the Seventeenth Century. Cambridge: Cambridge University Press.

Sloan, P. R. 1972. John Locke, John Ray, and the problem of the natural system. *Journal of the History of Biology* 5(1): 1-53.

Smith, A. B. 1994. *Systematics and the Fossil Record: Documenting Evolutionary Patterns.* Oxford: Blackwell.

Smith, H. 1962. Commentary on the 1961 Code of Zoological Nomenclature. *Systematic Zoology* 11(2): 85-91.

Sober, E. 1980. Evolution, population, and essentialism. In Ereshefsky 1992, ed., *The Units of Evolution*, pp. 247-278. Cambridge, Mass.: MIT Press.

_____. 1988. *Reconstructing the Past: Parsimony, Evolution, and Inference.* Cambridge, Mass.: MIT Press.

_____. 1993. *Philosophy of Biology.* Boulder: Westview Press.

Sprague, M. L. (= M. L. Green) 1944. A discussion on the differences in observance between zoological and botanical nomenclature. 1. The case for the botanists. *Proceedings of the Linnean Society of London* 156:126-134.

Sprague, T. A. 1950. The evolution of botanical taxonomy from Theophrastus to Linnaeus. In Linnean Society of London, *Lectures on the Development of Taxonomy*, pp. 1-23. London: Linnaean Society.

Stafleu, F. A. 1969. A historical review of systematic biology. In International Conference of Systematic Biology, University of Michigan, 1967, *Systematic Biology: Proceedings of an International Conference*, pp. 16-44. Washington, D.C.: National Academy of Sciences, Publication 1692.

_____. 1971. *Linnaeus and the Linnaeans: The Spreading of Their Ideas in Systematic Botany, 1735-1789.* Regnum Vegetabile 79. Utrecht: Oosthoek's Uitgeversmaatschappij.

Stafleu, F. A., C. E. B. Bonner, R. McVaugh, R. D. Meikle, R. C. Rollins, R. Ross, J. M. Schopf, G. M. Schulze, R. de Vilmorin, and E. G. Voss, eds. 1972. *International Code of Botanical Nomenclature.* Utrecht: International Bureau of Plant Taxonomy and Nomenclature.

Stafleu, F. A. and E. G. Voss, eds. 1978. *International Code of Botanical Nomenclature adopted Leningrad 1975.* Regnum Vegetabile 97. Utrecht: Bohn, Scheltema and Holkema.

Starobogatov, Y. I. 1991. Problems in the nomenclature of higher taxonomic categories. *Bulletin of Zoological Nomenclature* 48(1): 6-18.

Stearn, W. T., ed. 1953. *International Code of Nomenclature for Cultivated Plants.* London: Royal Horticultural Society.

_____. 1955. Linnaeus's 'Species Plantarum' and the language of botany. *Proceedings of the Linnean Society of London* 165(2): 158-164.

_____. 1959. The background of Linnaeus's contributions to the nomenclature and methods of systematic biology. *Systematic Zoology* 8(1): 4-22.

_____. 1960. Notes on Linnaeus's 'Genera Plantarum'. In C. Linnaeus, *Genera Plantarum*, pp. v-xxiv. Facsimile reprint of 5th ed., 1754, with an introduction by W. T. Stearn. Weinheim: H. R. Engelmann.

Stehli, F. G. and S. D. Webb, eds. 1985. *The Great American Biotic Interchange.* Topics in Geobiology 4. New York: Plenum.

Stejneger, L. 1924. A chapter in the history of zoological nomenclature. *Smithsonian Miscellaneous Collections* 77(1): 1-21.

Stenzel, H. B. 1950. Proposed uniform endings for names of higher categories in zoological systematics. *Science* 112:94.

Stevens, P. F. 1983. Augustun Augier's "Arbre botanique" (1801), a remarkably early botanical representation of the natural system. *Taxon* 32(2): 203-211.

_____. 1994. *The Development of Biological Systematics: Antoine-Laurent de Jussieu, Nature, and the Natural System.* New York: Columbia University Press.

Stevenson, B. E., ed. 1948. *The Home Book of Proverbs, Maxims and Familiar Phrases.* New York: Macmillan.

Steyskal, G. C. 1988. On the naming of higher taxa. *Systematic Zoology* 36(4): 400. [Dated 1987 but mailed 10 March 1988.]

Stiles, C. W. 1905. *The International Code of Zoological Nomenclature as applied to Medicine.* U.S. Hygienic Laboratory Bulletin 24. Washington, D.C.: Government Printing Office.

Stoll, N. R. 1961. Introduction. In International Commission on Zoological Nomenclature, *International Code of Zoological Nomenclature adopted by the XV International Congress of Zoology, London, July 1958*, pp. vii-xvii. London: International Trust for Zoological Nomenclature.

_____. 1964. Introduction. In International Commission on Zoological Nomenclature, *International Code of Zoological Nomenclature adopted by the XV International Congress of Zoology, London, July 1958*, pp. ix-xix. 2nd ed. London: International Trust for Zoological Nomenclature.

Storr, G. C. C. 1780. *Prodromvs methodi Mammalivm.* Tübingen: Reissianis.

Strickland, H. E. 1834. Observations on classification, in reference to the essays of Messrs. Jenyns (VI. 385.), Newman (481.), and Blyth (485.). *Magazine of Natural History* 7:62-64.

_____. 1837. Rules for zoological nomenclature. *The Magazine of Natural History and Journal of Zoology, Botany, Mineralogy, Geology, and Meteorology*, N.S., 1:173-176.

_____. 1841. On the true method of discovering the

natural system in zoology and botany. *Annals and Magazine of Natural History* 6(36): 184-194.

_____. 1843. Report of a committee appointed "to consider of the rules by which the nomenclature of zoology may be established on a uniform and permanent basis." *Report of the [Twelfth Meeting of the] British Association for the Advancement of Science [held at Manchester in June]* 1842:105-121.

Strickland, H. E., J. Phillips, J. Richardson, R. Owen, L. Jenyns, W. J. Broderip, J. S. Henslow, W. E. Shuckard, G. R. Waterhouse, W. Yarrell, C. Darwin, and J. O. Westwood. 1843. Series of propositions for rendering the nomenclature of zoology uniform and permanent, being the report of a committee for the consideration of the subject appointed by the British Association for the Advancement of Science. *Annals and Magazine of Natural History, including Zoology, Botany, and Geology* 11:259-275.

Strother, J. L. 1995. The emperors' new Code? *Taxon* 44(1): 83-84.

Stuessy, T. F. 1983. Phylogenetic trees in plant systematics. *Sida* 10:1-13.

Stümpke, H. 1957. *Bau und Leben der Rhinogradentia*. Stuttgart: Gustav Fischer. [Reprinted in translation in 1967 as *The Snouters, Form and Life of the Rhinogrades*. New York: Doubleday (Natural History Press).]

Stys, P. and I. Kerzhner. 1975. The rank and nomenclature of higher taxa in Recent Heteroptera. *Acta Entomologica Bohemoslovaka* 72(2): 65-79.

Swingle, D. B. 1946. *A Textbook of Systematic Botany*. 3rd ed. New York and London: McGraw-Hill.

Szalay, F. S. 1994. *Evolutionary History of the Marsupials and an Analysis of Osteological Characters*. Cambridge and New York: Cambridge University Press.

Tassy, P. 1988. The classification of Proboscidea: How many cladistic classifications? *Cladistics* 4(1): 43-57.

Telford, M. and R. Mooi. 1986. Echinoderms, Babel and the confusion of nomenclature. *Systematic Zoology* 35(2): 254-255.

Thompson, W. R. 1952. The philosophical foundation of systematics. *Canadian Entomologist* 84:1-16.

Thomson, K. S. 1995. By any other name. *American Scientist* 83:514-517.

Thorell, T. 1869-1870. On European Spiders. Part. 1. Review of the European genera of spiders, preceded by some observations on zoological nomenclature. *Nova Acta Regiae Societatis Scientiarum Upsaliensis*, ser. 3, 7: 1-242.

Tong, Y., S. Zheng, and Z. Qiu. 1995. Cenozoic mammal ages of China. *Vertebrata PalAsiatica* 33(4): 290-314.

Tournefort, J. P. de. 1694. *Éléments de Botanique, ou Méthode pour connoître les Plantes*. 3 vols. Paris: Imprimerie Royale.

_____. 1700-1703. *Institutiones Rei Herbariae*. 4 vols. Paris: Imprimerie Royale.

Trouessart, E.-L. 1897-1905. *Catalogus mammalium tam viventium quam fossilium*. Nova editio (prima completa). 6 pts. and 4 pt. supplement. Berlin: R. Friedländer.

Tubbs, P. K. 1992. The International Commission on Zoological Nomenclature: What it is and how it operates. *Systematic Biology* 41(1): 135-137.

Tuomikoski. R. 1967. Notes on some principles of phylogenetic systematics. *Annales Entomologici Fennici* 33(3): 137-147.

Ussher, J. 1650. *Annales veteris testamenti, a prima mundi origine deducti*. [Quotation in H. W. Smith 1952, Man and his Gods. Boston: Little Brown. See also Gould 1991.]

_____. 1658. *The Annals of the World, IV*. [Reprinted excerpt in G. Y. Craig and E. J. Jones, eds. 1982, A Geological Miscellany, pp. 2, 3. Oxford: Orbital Press.]

Valentine, J. W. and C. L. May. 1996. Hierarchies in biology and paleontology. *Paleobiology* 22(1): 23-33.

Van Valen, L. 1960. Therapsids as mammals. *Evolution* 14(3): 304-313.

_____. 1973a. Are categories in different phyla comparable? *Taxon* 22(4): 333-373.

_____. 1973b. A new evolutionary law. *Evolutionary Theory* 1:1-30.

Voss, E. 1952. The history of keys and phylogenetic trees in systematic biology. *Journal of the Scientific Laboratories of Denison University* 43:1-25.

Walker, E. P., F. Warnick, S. E. Hamlet, K. L. Lange, M. A. Davis. H. E. Uible, and P. F. Wright. 1964. *Mammals of the World*. 3 vols. Baltimore: Johns Hopkins Press. [2nd edition published in 1968; 3rd edition, 1975; 4th edition by R. M. Nowak and J. L. Paradiso 1983; 5th edition by R. M. Nowak, 1991.]

Wallace, A. R. 1856. Attempts at a natural arrangement of birds. *The Annals and Magazine of Natural History*, ser. 2, 18:193-216.

Warburton, F. E. 1967. The purposes of classification. *Systematic Zoology* 6(3): 241-245.

Watt, J. C. 1968. Grades, clades, phenetics, and phylogeny. *Systematic Zoology* 17(3): 350-353.

Webb, S. D. 1978. A history of savanna vertebrates in the New World. Part 2. South America and the Great American Interchange. *Annual Reviews of Ecology and Systematics* 9:393-426.

Werdelin, L. and N. Solounias. 1991. The Hyaenidae: Taxonomy, systematics and evolution. *Fossils and Strata* 30:1-104.

Wheeler, Q. D. 1986. Character weighting and cladistic analysis. *Systematic Zoology* 35(1): 102-109.

Whewell, W. 1837. *History of the Inductive Sciences, from the Earliest to the Present Times*. 3 vols.

London: John W. Parker. [Three editions, many reprintings.]

Whitehead, P. J. P. 1967. The dating of the 1st edition of Cuvier's *Le Règne Animal Distribué d'Apres son Organisation. Journal of the Society for the Bibliography of Natural History* 4(6): 300-301.

Wible, J. R. 1991. Origin of Mammalia: The craniodental evidence re-examined. *Journal of Vertebrate Paleontology* 11(1): 1-28.

Wible, J. R. and J. A. Hopson. 1993. Basicranial evidence for early mammal phylogeny. In F. S. Szalay, M. J. Novacek, and M. C. McKenna, eds., *Mammal Phylogeny.* Vol. 1: *Mesozoic Differentiation, Multituberculates, Monotremes, Early Therians, and Marsupials*, pp. 45-62. New York: Springer-Verlag.

Wiley, E. O. 1975. Karl R. Popper, systematics, and classification: A reply to Walter Bock and other evolutionary taxonomists. *Systematic Zoology* 24(2): 233-243.

_____. 1977. Are monotypic genera paraphyletic?: A response to Norman Platnick. *Systematic Zoology* 26(3): 352-355.

_____. 1979a. Ancestors, species, and cladograms: Remarks on the symposium. In J. Cracraft and N. Eldredge, eds., *Phylogenetic Analysis and Paleontology*, pp. 211-225. New York: Columbia University Press.

_____. 1979b. An annotated Linnaean hierarchy, with comments on natural taxa and competing systems. *Systematic Zoology* 28(3): 308-337.

_____. 1980. Is the evolutionary species fiction?: A consideration of classes, individuals and historical entities. *Systematic Zoology* 29(1): 76-80.

_____. 1981. *Phylogenetics: The Theory and Practice of Phylogenetic Systematics.* New York: Wiley.

_____. 1986. La Sistematica en la revolucion Darwiniana. *Anales del Museo de Historia Natural de Valparaíso* 17:25-31.

Wilkins, J. 1668. *An Essay Towards a Real Character and a Philosophical Language.* London: Royal Society.

Williams, C. B. 1950. A note on the relative sizes of genera in the classification of animals and plants. *Proceedings of the Linnean Society of London* 162:171-178.

Willmann, R. 1987. Phylogenetic systematics, classification and the plesion concept. *Verhandlungen des Naturwissenschaftlichen Vereins in Hamburg* 29:221-233.

Wilson, D. E. and D. M. Reeder, eds. 1993. *Mammal Species of the World: A Taxonomic and Geographic Reference.* Washington, D.C.: Smithsonian Institution Press.

Wood, A. E. 1965. Grades and clades among rodents. *Evolution* 19(1): 115-130.

Woodburne, M. O., ed. 1987. *Cenozoic Mammals of North America: Geochronology and Biostratigraphy.* Berkeley: University of California Press.

Woodburne, M. O. and C. C. Swisher, III. 1995. Land mammal high resolution geochronology, intercontinental overland dispersals, sea-level, climate, and vicariance. In W. A. Berggren, D. V. Kent, M.-P. Aubry, and J. Hardenbol, eds., *Geochronology, Time Scales and Global Stratigraphic Correlation*, pp. 335-364. Tulsa: SEPM (Society for Sedimentary Geology), Special Publication 54.

Woodger, J. H. 1937. *The Axiomatic Method in Biology.* Cambridge: Cambridge University Press.

Wyss, A. R. and J. J. Flynn. 1993. A phylogenetic analysis and definition of the Carnivora. In F. S. Szalay, M. J. Novacek, and M. C. McKenna, eds., *Mammal Phylogeny.* Vol. 2: *Placentals*, pp. 32-52. New York: Springer-Verlag.

Wyss, A. R. and J. Meng. 1996. Application of phylogenetic taxonomy to poorly resolved crown clades: A stem-modified node-based definition of Rodentia. *Systematic Biology* 45(4): 559-568.

Yates, F. A. 1982. *Lull and Bruno, collected essays.* London: Routledge and Kegan.

Zeller, U. 1988. The lamina cribrosa of *Ornithorhynchus* (Monotremata, Mammalia). *Anatomy and Embryology* 178:513-519.

Zimmermann, W. 1937. Arbeitsweise der botanischen Phylogenetik. *Handbuch der biologischen Arbeitsmethoden* 9(3): 941-1053.

Zoological Record. 1870-1995/1996. 1871-1996. Edited by various editors and published in London first by John van Voorst and most recently by the Zoological Society of London and BIOSIS. [For its predecessor, see Günther 1865-1870.]

INDEX OF COMMON NAMES

Boldface numbers indicate pages in the classification where the name is applied to a taxon. Page numbers followed by "*n*" indicate that the name appears in a note.

INDEX OF SCIENTIFIC NAMES

A boldface number indicates the page in the classification where the name appears as a currently used taxon. A page number followed by "*n*" indicates that the name appears in a note.

Aivukus **256**
Akmaiomys 183
Akodon **146**, 147*n*
Akodontini **146**
Aksyiria **392**
Aksyiromys **132**
Aktashmys **114**
Akzharomys **190**
Alabamornis 369
Alachterium **256**
Alachtherium 256
Alacodon 70
Alactaga 134
Alactaginae 133
Alactagulus 134
Alagomyidae **114**
Alagomys **114**
Alastor 301
Alastorius **301**
Alaymys **187**
Albanensia **127**
Albanohyus 398
Albanosmilus 229
Albertatherium **69**
Albertogaudrya **467**
Albertogaudryidae 467
Albertogaudryinae 467
Albertona **109**
Albionbaatar **37**
Albionbaataridae **37**
Albireo **391**
Albireonidae **391**
Alcadae 425, 430
Alce 430
Alcedae 425, 429, 430
Alceidae 425, 430
Alceini **430**
Alcelaphidae 435, 449
Alcelaphinae **449**
Alcelaphini **449**, 450*n*
Alcelaphus **449**
Alces **430**
Alcicephalus 433
Alcidedorbignya **215**
Alcidedorbignyidae 215
Alcinae 429, 430
Alcini 430
Aleamastodon 502
Alephis **445**
Aletesciurus 126
Aletocyon **244**
Aletodon **361**
Aletomerycini **424**
Aletomeryx **424**

Alexandromys 152
Alforjas **414**
Algarolutra **262**
Algeripithecus **342**
Aliama 382
Alicornops 482
Alienetherium **224**
Alilepini 110
Alilepus **111**
Alionycteris **299**
Alitoxodon 460
Aliveria **127**
Alkwertatherium **59**
Allaceropinae 480
Allacerops 480
Allacodon 40
Allactaga **134**
Allactagidae 132, 133
Allactaginae **133**
Allactodipus **134**
Allalmeia **457**
Allenopithecina 346
Allenopithecus **346**
Allictops **106**
Alloblairinella **289**
Allocaenelaphus **428**
Allocavia **200**
Allocebus **335**
Allochrocebus 346
Allocricetodon **138**
Allocricetulus **151**
Allocricetus 150
Allocyon **252**
Allodelphis **383**
Allodesmidae 255
Allodesmus **255**
Allodon 37
Allodontidae 36
Alloeodectes **69**
Allohippus 472
Allohyaena **242**
Alloiomys **197**
Allolagus 111, 112
Allomeryx 419
Allomops 315
Allomyidae **117**
Allomys **117**
Allopachyura **290**
Allophaiomys 152
Allops 476
Allopterodon **224**
Alloptox **110**
Alloscapanus 281
Allosciurus 126

Allosiphneus 172
Allosminthus **132**
Allosorex **293**
Allosoricinae **293**
Allosoricini 293
Allospalax 173
Allostylops **365**
ALLOTHERIA **36**
Allozebra 472
Allqokirus **78**
Alluvisorex **289**
Almogaver 363
Almogaveridae 363
Alobus 319
Alocodon 220
Alocodontulum **220**
Alopecocyon **272**
Alopecodon 272
Alopedon 247
Alopex 245
Alopsis 247
Alostera **211**
Alouatinae 354, 355
Alouatta **356**
Alouattidae 352, 354
Alouattinae 354, 355
Alouattini 354, **355**
ALPHADELPHIA 67
Alphadon **69**
ALPHADONTIA 68
Alphadontidae 68
Alphadontinae **68**
Alsatia 330
Alsaticopithecus 275
Alsomys 163
Altaniinae 339
Altanius **339**
Alterodon 209
Altiatlasius **337**
Alticamelinae 415
Alticamelus 415
Alticola **154**
Alticoli 154
Alticonodon **43**
Alticonodontinae **43**
Altilambda **216**
Altilemur 335
Altililemur 335
Altippus 470
Altomeryx 415
Altomiramys **177**
ALTUNGULATA **468**
Aluralagus **113**
Alveojunctus **324**

Haldanodon 32, 507, 508, 512
Halianassa 495
Halianassinae 494
Halibalaena 378
Halicetus **376**
Halichoerina 257
Halichoerus **258**
Halicore 496
Halicoridae 493, 496
Halicyon 257
Haligyna 495
Halipaedisca 496
Haliphilus 257
Halitherida 493, 494
Halitheriidae 493
Halitheriinae **494**
Halitherium 493*n*, **494**
Hallautherium **42**
Hallebune **403**
Hallelemur **323**
Hallensia **469**
Hallomys 157
Halmadromus 75
Halmarhiphus 74
Halmaselus 75
Halmatorhiphus 74
Halmaturida 62
Halmaturidae 62
Halmaturina 63
Halmaturini 62, 63
Halmaturus 62*n*, 64
HALOBIOIDEA 493
Halodon 41
Haltictops **106**
Halticus 135
Haltomys 135
Halychoerus 258
Halysiren 495
Hamadryas 346
Hamster 149, 150
Hanhaicerus **435**
Hanno 344
Hansdebruijnia 163
Hapale 352
Hapalemur **332**
Hapales 352
Hapalida 351
Hapalidae 351
Hapalina 351
Hapalineae 351
Hapalodectes **368**
Hapalodectidae **368**
Hapalodectinae 368
Hapaloides **98**

Hapalolemur 332
Hapalomys **163**
Hapalops **98**
Hapalorestes **368**
Hapalotis 170
Hapanella 352
Haplaletes **361**
Haplobunodon **403**
Haplobunodontidae **403**
Haplobunodontinae 403
Haploceros 441
Haplocerus 441
Haploconus **364**
Haplocyon 248
Haplocyoninae 248
Haplocyonoides 248
Haplocyonopsis 248
Haplodon 118
Haplodontheriidae 459, 461
Haplodontheriinae **461**
Haplodontherium **461**
Haplodontidae 118, 459, 461
Haplodontotherium 461
Haplodus 118
Haplohippus **470**
Haplolambda **216**
Haplomastodon **502**
Haplomeryx **412**
Haplomylomys 143
Haplomylus **362**
Haplomys **117**
Haplonycteris **299**
Haploodon 118
Haploodontidae 118
Haploodontini 117, 118
Haploodontoidea 117
Haploodus 118
HAPLORHINI **336**
Haplosiren 495
Haplostropha **207**
Haploudon 118
Haploudontia 118
Haploudus 118
Hapludon 118
Haptomys 205
Haqueina **392**
Haramiya 512
Haramiyavia 512
HARAMIYIDA 512
Haramiyidae 512
HARAMIYOIDEA 512
HARAMYOIDEA 512*n*
Harana 428
Harlanus 446

Harpagmotherium 500
Harpagocyon **248**
Harpagolestes **367**
Harpagophagus **249**
Harpaladae 351
Harpale 352
Harpaleocyon 251
Harpalodon 242
Harpiana 299
Harpiocephalus **322**
Harpiola 322
Harpyia 299
Harpyidae 299
Harpyiocephalus 322
Harpyionycterinae 296, 297
Harpyionycterini **297**
Harpyionycteris **297**
Harpyodidae **215**
Harpyodus **215**
Harrymyinae 183
Harrymyini **183**
Harrymys **183**
Hassianycterididae **301**
Hassianycteris **302**
Hathliacynidae 78
Hathliacyninae **78**
Hathliacynus 78
Hathlyacynidae 78
Hathlyacyninae 78
Hathlyacynus 78
Hattomys **150**
Hayoceros **422**
HEBETIDENTATA 464
Hebetotherium 97
Hecubides 249
Hedralophus **457**
HEGETOTHERIA **465**, 466*n*
Hegetotheridae 466
Hegetotheriidae **466**
Hegetotheriinae **466**
Hegetotherinae 466
Hegetotherium **466**
Hegetotheroidea 466
Heimyscus **166**
Helacytidae 347
Helacyton 350
Helaletes **487**
Helaletidae **487**
Helaletinae 487
Helamis 185
Helamyina 185
Helamys 185
Helarctos **251**
Helarctus 251

Xerinae 122, 123
Xerini **123**
Xeromys **170**
Xeroscapheus 281
Xerospermophilus 124
Xerus **123**
Xesmodon **454**
Xinjiangmeryx **420**
Xinyuictis **243**
Xiphacodon 243
Xiphias 382, 383
Xiphius 382
Xiphodon **412**
Xiphodonta 412
Xiphodontidae **412**
Xiphodontoidea 412
Xiphonycteris 315
Xiphuroides **90**
Xotodon 460
Xotodontidae 459, 460
Xotodontinae 460
Xotoprodon 460
Xylechimys **207**
Xylomys 184
Xyophorus **98**
Xyronomys **39**
Yakopsis **445**
Yalkaparidon **54**
YALKAPARIDONTIA **54**
Yalkaparidontidae **54**
YANGOCHIROPTERA **307**
YANGOTHERIA 43
Yanshuella **281**
Yantanglestes **367**
Yaquius **338**
Yarala **56**
Yatkolamys **138**
Yerbua 185
Yimengia **480**
Yindirtemys **190**
Yingabalanara **52**
Yingabalanaridae **52**
YINOCHIROPTERA **304**
YINOTHERIA 44
Yoderimyinae **180**
Yoderimys **180**
Youngia 172
Youngofiber **130**
Ysengrinia **249**
Yudinia 289
Yuelophus 492
Yumaceras **424**
Yumaceratinae 424

Yumanius 338
Yunnanochoerus **393**
Yunnanotherium **419**
Yunomys **164**
Yunoscaptor **281**
Yuodon **105**
Yuomyidae 188
Yuomyinae **188**
Yuomys **188**
Yuorlovia 433
Zaedius 85
Zaedypus 85
Zaedyus 85
Zaglossus **36**
Zagmys **113**
Zaisanamynodon **487**
Zaisaneomys **180**
Zalambdalestes **105**
Zalambdalestidae **105**
Zalambdodonta 49, 293
ZALAMBDOLESTA 104
Zalambdolestoidea 105
Zalophus **254**
Zamolxifiber **130**
Zanycteris **326**
Zaphilus **90**
Zapodidae 132, 133
Zapodinae **133**
Zapodini 133
Zapus **133**
Zarachis 383
Zarafa 432
Zaragocyon **251**
Zarhachis **383**
ZATHERIA **48**
Zati 345
Zazamys **208**
Zebra 472
Zegdoumyidae **185**
Zegdoumys **185**
Zelceina **288**
Zelomys **187**
Zelotomys **166**
Zemiodontomys **180**
Zenkerella **186**
Zenkerellinae **186**
Zetamys **179**
Zetis 125
Zetodon 364
Zeuctherium **103**
ZEUGLOCETA 368
Zeuglodon 368, 369
ZEUGLODONTA 368

ZEUGLODONTIA 368
Zeuglodontida 369
Zeuglodontidae 369
Zeuglodontinae 369
Zeusdelphys **67**
Zhelestes **103**
Zhelestinae 103
Zhujegale **359**
Ziamys **182**
Zibellina 265
Zibethailurus 231
Zindapiria **189**
Zinjanthropus 349
Ziphacodon **243**
Ziphiidae 381
Ziphiina 381
Ziphiinae 381
Ziphila 86
Ziphini 381
Ziphiodelphis **374**
Ziphioidea 381
Ziphioides **382**
Ziphiorhynchus 382
Ziphiorrhynchus 382
Ziphirostris 382
Ziphirostrum **382**
Ziphius **382**
Zodiolestes **261**
Zofiabaatar **37**
Zokor 172
Zonoplites 83
ZOOTOKA 35
Zorilla 267
Zorillina 265
Zorillinae 265
Zotodon 460
Zramys **141**
Zygaenocephalus 298
Zygiocuspis **49**
Zygodon 369
Zygodontomys **145**
Zygogeomys **183**
Zygolestes **71**
Zygolestina 71
Zygolestini 70, 71
Zygolophodon **500**
Zygolophodontinae 500
Zygomaturinae **59**
Zygomaturus **59**
Zygorhiza 369
Zyzomys **170**

CPSIA information can be obtained
at www.ICGtesting.com
Printed in the USA
BVHW061717030220
571274BV00005B/86

9 780231 110136